Emmanuil G. Sinaiski
and Eugeniy J. Lapiga
Separation of Multiphase, Multicomponent Systems

1807–2007 Knowledge for Generations

Each generation has its unique needs and aspirations. When Charles Wiley first opened his small printing shop in lower Manhattan in 1807, it was a generation of boundless potential searching for an identity. And we were there, helping to define a new American literary tradition. Over half a century later, in the midst of the Second Industrial Revolution, it was a generation focused on building the future. Once again, we were there, supplying the critical scientific, technical, and engineering knowledge that helped frame the world. Throughout the 20th Century, and into the new millennium, nations began to reach out beyond their own borders and a new international community was born. Wiley was there, expanding its operations around the world to enable a global exchange of ideas, opinions, and know-how.

For 200 years, Wiley has been an integral part of each generation's journey, enabling the flow of information and understanding necessary to meet their needs and fulfill their aspirations. Today, bold new technologies are changing the way we live and learn. Wiley will be there, providing you the must-have knowledge you need to imagine new worlds, new possibilities, and new opportunities.

Generations come and go, but you can always count on Wiley to provide you the knowledge you need, when and where you need it!

William J. Pesce
President and Chief Executive Officer

Peter Booth Wiley
Chairman of the Board

Emmanuil G. Sinaiski and Eugeniy J. Lapiga

Separation of Multiphase, Multicomponent Systems

BICENTENNIAL 1807 WILEY 2007 BICENTENNIAL

WILEY-VCH Verlag GmbH & Co. KGaA

The Authors

E. G. Sinaiski
Leipzig, Germany
e-mail: egsin@t-online.de

E. J. Lapiga
Moscow, Russia
e-mail: npfeitek@mtu-net.ru

Cover
E. J. Lapiga: Oil rig, developed by EITEK

Library of Congress Card No.: applied for

British Library Cataloguing-in-Publication Data
A catalogue record for this book is available from the British Library.

Bibliographic information published by the Deutsche Nationalbibliothek
The Deutsche Nationalbibliothek lists this publication in the Deutsche Nationalbibliografie; detailed bibliographic data is available in the Internet at ⟨http://dnb.d-nb.de⟩.

Printed in the Federal Republic of Germany
Printed on acid-free paper

Composition Asco Typesetters, Hong Kong
Printing strauss GmbH, Mörlenbach
Bookbinding Litges & Dopf GmbH, Heppenheim
Cover Design aktivComm GmbH, Weinheim
Wiley Bicentennial Logo Richard J. Pacifico

ISBN 978-3-527-40612-8

Contents

Separation of Multiphase, Multicomponent Systems. E. G. Sinaiski and E. J. Lapiga

ISBN: 978-3-527-40612-8

Preface

This book sets out the theoretical basis underpinning the separation of multiphase, multi-component systems with application to the processes used to prepare hydrocarbon mixtures (oil, natural gas, and gas condensate) for transportation. The text is divided into seven sections.

Section I provides an introduction to the basic processes, the technological schemes, and the components of the equipment employed in systems for the field preparation of oil, natural gas, and gas condensate. The emphasis is on the designs and the principles of operation of separators, absorbers, and cooling devices. Mathematical modeling of the processes in these devices is covered in subsequent sections of the book.

The media with which one has to deal when investigating preparation processes of hydrocarbon systems are invariably multi-phase and multi-component mixtures. Section II thus covers the aspects of the hydromechanics of physical and chemical processes necessary for an understanding of the more specialized material contained in following sections. Among these are transfer phenomena of momentum, heat, mass, and electrical charge; conservation equations for isothermal and non-isothermal processes for multi-component and multi-phase mixtures; equations of state, and basic phenomenological relationships.

Natural hydrocarbon systems exist as solutions, suspensions, colloidal systems, emulsions, gas-liquid and liquid-gas mixtures. Accordingly, Sections III–VII are devoted to each of the aforementioned kinds of systems.

Section III covers the theory and methods for investigating the behavior of multi-component charged and uncharged solutions. Considering non-charged solutions, the main focuses of attention are on diffusion processes with and without the possibility of chemical reactions, the flow of solutions in channels and pipes, processes on semi-permeable membranes (return osmosis), and mass exchange of particles, drops, and bubbles with the ambient media. For charged solutions, consideration is given to processes in electrolytic cells, electrodialysis, the structure of electrical double layers, electrokinetic phenomena, and electro-osmosis.

The behavior and stability of suspensions and colloidal systems, including non-charged and charged suspensions, along with the coagulation and sedimentation of particles and their deposition on obstacles, are considered in Section IV. Chap-

Separation of Multiphase, Multicomponent Systems. E. G. Sinaiski and E. J. Lapiga

ISBN: 978-3-527-40612-8

ter 8 (devoted to non-charged suspensions) provides an introduction to the micro-hydrodynamics of particles, covering the fundamentals of Brownian motion, the viscosity of dilute suspensions, and the separation of suspensions in a gravitational field or under centrifugal forces. Chapter 9, devoted to charged suspensions, deals with the definition of particle charge, electrophoretic effects, the motion of conductive drops in an electric field, and sedimentation potential. Chapter 10 deals with the problem of colloidal system stability, various mechanisms of particle coagulation, and the capture of particles by obstacles when a suspension is passed through a filter.

The behavior of emulsions is considered in Section V in connection with the process of oil dehydration. Actual problems of drop integration in emulsions are discussed. It is shown that this process occurs most effectively if the emulsion is subjected to an electric field. In this context, the behavior of conducting drops in emulsions, the interaction of drops in an electric field, and the coalescence of drops in emulsions are examined in detail. In terms of applications, processes of emulsion separation in settling tanks, electro dehydrators, and electric filters are considered.

Separation processes of gas-liquid (gas-condensate) mixtures are considered in Section VI. The following processes are described: formation of a liquid phase in a gas flow within a pipe; coalescence of drops in a turbulent gas flow; condensation of liquid in throttles, heat-exchangers, and turboexpanders; the phenomena related to surface tension; efficiency of division of the gas-liquid mixtures in gas separators; separation efficiency of gas-condensate mixtures in separators equipped with spray-catcher nozzles of various designs – louver, centrifugal, string, and mesh nozzles; absorbtive extraction of moisture and heavy hydrocarbons from gas; prevention of hydrate formation in natural gas.

Section VII is devoted to liquid-gas (oil-gas) mixtures. The topics discussed are the dynamics of gas bubbles in multi-component solutions; the separation of liquid-gas mixtures in oil separators both neglecting and taking into account the hindrance due to the floating-up of bubbles; and the coagulation of bubbles in liquids.

A list of literature is given at the end of each section.

All of the considered processes relate to the separation of multi-phase, multi-component media, hence the title of the book. It should be noted that in the preparation technology for the transportation of oil, natural gas, and gas condensates, the term separation is traditionally understood only as the process of segregation of either a condensate and water drops or of gas and gas bubbles (occluded gas) from an oil. The concept of separation used herein can mean any segregation of components in multi-component mixtures or of phases in multi-phase systems.

List of Symbols

Symbol	Definition	Dimension, SI
a	Sound velocity	$m \cdot s^{-1}$
a_i	Activity of i-th component	
a	Radius of tube, pipe, capillary, particle	m
a	Semi-axis of ellipsoid	m
a	Parameter of repulsive electrostatic force	
a	Specific surface of grid (mesh)	m^{-1}
a_t	Radius of particle	m
A	Dimensionless parameter	
A^*	Reduced gas constant	$J \cdot kg^{-1} \cdot K^{-1}$
A_{cyl}	Parameter of stream function at flow around cylinder	
A_i	Chemical affinity of reaction	$J \cdot mole^{-1}$
A_i	Dimensionless parameters of charged particles, of jalousie separator	
A_s	Parameter of stream function at flow around sphere	
Ar	Archimedean number	
Ar_{av}	Archimedean number calculated by average radius of particles	
b	Adsorption constant	
b	Ellipsoid semi-axis; radius of cell boundary, of collision section of particles with cylinder	m
b	Dimensionless parameter	
B	Constant of reaction of ν-th order	$mole^{1-\nu} \cdot m^{3\nu-2} \cdot s^{-1}$
B_i	Henry constant of i-th component	Pa
Bo	Bond number	
c	Specific heat capacity	$J \cdot kg^{-1} \cdot K^{-1}$
c	Wave velocity	$m \cdot s^{-1}$

Separation of Multiphase, Multicomponent Systems. E. G. Sinaiski and E. J. Lapiga

ISBN: 978-3-527-40612-8

c_{cap}	Capillary wave velocity	$m \cdot s^{-1}$
c_i	Inflow of energy to i-th phase due to work of external forces	$J \cdot m^{-2} \cdot s^{-1}$
c_i^n	Work of external surface forces	$J \cdot m^{-2} \cdot s^{-1}$
c_p	Specific heat capacity at constant pressure	$J \cdot kg^{-1} \cdot K^{-1}$
c_v	Specific heat capacity at constant temperature	$J \cdot kg^{-1} \cdot K^{-1}$
C	Molar concentration	$mole \cdot m^{-3}$
C	Reduced concentration of ions	$mole \cdot m^{-3}$
C	Euler constant	
C_{cr}	Critical concentration of electrolyte	$mole \cdot m^{-3}$
C_D	Resistance factor	
C_{ij}	Pair interaction factor of molecules of i-th and j-th components	
C_0	Initial concentration	$mole \cdot m^{-3}$
C_s	Saturation concentration of dissolved substance	$mole \cdot m^{-3}$
Ca	Capillary number	
d	Diameter of pipeline	m
d	Dimensionless parameter	
d_e	Hydraulic diameter of microchannel in porous environment medium	m
d_w	Wire diameter	m
D	Diffusion factor	$m^2 \cdot s^{-1}$
D^0	Diffusion factor of non-hindered (free) particle	$m^2 \cdot s^{-1}$
D^2	Variance distribution	m^2
D	Diameter of separator	m
D_{av}	Average diameter	m
D_{cr}	Critical diameter of drop to be broken	m
D_{br}	Factor of Brownian diffusion	$m^2 \cdot s^{-1}$
D_{eff}	Effective diffusion factor	$m^2 \cdot s^{-1}$
D_{ij}	Binary diffusion factor	$m^2 \cdot s^{-1}$
D_{max}	Maximal diameter of stable drop	m
D_{max}	Maximal drop diameter behind atomizer	m
D_{rot}	Rotation diffusion factor	s^{-1}
D_T	Turbulent diffusion factor	$m^2 \cdot s^{-1}$
Da	Damköhler number	
e	Specific internal energy	$J \cdot kg^{-1}$
E	Internal energy	J
E	Total energy	J
E	Strain rate tensor	$m \cdot s^{-2}$
E	Electric field strength	$W \cdot m^{-1}$

E	Activation energy	$J \cdot mole^{-1}$
E	Dimensionless parameter	
E_{cr}	Critical strength of electric field	$W \cdot m^{-1}$
E_{cyl}	Capture efficiency of particles by cylinder	
E_n	Normal component of electric field strength	$W \cdot m^{-1}$
E_s	Capture efficiency of particles by sphere	
f	Friction factor	$kg \cdot s^{-1}$
f	Resistance factor	
$f(W)$	Hinderness factor	
F	Stability factor	
f_i	Molar density of free energy	$J \cdot mole^{-1}$
f_i	Fugacity of *i*-th component	Pa
f_k	Dimensionless parameter of ***k***-th component of electric force of interaction between two charged particles	
$\mathbf{f}_k^0$	Dimensionless parameter of ***k***-th component of electric force of interaction between two far-spaced charged particles	
$\mathbf{f}_k^1$	Dimensionless parameter of ***k***-th component of electric force of interaction between two far-spaced charged particles, found with greater accuracy	
$\tilde{\mathbf{f}}_k$	Dimensionless parameter of ***k***-th component of electric force of interaction between two touching charged particles	
f_{ij}	Components of friction tensor	$kg \cdot s^{-1}$
f_{sr}, $f_{s\theta}$, f_{er}, $f_{e\theta}$, $f_{e\theta 1}$	Correction factors of hydrodynamic forces	
$f(D)$	Distribution of drops over diameters at jet disintegration	
$f(V)$	Breakage frequency of drop of volume V	$m^{-3} \cdot s^{-1}$
$\mathbf{f}$	Density of mass force	$N \cdot m^{-3}$
$\mathbf{f}$	Friction tensor	$kg \cdot s^{-1}$
$\mathbf{f}_E$	Density of electric force	$N \cdot m^{-3}$
$\mathbf{f}^r$	Rotation friction tensor	$m^2 \cdot kg \cdot s$
F	Free energy	J
F_{cap}	Capillary force	N

F_e	Electric force	N
F_{fr}	Friction force	N
F_h	Hydrodynamic force	N
F^{hyd}	Hydrodynamic force	N
F^{el}	Electric force	N
F_i	Component of *i*-th force	N
F^{mol}	Molecular force	N
F_n	Normal component of force	N
F_τ	Tangential component of force	N
F_{th}	Thermodynamic force	N
F_v	Viscous friction force	N
F_w	Resistance force	N
$\boldsymbol{F}$	Force	N
$\boldsymbol{F_\alpha}$	Molecular attraction force	N
$\boldsymbol{F}_\alpha^s$	Molecular attraction force between two spherical particles	N
$\boldsymbol{F_e}$	Force induced by particle own motion	N
$\tilde{\boldsymbol{F}}_i$	Electric force acting on *i*-th resting charged particle	N
$\boldsymbol{F_n}$	Normal to particle surface force component	
F_R^s	Electrostatic repulsion force between two spherical particles	N
$\boldsymbol{F_s}$	Stokes force	N
Fr	Froude number	
g_{eff}	Effective gravity acceleration at wave motion	$m \cdot s^{-2}$
G	Free energy (Gibbs energy)	J
G	Absolute value of vorticity vector	$m \cdot s^{-2}$
G	Dimensionless parameter	
G	Mass flow rate	$kg \cdot s^{-1}$
G	Capture (collision) section	m^2
$G(t)$	Random force	N
Gr	Grashoff number	
h	Specific enthalpy	$J \cdot kg^{-1}$
h	Half the channel height	m
h	Distance between particle centre and wall	m
h	Hydrodynamic resistance factor	$kg \cdot s^{-1}$
h	Dimensionless vorticity	
h	Distance between mash layers	m
h^0	Factor of hydrodynamic resistance at motion of non-hindered (free) particle	$kg \cdot s^{-1}$

h_{cr}	Critical thickness of liquid film on cylindrical string	m
h_N	Height of deposit layer	m
H	Height	m
H	Enthalpy	J
i_m	Limiting density of electric current	A
$\mathbf{i}$	Density of electric current	$A{\cdot}m^{-2}$
I	Nucleation rate in a unit volume	$m^{-3}{\cdot}s^{-1}$
I	Total mass flux	$mole{\cdot}s^{-1}$, $kg{\cdot}s^{-1}$
I	Electric current	A
$I(R_0)$	Correction factor for condensate growth of drop	
I_a	Rate of distribution change due to drop sedimentation	$m^{-3}{\cdot}s^{-1}$
I_b	Rate of distribution change due drop breakage	$m^{-3}{\cdot}s^{-1}$
I_D	Diffusion flux	$kg{\cdot}m^{-2}{\cdot}s^{-1}$
I_k	Rate of distribution change due to drop coagulation	$m^{-3}{\cdot}s^{-1}$
I_m	Rate of distribution change due to drop ablation	$m^{-3}{\cdot}s^{-1}$
I_n	Intensity of particle generation in a unit volume	$m^{-3}{\cdot}s^{-1}$
j	Mass flux through a unit surface	$kg{\cdot}m^{-2}{\cdot}s^{-1}$
j	Diffusion flux of particles	$m^{-3}{\cdot}s^{-1}$
j_0	Non-hindered (free) diffusion flux of particles	$m^{-3}{\cdot}s^{-1}$
j_{rw}	Diffusion flux of particles through a unit surface of solid angle	$m^{-2}{\cdot}s^{-1}$
j_s	Entropy flux through a unit surface	$J{\cdot}m^{-2}{\cdot}s^{-1}$
j_i	Individual mass flux of *i*-th component	$kg{\cdot}m^{-2}{\cdot}s^{-1}$
j_i^*	Individual mole flux of *i*-th component	$mole{\cdot}m^{-2}{\cdot}s^{-1}$
J	Diffusion flux of drops	s^{-1}
J	Mass flux	$kg{\cdot}s^{-1}$
J_0	Non-hindered (free) diffusion flux of drops	s^{-1}
J_A	Diffusion flux of drops with regard to molecular attraction force	s^{-1}
J_{A+R}	Diffusion flux of drops with regard to both molecular attraction force and electrostatic repulsion force	s^{-1}
J_{br}	Diffusion flux of drops at Brownian coagulations	s^{-1}

J_i	Moment of inertia of *i*-th particle	kg·m^2
J_i	Mass flux *i*-th component	kg·s^{-1}
$J_i^{(r)}$	Rate of *i*-th chemical reaction	mole·m^{-3}·s^{-1}
J_{ji}	Mass-exchange rate between *j*-th and *i*-th phases in a unit volume	kg·m^{-3}·s^{-1}
J_g	Diffusion flux of drops at gradient coagulation	s^{-1}
J_G	Gas flux from a unit surface of solution	kg·m^{-2}·s^{-1}
J_t	Diffusion flux of drops at turbulent coagulation	s^{-1}
J_i	Relative mass flux of *i*-th component	kg·m^{-2}·s^{-1}
J_i^*	Relative mole flux of *i*-th component	mole·m^{-2}·s^{-1}
J_q	Heat flux to a unit surface	W·m^{-2}
J_s	Entropy flux to a unit surface	W·m^{-2}
k	Heat conductivity factor	W·m^{-1}·K^{-1}
k	Specific kinetic energy	J·m^{-3}
k	Permeability of porous medium	m^2
k	Wave number	m^{-1}
k_1	Equilibrium constant	
k_i	Constant of *i*-th heterogeneous reaction of v_i-th order	mole$^{1-v_i}$·m$^{3v_i-2}$ · s^{-1}
k	Ratio of particle radiuses	
k	Adiabat constant	
k	Wetting factor	
k	Energy density	
k_T	Heat exchange factor	W·m^{-2}·K^{-1}
K	Kinetic energy	J
K	Kozeny factor	
K	Ablation factor of separator	
K	Dimensionless parameter	
$K(V, u)$	Kernel of kinetic equation (coagulation constant)	s^{-1}
K_i	Equilibrium constant of *i*-th component	
l	Characteristic linear size	m
l	Mean free path	m
l	Distance between centres of particles	m
l	Specific heat of evaporation	J·kg^{-1}
l	Step of particle random walk	m
$l \times l$	Average size of mesh cell	m × m
l	Radius of capture section	m
L	Length	m
L	Characteristic linear size	m

L	Mole fraction of liquid phase	
L	Work of friction forces	J
L	Work done on a unit mass of gas	$J \cdot kg^{-1}$
L_0	Distance between device of preliminary condensation (DPC) and separator	m
L_B	Distance from the point of jet outflow up to the place of jet disintegration	m
L_d	Throttle length	m
L_e	Height of separation contact element	m
L_{ik}	Phenomenological factor	
L_k	Length of absorber contact zone	m
L_c	Cyclone length	m
L_D	Length of entrance concentration region	m
L_{eq}	Length of equilibrium establishment	m
L_U	Length of entrance dynamic region	m
Le	Lewis number	
m	Mass	kg
$m_{C_{k+}}$	Relative amount of extracted components of fraction C_{k+}	
m_i	Mass of i-th component	kg
m_k	Distribution moment of k-th order	m^{3k-3}
$\hat{m}_k$	The dimensionless moment of k-th order	
M	Molecular mass	$kg \cdot mole^{-1}$
$\bar{M}$	Average molecular mass	$kg \cdot mole^{-1}$
M	Mach number	
Me^{z+}	Metal cation of charge z	
n	Number of moles	
n	Numerical concentration	m^{-3}
n	Number of absorbent recirculations in separation-contact element	
$\mathbf{n}$	Vector of a normal	
$n(R, t, P)$	Distribution of drops over radiuses	m^{-4}
$n(m, t, P)$	Distribution of bubbles over mass	$kg^{-1} \cdot m^3$
$n(V, t, P)$	Distribution of drops over volumes	m^{-6}
$n_d(D)$	Distribution of drops over diameters behind atomizer	m^{-4}
n_i	Components of normal vector	
n_i	Number of moles of i-th component	mole
nm^3	Cubic metre of gas under normal conditions	m^3
N	Number of moles	mole

N	Numerical concentration of particles	м$^{-3}$
N	Number of mesh layers	
N	Number of plates in absorber	
N_{ad}	Adhesion parameter of cylinder	
N_{ad}^{sph}	Adhesion parameter of sphere	
N_d	Number of moles in drop	mole
N_d	Numerical concentration of drops behind atomizer	m^{-3}
N_e	Number of separation-contact elements on the plate of absorber	
N_i	Dimensionless parameter	
$N_n(x_0, t)$	Rate of bubble nucleation at depth x at moment t	s^{-1}
Nu_D	Diffusion Nusselt number	
Nu_T	Temperature Nusselt number	
Oh	Ohnesorge number	
p	Pressure	Pa
p	Parameter of electromagnetic retardation	
p_a	Atmospheric pressure	Pa
p_c	Critical pressure	Pa
p_∞	Pressure above solution surface	Pa
$p_\infty^{(eq)}$	Established pressure above solution surface	Pa
p_e	Additional pressure at wave motion of liquid	Pa
p_i	Partial pressure	Pa
p_{iv}	Partial pressure of i-th solution component vapor	Pa
p_r	Reduced pressure	
p_s	Saturation pressure	Pa
p_v	Partial pressure of vapor	Pa
p_{vt}	Partial pressure of saturated vapor above drop surface	Pa
$p_{v\infty}$	Partial pressure of saturated vapor above flat surface	Pa
p_σ	Capillary pressure	Pa
$\boldsymbol{p}$	Unit vector	
P	Point of volume	
P	Probability of particle displacement	
$P(V, \omega)$	Probability of drop formation	
$\boldsymbol{P}_{ji}$	Intensity of momentum exchange between j-th and i-th phases	kg·m^{-2}·s^{-2}
Pe_D	Diffusion Peclet number	

Pe_T	Temperature Peclet number	
Pr	Prandtl number	
q	Specific quantity of heat	$\mathrm{J{\cdot}kg^{-1}}$
q	Electric charge	C
q	Specific heat flux	$\mathrm{J{\cdot}m^{-2}{\cdot}s^{-2}}$
q	Dimensionless parameter	
q_a	Specific flow rate of absorbent	$\mathrm{10^{-3}{\cdot}kg{\cdot}m^{-3}}$
q_i	Electric charge of *i*-th component	C
q_i^n	Normal component of heat flux of *i*-th component	$\mathrm{W{\cdot}m^{-2}}$
q_s	Density of surface charge	$\mathrm{C{\cdot}m^{-2}}$
$\mathbf{q}$	Heat flux	$\mathrm{W{\cdot}m^{-2}}$
$\mathbf{q}_i$	Heat flux of *i*-th component	$\mathrm{W{\cdot}m^{-2}}$
Q	Mole mass flux	$\mathrm{mole{\cdot}s^{-1}}$
Q	Volume flow rate	$\mathrm{m^3{\cdot}s^{-1}}$
Q	Dynamic pressure	Pa
Q	Total charge	C
Q	Heat brought to a unit mass of gas	$\mathrm{J{\cdot}kg^{-1}}$
Q_a	Absorbent flow rate	Tonne/day
Q_{cr}	Critical gas flow rate	$\mathrm{Mill.m^3/day}$
Q_G	Gas flow rate	$\mathrm{Mill.m^3/day}$
Q_h	Amount of hydrocarbons extracted from gas	Tonne/day
Q_i	Specific heat released due to work of friction forces	$\mathrm{J{\cdot}K\Gamma^{-1}}$
$\mathbf{Q}_i$	Mass flux of *i*-th component	$\mathrm{kg{\cdot}m{\cdot}s^{-1}}$
Q_{in}	Specific heat released by condensation	$\mathrm{J{\cdot}m^{-3}{\cdot}s^{-1}}$
Q_s	Specific heat due to heat transfer through pipe wall	$\mathrm{J{\cdot}kg^{-1}}$
Q_w	Heat transfer from pipe wall	J
r_c	Radius of wire	m
r_i	Rate of mass formation of *i*-th component in a unit volume	$\mathrm{kg{\cdot}m^{-3}{\cdot}s^{-1}}$
R_{av}	Average drop radius	m
R_{av}^0	Initial average radius of drop	m
R_c	Coagulation radius	m
R_c	Radius of cyclone	m
R_{cr}	Critical radius	m
R_i	Factors of resistance (components of resistance tensor) along principal axes of ellipsoid	m
R_i	Radius of *i*-th particles	m
R_{ij}	Components of resistance tensor	m

$R_i^{(s)}$	Specific mole rate of heterogeneous chemical reaction with formation of *i*-th component	mole·m^{-2}·s^{-1}
$R_i^{(v)}$	Specific mole rate of homogeneous chemical reaction with formation of *i*-th component	mole·m^{-2}·s^{-1}
R_m	Minimal radius of drops	m
R_{ms}	Minimal radius of drops settling with Stokesian velocity	m
R_s	Impede factor of membrane	
R_z	Radius of cell	m
Re	Reynolds number	
R	Dimensionless radius of cylinder capture section	
$\boldsymbol{R}$	Resistance tensor (Translation tensor)	m
s	Specific entropy	J·kg^{-1} K^{-1}
s	Specific surface	m^{-1}
s	Relative distance between particles	
s	Sedimentation factor	s
s	Random displacement	m
s	Supersaturation degree	
s_{cr}	Critical supersaturation degree	
S	Entropy	J·K^{-1}
S	Area	m^2
S	Surface	
S	Stokes number	
S	Spread factor	N·m^{-1}
S_A	Parameter of molecular interaction	
S_{av}	Average area of interface	m^2
S_{cr}	Critical Stokes number	
S_E	Parameter of electrohydrodynamic interaction	
S_f	Total area of microchannel sections	m^2
S_i	Dimensionless parameter	
S_m	Dimensionless parameter	
S_m	Minimal Stokes number	
S_R	Parameter of electrostatic interaction	
Sc	Shmidt number	
St	Strouhal number	
t	Time	s
t_b	Absorbent residence time on absorber plate	s
t_{br}	Characteristic time of Brownian coagulations	s

t_e	Residence time in separation-contact element	s
t_{ik}	Maxwell stress tensor	N·m^{-2}
t_{in}	Characteristic time of inertial coagulation	s
t_k	Characteristic time of drop coagulation (coalescence)	s
$\bar{t}_l$	Average life time of drops in turbulent flow	s
t_m	Characteristic time of drops mass exchange with gas	s
t_{mono}	Characteristic time of coagulation (coalescence) in monodisperse emulsion	s
t_{poly}	Characteristic time of drop coagulation (coalescence) in polydisperse emulsion	s
t_r	Time of drop relaxation in turbulent flow	m·s^{-1}
t_t	Characteristic time of drop turbulent coagulation	s
t_v	Characteristic time of velocity profile development in channel	s
$t_{s\varphi}$, $t_{e\varphi}$, $t_{e\varphi 1}$	Correction factors for hydrodynamic moments	
$\boldsymbol{t}$	Stress	N·m^{-2}
T	Absolute temperature	K
T	Characteristic time	s
T	Period of turbulent pulsations	s
T_G	Temperature of gas	K
T_L	Temperature of liquid	K
T_c	Critical temperature	K
$T_{cr}^{(k)}$	Temperature of condensation beginning	K
T_r	Reduced temperature	
T_t	Temperature of dew-point	K
$\boldsymbol{T}_e$	Moment caused by particle own motion	N·m
$\boldsymbol{T}_s$	Moment caused by Stokesion flow around particle	N·m
$\boldsymbol{T}$	Stress tensor	N·m^{-2}
$\boldsymbol{T}$	Moment vector	N·m
u	Velocity component	m·s^{-1}
u_{cr}	Critical velocity	m·s^{-1}

u_d	Dynamic velocity of gas	$m \cdot s^{-1}$
u_s	Stokesian velocity	$m \cdot s^{-1}$
u_s	Velocity of particle cross drift in turbulent flow	$m \cdot s^{-1}$
$\bar{u}$	Average velocity	$m \cdot s^{-1}$
u_e	Gas velocity in separation-contact element	$m \cdot s^{-1}$
u_i^n	Normal velocity component of *i*-th phase	$m \cdot s^{-1}$
u_m	Drop velocity near the wall in turbulent flow	$m \cdot s^{-1}$
u_{max}	Maximal velocity	$m \cdot s^{-1}$
u_λ	Velocity of turbulent pulsation of scale λ	$m \cdot s^{-1}$
$\mathbf{u}$	Velocity vector, mean-flow-rate velocity vector	$m \cdot s^{-1}$
$\mathbf{u}^*$	Mean-mole velocity vector	$m \cdot s^{-1}$
$\mathbf{u}_{st}$	Stokesian velocity	$m \cdot s^{-1}$
$\mathbf{u}_{sh}$	Velocity of shear flow	$m \cdot s^{-1}$
U	Characteristic velocity	$m \cdot s^{-1}$
U_e	Rate of filtration through porous medium	$m \cdot s^{-1}$
U_G	Velocity of motion of interface border	$m \cdot s^{-1}$
U_s	Sedimentation velocity	$m \cdot s^{-1}$
v	Velocity component	$m \cdot s^{-1}$
v	Specific volume	$m^3\ kg^{-1}$
v	Dimensionless velocity	
$\mathbf{v}_i$	Mobility of particles of *i*-th solution component	$mole \cdot s \cdot kg^{-1}$
v_i	Mobility factor of a body along *i*-th principal axis	$s \cdot kg^{-1}$
v_φ	Tangential component of velocity in cyclone	$m \cdot s^{-1}$
v_z	Longitudinal component of velocity in cyclone	$m \cdot s^{-1}$
V	Volume	m^3
V	Mole fraction of gas phase	
V	Electromotive force (emf)	V
V	Total potential energy of interaction between two particles	J
$\mathbf{V}$	Mobility tensor	$s \cdot kg^{-1}$
V_A^S	Potential of molecular attraction force between two spherical particles	J
V_A^P	Potential of molecular attraction force between two infinite parallel planes	$J \cdot m^{-2}$

V_{av}	Average volume of drops	m^3
V_{cr}	Critical volume	m^3
V_d	Volume of drop	m^3
V_e	Volume of eluent	m^3
V_i	Mole concentration of *i*-th component	$mole{\cdot}m^{-3}$
V_i	Potential of *i*-th particle surface	J
V_k	Volume of germ	m^3
V_{mol}	Volume of molecule	m^3
V_R^S	Potential of electrostatic repulsion force between two particles	J
V_t	Volume of particle	m^3
V_v	Volume of voids between particles of permeable medium	m^3
w	Velocity component	$m{\cdot}s^{-1}$
w	Specific work	$J{\cdot}kg^{-1}$
$\mathbf{w}_i$	Velocity of *i*-th phase relative medium as a whole	$m{\cdot}s^{-1}$
W	Volume concentration (volume content)	m^3/m^3
W	Work of drop done on the change of volume in a unit time	W
W	Energy of one mole	J
W	Stability factor	
W_0	Volume concentration of drops at the entrance of separator	m^3/m^3
W_1	Volume concentration of drops at the exit of separator	m^3/m^3
We	Weber number	
We_{cr}	Critical Weber number	
x_{cr}	Critical distance from top end of the string up to the point of liquid film detachment	m
x_i	Mole fraction of *i*-th component of liquid phase	
x_{Ir}	Mole fraction of hydrate inhibitore in solution	
x_M	Mole fraction of methanol in hydrate inhibitore	
x_{wr}	Mole fraction of water in solution	
$\mathbf{x}$	Radius-vector of point $P(x, y, z)$	
X	Thermodynamic force	
X_L	Length at which the flowing jet reaches wall	m
X^{z-}	Anion with charge *z*	

y_i	Mole fraction of *i*-th gas phase component	
y_M	Mole fraction of methanol vapor in gas	
z	Compressibility of gas	
Z	Dimensionless parameter	
z_m	Dimensionless minimal radius	
z_i	Charge of ion of *i*-th component	
α	Thermal diffusivity	$m^2 \cdot s^{-1}$
α	Thermal expansion factor	K^{-1}
α	Heat exchange factor	$W \cdot m^{-2} \cdot K^{-1}$
α	Dimensionless parameter	
α	Effective section	
α	Condensation factor	
α	Correction multiplier on microchannel curvature of porous medium	
α	Mass fraction of glycol in absorbent solution	
α	Slope of inclined wall	
β	Volume expansion factor	$Pa^{-1} \cdot s^{-1}$
β	Dimensionless parameter	
β	Coalescence parameter	s^{-1}
β	Design parameter of atomizer	
β_1	Asymmetry square of distribution	
β_2	Excess of distribution	
β_{ij}	Collisions frequency of particles *i* and *j*	s^{-1}
γ	Activity factor	
γ	Dimensionless parameter of repulsion force energy	
γ_I	Activity factor of inhibitor	
γ_w	Activity factor of water	
γ_φ	Dimensionless parameter of cyclone	
$\dot{\gamma}$	Shear rate	s^{-1}
$\bar{\dot{\gamma}}$	Dimensionless shear rate	
Γ	Hamaker constant	J
Γ	Surface concentration of surfactant	$mole \cdot m^{-2}$
Γ_∞	Limiting surface concentration of surfactant	$mole \cdot m^{-2}$
δ	Thickness of gap between two spherical particles	m
δ	Thickness of a boundary layer	m
δ_v	Thickness of viscous boundary layer	m
δ_f	Thickness of liquid film	m

δ_D	Thickness of diffusion boundary layer	m
Δ	Dimensionless thickness of gap between two spherical particles	
Δ_i	Dimensionless parameter	
Δ_k	Capillary length	m
Δy_i	Difference between mole fractions of *i*-th component at the interface and in gas bulk flow	
$\Delta\rho$	Difference of densities of bordering phases	$kg{\cdot}m^{-3}$
$\Delta\varphi_{om}$	Ohmic drop of potential	B
ε	Dielectric permittivity	$C{\cdot}V^{-1}{\cdot}m^{-1}$
ε	Void fraction of porous medium (porosity)	
ε	Dimensionless parameter	
ε_0	Dielectric permittivity in vacuum	$C{\cdot}V^{-1}{\cdot}m^{-1}$
ε_0	Specific energy dissipation of turbulent flow	$J{\cdot}kg^{-1}{\cdot}s^{-1}$
ε_{cr}	Critical specific energy dissipation of turbulent flow	$J{\cdot}kg^{-1}{\cdot}s^{-1}$
ε_r	Relative dielectric permittivity	
ε_{ij}	Components of strain rate tensor	$m{\cdot}s^{-2}$
ε_v	Void fraction of mesh layer	
ζ	ζ-potential	V
ζ	Vertical perturbation of interface	m
ζ	Dimensionless variable	
η	Dimensionless variable	
η	Separation efficiency, coefficient of effectiveness (CE)	
η_f	Capture efficiency of filter	
η_G	Effectiveness coefficient of mesh droplet capture	
η_h	Effectiveness coefficient of horizontal separator	
η_i	Mole fraction of *i*-th component	
η_k	Effectiveness coefficient of horizontal separator with regard to coagulation of drops	
η_s	Effectiveness coefficient of separator with string droplet capture	
η_t	Dehydration factor	
η_v	Effectiveness coefficient of vertical separator	
η_z	Effectiveness coefficient of cyclone	
Θ	Velocity divergence	

θ	Fraction of surface occupied by molecules of adsorbed substance	
θ	Dimensionless temperature	
λ	Heat conductivity factor	$W \cdot m^{-1} \cdot K^{-1}$
λ	Particle resistance factor	
λ	Scale of turbulent pulsation	m
λ	Ablation factor	
λ	Wave length	m
λ	Correction to minimal radius of drop on condensation growth of drops	
λ_0	Inner scale of turbulence	m
λ_D	Thickness of electric double layer	m
λ_G	Heat conductivity factor of gas	$W \cdot m^{-1} \cdot K^{-1}$
λ_h	Ablation factor of horizontal separator or settler	
λ_v	Ablation factor of vertical separator or settler	
λ_L	London wave length	Å
Λ	Mole conductivity	$S \cdot m^2 \cdot mole^{-1}$
Λ	Dimensionless parameter	
μ	Dynamic viscosity factor	Pa·c
μ	Chemical potential	$J \cdot mole^{-1}$ or $J \cdot kg^{-1}$
$\mu_i^{(0)}$	Chemical potential of pure *i*-th component	$J \cdot mole^{-1}$
$\bar{\mu}$	Ratio of viscosities of internal and external liquids	
ν	Kinematic viscosity factor	$m^2 \cdot s^{-1}$
ν	Stoichiometric factor	
ν_{ki}	Stoichiometric factor of *k*-th component in *j*-th reaction	
ν_+, ν_-	Number of ions	
ξ	Degree of completeness of reaction	$mole \cdot m^{-3}$
ξ	Dimensionless variable, dimensionless parameter	
ξ_i	Dimensionless parameter	
Ξ	Osmotic factor	
π	Pressure drop in reverse osmosis	Pa
π	Dimensionless parameter	
π_0	Osmotic pressure	Pa
Π_i	Mass percentage of *i*-th component	
$\mathbf{\Pi}$	Viscous stress tensor	$N \cdot m^{-2}$
χ	Debye reverse radius	m^{-1}
χ	Dimensionless parameter	
χ	Ratio of drop charges	

ρ	Mass density	$kg \cdot m^{-3}$
ρ_E	Electric charge density	$C \cdot m^{-3}$
ρ_G	Gas density	$kg \cdot m^{-3}$
ρ_i	Mass concentration of *i*-th a component	$kg \cdot m^{-3}$
ρ_i^0	True mass density	$kg \cdot m^{-3}$
ρ_{vG}	Mass concentration of water vapor in gas	$kg \cdot m^{-3}$
ρ_{vG}	Mass concentration of water in drops of diethyleneglycol (DEG)	$kg \cdot m^{-3}$
ρ_{wG}	Equilibrium mass concentration of water vapor in gas above water solution of DEG	$g \cdot m^{-3}$
σ	Electric conductivity	$S \cdot m^{-1}$
σ	Rate entropy generation	$J \cdot K^{-1} \cdot m^{-3} \cdot s^{-1}$
σ	Surface charge density	$C \cdot m^{-2}$
σ	Separation factor of gel – chromatography	
σ^2	Variance of distribution	m^2
σ_i	Relative deviation of exact value f_k from approximate values f_k^0	
σ_s	Surface conductivity	S
Σ	Coefficient of surface tension	$N \cdot m^{-1}$
Σ	Surface	
Σ_{SG}	Coefficient of surface tension of interface solid body – gas	$N \cdot m^{-1}$
Σ_{SL}	Coefficient of surface tension of interface solid body – liquid	$N \cdot m^{-1}$
τ	Dimensionless time	
τ_0	Shear stress	$N \cdot m^{-2}$
τ_+	Dimensionless time of drop relaxation	
$\tau^{(v)}$	Viscous stress	$N \cdot m^{-2}$
$\tau^{(\sigma)}$	Capillary stress	$N \cdot m^{-2}$
τ_i^{kl}	Stress components of *i*-th phase	$N \cdot m^{-2}$
τ_{ij}	Components of stress tensor	$N \cdot m^{-2}$
V	Specific energy	$J \cdot kg^{-1}$
V	Mole volume	$m^3 \cdot mole^{-1}$
V_c	Critical mole volume	$m^3 \cdot mole^{-1}$
V_i	Mole volume of *i*-th component	$m^3 \cdot mole^{-1}$
V_{ji}	Rate of energy exchange between *j*-th and *i*-th phases	$J \cdot m^{-3} \cdot s^{-1}$
V_m	Mole volume averaged over solution composition	$m^3 \cdot mole^{-1}$

V_r	Reduced mole volume	
ϕ	Electric potential	V
ϕ_{ij}	Equilibrium value of electric potential in solution	V
ϕ_{sed}	Sedimentation potential	V
φ	Dimensionless parameter	
φ	Velocity potential	$m^2{\cdot}s^{-1}$
φ	Dehydration factor	
φ_{conc}	Concentration overvoltage	V
φ_i	Central angle of *i*-th corrugation of jalousie droplet capture	
$\varphi_{m,i}$	Solution of Laplace equation in confluent bispherical coordinates	
$\varphi_{m,n}$	Solution of Laplace equation in bispherical coordinates	
Φ	Viscous dissipation, dissipation function	$W{\cdot}m^{-3}$
Φ_{ij}	Potential of molecular interaction between *i*-th and *j*-th particles	J
$\Phi(p)$	Correction factor of molecular interaction	
$\Phi(V)$	Transfer function	
$\Phi_h(V)$	Transfer function of horizontal separator or settler	
$\Phi_v(V)$	Transfer function of vertical separator or settler	
ψ	Stream function	$m^3{\cdot}s^{-1}$
ψ	Ratio of ablation factors of horizontal and vertical settler	
ψ_l	Stream function on limiting trajectory	$m^3{\cdot}s^{-1}$
ω	Non-centricity factor	
ω	Dimensionless parameter	
ω_i	Mass fraction of *i*-th component	
ω	Angular velocity	s^{-1}
ω	Dimensionless angular velocity	
ω	Frequency	s^{-1}
ω_+	Dimensionless frequency of large-scale pulsations	
ω_{ijl}	Collision frequency of collisions of particles *i* with particles *j*	s^{-1}
$\boldsymbol{\omega}$	Velocity vector	s^{-1}
$\boldsymbol{\Omega}$	Rotation tensor	m^3
Ω_{ij}	Rotation tensor components	m^3

Bottom Indexes

0	Initial value
$\pm$	Parameter appropriate to positive or negative ion
∞	Value of parameter far from boundary or in the infinity
a	Anion parameter
a	Value of parameter at anode
a	Parameter of particle a
av	Average value
b	Parameter of particle b
c	Value of parameter at cathode
e	Parameter of external medium
eff	Effective parameter
eq	Equilibrium value
f	Parameter of liquid phase
G	Parameter of gas phase
i	Parameter of internal medium
i, j	Parameter components
in	Parameter of gel pore
L	Parameter of liquid phase
m	Maximal value; value averaged over composition
Me	Parameter of metal
n	Normal component
n	Value of parameter under normal conditions
om	Ohmic drop
r, R	Radial component
s	Parameter of solution
s, B	Value of parameter at boundary
w	Boundary value

Top Indexe

$\sim$	Dimensionless parameter
$\cdot$	Ordinary time derivative
$'$	Dimensionless size, parameter of internal medium, transformed variable
∞	Value of parameter far from boundary or in the infinity
$\pm$	Charge sign

Mathematical Designations

$\langle\ \rangle$	Averaging
$\mathbf{e}_i$	Vectors of basis

$(\boldsymbol{i}, \boldsymbol{j}, \boldsymbol{k})$	Basis vectors in cartesian system of coordinates
(x, y, z)	Cartesian coordinates
(r, θ, z)	Cylindrical coordinates
$(\boldsymbol{i}_r, \boldsymbol{i}_\theta, \boldsymbol{i}_z)$	Basis vectors in cylindrical system of coordinates
(ρ, Φ, z)	Cylindrical coordinates in narrow gab between particles
(r, θ, φ)	Spherical coordinates
(ξ, η)	Bipolar coordinates
(η, μ, Φ)	Bispherical coordinates
$\delta(x)$	Delta-function
$\{v_{ij}\}$	Matrix
$\mathrm{d}\Sigma$	Oriented surface element
$Ei(x)$	Integral exponential function
$G_{ij}(\boldsymbol{x} - \boldsymbol{y})$	Green Function
$\boldsymbol{I}$	Unit tensor with components δ_{ij}.
$J_m(x)$	Bessel function of m-th order
$\mathrm{I}_n(x)$	Modified Bessel function of 1-st order
$\mathrm{K}_n(x)$	Modified Bessel function of 2-nd order
L_{ki}	Laguerre associated polinomial
$L_q^{(r)}$	Lagrange polynomial factor
$p(x, y)$	Joint distribution density of random variables x and y
$P(\boldsymbol{r})$	Binary distribution function
P_n^m	Legendre associated polinomial
$\gamma(\alpha, x)$	Incomplete gamma – function
$\Gamma(x)$	Complete gamma – function
δ_{ij}	Kronecker symbols
ε_{ijk}	Permutation symbols
$\overset{o}{\boldsymbol{\Pi}}_{ij}$	Tensor with zero-trace matrix, that is $\overset{o}{\boldsymbol{\Pi}}_{ii} = 0$
$\boldsymbol{\Pi}^{(\alpha)}$	Antisymmetric part of tensor
$\boldsymbol{\Pi}^{(s)}$	Symmetric part of tensor
σ^2	Variance of distribution
$\Phi(\alpha; \beta; \gamma; z)$	Hypergeometrical function
$\psi(x)$	Euler psi-function
$(\ )^{\mathrm{T}}$	Transposition
$\boldsymbol{a} \cdot \boldsymbol{b} = a_i b_i$	Scalar product of vectors
$\boldsymbol{a} \times \boldsymbol{b} = \varepsilon_{\mathrm{ijk}} a_j b_j$	Vector product of vectors
$\boldsymbol{T} \cdot \boldsymbol{n} = (T_{ji} n_i)\mathbf{e}_j$	Scalar product of tensor and vector
$\boldsymbol{T} : \boldsymbol{E} = T_{ij} E_{ij}$	Full scalar product of tensors
$D/Dt = \partial/\partial t + u_i \partial/\partial x_i$	Substancial derivative
$\nabla\varphi = (\partial\varphi/\partial x_i)\mathrm{e}_i$	Gradient of scalar
$\nabla \cdot \boldsymbol{u} = \partial u_i/\partial x_i$	Divergence of vector
$\nabla \cdot \boldsymbol{T} = (\partial T_{ij}/\partial x_j)\mathbf{e}_i$	Divergence of tensor
$\nabla \boldsymbol{u} = \partial u_i/\partial x_k$	Gradient of vector
Δ	Laplace operator
∇_θ	Gradient in tangential direction to axisymmetric body

$\binom{i}{k}$	Binomial factors
∇_S	Surface gradient

Physical Constants

$N_A = 6.022 \cdot 10^{23}$	Avogadro number constant	$mole^{-1}$
$k = 1.381 \cdot 10^{-23}$	Bolzmann number	$J{\cdot}K^{-1}$
$e = 1.602 \cdot 10^{-19}$	Elementary electric charge	C
$F = 9.648 \cdot 10^{4}$	Faraday constant	$C{\cdot}mole^{-1}$
$A = 8.314$	Gas constant	$J{\cdot}mole^{-1}{\cdot}K^{-1}$
$\varepsilon_0 = 8.854 \cdot 10^{-12}$	Dielectric permittivity in vacuum	$C{\cdot}V^{-1}{\cdot}m^{-1}$
$g = 9.807$	Acceleration of gravity	$m{\cdot}s^{-2}$
$Atm = 1.013 \cdot 10^{5}$	Atmospheric pressure	Pa
$T_0 = 273.15$	Absolute temperature 0 °C	K

I
Technological Fundamentals of Preparation of Natural Hydrocarbons for Transportation

Separation of Multiphase, Multicomponent Systems. E. G. Sinaiski and E. J. Lapiga

ISBN: 978-3-527-40612-8

Introduction

The product obtained from wells on petroleum, natural gas and gas-condensate fields is invariably a multi-phase, multi-component mixture. The raw hydrocarbon material produced needs to be processed before it can be transported by pipeline and delivered to gasoline plants, oil refineries, and fractionating plants. In this context, engineers widely employ technological processes based on the principle of division (separation) of the native mixture into liquid and gaseous phases as a result of the action of intrinsic forces such as gravity or inertia.

Gas-oil and gas-condensate systems consist of petroleum and gas or gas and condensate, respectively. The state and properties of these systems are determined by various parameters, the most important of which are pressure, temperature, specific volumes and composition of the phases. The pressure and temperature change continuously during movement of the hydrocarbon system throughout the production chain, i.e. from bed to well to the system of gathering and preparation and thence to the pipeline. As a result, the phase condition of system as well as the composition of the phases change accordingly. Besides, some components of the mixture (liquid, gas, or solid phases) may be introduced into or removed from the system. This also results in a change of both the phase condition and the composition of the mixture.

Natural gas contains hydrocarbon components – methane, ethane, propane, butane, and heavier components (which are designated C_1, C_2, etc.), sour gases – carbon dioxide, hydrogen sulfides, thiols, and other components. Besides the listed components, natural gas also contains water vapor and inorganic admixtures that are removed together with the extracted from wells. The composition of natural gas from a given source does not remain constant, but changes in the course of exploitation as the reservoir pressure falls. Table I.1 gives an overview concerning rates of change of reservoir pressure through data relating to a real gas-condensate field. The numbers in the third row are the projected values of reservoir pressure.

Natural gas and the products of its processing, namely ethane, propane, butane or the wide fraction of light hydrocarbons (WFLH), as well as the condensate, represent fuels for industry and household consumers and initial raw materials for gas-processing plants. If natural gas and the products of its processing are to be used as fuels and raw materials, then they must meet certain requirements, con-

Separation of Multiphase, Multicomponent Systems. E. G. Sinaiski and E. J. Lapiga

ISBN: 978-3-527-40612-8

Table I.1

Years after beginning of the exploitation	1	5	10	15	25
Annual gas extraction, million m^3	3.9	7	7	7	6.38
Reservoir pressure, MPa	26.6–30.1	23.2–25.6	19.6–21.5	16.3–17.8	10–11.4
Wellhead pressure, MPa	16.67	12.37	9.09	6.24	1.86

cerning on the one hand the quality of commodity output and, on the other hand, the restriction of levels of possible environmental contaminations. Specifications and standards concerning natural gas are dependent on where it is delivered. Basic requirements concerning natural gas supplied through a pipeline, and salient quantitative data are presented in Table I.2.

The restriction on the dew point for hydrocarbons is stipulated for natural gas with contents of hydrocarbons C_{5+} not less than 1 g m^{-3}.

Limitations on the dew point for moisture and hydrocarbons are caused by the requirement that hydrates do not form and the condensate does not precipitate as the gas is transported at low temperature. The moisture content in a gas is determined by the given values of dew point temperature and pressure, using nomo-

Table I.2

Attributes	**Climatic area**			
	Moderate		**cold**	
	1.V–30.IX	**1.X–30.IV**	**1.V–30.IX**	**1.X–30.IX**
Dew point, °C:				
at moisture, no more	0	−5	−10	−20
at hydrocarbons, no more	0	0	−5	−10
Mass of mechanical admixtures in 1 m^3 of gas, no more (g)	0.003	0.003	0.003	0.003
Mass of hydrogen sulfide in 1 m^3 of gas, no more (g)	0.02	0.02	0.02	0.02
Mass of a sour sulfur in 1 m^3 of gas, no more (g)	0.035	0.035	0.035	0.035
Volume fraction of oxygen, no more (%)	0.1	0.1	0.1	0.1

grams or empirically-obtained calculating formulas [1]. The dew point for hydrocarbons depends not only on the pressure of the gas, but also on its composition. To find the dew point, we either use special tables or carry out the vapor-liquid balance calculation for our multi-component system [2].

Heavy hydrocarbons condensed from a gas during its extraction form a condensate enriched with a group of hydrocarbons C_{5+}. This by-product of gas-producing and gas-processing industries is an important commodity. The condensate is used as a raw material by oil refineries and in the production of natural gasoline. The fractional composition of condensates varies from one gas field to another. One should make a distinction between a stable condensate containing C_{5+}, and an unstable one, containing the lighter components excluding C_{5+}. The condensate grade is characterized by the vapor pressure p_{sat} and by the evaporation of up to 25–85% at the temperature of 50 °C and the atmospheric pressure. The vapor pressure of the stable condensate should be such as to assure its storage in the liquid state at 37.8 °C.

Thus, both natural gas and gas condensate must be thoroughly processed in order to achieve the required conditions before we can feed it into a pipeline, whether on a gas-transporting factory or in the communal distribution network. This processing includes the following procedures:

1) Isolation of mechanical admixtures from the gas, precipitation of moisture and condensate. This process is called the separation;
2) Removal of water vapors from the gas. This process is called gas dewatering (dehydration). Since dehydration causes a decrease in the threshold temperature of hydrate formation, this procedure often includes additional steps intended to prevent the formation of hydrates.
3) Extraction of heavy hydrocarbons from the gas.

The procedures detailed here are performed by special devices for the complex preparation of gas (DCPG), which are located at the gas field. In their heavy concentration of machinery and the complexity of technological processes involved, DCPG's come close to industrial plants.

Petroleum is a heavier liquid than gas condensate because it contains much more oils, paraffin and other high-molecular compounds. Many types of petroleum are more than 99% hydrocarbons, of which paraffin and the naphthenic series are most widely submitted. Other classes of organic compounds – oxygen compounds, sulphurous compounds, asphalt-tars and others – are also present in small amounts. The majority of sulphurous and oxygen-containing compounds are surface-active compounds. They are aggressive with respect to metal and cause heavy corrosion. Yet another common admixture in petroleum is mineralized water, which causes significant complications for its collection and transportation. The harmful feature of oil-field brines is their ability to form water-oil emulsion that complicates preparation and refining of oil, as well as the movement of petroleum in pipelines (water can accumulate in bends and then freeze, resulting in pipeline breakage). The surface-active agents (surfactants) promote formation of emulsions and therefore are called emulsifiers. The presence of surfactants in oil facilitates formation of emulsions and increases their stability (the property to keep emulsion for a long time). Oil also contains low-molecular components, which are especially abundant in light oil. These components can be present in liquid as well as in gaseous phase. The change of pressure and temperatures during the movement of petroleum throughout the chain: bed–well–system of gathering and preparation-pipeline results in intense separation of the lighter components, accompanied by an increase in the gas factor (gas volume per unit volume of oil mixture, m^3/m^3).

The presence of free gas in oil (oil gas) also causes complications in production, gathering, preparation, and transportation of oil. A gas breakthrough in productive wells from the gas cap of the geological horizon or from gas-bearing horizons is sometimes observed. It leads to an increase of the gas factor of produced oil.

In Russia, the government-imposed standards (SAST) define petroleum as fit for delivery to oil refineries if it contains no more than 0.1% of water and no more than 40 mg $liter^{-1}$ of chlorine salts. Besides these two requirements there are also others. Therefore, before petroleum is pumped into the pipeline, it must be subjected to processing that includes the following operations:

1) Removal of the light gases contained in the oil in free or dissolved state. This process is called separation;
2) Removal of water from the oil. This process is called oil dehydration or dewatering;
3) Extraction of salts dissolved in the oil. This process is called oil desalting.

More information about the properties and processes of gathering and preparation of natural gases and oils can be found in Refs. [3, 4].

1
Technological Schemes of Complex Oil, Gas and Condensate Processing Plants

The product obtained from wells on gas and gas condensate fields is always a complex heterogeneous mixture that contains a mixture of gases (saturated with water vapor and heavy hydrocarbons), liquid hydrocarbons (oil or gas condensate) and water, solid particles of rock, and other components. In order for the gas delivered to the consumer to meet all requirements placed upon it, it is necessary to eliminate the solid and liquid phases and also a part of the water vapor and heavy hydrocarbons before feeding it in the gas-transmission pipeline. These processes are carried out in special field devices of the complex preparation of a gas and condensate (DCPG). The typical design of the DCPG apparatus at a gas-condensate field is shown schematically in Fig. 1.1. The preparation of the gas is carried out by the method of low-temperature separations (LTS).

In Fig. 1.1 the following designations are used: S1, S2, S3 are separators used in the I, II, and III step, respectively; D1 and D2 are the three-phase dividers for the I and II steps; T1, T2, T3 are throttles; H1, H2, H3 are heat exchangers; TEA is a turbo-expander set (or evaporative refrigerator); F is flare.

Table 1.1 shows the approximate values of pressure and temperatures at the relevant points of the technological scheme.

Gas enters the DCPG with the temperature of 35 °C at the pressure of 15.5 MPa. Before the Step I separator a throttle (T1) is located. Here gas pressure is reduced to 13.1 Mpa, and the temperature drops to 21 °C. In the Step I separator (S1) drops of liquid are separated from the gas. The dehydrated gas is cooled in the heat exchanger H1 to 0 °C, and the condensate that forms as a result of cooling is separated from the gas in the Step II separator S3. The condensate from the first two separators gets collected and, passing through throttle T2, goes to the three-phase divider D1, where gas containing light hydrocarbons is separated from the condensate.

Condensate stabilization occurs in the divider. Deep cooling of gas (up to −30 °C) occurs as the result of consecutive cooling in the heat exchanger H2 and throttling in T3. As the wellhead pressure decreases with time, the pressure in DCPG also falls. Therefore after a while, the pressure in T3 might no longer be sufficient to achieve the required temperature of −30 °C. To overcome this predicament, the scheme envisages the possibility of adding a turboexpander TEA for additional cooling of gas. Cooled gas with high flow rate interfuses in the ejec-

Separation of Multiphase, Multicomponent Systems. E. G. Sinaiski and E. J. Lapiga

ISBN: 978-3-527-40612-8

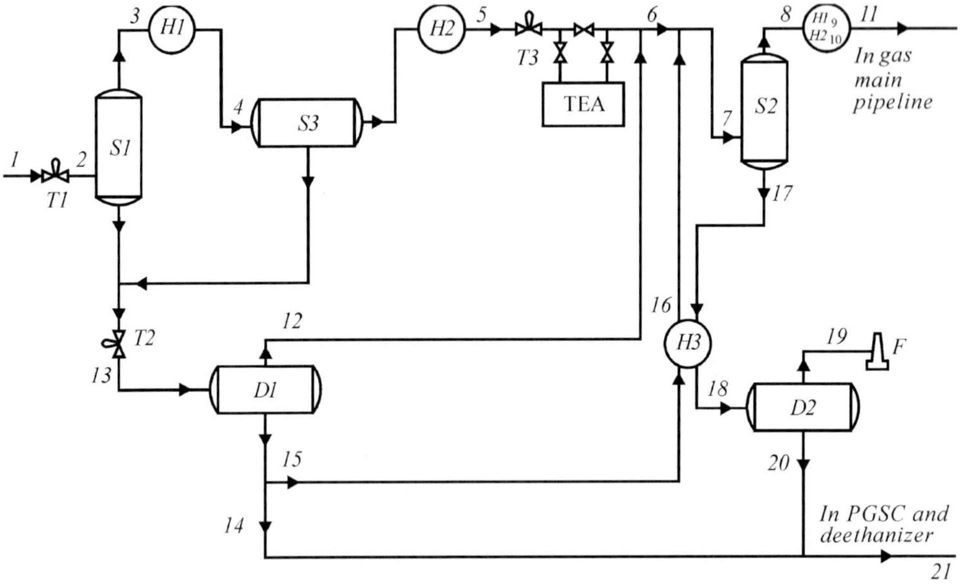

Fig. 1.1 Scheme of the low-temperature separation (LTS).

Table 1.1

Point	**1**	**2**	**3**	**4**	**5**	**6**	**7**	**8**	**9**	**10**	**11**
Pressure, Mpa	17	13	13	13	13	8	7.8	7.8	7.7	7.5	7.4
Temperature, °C	35	21	21	0	−15	−30	−30	−30	−9	−12	−10
Point	**12**	**13**	**14**	**15**	**16**	**17**	**18**	**19**	**20**	**21**	
Pressure, Mpa	7.8	8	7.2	7.9	7.8	7.8	7.8	7.6	7.2	7.2	
Temperature, °C	13	13	13	13	−22	−30	−25	−25	−25	−1	

tor with the gas that comes from divider D1 with a relatively low flow rate. Step 3 separator (a.k.a. the end separator) is responsible for the final separation of condensate from the gas. The separated gas gets heated in the heat exchangers H1 and H2 and goes into the gas pipeline, while the condensate is channelled to the three-phase divider D2. The free gas from D2 goes toward the flare F, or to the ejector where it interfuses with the basic stream of the separation gas. The condensate from D1 and D2 goes into the condensate pipeline and next to the plant for gathering and stabilization of condensate (PGSC) or to a condensate deethanizer.

Let's note some more features of the above-described technological scheme of DCPG. The heat exchangers used in the scheme (except for H1) cool or warm up the moving medium (gas or liquid), employing the already cooled (or heated) streams. So, for example, the cooling of gas in H3 is carried out with the help of the cold condensate that comes from the end separator S3. As a rule, engineers use the shell-and-tube heat exchanger where the previously cooled (or heated) gas or liquid moves in internal tubes. The actuating medium (gas or liquid) is moving in the intertubular space and is cooled (or heated) by contact with the tube surface. The second peculiarity of the scheme is the opportunity to feed a part of the condensate gathered in divider D1 through the ejector into the stream of separated gas. Since the temperature of the condensate is higher than the temperature of the separated gas (compare conditions in points 6 and 16), the condensate injected into the gas can actively absorb the heavy hydrocarbons (C_{3+} and C_{5+}). Thus, it helps achieve one of the purposes of gas preparation – to extract from the gas as many heavy hydrocarbons as possible.

In the scheme represented in Fig. 1.1, a plant for gas dehydration is absent. Dehydration is carried out by injecting highly concentrated methanol solution into the gas stream before the DCPG. In this way we achieve two goals at once. First, water vapors are eliminated from the gaseous phase; secondly, the temperature of hydrate formation is reduced. Some technological schemes call for the installation of special absorbing units that utilize liquid absorbents for gas dehydration. Diethyleneglycol (DEG) and triethylene glycol (TEG) are the most common absorbents. The use of glycols results in the necessity to establish additional equipment for capture, removal, and regeneration of exhausted absorbent.

For a natural gas field with methane content of more than 95% and small contents of heavy hydrocarbons, there is no need for such complex DCPGs. For such gas fields, the gas processing scheme needs only to ensure the separation and dehydration of gas.

Another scheme of gas preparation on the gas-condensate field is presented in Fig. 1.2. In this scheme the process of low-temperature absorption (LTA) is used.

In Fig. 1.2, we introduce the following designations: T1, T2, T3, T4 are throttles; H1, H2, H3 are heat exchangers; A1 is the absorber for gas dehydration; A2 is the absorber for the extraction of heavy hydrocarbons; T is the turbine; S1, S2, S3 are separators; D1, D2 are three-phase dividers; DA is the deflationer of the absorber; DPC is a complex that includes a compressor of the turboexpander, an air refrigerator and a heat exchanger. The gas at the entrance to DPC has the pressure of 15 MPa and the temperature of 26.8 °C (Table 1.2).

Before the entrance to separator S1, gas is throttled through T1, achieving the pressure of 10.5 MPa and the temperatures of 15 °C. The separated gas leaving S1 is then dehydrated in the absorber A1, and then, after consecutive compression in the compressor and cooling in the heat exchanger, enters the intermediate separator S2. The separated gas is further cooled in TEA as a result of performing work on the turbine and then moves to the bottom part of the contact chamber of

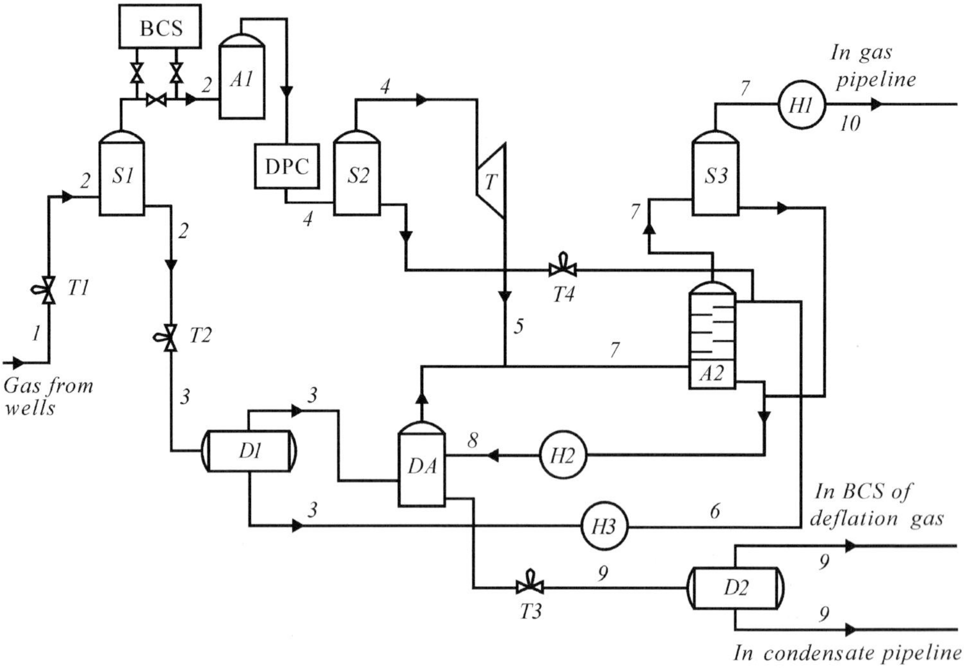

Fig. 1.2 Scheme of the low-temperature absorption (LTA).

Table 1.2

point	1	2	3	4	5	6	7	8	9	10
p, Mpa	15	10.5	7.7	12	7.65	7.7	7.65	7.7	3.5	7.4
T, °C	26.8	15	11.56	−3.4	−25.1	−15.1	−23.9	5.8	−3.9	−2.1

the hydrocarbon absorber A2. The condensate, which comes from the divider D1 and from Step 2 separator S2, enters the top part of the contact chamber. As a result of contact between the gas and the condensate, where the latter serves as an absorbent, heavy hydrocarbons are extracted from the gas. Afterwards, the gas from A2 is separated in the end separator S3, undergoes heating in H1, and moves into the gas pipeline. The used condensate is removed from absorber A2, undergoes heating in H2, and enters the deflationer DA, where light hydrocarbons evaporate from the condensate.

The obtained vaporous phase is pushed again into the contact chamber of the absorber to eliminate from it any remaining heavy hydrocarbons. The condensate formed (after aeration) in DA is throttled through T3 and goes to the divider D2 where its final stabilization occurs. From D2, the condensate goes into the condensate pipeline, and the vaporous phase goes to the booster compressor station BSC.

The scheme considered above calls for the installation of the booster compressor station (BCS), which starts to operate when the pressure at the entrance of DCPG will be less than the required pressure in the entrance separator.

As you can see, the basic distinction of the scheme in Fig. 1.2 from the scheme in Fig. 1.1 consists in the following:

1. The process of gas dehydration occurs in the special absorber A1, which ensures a higher quality of dehydration.
2. The extraction of heavy hydrocarbons from the gas is conducted in a special multistage absorber equipped with devices for condensate deflation. A higher efficiency of extraction of heavy hydrocarbons from the separated gas and a higher quality of condensate stabilization is achieved as the result.
3. The employment of turboexpander (TDA) for gas cooling allows us to decrease pressure by 4 MPa and to cool the gas by more than 20 °C. It should be noted that the same pressure drop on a throttle would cool the gas by only 12 °C.

It is worthwhile to note that the blocks A2, S3, and DA are installed in one three-stage vertical apparatus, which allows for a considerable reduction of the area set aside for the DCPG installation.

Modern on-site plants that perform complex preparation of gas and condensate come close to specialized factories in terms of complexity, multifunctionality, and quality of treatment of natural hydrocarbons. In Russia, the basic creator of the equipment for DCPGs is the central engineering office of the oil-production machinery company (CEOM).

Thus, in the process of gas and condensate preparation for transportation the most common elements of equipment are separators, dividers, absorbers, heat exchangers and turboexpander aggregates.

The oil preparation scheme should also provide for the arrangement of processes of oil degassing, dehydration, and demineralization (desalting). Figure 1.3

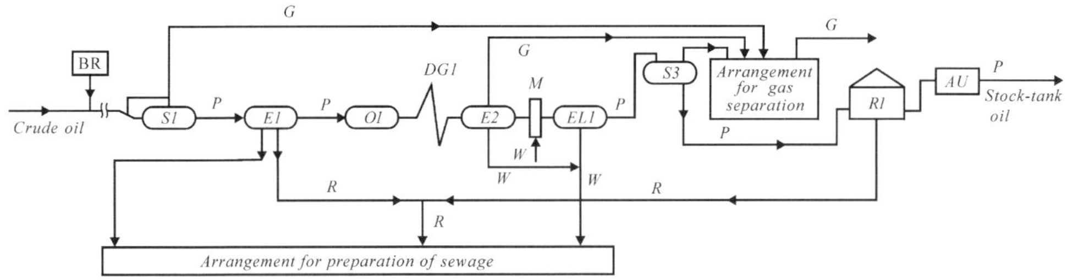

Fig. 1.3 Scheme of oil preparation: *P* – oil, *W* – water, *R* – residuum, *G* – gas.

shows just such a typical technological scheme for the preparation of oil. This scheme includes the following elements: BR is the dosage block for the reagent that destroys the water-oil emulsion in the gathering collector; S1 is the 1st stage separator; DG1 is the drop-generator; E1 is the settler for preliminary oil dehydration; O1 is the furnace for emulsion heating; E2 is settler for deep dehydration; M is the agitator for mixing fresh water W with dewatered oil for its preliminary demineralization; EL1 is the electrodehydrator for deep oil demineralization; S3 is hot third stage separator; R1 is the reception tank for stock-tank oil; AU is the automatic device that measures the quality and quantity of oil. A gas separation arrangement is provided as well. There is also an arrangement for the preparation of sewage, i.e. water formed during dewatering and used in the course of demineralization.

A few words must be said about some special features of oil preparation process. Degassing of oil (oil separation) is usually carried out in several separators connected in series. In each separator, a pressure drop occurs, as a result of successive extraction of a certain fraction of light hydrocarbon components from oil. If degassing occurs without gas being siphoned off from the system then such separation is called a "contact separation". If in the course of degassing the separated gas is siphoned off, then the procedure is referred to as "differential separations". The other feature is the organization of oil dehydration and oil demineralization processes. The main challenge here is to eliminate small water droplets from oil. For this purpose, one can reduce the viscosity of oil by warming it. Another method is cohesion of water drops into a liquid emulsion, with the help of an electric field in an electrical dehydrator.

Analysis and mathematical modeling of processes occurring in the abovementioned technological schemes are the subject of the book. They are successively considered in their corresponding sections. Further on, we shall consider the basic mechanisms of separation processes for multi-phase, multi-component hydrocarbonic systems to which natural gases and oil belong. In the next chapter, we consider constructional features of some elements of technological schemes: separators, dividers, settlers, absorbers, heat exchangers, and turboexpanders.

2
Construction of Typical Apparatuses

2.1
Separators, Dividers, and Settlers

Separators appear among the most indispensable elements of any technological scheme of on-site preparation of oil and natural gas on oil and gas-condensate fields. Separators are the necessary pieces of equipment in processes of gas condensate treatment, gas compression and gas cooling at the final stage of field exploitation, in cycling-processing plants [5–9].

Depending on the kind of the substance they are designed to process, separators can be labeled as "gas-oil separators" and "gas separators" (Fig. 2.1). Gas-oil separators are used for separation of oil from oil gas, and gas separators – for separation of natural gas from droplets of condensate, water, and solid particles. Gas separators, as a rule, process gas-liquid mixtures with a rather small content of the liquid phase. But they can also operate in the so-called flood regime (when a large volume of liquid enters the separator as a result of accidental injection of water or accumulation of condensate in pipes).

Separators consist of several sections, with each section assigned its own function. The input section for gas-liquid mixture is assigned the task of scattering apart the coarsely-dispersed phase to the maximum possible degree (especially in case of high initial content of the liquid phase), and ensuring a uniform input of the gas-liquid mixture into the apparatus (including, but not limited to, the final clearing section). The section for small droplet coagulation is located in the sedimentation zone before the settling section. It is intended for the integration of fine droplets, splitting of large drops and for smoothing the flow of gas that is being fed into the final clearing section. The final clearing section provides the required efficiency of separation in the project range of loads both on gas, and on liquid. The section of gathering of separated liquid ensures steady work of regulating devices and the signaling system that detects the top and bottom threshold levels. This section also performs degassing of the gathered liquid, prevents funnel formation at the entrance, and, whenever necessary, also separates the liquid into individual fractions.

Depending on their configuration, separators can be classified as spherical or cylindrical, and depending on their spatial arrangement – as horizontal or vertical.

Separation of Multiphase, Multicomponent Systems. E. G. Sinaiski and E. J. Lapiga

ISBN: 978-3-527-40612-8

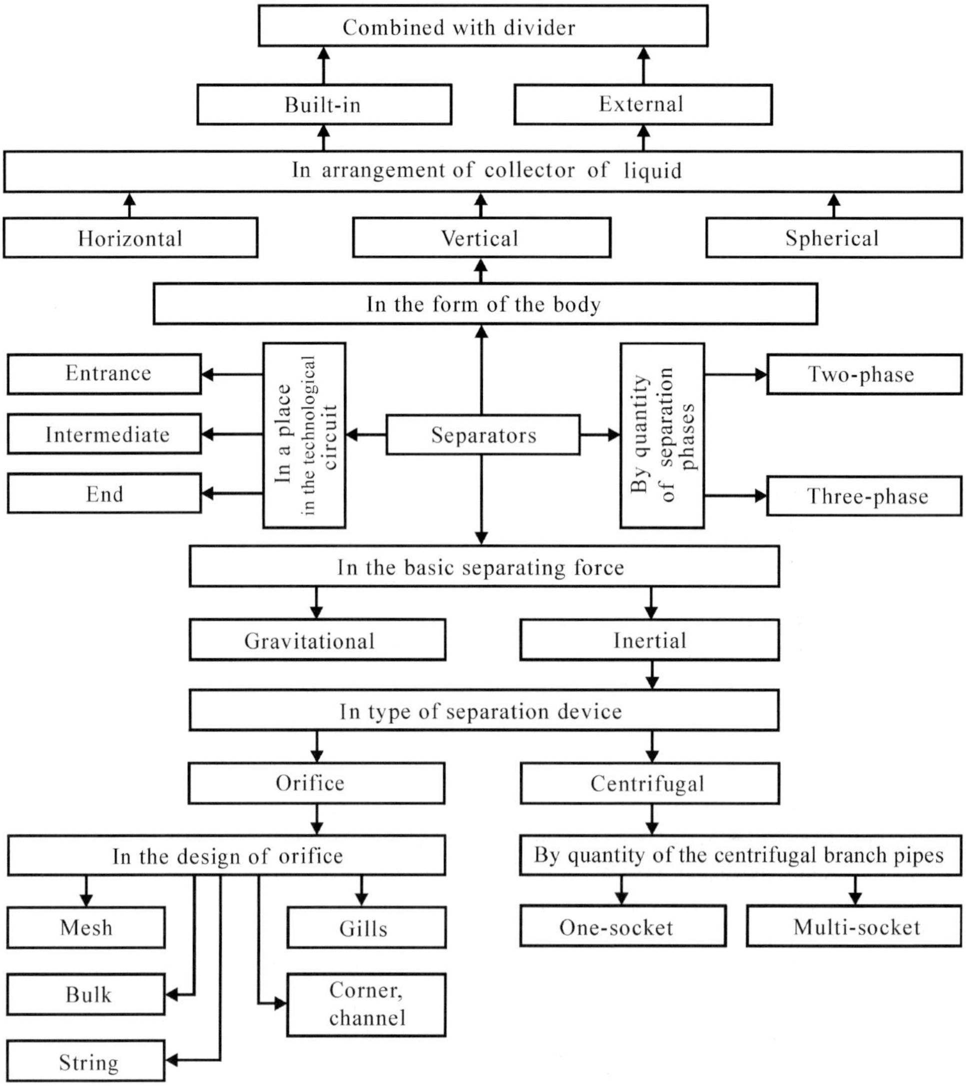

Fig. 2.1 Classification of separators.

Spherical separators are more compact and require less metal to manufacture. The higher the pressure load and the greater the production output of the separator, the more manifest these advantages become. The basic disadvantage of spherical separators is their awkward spatial configuration that excludes the possibility to arrange coagulation and gathering sections in the frame. As a result, the efficiency of processing of gas streams with finely-dispersed liquid phase or with the initial liquid content of more than 100 sm^3/m^3 in spherical separators is minimal.

The basic advantage of horizontal cylindrical separators is their high productivity. They are intended for separation of gas-liquid mixtures with high liquid content and for separation of liquids that tend to produce foam. The shortcoming of horizontal separators is the difficulty of removal of solid admixtures. This drawback is absent in vertical cylindrical separators. The elliptic bottom of these separators provides for the accumulation of liquid and solid impurities in the bottom part of the device and their removal via the drainage system. The liquid gathering section can be taken outside the frame of the separator and its shape can be either spherical or cylindrical (usually a horizontal cylinder).

Spherical collectors are mostly used when high pressures are involved.

Sedimentation of drops in a gravitational separator occurs chiefly due to the force of gravity. The larger the size of liquid drops in a gas stream and the lower the stream's velocity inside the separator, the greater the efficiency of separation. Therefore, it may be difficult to achieve high efficiency in a gravitational separator at high gas flow rates or if there is too much breakage of large drops as the gas moves inside the main field gathering pipeline.

In inertial separators, the separation of liquid from gas occurs mainly due to the action of inertial (most often, centrifugal) forces. Basic elements of orifice inertial separators are orifices of various constructions, which are installed in the final gas clearing section. Separators installed at oil fields are increasingly using, among others, the so-called string orifices, which are essentially sets of frames with 0.3–0.5 mm wire coiled around them.

Efficiency of orifice inertial separators mostly depends on the orifice design and on spatial arrangement of orifices inside the separator and can reach 99.5%–99.8% at gas velocities 3–5 times above the gas velocity in gravitational separators. High efficiency of these separators is predetermined by large area of contact of separating elements with the gas-liquid flow, which ensures separation of drops of diameter above 3–5 μ for meshes and above 10–20 μ for gills.

Centrifugal separators use swirlers to transform translational motion into rotational one. The chief advantage of centrifugal separators is high operating velocity of the gas in the body of the centrifugal element. Owing to the action of centrifugal forces, it is possible to perform separation for gas drops of diameter above 10–20 microns. The efficiency of centrifugal separators at high pressure varies from 80% to 99%. To improve efficiency, separators are equipped with centrifugal elements of small diameter – cyclones.

Since practically all separators are equipped with droplet catchers of different construction, let us consider them in more detail.

Gravitational gas separators can be supplied with Dickson's plane-parallel plates (Fig. 2.2, a), allowing for a slight increase in the operating velocity and efficiency of sedimentation of droplets, and with gills orifices that enlarge the operating area in comparison with the previous construction, increasing the gas flow rate and the efficiency of separation. Bent surface of gills helps to improve efficiency by utilizing the inertia of droplets. In addition to maximizing droplet sedimentation, it is necessary to reduce the clearance between surfaces of gills facing downstream. In horizontal-body separators, gills orifices are arranged in the cross section (Fig. 2.2, b) or bilaterally along the main axis of the device (Fig. 2.2, c).

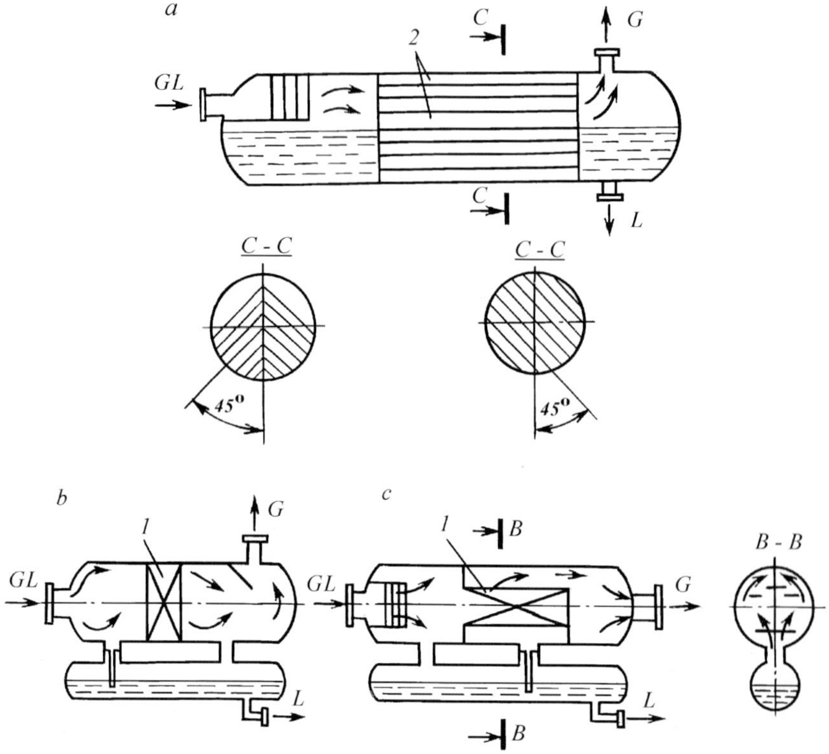

Fig. 2.2 Gas-liquid gravitational separators: *1* – Dickson's plates, *2* – gill orifices.

Vertical gas separators with gill orifices (Fig. 2.3) differ from each other mainly by the arrangement of gills stacks and gas inlet and outlet branch pipes relative to the orifices. For the purpose of increasing the flow rate, gills orifices can be made either one- or two-sectional (Fig. 2.3, a, b). The separator in Fig. 2.3, c differs from the previous ones by that the orifices are arranged in a ring, which improves separation efficiency. Gill separators are sometimes used in plants of LTS as entrance, intermediate and end stages, but their basic application is preliminary separation of liquid from gas. This is why gill separators are mostly used to clean field or factory gas from liquid admixtures. In some cases, the use of such separators represents a preliminary step when separating liquid from gas at gas compressor stations on the main pipeline.

Separators with horizontal mesh orifices (Fig. 2.4) differ from each other mostly by the structure of the inlet section: a baffle plate with incoming stream making a 90° turn (Fig. 2.4, a); a conic shell, with the stream turning 180° (Fig. 2.4, b); a cylindrical shell, stream swirler placed between the separator body and the shell, and a 180° turn of the stream (Fig. 2.4, c).

Vertical mesh separators are used at gas fields, as the end separators in LTS plants, as intermediate and end separators at gas-processing factories (GPF), and for clearing the gas from liquid before its delivery to the flare. They can also be

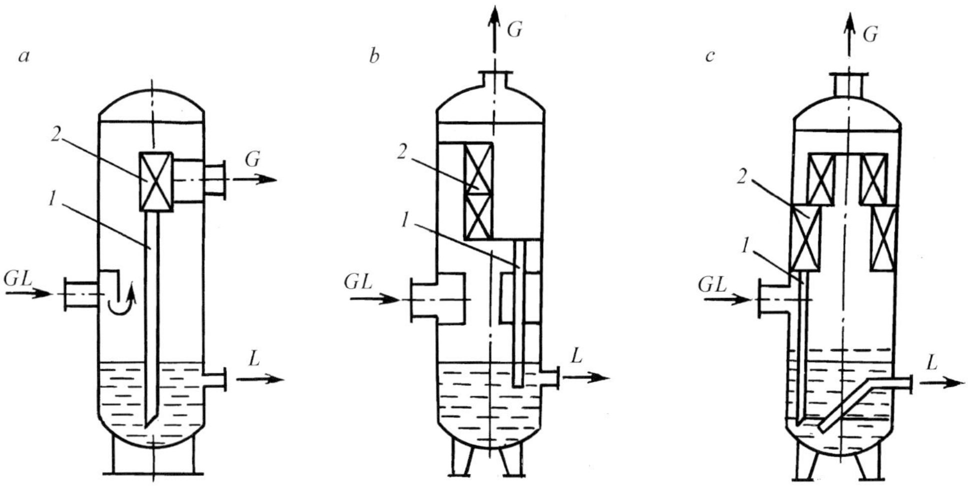

Fig. 2.3 Vertical gas separators with gill (louver) orifice: 1 – entrainment trap, 2 – gills orifice.

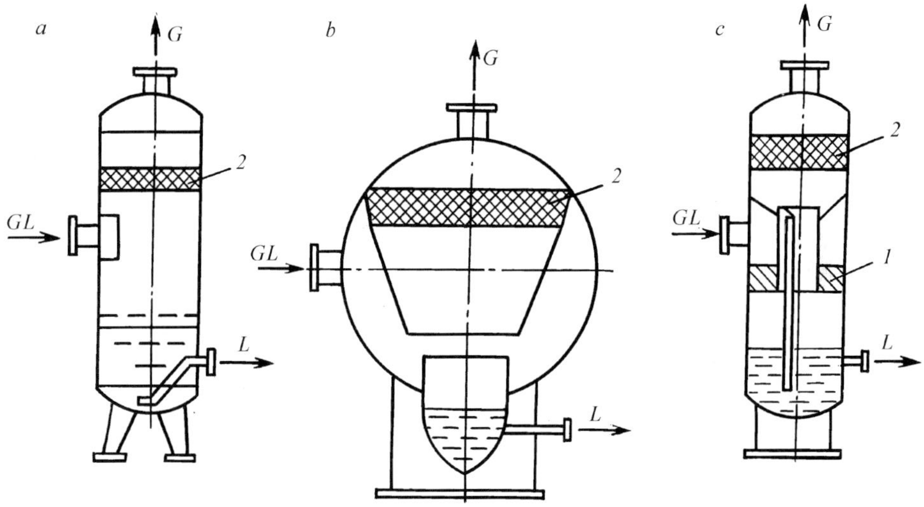

Fig. 2.4 Separators with horizontal grid orifice: 1 – swirler, 2 – grid orifice.

used as entrance separators whenever highly efficient clearing of gas is needed before the gas is taken to the absorber for dehydration.

Sketches of one-socket centrifugal direct-flow gas separators without the device for preliminary separation of liquid are shown in Fig. 2.5. Among them are separators in which the swirling stream is moving horizontally (Fig. 2.5, a, b) and separators in which clearing of the gas from liquid occurs in a swirling stream that moves downward (Fig. 2.5, b). The controlled swirler 1 is designed to maintain a constant effective rate of separation as the gas flow rate and pressure change; and

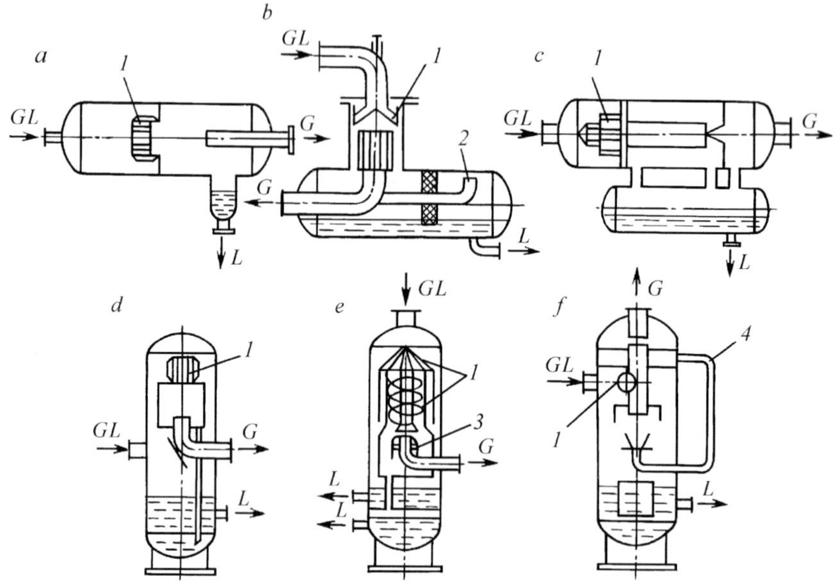

Fig. 2.5 Centrifugal one-socket separators.

the suction channel 2 works to increase the efficiency and the flow rate of the separator. A centrifugal element of the cascade type (Fig. 2.5, c) allows to perform efficient clearing of gas by lowering the hydraulic resistance of the device. Separators depicted in Fig. 2.5, d, e, f, on the other hand, are equipped with devices for preliminary separation of liquid-baffle plates causing a 90° turn of the stream. The presence of apertures 3 (Fig. 2.5, e) and the gas recycling pipe (Fig. 2.5, c) helps increase the flow rate and efficiency of separation. Separators can be made with a spherical collector for liquid (Fig. 2.5, b, e, c), which allows to increase the flow rate and to reduce the weight of devices.

Separators shown in Fig. 2.6 are centrifugal multisocket (multicyclone) constructions, from which the first two are not equipped with any devices for preliminary separation of admixtures. In the first design (Fig. 2.6, a), a set of cyclones is used; the gas is fed into each cyclone directly from the gas entrance pipe. Gas purifiers in the device in Fig. 2.6, b are cyclone elements (of the counterflow type) mounted on a grate plate that is located below the branch pipe for gas input. Output tubes of these elements are fixed at the grate plate located above the gas input pipe. Because practice has shown that swirlers easily get choked with various impurities, some constructions (Fig. 2.6, c) are equipped with a special chamber for preliminary separation of impurities and disposition of cyclone swirlers at the top grate plate. However, even with these devices, cases of blocking of cyclone swirlers with the resulting decrease of separation efficiency are still observed. In devices where the basic separation unit is designed in the form of co-current centrifugal elements, (Fig. 2.6, d, e, f, g), suction of a part of the gas (Fig. 2.6, d, e)

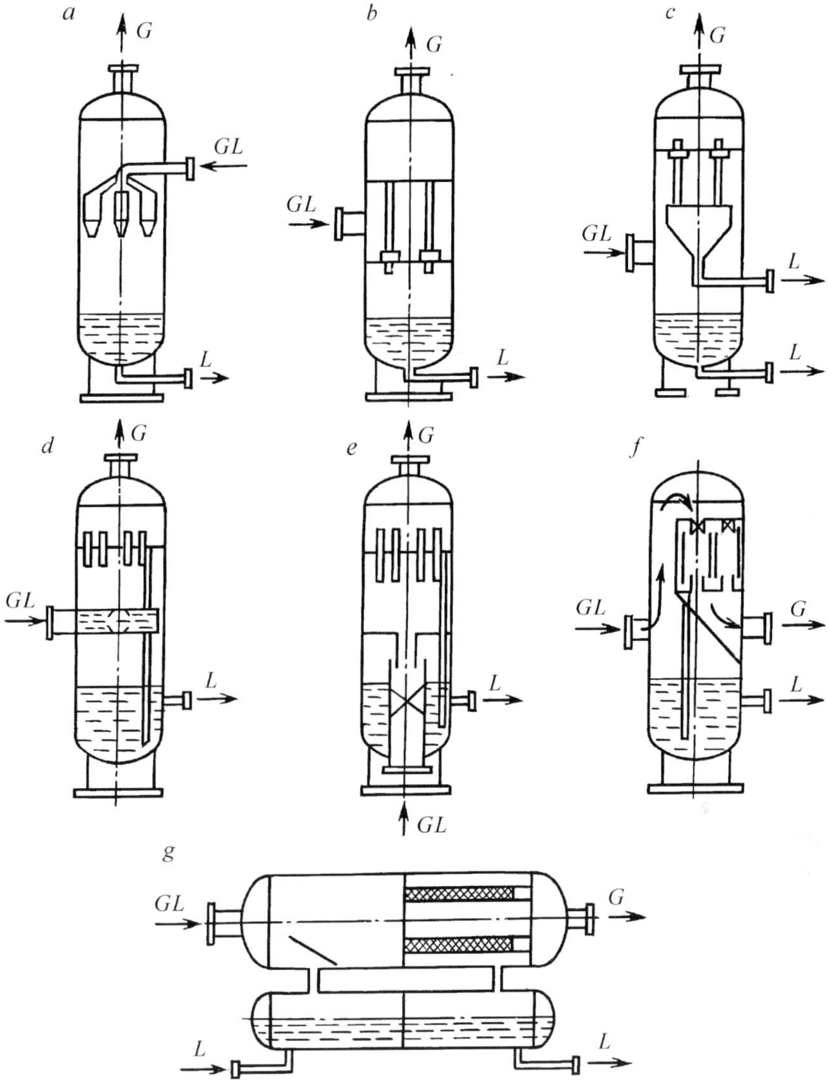

Fig. 2.6 Centrifugal multisocket separators.

and its recirculation (Fig. 2.6, f, g) is a part of the scheme. Separators in Fig. 2.6, d use branch pipes with axial and tangential swirlers.

Preliminary clearing of gas is carried out by using a radial-aperture input device, through which the gas passes into the free volume of the separator. In separators with centrifugal elements of the cascade type (Fig. 2.6, e), preliminary clearing of gas is carried out with the help of centrifugal forces in the gas supply pipe, which serves as an independent centrifugal separation element. Some separators employ direct-flow elements placed between the two grates (Fig. 2.6, f, g).

These elements are supplied with recirculation channels and can provide efficient gas clearing; their arrangement can be either horizontal or vertical, with downward or upward gas flow in the latter case. Preliminary clearing of the gas also occurs in the free volume due to the impact of the stream on the membrane that causes the incoming stream to make a 90° turn (Fig. 2.6, f).

Centrifugal separators are mostly used as preparatory and intermediate clearing stages at plants for complex preparation of gas, and also in major gas pipelines. Sometimes, however, centrifugal separators are used in the final stage of gas clearing.

A more thorough separation of liquid from gas is carried out in separators supplied with a mesh coagulator (Fig. 2.7) or a glass-fiber coagulator (Fig. 2.8). A horizontal mesh coagulator (Fig. 2.7, a, b) is placed inside the separator shell, coaxial

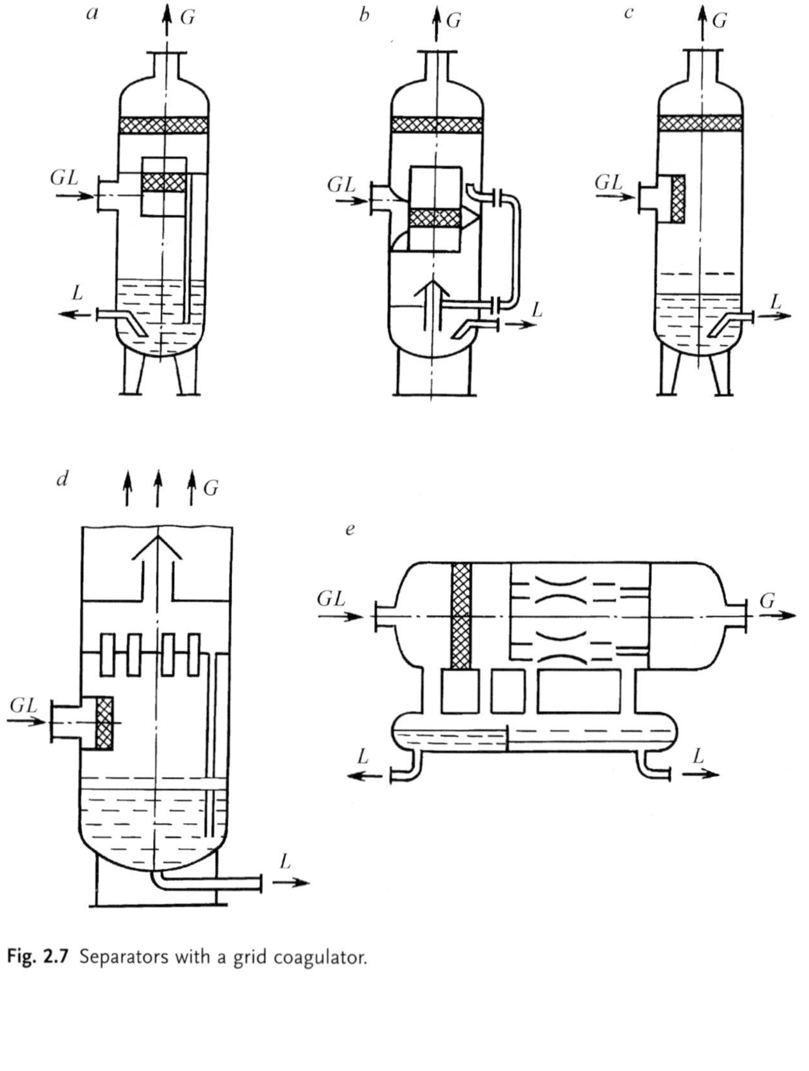

Fig. 2.7 Separators with a grid coagulator.

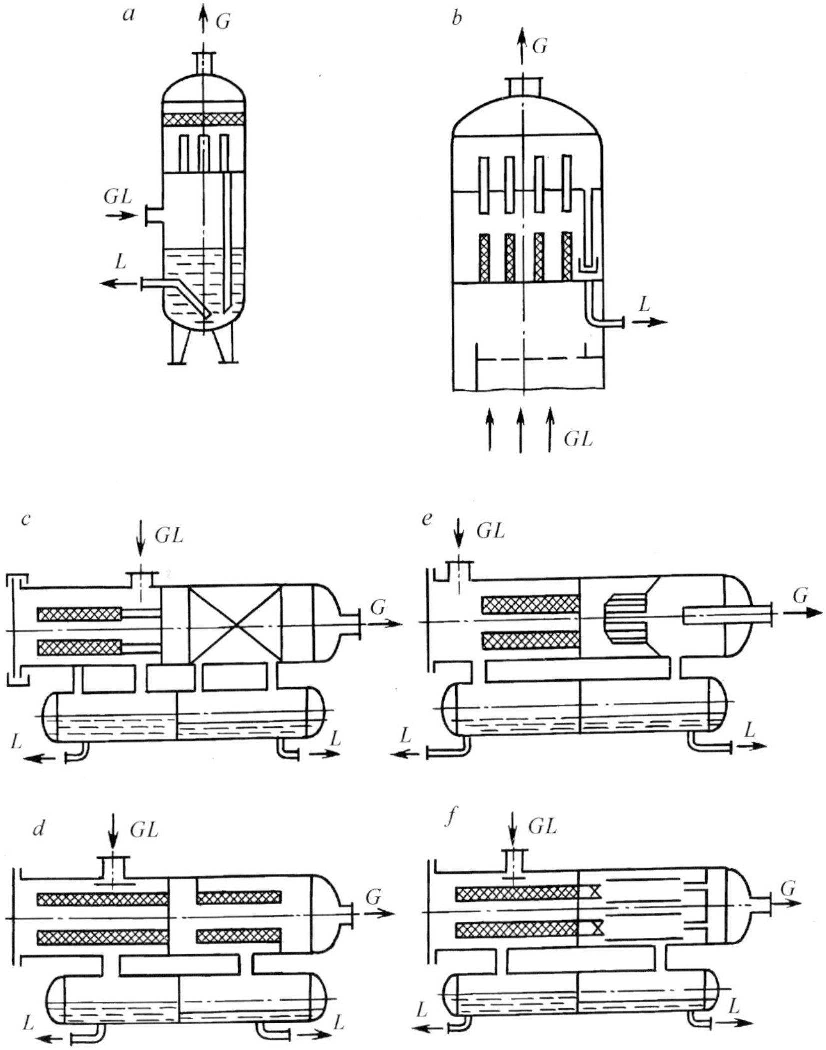

Fig. 2.8 Separator with the glass-fibred coagulator.

to the body, and acts as a device for preliminary separation of liquid from gas, while also accelerating the integration of fine droplets. At the entrance to the separator in Fig. 2.7, b, a screw swirler is located.

The section for final treatment of gas is made in the form of a horizontal grid orifice. Vertical grid coagulators (Fig. 2.7, c, d, e) are located in separators behind the gas input branch pipe in the apparatus. Preliminary separation of liquid from gas is carried out in the free volume (Fig. 2.7, e). The final separation of liquid is performed in a grid orifice or in centrifugal direct-flow elements. The final gas clearing section is made in the form of a horizontal grid orifice.

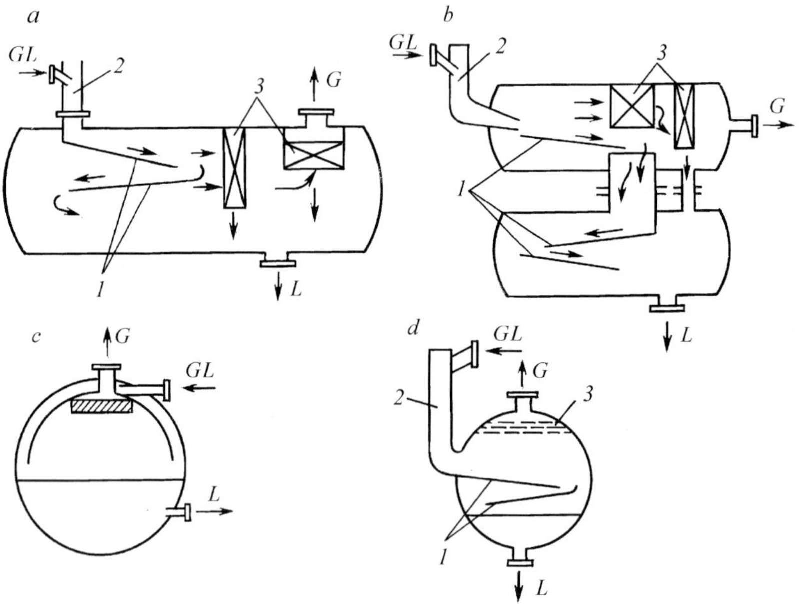

Fig. 2.9 Oil-and-gas hydrocyclone separators: *1* – inclined planes (shelves), *2* – hydrocyclone, *3* – entrainment trap orifice.

When extremely small droplets are present in the gas, we can use separators with a glass-fiber coagulator that promotes droplet integration (Fig. 2.8). The final separation of enlarged drops occurs in the base separation section that can be made in form of a mesh stack (Fig. 2.8, a, d), a centrifugal construction (Fig. 2.8, b, e, c), or gills (Fig. 2.8, c).

The idea of utilizing the centrifugal force for separation of liquid from gas also finds application in gas-oil hydrocyclone separators (Fig. 2.9). In order to create a centrifugal stream inside these separators, an attached unit – a hydrocyclone – is installed on their lateral surface (Fig. 2.9, a, b, c). A hydrocyclone is usually a vertical device with a flat tangential input and directing branch pipe for gas flow reversal in the top part, and a liquid cross-flow section in the bottom part. The technological reservoir is made in the form of a horizontal separator with various devices for additional separation of liquid from gas that are typical for oil-gas separators.

The constructions of some centrifugal elements used in gas separators are shown in Fig. 2.10. Branch pipes with axial and tangential swirlers are employed in multisocket separators shown in Fig. 2.6.

Oil-gas separators have to process a gas-liquid mixture with a rather small content of gas; it is therefore appropriate to call it a liquid-gas mixture. Gas is present in it in the form of bubbles. The separation of oil from gas occurs mostly due to the gravitational force. The scheme of gas-oil separators usually includes a deflector, baffles, a tangentially located branch entrance pipe, and other devices that

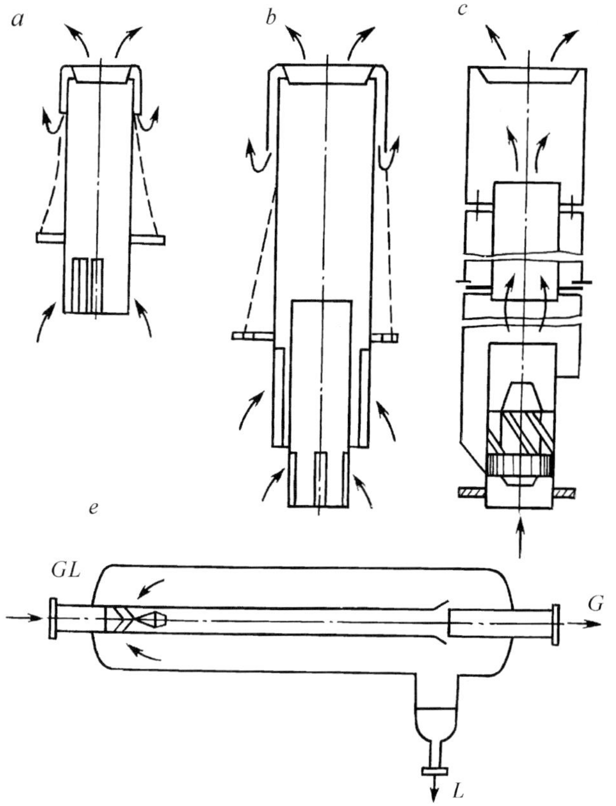

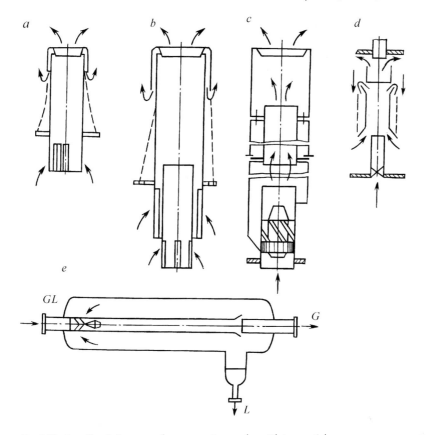

Fig. 2.10 Centrifugal elements of gas separators: *a*, *b* – with tangential swirler; *c*, *d*, *e* – with axial swirler; *d*, *e* – with recirculation channel.

promote increased efficiency of separation of gas from oil by employing inertia and adhesion forces.

Electromagnetic, pneumatic (vibratory), and ultrasonic sources are used to intensify coalescence of gas bubbles, with their subsequent integration and fast extraction from oil. The sedimentation section of a gas-oil separator is often equipped with inclined planes (plates, shelves) in order to reduce the oil layer thickness and promote a faster release of gas. Liquid drops captured by the gas that is being extracted from oil are swept away by the gas stream. That's why the top part of the separator features entrainment traps (orifices in the form of grids, box irons, jalousies, etc., designed to separate the liquid drops entrained by the gas stream) that are installed before the gas branch pipe. The arrival of the oil-gas mixture in the reception section is frequently accompanied by the formation of foam. The foam complicates the process of extracting gas from oil. To avoid this problem, separators are equipped with special partitions, grates, and other similar devices to suppress foam formation, and also to decrease mixing.

Basic diagrams of gravitational vertical separators with radial-aperture input, and of horizontal plate-shaped separators, are presented on Fig. 2.11. In order to

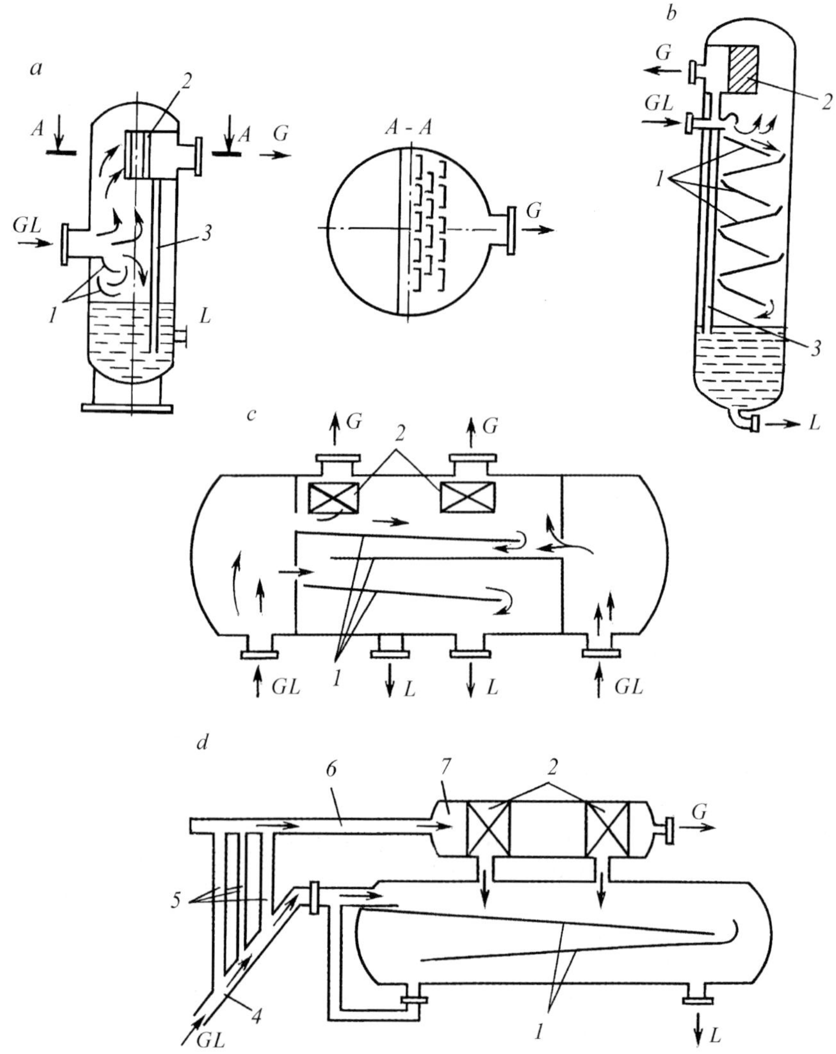

Fig. 2.11 Gravitational oil-gas separators: *a*, *b* – vertical, with radial-aperture input; *c*, *d* – horizontal, disk type; *1* – inclined planes; *2* – entrainment trap orifices; *3* – liquid drainage; *4* – supply pipeline; *5* – gas-bleeding tubes; *6* – built-up collector (depulser); *7* – spray catcher body.

increase operating efficiency of oil-gas separators intended for processing of oil with high content of gas, the gas released from oil in the supply pipeline is removed just before the separator entrance (Fig. 2.11, d). In this case the gas containing droplets of liquid goes to the entrainment trap placed above the separator. The device for preliminary removal of gas also allows us to reduce pulsations in the stream and to provide a more uniform input of the oil-gas mixture into the

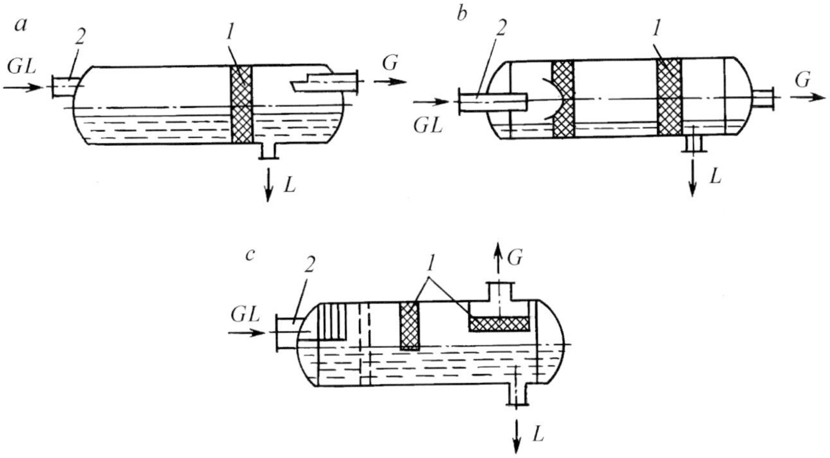

Fig. 2.12 Entrainment traps with mesh orifices: *a*, *b* – for final separation of liquid; *c* – for preliminary separation and drop coagulation; *1* – mesh orifice; *2* – branch pipe of mixture input.

separator. This is why such a device is called the depulser. A dispenser, where fragmentation of the mixture takes place, is frequently installed before the separator entrance. This increases the area of oil-gas contact surface. Entrainment traps often rely on mesh orifices (Fig. 2.12) to separate oil droplets from gas.

Nowadays, entrainment traps with string orifices are adopted. The vertical mesh orifice can be used as the section for final separation of liquid drops (Fig. 2.12, a), or to carry out preliminary separation and droplet coagulation (Fig. 2.12, b). In the latter case, the final separation of droplets is achieved in a horizontal mesh installed before the gas output branch pipe. In some gravitational gas-oil separators, the gas-liquid stream is directed toward a special reflector called a deflector (or baffle). The wall of the vertical separator can also act as a baffle. As a result of an impact of the liquid-gas mixture against the obstacle, the initial separation of gas from liquid occurs. To ensure smooth flow of liquid without foaming, separators are provided with a series of horizontal and inclined surfaces – shelves, cones, or hemispheres.

Dividers used in plants for complex preparation of gas and condensate are intended for separation of gas from condensate. Their construction is basically the same as that of oil separators.

Oil dehydration occurs in devices that separate water-oil emulsion – gravitational settlers, i.e. tanks where emulsion separation occurs due to gravity. Small sizes of water droplets and small density difference between oil and water necessitate the use of a large apparatus. Therefore, the main challenge in designing a settler is to provide for the enlargement of drops. This goal can be achieved by thermochemical methods or by applying electric field to the emulsion. Apparatuses operating on these principles are called, respectively, thermochemical devices and electrodehydrators.

The main operating principle of thermochemical settling devices is emulsion heating that reduces viscosity of oil, thus accelerating sedimentation of water droplets. Addition of chemical reagents (de-emulsifiers) to the emulsion promotes destabilization of the emulsion and increases the rate of droplet coalescence. In their design, thermochemical settlers do not differ much from gravitational gas separators. Settlers are differentiated by their geometry, by the design of input and output devices, and also by specific details that have to do with organization of hydrodynamic regime inside settlers. The most common type at the present moment is a horizontal settling device, with length-to-diameter ratio close to 6. One distinctive feature of settling tanks is utilization of special input-output devices called "mother liquor" whose function is to ensure a uniform emulsion distribution in the cross section. Distributors for the emulsion entering the apparatus can differ according to whether the emulsion is fed under the drainage water layer or directly into the oil phase. If the water-oil emulsion is fed under the drainage water layer gathered in the bottom part of the apparatus, then to accelerate breakup of water-oil emulsion jets issuing from orifices of the tubular mother liquor, the orifices are made in the bottom or lateral part of the mother liquor. To achieve a uniform distribution of the emulsion over the apparatus cross section, the tubular mother liquors are aligned with the height of the device. Such an arrangement is not always convenient. Another possibility is the mother liquor in the form of a duct with apertures in the upper parts and opened from below. These ducts are put at some distance from each other on two distributive tubes with apertures just under the ducts. The ducts is where the spontaneous separation of oil and water occurs. Oil flows out from apertures of the duct from above, while water remains in the bottom part. In the case when emulsion is injected into the oil layer, we use tubular mother liquors with apertures on the top. In this context, there arises the problem of how to distribute the apertures over the length of the tube to achieve a uniform liquid flow rate. A non-uniform flow rate would cause an undesirable mixing of emulsion in the settler.

In the volume of the settling tank, water-oil emulsion is separated into water, which accumulates in the bottom part, and oil. There is no precise boundary between oil and water, because there is an intermediate emulsion layer between the two substances. This layer exists in any settling tank, and it has important technological functions. As water descends to the bottom, it has to pass through this layer. This promotes the coalescence process on the interface. In the layer itself, an inter-drop coalescence can occur; a finely dispersed component of the emulsion can be filtered out when crude oil passes through the intermediate layer. This intermediate layer can increase or decrease in size, or exist in the state of dynamic equilibrium, when its thickness does not change. Stabilization of the emulsion in the intermediate layer leads to a delay in drop coalescence, with growth of the layer thickness and its removal from the device, which worsens the quality of oil dehydration. Similar phenomena are frequently observed in the process of dehydration of high-paraffin crude oil at low temperature.

The presence of salts in brine water causes high electric conductivity of water in the water-oil emulsion, and this can be used to enhance water droplet integra-

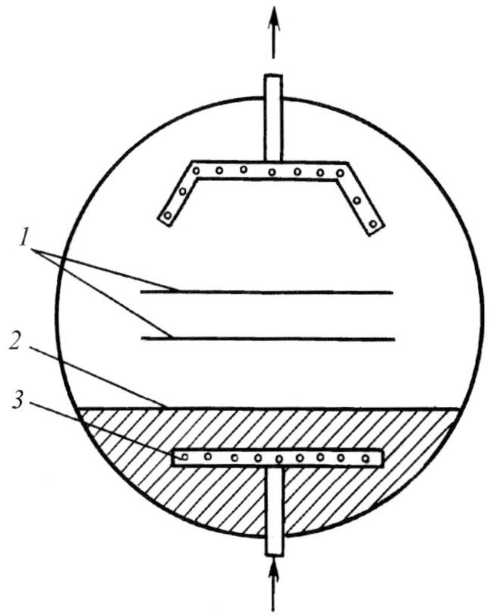

Fig. 2.13 Electrodehydrator: *1* – electrodes; *2* – interface, *3* – distributor (mother liquor).

tion in the presence of electric field. Apparatuses that separate water-oil emulsion by applying electric field are called electric dehydrators. Depending on the voltage used, they can be classified as electric dehydrators, operated by industrial-frequency voltage, and electrostatic dehydrators, operated by constant electric current. Electrostatic dehydrators have the widest range of applications. They are created on the basis of settling tanks of all types: spherical, cylindrical, vertical, and horizontal. In all types of electric dehydrators used in industry, the mother liquors are arranged to provide vertical ascending flow of liquid (Fig. 2.13). Devices in which the emulsion is subjected to preliminary integration in the electric field of an electric coalescencer before arriving in the settling tank are also possible.

2.2 Absorbers

Gas dewatering, treatment, separation by low-temperature absorption, low-temperature condensation, and rectification methods, and stabilization of condensate are common stages of processing of natural and associated gases and are therefore widespread in oil and gas industry. In the past, gas preparation at gas-condensate fields was limited to gas dewatering and separation of the condensate. In the resent years, due to the discovery and beginning of industrial exploitation of large deposits of natural gas, whose composition includes (besides light hydrocarbons) plenty of heavy hydrocarbons, hydrogen sulfide, carbon dioxide, mercaptans, and heavy paraffin hydrocarbons, field preparation of gas in terms of its objectives and processes involved has come close to the procedures that had hitherto been employed in cleaning and processing of gas at gas processing plants and oil refineries [10].

The process of mass exchange between the gas and the liquid in dewatering absorbers and treatment absorbers has the following characteristic features: high pressure (4–12 MPa); rather high energy of gas inside field equipment; requirement of minimal ablation of the expensive absorbent (up to 10–15 g on 1000 m^3 of gas); high flow rates of gas (up to 5–10 million m^3 of gas per day in one apparatus); low flow rates of liquid (15–25 kg per 1000 m^3 of gas) in dewatering absorbers and high flow rates of both liquid and gas in absorbers designed for clearing of gas from admixtures and heavy hydrocarbons [11]. Specific conditions encountered in some gas preparation procedures; emphasis on improving productivity and efficiency of the equipment; and also the difficulty of moving, installing and assembling equipment in remote, isolated and underpopulated areas – all of this has caused an increased interest in the creation of the technological devices with minimum specific metal capacity. In this context, much attention is given to the problem of intensification of processes in column devices by utilizing progressive contact devices that allow for a significant increase of specific loads on the vapor and liquid phases while sustaining high efficiency of separation. The variety of processes utilized in field preparation and the diversity of natural and operating conditions means that one can not come up with a universal construction of the absorber.

In plants for complex preparation of gas, absorbers are used for gas dehydration and extraction of heavy hydrocarbons from the gas.

There exist two ways of extracting a desired product from natural gas: adsorption using rigid absorbent (zeolites) as the absorber, and absorption by a liquid absorbent. The first method is periodical because it demands a periodic change of the solid absorbent. For this purpose, it is often necessary to swap different gas streams and accordingly to heat and cool the equipment. The second method can be run continuously, it involves less operational expenses and cheaper equipment, it helps improve reliability and is simple to control. Therefore the absorption method of gas dewatering and extraction of heavy carbohydrates is the method of choice for natural gas fields.

According to the data [12–14], the whole variety of absorbers can be divided into three classes: barbotage (bubble), surface, and atomized-spray devices.

In barbotage absorbers, gas in the form of jets or bubbles passes through a liquid layer. Barbotage absorbers are classified into the following basic groups:

1. Absorbers with a continuous barbotage layer, in which continuous contact between phases takes place.
2. Absorbers of the plate type with step contact between phases.
3. Absorbers with a mobile (floating) orifice.
4. Absorbers with mechanical mixing of the liquid.

Surface absorbers are the devices in which the surface of phase contact is determined by the geometry of absorber elements. These devices are subdivided into the following groups:

1. Surface absorbers formed by a horizontal surface of a liquid.
2. Film absorbers.
3. Packed (column) absorbers (with motionless orifice).
4. Mechanical film absorbers.

In atomized-spray absorbers, the surface of phases contact is formed by spraying the liquid into the gas. This class of absorbers is subdivided into:

1. Absorbers, in which the liquid is sprayed through nozzles.
2. High-velocity direct-flow spray absorbers, in which the liquid is atomized by a high-velocity stream of gas (i.e. the work necessary for atomization comes from the kinetic energy of the stream).
3. Mechanical spray absorbers, in which the liquid is atomized by rotating elements.

However, there is no clear distinction between the above-mentioned classes of devices, because under certain hydrodynamic conditions, the character and method of interface formation may change. For example, in plate absorbers, at high velocities of the gas stream, inversion of phases occurs resulting in ablation of liquid from the plates results in direct flow and the absorbers begin to operate as high-velocity, direct-flow spray absorbers. Operation of absorbers with mobile orifices at low gas velocities is not different from operation of devices with motionless orifices. It is possible to give many other similar examples.

For every type devices listed above, we can further distinguish between vertical and horizontal constructions.

Bubble absorbers (of the plate type), surface absorbers (film and column), and also spray absorbers (nozzle and high-velocity direct-flow absorbers) are widely used in plants for complex preparation of gas to perform absorption dehydration, or absorption extraction of heavy hydrocarbons from the gas.

In Fig. 2.14, a, a vertical multistage plate absorber is shown. The apparatus consists of three sections. The first separation section encountered by the stream of gas consists of a net-shaped impingement plate located directly at the entrance, and a separation plate with 178 centrifugal separation elements whose diameter is 60 mm (Fig. 2.14, b). This section is followed by the mass-exchange section, which has five contact stages. Each stage consists of a perforated plate with apertures of 6.3 mm diameter (Fig. 2.14, c), where the mass exchange occurs, and a separation plate equipped with centrifugal elements of 60 mm diameter. The last section is assigned the task of catching the absorbent (for instance, glycol). It consists of a partition plate, to which a cartridge filter of 1100 mm length and 100 mm diameter is attached, and a separation plate identical to the one used in the separation section. Cartridge filters consist of a perforated cylindrical core wound around by 10–15 layers of glass-canvas or technological linen made from synthetic fibres. From the inside and outside, the layer of filtering material is fixed by two or three layers of bag grids.

The absorbent (i.e. regenerated glycol is used for gas dewatering) is fed into the device through the branch pipe of 80 mm diameter on the top of the perforated

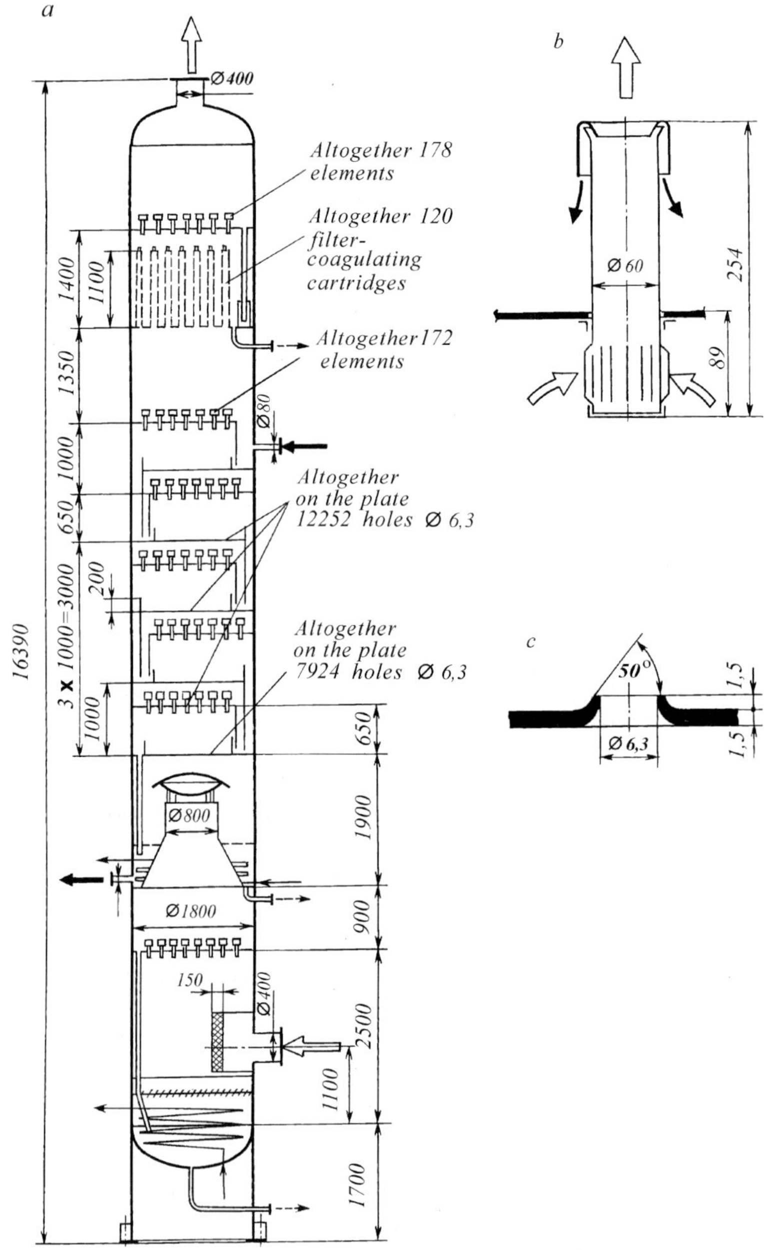

Fig. 2.14 A vertical multistage plate absorber.

plate, and the saturated absorbent is bled through the branch pipe on the semi-dead plate. Crude gas from the gathering system enters the separation part of the multipurpose apparatus through the entrance branch pipe. A large part of liquid drops contained in the gas is separated at the grid impingement plate and in the inter-plate space by the force of gravity. The separated liquid and mechanical impurities accumulate in the bottom part of the apparatus, protected from disturbance by the gas stream by a bubble made from cutting-out sheet. Partly, refined gas arrives at the separation plate, where finely-dispersed drops are separated from it under the action of the centrifugal force. The separated drops merge into a liquid film that flows down over the plate and then through the drain pipe into the collecting unit of the separation section. The gas cleared of drops is directed through the cone-shaped pipe branching off the semi-dumb plate to the mass-exchange section. The regenerated absorbent is reentered into the top part of the mass-exchange section. There it comes into contact with the gas stream and dewaters it, that is, removes the targeted component (water vapor in case of gas dehydration, or heavy hydrocarbons in the case of gas processing) from the gas. Note that in course of gas dehydration, glycol is used as absorbent, while transformer oil, petroleum, weathered condensate, etc. are used for the extraction of heavy hydrocarbons. Intensive mass exchange between the phases is achieved by utilizing the barbotage of gas through the absorbent layer at the perforated plate operating in the ablation regime. Drops of absorbent carried off by the gas stream from the perforated plate are caught by a separation plate located above and return (through the hydraulic hitch) to the perforated plate where they again come into contact with the gas. Thus, there is a constant circulation of absorbent inside the contact stage. Dewatered gas from the mass-exchange section goes to the collecting section which catches the remaining drops of absorbent. The used absorbent saturated by target components (water vapor or heavy hydrocarbons) extracted from the gas is directed towards the regeneration device, where it is brought back into working condition.

The same apparatus, but in the horizontal variant, is shown in Fig. 2.15. Here the section of preliminary separation of gas from liquid consists of a mesh coagulator mounted at the entrance to the apparatus, and one separation plate. The

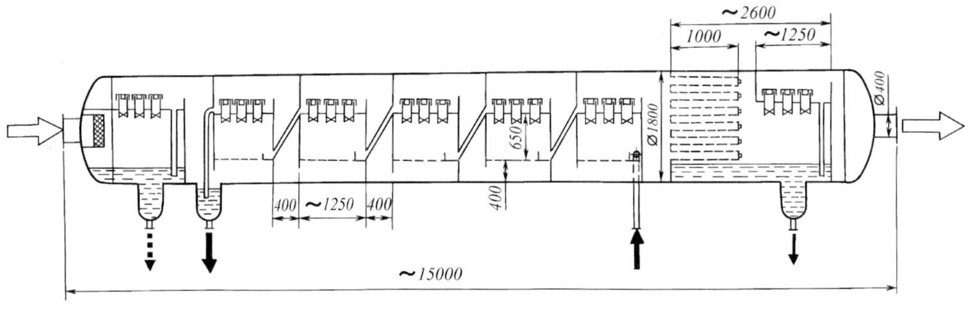

Fig. 2.15 Horizontal multistage plate absorber.

mass-exchange section consists of five contact stages, each stage consisting of a perforated plate and a separation plate placed 650 mm above it. The section of final gas separation consists of a coalescing orifice and a separation plate preventing the ablation of absorbent from the apparatus. The same type of separating elements is used in all sections: direct-flow centrifugal elements with tangential swirlers. Operation of the horizontal absorber is based on the principle of counterflow of gas and liquid, with the cross-flow of liquid from one contact stage to another driven by the kinetic energy of gas.

In gas industry, absorbers with ascending motion of a film are the most widespread. The operating principle of these absorbers is based on the fact that at sufficiently high velocities (more than 10 m/s), the gas moving upwards entrains the liquid film of the absorbent along the direction of its motion. Thus, an ascending co-current flow is formed. Absorption in such apparatuses takes place at high velocities (up to 40 m/s), resulting in a high mass-exchange factor between the phases. One construction of an absorber with ascending film flow is shown in Fig. 2.16. The apparatus is a box with plates and injection elements built in it. The box is divided into chambers by vertical partitions that just fall short of reaching its walls, thus providing channels for the passage of gas. The plates of each chamber are equipped with injection elements. Here the top-part partitions are placed to collect the separated liquid and feed it into the neighboring chamber. Such a design allows for a counter-flow motion of gas and liquid over the apparatus. The phase contact at each plate occurs in the ascending overflow regime.

The film device category also embraces absorbers with centrifugal contact elements, where the gas entering from below, passes through a swirler, involving the liquid in its upward motion. Under the action of the centrifugal force, the liquid is flung to the pipe wall, forming there a spiral film moving upwards.

Such absorbers with centrifugal contact elements have advantage over other film absorber designs. There are a few different constructions of centrifugal contact elements, for instance, elements with axial swirlers (of vertical or horizontal arrangement), or with tangential swirlers. The diameter of the elements is 100 mm, and all of them are equipped with devices for recirculation and suction of gas, which promotes faster operation, more efficient separation, and broader range of efficient operation. Elements with axial swirlers are equipped with axial shovel swirlers with the twist angle of about 45°. The tangential swirler construction is different in that it is not a "real" swirler; rather, the gas stream just enters

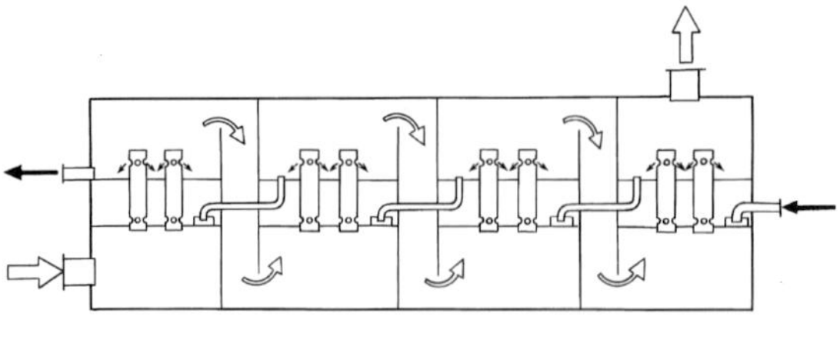

Fig. 2.16 Absorber with ascending motion of a film.

the element at a tangent to the surface through special slits in the wall. The length of elements is equal to three diameters. Tests performed on tangential elements have shown that their efficiency is greater than that of axial elements. Shown in Fig. 2.17 is one of the designs of a multipurpose absorber. Here both

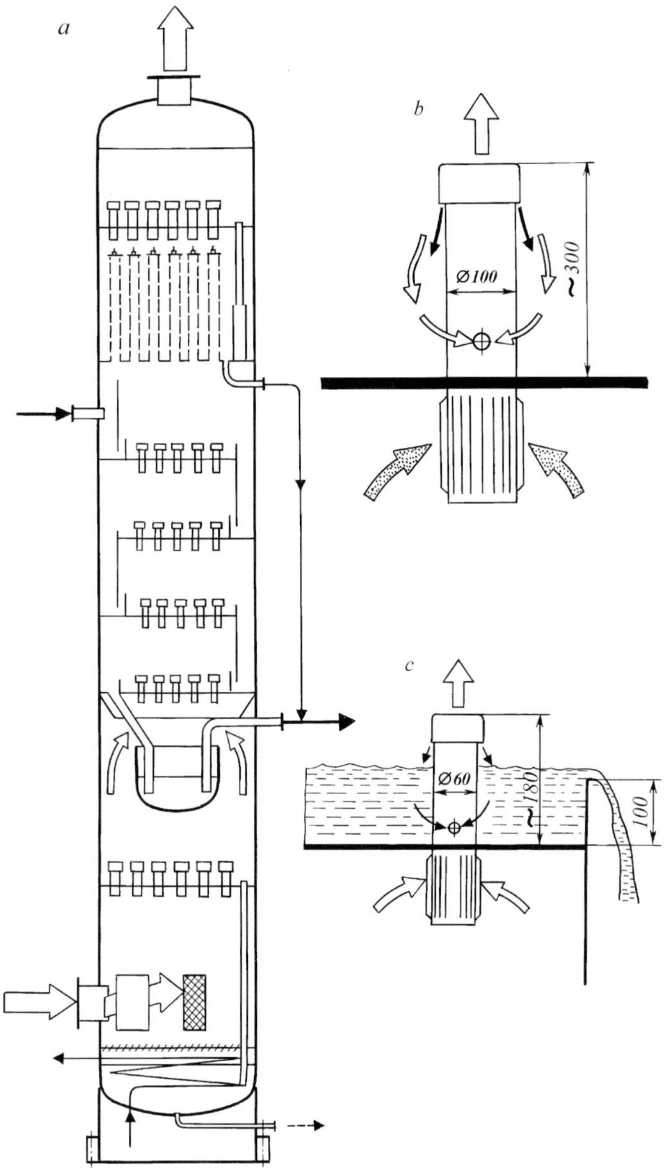

Fig. 2.17 Vertical absorber of the film type.

the entrance and the exit section are equipped with tangential centrifugal elements with devices for recirculation and suction of gas (Fig. 2.17, b, c). Contact between the gas and the absorbent takes place at four contact plates equipped with the same centrifugal elements.

Packed absorbers are columns whose contact zones are filled with bodies of various shapes. For the most part, the contact of gas with liquid takes place at the wetted surface of the orifice, where the sprinkled liquid flows down. The large surface area of the orifice per unit volume of the device makes it possible to create a sufficiently large contact surface that is required for efficient mass exchange of phases, in relatively small volumes. One of the constructions of such an absorber is shown in Fig. 2.18. The absorber consists of an entrance separator 1, a contact zone filled with orifice 2, and a filter-separator 3. In the same figure, a few kinds of orifices are shown.

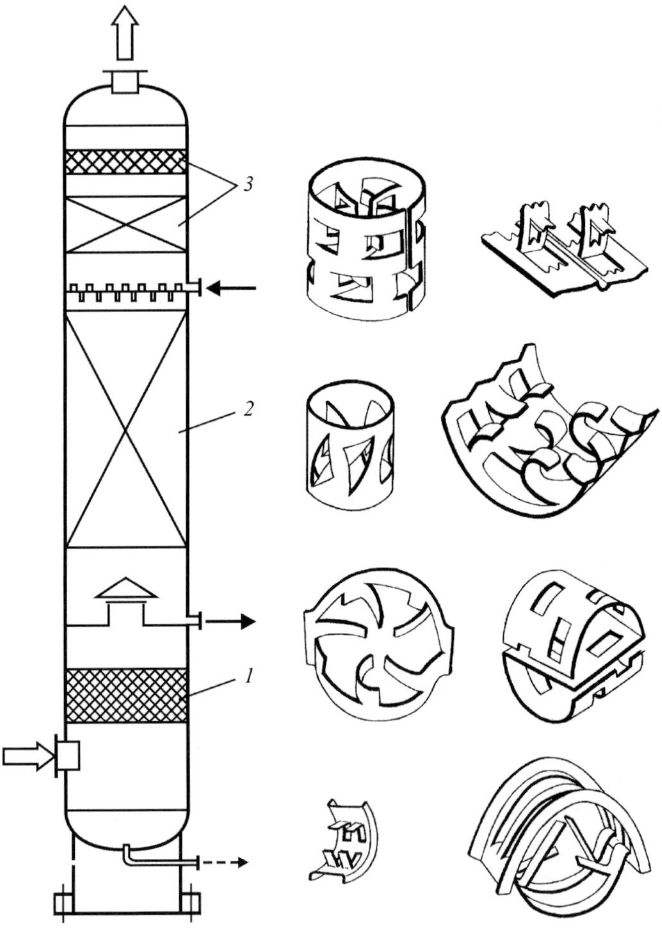

Fig. 2.18 Multipurpose packed absorber.

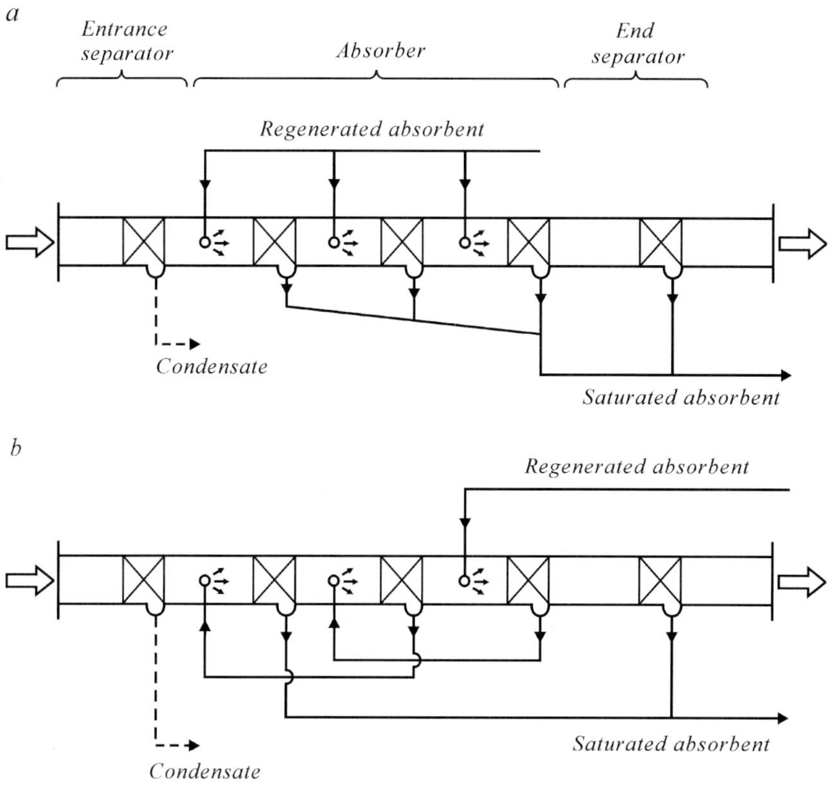

Fig. 2.19 The circuit of binding of spray absorbers: *a* – direct-flow multistage absorption; *b* – step directflow-counterflow absorption.

Spray absorbers are usually apparatuses of the horizontal type and are employed in plants for preparation of gas having a small flow rate. Their schemes are shown on Fig. 2.19, a, b. These devices are equipped with entrance and exit separators. The contact zone consists of several identical stages. In each contact stage, the regenerated absorbent is entered into the atomizer parallel to, or against the stream. Thanks to its large contact surface, the sprayed absorbent quickly absorbs the targeted component from the gas. The exhausted absorbent is separated from the gas in a separation unit after each contact stage. The distinction between absorbers depicted in Fig. 2.19, a and 2.19, b consists in the absorbent supply arrangement: the first figure shows direct-flow multistage absorption, while the second figure shows step-by-step, direct flow – counterflow absorption. The second method of absorbent supply provides for a lower flow rate of absorbent and higher quality of gas clearing as compared to the first one. Its disadvantage is a rather complex arrangement for pumping the absorbent from one stage into another.

2.3 Cooling Devices

The main cooling field devices are throttles, heat exchangers, turboexpanders, air cooling apparatuses, and artificial cold mechanisms.

Throttling of gas is performed in throttles (orifice plate, snouts) whose operation is based on narrowing the cross section of the pipe in which the gas is flowing. As a result, gas velocity increases, while both pressure and temperature drop. The drop of temperature is approximately proportional to the drop of pressure. The proportionality factor is determined from thermodynamic formulas describing the Joule–Thomson effect. For hydrocarbonic natural gases, it has the order of 0.3 degree/standard atmosphere. So, a 3 Mpa pressure drop is needed to cool the gas by ten degrees. Low thermodynamic efficiency of throttling limits the duration of its useful employment, because in the course of exploitation of a gas field, the reservoir pressure falls, and consequently, there is a decrease of pressure at the entrance to the plant for complex gas preparation. After approximately ten years of exploitation of the gas field (at the typical gas production rate), throttling can no longer provide the required degree cooling, and it becomes necessary to either increase the pressure with the help of a booster compressor station, or use other cooling devices.

An efficient mechanism for cold generation is the turboexpander aggregate (TEA), [15]. Cooling of gas is achieved by arranging the expansion process in such a way that the gas flowing through the TEA does external work. This causes the pressure and temperatures of the gas to decrease. The meridian cross section of the setting part of the turbo-expander is shown schematically in Fig. 2.20. The elements of the turbo-expander, in which the energy of the gas will be transformed into work, are: the motionless nozzle apparatus 1 with nozzle blades 2, and the rotating wheel 4 with runner blades 3. An "unfolded" view of the cylindrical cross section of the turbo-expander's blade set is shown in Fig. 2.21. Mapped on the same figure are characteristic gas velocities and forces arising from gas interaction with runner blades of the turbine wheel. The rotating part

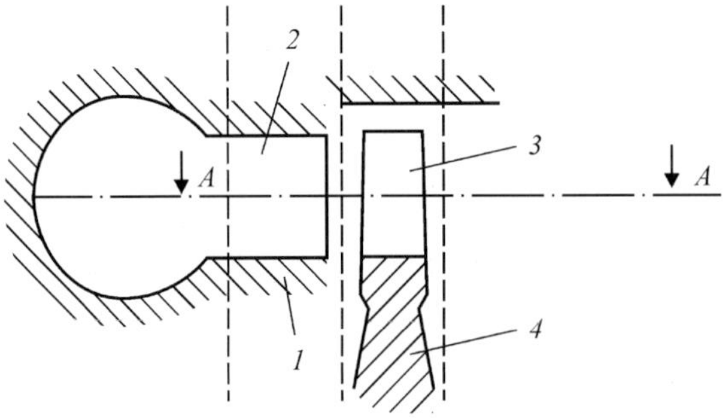

Fig. 2.20 Scheme of the meridian section of the setting part of turboexpander.

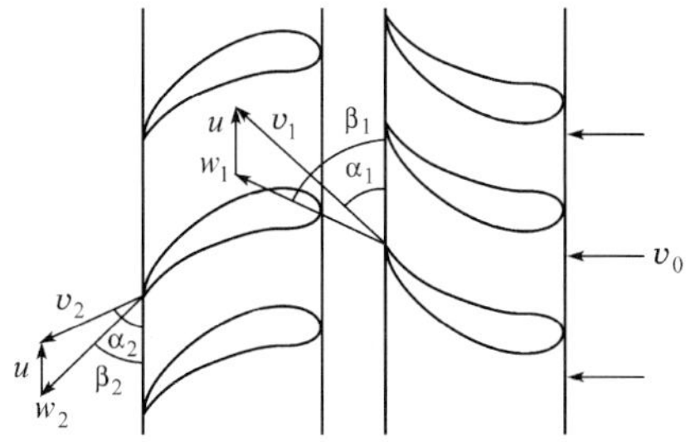

Fig. 2.21 The "unfolded" view of the cylindrical section A – A of the turboexpander (see Fig. 2.20).

of the turbo-expander, which consists of a wheel with runner blades and a shaft with bearings, is called the rotor. The immobile part of the turbo-expander, which includes the body, the nozzle block, and other details, is called the stator. The principle of operation of the turbo-expander is as follows. The gas with the velocity v_0 enters the blade channels of the nozzle guide apparatus and expands so that its initial pressure p_0 drops down to pressure $p_1 < p_0$. As a result, the gas velocity increases up to v_1, and its temperature decreases. The gas jet leaves the nozzle guide apparatus at the angle α_1 with the plane of the wheel's rotation and then enters the channels between the blades, where it further expands so that its pressure drops to $p_2 < p_1$, and accordingly, experiences additional cooling. At the same time, the gas jet changes its direction significantly while in contact with the working blades. The change of momentum gives rise to a force on the blades, thus forcing them to rotate.

Depending on direction of gas flow, we distinguish between the axial and radial turbo-expanders. In the first case, the stream moves basically along the coaxial cylindrical surfaces whose symmetry axis coincides with the axis of the wheel's rotation. In radial centrifugal turbo expanders, gas moves from the driving wheel's axis to the periphery, and in radial centripetal ones – in the reverse direction. One operational feature of radial turbo expanders is that, while in the driving wheel, the expanding gas stream does work in the field of centrifugal forces. Other conditions equal, this results in reduction of the gas velocity at the outlet of the radial turboexpander's wheel as compared with that in the axial turbo-expander. Because of this effect in centripetal turbo-expander, an increased pressure drop of gas is applied on the driving wheels.

The axial and radial turbo-expanders may be single- or multi-staged. The application of multi-staged turbo-expanders is expedient when a significant decrease of gas pressure and temperature is needed. Radial turbo-expanders are used mostly in plants operating at a low gas flow rate. Axial turbo-expanders are used for cooling in apparatuses with a high gas flow rate.

Since high power is applied to the turbo-expander's arbor, it is used to drive generators, pumps, and compressors. Therein lies the basic advantage of turbo-expanders as compared to throttles, where the energy released at pressure drop is just wasted.

Heat exchangers are used at oil-gas fields for heating, cooling, condensation and evaporation of liquid, gas, vapor, and their mixtures. Heat exchangers of the double-pipe and shell-and-tube types and apparatuses of air cooling (AAC) are the most widespread. The double-pipe heat exchanger consists of two coaxial pipes. The medium (gas, liquid) to be heated (or cooled) moves in the internal pipe, and the heating or cooling agent (gas, liquid) moves in the inter-tube space. The agent's and the medium's flows can be co-currrent or countercurrent. This kind of heat exchangers usually has several sections. Simplicity of design and of the system for feeding hydrate inhibitor motivates the use of such heat exchangers in LTS plants. The gaseous coolant is a separated cold gas, which comes from the low temperature separator into the inter-tube space of the heat exchanger. The shell-and-tube heat exchanger enjoys wider application than the double-pipe heat exchanger, because it needs less metallic material. But because of the absence of reliable hydrate inhibitor feeding system, these heat exchangers are applied only for gas cooling down to temperature not below gas hydrate formation point, or else when the gas has already been dewatered before the heat exchanger.

Depending on the kind of cooling agent and the medium to be cooled, we distinguish between gas-gas, gas-water, and condensate-condensate types of heat exchangers.

Devices for air cooling are intended to operate on open air, in areas with moderate or cold climate. There are various designs for cooling natural gas, condensate, and water.

The reader can find more detailed information about other equipment (pumps, compressors etc.) in Refs. [16, 17].

3
Basic Processes of Separation of Multi-phase, Multi-component Hydrocarbon Mixtures

Extracted hydrocarbon raw material is a multi-phase multi-component mixture. Formation of the multi-phase mixture (water-oil emulsion included), begins in a reservoir, and continues as the liquid is moving in the well, in elements of gathering and preparation equipment, and finally, in the main pipeline as a result of change of thermobaric conditions and geometrical sizes of regions through which the mixture flows. If these changes occur slowly enough compared to the characteristic relaxation time (the time it takes for the system to come to phase equilibrium), then it is possible to assume that the mixture motion occurs under conditions of thermodynamic and dynamic equilibrium (this is the case for mixture motion in beds and wells). It means that once the initial component composition of the mixture and the initial pressure and temperature at the point of interest are given, we can, in principle, determine specific volumes and component structure of phases, using equations of phase equilibrium. There currently exists a well-advanced semi-empirical theory of calculation of liquid-vapor equilibrium for systems of natural hydrocarbons [2], which allows to perform such calculations with a sufficient accuracy (this issue will be discussed further in Section 5.7). However, the problem of calculating phase equilibrium for systems containing water turns out to be more difficult. This issue is addressed in Refs. [18–22]. The results established in these papers allow us (to a certain degree) to estimate the equilibrium concentration of moisture in a hydrocarbon gas and, in particular, the temperature of hydrate formation. As of today, we don't have enough data to perform calculations of three-phase systems, for example, oil-gas-water or condensate-gas-water systems.

An absence of phase and dynamic balance in the system makes it necessity to take into account process dynamics. This is the case for mixture motion in regions with rapidly varying external conditions, as, for instance, in throttles, heat exchangers, turbo-expanders, separators, settlers, absorbers, and other devices. Violation of thermodynamic and dynamic balance may cause intense nucleation of one of the phases (liquid or gaseous) with formation of drops and bubbles, and their further growth due to inter-phase mass exchange (condensation, evaporation); this process is accompanied by mutual interaction of drops, bubbles, and other formations, which results in their coagulation, coalescence, and breakup.

Separation of Multiphase, Multicomponent Systems. E. G. Sinaiski and E. J. Lapiga

ISBN: 978-3-527-40612-8

The analysis of physical processes occurring in plants for oil, gas, and condensate preparation leads us identify the most important processes: separation of phases (liquid from gas, gas from liquid, liquid from liquid, solid particles of an impurity from gas or from liquid), and extraction of a certain component from gas or liquid mixture. In special literature devoted to these processes, each process has its own name. Thus, the term "separation" implies separation of liquid from gas or gas from liquid. Separation of liquid from liquid is called demulsification, and separation of suspensions (liquids or gases mixed with solid particles) is known as sedimentation. From the physical viewpoint, any of these processes occurs under the action of certain driving forces, which cause separation of phases or of components of one phase. For heterogeneous mixtures, such driving forces are: gravity, inertia, surface, hydrodynamic, electromagnetic and thermodynamic forces. For homogeneous mixtures, for example, mixtures of gases or solutions, the driving forces are gradients of concentration, temperature, pressure, or chemical potential. Mathematical simulation of these processes is based on common physical laws of conservation of mass, momentum, angular momentum, and energy, complemented by phenomenological correlations that particularize the model of the considered medium, and also by initial and boundary conditions. This makes it possible to incorporate the whole variety of physical processes involved in separation of multi-phase multi-component systems into a common theoretical framework. To facilitate understanding of special topics in Parts II–VII, each chapter will provide a summary of physicochemical foundations of relevant processes.

References

1 Zhdanova H. B., Haliph A. L., Natural gas dewatering, Nedra, Moscow, 1975 (in Russian).
2 Batalin O. J., Brusilovskiy A. I., Zaharov M. J., Phase equilibria in systems of natural hydrocarbons, Nedra, Moscow, 1992 (in Russian).
3 Guzhov A. I., Joint gathering and transport of oil and gas, Nedra, Moscow, 1973 (in Russian).
4 Loginov V. I., Dewatering and desalting of oil, Chemistry, Moscow, 1979 (in Russian).
5 Berlin M. A., Gorichenkov V. G., Volkov N. P., Treatment of oil and natural gases, Chemistry, Moscow, 1981 (in Russian).
6 Gritsenko A. I., Aleksandrov I. A., Galanin I. A., Physical bases of natural gas treatment and utilization, Nedra, Moscow, 1981 (in Russian).
7 Marinin N. S., Savvateev J. N., Oil degassing and preliminary dehydration in gathering systems, Nedra, Moscow, 1982 (in Russian).
8 Baras V. I., Production of the oil gas, Nedra, Moscow, 1982 (in Russian).
9 Sinaiski E. G., Gurevich G. R., Kashitskiy J. A., et al., Efficiency of separation equipment in field plants of gas preparation, VNIIEGASPROM Review (Moscow), 1986, No. 6, p. 41 (in Russian).
10 Mil'shtein L. M., Boiko S. I., Zaporogez E. P., Oil-gas-field separation techniques (Handbook) (Ed.: Mil'shtein L. M.), Nedra, Moscow, 1992 (in Russian).
11 Gorechenkov V. G., et al., The condition and perspectives of development of absorption processes at gas clearing, Gas Industry, Survey

information. Series: Preparation and treatment of gas and gas condensate, VNIIEGASPROM, Moscow, 1984, No. 7, p. 1–4 (in Russian).

12 Zibert G. K., Alexandrov I. A., Contact devices of mass-exchange apparatuses for processes of preparation and treatment of natural and associated gases, Express information, CINTEHIMNEFTEMASH, Moscow, 1980, No. 6 (in Russian).

13 Skoblo A. I., Molokanov J. K., Vladimirov A. I., Schelkunov V. A., Processes and devices of oil and gas treatment and petrochemistry, Nedra, Moscow, 2000 (in Russian).

14 Ramm V. M., Gas absorption, Chemistry, Moscow, 1976 (in Russian).

15 Yasik A. V., Turbo-expanders in systems of field preparation of natural gas, Nedra, Moscow, 1977 (in Russian).

16 Gas equipment, devices, and armature (Handbook) (Ed.: Riabtsev N. I.), Nedra, Moscow, 1988 (in Russian).

17 Gvozdev B. V., Gritsenko A. I., Kornilov A. E., Exploitation of gas and gas-condensate fields, Nedra, Moscow, 1988 (in Russian).

18 Korotaev J. P., Margulov R. D. (Eds.), Production, preparation and transport of natural gas and condensate (Handbook in 2 Volumes), Nedra, Moscow, 1984 (in Russian).

19 Namiot A. J., Phase equilibria in oil production, Nedra, Moscow, 1976 (in Russian).

20 Namiot A. J., Solubility of gases in water (Handbook), Nedra, Moscow, 1991 (in Russian).

21 Anderson F. E., Prausnitz J. M., Inhibition of gas hydrates by methanol, AICHE J. 1986, Vol. 32, No. 8, p. 1321–1333 (in Russian).

22 Istomin V. A., Yakushev V. S., Gas hydrates at natural conditions, Nedra, Moscow, 1992 (in Russian).

II
Physical and Chemical Bases of Technological Processes

Separation of Multiphase, Multicomponent Systems. E. G. Sinaiski and E. J. Lapiga

ISBN: 978-3-527-40612-8

4
The Transfer Phenomena

4.1
Phenomenological Models

The behavior of continuous medium is described by equations that follow from the laws of conservation of mass, charge, momentum, angular momentum, and energy. These equations should be completed by correlations reflecting the accepted model of a continuous medium. Such correlations are called constitutive equations or phenomenological relationships. Examples of constitutive relations are the Navier-Stokes law which establishes the linear dependence of a stress tensor on the rate of a strain tensor; the Fourier law, according to which heat flux is proportional to a temperature gradient; Fick's law, according to which mass flux is proportional to the gradient of a substance concentration; and Ohm's law, which states that the force of a current in a conducting medium is proportional to the applied electric field strength or to the potential gradient. These constitutive equations have been derived experimentally. The coefficients of proportionality, that is, coefficients of viscosity, heat conductivity, diffusion, and electrical conductivity, referred to as transfer coefficients, can be derived experimentally, and in some cases theoretically through the use of kinetic theory [1].

Conservation equations together with constitutive equations describe a phenomenological model of the continuous medium (continuum).

If one considers a liquid containing particles, then the disperse phase may be treated as a continuum, its description requiring a special phenomenological model with constitutive equations that differ from those relating to the continuous phase.

The flux of mass and energy in a continuum occurs in the presence of spatial gradients of state parameters, such as temperature, pressure, or electrical potential. Variables such as the volume of a system, its mass, or the number of moles are called extensive variables because their values depend on the total quantity of substance in the system. On the other hand, variables such as temperature, pressure, mole fraction of components, or electrical potential constitute intensive variables because they have certain values at each point in the system. Therefore, constitutive equations express the connection between fluxes and gradients of the intensive parameters. In the constitutive equations listed above, the flux de-

Separation of Multiphase, Multicomponent Systems. E. G. Sinaiski and E. J. Lapiga

ISBN: 978-3-527-40612-8

pends only on the gradient of a single parameter. Cases in which the flux is a superposition of gradients of several parameters are also possible. For example, mass flux is determined by the gradients of concentration, pressure, and electrical potential, and therefore the constitutive equation connects mass flux to all of the listed gradients. Generally, these connections have tensor character. Note that scalars and vectors are tensors of zero and first rank, respectively. In describing the mechanics of a continuum, tensors of second rank are also used, in particular tensors of stresses and of rate of strain. For example, the Fourier, Fick, and Ohm laws for an isotropic medium are vector laws, whereas the Navier–Stokes law is a tensor one.

4.2 Momentum Transfer

Real liquids are viscous and are characterized by internal shear stresses and viscous dissipation of energy. The processes occurring in a viscous liquid are thermodynamically irreversible and have spatial heterogeneity.

Consider a liquid conforming to the Newtonian law of proportionality between the shear stress and the shear velocity, giving rise to so-called Couette flow (Fig. 4.1).

Couette flow represents a stationary shear flow between two infinite plates separated from each other by a distance h. One of the plates is immovable, while the other moves translationally with a constant velocity U along the x-axis. The pressure p in the liquid is constant. The velocity of the liquid has one component $u(y)$ along the x-axis, and satisfies the conditions of adherence of liquid to surfaces, $y = 0$ and $y = h$, that is, $u(0) = 0$ and $u(h) = U$. Consider an arbitrary in-

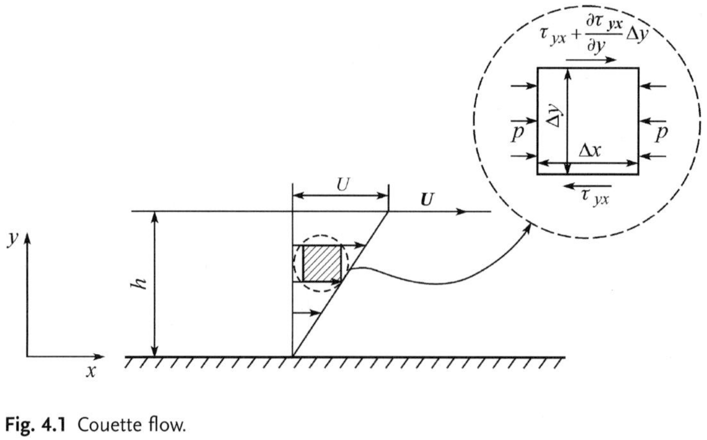

Fig. 4.1 Couette flow.

finitesimal element of liquid (see Fig. 4.1). The balance of forces acting on this element of the liquid gives

$$\sum F_x = \frac{\partial \tau_{yx}}{\partial y} \Delta y \Delta x \cdot 1, \tag{4.1}$$

where τ_{yx} is the projection on the x-axis of the shear stress at the area element perpendicular to the y-axis. Hereinafter, τ_{yx} is taken to be positive for an area element with exterior normal directed along the y-axis and negative otherwise.

It follows from Newton's second law that for the considered liquid element:

$$\sum F_x = \rho \frac{Du}{Dt} \Delta y \Delta x \cdot 1, \tag{4.2}$$

where ρ is the density of the liquid; Du/Dt is the substantial derivative:

$$\frac{Du}{Dt} = \frac{\partial u}{\partial t} + u \frac{\partial u}{\partial x}. \tag{4.3}$$

In the considered case, $\sum \mathrm{F_x} = 0$ and, according to (4.1):

$$\tau_{yx} = \text{const} \tag{4.4}$$

In compliance with Newton's constitutive equation

$$\tau_{yx} = \mu \frac{du}{dy}, \tag{4.5}$$

Here, μ is the coefficient of dynamic viscosity of the liquid, which depends on temperature and to a lesser degree on pressure.

A liquid that obeys Eq. (4.5) is called a Newtonian liquid. All gases and the majority of liquids can be taken to be Newtonian. However, there are liquids, for example solutions of polymers, or suspensions with high particle concentrations of the disperse phase, which do not obey Eq. (4.5). Such liquids are referred to as non-Newtonian.

From Eqs. (4.5) and (4.4), taking into account the boundary conditions, one obtains the velocity distribution

$$u = U \frac{y}{h} \tag{4.6}$$

and the shear stress

$$\tau_{yx} = \mu \frac{U}{h}. \tag{4.7}$$

The expression for the shear stress can be treated [2] as the flux of the x-th component of the momentum of the viscous liquid in the direction opposite to the y-axis. This flux is induced by the motion of the top plane, and since the bottom plane is at rest the flux of momentum causes liquid motion in the direction of the x-axis.

Thus, the flux of the momentum is directed against the direction of the velocity gradient. Hence, the velocity gradient may be considered as the driving force for the process of momentum transfer.

In parallel with the dynamic viscosity, one also needs to consider the kinematic viscosity:

$$\nu = \mu/\rho \tag{4.8}$$

The kinematic viscosity has dimensions of $m^2\ s^{-1}$. The diffusion coefficient has the same dimensions, and so the value ν can be viewed as the diffusion coefficient of the momentum in a viscous liquid.

As has already been noted, the coefficient of viscosity depends on temperature, but is practically independent of pressure. With increasing temperature, the viscosity of a gas increases, whereas the viscosity of a liquid decreases. This distinction arises because of various mechanisms of momentum transfer in a gas and in a liquid. In a gas, the molecules are relatively far from one another and are characterized by their average path length, l. Therefore, in a gas

$$\nu = \bar{u}l, \tag{4.9}$$

where $\bar{u}$ is the average velocity of molecules and is proportional to $(T)^{1/2}$.

In a liquid, the situation is different. Since molecules in a liquid are closely spaced, they need a much higher activation energy, ΔG, to move into the next vacant site. The velocity gradient in the direction perpendicular to the direction of liquid flow (du/dy) is equal to the shear stress multiplied by $\exp(-\Delta G/AT)$, where A is the gas constant, giving an indication of the probability of a molecule moving into the next vacant site. If the liquid flows in the direction of molecular motion, then this probability is proportional to τ_{yx}, since the shear stress forces the molecule to do additional work in the direction of the flow. By virtue of the correlation according to Eq. (4.5), it may be concluded that $v = \exp(\Delta G/AT)$, that is, the viscosity decreases with increasing temperature.

We now generalize Eq. (4.5) for a three-dimensional case. The stress at an area element in a liquid depends on the orientation of this area element, that is, on the direction of the external normal $\boldsymbol{n}$ (Fig. 4.2).

Consider a point in a liquid $\mathbf{x}$ (x, y, z) at some moment in time, and in the vicinity of this point three mutually orthogonal area elements perpendicular to the vectors $\boldsymbol{i}$, $\boldsymbol{j}$, $\boldsymbol{k}$. Through $\boldsymbol{t}(\boldsymbol{i})$, $\boldsymbol{t}(\boldsymbol{j})$, and $\boldsymbol{t}(\boldsymbol{k})$, we designate the stresses at these elements. Each of these stresses may be dissected with respect to the coordinate axes (Fig. 4.3). A projection of the stress t on the area element of arbitrary orientation with normal $\boldsymbol{n}$ can then be written as:

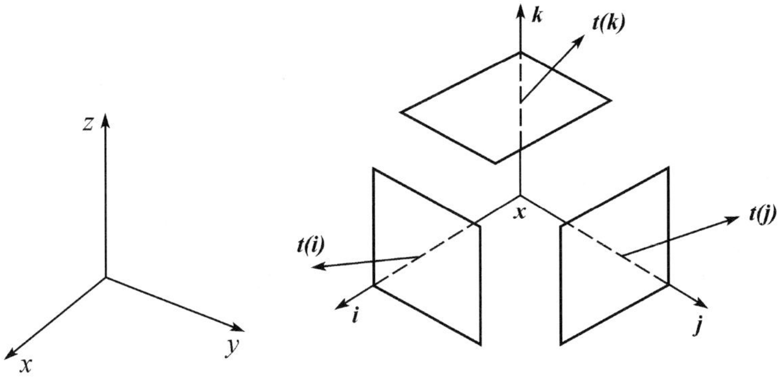

Fig. 4.2 Forces at the area elements perpendicular to the respective coordinate axes.

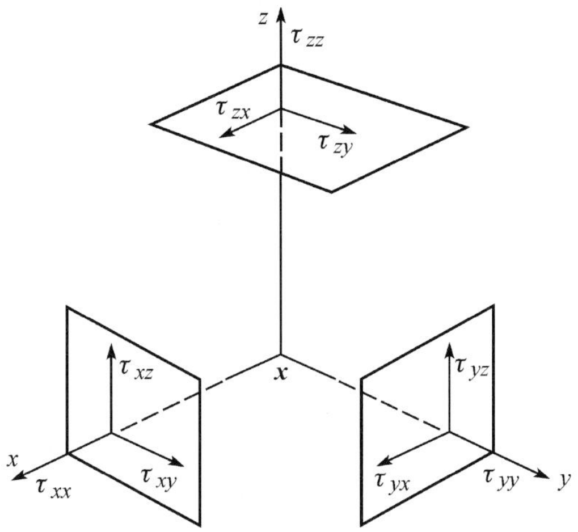

Fig. 4.3 Stresses on the area elements that are perpendicular to coordinate axes.

$$
\begin{aligned}
\boldsymbol{t}_x &= \tau_{xx}\boldsymbol{n}_x + \tau_{yx}\boldsymbol{n}_y + \tau_{zx}\boldsymbol{n}_z \\
\boldsymbol{t}_y &= \tau_{xy}\boldsymbol{n}_x + \tau_{yy}\boldsymbol{n}_y + \tau_{zy}\boldsymbol{n}_z \\
\boldsymbol{t}_z &= \tau_{xz}\boldsymbol{n}_x + \tau_{yz}\boldsymbol{n}_y + \tau_{zz}\boldsymbol{n}_z
\end{aligned} \tag{4.10}
$$

In τ_{ij}, the first index designates the axis to which the considered area element is perpendicular, and the second index designates a projection of the stress on the appropriate axis. Thus, τ_{ij} comprises nine components depending on x and t. These are components of the tensor of second rank, $\boldsymbol{T}$, called the stress tensor. In abbreviated form, the correlations of Eq. (4.10) can be written as:

$$
\mathbf{t}_i = \tau_{ij}\mathbf{n}_j, \tag{4.11}
$$

bearing in mind that in the right-hand part the summation is over repeating indices.

The stress tensor thus obtained characterizes the stress state of the liquid at point $\boldsymbol{x}$. If in the liquid there are no internally distributed moments (pairs), then the stress tensor is symmetric, that is, $\tau_{ij} = \tau_{ji}$.

When it is said that a liquid is Newtonian, one may assume that:

1) the liquid is isotropic, that is, its properties are identical in all directions;
2) if the liquid is non-viscous or quiescent, then

$$\boldsymbol{T} = -p\boldsymbol{I}, \tag{4.12}$$

where $\boldsymbol{I}$ is the unit tensor with components

$$\delta_{ij} = \begin{cases} 0 & \textit{if } \boldsymbol{i} \neq \boldsymbol{j}, \\ 1 & \textit{if } \boldsymbol{i} = \boldsymbol{j}; \end{cases}$$

3) if the liquid is viscous, then the stress tensor is a linear function of the strain rate tensor (Navier–Stokes law)

$$\boldsymbol{T} = \left(-p + \frac{2\mu}{3}\Theta\right)\boldsymbol{I} + 2\mu\boldsymbol{E} = -p\boldsymbol{I} + \boldsymbol{\Pi}. \tag{4.13}$$

Components of the strain rate tensor are equal to:

$$\varepsilon_{ij} = \frac{1}{2}\left(\frac{\partial u_i}{\partial x_j} + \frac{\partial u_j}{\partial x_i}\right). \tag{4.14}$$

In Eq. (4.13), the value $\Theta = \varepsilon_{kk} = \nabla \cdot \boldsymbol{u}$ is the first invariant of the strain rate tensor.

If a Newtonian liquid is incompressible, then $\nabla \cdot \boldsymbol{u} = 0$ and in the Cartesian system of coordinates it follows from Eq. (4.14) that:

$$\begin{aligned} &\tau_{xx} = -p + 2\mu\frac{\partial u}{\partial x}, \quad \tau_{yy} = -p + 2\mu\frac{\partial v}{\partial y}, \quad \tau_{zz} = -p + 2\mu\frac{\partial w}{\partial z}, \\ &\tau_{xy} = \tau_{yx} = \mu\left(\frac{\partial u}{\partial y} + \frac{\partial v}{\partial x}\right), \quad \tau_{xz} = \tau_{zx} = \mu\left(\frac{\partial u}{\partial z} + \frac{\partial w}{\partial x}\right), \\ &\tau_{yz} = \tau_{zy} = \mu\left(\frac{\partial v}{\partial z} + \frac{\partial w}{\partial y}\right), \end{aligned} \tag{4.15}$$

where u, v, and w are components of the velocity $\boldsymbol{u}$.

4.3 Thermal Conduction and Heat Transfer

The transfer of heat in a liquid may occur by means of conduction, convection, diffusion, or radiation. Heat transport by way of convection and diffusion will be considered in the appropriate sections, and since heat transport through radiation is small (essentially only occurring at very high temperatures), we consider here only heat transport due to conduction.

If in a liquid there is a temperature gradient $\nabla \boldsymbol{T}$, then according to the Fourier law, the heat transport is proportional to $\nabla \boldsymbol{T}$ and in the opposite direction to $\nabla \boldsymbol{T}$:

$$\boldsymbol{q} = -k\nabla \boldsymbol{T}. \tag{4.16}$$

The coefficient k is called the heat conductivity factor. In parallel with this, the factor of thermal diffusivity is frequently used in practice (sometimes also called the coefficient of heat conductivity)

$$\alpha = k/C_p\rho, \tag{4.17}$$

where C_p is the specific heat at constant pressure. Note that for an incompressible liquid $C_p = C_v = C$.

The dimensions of α are the same as those of the diffusion coefficient and of the kinematic viscosity, therefore the process of heat transport due to conduction can be treated as the diffusion of heat with the diffusion coefficient α, bearing in mind that the transport mechanisms of diffusion and heat conductivities are identical. The coefficient of heat conductivity of gases increases with temperature. For the majority of liquids the value of k decreases with increasing T. Polar liquids, such as water, are an exception. For these, the dependence $k(T)$ shows a maximum value. As well as the coefficient of viscosity, the coefficient of heat conductivity also shows a weak pressure-dependence.

Heat flux may also occur as a result of a concentration gradient (the Dufour effect). However, as a rule, it is negligibly small, and so hereafter we shall neglect this effect.

4.4 Diffusion and Mass Transfer

The diffusion or transfer of mass of an i-th component in a multicomponent mixture is possible if there is a spatial concentration gradient of this component (ordinary diffusion), a pressure gradient (barodiffusion), a temperature gradient (thermodiffusion), or if there are external forces that act selectively on the considered component (forced diffusion) [2]. Therefore, concentration, pressure, and temperature gradients, along with external forces, may be considered as the driving forces of the diffusion process. In this section, we focus on diffusion due to concentration gradients.

Consider first a diffusion process in a binary mixture, for example, the diffusion of a colored dye in water. If the dye is injected at any point in the liquid, it spreads throughout the liquid. The dye flows from the area in which it has high concentration to areas in which its concentration is lower. Simultaneously, one can observe the transfer of molecules of liquid in the opposite direction. In time, the process reaches equilibrium, diffusion ceases, and the solution becomes homogeneous. The cited case is a typical example of binary diffusion; moreover, the coefficient of diffusion, D_{12}, of the dye relative to water is equal to the coefficient of diffusion, D_{21}, of water relative to the dye.

There are several ways to specify the concentration of a substance [2]:

1) Mass concentrations:

$$\rho_i = \frac{m_i}{V} = \frac{\text{mass of } i\text{-th component}}{\text{volume of mixture}} \left(\frac{kg}{m^3}\right); \tag{4.18}$$

2) mole concentrations:

$$C_i = \frac{n_i}{V} = \frac{\text{number of } i\text{-th component moles}}{\text{volume of mixture}} \left(\frac{mole}{m^3}\right); \tag{4.19}$$

3) mass fractions (dimensionless):

$$\omega_i = \frac{\rho_i}{\rho} = \frac{\text{mass concentration of } i\text{-th component}}{\text{density of mixture}}; \tag{4.20}$$

4) mole fractions:

$$x_i = \frac{C_i}{C} = \frac{\text{mole concentration of } i\text{-th component}}{\text{mole density of mixture}}. \tag{4.21}$$

The average mole mass of a mixture may be determined by the formula:

$$\bar{M} = \rho / C. \tag{4.22}$$

Table 4.1 shows the formulae that connect the concentrations defined by the relationships of Eqs. (4.18)–(4.22).

There are several definitions of the average velocity of motion of multicomponent mixture in which components move with different velocities $\boldsymbol{u}_i$ [2]. The mass average velocity of a mixture is given by

$$\boldsymbol{u} = \frac{1}{\rho} \sum \rho_i \boldsymbol{u}_i. \tag{4.23}$$

Table 4.1

Concentration	
Mass	**Mole**
$m_i = n_i M_i;\ \rho_i = C_i M_i;\ \sum \omega_i = 1$	$n_i = \frac{m_i}{M_i};\ C_i = \frac{\rho_i}{M_i};\ \sum x_i = 1$
$m = \sum m_i;\ \rho = \sum \rho_i;\ \left(\sum \frac{\omega_i}{M_i}\right)^{-1} = \bar{M}$	$n = \sum n_i;\ C = \sum C_i;\ \sum x_i M_i = \bar{M}$
$\omega_i = \frac{m_i}{m};\ \omega_i = \frac{\rho_i}{\rho};\ \omega_i = \frac{x_i M_i}{\sum x_i M_i}$	$x_i = \frac{n_i}{n} = \frac{C_i}{C} = \frac{\omega_i/M_i}{\sum \omega_i/M_i}$

The mole average velocity is determined by

$$\boldsymbol{u}^* = \frac{1}{C}\sum C_i \boldsymbol{u}_i. \tag{4.24}$$

In the right-hand sides of Eqs. (4.23) and (4.24), contain the individual fluxes

$$\boldsymbol{j}_i = \rho_i \boldsymbol{u}_i, \quad \boldsymbol{j}_i^* = C_i \boldsymbol{u}_i. \tag{4.25}$$

Note that $\rho_i \boldsymbol{u}_i$, $C_i \boldsymbol{u}_i$, $\rho \boldsymbol{u}^*$ and $C\boldsymbol{u}^*$ represent the fluxes of the i-th component and of the entire mixture with reference to some fixed coordinate system. For infinitely dilute solutions with $M_1 = M_2 = \cdots = M_n$, $\boldsymbol{u} = \boldsymbol{u}^*$.

In multicomponent systems it is not individual flux streams that are of interest, but streams relating to the entire mixture:

$$\boldsymbol{J}_i = \rho_i(\boldsymbol{u}_i - \boldsymbol{u}), \quad \boldsymbol{J}_i^* = C_i(\boldsymbol{u}_i - \boldsymbol{u}^*). \tag{4.26}$$

Consider first a binary mixture. According to Fick's law

$$\boldsymbol{J}_1 = \boldsymbol{j}_1 - \rho_1 \boldsymbol{u} = -\rho D_{12}\nabla\omega_1, \quad \boldsymbol{J}_1^* = \boldsymbol{j}_1^* - C_1\boldsymbol{u}^* = -CD_{12}\nabla x_1, \tag{4.27}$$

and $D_{12} = D_{21}$, $\boldsymbol{J}_2 = -\boldsymbol{J}_1$, $\boldsymbol{J}_2^* = -\boldsymbol{J}_1^*$.

In the specific case of constant density of a liquid, ρ, one may write

$$\boldsymbol{J}_1 = -D_{12}\nabla\rho_1. \tag{4.28}$$

The coefficient D_{12} is called the coefficient of binary diffusion, though it is frequently designated simply as D. In gases, D_{12} is practically independent of composition, it increases with temperature, and is inversely proportional to pressure.

In liquids, D_{12} strongly depends on concentration of components and increases with temperature. Therefore, in multi-component systems, the flux of the *i*-th component depends on the concentration gradients of all components.

In multicomponent mixtures, fluxes with respect to a fixed coordinate system and relative fluxes are connected by relationships that follow from Eqs. (4.25) and (4.26)

$$J_i = j_i - \omega_i \sum j_j, \quad J_i^* = j_i^* - x_i \sum j_j^*, \tag{4.29}$$

from which one obtains

$$\sum J_i = \sum J_i^* = 0. \tag{4.30}$$

For infinitely dilute solutions, the diffusion coefficient of each component may be considered as the coefficient of binary diffusion of this component relative to entire mixture. Therefore, for each infinitely dilute component, Fick's law in the form of Eq. (4.27) applies. Moreover, the approach of considering an infinitely dilute solution allows estimation of the coefficient of binary diffusion, using simple thermodynamic reasoning. Consider the motion of a molecule of a dissolved substance as Brownian motion with a kinetic energy of thermal movement $\mathsf{k}T$ (k is Boltzmann constant). The viscosity of a liquid provides resistance to motion with a force that may be estimated by means of Stokes' formula $\mu_2 u_1 d_1$ (d_1 is the average diameter and u_1 is the average velocity of a molecule, μ_2 is the viscosity of the liquid). The work done by the molecule against the resistance of the liquid over distance l, is equal to $\mu_2 u_1 d_1 l$. Equating the work and the kinetic energy and taking $D_{12} \sim u_1 l$, one obtains

$$D_{12} = \mathsf{k}T/\mu_2 d_1. \tag{4.31}$$

Using the further dependence of viscosity on temperature, it is found that

$$D_{12} = \frac{\mathsf{k}T}{d_1} \exp(-\Delta G/AT).$$

The formula thus obtained is applicable to both the molecules and the particles suspended in a liquid. In the case of molecules, d_1 should be taken as the linear dimension δ, of the cubic cell containing a molecule. With the assumption of closely packed molecules, this dimension is equal to $(\bar{V}_1/N_A)^{1/3}$, where N_A is the Avogadro constant, and V_1 is the mole volume of particles of type 1.

In conclusion, it should be noted that transfer of mass is also possible due to a temperature gradient (Soret effect).

4.5 Electro-Conductivity and Charge Transfer

Consider mass transfer in a mixture containing components that carry different charges. If an imposed external electric field is applied, different electric forces act on these charged components accelerating them in the two opposite directions depending on their charge. The resulting motion of components in the electric field is called their migration, although in some sense it may also be considered as diffusion in a selective direction, because, on the one hand, charged particles are accelerated by an external electric field, while on the other hand, they collide with particles of the solution.

For simplicity, we consider only infinitely dilute solutions, for which it is possible to neglect the interaction of the charged components. If a solution containing the charged components, for example an electrolyte, is exposed to an electric field $\boldsymbol{E} = -\nabla\varphi$, then the force acting on one mole of the i-th dissolved component, is equal to $-z_i F \nabla\varphi$, where z_i is the charge on ions of the i-th component, $F = N_A e = 9.65 \times 10^4$ Kmol^{-1} is the Faraday constant, and e is the charge of an electron. The governing equation for the process of charge transport is similar to the equations for the processes of diffusion and heat conductivity. Therefore, the mole flux of the charged substance in an electric field is proportional to the force acting on the particles, and also to their concentration:

$$j_i^* = -\upsilon_i z_i F C_i \nabla\phi = \upsilon_i z_i F C_i \boldsymbol{E} \ (\text{mol/m}^{-2}\text{s}^{-1}) \tag{4.32}$$

The proportionality factor υ_i, as well as the heat-conductivity coefficient and diffusion coefficient, is a transport factor and is called mobility, since it characterizes the ability of charged particle to move in the medium under action of driving force (in this case – the electric field). This coefficient can be interpreted as average velocity of motion of charged particle in solution under action of a force equal to 1 N/mole. The dimensionality of υ_i is $\text{N}^{-1}\cdot\text{m}\cdot\text{s}^{-1}\cdot\text{mole}$.

The concept of mobility, which we shall frequently encounter with further, is a generalized concept, because the mobility characterizes the particle's capacity of drift in liquid under action of any force (electric, magnetic, gravitational, centrifugal, etc.).

Together with molar flux (4.32), we can also introduce the mass flux

$$j_i = -\upsilon_i z_i F \rho_i \nabla\phi = \upsilon_i z_i F \rho_i \boldsymbol{E} \ (\text{kg/m}^2\text{s}). \tag{4.33}$$

The mole flux is commonly used in research of electrochemical processes.

The diffusion coefficient and mobility of a particle are linked together. As has been shown in Section 4.4, an estimation $D_i \sim u_i l_i$ holds for extremely diluted solutions and gases. Multiplying and dividing the right part of this equality by $N_A F_{\upsilon i}$, where $F_{\upsilon i}$ is the viscous resistance force on a particle, one obtains

$$D_i \sim \frac{u_i}{N_A F_{\upsilon i}} F_{\upsilon i} l_i N_A = \upsilon_i F_{\upsilon i} l_i N_A = \upsilon_i \mathbf{k} T N_A = \upsilon_i T A. \tag{4.34}$$

Here the definition of gas constant $A = \mathbf{k} N_A$ is taken into account.

The correlation (4.34) is called the Nernst-Einstein equation.

Motion of charged particles in an external electric field can be considered as the electric current with density

$$\boldsymbol{i} = F \sum z_i \boldsymbol{j}_i^* \ (\mathrm{A/m^2}). \tag{4.35}$$

If there are no other driving forces in the system except for the electric field, then

$$\boldsymbol{i} = \sigma \mathbf{E} = -\sigma \nabla \phi, \tag{4.36}$$

where the quantity

$$\sigma = F^2 \sum z_i^2 \upsilon_i C_i, \tag{4.37}$$

is called the electroconductivity of solution. Equation (4.36) is the Ohm's law for a solution containing the charged components.

In parallel with ordinary electroconductivity determined by (4.37), the molar electroconductivity Λ_i can be introduced:

$$\Lambda_i = \frac{\sigma_i}{C_i} = F^2 \upsilon_i z_i^2. \tag{4.38}$$

Similar to diffusion coefficient, the molar electroconductivity exponentially increases with temperature. It also depends on electrolyte concentration, decreasing with concentration increase.

5
Conservation Laws and Equations of State

5.1
Isothermal Processes

Consider conservation laws for mass and momentum of viscous Newtonian liquids. In case of isothermal flow of incompressible liquid, it is necessary to add to these laws the basic equation (4.13) together with appropriate initial and boundary conditions. It is sufficient to determine the velocity distribution and stresses at any point of space filled with the liquid, and at any moment of time. If the flow is not isothermal, then in order to find the temperature distribution in liquid, we need to use the energy conservation. If, besides, the liquid is compressible, it is necessary to add the equation of state.

The derivation of equations expressing the conservation laws could be found, for example, in [3].

Recall the two important equations before we start to discuss the conservation laws. The first one is the Gauss-Ostrogradsky theorem[1]. It relates the surface and volume integrals of the type

$$\int_{\Sigma} F n_i \, ds = \int_V \frac{\partial F}{\partial x_i} dV. \tag{5.1}$$

The formula (5.1) is true for a scalar, vector or tensor function F at a fixed index i; the summation over i is also allowed. The function F is supposed to be continuously differentiable in volume V enclosed by a sufficiently smooth surface Σ. Symbol n_i designates the i-th component of the normal $\boldsymbol{n}$ external to V (the direction cosine). If $F = u_i$ are the velocity components, then performing in (5.1) the summation over i, one obtains

$$\int_{\Sigma} \boldsymbol{u} \cdot \boldsymbol{n} \, ds = \int_V \nabla \cdot \boldsymbol{u} \, dV. \tag{5.2}$$

1) I retain the original terminology here. In Russian scientific literature, the Gauss's theorem is called the Gauss-Ostrogradsky theorem. – VF

Separation of Multiphase, Multicomponent Systems. E. G. Sinaiski and E. J. Lapiga

ISBN: 978-3-527-40612-8

Let's take $F = \varepsilon_{ijk} u_i$ where ε_{ijk} is the Levi-Civitta permutation symbol equal to 1 when indexes ijk form a cyclic permutation of numbers 123, equal to 0 when among ijk are two (or more) identical numbers from the set 123, and to -1 otherwise. Using this tensor, the components of a cross product can be written as $(\boldsymbol{b} \times \boldsymbol{c})_i = \varepsilon_{ijk} b_i c_i$. Then it follows from (5.1)

$$\int_{\Sigma} \boldsymbol{n} \times \boldsymbol{u}\, ds = \int_{V} \nabla \times \boldsymbol{u}\, dV. \tag{5.3}$$

Surface integrals in (5.2) and (5.3) can be rewritten in terms of oriented area elements of the surface $\boldsymbol{ds} = \boldsymbol{n} \cdot ds$:

$$\int_{\Sigma} \boldsymbol{u} \cdot \boldsymbol{n}\, ds = \int_{\Sigma} \boldsymbol{u} \cdot \boldsymbol{ds}, \quad \int_{\Sigma} \boldsymbol{n} \times \boldsymbol{u}\, ds = \int_{\Sigma} \boldsymbol{ds} \times \mathbf{u}. \tag{5.4}$$

If F is a tensor $\boldsymbol{T}$, then the equations (5.2) can be written as

$$\int_{\Sigma} \boldsymbol{T} \cdot \boldsymbol{n}\, ds = \int_{V} \nabla \cdot \boldsymbol{T}\, dV. \tag{5.5}$$

Here we have on the left a scalar product of tensor and vector, and on the right the divergence of tensor. By definition, they are both vectors with coordinates

$$(\boldsymbol{T} \cdot \boldsymbol{n})_i = T_{ji} n_i, \quad (\nabla \cdot \boldsymbol{T})_i = \partial T_{ij} / \partial x_j. \tag{5.6}$$

The second equation represents a formula for differentiation of the volume integral over time when the volume itself is a variable $\mathrm{V} \rightarrow \mathrm{V}^*$

$$\frac{\mathrm{D}}{Dt} \int_{V^*} F\, dV = \int_{V} \left(\frac{DF}{Dt} + F \nabla \cdot \boldsymbol{u} \right) dV. \tag{5.7}$$

Using the expression (4.3) for the substantial derivative, it is possible to transform the integrand in the right part (5.7)

$$\frac{DF}{Dt} + F \nabla \cdot \boldsymbol{u} = \frac{\partial F}{\partial t} + \nabla \cdot (F \boldsymbol{u}). \tag{5.8}$$

Applying the Gauss–Ostrogradsky theorem (5.1) to the integral in the right part of (5.7) and using (5.8), we obtain

$$\frac{\mathrm{D}}{Dt} \int_{V^*} F\, dV = \int_{V} \frac{\partial F}{\partial t} dV + \int_{\Sigma} F \boldsymbol{u} \cdot \boldsymbol{n}\, ds. \tag{5.9}$$

Employing the relations (5.1)–(5.9), it is easy to obtain the equations for mass and momentum conservation.

Consider now a moving volume V^*. Denote as m mass within this volume. Then the condition of mass conservation takes the form

$$\frac{Dm}{Dt} = \frac{D}{Dt}\int_{V^*} \rho \, dV = 0. \tag{5.10}$$

Use the equation (5.7)

$$\frac{D}{Dt}\int_{V^*} \rho \, dV = \int_V \left(\frac{D\rho}{Dt} + \rho \nabla \cdot \boldsymbol{u}\right) dV = 0. \tag{5.11}$$

Since the formula (5.11) is true for any volume, it reduces to

$$\frac{D\rho}{Dt} + \rho \nabla \cdot \boldsymbol{u} = \frac{\partial \rho}{\partial t} + \nabla \cdot (\rho \boldsymbol{u}) = 0. \tag{5.12}$$

The equation (5.12) is called the continuity equation. If liquid is incompressible, then $D\rho/Dt = 0$ and it follows from (5.12)

$$\nabla \cdot \boldsymbol{u} = 0. \tag{5.13}$$

The concept of incompressibility should not be confused with homogeneity of liquid. A liquid is called homogeneous if its density ρ is the same in all parts of the liquid, that is ρ does not depend on spatial coordinates x, y, z. Otherwise the liquid refers to as non-homogeneous. The condition of the incompressibility $D\rho/Dt = 0$ means, that the local density of a liquid (or an isolated liquid particle) remains constant during its motion, but can be different for different particles. Therefore the equation (5.13) is valid both for homogeneous and non-homogeneous liquids. Hence, if the liquid is incompressible and in addition homogeneous, then $\rho = \text{const}$ in space, while for the incompressible non-homogeneous liquid the following conditions should be satisfied

$$\nabla \cdot \boldsymbol{u} = 0, \quad \frac{D\rho}{Dt} = \frac{\partial \rho}{\partial t} + \boldsymbol{u} \cdot \nabla \rho = 0. \tag{5.14}$$

Let us turn now to derivation of the momentum conservation equation. We will need a formula following from (5.7) and (5.12):

$$\frac{D}{Dt}\int_{V^*} \rho F \, dV = \int_V \rho \frac{DF}{Dt} dV. \tag{5.15}$$

According to a postulate called the Cauchy's stress principle, for any closed surface Σ there is a distribution of the stress vector $\boldsymbol{t}$ with resultant and moment, equivalent to a force field acting on the continuum in volume V enclosed by a surface Σ as opposed to the medium outside of Σ. It is assumed that $\boldsymbol{t}$ depends only on position and orientation of the surface element ds, that is $\boldsymbol{t} = \boldsymbol{t}(\boldsymbol{x}, t, \boldsymbol{n})$. According to the other postulate, for a volume V of liquid enclosed by a surface S, by analogy with a system of material points, the following conservation equation of momentum takes place

$$\frac{D}{Dt}\int_V \rho\boldsymbol{u}\, dV = V\int_V \rho\boldsymbol{f}\, dV + \int_\Sigma \boldsymbol{t}\, ds, \tag{5.16}$$

where $\boldsymbol{f}$ is the density of mass forces. It can be gravity, electric, or magnetic forces. In particular, for gravity force, we have $\boldsymbol{f} = \boldsymbol{g}$, where $\boldsymbol{g}$ is the acceleration due to gravity.

Consider the surface integral in (5.16). Putting there the expression (4.11) and transforming the surface integral into volume integral according to (5.5), one obtains

$$\int_\Sigma \boldsymbol{t}\, ds = \int_\Sigma \boldsymbol{T}\cdot\boldsymbol{n}\, ds = \int_V \nabla\cdot\boldsymbol{T}\, dV. \tag{5.17}$$

Putting it back to (5.16) and using (5.15), one finds

$$\int_V \left(\rho\frac{D\boldsymbol{u}}{Dt} - \rho\boldsymbol{f} - \nabla\cdot\boldsymbol{T}\right) dV = 0. \tag{5.18}$$

Since the last equation holds for any volume, the following differential equation must hold as well

$$\rho\frac{D\boldsymbol{u}}{Dt} = \rho\boldsymbol{f} + \nabla\cdot\boldsymbol{T}. \tag{5.19}$$

If the liquid is Newtonian, then the stress tensor $\boldsymbol{T}$ is related to the strain rate tensor by the equation (4.13). In particular, in the Cartesian system of coordinates the equation of motion for a Newtonian liquid is

$$\rho\left(\frac{\partial u_i}{\partial t} + u_j\frac{\partial u_i}{\partial x_j}\right) = \rho f_i - \frac{\partial p}{\partial x_i} + \frac{\partial}{\partial x_j}\left(\mu\left(\frac{\partial u_i}{\partial x_j} + \frac{\partial u_j}{\partial x_i} - \frac{2}{3}\delta_{ij}\frac{\partial u_k}{\partial x_k}\right)\right). \tag{5.20}$$

Here, as before, $i, j = 1, 2, 3$, where 1, 2, 3 stand for the axes x, y, z respectively, and the summation over repeating indexes is assumed.

If the liquid is incompressible, then $\partial u_k/\partial x_k = \nabla\cdot\boldsymbol{u} = 0$ and (5.20) can be simplified. In a vector form, the equation of motion of incompressible liquid is

$$\rho\frac{D\boldsymbol{u}}{Dt} = \rho\boldsymbol{f} - \nabla p + \mu\Delta\boldsymbol{u}. \tag{5.21}$$

The isothermal flow of incompressible liquid is described by equations (5.13) and (5.21), and the viscosity coefficient $\mu =$ const. Hence, there are four equations for four unknowns – the pressure p and three velocity components u, v, and w. Thus, the system of equations is a closed one. For its solution it is necessary to formulate the initial and boundary conditions. Let us discuss now possible boundary conditions. Consider conditions at an interface between two mediums denoted as 1 and 2. The form and number of boundary conditions depends on whether the boundary surface is given or it should be found in the course of solution, and also from the accepted model of the continuum. Consider first the boundary between a non-viscous liquid and a solid body. Since the equations of motion of non-viscous liquid contain only first derivatives of the velocity, it is necessary to give one condition of the impermeability: $u_{n1} = u_{n2}$ at the boundary Σ, where u_n is the normal component of the velocity. The equations of motion of viscous liquid include the second-order derivatives, therefore at the boundary with a solid body it is necessary to assign two conditions following from the condition of sticking: $u_{n1} = u_{n2}$, $u_{\tau 1} = u_{\tau 2}$ where u_τ is the tangential to Σ component of the velocity. If the boundary Σ is an interface between two different liquids or a liquid and a gas, then it is necessary to add the kinematic condition $u_1 = u_2 = u_\sigma$ (where u_σ is the surface velocity) and the condition for the normal and tangential stresses $\tau_{nn}^{(1)} = \tau_{nn}^{(2)}$, $\tau_{n\tau}^{(1)} = \tau_{n\tau}^{(2)}$ on Σ. In case of incompressible non-viscous liquid, the last conditions degenerate to equality of pressures at Σ. The curvature of the interface gives rise to an additional pressure, called capillary pressure, in the phase where the centre of the surface curvature lies. Therefore there must be a jump $\{\tau_{mn}\} = p_{cap}$ on Σ in such cases. This question will be discussed in more detail in the chapter devoted to the surface tension.

5.2
Non-isothermal Processes

Processes accompanied by temperature change are called non-isothermal. The change of temperature may be caused by heat transfer through the medium's boundary Σ, or heat production within the volume V of a liquid; it may be the Joule heat from chemical reactions, viscous energy dissipation in the medium's motion, the change of medium's phase state (evaporation, condensation, fusion etc.).

The thermal energy of a liquid particle is defined by the internal energy density, which depends on parameters of local thermodynamic condition. According to the first law of thermodynamics,

$$dq = de - dw, \tag{5.22}$$

where dq is the amount of heat delivered from the outside to a unit of liquid mass; de is the change of the internal energy; dw is work, performed on a unit of liquid mass.

Here we consider heat transport in the system due only to heat conductivity, without taking into account internal heat production due to Joule heat and/or the chemical reactions.

Consider a liquid element of volume V enclosed by a surface Σ. Determine the full energy of the considered volume as sum of kinetic K and internal E energy, where

$$\mathsf{K} = \frac{1}{2}\int_V \rho u^2\, dV, \quad \mathsf{E} = \int_V \rho e\, dV. \tag{5.23}$$

Note that thermodynamic interpretation of e is essentially different for compressible and incompressible fluid. Compressible fluid can be considered as a two-parametric system. According to the second law of thermodynamics, there is a state function – entropy s, playing the role of a thermodynamic potential. If e and specific volume $1/\rho$ are taken as independent parameters, then the equation of state of compressible gas will be $s = s(e, 1/\rho)$, and the perfect differential of the entropy is

$$ds = \frac{1}{T}de + \frac{p}{T}d\left(\frac{1}{\rho}\right). \tag{5.24}$$

It follows from (5.24) that $p = p(\rho, s)$, that is, the pressure is a function of the state parameters.

For incompressible fluid we have, instead of (5.24):

$$ds = \frac{1}{T}de, \tag{5.25}$$

and $e = e(s)$. Since pressure does not figure in (5.25), p is an independent quantity.

Accept as an axiom the following statement (conservation law for the total energy of a fluid particle): the rate of total energy change in a volume moving with a fluid is equal to the sum of the corresponding force powers and the amount of heat received by the volume per unit time

$$\frac{D}{Dt}(\mathsf{K} + \mathsf{E}) = \int_V \rho \boldsymbol{f} \cdot \boldsymbol{u}\, dV + \int_\Sigma \boldsymbol{t} \cdot \boldsymbol{u}\, dV - \int_\Sigma \boldsymbol{q} \cdot \boldsymbol{n}\, ds. \tag{5.26}$$

Transform the left part of (5.26) by using (5.15) to

$$\frac{D}{Dt}(\mathsf{K} + \mathsf{E}) = \int_V \rho \frac{D}{Dt}\left(\frac{u^2}{2} + e\right) dV,$$

and the surface integral on the right of (5.26) into the volume integral by applying the Gauss–Ostrogradsky theorem

$$\int_{\Sigma} \boldsymbol{t} \cdot \boldsymbol{u}\, ds = \int_{\Sigma} (\boldsymbol{T} \cdot \boldsymbol{n}) \cdot \boldsymbol{u}\, ds = \int_{V} \nabla \cdot (\boldsymbol{T} \cdot \boldsymbol{u})\, dV,$$

$$\int_{\Sigma} \boldsymbol{q} \cdot \boldsymbol{n}\, ds = \int_{V} \nabla \cdot \boldsymbol{q}\, dV,$$

one obtains

$$\int_{V} \left(\rho \frac{D}{Dt} \left(\frac{u^2}{2} + e \right) - \rho \boldsymbol{f} \cdot \boldsymbol{u} - \nabla \cdot (\boldsymbol{T} \cdot \boldsymbol{u}) + \nabla \cdot \boldsymbol{q} \right) dV = 0. \tag{5.27}$$

Since equality (5.27) is valid for any volume V, then

$$\rho \frac{D}{Dt} \left(\frac{u^2}{2} + e \right) = \rho \boldsymbol{f} \cdot \boldsymbol{u} + \nabla \cdot (\boldsymbol{T} \cdot \boldsymbol{u}) - \nabla \cdot \boldsymbol{q}, \tag{5.28}$$

or, opening the scalar products,

$$\rho \frac{D}{Dt} \left(\frac{u^2}{2} + e \right) = \rho f_i u_i - \frac{\partial q_i}{\partial x_i} + \frac{\partial}{\partial x_j} (\tau_{ji} u_i). \tag{5.29}$$

Further, substituting the heat flux $\boldsymbol{q}$ in (5.28) with the expression (4.16), expressing $\nabla \cdot (\boldsymbol{T} \cdot \boldsymbol{u})$ according to definition (4.13) of the strain rate tensor, and taking into account the equation of motion (5.19), one gets

$$\rho \frac{De}{Dt} = \boldsymbol{T} : \boldsymbol{E} + \nabla \cdot (k \nabla T). \tag{5.30}$$

Here denoted as $\boldsymbol{T} : \boldsymbol{E}$ is the scalar product of tensors, which by definition is equal to

$$\boldsymbol{T} : \boldsymbol{E} = T_{ij} E_{ij} = -p \frac{\partial u_i}{\partial x_i} + 2\mu \left(\varepsilon_{ij} \varepsilon_{ij} - \frac{1}{3} \varepsilon_{ii}^2 \right). \tag{5.31}$$

The second term in the right part (5.31) is called dissipative function Φ, because it characterizes the rate of viscous dissipation of energy in unit volume of fluid.

In Cartesian coordinates

$$\begin{aligned} \Phi = {} & 2\mu \left(\left(\frac{\partial u}{\partial x} \right)^2 + \left(\frac{\partial v}{\partial y} \right)^2 + \left(\frac{\partial w}{\partial z} \right)^2 \right) \\ & + \mu \left(\left(\frac{\partial v}{\partial x} + \frac{\partial u}{\partial y} \right)^2 + \left(\frac{\partial w}{\partial y} + \frac{\partial v}{\partial z} \right)^2 + \left(\frac{\partial u}{\partial z} + \frac{\partial w}{\partial x} \right)^2 \right) \\ & - \frac{2}{3} \mu \left(\frac{\partial u}{\partial x} + \frac{\partial v}{\partial y} + \frac{\partial w}{\partial z} \right)^2. \end{aligned} \tag{5.32}$$

In case of an incompressible fluid, the last term in (5.32) is equal to zero. Note also that for the viscous fluid $\Phi > 0$.

Thus, in a viscous flow, the mechanical energy associated with the work of internal stress forces, partially dissipates into heat (the term Φ, irreversible process) and partially converts into the internal energy (the term $-p\partial u_i/\partial x_i$, reversible process).

The energy conservation law (5.30) can be represented in other forms as well. If we introduce instead of e the quantities h (specific enthalpy) and s (specific entropy) and use known thermodynamic equalities

$$dh = de + d\left(\frac{p}{\rho}\right), \quad T\,ds = dh = \frac{dp}{\rho}, \tag{5.33}$$

then instead of (5.30) we obtain

$$\rho\frac{Dh}{Dt} = \frac{D\rho}{Dt} + \nabla\cdot(k\nabla T) + \Phi, \tag{5.34}$$

$$\rho T\frac{Ds}{Dt} = \nabla\cdot(k\nabla T) + \Phi. \tag{5.35}$$

In equations (5.30), (5.34) and (5.35) only the heat conductivity and the viscous dissipation of energy are taken into account. It is easy to take account of other kinds of energy, for example, the energy associated with chemical reactions, or Joule heat, by adding the corresponding terms in the right part. The case of an incompressible fluid and small temperature and pressure variations is of special interest, since it occurs in many situations. In this case the density and transfer coefficients can be assumed to be independent of p and T, and equations of motion can be separated from the energy conservation law. This means that distributions for u and p may be found without use of energy conservation, and the latter will determine the temperature distribution. The liquid or gas can be considered as incompressible if the flow velocity u is small in comparison with the sound velocity a. Therefore the criterion of incompressibility is the smallness of the Mach number $M = u/a$.

For incompressible liquid the following thermodynamic equalities [4] are obeyed

$$\frac{\partial s}{\partial t} = \left(\frac{\partial s}{\partial T}\right)_p \frac{\partial T}{\partial t}, \quad \nabla s = \left(\frac{\partial s}{\partial T}\right)_p \nabla T. \tag{5.36}$$

It follows from (5.35) and (5.36) that in case of an incompressible fluid and small velocity, pressure, and temperature variations, when the viscous dissipation can be neglected, the equation for energy reduces to the heat conduction equation

$$\frac{DT}{Dt} = \alpha\Delta T. \tag{5.37}$$

In the case of compressible fluid (gas), it is necessary to add the state equation to equations of continuity, motion and energy. At pressure and temperature close to normal, the gaseous state is well described by the state equation for an ideal gas

$$p = \frac{\rho A T}{M} \quad \text{or} \quad pV = nAT, \tag{5.38}$$

where $A = 8.3$ J/K·mole is the gas constant; M is the molecular mass of gas; $\rho = nM/V$.

Accordingly, for a liquid

$$\rho = \rho_0(1 - \alpha(T - T_0) + \beta(p - p_0)). \tag{5.39}$$

Here T_0 and p_0 are reference values of the temperature and pressure, α and β are coefficients of thermal and volume expansion of liquid. In particular, for water under normal conditions $\alpha = 2.1 \cdot 10^{-4}\ \text{K}^{-1}$, $\beta = 4.6 \cdot 10^{-10}\ \text{Pa}^{-1}$.

The state equation for a real gas and the unified state equations for gas and liquid will be consider in the section 5.6.

5.3 Multi-Component Mixtures

Hitherto, only homogeneous mediums have been considered in our formulation of conservation laws. Consider now these laws for multi-component systems in which chemical reactions can run. In homogeneous mediums the mass transfer takes place, if the velocity of the medium is non-zero or/and an external mass force acts on the medium. In a multi-component system mass transfer can also occur due to concentration gradients. Simultaneously with the transfer of mass, the momentum and heat transfer takes place in such systems as well.

For multi-component systems it is possible to derive the continuity equation for each component. Let u_i and ρ_i be the velocity and mass density of i-th component. If this component is not formed in volume of the mixture, then the continuity equation for this component has a form (5.12) with u and ρ replaced by u_i and ρ_i. If a chemical reaction runs in the volume, generating the i-th component at a rate r_i, then the mass conservation equation can be written as

$$\frac{Dm_i}{Dt} = \int_V r_i \, dV. \tag{5.40}$$

By reiterating the procedure of deriving equation (5.12), it is easy to obtain the continuity equation for the i-th component

$$\frac{D\rho_i}{Dt} + \nabla \cdot (\rho_i \mathbf{u}_i) = r_i. \tag{5.41}$$

Note that $r_i > 0$ in reactions with the formation of i-th component, and $r_i < 0$ in case of its absorption. The values r_i are determined by the kinetics of chemical reactions, local thermodynamic state and the stoichiometric coefficients characteristic for given chemical reactions. According to chemical thermodynamics [5], the rate of mass exchange in a chemical reaction for the i-th component is determined by

$$r_i = \frac{dm_i}{dt} = \nu_i M_i \frac{d\xi}{dt}, \tag{5.42}$$

where ν_i is the corresponding stoichiometric coefficient; ξ is degree of completeness of the reaction or simply the reaction coordinate. Since the total mass in the run of chemical reactions remains constant, summation of (5.42) over components gives

$$\sum r_i = 0 \quad \text{or} \quad \sum \nu_i M_i = 0. \tag{5.43}$$

Earlier in section 4.4 the weight-average velocities $\boldsymbol{u}$ were introduced (see (4.23)), as well as the mass fluxes $\boldsymbol{J}_i$ relative to the average velocity (see (4.26)). In view of these equations, (5.41) may be rewritten as

$$\frac{\partial \rho_i}{\partial t} + \nabla \cdot (\rho_i \boldsymbol{u}) = -\nabla \cdot \boldsymbol{J}_i + r_i \tag{5.44}$$

or

$$\frac{D\rho_i}{Dt} + \rho_i \nabla \cdot \boldsymbol{u} = -\nabla \cdot \boldsymbol{J}_i + r_i. \tag{5.45}$$

The continuity equation for all mixture can be obtained from (5.44) by summing over all components

$$\frac{\partial \sum \rho_i}{\partial t} + \nabla \cdot \left(\sum \rho_i \boldsymbol{u}\right) = -\nabla \cdot \sum \boldsymbol{J}_i + \sum r_i. \tag{5.46}$$

Since $\rho = \sum \rho_i$ is the net density of mixture, and in accord with (4.30) and (5.43) $\sum \boldsymbol{J}_i = 0$ and $\sum r_i = 0$, the equation (5.46) reduces to the continuity equation (5.12) for pure liquid.

The equation of motion of viscous fluid can be applied also to multi-component mixtures so far as mass forces act equally on all components of the mixture. Such a force, for example, is the gravity force. The electric force can act selectively on some components, for example, on the electrolyte mixed with electrically neutral liquid. The principal cause of this fact is, that the phenomenological equation of the viscous fluid (4.13), determining the form of stress tensor, does not depend on concentration gradients of the components. Since the equation (4.13) is tensor equation containing the second-rank tensors, such dependence could contain only the products $\nabla\rho_i\nabla\rho_j$, which form a tensor of the second rank. However, the terms $\nabla\rho_i\nabla\rho_j$ would be of the second order of smallness in comparison with the strain rate tensor. Recall that the law (4.13) is true only for small values of strain rate. Hence, in this approximation the stress tensor cannot depend on concentration gradients.

We have quite different situation with the energy equation. If only the gravity force acts on the medium, then the energy equation for multi-component mixture is the same as for a pure fluid (5.28), but in the expression for a thermal flux one has to take into account the difference between components. According to [2], the main equation generalizing the Fourier law for a multi-component mixture, is

$$\boldsymbol{q} = \sum h_i \boldsymbol{J}_i - k\nabla T, \tag{5.47}$$

where h_i is partial specific enthalpy of the i-th component.

Fluxes $\boldsymbol{J}_i$ relative to average mass velocity enter the continuity and energy equations for the multi-component mixture, therefore these expressions have to be added to the corresponding equations. They connect fluxes $\boldsymbol{J}_i$ with driving forces causing the corresponding component mass transfer. In section 4.4 it was shown, that such driving forces are gradients of concentration and the external electric field. Influence of the electric field will be considered in the next section. The flux is also possible due to the gradient of pressure (baro-diffusion), however it is usually small, except for cases of big pressure gradients, for example, in the centrifugation. Finally, a flux induced by a temperature gradient is possible (the Soret effect). A detailed discussion of these effects and the estimation of their contribution to mass transfer can be found in [1, 2]. The presentation here will be restricted to the case of ordinary diffusion.

At the core of thermodynamics with ordinary diffusion lies the statement that a driving force is gradient of chemical potential. The chemical potential is defined in the following way [5].

Consider a multi-component system determined by the following parameters: entropy S, molar volume V, and the molar quantities of components $n_1, n_2, \ldots, n_N$. Then the internal energy of the system may be written as

$$E = E(S, V, n_1, n_2, \ldots, n_N). \tag{5.48}$$

From the Gibbs fundamental equation it follows that

$$dE = T\,dS - p\,dV + \sum \left(\frac{\partial E}{\partial n_i}\right)_{S,\,V,\,n_j} \partial n_i. \tag{5.49}$$

The coefficients at dn_i are called chemical potentials:

$$\mu_i = \left(\frac{\partial E}{\partial n_i}\right)_{S,V,n_j}. \tag{5.50}$$

The chemical potential is a partial molar quantity, and therefore intensive quantity, and accordingly it can be treated as concentration or molar fraction. If the chemical reactions take place in the system, the condition of chemical balance is

$$\sum v_i \mu_i = 0. \tag{5.51}$$

Now we also define the ideal solution. The solution is termed ideal, if the corresponding chemical potentials are

$$\mu_i = \mu_i^0(T, p) + AT \ln x_i. \tag{5.52}$$

It has been found experimentally that all sufficiently diluted solutions, which were earlier named "extremely diluted", behave as ideal solutions.

The chemical potential for a non-ideal solution is also expressed similar to (5.52), but instead of molar fractions x_i, the component activities a_i enter the expression. The activities are related to x_i by the equations $a_i = \gamma_i x_i$, where γ_i are the activity coefficients. Thus, for a non-ideal solution

$$\mu_i = \mu_i^0(T, p) + AT \ln a_i. \tag{5.53}$$

Notice that for ideal solution $\gamma_i = 1$ and $a_i = \gamma_i x_i$.

To characterize the solution imperfection relative to the solvent it is frequently more preferable to use osmotic coefficient Ξ instead of the activity coefficients

$$\mu_i = \mu_i^0(T, p) + \Xi AT \ln x_1. \tag{5.54}$$

where x_i is the molar fraction of the solvent.

Comparing (5.53) with (5.54) we obtain

$$\Xi = 1 + \frac{\ln \gamma_1}{\ln x_1}. \tag{5.55}$$

According to principles of thermodynamics of non-equilibrious processes, the relative fluxes are

$$J_i = \frac{C^2}{\rho} \sum_j M_i M_j D_{ij} \left(\frac{x_j}{AT} \nabla \mu_j\right). \tag{5.56}$$

Here the expression in parentheses is the dimensionless driving force of ordinary diffusion for j-th component, D_{ij} are coefficients of binary diffusion.

Putting here the expressions for μ_i from (5.53) yields

$$J_i = \frac{C^2}{\rho}\sum_j M_i M_j D_{ij}\left(\frac{\nabla \ln a_j}{\nabla \ln x_j}\right)_{T,p} \nabla x_j. \tag{5.57}$$

In case of an extremely diluted solution $a_i = x_i$, and it follows from (5.57)

$$J_i = \frac{C^2}{\rho}\sum_j M_i M_j D_{ij} \nabla x_j. \tag{5.58}$$

In particular, for a binary solution $x_2 = 1 - x_1$, $D_{12} = D_{21}$ and

$$J_1 = \frac{C^2}{\rho}\sum_j M_1 M_2 D_{12} \nabla x_1. \tag{5.59}$$

Let's show that the Fick's law (4.27) follows from (5.59). Express the molar fraction x_1 in terms of mass fraction

$$x_1 = \frac{\omega_1/M_1}{\omega_1/M_1 + \omega_2/M_2},$$

$$\nabla x_1 = \frac{\nabla\omega_1}{M_1 M_2 (\omega_1/M_1 + \omega_2/M_2)^2} = \frac{\overline{M}^2}{M_1 M_2}\nabla\omega_1 \tag{5.60}$$

and substitute in (5.59). Then, in view of $\overline{M} = \rho/C$, we obtain

$$J_1 = -\rho D_{12} \nabla\omega_1. \tag{5.61}$$

Getting back to (5.44), one obtains the continuity equation for binary mixture

$$\frac{\partial \rho_1}{\partial t} + \nabla\cdot(\rho_1 \boldsymbol{u}) = D_{12}\Delta\rho_1 + r_1 \tag{5.62}$$

or

$$\frac{\partial C_1}{\partial t} + \nabla\cdot(C_1 \boldsymbol{u}^*) = D_{12}\Delta C_1 + R_1. \tag{5.63}$$

In the absence of chemical reactions $r_1 = R_1 = 0$ and the equation (5.63) converts to the known equation of convective diffusion.

5.4
Multi-Phase Mixtures

Multiphase mixtures, in contrast to homogeneous mixtures (solutions, mixtures of gases) are characterized by the presence of macroscopic heterogeneities or inclusions (solid particles, bubbles, drops, macromolecules). Multiphase mixtures are also named heterogeneous. Among these are mixtures of gas with particles (aerosols), of liquid with solid particles (suspensions), of gas with drops (gas-liquid mixtures), of liquid with bubbles (liquid-gas mixtures), and also mixtures of liquid with drops of another liquid (emulsions). Colloidal mixtures, which are frequently called colloids, and also micellar solutions, fill in an intermediate domain between heterogeneous and homogeneous mixtures.

The listed types of multiphase mixtures form the class of disperse mediums consisting of two phases. Inclusions in a continuous or disperse phase are in the suspended state. The disperse phase consists of particles of various nature, in particular drops, solid particles, bubbles.

In studying various processes in multiphase mixtures, the scientists usually assume that the size of inclusions in a mixture (particles, drops, bubbles, the pores in the porous mediums) is much greater than the size of the molecules. This assumption named the continuity hypothesis, allows us to use the mechanics of continuous mediums for description of processes occurring inside or near the separate inclusions. For description of physical properties of phases, such as viscosity, heat conductivity etc., it is possible to use equations and parameters of an appropriate single-phase medium.

If the average distance between the inclusions is much less than the characteristic size of a continuous phase over which the macroscopic parameters (velocity, pressure, temperature etc.) change, then it is possible to describe the macroscopic processes in the mixture by the methods of mechanics of continuous mediums as well. To this end, the averaged or macroscopic parameters are introduced. At this point, the concept of multi-velocity continuum [6], representing a set of N continuums, can also be introduced. The number N is the number of considered phases, each of which corresponds to a certain constituent phase of the mixture and fills one and the same volume. For each of these constituent continuums we define in the usual manner the local density, which is called the reduced density

$$\rho_i = \frac{m_i}{V} = \frac{\text{mass of } i\text{-th phase}}{\text{volume of mixture}}, \tag{5.64}$$

and also the velocity $\boldsymbol{u}_i$ and other parameters. Thus, to each point of space occupied by the mixture, N densities, N velocities etc., are assigned. As well as for homogeneous mixtures (solutions), it is possible to define parameters describing the mixture as a whole. The mixture's mean mass density and mass (bari-centric) velocity are

$$\rho = \sum_{i=1}^{N} \rho_i, \quad \boldsymbol{u} = \frac{1}{\rho} \sum_{i=1}^{N} \rho_i \boldsymbol{u}_i. \tag{5.65}$$

In the same manner as for solutions, one can introduce the phase velocity relative to the mass center of the mixture or medium as a whole

$$\mathbf{w}_i = \boldsymbol{u}_i - \boldsymbol{u}. \tag{5.66}$$

From (5.65) and (5.66) it follows that

$$\sum_{i=1}^{N} \rho_i \mathbf{w}_i = 0. \tag{5.67}$$

The distinguishing feature of multiphase medium is that here it is necessary to introduce two substantial derivatives

$$\frac{d_i}{dt} = \frac{\partial}{\partial t} + \boldsymbol{u}_i \cdot \nabla = \frac{\partial}{\partial t} + u_i^k \nabla^k = \frac{\partial}{\partial t} + u_i^k \frac{\partial}{\partial x^k},$$

$$\frac{d}{dt} = \frac{\partial}{\partial t} + \boldsymbol{u} \cdot \nabla = \frac{\partial}{\partial t} + u^k \nabla^k = \frac{\partial}{\partial t} + u^k \frac{\partial}{\partial x^k}. \tag{5.68}$$

In (5.68) summation over repeating indexes is carried out.

The mechanics of multiphase mixtures can be constructed on the basis of conservation laws for mass, momentum and energy in each phase. The main difference from homogeneous solutions is in that the mass, momentum and energy exchange occurs between the phases.

Let J_{ji} be the rate of mass transfer (evaporation, condensation, chemical reactions) from j-th to i-th phase, or the reverse. In the latter case, $J_{ji} < 0$. Then the continuity equation for i-th phase in the integral form is

$$\int_V \frac{\partial \rho_i}{\partial t} dV = -\int_S \rho_i u_i^n \, ds + \int_V \sum_{j=1}^{N} J_{ji} \, dV. \tag{5.69}$$

Here is formally set $J_{ii} = 0$. It follows from the mass conservation in physicochemical processes that $J_{ji} = J_{ij}$ [2].

2) In conventional (and more convenient) notation one distinguishes between $\boldsymbol{J}_{ij}$ and $\boldsymbol{J}_{ji}$, namely, $\boldsymbol{J}_{ij} = -\boldsymbol{J}_{ii}$. This has a simple physical meaning: a mass transferred from j to i-component in a certain process, is negative of the mass transferred from i to j-component in this process. In this notation, $\boldsymbol{J}_{ij}$ can be considered as an antisymmetric second-rank tensor, and the condition $\boldsymbol{J}_{ii} = 0$ for its diagonal elements (no summation assumed here!) follows automatically. – VF

The equation (5.69), with the use of the Gauss formula (5.1), transforms to the mass conservation law for each phase in differential form

$$\frac{\partial \rho_i}{\partial t} + \nabla \cdot (\rho_i \mathbf{u}_i) = \sum_{j=1}^{N} J_{ji} \quad (i = 1, 2, \ldots, N). \tag{5.70}$$

After summation of (5.70) over i, one obtains the continuity equation for a mixture as a whole. It has the same form as the corresponding equation for a homogeneous medium.

The momentum conservation for each phase in the integral form is expressed by

$$\int_V \frac{\partial \rho_i \mathbf{u}_i}{\partial t} dV = -\int_S \rho_i \mathbf{u}_i u_i^n \, ds + \int_S \mathbf{t}_i^n \, ds + \int_V \rho_i \mathbf{f}_i \, dV + \int_V \sum_{j=1}^{N} \mathbf{P}_{ji} \, dV. \tag{5.71}$$

Here the first term on the right corresponds to momentum inflow of the i-th phase through the surface S; second and third terms describe the influence of the external forces, which act on i-th phase and are described by the stress tensor τ_i^{kl} and vector $\mathbf{f}_i$ of mass forces; $\mathbf{P}_{ji}$ is the momentum exchange rate between j-th and i-th phases. It follows from momentum conservation that $\mathbf{P}_{ji} = -\mathbf{P}_{ij}$, $\mathbf{P}_{ii} = 0$. Applying the Gauss-Ostrogradskiy theorem to (5.71), one obtains the equation of motion for the i-th phase, which by virtue of (5.70) and (5.67) transforms to

$$\rho_i \frac{D_i \mathbf{u}_i}{Dt} = \nabla^{\mathrm{k}} \mathbf{t}_i^{\mathrm{k}} + \rho_i \mathbf{f}_i + \sum_{j=1}^{N} (\mathbf{P}_{ji} - J_{ji} \mathbf{u}_i); \quad (i = 1, 2, \ldots, N). \tag{5.72}$$

If we determine the stress tensor of the surface-forces and the vector of mass-forces for the entire medium as

$$\mathbf{t} = \sum_{i=1}^{N} \mathbf{t}_i, \quad \rho \mathbf{f} = \sum_{i=1}^{N} \rho_i \mathbf{f}_i, \tag{5.73}$$

sum (5.72) over i and then use (5.66), we obtain

$$\rho \frac{d\mathbf{u}}{dt} = \nabla^{\mathrm{k}} \mathbf{t} + \rho \mathbf{f} - \sum_{i=1}^{N} \nabla^{\mathrm{k}} \cdot (\rho_i w_i^{\mathrm{k}} \mathbf{w}). \tag{5.74}$$

This equation differs from the equation of motion for a homogeneous medium, since it contains the relative rather than absolute, velocities.

Let us now turn to the energy-equation for multiphase mixture and introduce specific energy as a sum of specific kinetic and internal energies:

$$\mathrm{u} = k + e, \tag{5.75}$$

where it is taken that the internal energy is additive over components involved

$$e = \frac{1}{\rho} \sum_{i=1}^{N} \rho_i e_i, \tag{5.76}$$

and the kinetic energy is determined by

$$k = \frac{1}{\rho} \sum_{i=1}^{N} \frac{\rho_i u_i^2}{2}. \tag{5.77}$$

Then the energy of the mixture is

$$u = \sum_{i=1}^{N} \rho_i \left(e_i + \frac{u_i^2}{2} \right) = \sum_{i=1}^{N} \rho_i u_i. \tag{5.78}$$

The energy balance equation for i-th phase can be written in the integral form

$$\int_V \frac{\partial \rho_i \mathsf{u}_i}{\partial t} dV = - \int_S \rho_i \mathsf{u}_i u_i^n \, ds + \int_S c_i^n \, ds + \int_V \rho_i \boldsymbol{f} \cdot \boldsymbol{u}_i \, dV$$
$$+ \int_V \sum_{j=1}^{N} \mathsf{u}_{ji} \, dV - \int_S q_i^n \, ds, \tag{5.79}$$

where the first term on the right describes the energy inflow of i-th phase through the surface S, the second and the third terms represent the work of the external surface ($c_i^n = \boldsymbol{t}_i^n \cdot \boldsymbol{u}_i$) – and mass-forces acting on the i-th phase, u_{ji} is the energy exchange rate between i-th and j-th phases, with $\mathsf{u}_{ji} = -\mathsf{u}_{ij}$ and $\mathsf{u}_{ji} = 0$, the fifth term represents heat inflow through the surface S characterized by the heat flux vector $\boldsymbol{q}$.

Applying to (5.79) the Gauss-Ostrogradsky theorem and using expressions (5.68), (5.70), one obtains

$$\rho_i \frac{d_i}{dt} \left(e_i + \frac{u_i^2}{2} \right) = \nabla \cdot (\boldsymbol{c}_i - \boldsymbol{q}_i) + \rho_i \boldsymbol{f}_i \cdot \boldsymbol{u}_i + \sum_{i=1}^{N} \left(\mathsf{u}_{ji} - J_{ji} \left(e_i + \frac{u_i^2}{2} \right) \right) \tag{5.80}$$

Thus, equations (5.70), (5.72) and (5.80) describe the behavior of multiphase mixture if the expressions for J_{ji}, $\boldsymbol{P}_{ji}$, u_{ji}, e_i, τ_1^{kl} and $\boldsymbol{q}_i$ are known. A more detailed analysis of this problem can be found in [6].

In contrast to homogeneous mixtures, the heterogeneous mixtures are generally described by multi-velocity model with regard to dynamic effects, because of discrepancies between the phase velocities. The relative velocities $\mathbf{w}_i$ of phases can be close (by order of magnitude) to velocities $\boldsymbol{u}_i$ of their absolute motion, or to mass-mean velocities $\boldsymbol{u}$. Another difference between a heterogeneous mixture and a homogeneous one is, that each component of the homogeneous mixture is considered as effectively occupying the whole volume together with other components, therefore $V_1 = V_2 = \cdots = V_N = V$, while in heterogeneous mixture each phase occupies only a part of the volume, that is $V_1 + V_2 + \cdots + V_N = V$. Therefore in the theory of heterogeneous mixtures the volume fractions W_i occupied by phases, are introduced, so that

$$\sum_{i=1}^{N} W_i = 1. \tag{5.81}$$

These parameters are also known as volume concentration or volume content of a phase.

Besides the reduced density ρ_i, the true densities of components are defined

$$\rho_i^0 = \frac{m_i}{V_i} = \frac{\text{mass of } i\text{-th phase}}{\text{volume of } i\text{-th phase}} = \frac{\rho_i}{W_i}. \tag{5.82}$$

The laws describing relative motion of phases in heterogeneous mediums are complicated, because these motions are determined by processes of phase interactions, for example, liquid-carrier flow over the particles or direct interactions between the particles.

Sometimes, when the inertial effects of relative motion of phases are insignificant, a one-liquid approximation for description of motion of a heterogeneous medium can be to used as well. As an example we can name an "inertia-less" description of phases' motion in a porous medium, known as Darcy's law

$$\rho_i w_i = -\chi_i \nabla p, \tag{5.83}$$

which is used in models of saturated porous mediums and sometimes in models of gas-liquid flows.

Of particular interest is the case of a heterogeneous mixture with small volume content of the disperse phase $W_i \ll 1$. In this case we can assume that the disperse phase exerts a weak influence on the continuous phase. Then fields of velocity, pressure, temperature and other parameters of the continuous phase could be determined by using one-velocity model, and then, for given distributions of parameters, one can determine the behavior of disperse phase. If the disperse phase represents a discrete system of inclusions (solid particles, drops, bubbles, macromolecules), it can be characterized by a distribution $n(V, t, P)$ of inclusions over volumes V at a point of space P. Inclusions can exchange mass (due to evaporation, condensation, fusion etc.) with the continuous phase, and also interact between themselves, – they can collide, coagulate, coalesce, break, form the inclusions of various size and shape. In addition, a phase can nucleate in conditions of mixture super-saturation and then increase in size due to a phase transition. The equation describing change of inclusion distribution over sizes (with account of only pair collisions) takes the form

$$\frac{\partial n}{\partial t} + \nabla(\mathbf{u}n) + \frac{\partial}{\partial V}\left(\frac{\partial V}{\partial t}n\right) = I_c + I_b + I_n + I_a, \tag{5.84}$$

where $\mathbf{u}$ is the velocity of particle motion; V is the particle volume; dV/dt is the volume change rate due to phase transitions; I_c, I_b, I_n and I_a are rates of formation of inclusions of volume V due to processes of coagulation, breakage, nucleation and deposition at the boundary of flow volume.

The majority of processes to be considered further is characterized by small volume content of the dispersed phase, therefore the main attention will be given to behavior of dispersed phase.

5.5 Charged Mixtures

In section 4.5 it was shown that the presence of an external electric field results in migration of a charged component in a mixture. At the same time, the mixture transport occurs due to convection and diffusion. Consider how the transport equations change when a charged component is in an external electric field. We shall be limited to a case of extremely diluted mixtures.

Notice first of all that equations (5.56) for the relative fluxes hold for the charged components as well, but for these components, the quantities μ_i should be understood as the electrochemical potentials depending on pressure, temperature, chemical composition and electric parameters. The last are, in turn, described by independent equations. As to phenomenological equations for concentrated mixtures, they are cumbersome and demand a lot of empirical parameters.

In the approximation of an extremely diluted solution, the molar flux of i-th component is a sum of fluxes caused by migration, diffusion, and convection, and in view of the expressions (4.27) and (4.32), is equal to

$$\mathbf{j}_i^* = -\upsilon_i z_i F C_i \nabla\phi - D_i \nabla C_i + C_i \mathbf{u}. \tag{5.85}$$

Here $\mathbf{u}^*$ is replaced by $\mathbf{u}$, because in an extremely diluted solution $\mathbf{u}^* = \mathbf{u}$. It is a common practice in electrochemistry to use the molar fluxes $\mathbf{j}_i^*$, but mass mean velocity $\mathbf{u}$. A similar equation may be written for the mass flux

$$\mathbf{j}_i^* = -\upsilon_i z_i F \rho_i \nabla\phi - D_i \nabla \rho_i + \rho_i \mathbf{u}. \tag{5.86}$$

Recall that υ_i is the mobility, D_i is the coefficient of binary diffusion, related to with mobility by the equation $D_i = AT\upsilon_i$. Equations (5.85) and (5.86) are known as the Nernst-Planck equations.

The motion of charged component produces the electric current. Its density is, in accord with (4.35), given by

$$\boldsymbol{i} = F\sum z_i \boldsymbol{j}_i^* = -\sigma\nabla\phi - F\sum z_i D_i \nabla C_i + F\boldsymbol{u}\sum z_i C_i, \tag{5.87}$$

where $\sigma = F^2 \sum z_i^2 \upsilon_i C_i$ is the electric conductivity of the solution.

In order to determine fluxes $\boldsymbol{j}_i$ and current density $\boldsymbol{i}$, it is necessary to know $\nabla\varphi$ or $\boldsymbol{E}$. For their definition it is necessary to use the Maxwell equations. In general case the external electric field induces secondary electric and magnetic fields (the medium's response), which in turn influence the external field. However, if the external magnetic field is absent, and the external electric field is quasi-stationary, then the electrodynamical problem reduces to electrostatic one, namely, to determining of the electric potential distribution in liquid, described by Poisson equation

$$\Delta\phi = -\frac{\rho_E}{\varepsilon}, \quad \varepsilon = \varepsilon_0\varepsilon_r. \tag{5.88}$$

Here $\rho_E = F\sum z_i C_i$ is the electric charge density, $\varepsilon_0 = 8.85 \cdot 10^{-12}\ \mathrm{K}^2\ \mathrm{m}^2/\mathrm{N}$ is the permittivity of vacuum, ε_r is the relative permittivity. For water $\varepsilon_r = 78.3$.

In compliance with the Lorentz law, the mass electric force acting on liquid bearing the charge ρ_E, is

$$\boldsymbol{f}_\mathrm{E} = \rho_E \boldsymbol{E}. \tag{5.89}$$

A practically important case is the case of an electrically neutral mixture. In this approximation $\rho_E = 0$ and

$$\sum z_i C_i = 0 \tag{5.90}$$

As, a rule this condition is approximately satisfied. In the absence of processes resulting in charge separation, a stream remains electrically neutral except for the areas adjoining the charged surfaces. These areas are known as double electric layers or Debye's layers. Their thickness is small (from 1 up to 10 nanometers). The account of these layers is important when interactions between the small particles, and processes in the immediate proximity of the charged surfaces are considered.

Consider a binary solution of the electrolyte consisting of a non-ionized solvent and completely ionized dissolved substance, for example, a salt in very small concentration. The dissolved substance consists of positively and negatively charged ions (formed by molecule dissociation) with concentrations υ_+ and υ_-, and the molar concentrations C_+ and C_- respectively. Consider the case when chemical reactions are absent, that is $r_i = 0$. From the condition (5.90) of the electroneutrality (5.90) it follows that

$$z_+ C_+ + z_- C_- = 0 \tag{5.91}$$

Introduce reduced concentration of ions

$$C = \frac{C_+}{\upsilon_+} = \frac{C_-}{\upsilon_-}. \tag{5.92}$$

The concentration C so determined satisfies equation (5.91). From the continuity equation (5.44), under the condition of solvent incompressibility ($\nabla \cdot \boldsymbol{u} = 0$) and absence of the chemical reactions ($r_i = 0$) it follows that

$$\frac{\partial C}{\partial t} + \boldsymbol{u} \cdot \nabla C = z_\pm \upsilon_\pm F \nabla \cdot (C \nabla \phi) + D_\pm \Delta C. \tag{5.93}$$

We assume that $\upsilon_\pm$ and $D_\pm$ are constants. Subtract from the equation (5.93) for the positive ions the same equation for the negative ions. The result is

$$(z_+ \upsilon_+ - z_- \upsilon_-) F \nabla \cdot (C \nabla \phi) + (D_+ - D_-) \Delta C = 0. \tag{5.94}$$

Solving this for the term with the electric potential and putting it into (5.93), one obtains the equation of convective diffusion

$$\frac{\partial C}{\partial t} + \boldsymbol{u} \cdot \nabla C = D \Delta C \tag{5.95}$$

with the effective coefficient of diffusion D

$$D = \frac{z_+ \upsilon_+ D_- - z_- \upsilon_- D_+}{z_+ \upsilon_+ - z_- \upsilon_-}. \tag{5.96}$$

Since $D_\pm = AT\upsilon_\pm$, D can be transformed to

$$D = D_+ \frac{1 - z_+/z_-}{1 - z_+ D_+ / z_- D_-}. \tag{5.97}$$

In particular, if $D_+ = D_-$, then $D = D_+ = D_-$.

Thus, the reduced-concentration distribution in a binary extremely diluted solution of electrolyte is described by the same equation (5.95) as for uncharged components, but with the different diffusion coefficient.

The electric potential distribution is given by the equation (5.94), which is essentially a continuity equation for the electric current density. Indeed, taking into account the condition (5.91) of electric neutrality, one can rewrite (5.87) as

$$-\frac{i}{z_+ \upsilon_+ F} = (z_+ \upsilon_+ - z_- \upsilon_-) F C \nabla \phi + (D_+ - D_-) \nabla C. \tag{5.98}$$

Taking the divergence of the Eq. (5.98) and noting that the divergence of the right part is zero in view of (5.94), we obtain $\nabla \cdot \boldsymbol{i} = 0$. It is the continuity equation for a stationary electric current density $\boldsymbol{i}$.

5.6
The Criteria of Similarity

In the previous section we have derived the transfer equations for pure and multicomponent liquids, and also for mixtures containing charged components. The initial and boundary conditions for these equations for specific applications will be considered later. In this section we consider the most frequently used system of equations and describe the parameters of similarity.

Until the further applications, the treatment here will be limited to binary extremely diluted incompressible solutions without chemical reactions. We assume that the solution remains electrically neutral in the presence of an electrolyte. It is also assumed that all physical and chemical parameters and transfer coefficients are constants. Out of all possible mass forces, only the gravity force operates.

These assumptions allow us to write equations of continuity, momentum, energy and diffusion as

$$\nabla \cdot \boldsymbol{u} = 0, \tag{5.99}$$

$$\frac{D\boldsymbol{u}}{Dt} = -\frac{1}{\rho}\nabla p + \nu \Delta \boldsymbol{u} + \mathbf{g}. \tag{5.100}$$

$$\frac{DT}{Dt} = \alpha \Delta T, \tag{5.101}$$

$$\frac{DC}{Dt} = D \Delta C. \tag{5.102}$$

In the energy equation, the terms with heat transfer due to transfer of mass, viscous dissipation, and internal energy extraction, are omitted.

Now transform the system (5.99)–(5.102) to a dimensionless form. To this end, introduce the characteristic values of length L, velocity U and time τ. Notice that the characteristic time is natural in problems, in which processes under the external, time-varying actions are considered. We denote as τ the characteristic time of such an action. In many problems τ is not an independent value, and is determined as $\tau = L/U$.

Introduce the following dimensionless variables

$$\mathbf{x}' = \mathbf{x}/L, \quad t' = t/\tau, \quad \mathbf{u}' = \mathbf{u}/U. \tag{5.103}$$

We define the quantity ρU^2 as the characteristic pressure (that is, characteristic dynamic pressure). Then the dimensionless pressure

$$p' = \frac{p - p_0}{\rho U^2}. \tag{5.104}$$

Here p_0 is given pressure, for example, the pressure in a liquid in the unperturbed flow.

As dimensionless values of temperature and concentration are taken the quantities

$$T' = \frac{T - T_0}{T_0 - T_w}, \quad C' = \frac{C - C_0}{C_0 - C_w}, \tag{5.105}$$

where the index 0 denotes, as a rule, the initial value, and w is the value at the surface bounding the flow region.

Switching in (5.99) to (5.102) to dimensionless variables, one obtains

$$\nabla \cdot \mathbf{u}' = 0, \tag{5.106}$$

$$St\frac{\partial \mathbf{u}'}{\partial t} + (\mathbf{u}' \cdot \nabla)\mathbf{u}' = -\nabla p' + \frac{1}{\mathrm{Re}}\Delta \mathbf{u}' + \frac{1}{Fr}\frac{\mathsf{g}}{g}. \tag{5.107}$$

$$St\frac{\partial T'}{\partial t} + (\mathbf{u}' \cdot \nabla)T' = \frac{1}{\mathrm{Pe}_T}\Delta T', \tag{5.108}$$

$$St\frac{\partial C'}{\partial t} + (\mathbf{u}' \cdot \nabla)C' = \frac{1}{\mathrm{Pe}_D}\Delta C'. \tag{5.109}$$

Here the following dimensionless parameters are introduced:

1. **The Strouhal number**

$$St = \frac{L}{\tau U} = \frac{L/U}{\tau} = \frac{\text{characteristic time of flow}}{\text{characteristic time of action}}. \tag{5.110}$$

2. **The Froude number**

$$\mathrm{Fr} = \frac{U^2}{\mathsf{g}L} = \frac{\rho U^2/L}{\rho \mathsf{g}} = \frac{\text{force of inertia}}{\text{gravity force}}. \tag{5.111}$$

3. **The Reynolds number**

$$\mathrm{Re} = \frac{\rho UL}{\mu} = \frac{\rho U^2/L}{\mu U/L^2} = \frac{\text{force of inertia}}{\text{force of viscosity}}. \tag{5.112}$$

4. The Peclet number

$$\mathrm{Pe}_T = \frac{UL}{\alpha} = \frac{\rho C_p U(T_0 - T_w)/L}{k(T_0 - T_w)/L^2}$$

$$= \frac{\text{heat due to convection}}{\text{heat due to heat conductivity}}. \qquad (5.113)$$

$$\mathrm{Pe}_D = \frac{UL}{D} = \frac{U(C_0 - C_w)/L}{D(C_0 - C_w)/L^2}$$

$$= \frac{\text{mass due to convection}}{\text{mass due to diffusion}}. \qquad (5.114)$$

The first parameter, the Strouhal's number, is a measure of non-stationarity of the process. At $\mathrm{St} \ll 1$ or $\tau \gg L/U$ the first term in the left part of (5.107)–(5.109) can be neglected. At that, the flow can be considered as stationary. In this connection it is necessary to notice, that there are problems in which the time-dependence can exist at the boundary, for example in problems with time-varying interface (drops or bubbles deformation, surface waves in liquid etc.). In this case the bulk flow can be assumed stationary everywhere except for a thin layer near a surface. This kind of a problem is called sometimes the quasi-stationary problem.

The Froude's number characterizes the influence of gravity on liquid's motion. In many problems $\mathrm{Fr} \gg 1$, that is, the gravity exerts weak influence on the flow, therefore the last term in the right part (5.107) can be neglected.

An important parameter is the Reynolds number. At $\mathrm{Re} \gg 1$ the viscous term in (5.107) is small in comparison with the inertial one. Neglecting it, one obtains the equations of motion of an ideal liquid (Euler's equations). These equations describe flow of liquid in a volume, with the exception of small regions, adjoining the surface of an immersed body. Near such surfaces, the viscosity force can be comparable with inertial force, which results in formation of a viscous boundary layer with thickness $\delta \sim L/(\mathrm{Re})^{1/2}$, where L is the characteristic size of the body. Approximation $\mathrm{Re} \ll 1$ leads to an "inertialess" flow described by Stokes' equations. These equations follow from (5.107), in which the inertial terms are omitted. Such equations describe the problems of micro-hydrodynamics, for example, problems of the small particles' motion in a liquid.

The Peclet's number is defined as thermal Peclet's number (5.113) and the diffusion one (5.114). At $\mathrm{Pe}_T \gg 1$ the basic contribution to the heat transfer comes from convection, and at $\mathrm{Pe}_T \ll 1$ – from the thermal conductivity. Similar conclusions refer to Pe_D.

In problems of heat-mass exchange, the following dimensionless quantities are used apart from Peclet's number:

5. The Prandtl number

$$\mathrm{Pr} = \nu/\alpha, \quad \mathrm{Pe}_T = \mathrm{Re}\ \mathrm{Pr}. \qquad (5.115)$$

6. The Schmidt number

$$\mathrm{Sc} = \nu/D, \quad \mathrm{Pe}_D = \mathrm{Re}\ \mathrm{Sc}. \tag{5.116}$$

7. The Lewis number

$$\mathrm{Le} = \alpha/D. \tag{5.117}$$

The Prandtl's number, the Schmidt's number, and the Lewis's number characterize different properties of liquid. In table 5.1 and 5.2 some values of these numbers for various substances [7] are given.

Knowing properties of liquids, it is possible to find Pr, Sc, Le, and consequently Peclet's number for various values of Re. Such estimations are of great importance, because they allow us to simplify the system of equations (5.106) to (5.109).

Since for liquids $\mathrm{Sc} \gg 1$, it follows from (5.116) that $\mathrm{Pe_D} \gg 1$ for values $\mathrm{Re} \geq 1$, that is the convective flux dominates over the diffusion in liquids at finite (and sometimes at small) Reynolds numbers. In high-viscous liquids, $\mathrm{Pr} \gg 1$ and from (5.115) it follows that $\mathrm{Pe_T} \gg 1$ at not very small numbers Re. Hence, in this case the heat transfer is basically due to convection.

The Lewis's number appears in the problems, in which the substance- and heat transfers are considered simultaneously.

Table 5.1

Matter	Temperature °C	The Prandtl number $\mathrm{Pr} = \nu/\alpha$
Mercury	27	2.72×10^{-2}
Air	27	7.12×10^{-1}
Water	27	5.65
Glycerine	20	1.16×10^{4}

Table 5.2

Matter	Temperature °C	The Schmidt number $\mathrm{Sc} = \nu/D$	The Lewis number $\mathrm{Le} = \alpha/D$
O_2–N_2	0	7.3×10^{-1}	1
Infinite diluted gas mixture	20	-1	-1
Water solution of NaCl	20	70×10^{2}	10^{2}
Infinite diluted solution	20	$\sim 10^{3}$	$\sim 10^{2}$

As seen from the data in table 5.2, for gases Le ~ 1, that is, both processes are equally important, while in liquid Le $\gg 1$, that is the heat transfer dominates over diffusion transfer of mass.

In conclusion, it is worthwhile to cite two additional dimensionless parameters describing process of heat and mass transfer at convective flow of a medium. Determine first the heat-transfer coefficient α as ratio of thermal flux to the temperature difference, that is $q = \alpha\Delta T$, and then the mass-transfer coefficient β as the ratio of the diffusion flux to difference of concentrations $j = \Delta C$. Then one can introduce two dimensionless Nusselt's numbers

$$\mathrm{Nu}_T = \frac{\alpha d}{k}, \quad \mathrm{Nu}_D = \frac{\beta d}{D}. \tag{5.118}$$

Nusselt's numbers, which are functions of numbers Re and Pr, are used for definitions of heat transfer and mass transfer coefficients.

It should be noted that the set of the meaningful characteristics is not limited to the cited dimensionless parameters. Other parameters will be specified when we come to specific relevant problems.

5.7 The State Equations

5.7.1 The State Equation for an Ideal Gas and an Ideal Gas Mixture

Ideal gas is determined as gas, satisfying the following conditions:

1) The Joule's law, according to which the internal energy e of one mole of gas depends only on absolute temperature T. If gas consists of n moles then its internal energy is

$$\mathsf{E} = ne(T); \tag{5.119}$$

2) The Boyle's law, according to which the volume V occupied by n moles of gas at constant temperature, is inversely proportional to pressure p:

$$V = n\frac{f(T)}{p}, \tag{5.120}$$

where $f(T)$ is a universal function independent of nature of gas.

The change of internal energy in a system under a reversible process is determined by

$$dE = T\,dS - p\,dV, \tag{5.121}$$

where S is the entropy of the system.

Dividing (5.121) by number of moles n yields

$$ds = \frac{de + p\,dv}{T} = \frac{c_v}{T}dT + \frac{p}{T}dv, \tag{5.122}$$

where we have introduced the molar volume $v = V/n$ and heat capacity at constant volume $c_v = de/dT$. Since $e = e(T)$, we have $c_v = c_v(T)$. Considering (5.122) as a perfect differential of function $s(T, v)$, one obtains

$$\frac{c_v}{T} = \left(\frac{\partial s}{\partial T}\right)_v, \quad \frac{p}{T} = \left(\frac{\partial s}{\partial v}\right)_T. \tag{5.123}$$

Differentiating first equation (5.123) with respect to v, second equation – with respect to T, and taking into account that according to (5.120) $p = f(T)/v$, we get

$$\frac{\partial}{\partial v}\left(\frac{c_v}{T}\right) = \frac{\partial}{\partial T}\left(\frac{p}{T}\right) = \frac{\partial}{\partial T}\left(\frac{f(T)}{Tv}\right).$$

As far as c_v/T depends only on T, we have

$$\frac{\partial}{\partial T}\left(\frac{f(T)}{T}\right) = 0 \quad \text{or} \quad \frac{f(T)}{T} = \text{const}. \tag{5.124}$$

This constant is called the gas constant A. Its substitution into (5.120) results in the following equation of state for an ideal gas

$$pv = AT. \tag{5.125}$$

The value of the gas constant $A = 8.314\ \text{J} \cdot \text{mole}^{-1} \cdot \text{K}$.

Recall some simple properties of an ideal gas. The equation relating molar thermal capacity at constant pressure to that at constant volume is

$$c_p - c_v = A. \tag{5.126}$$

Using expression $c_v(T) = de/dT$, one can find the change of internal energy of ideal gas with change of its temperature from T_0 to T:

$$e(T) - e(T_0) = \int_{T_0}^{T} c_v(T)\,dT. \tag{5.127}$$

Similarly, for molar enthalpy $h = e + pv = e + AT$ we have $dh = de + A\,dT = c_p\,dT$ and

$$h(T) - h(T_0) = \int_{T_0}^{T} c_p(T)\, dT. \tag{5.128}$$

To determine the molar entropy we need to integrate (5.122) from (T_0, υ_0) to (T, υ). In a result we obtain

$$\begin{aligned} s(T,\upsilon) &= \left(s(T_0,\upsilon_0) + \int_{T_0}^{T} \frac{c_\upsilon(T)}{T} dT - A \ln \upsilon_0 \right) + A \ln \upsilon \\ &= \left(s(T_0,\upsilon_0) + \int_{T_0}^{T} \frac{c_p(T)}{T} dT + A \ln p_0 \right) - A \ln p \\ &= s^{(0)}(T) - A \ln p. \end{aligned} \tag{5.129}$$

Using the expression for chemical potential

$$\mu = h - Ts, \tag{5.130}$$

one can obtain the algebraic expression for chemical potential of an ideal gas

$$\mu(T, p) = \mu^{(0)}(T) + AT \ln p. \tag{5.131}$$

Thus thermodynamic functions e, h, s and μ of the gas are determined from the relation between the heat capacity (c_p or c_υ) and temperature depending on molecular composition of gas. The reader can find the formulas for heat capacities of ideal gas in the book [1].

Consider now the gas mixtures. Introduce a concept of partial pressure

$$p_i = p x_i, \tag{5.132}$$

where x_i is the molar fraction of the i-th component.

It follows from this definition that the full pressure of the mixture is

$$p_i = \sum_i p_i. \tag{5.133}$$

The gas mixture in a volume V at temperature T is called ideal if its free energy $F = \mathrm{E} - TS$ is equal to the sum of free energies of the individual components if each of them occupied the same volume at the same temperature. This definition implies that for the mixture of ideal gases the following equation of a state must hold

$$p\upsilon = AT, \tag{5.134}$$

where p is determined by the equation (5.133); $\upsilon = V/n = V/\sum n_i$; n_i is the number of moles of i-th component in volume V.

Accordingly, we have for each component

$$p_i \upsilon_i = AT, \quad \upsilon_i = V/n_i. \tag{5.135}$$

If we introduce the molar concentration $C_i = 1/\upsilon_i$, then the equation of state (5.135) can be rewritten as

$$p_i = C_i AT. \tag{5.136}$$

Thermodynamic functions for an ideal gas mixture are [5]:

the internal energy

$$\mathsf{E} = \sum_i n_i e_i(T) = \sum_i \mathsf{E}_i(T, n_j),$$

the enthalpy

$$H = \mathsf{E} + nAT,$$

the free energy

$$F = \sum_i n_i f_i = \sum_i F_i(T, V, n_j),$$

the entropy

$$S = \sum_i n_i s_i = \sum_i n_i (s^{(0)}(T, p) - A \ln(px_i)).$$

The chemical potential for each component is determined in accordance with (5.131)

$$\begin{aligned} \mu_i &= \mu_i^{(0)}(T) + AT \ln p_i = \mu_i^{(0)}(T) + AT \ln(px_i) \\ &= \mu_i^{(0)}(T, p) + AT \ln(x_i), \end{aligned} \tag{5.137}$$

where $\mu_i^{(0)}(T, p)$ is the chemical potential of a pure gas at temperature T and pressure p.

Note that the expression for chemical potential of components of ideal gas mixture is similar to that of the ideal solution. The distinction is in the form of $\mu_i^{(0)}(T, p)$.

5.7.2
The State Equation for a Real Gas and a Real Gas Mixture

At pressure and temperature slightly deviating from the normal ones, the majority of gases and real mixtures behave as ideal gases and mixtures. However at a sufficiently large departure of pressure and temperature from the normal ones, gases and gas mixtures do not satisfy the equations (5.125) and (5.135). The natural gas, representing a multi-component system consisting of hydrocarbonic and other components, can serve as an example. The typical conditions in its natural reservoirs amount to tens MPa of pressure and about 600 K of temeperature. Under these conditions the natural gas, as a rule, remains in a gaseous state. During production, preparation, processing and transportation of gas, its pressure changes from the reservoir pressure to atmospheric pressure and the temperature may lower down to cryogenic temperatures. The change of thermo-baric conditions results in the change of the mixture's state from one-phase state in reservoir to binary-phase state (liquid-vapors). In order to construct a mathematical model of motion of hydrocarbon mixture from a natural reservoir to the ground surface, and then through the units of technological chain – the preparation, processing and transport, it is necessary to describe the phase state of hydrocarbon system in conditions of thermodynamic equilibrium. The state equations of a multi-component multi-phase system for a big range of pressure and temperature are the basis for thermodynamic calculations of the phase balance and physical properties of the system. The equation of state of ideal gas and ideal mixture of gases are not suitable for these purposes, and there is a need in equations of state for a real gas and real gas mixture.

The sought-for equations of state must describe the coexistence of both – gaseous and liquid – phases of the system for a large range of pressures and temperatures. Therefore the versions of the state equations suggested recently are called the unified equations of state.

Up to the present time, many versions of state equations [8, 9] have been suggested. The majority of them aim at the description of behavior of hydrocarbon systems. These equations can be conventionally divided into two basic types: the multi-coefficient- and cubic equations of state.

Consider first the state equations for pure substances. The representation of a state equation in a virial form lies at the basis of multi-coefficient equations. The gas pressure can be represented as a power series in $1/v$, where v is the molar volume of the substance

$$p = \frac{AT}{v}\left(1 + \frac{B}{v} + \frac{C}{v^2} + \cdots\right). \tag{5.138}$$

Coefficients $B, C, \ldots,$ are called the virial coefficients. For pure substances they depend only on temperature and characterize molecular interactions of the substance: B corresponds to pair interactions, C to triple interactions etc. Virial coefficients can be calculated for gases of small density, and consequently, low pressure, by methods of statistical physics. Most often calculations are limited to the second term of the sum in (5.138), because the account of multiple interactions makes the calculations of higher-order terms increasingly difficult. Therefore for gases slightly deviating from ideal gases we take

$$p = \frac{AT}{\upsilon}\left(1 + \frac{B}{\upsilon}\right). \tag{5.139}$$

To obtain a virial form of state equation suitable for a real gas, engineers used numerous experimental data. They have been used to formulate empirical equations of state. In these equations pressure is presented as polynomial of substance molar density $\rho = 1/\upsilon$ with coefficients dependent on temperature. Out of many suggested equations of state, the most widely accepted is the Benedict–Weber–Rubin equation (BWR) and its modifications. One of successful (from the viewpoint of computational accuracy) modifications of empirical equations of state is the Starling–Khan equation

$$p = AT\rho + \left(B_0AT - A_0 - \frac{C_0}{T^2} + \frac{D_0}{T^3} - \frac{E_0}{T^4}\right)\rho^2 + \left(ATb - a - \frac{d}{T}\right)\rho^3$$
$$+ \alpha\left(a + \frac{d}{T}\right)\rho^6 + \frac{c\rho^3}{T^2}(1 + \gamma\rho^2)\exp(-\gamma\rho^2), \tag{5.140}$$

where A_0, B_0, C_0, D_0, E_0, a, b, c, d, α, γ are empirical coefficients.

Note that the series (5.140) is analogous to (5.138). The last term with the exponential factor is chosen in order to compensate for the rest of the series.

Thus, multi-coefficient equations represent a state with molar density ρ at given values p and T. To find the roots of this equation, one has to solve the nonlinear algebraic equation.

Cubic equations of state are more convenient for calculations in practical cases. The simplest cubic equation is Van der Waals equation of state

$$p = \frac{AT}{\upsilon - b} - \frac{a}{\upsilon^2}, \tag{5.141}$$

where a and b are constant coefficients.

In deriving this equation we consider the inter-molecular interactions, modeling the molecules as solid spheres: the first term in the right part of (5.141) corresponds to the direct-contact repulsion of the rigid spheres, and the second term describes the long-range attraction force. If p and T are given, then the equation (5.141) represents cubic equation for molar volume υ or molar density $\rho = 1/\upsilon$. Note that for weakly non-ideal gas we have $b/\upsilon \ll 1$ and, expanding the right

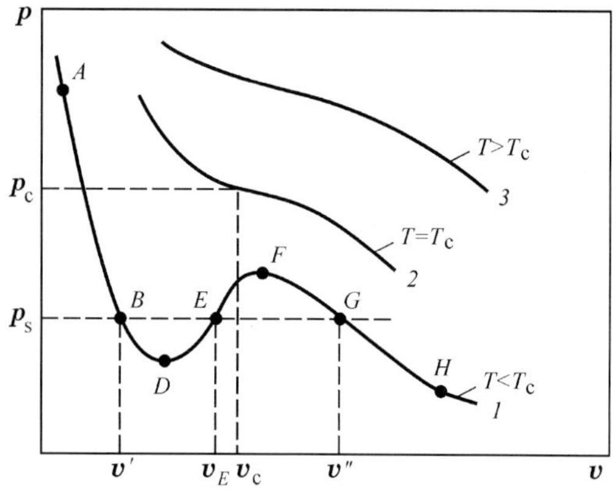

Fig. 5.1 Isotherms of a real gas' states.

part of (5.141) in the power series in $1/\upsilon$, one obtains the equation of state in the virial form with the second virial coefficient

$$B(T) = b - \frac{a}{AT}. \tag{5.142}$$

The function $p(\upsilon)$ at fixed temperature (the isotherm) is shown in Fig. 5.1. The curves *1, 2, 3* correspond to different temperatures. The curve 3 corresponds to a temperature above the critical temperature ($T > T_c$). In this state the curve changes smoothly, pressure falls with increase of υ, and the substance can be in equilibrium only in the gaseous form. The second curve corresponds to the critical temperature T_c. It is the highest temperature at which liquid and vapor states can coexist in balance with each other. At temperature $T < T_1$ (curve 1) the dependence $p(\upsilon)$ is non-monotonous. To the left of the point B (line AB) the substance is in the mono-phase liquid state, to the right of point G (line GH) the substance is in the mono-phase vapor state. The region between points B and G corresponds to the equilibrium the bi-phase state liquid – vapor. In accordance with the Maxwell's rule, squares of areas BDE and EFG are equal. From the form of isotherms it follows that in pre-critical area ($T < T_c$) the cubic equation (5.141) has three real positive roots υ', υ_E and υ'', with the least root corresponding to the molar volume of the liquid phase, and the largest root – to the molar volume of the gas phase. The root υ_E does not have a physical meaning. In the critical and super-critical area there is one real positive root of the equation (5.141). Besides, at the critical point (the point of inflection) the following conditions are satisfied

$$\frac{\partial p}{\partial \upsilon} = \frac{\partial^2 p}{\partial \upsilon^2} = 0 \quad \text{at } T = T_c;\ \upsilon = \upsilon_c, \tag{5.143}$$

Substitution (5.141) into conditions (5.143) allows us to determine coefficients a and b

$$a = 3p_c\upsilon_c^2 = \frac{27A^2T_c^2}{64p_c}, \quad b = \frac{\upsilon_c}{3} = \frac{AT_c}{8p_c}. \tag{5.144}$$

Introduce dimensionless pressure p_r, temperature T_r and molar volume υ_r:

$$p_r = \frac{p}{p_c}, \quad T_r = \frac{T}{T_c}, \quad \upsilon_r = \frac{\upsilon}{\upsilon_c}. \tag{5.145}$$

The dimentionless variables defined by (5.145) are called the reduced parameters. In these variables the Van der Waals equation can be written as

$$\left(p_r + \frac{3}{\upsilon_r^2}\right)(3\upsilon_r - 1) = 8T_r. \tag{5.146}$$

The equation (5.146) contains only the reduced parameters, therefore if two different substances are described by the Van der Waals equation and have identical values of p_r and T_r, their reduced molar volumes υ_r must also be the same. This law is known as the principle of homologous states. It lies at the basis of the theory of thermodynamic similarity.

The majority of modifications of the cubic equations is based on the Van der Waals equation. Up to the present time many cubic equations of state [8, 9] have been proposed. Below are given some, most frequently used, equations.

1. The Redlich-Kvong equation

$$p = \frac{AT}{\upsilon - b} - \frac{a}{T^{0.5}\upsilon(\upsilon + b)}. \tag{5.147}$$

From conditions (5.143) at the critical point one can easily find the coefficients a and b:

$$a = 0.42747\frac{A^2T_c^{2.5}}{p_c}; \quad b = 0.08664\frac{AT_c}{p_c}. \tag{5.148}$$

The Redlich-Kvong equation does not have any theoretical grounds, – it was obtained empirically. This equation has been widely used for calculations of properties of the gas phase. However, it leads to big errors in calculations of properties of the liquid phase. Wilson has proposed a modification that allows us to calculate the properties of both – gas and liquid phases. The coefficient a in the equation (5.147) has been modified to:

$$a = 4.934(1 + (1.45 + 1.62\omega)(T_r^{-1} - 1)T_r^{0.12})bAT^{1.5}. \tag{5.149}$$

Here an additional parameter ω named the non-centricity (or eccentricity) factor has been introduced. It characterizes the degree of deviation of molecular shape from the spherical. The greater this deviation, the greater is ω-factor.

2. The Soav-Redlich-Kvong equation

$$p = \frac{AT}{v-b} - \frac{a}{v(v+b)}. \tag{5.150}$$

The coefficient b in this equation is determined by the equation (5.148); the coefficient a is given by

$$a = a_c \alpha(T_r, \omega), \tag{5.151}$$

where

$$a = 0.42747 \frac{A^2 T_c^2}{p_c}; \quad \alpha(T_r, \omega) = (1 + m(1 - T_r^{0.5}))^2;$$

$$m = 0.48 + 1.574\omega - 0.17\omega^2.$$

3. The Peng-Robinson equation

$$p = \frac{AT}{v-b} - \frac{a}{v(v+b) + b(v-b)}. \tag{5.152}$$

The coefficient a is determined by the equation (5.151), in which

$$a_c = 0.457 \frac{A^2 T_c^2}{p_c}; \quad m = 0.375 + 1.542\omega - 0.27\omega^2,$$

and coefficient $b = 0.078AT_c/p_c$.

The Redlich-Kvong, Soav-Redlich-Kvong and Peng-Robinson equations may be used also for the multi-component mixtures. For first two equations in this case the coefficients a and b are determined by

$$a = \left(\sum_{i=1}^{N} y_i a_i^{0.5} \right)^2, \quad b = \sum_{i=1}^{N} y_i b_i, \tag{5.153}$$

where N is the number of components in the mixture; a and b are coefficients of the pure i-th substance; y_i is the molar fraction of the i-th component.

At calculations of multi-component systems by means of the Peng-Robinson equation, the coefficients a and b are calculated using the formulas

$$a = \sum_{i,j=1}^{N} (1 - C_{ij}) y_i y_j (a_i a_j)^{0.5}, \quad b = \sum_{i=1}^{N} y_i b_i, \tag{5.154}$$

where C_{ij} are the empirically determined coefficients of pair interactions between the molecules of i-th and j-th components.

Knowing the equation of state for a multi-component system of interest, one can determine thermodynamic functions consistent with basic thermodynamic relations [5]. In view of applications to calculation of the liquid-vapor equilibrium, we now turn to considering the definition of chemical potentials of components. For a real gas mixture, the chemical potential of an i-th component is

$$\mu_i(T, p) = \mu_i^{(0)}(T) + AT \ln f_i, \tag{5.155}$$

where $\mu_i^{(0)}(T)$ is the chemical potential of ideal i-th component at temperature T and the unit pressure in the accepted system of units; f_i is the fugacity of i-th component.

Comparing (5.155) with the similar expression (5.137) for an ideal gas mixture shows that for the ideal mixture of gases $f_i = p_i$. On the other hand, the relation (5.155) can be written in somewhat different form:

$$\mu_i(T, p) = \mu_i^{(0)}(T, p) + AT \ln a_i, \tag{5.156}$$

where a is the activity of the i-th component.

At the heart of calculation of μ_i or f_i lies the known thermodynamic relation for the derivative of chemical potential

$$\left(\frac{\partial \mu_i}{\partial p}\right)_{T, n_j} = \left(\frac{\partial V}{\partial n_i}\right)_{T, p, n_j \neq n_i} = v_i. \tag{5.157}$$

The derivation of the equation for μ_i is mathematically involved and is not presented here. The details can be found in work [9].

5.7.3 Methods of Calculation of Liquid–Vapor Equilibrium

The primary problem in the engineering practice is the calculation of a phase state of multi-component systems, for example, the hydro-carbonic mixtures. Under the term "calculation of phase state" we mean the determining of structure of vapor and liquid phases at the given values of pressure and temperature, and the overall structure of mixture.

Consider a multi-component system, whose structure is given by molar fractions η_i of components included in it. At a pressure p and temperature T the system is divided into vapor and liquid phases. Denote as x_i and y_i the molar fractions of the i-th component in liquid and vapor phases respectively, and let V and L be the molar fractions of vapor and liquid phases. The equilibrium constant K_i of the i-th component is defined as the ratio of this component's molar fractions

$$K_i = y_i / x_i \quad (i = \overline{1, N}). \tag{5.158}$$

The condition of equilibrium of phases is the equality of chemical potentials of components in both phases:

$$\mu_i^{(L)} = \mu_i^{(V)} \quad (i = \overline{1, N}). \tag{5.159}$$

By virtue of relations (5.155), the conditions (5.159) are equivalent to equalities of the fugacities

$$f_i^{(L)} = f_i^{(V)} \quad (i = \overline{1, N}). \tag{5.160}$$

Let's add to the equations (5.160) the equations of material balance for each component

$$\eta_i = y_i V + x_i L \quad (i = \overline{1, N}). \tag{5.161}$$

Since in accord with (5.158), $V + L = 1$, and $y_i = K_i x_i$, it is possible to transform (5.161) to

$$x_i = \frac{\eta_i}{V(K_i - 1) + 1}, \qquad y_i = \frac{\eta_i K_i}{V(K_i - 1) + 1}. \tag{5.162}$$

Equations (5.162) are known as the equations for the phase concentrations of components of mixture. They allow us to find molar fractions of components in phases if the structure of mixture η_i, the equilibrium constants K_i, and the molar fractions of the vapor phase are known. Summing (5.162) over i and taking into account the obvious equality $\sum_i x_i = \sum_i y_i = 1$, one obtains the equation determining V for the given values η_i and K_i:

$$F(V) = \sum_{i=1}^{N} \frac{\eta_i (K_i - 1)}{V(K_i - 1) + 1} = 0. \tag{5.163}$$

The function $F(V)$ is monotonously decreasing because $F'(V) < 0$. This property allows us to obtain the following criteria for the mixture's phase state.

1. $V < 0$. This case corresponds to a one-phase non-saturated liquid state. Since $F(0) < 0$, the necessary criterion for the mixture to be in one-phase liquid condition at preset values of K_i and η_i, is transformed to inequality

$$\sum_{i=1}^{N} \eta_i K_i < 1. \tag{5.164}$$

2. $V = 0$. This case corresponds to the saturated liquid state (boiling point). From condition $F(0) = 0$ it follows that necessary condition of the single-phase saturated liquid state is

$$\sum_{i=1}^{N} \eta_i K_i = 1. \tag{5.165}$$

3. $0 < V < 1$. This case corresponds to a two-phase vapor-liquid state. From conditions $F(0) > 0$ and $F(1) < 0$ it follows that a necessary condition of biphasic state is

$$\sum_{i=1}^{N} \eta_i K_i > 1 \quad \text{and} \quad \sum_{i=1}^{N} \frac{\eta_i}{K_i} > 1. \tag{5.166}$$

4. $V = 1$. This case corresponds to a one-phase saturated state (condensation point). From condition $F(1) = 0$ there follows the necessary condition of existence of mixture in a saturated gas state

$$\sum_{i=1}^{N} \frac{\eta_i}{K_i} = 1. \tag{5.167}$$

5. $V > 1$. This case corresponds to a one-phase non-saturated gas state. From condition $F(1) > 0$ there follows the necessary condition

$$\sum_{i=1}^{N} \frac{\eta_i}{K_i} < 1. \tag{5.168}$$

Note that criteria 1 and 5 are not physically realizable, therefore they may be considered only as the auxiliary criteria helpful in numerical solutions of appropriate problems.

Thus, we have the following system of $2N + 2$ equations for $2N + 2$ unknowns x_i, y_i, V and L:

$$f_i^{(L)} = f_i^{(V)} \quad (i = \overline{1, N}), \tag{5.169}$$

$$\eta_i = y_i V + x_i L \quad (i = \overline{1, N}), \tag{5.170}$$

$$\sum_i y_i = 1, \tag{5.171}$$

$$V + L = 1. \tag{5.172}$$

Note that fugacities are calculated on the basis of thermodynamic equations with use of one of the state equations. The system of equations (5.169)–(5.172) represents nonlinear system of algebraic equations. Its solution can be found by using appropriate numerical methods.

5.8
Balance of Entropy – The Onsager Reciprocal Relations

Phenomenological relations, described in the subsection 4.1, play important role in the thermodynamics of irreversible processes. The general basis of the macroscopic description of irreversible processes is the non-equilibrium thermodynamics, which is developing as a general theory of continuous media. Its parameters, in contrast to the equilibrium thermodynamics, are functions of spatial coordinates and time. The central role in the non-equilibrium thermodynamics plays the equation of entropy balance [10]. This equation expresses the fact that the entropy in a volume element of a continuous medium changes in time due to the entropy flux into considered element from the outside, and also due to irreversible processes leading to entropy production within the element. If the processes are reversible, we can say that sources of entropy are absent. The above two statements constitute the local formulation of the second law of thermodynamics.

The primary goal of the theory of irreversible processes is to obtain an expression for the intensity of the corresponding entropy source. For this purpose it is necessary to use conservation laws for mass, momentum and energy in differential form. The diffusion and heat fluxes and also stress tensor enter the conservation equations describing the mass, energy and momentum transfer. The Gibbs thermodynamic equation (5.49), connecting the rate of entropy change with rates of energy and a mixture's structure changes, plays an important role. The expression for intensity of entropy source is a sum of terms, each being a product of a flux describing the corresponding irreversible process, and a quantity known as a thermodynamic force. A thermodynamic force causes heterogeneity of the system or deviation of its parameter from equilibrium values. Fluxes, in first approximation, linearly depend on thermodynamic forces, in accord with phenomenological relations. These linear laws reflect the flux dependence on all thermodynamic forces, that is, they take into account cross-interaction effects. Thus, for example, the flux of a substance depends not only on its concentration gradient, but also on gradients of pressure, temperature, electric potential etc. The non-equilibrium thermodynamics is limited basically to study of linear phenomenological dependencies.

In this framework, the intensity of an entropy source is represented by a quadratic form of thermodynamic forces. The corresponding phenomenological coefficients form a matrix with remarkable properties. These properties, formulated as the Onsager reciprocity theorem, allow to reduce the number of independent quantities and to find relations between various physical effects.

The change of a system's entropy is

$$dS = d_e S + d_i S. \tag{5.173}$$

Here $d_e S$ is the entropy entering the system from the outside, $d_i S$ is the entropy produced within the system.

According to the second law of thermodynamics, $d_i S = 0$ for the equilibrium processes, and for non-equilibrium processes $d_i S \geq 0$. For an adiabatic isolated system, that is for a system, which can not exchange its heat or substance with the environment, $d_e S = 0$, and it follows from (5.119)

$$dS = d_i S \geq 0 \tag{5.174}$$

For a closed system, that is a system, which can exchange with environment only its thermal energy, the Carnot-Clausius equation holds

$$d_e S = \frac{dQ}{T}, \tag{5.175}$$

where dQ is the amount of heat entering the system from the outside.

It follows from (5.173) and (5.175) that for a closed system

$$dS = d_i S + \frac{dQ}{T} \geq \frac{dQ}{T}. \tag{5.176}$$

The equations (5.174) and (5.176) represent different forms of the second law of thermodynamics, corresponding to an adiabatic isolated and a closed system, respectively.

Of the greatest interest are the open systems, that is systems, which can exchange with the environment both – their thermal energy and substance.

Introduce the entropy density s by the expression

$$S = \int_V \rho s \, dV. \tag{5.177}$$

Since $d_e S$ is the entropy that enters the system from the outside through the surface Σ enclosing volume V, its change in time dt is

$$d_e S = -\int_\Sigma \boldsymbol{j}_s \cdot d\boldsymbol{\Sigma} \, dt, \tag{5.178}$$

where $\boldsymbol{j}_s$ is the entropy flux through the unit surface; $d\boldsymbol{\Sigma}$ is the oriented element of the surface.

Denote as σ the entropy production rate in the unit volume. Then the entropy change in time dt in volume V due to a non-equilibrium process is

$$d_i S = \int_V \sigma \, dV \, dt, \quad \sigma \geq 0. \tag{5.179}$$

Substituting (5.177)–(5.179) into (5.173) and transforming the surface integral into the volume integral according to the Gauss-Ostrogradsky formula (5.1), one obtains

$$\int_V \left(\frac{D\rho s}{Dt} + \nabla \cdot \boldsymbol{j}_s - \sigma \right) dV = 0. \tag{5.180}$$

Since the equation (5.180) holds for an arbitrary volume, it implies the corresponding differential equation, representing the local formulation of the second law of thermodynamics for an open system:

$$\frac{D\rho s}{Dt} = -\nabla \cdot \boldsymbol{j}_s + \sigma, \quad \sigma \geq 0. \tag{5.181}$$

Using the continuity equation (5.12), on can transform the equation (5.181) to

$$\rho \frac{Ds}{Dt} = -\nabla \cdot \boldsymbol{J}_s + \sigma, \quad \sigma \geq 0. \tag{5.182}$$

where $\boldsymbol{J}_s = \boldsymbol{j}_s - \rho s \boldsymbol{U}$ is the relative flux.

Use now the Gibbs formula (5.49), rewriting it as

$$T \frac{Ds}{Dt} = \frac{De}{Dt} + p \frac{D\upsilon}{Dt} - \sum_{k=1}^{n} \mu_k \frac{D\omega_k}{Dt}, \tag{5.183}$$

where ω_k is the mass concentration of k-th component and μ_k is chemical potential of k-th component.

To determine of De/Dt and $D\omega/Dt$ it is necessary to use equation of energy (5.30) and diffusion (5.45). Note that equation (5.30) is written without regard to multicomponentness medium and chemical reactions. In view of these phenomena it will be rewritten as [10]

$$\rho \frac{De}{Dt} = \boldsymbol{T} : \boldsymbol{E} - \nabla \cdot \boldsymbol{J}_q + \sum_{k=1}^{n} \boldsymbol{J}_k \cdot \boldsymbol{F}_k, \tag{5.184}$$

where $\boldsymbol{J}_q$ is heat flux (in section 5.2 it was designated as $\boldsymbol{q}$); $\boldsymbol{J}_k$ is flux diffusion of k-th component (see (4.26)); $\boldsymbol{F}_k$ is external force acting on k-th component.

The scalar product of a stress tensor and a strain rate tensor $\boldsymbol{T} : \boldsymbol{E}$ is determined by equation (5.31). Then (5.184) takes the form

$$\rho \frac{De}{Dt} = -\nabla \cdot \boldsymbol{J}_q - p\nabla \cdot \boldsymbol{u} + \boldsymbol{\Pi} : \nabla \boldsymbol{u} + \sum_{k=1}^{n} \boldsymbol{J}_k \cdot \boldsymbol{F}_k, \tag{5.185}$$

where $\nabla \boldsymbol{u} = \partial u_i / \partial x_k$ is the vector gradient, that is, a second rank tensor.

Write down the diffusion equation in the form

$$\rho \frac{D\omega_k}{Dt} = -\nabla \cdot \boldsymbol{J}_k + r_k, \tag{5.186}$$

where r_k is the formation rate per unit volume of the k-th component, resulting from a chemical reaction; ω_k is the mass fraction of k-th component.

Formation of the k-th component may occur as a result of several simultaneous reactions. If reactions are not accompanied by change of volume, then the formation rate for each formed component is determined by the equation

$$r_k = \sum_{j=1}^{r} \nu_{kj} M_k J_j^{(r)}, \tag{5.187}$$

where ν_{kj} is the stoichiometric coefficient of the k-th component in the j-th reaction; M is molecular mass; $J_j^{(r)} = \rho \partial \xi_j / \partial t$ is a rate of the j-th chemical reaction in unit volume. From the condition of mass conservation in every chemical reaction there follows

$$\sum_{j=1}^{r} M_k \nu_{kj} = 0. \tag{5.188}$$

The equation (5.186), with regard to formulas (5.45), (5.46), and (5.187), can be rewritten in the form

$$\rho \frac{D\omega_k}{Dt} = -\nabla \cdot \boldsymbol{J}_k + \sum_{j=1}^{r} \nu_{kj} M_k J_j^{(r)}. \tag{5.189}$$

Substitution of (5.185) and (5.189) into (5.183) yields

$$\rho \frac{Ds}{Dt} = -\frac{1}{T} \nabla \cdot \boldsymbol{J}_q + \frac{1}{T} \boldsymbol{\Pi} : \nabla \boldsymbol{u} + \frac{1}{T} \sum_{k=1}^{n} \boldsymbol{J}_k \cdot \boldsymbol{F}_k$$

$$+ \frac{1}{T} \sum_{k=1}^{n} \mu_k \nabla \cdot \boldsymbol{J}_k - \frac{1}{T} \sum_{j=1}^{r} A_j J_j^{(r)}. \tag{5.190}$$

Here $A_j = \sum_{k=1}^{n} M_k \nu_{kj} \mu_k$ are chemical affinities of reactions [5].

The following step is the reduction of equation (5.190) to the form (5.182), for which purpose one has to collect terms containing divergence. After simple algebra one obtains the expression for entropy flux $\boldsymbol{J}_s$ and entropy source σ, known as dissipative function

$$\boldsymbol{J}_s = \frac{1}{T} \left(\boldsymbol{J}_q - \sum_{k=1}^{n} \mu_k \boldsymbol{J}_k \right), \tag{5.191}$$

$$\sigma = \frac{1}{T^2} \boldsymbol{J}_q \cdot \nabla T - \frac{1}{T} \sum_{k=1}^{n} \boldsymbol{J}_k \cdot \left(T \nabla \left(\frac{\mu_k}{T} \right) - \boldsymbol{F}_k \right)$$

$$+ \frac{1}{T} \boldsymbol{\Pi} : \nabla \boldsymbol{u} - \frac{1}{T} \sum_{j=1}^{r} A_j J_j^{(r)} \geq 0. \tag{5.192}$$

The expression for $\boldsymbol{J}_s$ shows, that the entropy flux for open systems consists of two parts: the thermal flux associated with the heat transfer, and the flux due to diffusion. The second expression consists of four terms associated with, respectively, the heat transfer, diffusion, viscosity, and chemical reactions. The expression for the dissipative function σ has quadratic form. It represents the sum of products of two factors: a flux (specifically, the heat flux $\boldsymbol{J}_q$, diffusion flux $\boldsymbol{J}_k$, momentum flux $\boldsymbol{\Pi}$, and the rate of a chemical reaction $J_j^{(r)}$) and a thermodynamic force, proportional to gradient of some intensive variable of state (temperature, chemical potential, or velocity). The second factor can also include external force $\boldsymbol{F}_k$ and chemical affinity A_j.

For a wide class of irreversible processes, the fluxes are linear functions of thermodynamic forces determined by phenomenological laws. Some of these laws were briefly discussed earlier, for example, the Fourier and Fick's laws. These laws embrace also the mixed or cross-phenomena, for example, thermo-diffusion, when the diffusion flux linearly depends on both – concentration- and temperature gradients. In general case linear phenomenological relations can be written as

$$J_i = \sum_k L_{ik} X_k, \tag{5.193}$$

where J_i and X_i are the Cartesian components of the independent fluxes and thermodynamic forces entering the formula for a dissipative function

$$\sigma = \sum_i J_i X_i. \tag{5.194}$$

Quantities L_{ik} are called the phenomenological or kinetic coefficients. Combining (5.193) and (5.194), one obtains

$$\sigma = \sum_i \sum_j L_{ij} X_i X_j. \tag{5.195}$$

By virtue of condition $\sigma \geq 0$, the quadratic form (5.195) should be a positively determined form or, at the least, a non-negative one. The sufficient condition for this is the positiveness of diagonal elements of the matrix $\{L_{ij}\}$ and the inequality $L_{ii} \cdot L_{kk} \geq (L_{ik} + L_{ki})$ for non-diagonal elements.

Before formulating the basic result of the Onsager theory one should transform the expression for the dissipative function σ. Use the thermodynamic equation [5]

$$Td\left(\frac{\mu_k}{T}\right) = (d\mu_k)_T - \frac{h_k}{T} dT, \tag{5.196}$$

where h_k is specific partial enthalpy, and the index T indicates that the differential is taken at constant temperature.

Introduce now, in accordance with (5.47), the thermal flux in a heat transfer due to heat conductivity

$$\boldsymbol{J}'_q = \boldsymbol{J}_q - \sum_{k=1}^{n} h_k \boldsymbol{J}_k. \tag{5.197}$$

Then the equation (5.192) can be rewritten in terms of $\boldsymbol{J}_k$ and $\boldsymbol{J}'_q$ as

$$\sigma = -\frac{1}{T^2}\boldsymbol{J}'_q \cdot \nabla T - \frac{1}{T}\sum_{k=1}^{n} \boldsymbol{J}_k \cdot ((\nabla \mu_k)_T - \boldsymbol{F}_k) + \frac{1}{T}\boldsymbol{\Pi} : \nabla \boldsymbol{u} - \frac{1}{T}\sum_{j=1}^{r} A_j J_j^{(r)}. \tag{5.198}$$

Since the specific partial entropy of a component is $s_k = -(\mu_k - h_k)/T$, the entropy flux can be written, using (5.197), in the same terms:

$$\boldsymbol{J}_s = \frac{1}{T}\boldsymbol{J}'_q + \sum_{k=1}^{n} s_k \boldsymbol{J}_k. \tag{5.199}$$

Present the viscous stress tensor $\boldsymbol{\Pi}$ and the tensor $\nabla \boldsymbol{u}$ in the form

$$\boldsymbol{\Pi} = \Pi \boldsymbol{I} + \overset{\circ}{\boldsymbol{\Pi}}, \tag{5.200}$$

$$\nabla \boldsymbol{u} = \frac{1}{3}(\nabla \cdot \boldsymbol{u})\boldsymbol{I} + \overset{\circ}{\nabla \boldsymbol{u}}, \tag{5.201}$$

where $\Pi = \frac{1}{3}\boldsymbol{\Pi} : \boldsymbol{I} = \frac{1}{3}\Pi_{\alpha\alpha} = \frac{1}{3}\, Tr(\boldsymbol{\Pi})$, $Tr(\boldsymbol{\Pi})$ is the trace of matrix $\boldsymbol{\Pi}$. Notice that $\nabla \cdot \boldsymbol{u} = \nabla \boldsymbol{u} : \boldsymbol{I} = \partial u_i/\partial x_i = Tr(\nabla \boldsymbol{u})$. By virtue of this, the tensors $\overset{\circ}{\boldsymbol{\Pi}}$ and $\overset{\circ}{\nabla \boldsymbol{u}}$ are traceless, since $\overset{\circ}{\boldsymbol{\Pi}} : \boldsymbol{I} = 0$ and $\overset{\circ}{\nabla \boldsymbol{u}} = 0$. Now make a scalar product $\boldsymbol{\Pi} : \nabla \boldsymbol{u}$. Using (5.200), (5.201) and above mentioned properties of $\overset{\circ}{\boldsymbol{\Pi}}$ and $\overset{\circ}{\nabla \boldsymbol{u}}$ we get

$$\boldsymbol{\Pi} : \nabla \boldsymbol{u} = \overset{\circ}{\boldsymbol{\Pi}} : \overset{\circ}{\nabla \boldsymbol{u}} + \Pi \nabla \cdot \boldsymbol{u} \tag{5.202}$$

Separate the tensor into symmetric and anti-symmetric parts

$$\overset{\circ}{\nabla \boldsymbol{u}} = (\overset{\circ}{\nabla \boldsymbol{u}})^s + (\overset{\circ}{\nabla \boldsymbol{u}})^a, \tag{5.203}$$

where $(\overset{\circ}{\nabla \boldsymbol{u}})^s = \frac{1}{2}(\overset{\circ}{\nabla \boldsymbol{u}} + (\overset{\circ}{\nabla \boldsymbol{u}})^{\mathrm{T}})$, $(\overset{\circ}{\nabla \boldsymbol{u}})^a = \frac{1}{2}(\overset{\circ}{\nabla \boldsymbol{u}} + (\overset{\circ}{\nabla \boldsymbol{u}})^{\mathrm{T}})$ The index "T" denotes the transposition operation.

Components of these tensors are

$$(\overset{\circ}{\nabla \boldsymbol{u}})^s_{\alpha\beta} = \frac{1}{2}\left(\frac{\partial u_\beta}{\partial x_\alpha} + \frac{\partial u_\alpha}{\partial x_\beta}\right) - \frac{1}{3}\delta_{\alpha\beta}\frac{\partial u_\gamma}{\partial x_\gamma},$$

$$(\overset{\circ}{\nabla \boldsymbol{u}})^a_{\alpha\beta} = \frac{1}{2}\left(\frac{\partial u_\beta}{\partial x_\alpha} - \frac{\partial u_\alpha}{\partial x_\beta}\right).$$

Here the summation is to be performed over repeating indexes. Substitution of (5.203) into (5.202) leads to

$$\mathbf{\Pi} : \nabla \boldsymbol{u} = \overset{\circ}{\mathbf{\Pi}} : (\overset{\circ}{\nabla \boldsymbol{u}})^s + \Pi \nabla \cdot \boldsymbol{u}. \tag{5.204}$$

Here a known rule has been used: the scalar product of symmetric and antisymmetric tensors vanishes.

Substituting the expression (5.204) into (5.198) yields

$$\sigma = -\frac{1}{T^2} \boldsymbol{J}'_q \cdot \nabla T - \frac{1}{T} \sum_{k=1}^{n} \boldsymbol{J}_k \cdot ((\nabla \mu_k)_T - \boldsymbol{F}_k)$$
$$+ \frac{1}{T} \overset{\circ}{\mathbf{\Pi}} : (\overset{\circ}{\nabla \boldsymbol{u}})^s - \frac{1}{T} \Pi \nabla \cdot \boldsymbol{u} - \frac{1}{T} \sum_{j=1}^{r} A_j J_j^{(r)}. \tag{5.205}$$

Let us now derive phenomenological equations of the kind (5.193) corresponding to the expression (5.205). As has been mentioned before, each flux is a linear function of all thermodynamic forces. However the fluxes and thermodynamic forces that are included in the expression (5.205) for the dissipative function, have different tensor properties. Some fluxes are scalars, others are vectors, and the third one represents a second rank tensor. This means that their components transform in different ways under the coordinate transformations. As a result, it can be proven that if a given material possesses some symmetry, the flux components cannot depend on all components of thermodynamic forces. This fact is known as Curie's symmetry principle. The most widespread and simple medium is isotropic medium, that is, a medium, whose properties in the equilibrium conditions are identical for all directions. For such a medium the fluxes and thermodynamic forces represented by tensors of different ranks, cannot be linearly related to each other. Rather, a vector flux should be linearly expressed only through vectors of thermodynamic forces, a tensor flux can be a liner function only of tensor forces, and a scalar flux – only a scalar function of thermodynamic forces. The said allows us to write phenomenological equations in general form

$$\boldsymbol{J}'_q = -L_{qq} \frac{\nabla T}{T^2} - \sum_{k=1}^{n} L_{qk} \frac{(\nabla \mu_k)_T - \boldsymbol{F}_k}{T}, \tag{5.206}$$

$$\boldsymbol{J}_i = -L_{iq} \frac{\nabla T}{T^2} - \sum_{k=1}^{n} L_{ik} \frac{(\nabla \mu_k)_T - \boldsymbol{F}_k}{T}, \quad (i = 1, 2, \ldots, n), \tag{5.207}$$

$$\overset{\circ}{\mathbf{\Pi}}_{\alpha\beta} = -\frac{L}{T} (\nabla \boldsymbol{u})^s_{\alpha\beta}, \quad (\alpha, \beta = 1, 2, 3) \tag{5.208}$$

$$\Pi = -l_{vv}\frac{\nabla \cdot \boldsymbol{u}}{T} - \sum_{m=1}^{r} l_{vm}\frac{A_m}{T}, \tag{5.209}$$

$$J_i^{(r)} = -l_{iv}\frac{\nabla \cdot \boldsymbol{u}}{T} - \sum_{m=1}^{r} l_{im}\frac{A_m}{T}, \quad (i = 1, 2, \ldots, n). \tag{5.210}$$

Equations (5.206) and (5.207) describe the vector phenomena of heat conductivity, diffusion, and cross-effects. Coefficients L_{qq}, L_{qk}, L_{iq} and L_{ik} are scalars. Equations (5.208) relate components of the stress tensor to components of a symmetric tensor. Equations (5.209) and (5.210) describe the scalar processes of the chemical character associated with the phenomena of volume viscosity and the cross-phenomena.

In case of the anisotropic media, for example, anisotropic crystals, the phenomenological equations in the absence of chemical reactions are similar to (5.206) and (5.207), but the quantities L_{qq}, L_{qk}, L_{iq} and L_{ik} are tensors. In particular the tensor L_{qq} is proportional to heat conductivity tensor.

The phenomenological equations (5.206)–(5.210) have been obtained under the condition of spatial symmetry of the medium. Another feature of the physical laws is the invariance of equations of motion of particles constituting the medium with respect to the time reversal. This property means that the equations of motion are symmetric with respect to time, that is, at the change of the sign of all velocities, the particles will pass their respective trajectories in the opposite direction. The Onsager theorem is based on this principle (its discussion can be found in [10]): for an isotropic liquid or gas in the absence of magnetic field, the phenomenological coefficients obey the following conditions:

$$\begin{aligned} &L_{qi} = L_{iq}, \quad L_{ik} = L_{ki}, \\ &L_{vi} = -L_{iv}, \quad L_{jm} = L_{mj}. \end{aligned} \tag{5.211}$$

As an example, consider a special case of a process, in which only heat conductivity and diffusion are taken into account. This and other cases are considered in detail in work [10].

Let concentration and temperature in a mixture are non-uniform (that is, position-dependent). The mixture is considered as isotropic, the viscosity is neglected, chemical reactions and external forces are absent. Under these conditions, the pressure in the system can be taken a constant. Then the dissipative function (5.205) will become

$$\sigma = -\frac{1}{T^2}\boldsymbol{J}_q' \cdot \nabla T - \frac{1}{T}\sum_{k=1}^{n} \boldsymbol{J}_k \cdot (\nabla \mu_k)_{T,p}. \tag{5.212}$$

Consider the sum in the right part of this expression. From the relation for diffusion fluxes $\sum_{k=1}^{n} \boldsymbol{J}_k = 0$ (see (4.30)) it follows that $\boldsymbol{J}_n = -\sum_{k=1}^{n-1} \boldsymbol{J}_k$. After substitution in (5.212) one obtains

$$\sigma = -\frac{1}{T^2} J'_q \cdot \nabla T - \frac{1}{T} \sum_{k=1}^{n-1} J_k \cdot (\nabla(\mu_k - \mu_n))_{T,p}. \tag{5.213}$$

Now we can apply the Gibbs-Duhem thermodynamic relation, according to which the changes of chemical potentials of components $\delta\mu_k$ in an n-component system at constant p and T satisfy the condition

$$\sum_{k=1}^{n} \omega_k \delta\mu_k = 0. \tag{5.214}$$

This equation allows us to eliminate μ_k from (5.213) and to get

$$\sigma = -\frac{1}{T^2} J'_q \cdot \nabla T - \frac{1}{T} \sum_{k=1}^{n-1} \sum_{m=1}^{n-1} J_k A_{km} \cdot (\nabla\mu_k)_{T,p}. \tag{5.215}$$

where $A_{km} = \delta_{km} + \omega_m/\omega_n$.

Since the chemical potential $\mu_m = \mu_m(p, T, \omega_m)$, we have

$$(\nabla\mu_m)_{p,T} = \sum_{i=1}^{n} \left(\frac{\partial\mu_m}{\partial\omega_i}\right)_{p,T,\omega_j} \nabla\omega_i. \tag{5.216}$$

Denote $\mu^{\omega}_{mi} = \left(\frac{\partial\mu_m}{\partial\omega_i}\right)_{p,T,\omega_j}$. Then the phenomenological equations (5.206) and (5.207) take the form

$$J'_q = -L_{qq} \frac{\nabla T}{T^2} - \sum_{k=1}^{n-1} \sum_{m=1}^{n-1} \sum_{j=1}^{n-1} L_{qk} \frac{A_{km}\mu^{\omega}_{mj} \nabla\omega_j}{T}, \tag{5.217}$$

$$J_i = -L_{iq} \frac{\nabla T}{T^2} - \sum_{k=1}^{n-1} \sum_{m=1}^{n-1} \sum_{j=1}^{n-1} L_{ik} \frac{A_{km}\mu^{\omega}_{mj} \nabla\omega_j}{T}, \tag{5.218}$$

The Onsager relations for phenomenological coefficients in these equations are

$$L_{iq} = L_{qi}, \quad L_{ik} = L_{ki}, \quad (i, k = 1, 2, \ldots, n-1). \tag{5.219}$$

The coefficients L_{qq} and L_{ik} are related to coefficients of heat conductivity and diffusion. Coefficients L_{iq} characterize the phenomenon of thermo-diffusion (the Soret effect). Coefficients L_{qk} describe the reverse phenomenon consisting in the occurrence of heat flux due to the concentration gradient (the Dufour effect).

In special case of a binary system ($n = 2$) we have $A_{11} = 1 + \omega_1/\omega_2 = 1/\omega_2$ because $\omega_1 + \omega_2 = 1$, and phenomenological equations reduce to

$$J_q' = -L_{qq}\frac{\nabla T}{T^2} - L_{q1}\frac{\mu_{11}^{\omega}\nabla\omega_1}{\omega_2 T}, \tag{5.220}$$

$$J_1 = -L_{1q}\frac{\nabla T}{T^2} - L_{11}\frac{\mu_{11}^{\omega}\nabla\omega_1}{\omega_2 T}. \tag{5.221}$$

The Onsager relations for $L_{1q} = L_{q1}$, and the sufficient conditions for the quadratic form σ to be non-negative yield the inequalities

$$L_{qq} \geq 0, \quad L_{11} \geq 0, \quad L_{qq}L_{11} - (L_{1q} + L_{q1})^2/4 = L_{qq}L_{11} - L_{1q}^2 \geq 0. \tag{5.222}$$

Usually, instead of phenomenological coefficients, other coefficients are introduced: the coefficient of heat conductivity $\lambda = L_{qq}/T^2$; the Dufour coefficient $D'' = L_{q1}/\rho\omega_1\omega_2 T^2$; coefficient of thermo-diffusion $D' = L_{1q}/\rho\omega_1\omega_2 T^2$; the diffusion coefficient $D = L_{11}\mu_{11}^{\omega}/\rho\omega_2 T$.

The coefficient of thermo-diffusion D' has the order of magnitude 10^{-12}–10^{-14} (m^2/s·degree) in a liquid and 10^{-8}–10^{-10} (m^2/s·degree) in a gas.

When expressed in terms of these coefficients, the phenomenological equations take the following form

$$J_q' = -\lambda\nabla T - \rho_1\mu_{11}^{\omega}TD''\nabla\omega_1, \tag{5.223}$$

$$J_q = -\rho_1\omega_1\omega_2 D'\nabla T - \rho D\nabla\omega_1. \tag{5.224}$$

The appropriate equations of heat and substance transfer follow from the conservation laws for energy and mass, and from the phenomenological equations (5.223) and (5.224):

$$\rho C_p\frac{\partial T}{\partial t} = -\nabla\cdot J_q' = \nabla\cdot(\lambda\nabla T + \rho_1\mu_{11}^{\omega}TD''\nabla\omega_1), \tag{5.225}$$

$$\frac{\partial\omega_1}{\partial t} = -\nabla\cdot(J_q'/\rho) = \nabla\cdot(\omega_1\omega_2 D'\nabla T + D\nabla\omega_1). \tag{5.226}$$

Note that equations (5.225) and (5.226) differ from equations of diffusion and heat conductivity obtained above (see (5.37) and (5.62)) by the account taken of the cross-effects.

References

1 Hirschfelder J. O., Curtiss C. F., Bird R. B., *Molecular theory of gases and liquids*, Wiley, New York, 1954.

2 Bird R., Stuart V., Lightfoot E., *Transport phenomena*, Wiley, New York, 1965.

3 Serrin J., *Mathematical principles of classical fluid mechanics*, Handbuch der Physik Band VIII/I Strömungsmechanik Berlin-Göttingen-Heidelberg, 1959.

4 Landau L. D., Lifshitz E. M., *Theoretical physics*. Volume VI:

Hydrodynamics, Nauka, Moscow, 1988, p. 736 (in Russian).
5 Prigogine I., Defay R., *Chemical thermodynamics*, Longmans Green and Co., London, New York, Toronto, 1954.
6 Nigmatullin R. I., *Dynamics of multi-phase media* (in 2 vols), Nauka, Moscow, 1987 (in Russian).
7 Probstein R. F., *Physicochemical hydrodynamics*, Butterworths, 1989.
8 Walas S., *Phase equilibria in chemical technology* (in 2 vols), Butterworths, Boston, 1985.
9 Batalin O. Yu., Brusilovskiy A. I., Zaharov M. Yu., *Phase equilibria in systems of natural hydrocarbons*, Nedra, Moscow, 1992 (in Russian)
10 de Groot S. R., Mazur P., *Non-equilibrium thermodynamics*, Elsevier, Amsterdam, 1962.

III
Solutions

Separation of Multiphase, Multicomponent Systems. E. G. Sinaiski and E. J. Lapiga

ISBN: 978-3-527-40612-8

6 Solutions Containing Non-charged Components

6.1 Diffusion and Kinetics of Chemical Reactions

Consider the process of mass transfer taking place in a solution as a result of convective diffusion, given that the solution is isothermal and contains only uncharged components [1]. Chemical, physicochemical, and biochemical reactions are possible in the solution. If reactions occur throughout the entire expanse of the solution, they are called homogeneous, or volume reactions. If, on the other hand, reactions occur in a limited region, for example, at the boundary of a region occupied by the solution, or at some fragment of this boundary, such reactions are said to be heterogeneous, or surface reactions. In the former case, production of components during the reaction is described with regard to the source type in transfer equations, and in the latter case – with regard to boundary conditions at the reacting surfaces.

Homogeneous reactions are sometimes accompanied by significant calorification, hence the energy equation should take into account heat evolution in the volume. In the present chapter, we consider processes with no volume calorification, so the assumption of an isothermal solution will be valid.

A wide range of problems involve flows accompanied by heterogeneous reactions on solid surfaces or at interfaces, in particular, at membranes, or processes of dissolution and deposition of substances from solutions or melts, etc. Heterogeneous reactions consist of several stages. The first stage, namely the transport stage, involves delivery of the reacting components to the relevant surface. The second stage involves the actual chemical reaction at the surface. This stage can, in turn, consist of several stages, including diffusion of the reacting substance through a wall or surface layer, adsorption of a substance at the surface, chemical reaction, desorption of the reaction products, and their diffusion from the wall or surface layer. The third stage involves transfer of reaction products into the bulk flow. Each stage is associated with its own characteristic time. The stage having the longest characteristic time is called the limiting stage, and the associated process is referred to as controlling process. In cases where the first or the third stage (as listed above) is the limiting one, the relevant process is referred to as diffusion-controlled. Equations describing this stage are thus called diffusion

Separation of Multiphase, Multicomponent Systems. E. G. Sinaiski and E. J. Lapiga

ISBN: 978-3-527-40612-8

equations. If the second stage takes the longest time, then we can speak of a process controlled by reaction kinetics, or a kinetically-controlled process. Since this stage consists of several substages, then within this stage it is possible to distinguish between diffusion-controlled and kinetically-controlled substages. If characteristic times of stages or substages have the same order, then such a reaction is called a mixed reaction.

The specific molar production rate of an *i*-th component in the course of a homogeneous reaction is determined by the expression:

$$R_i = \frac{1}{V}\frac{dn_i}{dt}, \tag{6.1}$$

where V is the volume of the mixture and $n_i = C_i V$ is the number of moles of *i*-th component having the concentration C_i.

Generally, R_i^0 depends on T, p, and the concentration of a component participating in the reaction, for example, a catalyst or inhibitor.

Heterogeneous reactions occur at a surface, and therefore the specific molar rate is determined by:

$$R_i^S = \frac{1}{S}\frac{dn_i}{dt}, \tag{6.2}$$

where S is the area of the reacting surface.

Theoretical descriptions of homogeneous and heterogeneous chemical reactions are similar. Transition from one reaction to another can be carried out by using the formula:

$$R_i^{(v)} V = R_i^{(S)} S. \tag{6.3}$$

A reaction is termed simple, if its rate at a given temperature depends only on the collision rate of reacting molecules. In accordance with the law of mass action, the rate of a simple reaction is proportional to the concentrations of reacting components:

$$R_i = \mathrm{k}_i \prod_j C_j^{\nu_j}; \quad \nu = \sum_j \nu_j, \tag{6.4}$$

where $\boldsymbol{k}$ is a reaction constant; ν_i is the reaction order with respect to *i*-th reactant; ν is the total reaction order.

In the case of reversible, or two-way reaction, the term “reaction rate” is used for the difference of direct and reverse reaction rates. At the moment when equilibrium is established between these reactions, $R_i = 0$, and the constants of direct (f) and reverse (r) reactions are connected by the relation

$$\mathsf{k} = \mathsf{k}_f/\mathsf{k}_r, \tag{6.5}$$

where k is the equilibrium constant of the reverse reaction.

We define a complex reaction as a reaction whose rate depends on both the concentration of components entering the reaction, and on the concentration of intermediate and final products of reaction. Even though for complex reactions, the reaction order is frequently unknown, the law (6.4) can be applied for such reactions as well; but the values of the exponents may in this case be non-integral and need not be positive.

If the reaction mechanism is not known, an empirical dependence

$$R_i = \mathsf{k}_i C_i^{\nu}. \tag{6.6}$$

is often used to express the reaction rate. The reaction constant included in Eqs. (6.4) and (6.6), is, strictly speaking, not a constant. It depends on temperature in accordance with Arrhenius's law:

$$\mathsf{k} = B \exp\left(-\frac{E}{AT}\right), \tag{6.7}$$

where E is the activation energy, i.e. energy that is necessary for a molecule to react; B is the reaction constant; A is the gas constant.

The laws (6.4) and (6.6) are also valid for heterogeneous reactions, but the reaction constants have a different dimensionality, since the reaction rate is related to a surface unit, rather than a volume unit. In particular, the empirical law (6.6) can be written as

$$(R_i^{(S)})_{\mathrm{w}} = (\mathsf{k}_i' C_i^{\nu})_{\mathrm{w}}. \tag{6.8}$$

The index w means that the reaction is taking place at the surface.

Consider in detail what happens at the reacting surface, since the condition (6.8) is, in fact, a boundary condition for the solution of diffusion equations. In the course of heterogeneous reaction, the molecules of reacting substance get adsorbed at the surface. If the surface is homogeneous, and the adsorbed substance forms a monolayer on this surface, then the process of adsorption is described by a Langmuir's isotherm, which relates the quantity of substance adsorbed at the surface to concentration C of this substance in the solution near the surface:

$$\theta = C/(C + b). \tag{6.9}$$

Here θ is the fraction of the surface occupied by the molecules of absorbed substance; b is the adsorption constant.

If adsorption rate is high, and the rate of chemical reaction at the surface is low, then

$$R^{(S)} = \mathsf{k}'_\alpha \theta = \mathsf{k}'_\alpha \frac{C}{C+b}, \tag{6.10}$$

where k'_α is the surface reaction constant.

Such process is known as physical adsorption. Intermolecular forces similar to Van der Waals or dipole interaction forces operate in physical adsorption. Therefore physical adsorption does not require activation energy and occurs very quickly. In particular, at small adsorbed substance concentrations in the solution near the surface we have $R^{(S)} \sim C$, and at high concentrations – $R^{(S)} \sim 1$. Hence, physical adsorption can be regarded as a chemical reaction of the order $0 < \nu < 1$.

The other case is that of slow adsorption of the substance, with fast chemical reaction at the surface. Such process is known as chemical adsorption or chemisorption. Here adsorbed molecules are glued to the surface by chemical forces of the same type as the forces of valence bonds. In order for these forces to reveal themselves, a molecule should come into a deformed state and overcome the activation barrier. Therefore the process of chemical adsorption requires a certain activation energy. Sometimes chemical adsorption is called activated adsorption. Chemical adsorption is closely related to the process of heterogeneous catalysis. The reaction rate of chemical adsorption is given by the expression:

$$R^{(S)} = \mathsf{k}'_\beta (1-\theta) = \mathsf{k}'_\beta \frac{bC}{C+b}. \tag{6.11}$$

For an inhomogeneous surface, the adsorption process will be described by Freindlich's isotherm, rather than Eq. (6.9):

$$\theta = \mathsf{k}' C^{1/n}, \quad (n > 1). \tag{6.12}$$

The molar flux of uncharged i-th component at the surface is

$$(\boldsymbol{j}_i^*)_w = (C_i \boldsymbol{u} - D_i \nabla C_i)_w. \tag{6.13}$$

If the surface is impenetrable and the heterogeneous reaction is not accompanied by a local density change, then the condition of sticking, $\boldsymbol{u} = 0$ must be satisfied on the surface. Otherwise, near the surface, there arises a convective flux of reactant, which is normal to the surface and directed toward it. This flux is known as Stefan's flux. As a rule, it does not exert a noticeable influence on chemical and biochemical heterogeneous reactions, and can be disregarded. However, in problems involving intense melting, evaporation, or condensation of substance, Stefan's flux may not be small and thus should be taken into account.

If the surface at which the heterogeneous reaction occurs is permeable or semipermeable (e.g., a membrane), there is also a nonzero flux component in the direction of normal ward to the surface. The boundary conditions at such surface will be discussed later, in the context of filtration processes in membranes.

Go back to the case of an impenetrable surface. Take a local coordinate system attached to the surface, with the y-axis along the local normal, in the direction opposite to ∇C_i. Then it follows from conditions (6.13) that

$$D\left(\frac{\partial C}{\partial y}\right)_w = \mathsf{k}' C_w^\nu. \tag{6.14}$$

Here, for simplicity's sake, the index i is dropped.

The condition (6.14) suggests an additional dimensionless criterion. Indeed, let us introduce dimensionless parameters

$$y^* = y/L, \quad C^* = C/C_0,$$

where L is the characteristic linear size of the problem; C_0 is the characteristic reactant concentration, for example, concentration in the bulk of the solution far from the surface.

Considering D and k' as constants, we obtain from (6.14):

$$\frac{1}{\mathrm{Da}}\left(\frac{\partial C^*}{\partial y^*}\right)_w = (C_w^*)^\nu. \tag{6.15}$$

The dimensionless parameter

$$\mathrm{Da} = \frac{\mathsf{k}' C_0^{\upsilon-1}}{D/L} = \frac{\text{reaction rate}}{\text{diffusion rate}} \tag{6.16}$$

is known as Damköler's number. At $\mathrm{Da} \gg 1$, the production rate of the reaction at the surface is much higher than the rate of substance delivery to the surface due to diffusion. Then the condition (6.15) reduces to

$$C_w = 0. \tag{6.17}$$

The physical meaning of this condition is that the reaction rate at the surface is so high that particles (molecules) of reacting substance reaching the surface by diffusion instantly engage in a reaction with the surface and disappear from the solution. Since the flux of substance toward the surface is proportional to $C_0 - C_w$, this flux achieves its maximum value, called the limiting flux, at $C_w = 0$. The boundary condition (6.17) is formulated for many problems known as diffusion-controlled problems.

At $\mathrm{Da} \ll 1$, the rate of reaction products formation is small in comparison with the rate of supply of reacting substances to the surface via diffusion. Therefore, in the first approximation, it is reasonable to assume the diffusion flux at the surface to be zero:

$$\left(\frac{\partial C^*}{\partial y^*}\right)_w = 0. \tag{6.18}$$

In this approximation, it is possible to set $C \approx C_0$ at the surface, and to take the reaction rate equal to $k' C_0^{\nu}$.

6.2
Convective Diffusion

As was shown earlier, for an extremely diluted solution, one has $\mathrm{Sc} \gg 1$.

As a consequence, if Re is not too small, there follows the inequality $\mathrm{Pe}_D \gg 1$. In the limiting case ($\mathrm{Pe}_D \to \infty$), Eq. (5.109) becomes:

$$DC/Dt = 0. \tag{6.19}$$

It follows from (6.19) that C remains constant during the motion of a liquid particle. If $C = C_0$ in the bulk of the flow and $C_w = 0$ at the surface, no solution of Eq. (6.19) can be found that would satisfy these conditions. Therefore, by analogy with the theory of a viscous boundary layer, there should exist near the surface a thin diffusion boundary layer of thickness δ_D, in which the concentration changes from C_0 to C_w. Inside this layer, the partial y-derivatives (along the normal to the surface) would be much greater then the x-derivatives (along the tangent to the surface).

Recall that the thickness of a viscous boundary layer is estimated by the expression [2]

$$\frac{\delta_u}{L} \sim \frac{1}{\sqrt{\mathrm{Re}}} = \left(\frac{\nu}{UL}\right)^{1/2}. \tag{6.20}$$

The diffusion coefficient plays the same role as the coefficient of kinematic viscosity, and so far as $\mathrm{Sc} \gg 1$, i.e. $\nu \gg D$, we have $\delta_D \ll \delta_U$ (Fig. 6.1).

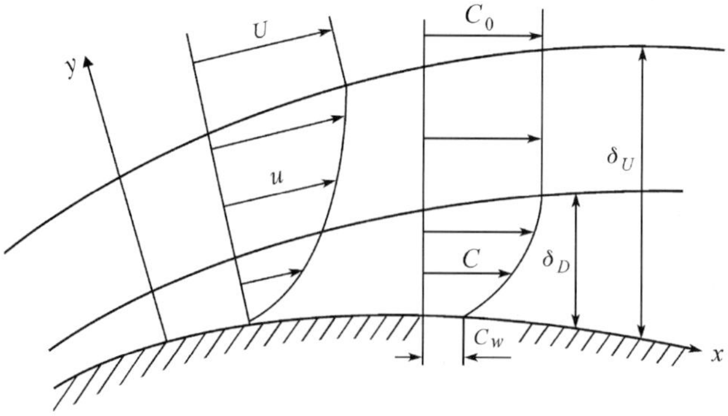

Fig. 6.1 Diffusion and viscous boundary layers at a body's surface.

In the diffusion boundary layer, we have $\partial^2 C/\partial x^2 \ll \partial^2 C/\partial y^2$, so the equation of convective diffusion in the layer becomes:

$$\frac{\partial C}{\partial t} + u\frac{\partial C}{\partial x} + \mathrm{v}\frac{\partial C}{\partial y} = D\frac{\partial^2 C}{\partial y^2}. \tag{6.21}$$

Consider the diffusion process at an infinite flat wall [3, 4]. Estimate the thickness of the diffusion boundary layer. For simplicity, assume the process to be stationary. It is known from the theory of a viscous boundary layer that $\mathrm{v}/u \sim \delta_D/L$, therefore $u\partial C/\partial x \sim \mathrm{v}\partial C/\partial y$. Since $\delta_D \ll \delta_U$, the structure of velocity in diffusion layer is equal to the velocity in the immediate vicinity of the wall

$$\frac{u}{U} \sim \frac{y}{\delta_U}. \tag{6.22}$$

Consequently,

$$u\frac{\partial C}{\partial x} \sim U\frac{y}{\delta_U}\frac{(C_0 - C_w)}{x} \sim \frac{\nu\delta_D(C_0 - C_w)}{\delta_U^3}. \tag{6.23}$$

Here we made use of the expressions $y \sim \delta_D/L$ and $(\delta_D)^2/x \sim \mathrm{vU}$.

Our thickness estimation requires the knowledge of the concentration C_w. If the process is diffusion-controlled, then $C_w = 0$. Another example is the surface of a material dissolvable in liquid. If dissolution occurs much faster than removal of dissolved substance into to the bulk of the liquid, then $C_w = C_{\text{sat}}$, where C_{sat} is the equilibrium concentration of dissolved substance near the surface. The values of C_w for mixed heterogeneous reactions and penetrable surfaces will be determined later on in the relevant sections.

Inside the boundary layer, we have $y = \delta_D$ and $D\partial^2 C/\partial y^2 \sim \mathrm{u}\partial C/\partial x$. Estimating the terms in the second equality, we get:

$$D\frac{(C_0 - C_{\mathrm{W}})}{\delta_U^2} \sim \frac{\nu\delta_D(C_0 - C_{\mathrm{W}})}{\delta_U^3},$$

which leads to

$$\delta_D \sim \left(\frac{D}{\nu}\right)^{1/3}\delta_U = \frac{\delta_U}{\mathrm{Sc}^{1/3}} \sim \frac{\sqrt{x\nu/U}}{\mathrm{Sc}^{1/3}}. \tag{6.24}$$

For infinite diluted solutions, $\mathrm{Sc} \sim 1000$, therefore $\delta_D \sim 0.1\delta_U$. Consequently, $\delta_D \ll \delta_U$, and the distribution of velocities in the diffusion boundary layer may be determined independently of the appropriate hydrodynamic problem. As an example, consider stationary laminar flow of viscous incompressible liquid in a flat channel. It is known that at some distance from the channel entrance, the velocity profile changes to a parabolic Poiseuille profile (Fig. 6.2) [5].

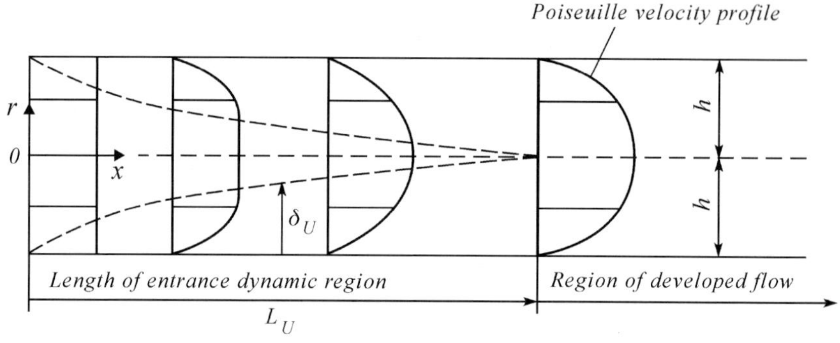

Fig. 6.2 Flow development in a channel.

By looking at Fig. 6.2, one can follow the evolution of the velocity profile from a homogeneous profile $u = U$ at the entrance to a parabolic one, $u = u(y)$, established at some distance L_U. This distance is called the length of the entrance dynamic region [5]. The dashed line shows how the thickness of the viscous boundary layer changes with length x. The characteristic time required for the layer to reach the channel axis ($\delta_U \sim h$) is estimated as $t_v \sim h^2/v$. In accordance with [4, 5], the length of the entrance region is

$$L_U \sim 0.16h\mathrm{Re}, \quad \mathrm{Re} = hU/v. \tag{6.25}$$

Thus, for Re = 1000, it equals $L_U \sim 80(2h)$.

A similar picture is observed for the flow of an infinite diluted solution accompanied by diffusion of the dissolved component away from the channel wall (Fig. 6.3).

Since $v \gg D$ for an infinite diluted solution, the entrance concentration region has the length $L_D \gg L_U$. Fig. 6.3 shows the case when dissolved substance gets into the flow from the soluble walls, with the boundary conditions $C = 0$ at the entrance and $C = C_w = C_{\mathrm{sat}}$ at the wall for $x > 0$. Dissolved substance diffuses from the wall to the channel axis.

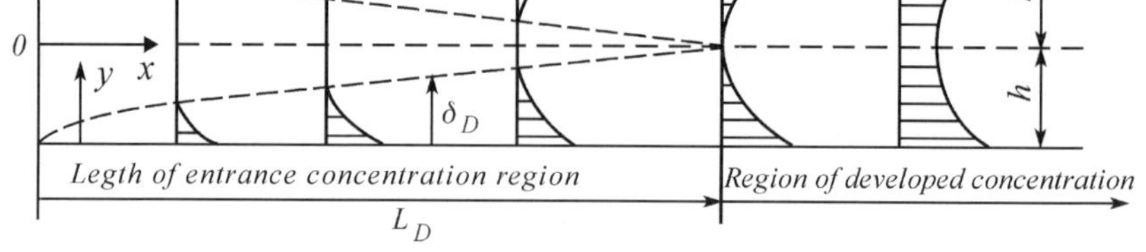

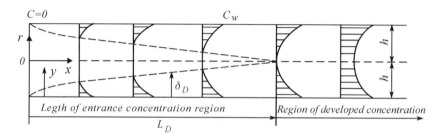

Fig. 6.3 Development of concentration profile in a channel.

There is yet another case – when the wall adsorbs the component dissolved in the liquid entering the channel, with a given concentration C_0 at the entrance. In this case, the boundary condition $C = 0$ at $x > 0$ is satisfied at the wall, and the picture of flow development is similar to Fig. 6.2 where U is replaced by C_0, δ_U – by δ_D, and L_U – by L_D. In contrast to the development of velocity profile, the concentration at the channel axis remains constant throughout the entire length of the entrance concentration region, and only then starts to decrease with x as the dissolved substance is being adsorbed on the wall surface.

The velocity profile in a developed channel flow has the form:

$$u = u_{\max}\left(-\frac{(y-h)^2}{h^2}\right), \qquad u_{\max} = \frac{3}{2}U = -\frac{h^2}{2\mu}\frac{dp}{dx}. \tag{6.26}$$

Accordingly, in the round pipe

$$u = u_{\max}\left(1 - \frac{r^2}{h^2}\right), \qquad u_{\max} = 2U. \tag{6.27}$$

Taking $y/h \ll 1$ near the wall and neglecting the term y^2/h^2 in (6.26), we obtain

$$u \approx u_{\max}\frac{2y}{h}. \tag{6.28}$$

Since $\delta_D \ll \delta_U$ in the diffusion boundary layer, it is possible to take the expression (6.28) for velocity. This helps us estimate the thickness δ_D more precisely. From

$$D\frac{\partial^2 C}{\partial y^2} \sim u\frac{\partial C}{\partial x},$$

there follows

$$D\frac{(C_0 - C_w)}{\delta_D^2} \sim u_{\max}\frac{2\delta_D}{h}\frac{(C_0 - C_w)}{x}$$

and

$$\delta_D \sim x\left(\frac{h}{x}\right)^{2/3}\left(\frac{D}{u_{\max}h}\right)^{1/3}. \tag{6.29}$$

Comparing (6.29) with (6.24), it is possible to conclude that at the infinite flat wall, $\delta_D \sim x^{1/2}$, whereas at the channel wall, $\delta_D \sim x^{1/3}$. To estimate the entrance concentration region length L_D, it is necessary to set $\delta_D \sim h$, $x \sim L_D$, and $u_{\max} \sim U$. The outcome is:

$$\frac{L_D}{h} \sim \frac{Uh}{D} = \mathrm{Pe}_D. \tag{6.30}$$

Thus, in the considered problem, the number Pe_D plays the same role as the number Re in the corresponding hydrodynamic problem.

In the above-considered problems involving a diffusion boundary layer at an infinite flat wall and in a flat channel, it is possible to determine the flux of matter at the wall, assuming a linear concentration profile near the wall:

$$j^* \sim D\frac{(C_0 - C_w)}{\delta_D^2}. \tag{6.31}$$

Substitution of expressions for δ_D from (6.29) or (6.24) into (6.31) gives us the diffusion flux at the wall as a function of length.

Summarizing, we should note that the methods presented in the present section can be applied without any modifications to heat exchange problems, because temperature distribution is described by an equation similar to the diffusion equation. The boundary conditions are also formulated in a similar way. One only has to replace D by the coefficient of thermal diffusivity, and the number Pe_D – by Pe_T. The corresponding boundary layer is known as the thermal layer. Detailed solutions of heat conductivity problems can be found in [6].

6.3
Flow in a Channel with a Reacting Wall

In the previous section, the thickness of the diffusion boundary layer at the channel wall was evaluated in the approximation $\mathrm{Sc} \gg 1$. Consider now the solution of the convective diffusion equation in the region of concentration development, where $\delta_D/h \ll 1$, that is, at small distances from the channel entrance. Suppose the channel wall is made from material that can dissolve in the liquid flow, so that the boundary conditions will be $C = C_{\mathrm{sat}}$ at the wall and $C = C_0$ far away from the wall. If the problem involves adsorption of substance from an infinite diluted binary solution on the wall, then $C = C_w = 0$ at the wall, and $C = C_0$ on a large distance from the wall, C_0 being the concentration of the dissolved component. Consider the diffusion happening in the region of hydrodynamic development of the flow. Then we should take $\mathrm{v} = 0$ and use (6.28) as the expression for the longitudinal velocity. Note that a similar situation takes place in a developed thin film of thickness d flowing under the influence of gravity along a flat vertical wall [3]. The velocity profile in the film is parabolic, with the maximum velocity

$$u_{\max} = g\delta^2/2\nu \tag{6.32}$$

achieved at the free surface of the film. The thickness of the film is

$$\delta = (3\nu\bar{Q}/g)^{1/3} \tag{6.33}$$

where $\bar{Q}$ is the volume flow rate through a unit cross section of the film.

The adopted assumptions allow us to write the equation of stationary convective diffusion:

$$2\frac{y}{h}u_{\max}\frac{\partial C}{\partial x} = D\frac{\partial^2 C}{\partial y^2} \tag{6.34}$$

with boundary conditions

$$C = C_{\text{sat}} \quad \text{at } y = 0; \qquad C = 0 \quad \text{at } y \to \infty. \tag{6.35}$$

Note that the second condition should, generally speaking, be formulated at $y = h$, but as long as we consider the region $y/h \ll 1$, this condition can be formulated at $y \to \infty$.

Eq. (6.34) with conditions (6.35) has a self-similar solution. To derive this solution, one should proceed as follows. From Eq. (6.34), it follows that

$$\frac{C}{C_{\text{sat}}} = f\left(x, y, \frac{2u_{\max}}{Dh}\right). \tag{6.36}$$

Since the left-hand side of (6.36) is dimensionless, the right-hand side must be dimensionless as well. Three parameters x, y, and $2u_{\max}/Dh$ can be used to construct one dimensionless complex parameter. It will be sought in the general form

$$\eta = x^{\alpha}y^{\beta}(2u_{\max}/Dh)^{\gamma}, \tag{6.37}$$

such that

$$\tilde{C} = \frac{C}{C_{\text{sat}}} = f(\eta) \tag{6.38}$$

obeys Eq. (6.34). The following equations should hold:

$$\frac{\partial C}{\partial x} = f'(\eta)\alpha x^{\alpha-1}y^{\beta}\left(\frac{2u_{\max}}{Dh}\right)^{\gamma} = \frac{f'(\eta)\alpha\eta}{x},$$

$$\frac{\partial C}{\partial y} = f'(\eta)x^{\alpha}\beta y^{\beta-1}\left(\frac{2u_{\max}}{Dh}\right)^{\gamma} = \frac{f'(\eta)\beta\eta}{y},$$

$$\frac{\partial^2 C}{\partial y^2} = \frac{\beta}{y}x^{\alpha}\beta y^{\beta-1}\left(\frac{2u_{\max}}{Dh}\right)^{\gamma}\frac{d}{d\eta}(\eta f') - \frac{f'(\eta)\beta\eta}{y^2} = \frac{\beta^2\eta}{y^2}\frac{d}{d\eta}(\eta f')' - \frac{f'(\eta)\beta\eta}{y^2}.$$

Substituting these expressions in (6.34), we find:

$$\left(\left(\frac{2u_{\max}}{Dhx}\right)^{1/3} y\right)^3 \alpha f' = \beta^2(\eta f')' - \beta f'. \qquad (6.39)$$

It follows from this equation that if

$$\eta = y\left(\frac{2u_{\max}}{Dhx}\right)^{1/3}, \qquad (6.40)$$

then the solution of Eq. (6.34) could be sought in the form (6.37), by taking $\alpha = -1/3$, $\beta = 1$, $\gamma = 1/3$, and $f(\eta)$ that satisfies the equation

$$f'' + \frac{1}{3}\eta^2 f' = 0. \qquad (6.41)$$

The boundary conditions (6.35) then transform to

$$f = 1 \quad \text{at } \eta = 0, \quad f = 1 \quad \text{at } \eta \to \infty. \qquad (6.42)$$

Eq. (6.41) can be rewritten as

$$\frac{d}{d\eta}(\ln f') = -\frac{\eta^2}{3}. \qquad (6.43)$$

After double integration, this equation gives the following solution:

$$f = A\int_0^{\eta} e^{-\eta^3/9}\, d\eta + B.$$

Using boundary conditions (6.42), we finally get:

$$f = 1 - \frac{\int_0^{\eta} e^{-\eta^3/9}\, d\eta}{\int_0^{\infty} e^{-\eta^3/9}\, d\eta}. \qquad (6.44)$$

The integral in the denominator is expressed through a full gamma function:

$$\Gamma(z) = \int_0^{\infty} z^{n-1} e^{-z}\, dz, \quad n > 0.$$

Indeed, making the replacement $\eta^3/9 = z$, we get:

$$\int_0^{\infty} e^{-\eta^3/9}\, d\eta = \frac{1}{3^{1/3}}\int_0^{\infty} z^{-2/3} e^{-z}\, dz = 1.858.$$

and

$$\frac{C}{C_{sat}} = f(\eta) = 1 - 0.538\int_0^{\eta} e^{-\eta^3/9}\,d\eta. \tag{6.45}$$

The diffusion flux at the wall is equal to

$$j^* = -D\left(\frac{\partial C}{\partial y}\right)_w = -C_{sat}D\left(\frac{2u_{max}}{Dhx}\right)^{1/3} f'(\eta)_w = 0.678C_{sat}D\left(\frac{2u_{max}}{Dhx}\right)^{1/3}. \tag{6.46}$$

Substitution of (6.46) into (6.31) gives us the expression for diffusion layer thickness:

$$\frac{\delta_D}{x} = 1.475\left(\frac{h}{x}\right)^{2/3}\left(\frac{D}{u_{max}h}\right)^{1/3}. \tag{6.47}$$

If we solve a similar problem of adsorption at a quickly reacting wall, then we should take the following boundary conditions:

$$C = C_w = 0 \quad \text{at } y = 0, \qquad C = C_0 \quad \text{at } y \to \infty. \tag{6.48}$$

Repeating the calculations performed earlier, we get:

$$\frac{C}{C_0} = 0.538\int_0^{\eta} e^{-\eta^3/9}\,d\eta. \tag{6.49}$$

We may then derive an expression for the diffusion flux, which turns out to be directed not from the wall, but toward it, and the thickness of the diffusion boundary layer. It is obvious that these characteristics will coincide with (6.46) and (6.47), but in these equations, C_{sat} should be replaced with C_0.

6.4
Reverse Osmosis

The method of reverse osmosis [7] is based on filtration of solutions under pressure through semi-permeable membranes, which let the solvent pass through while preventing (either totally or partially) the passage of molecules or ions of dissolved substances. The phenomenon of osmosis forms the physical core of this method. Osmosis is a spontaneous transition of the solvent through a semi-permeable membrane into the solution (Fig. 6.4, a) at a pressure drop ΔP lower than a certain value π. The pressure π at which the equilibrium is established, is known as osmotic pressure (Fig. 6.4, b). If the pressure drop exceeds π, i.e. pressure $p' > p'' + \pi$ is applied on the solution side, then the transfer of solvent will reverse its direction. Therefore, this process is known as reverse osmosis (Fig. 6.4,

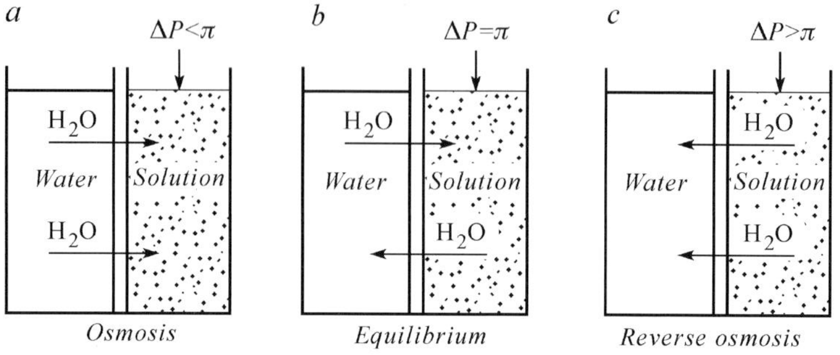

Fig. 6.4 Osmosis and reverse osmosis.

c). All of this implies that the driving force of reverse osmosis for an ideal semi-permeable membrane is

$$\Delta p = \Delta P - \pi, \tag{6.50}$$

where ΔP is the excess pressure (working pressure) above the solution; π is the osmotic pressure of the solution.

In practice, however, no membrane can act as an ideal semi-permeable membrane, and some transition of the dissolved substance through a real membrane always happens – to a larger or smaller extent. With this in mind, we rewrite Eq. (6.50) as

$$\Delta p = (p' - p'') - \Delta\pi, \quad \Delta\pi = \pi' - \pi'', \tag{6.51}$$

where π' and π'' are the respective osmotic pressures in the solution in front of the membrane and in the filtrate behind the membrane.

In chemical thermodynamics, osmotic pressure should be determined as follows [8]. Consider the equilibrium between a solution whose chemical potential is μ' and a solvent whose chemical potential is μ''. The solution and the solvent are separated by a semi-permeable membrane. The chemical (osmotic) equilibrium between them occurs under the condition

$$\mu_i'' = \mu_i', \tag{6.52}$$

where $i = 1$ for the solvent, $i = 2$ for the solute.

Used the expression (5.54) for chemical potential, we find that for a solvent in the solution, there should hold

$$\mu_1' = \mu_1^0(T, p') + \Xi AT \ln x_1 \tag{6.53}$$

and for a solvent in the filtrate, that is, in the liquid that has passed through an ideal membrane,

$$\mu_1'' = \mu_1^0(T, p''). \tag{6.54}$$

Eqs. (6.52)–(6.54) give us the following:

$$\mu_1^0(T, p'') - \mu_1^0(T, p') = \Xi AT \ln x_1. \tag{6.55}$$

One can see from the last equality that $p'' = p'$ only at $x_1 = 1$, and since $x_1 < 1$, it follows that $p'' \neq p'$. Using the expression for $\mu_1^0(T, p)$ [8], we establish that the following equality must hold at osmotic equilibrium:

$$\pi = p' - p'' = -\frac{\Xi AT \ln x_1}{\bar{v}_1^0}, \tag{6.56}$$

where $\bar{v}_1^0$ is the molar volume of pure solvent at the pressure $0.5(p' + p'')$.

We have noted in Section 5.3 that an infinite diluted solution behaves as an ideal one. The reverse statement, generally speaking, is wrong. Consider an infinite diluted solution, in which $x_1 \to 1$ and $\sum_{i\neq 1} x_i \to 0$. Then from (6.56), it follows that

$$\pi = -\frac{\Xi AT}{\bar{v}_1^0} \ln\left(1 - \sum_{i\neq 1} x_i\right) \approx \frac{\Xi AT}{\bar{v}_1^0} \sum_{i\neq 1} x_i \approx \frac{\Xi AT}{\bar{v}_1^0} \sum_{i\neq 1} C_i \bar{v}_1^0 = \Xi AT, \tag{6.57}$$

where C_i and C are molar concentrations of dissolved components.

Eq. (6.57) for osmotic pressure in an ideal infinite diluted solution is known as Van't Hoff equation. It is analogous to the equation of state for an ideal gas.

It follows from Eq. (6.57) that osmotic pressure is a property of the solution and is not dependent on the properties of membrane material.

The values of osmotic pressure π for some water solutions at a standard temperature are given in Tab. 6.1 for various dissolved substance concentrations [4]. One can see from these data that for a given component, the value of π decreases

Table 6.1

Dissolved component	Concentration, kg/m^3	Osmotic pressure, MPa
NaCl (M = 58.5)	50	4.609
	10	0.844
	5	0.421
Carbamide (M = 60)	50	2.127
	10	0.427
	5	0.213
Sucrose (M = 342)	50	0.380
	10	0.076
	5	0.038

with the reduction of component concentration. Osmotic pressure also decreases with the increase of the component's molecular weight at a fixed concentration. The flow rate of a solvent passing through a membrane is equal to

$$j_A{}^* = A(\Delta P - \Delta \pi), \tag{6.58}$$

where A is the penetration factor of the membrane with respect to the solvent. As it was already noted, the membrane does not impede the dissolved substance completely. Since the driving force of the dissolved substance's transfer through the membrane is the difference between its concentrations in front of, and behind the membrane, the flux of this substance is given by

$$j_A{}^* = B\Delta C_w \approx BC_w R_S. \tag{6.59}$$

Here B is the penetration factor of the membrane with respect to the dissolved substance, C_w is the dissolved substance concentration in the solution in front of the membrane (it is assumed that behind the membrane, $C_{w1} \ll C_w$), R_S is the hindrance factor of the membrane. According to (6.57), osmotic pressure of infinite diluted solutions is proportional to the concentration of dissolved substance. So, we can use formulas (6.57) and (6.58) to estimate the rate of solution filtration through the membrane [4]:

$$v_w = \frac{j}{\rho} = A'\Delta P\left(1 - R_S \frac{C_w \pi_0}{C_0 \Delta P}\right). \tag{6.60}$$

Here $A' = A/C_A$, C_A is the molar concentration of solvent in the solution, C_0 is the concentration of dissolved substance in the bulk flow before the membrane, π_0 is the osmotic pressure that corresponds to the concentration C_0.

Note that at $\Delta P \gg \pi_0$, the impede factor of the membrane is $R_S \to 1$.

A distinguishing feature of solution filtration through a membrane is that the concentration of dissolved substance at the surface before the membrane is higher than in the bulk of the solution. This effect is known as concentration polarization. We must mention some negative consequences of this phenomenon. First, the increase of surface concentration also increases the osmotic pressure before the membrane; therefore at a given hydrostatic pressure drop πP at the membrane, the flux of solvent through the membrane (see Eq. (6.58)) decreases. Second, according to Eq. (6.59), the flux of dissolved substance through the membrane increases, which is undesirable.

The degree of concentration polarization depends on the hydrodynamics of the flow before the membrane and on the geometry of the membrane surface. Therefore, if we want to reduce negative consequences of concentration polarization, we should be looking at possible ways to change the flow hydrodynamics or the surface geometry. Thus, an increase of flow velocity along the membrane surface, in particular, turbulization of the flow, promotes transfer of dissolved substance from the wall and weakens the effect of polarization.

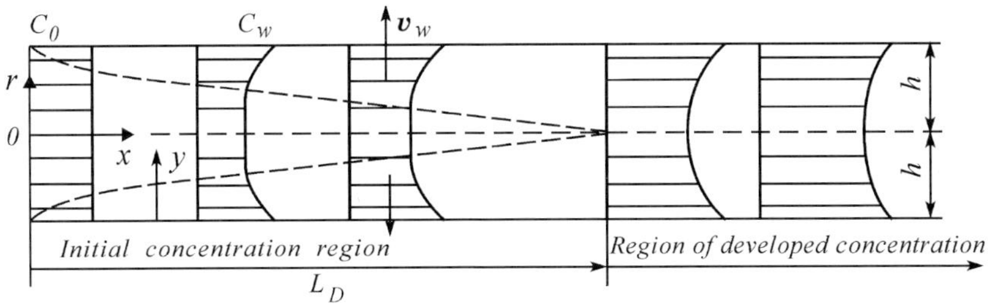

Fig. 6.5 Convective diffusion in a channel with semipermeable membrane walls.

Let us illustrate the effect of concentration polarization by solving a relevant stationary problem of convective diffusion of binary infinite diluted solution in a channel whose walls are semipermeable membranes [9, 10]. A qualitative picture of the distribution of dissolved substance concentration is shown in Fig. 6.5.

At the channel entrance, the profile of concentration is taken to be uniform in the channel cross section. When the solution flows into the channel, the solvent starts filtering through the wall. It means that in addition to the longitudinal convective transfer, there is a transfer of dissolved substance to the wall. Since for all practical purposes, dissolved substance does not pass through the membrane, its concentration at the wall increases, i.e. $C_w > C_0$.

The concentration difference $C_w - C_0$ is the driving force of diffusion flux of dissolved substance from the wall to the channel axis. As a result, the profile of concentration is deformed, and near the wall, a concentration boundary layer is formed, whose thickness increases along the length of the channel. The length at which the layer grows thick enough to reach the channel axis is called the length of the entrance region. Qualitatively, the picture looks similar to the one encountered in the diffusion problem that was considered in Section 5.3. Since $\mathrm{Sc} \gg 1$, diffusion layer thickness is much smaller than the thickness of the viscous boundary layer, and in consideration of convective diffusion, the velocity distribution in the developed flow in the channel can be taken as the flow velocity. However, in contrast to the problem of diffusion in a channel with a soluble or quickly reacting wall considered in Section 5.3, for a flow in a channel with semi-permeable walls, it is necessary to take into account the transverse component of velocity v, which on a wall is equal to the velocity of the solvent v_w. In the present case, the equation of diffusion has the form

$$u\frac{\partial C}{\partial x} + \mathrm{v}\frac{\partial C}{\partial y} = D\frac{\partial^2 C}{\partial y^2}. \tag{6.61}$$

We find the velocities by solving the hydrodynamic problem of developed flow in a channel with permeable walls. If the Reynolds number (which corresponds to the velocity of liquid filtration v_w through the wall) is small (since v_w is small, this condition is usually satisfied), the form of the longitudinal velocity of this flow is similar to that of a flow in the channel with impenetrable walls (Poiseuille flow):

$$u = u_{max}\left(1 - \frac{r^2}{h^2}\right) = \frac{3}{2}U\left(1 - \frac{r^2}{h^2}\right), \tag{6.62}$$

and the transverse velocity is equal to [11]

$$\mathrm{v} = \mathrm{v}_w \frac{r}{2h}\left(3 - \frac{r^2}{h^2}\right), \quad \frac{h\mathrm{v}_w}{\nu} \ll 1. \tag{6.63}$$

Let us adopt the same procedure as in Section 5.3. Instead of r, introduce the coordinate $y = r + h$ and, as long as the region adjacent to the entrance section of the channel is being considered, we can take $y/h \ll 1$. Then from (6.62) and (6.63) there follow approximate equalities

$$u = 2u_{max}\frac{y}{h} = 3U\frac{y}{h}, \tag{6.64}$$

$$\mathrm{v} = -\mathrm{v}_w. \tag{6.65}$$

Let us formulate the boundary conditions. The condition at the entrance is:

$$C = C_0 \quad \text{at } x = 0. \tag{6.66}$$

The boundary condition at the wall (membrane) can be formulated by using the condition of conservation of flow rate of dissolved substance through the membrane:

$$-\mathrm{v}_w C_w - D\left(\frac{\partial C}{\partial y}\right)_w = -(1 - R_S)\mathrm{v}_w C_w. \tag{6.67}$$

Here on the left-hand side is the difference between the convective flux of dissolved substance toward the wall and the diffusion flux from the wall, and at the right-hand side is the flux of the same substance through the wall.

The condition (6.67) immediately simplifies to

$$R_S \mathrm{v}_w C_w = -D\left(\frac{\partial C}{\partial y}\right)_w. \tag{6.68}$$

The relation (6.68) corresponds to the case when the membrane lets the dissolved substance pass through ($R_S < 1$). In the case of an ideal semi-permeable membrane, $R_S = 1$, and the condition (6.68) becomes

$$\mathrm{v}_w C_w = -D\left(\frac{\partial C}{\partial y}\right)_w. \tag{6.69}$$

Compare the obtained boundary condition to the condition for a mixed heterogeneous reaction,

$$k' C_w^\nu = -D \left(\frac{\partial C}{\partial y} \right)_w . \tag{6.70}$$

The condition (6.69), which implies $R_S = 1$, is similar to the condition for a mixed heterogeneous reaction of the first order ($\nu = 1$). In the case $R_S \sim C_w^m$, the condition at the wall corresponds to the condition for reaction of ($\nu = m + 1$)-th order.

The analogy between the problem of convective diffusion in a channel with semipermeable walls and the problem of mixed heterogeneous reaction makes it possible to consider these problems together. Conclusions that are true for one problem would be true for the other problem as well, even though these processes are distinct from both physical and chemical viewpoints. Besides, the analogy with chemical reactions allows us to introduce a dimensionless parameter of the problem (see (6.16)) – the Damköler number:

$$\mathrm{Da} = \frac{\mathrm{v}_w}{D/h} = \frac{\text{filtration rate}}{\text{diffusion rate}} . \tag{6.71}$$

On the other hand, this number can be interpreted as the Peclet diffusion number:

$$\mathrm{Pe}_D = \frac{\mathrm{v}_w h}{D} = \frac{\text{mass transfer in filtration}}{\text{mass transfer in diffusion}} . \tag{6.72}$$

In the region occupied by the diffusion boundary layer, we can use the Peclet number determined by the thickness of the boundary layer δ_D. The following inequality holds for this number:

$$\mathrm{Pe}_D = \frac{\mathrm{v}_w \delta_D}{D} \ll 1. \tag{6.73}$$

If we consider a diffusion process with the boundary condition at the wall that corresponds to a mixed heterogeneous reaction, then the inequality (6.73) is similar to the inequality $\mathrm{Da} \ll 1$. It means that the rate of formation of reaction products is small in comparison with the diffusion flux.

We can estimate by the order of magnitude the concentration of dissolved substance at the membrane surface, regarding it as an ideal semi-permeable membrane. From (6.69), it follows that for a relatively small change of concentration at the entrance region ($C_w/C_0 \sim 1$),

$$\frac{C_w - C_0}{C_0} \sim \frac{\mathrm{v}_w \delta_D}{D} \ll 1. \tag{6.74}$$

The thickness of the diffusion layer can be estimated by using the expression (6.29). Then from (6.74) there follows:

$$\frac{C_w - C_0}{C_0} \sim \xi^{1/3}, \quad \xi = \left(\frac{V_w h}{D}\right)^3 \left(\frac{D}{3Uh}\right)\left(\frac{x}{h}\right). \tag{6.75}$$

The parameter $(C_w - C_0)/C_0$ characterizes the degree of concentration polarization. Eq. (6.61) with the expressions (6.64) and (6.65) for velocities at the wall and the condition $C \to C_0$ at y (this condition corresponds to the constancy of concentration outside of the diffusion boundary layer in the bulk flow, and since $y/h \ll 1$, the condition is formulated at $y \to \infty$) has a self – similar solution, which can be derived in the same way as in Section 6.3. Consequently, the solution can be sought in the form

$$\frac{C - C_0}{C_w - C_0} = f(\eta), \quad \eta = y\left(\frac{3U}{hDx}\right)^{1/3} = \frac{y}{\xi^{1/3}}\left(\frac{V_w}{D}\right). \tag{6.76}$$

or, in view of (6.75), as

$$C^* = \frac{C_w - C_0}{C_0} = \xi^{1/3} f(\eta). \tag{6.77}$$

We further obtain:

$$\frac{\partial C}{\partial x} = C_0 \frac{\partial C^*}{\partial x} = C_0\left(\frac{\partial C^*}{\partial \xi}\frac{\partial \xi}{\partial x} + \frac{\partial C^*}{\partial \eta}\frac{\partial \eta}{\partial x}\right) = C_0 \frac{\xi^{1/3}}{3x}(f - \eta f'),$$

$$\frac{\partial C}{\partial y} = C_0 \frac{\partial C^*}{\partial y} = C_0\left(\frac{3U}{hDx}\right)^{1/3} \xi^{1/3} f'(\eta),$$

$$\frac{\partial^2 C}{\partial y^2} = C_0 \frac{\partial^2 C^*}{\partial y^2} = C_0\left(\frac{3U}{hDx}\right)^{1/3} \xi^{1/3} f''(\eta).$$

Substituting the obtained derivatives into the diffusion equation (6.61), we find that at $\delta_D \to 0$ and $\xi^{1/3} \to 0$,

$$f'' + \frac{\eta^2}{3} f' - \frac{\eta f}{3} = 0. \tag{6.78}$$

The boundary conditions are transformed accordingly to

$$f' = -1 \quad \text{at } \eta = 0, \quad f \to 0 \quad \text{at } \eta \to \infty. \tag{6.79}$$

The solution of Eq. (6.78) is sought in the form

$$f(\eta) = \eta q(\eta). \tag{6.80}$$

Substitution of (6.80) into (6.78) results in the equation

$$\eta q'' + \left(\frac{\eta^3}{9} + 2\right) q' = 0, \tag{6.81}$$

whose solution, subject to the second condition (6.79), has the form

$$q = -A\left(\frac{e^{-\eta^3/9}}{\eta} + \frac{1}{3}\int_{\infty}^{\eta} \eta e^{-\eta^3/9}\, d\eta\right). \tag{6.82}$$

The first boundary condition (6.79) allows to find A:

$$A = \left(\frac{1}{3}\int_{\infty}^{\eta} \eta e^{-\eta^3/9}\, d\eta\right)^{-1} = \frac{9^{1/3}}{\Gamma(2/3)} = 1.536.$$

As a result, we obtain:

$$f(\eta) = 1.536\left(e^{-\eta^3/9} - \frac{\eta}{3}\int_{\infty}^{\eta} \eta e^{-\eta^3/9}\, d\eta\right). \tag{6.83}$$

From (6.77) and (6.83), we find the concentration of dissolved substance in the solution before the membrane:

$$\frac{C_w}{C_0} = 1 + \xi^{1/3} f(0) = 1 + 1.536\xi^{1/3}. \tag{6.84}$$

As was already remarked, the solution (6.83) and (6.84) is applicable at $\xi^{1/3} \to 0$. A more detailed analysis shows that it is applicable at $\xi \le 0.02$.

Until now, we have considered only one of the processes related to the separation of mixtures by a membrane, namely, the reverse osmosis. Let us enumerate some other processes:

Dialysis. If in the process of osmosis, a part of the dissolved substance is transferred together with the solvent, then such a process is called dialysis.

Ultrafiltration. It is a separation process in which molecules or colloidal particles are filtered from the solution via membranes. The peculiarity of this process, distinguishing it from reverse osmosis, is that it achieves separation of systems in which the molecular mass of dissolved components is much greater than that of the solvent. Pressure gradient is the driving force in this process, as well as in reverse osmosis and dialysis.

Electrodialysis. It is a process in which ions of the dissolved substance are pushed through the membrane under the action of electric field. The gradient of electric potential is the driving force in this process.

Finally, it should be noted that reverse osmosis and ultrafiltration differ essentially from the usual filtration. In ordinary filtration, the product deposits at the surface of the filter, whereas in reverse osmosis or ultrafiltration, two solutions are formed at both sides of the membrane, one of which is enriched by the dissolved substance.

6.5 Diffusion Toward a Particle Moving in a Solution

Consider convection diffusion toward a spherical particle of radius R, which undergoes translational motion with constant velocity U in a binary infinite diluted solution [3]. Assume the particle is small enough so that the Reynolds number is $\mathrm{Re} = UR/\nu \ll 1$. Then the flow in the vicinity of the particle will be Stokesean and there will be no viscous boundary layer at the particle surface. The Peclet diffusion number is equal to $\mathrm{Pe}_D = \mathrm{Re\,Sc}$. Since for infinite diluted solutions, $\mathrm{Sc} \sim 10^3$ and the flow can be described as Stokesian for the Re up to $\mathrm{Re} \sim 0.5$, it is perfectly safe to assume $\mathrm{Pe}_D \gg 1$. Thus, a thin diffusion boundary layer exists at the surface. Assume that a fast heterogeneous reaction happens at the particle surface, i.e. the particle is dissolving in the liquid. The equation of convective diffusion in the boundary diffusion layer, in a spherical system of coordinates r, θ, φ, subject to the condition that concentration does not depend on the azimuthal angle φ, has the form:

$$u_r \frac{\partial C}{\partial r} + \frac{u_\theta}{r}\frac{\partial C}{\partial \theta} = D\left(\frac{\partial^2 C}{\partial r^2} + \frac{2}{r}\frac{\partial C}{\partial r}\right). \tag{6.85}$$

We dropped the Laplacian operator $\dfrac{1}{r^2 \sin\theta}\dfrac{\partial}{\partial\theta}\left(\sin\theta \dfrac{\partial C}{\partial \theta}\right)$ in the right-hand side of the diffusion equation, since in the boundary layer, derivatives in the tangential direction to the surface that is, perpendicular to the radius are small in comparison with normal derivatives. Stokesean flow past the sphere [12] is characterized by the stream function

$$\psi = -\frac{U}{2}\sin^2\theta\left(r^2 - \frac{3}{2}Rr + \frac{1}{2}\frac{R^3}{r}\right). \tag{6.86}$$

Consider diffusion in a boundary layer with thickness $\delta_D \ll R$. It is convenient to introduce the coordinate $y = r - R$ instead of r. The new coordinate has the meaning of distance from the spherical surface. Then, taking $y/R \ll 1$, we obtain:

$$\psi = -\frac{3}{4}Uy^2 \sin^2\theta. \tag{6.87}$$

The continuity equation in a spherical system of coordinates is written as

$$\frac{\partial}{\partial r}(\sin\theta r^2 u_r) + \frac{\partial}{\partial\theta}(r \sin\theta u_\theta) = 0. \tag{6.88}$$

The form of Eq. (6.88) enables us to express velocities through the stream function:

$$u_r = \frac{1}{r^2 \sin\theta} \frac{\partial \psi}{\partial \theta}, \quad u_\vartheta = -\frac{1}{r \sin\theta} \frac{\partial \psi}{\partial r}. \tag{6.89}$$

Switching from r to y and noticing that $y/R \ll 1$, we rewrite (6.89) as

$$u_r \approx \frac{1}{R^2 \sin\theta} \frac{\partial \psi}{\partial \theta}, \quad u_\vartheta \approx -\frac{1}{R \sin\theta} \frac{\partial \psi}{\partial y} = \frac{3}{2} U \frac{y}{R} \sin\theta. \tag{6.90}$$

Now, go back to the diffusion equation (6.85) and introduce new variables ψ and θ instead of r and θ. Then

$$\frac{\partial C(r,\theta)}{\partial \theta} = \frac{\partial C}{\partial \theta} + \frac{\partial C}{\partial \psi}\frac{\partial \psi}{\partial \theta} \approx \frac{\partial C}{\partial \theta} + R^2 \sin\theta u_r \frac{\partial C}{\partial \psi},$$

$$\frac{\partial C(r,\theta)}{\partial r} = \frac{\partial C}{\partial \psi}\frac{\partial \psi}{\partial r} \approx -R \sin\theta u_\theta \frac{\partial C}{\partial \psi},$$

$$\frac{\partial^2 C}{\partial r^2} = \frac{\partial}{\partial r}\frac{\partial C(r,\theta)}{\partial r} \approx \frac{\partial \psi}{\partial r}\frac{\partial}{\partial \psi} - \left(R \sin\theta u_\theta \frac{\partial C}{\partial \psi}\right) \approx R^2 \sin^2\theta u_\vartheta \frac{\partial}{\partial \psi}\left(u_\vartheta \frac{\partial C}{\partial \psi}\right),$$

We can show that $\frac{2}{r}\frac{\partial C}{\partial r} \ll \frac{\partial^2 C}{\partial r^2}$. Indeed,

$$\frac{2}{r}\frac{\partial C}{\partial r} \approx \frac{2}{R}\frac{\partial C}{\partial y},$$

$$\frac{\partial^2 C}{\partial r^2} = \frac{\partial^2 C}{\partial y^2} = \frac{\partial}{\partial y}\left(\frac{\partial C}{\partial y}\right) \sim \frac{1}{\delta_D}\frac{\partial C}{\partial r},$$

and since $\delta_D/R \ll 1$, we have:

$$\frac{2}{r}\frac{\partial C}{\partial r} : \frac{\partial^2 C}{\partial r^2} \sim \frac{2}{R}\frac{\partial C}{\partial y} : \frac{1}{\delta_D}\frac{\partial C}{\partial y} = \frac{2\delta_D}{R} \ll 1.$$

Substitution of the obtained expressions into (6.85) yields

$$\frac{\partial C}{\partial \theta} = DR^3 \sin^2\theta \frac{\partial}{\partial \psi}\left(u_\theta \frac{\partial C}{\partial \psi}\right). \tag{6.91}$$

Having expressed u_θ through ψ from (6.87) and (6.90), we finally obtain:

$$\frac{\partial C}{\partial \theta} = DR^2\sqrt{3U} \sin^2\theta \frac{\partial}{\partial \psi}\left(\sqrt{\psi}\frac{\partial C}{\partial \psi}\right). \tag{6.92}$$

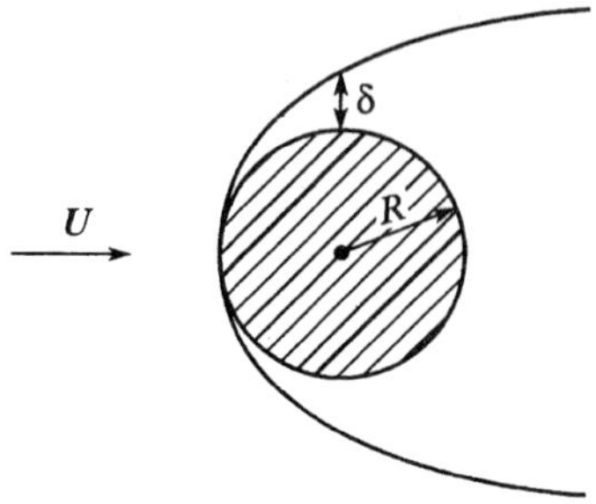

Fig. 6.6 Convective diffusion toward a solid particle in a solution.

The boundary conditions for Eq. (6.92) are:

$$\begin{aligned} &C = 0 \quad \text{at } \psi = 0 \text{ (surface of the particle)} \\ &C \to C_0 \quad \text{at } \psi \to \infty \text{ (far from the particle)}, \\ &C = C_0 \quad \text{at } \theta = 0 \text{ and } \psi = 0. \end{aligned} \tag{6.93}$$

Here the condition in the bulk flow is formulated at $r \to \infty$ (or at $\psi \to \infty$). The last condition (6.93) has a simple meaning. The point $\theta = 0$ and $\psi = 0$ corresponds to the point of flow run on the sphere. At this point (Fig. 6.6) the flow has not yet been depleted by diffusion, and the concentration of solute substances coincides with the concentration C_0 in the bulk flow.

To solve Eq. (6.92), a new variable should be introduced instead of θ:

$$t = DR^2\sqrt{3U}\int \sin^2\theta\, d\theta = \frac{DR^2\sqrt{3U}}{2}\left(\theta - \frac{\sin\theta}{2}\right) + A. \tag{6.94}$$

Then (6.92) may be reduced to the equation

$$\frac{\partial C}{\partial t} = \frac{\partial}{\partial \psi}\left(\sqrt{\psi}\frac{\partial C}{\partial \psi}\right), \tag{6.95}$$

which has a self-similar solution of the kind

$$C = f(\eta), \quad \eta = \frac{\psi}{t^{2/3}}. \tag{6.96}$$

Then

$$\frac{\partial C}{\partial t} = -\frac{2}{3}\frac{\psi}{t^{5/3}}f' = -\frac{2}{3}\frac{\eta}{t}f', \quad \frac{\partial C}{\partial \psi} = \frac{1}{t^{2/3}}f',$$

$$\frac{\partial}{\partial \psi}\left(\sqrt{\psi}\frac{\partial C}{\partial \psi}\right) = \frac{1}{t}\frac{d}{d\eta}(\sqrt{\eta}f').$$

Eq. (6.95) transforms to the ordinary differential equation

$$\frac{d}{d\eta}(\sqrt{\eta}f') = -\frac{2}{3}\eta f'. \tag{6.97}$$

Introducing $z = (\eta)^{1/2}$, we bring Eq. (6.97) to the form

$$\frac{d^2 f}{dz^2} + \frac{4}{3}z^2\frac{df}{dz} = 0.$$

Its solution is given by the expression

$$f(z) = B\int_0^z e^{-4z^3/9}\,dz + E, \tag{6.98}$$

where

$$z = \sqrt{\eta} = \left(\frac{\psi}{t^{2/3}}\right)^{1/2} = \frac{\sqrt{3U}}{2}\,\frac{y\sin\theta}{\left(DR^2\sqrt{\dfrac{3U}{4}}\left(\theta - \dfrac{\sin 2\theta}{2}\right) + A\right)^{1/3}}. \tag{6.99}$$

The constants A, B, E are obtained by using the boundary conditions (6.93). The first condition at $\psi = 0$ (or $z = 0$) results in $f(0) = E = 0$. The second condition reduces to $f(\infty) = C_0$, leading to

$$B = \left(\int_0^\infty e^{-4z^3/9}\,dz\right)^{-1} C_0 = C_0\left(\left(\frac{9}{4}\right)^{1/3}\Gamma\left(\frac{1}{3}\right)\right)^{-1} = \frac{C_0}{1.15}.$$

Consider now the last boundary condition. In the region close to the pole facing the flow, the angle θ is small. Therefore, expanding $\sin x$ in (6.99) as a power series in x, we obtain:

$$z \approx \sqrt{\frac{3U}{4}}y\theta\left(\frac{DR^2}{2}\frac{2\theta^3}{3}\sqrt{3U} + A\right)^{-1/3}. \tag{6.100}$$

For the condition $C = C_0$ to hold at $\theta = 0$, we need to have $A = 0$. Indeed, if $A \neq 0$, then one can see from (6.100) that $z \sim \theta$, and for $\theta \to 0$ we have $z \to 0$. Hence, $C = f(z) \to 0$ and the condition at the pole facing the flow is not obeyed. Thus, $A = 0$ and

$$z = \sqrt[3]{\frac{3U}{4DR^2}}\,\frac{y\sin\theta}{\left(\theta - \dfrac{\sin 2\theta}{2}\right)^{-1/3}},$$

$$C = \frac{C_0}{1.15}\int_0^z e^{-4z^3/9}\,dz. \tag{6.101}$$

Determine now the diffusion flux at the particle surface:

$$j_w = D\left(\frac{\partial C}{\partial y}\right)_{y=0} = \frac{DC_0}{1.15}\sqrt[3]{\frac{3U}{4DR^2}}\frac{y\sin\theta}{\left(\theta - \dfrac{\sin 2\theta}{2}\right)^{1/3}}. \tag{6.102}$$

The flux proves to be proportional to C_0, $U^{1/3}$, $D^{2/3}$, $R^{-2/3}$, and also depends on the angle θ. At $\theta = \pi/2$, this function of angle is equal to $(2/\pi)^{1/3}$; at $\theta = \pi$, it is equal to 0; and at the pole facing the flow, i.e. at $\theta = 0$, the function of angle is equal to 1. So, the flux decreases with the growth of θ, assuming the maximum value at the pole facing the flow, and the minimum value at the opposite pole. Diffusion boundary layer thickness is estimated by the formula

$$\left(\frac{\partial C}{\partial y}\right)_w \sim \frac{C_0}{\delta_D}:$$

$$\delta_D \sim \frac{1.15\left(\theta - \dfrac{\sin 2\theta}{2}\right)^{1/3}}{\sin\theta}\sqrt[3]{\frac{4DR^2}{3U}}. \tag{6.103}$$

The thickness of the boundary layer grows with θ and becomes infinite at $\theta = \pi$. Recall that the basic assumption was that $\delta_D \ll R$. However, starting from some angle θ, the value δ_D becomes comparable to the radius R. In this range of angles, the above-formulated theory is not applicable. However, since the flux quickly decreases with growth of θ, the range of angles corresponding to the rear part of the sphere does not make a noticeable contribution to the total flux toward the particle. Therefore, to determine the total flux I, one should integrate (6.102) over the entire sphere surface (i.e., over θ) from 0 to π. The result is:

$$I = \int j_w\, ds = 2\pi R^2 \int_0^{\pi} j_w \sin\theta\, d\theta$$

$$= 2\pi\frac{DC_0R^2}{1.15}\sqrt[3]{\frac{3U}{4DR^2}}\int_0^{\pi}\frac{\sin^2\theta\, d\theta}{\left(\theta - \dfrac{\sin^2\theta}{2}\right)^{1/3}} = 7.98C_0D^{2/3}U^{1/3}R^{4/3}. \tag{6.104}$$

Until now, we have been considering the case of $\mathrm{Pe}_D \gg 1$. If $\mathrm{Pe}_D \ll 1$, it follows from (5.114) that transport of substance to the sphere occurs mostly via diffusion, and the convective flux can be ignored. The diffusion equation then reduces to

$$\Delta C = 0. \tag{6.105}$$

Since this approach does not take into account the flow past the sphere, the distribution of concentration will depend only on r. Therefore Eq. (6.105) will become

$$\frac{1}{r^2}\frac{d}{dr}\left(r^2\frac{\partial C}{\partial r}\right) = 0. \tag{6.106}$$

The solution of this equation with the boundary conditions $C = 0$ at $r = R$, $C = C_0$ at $r \to \infty$ is

$$C = C_0\left(1 - \frac{R}{r}\right). \tag{6.107}$$

Accordingly, the flux onto the sphere's surface is

$$j_w = D\left(\frac{\partial C}{\partial y}\right)_{r=R} = \frac{DC_0}{R}. \tag{6.108}$$

Integration of j_w over the surface gives the total flux toward the particle for the case $\mathrm{Pe}_D \ll 1$:

$$I = 4\pi DRC_0. \tag{6.109}$$

Of special interest is the diffusion flux to the surface of a drop with internal viscosity μ'. This problem has been considered in [3]. The main difference from diffusion toward a rigid particle is in the necessity to take into account the possibility of liquid slippage over the drop surface due to the motion of liquid within the drop. For the case when the substance is adsorbed on the surface but does not penetrate it, the expression for the diffusion flux at the surface has the form:

$$I = 8\left(\frac{\pi}{3}\right)^{1/2}\left(\frac{D\mu}{2a(\mu+\mu')}\right)^{1/2} a^2 C_0 U^{1/2}. \tag{6.110}$$

Comparing the expressions for diffusion flux onto a solid (6.104) and a liquid (6.110) particle, one can conclude that, all other things equal, the flux toward a liquid particle exceeds the flux toward a solid one. This is a consequence of more favorable conditions for substance delivery to the surface of liquid, as mixing of solution takes place near the drop surface.

The solution of problems on convective diffusion toward particles at finite Re and Pe_D numbers, was well as for boundary conditions more general than (6.93), can be found in work [13].

6.6 Distribution of Matter Introduced Into a Fluid Flow

The measurement of the flow velocity by introduction of radioactive substance or electrolyte into a cross section, followed by measurement of the radioactivity or electric conductivity in cross sections further downstream, has gained wide acceptance in different branches of engineering and, in particular, in physiology (the measurement of flow velocity of blood in arteries). The main difficulty associated

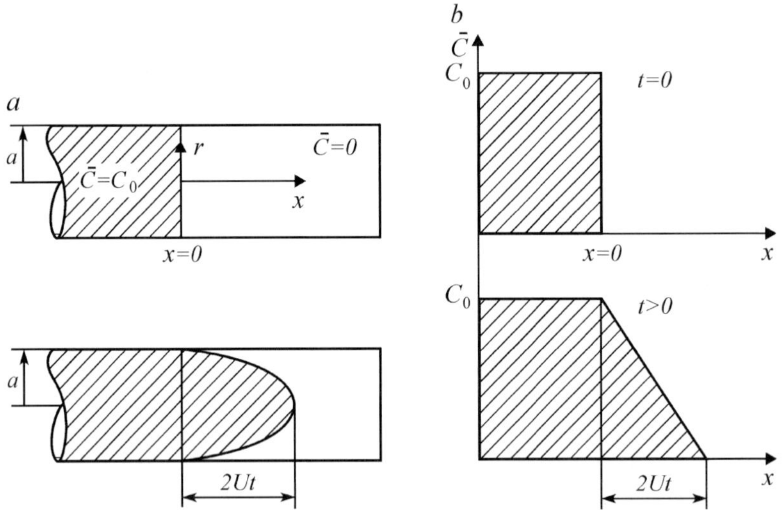

Fig. 6.7 Distribution of the average concentration during convective transfer of matter, which initially occupied a semi-infinite region.

with this method is that the introduced substance diffuses in all directions – upstream, downstream, and across the flow. If some portion of substance is introduced into the flow, then, due to convection and diffusion, the contours of the substance's volume deform and spread. As a result, in a radioactivity or electric conductivity measurement at some point further downstream, the measured characteristics will begin to increase at some moment of time, reach a maximum value, and then fall down to zero. The theory of this method for the case of fluid flow in a pipe has been developed by Taylor [14]. The qualitative picture of the average concentration distribution of a substance introduced into the flow in the absence of molecular diffusion is shown in Fig. 6.7a for the case when the substance initially occupies a semi-infinite region in the flow. Fig. 6.8a illustrates the case when the substance occupies a finite region.

Let us consider first the process of spreading of a semi-infinite layer due to convection only (see Fig. 6.7, b). We assume a developed flow, so the velocity has only the longitudinal component:

$$u(r) = 2U\left(1 - \frac{r^2}{R^2}\right), \tag{6.111}$$

where R is the tube radius.

The leading front of the layer changes according to the law

$$x = 2Ut\left(1 - \frac{r^2}{R^2}\right). \tag{6.112}$$

Now bring into consideration the average (over the cross section) concentration,

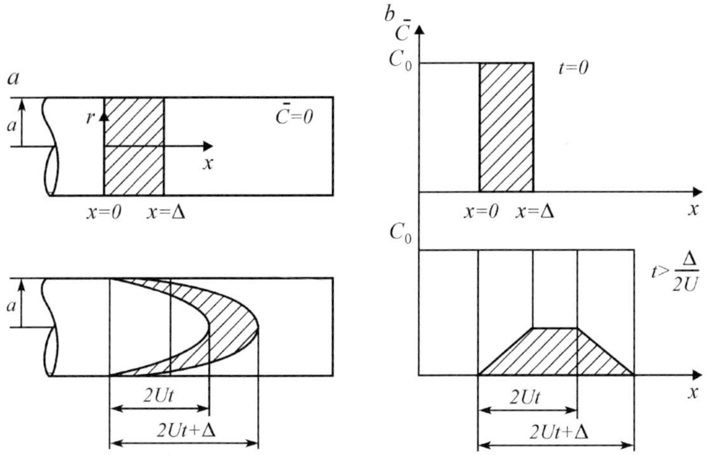

Fig. 6.8 Distribution of the average concentration during convective transfer of matter, which initially occupied a finite region.

$$\bar{C} = \frac{1}{\pi R^2} \int_0^R C(r) 2\pi r \, dr. \tag{6.113}$$

Using (6.112), one gets

$$\bar{C} = \begin{cases} C_0 & \text{at } x < 0, \\ C_0 \left(1 - \dfrac{x}{2Ut}\right) & \text{at } 0 < x < 2Ut. \end{cases} \tag{6.114}$$

In Fig. 6.7b, the distribution of $C(x)$ is shown.

In the case when a substance is introduced at the initial moment in the form of a layer of finite thickness Δ (see Fig. 6.8a), it is possible to obtain the average concentration distribution (Fig. 6.8b) as superposition of two methods of input:

$$\bar{C} = C_0 \quad \text{at } x < \Delta \quad \text{and} \quad \bar{C} = 0 \quad \text{at } x > \Delta, \tag{6.115}$$

$$\bar{C} = -C_0 \quad \text{at } x < 0 \quad \text{and} \quad \bar{C} = 0 \quad \text{at } x > 0. \tag{6.116}$$

Using (6.114) as applied to (6.115) and (6.116), and summing the obtained values, we obtain for the time $t > \Delta/2U$:

$$\bar{C} = \begin{cases} 0 & \text{at } x < 0, \\ C_0 \dfrac{x}{2Ut} & \text{at } 0 < x < \Delta, \\ C_0 \dfrac{\Delta}{2Ut} & \text{at } \Delta < x < 2Ut + \Delta, \\ C_0 \dfrac{(\Delta + 2Ut - x)}{2Ut} & \text{at } 2Ut < x < 2Ut + \Delta, \\ 0 & \text{at } x > \Delta + 2Ut. \end{cases} \tag{6.117}$$

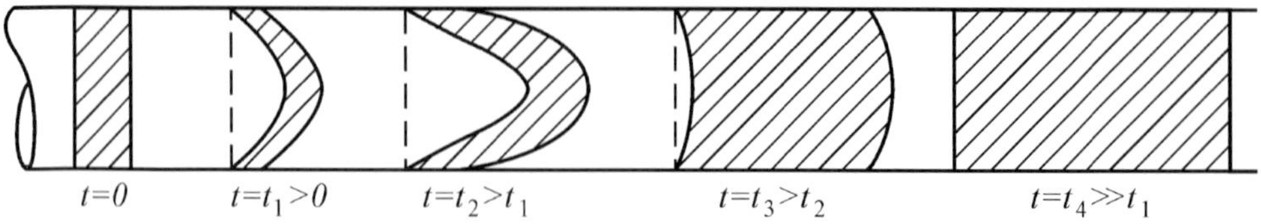

Fig. 6.9 The picture of time evolution of matter introduced in a flow.

The actual picture of the substance spread (Fig. 6.9) differs from that presented in Fig. 6.8.

The comparison of layer spread velocities, especially at later times, shows that it is impossible to explain the picture that is actually observed by convection mechanism only. Obviously, molecular diffusion needs to be taken into account as well.

So consider the equation of convective diffusion in a developed flow,

$$\frac{\partial C}{\partial t} + u(r)\frac{\partial C}{\partial x} = D\left(\frac{\partial^2 C}{\partial x^2} + \frac{1}{r}\frac{\partial}{\partial r}\left(\frac{\partial C}{\partial r}\right)\right). \tag{6.118}$$

The criterion of convective character of substance transfer is the smallness of diffusion terms in the right-hand side of (6.118) in comparison with the convective term on the left. The right-hand side contains two terms. The first term determines longitudinal diffusion with the characteristic time $t_{LD} \sim L^2/D$, where L is the characteristic length, while the second term determines transversal diffusion with the characteristic time $t_{RD} \sim R^2/D$. The characteristic time of convective transfer is equal to $t_C \sim L/U$. For the ratio of characteristic times of diffusion and convection, one gets:

$$\frac{t_{LD}}{t_C} \sim \frac{L^2 U}{DL} = \frac{RU}{D}\frac{L}{R} = \mathrm{Pe}_D\frac{L}{R}, \quad \frac{t_{RD}}{t_C} \sim \frac{R^2 U}{DL} = \mathrm{Pe}_D\frac{R}{L}.$$

If $t_{LD} \gg t_C$ or $\mathrm{Pe}_D \gg R/L$, longitudinal convection dominates over longitudinal diffusion, and at $t_{RD} \gg t_C$ or $\mathrm{Pe}_D \gg L/R$, longitudinal convection dominates over radial diffusion. In the case $L \gg R$, both conditions are true, and the substance transfer is purely convective. This is the case presented in Fig. 6.8. Thus, the profile of average concentration (6.117) can be considered as the averaged solution of the equation

$$\frac{\partial C}{\partial t} + u(r)\frac{\partial C}{\partial x} = 0. \tag{6.119}$$

In the other limiting case, $t_{RD} \ll t_C$ or $\mathrm{Pe}_D \ll L/R$, radial diffusion occurs very quickly and it quickly washes out the heterogeneity of concentration in the radial direction. If $t_{LD} \ll t_C$ or $\mathrm{Pe}_D \ll R/L$, then axial diffusion occurs much faster than convective transfer, and the axial heterogeneity of concentration spreads out very quickly. In the case of $L > R$, and under the condition $\mathrm{Pe}_D \ll R/L$, the condition $\mathrm{Pe}_D \ll L/R$ is automatically satisfied.

Compare the rates of radial and longitudinal diffusion. We have: $t_{LD}/t_{RD} \sim (L/R)^2$. For diffusion processes occurring in thin tubes, for example in capillaries, the condition $L \gg R$ is usually satisfied, therefore radial diffusion dominates over the axial one.

Thus, if the conditions $L/R \gg 1$ and $\mathrm{Pe}_D \gg 1$ are satisfied, substance transport in the axial direction happens as a result of convection, while transport in the radial directions occurs via molecular diffusion.

It is convenient to look for the solution in a coordinate system moving with velocity U relative to the pipe wall. In this system of coordinates, the equation of convective diffusion (6.118) subject to the inequalities $L/R \gg 1$ and $\mathrm{Pe}_D \gg 1$ takes the form

$$\frac{\partial C}{\partial t} + u(r)\frac{\partial C}{\partial x'} = D\frac{1}{r}\frac{\partial}{\partial r}\left(\frac{\partial C}{\partial r}\right); \quad x' = x - Ut. \tag{6.120}$$

The velocity figuring in (6.120) is determined by

$$u(r) = 2U\left(1 - \frac{r^2}{R^2}\right) - U = U\left(1 - \frac{2r^2}{R^2}\right). \tag{6.121}$$

The boundary conditions here are the requirements of absence of radial flux at the wall and at the pipe axis:

$$\frac{\partial C}{\partial t} = 0 \quad \text{at } r = R \text{ and at } r = 0. \tag{6.122}$$

Assume that in the first approximation, the flow is quasi-stationary with respect to the mobile system of coordinates. Actually, it corresponds to the asymptotic solution at $t \gg R^2/D$. Another assumption is that the longitudinal component of concentration gradient is constant, that is, $\partial C/\partial x' = \text{const}$. Then, in this approximation, concentration depends only on r and is described by the equation

$$\frac{\partial^2 C}{\partial r^2} + \frac{1}{r}\frac{\partial C}{\partial r} = \frac{U}{D}\left(1 - \frac{2r^2}{R^2}\right)\frac{\partial C}{\partial x'}. \tag{6.123}$$

The solution of this equation with conditions (6.122) has the form

$$C = C_0 + \frac{UR^2}{4D}\left(\frac{r^2}{R^2} - \frac{r^4}{2R^4}\right)\frac{\partial C}{\partial x'}. \tag{6.124}$$

Find now the concentration averaged over the flow's cross section:

$$\bar{C} = C_0 + \frac{UR^2}{12D}\frac{\partial C}{\partial x'}. \tag{6.125}$$

Comparing (6.124) with (6.125), one obtains

$$C = \bar{C} + \frac{UR^2}{4D}\left(-\frac{1}{3} + \frac{r^2}{R^2} - \frac{r^4}{2R^4}\right)\frac{\partial C}{\partial x'}. \tag{6.126}$$

Thus, concentration C is the sum of the average concentration and the concentration disturbance that depends on r. The condition $\partial C/\partial x' = \text{const}$ may be written as $\partial C/\partial x' = \partial \bar{C}/\partial x'$, from which it follows, in view of (6.126), that it is satisfied at the performance of condition $R^2 U/4DL \ll 1$ or

$$\mathrm{Pe}_D \ll 4L/R. \tag{6.127}$$

Hence, the Peclet's number in considered problem must satisfy the inequalities

$$1 \ll \mathrm{Pe}_D \ll 4L/R. \tag{6.128}$$

Let us now find the flux of substance through the pipe's cross section:

$$Q = 2\pi \int_0^R Cu(r)r\,dr = -\pi R^2 \frac{U^2 R^2}{48D}\frac{\partial \bar{C}}{\partial x'} \tag{6.129}$$

The corresponding flux density (i.e., flux through a unit area) is

$$J = \frac{Q}{\pi R^2} = -\left(\frac{U^2 R^2}{48D}\right)\frac{\partial \bar{C}}{\partial x'}. \tag{6.130}$$

In the moving system of coordinates, the flux J is similar to the relative flux introduced in Section II. Therefore the expression (6.130) can be considered as Fick's law with the effective coefficient of diffusion

$$D_{\mathit{eff}} = U^2 R^2/48D. \tag{6.131}$$

It is called Taylor's (dispersion) coefficient.

Introduction of the quantity D_{eff} allows us to write the diffusion equation for the average concentration in the stationary system of coordinates:

$$\frac{\partial \bar{C}}{\partial t} + U\frac{\partial \bar{C}}{\partial x} = D_{\mathit{eff}}\frac{\partial^2 \bar{C}}{\partial x^2}. \tag{6.132}$$

The condition for applicability of this equation follows from the condition that Taylor's longitudinal diffusion by far exceeds the longitudinal molecular diffusion, that is, $D_{\mathit{eff}} \gg D$. This condition, with account taken of (6.131) and (6.127), gives the final condition for applicability of the considered solution:

$$7 \ll \mathrm{Pe}_D \ll 4\frac{L}{R}. \tag{6.133}$$

The following problem illustrates the usage of Eq. (6.132). Some amount of a substance N_0 is introduced into the flow at the origin $x = 0$ at the moment $t = 0$ within a region that is small in comparison with the pipe's radius. The initial concentration is

$$\bar{C}_0 = \frac{N_0}{\pi R^2}\delta(x), \tag{6.134}$$

where $\delta(x)$ is the delta function.

The solution of Eq. (6.132) with the condition (6.134) is

$$\bar{C}_0 = \frac{N_0}{\pi R^2 (\pi D_{eff} t)^{1/2}} e^{-(x-Ut)^2/4D_{eff}t}. \tag{6.135}$$

The sole parameter D_{eff} entering (6.135) can be determined experimentally, by measuring the change of average concentration with time. Knowing D_{eff} it is possible to determine with the help of (6.135) the flow velocity U of dissolved substance, that is, to solve the problem formulated in the beginning of this section.

Another approximate method of solution of Eq. (6.118) that covers both longitudinal and radial diffusion of substance was proposed in [15]. The essence of the method is to reduce the diffusion equation to a system of equations for the moments, which have a simple physical meaning. Introduce the following dimensionless variables:

$$\xi = (x - Ut)/R,\ \eta = r/R,\ \tau = Dt/R^2,\ \mathrm{Pe}_D = UR/D,\ c = C/C_0 \tag{6.136}$$

and dimensionless moments of concentration distribution (which is also dimensionless):

$$c(\eta, \tau) = \int_{-\infty}^{+\infty} \xi c(\xi, \eta, \tau)\, d\xi, \quad m(\tau) = \bar{c} = 2\int_{-\infty}^{+\infty} c(\eta, \tau)\eta\, d\eta. \tag{6.137}$$

Since all values of dimensionless concentration lie in the interval [0, 1], c has the meaning of probability density distribution of molecules of dissolved substance. The first three moments of this distribution are: $m_0(\tau)$ – the dimensionless mass of dissolved substance in a unit volume, $m_1(\tau)$ – the dimensionless mean position of the center of mass of the dissolved substance, $D = m_2 - m_1^2$ – the dimensionless distribution variance, characterizing void-mean-square deviation of dissolved substance distribution relative moving coordinate system, and associated with the above-introduced effective diffusion coefficient D_{eff} through the relation

$$D_{eff} = \frac{1}{2}\frac{dD}{d\tau}D. \tag{6.138}$$

To derive the equations for the moments, we multiply both sides of (6.118) by x^P, then integrate over ξ from $-\infty$ to $+\infty$ and average over the pipe's cross section, taking into account the initial condition $c(\xi, \eta, 0) = 1$ and the corresponding boundary conditions following from (6.122):

$$\frac{\partial D}{\partial \tau} = \frac{1}{\eta}\frac{\partial}{\partial \eta}\left(\eta \frac{\partial c_p}{\partial \eta}\right) + (p-1)c_{p-2} + c_D p(1-2p^2)c_{p-1}.$$

$$\frac{dm_p}{d\tau} = p(p-1)m_{p-2} + 2pc_D \int_0^1 p(1-2p^2)c_{p-1}\, dp. \tag{6.139}$$

$$c_p(\eta, 0) = c_{p0}(\eta), \quad \frac{\partial c_p}{\partial \tau} = 0, \quad \eta = 0, \quad m_p(0) = m_{p0}.$$

At $p = 0$, we obtain $dm_0/dt = 0$, which leads to the obvious condition of total mass conservation for a dissolved substance. From the equation for the first moment one obtains $dm_1/d\tau \to 0$ at $\tau \to \infty$. It means that at $\tau \to \infty$, the center of mass of dissolved substance moves with the average flow velocity U. The second moment at $\tau \to \infty$ tends to

$$m_2 = 2(1 + c_D^2/48) + \text{const.} \tag{6.140}$$

Going back to (6.138), we find that at $\tau \to \infty$ or at $t \gg R^2/D$, the coefficient of effective diffusion tends to

$$D_{\mathit{eff}} = D + \frac{R^2U^2}{48D} = D\left(1 + \frac{c_D^2}{48}\right). \tag{6.141}$$

This coefficient is called the Taylor-Aris's dispersion factor. It follows from (6.141) that at $t \gg R^2/D$ the coefficient of effective diffusion is equal to the sum of the molecular diffusion coefficient and the effective diffusion coefficient given by (6.131).

At $\mathrm{Pe}_D \gg 1$, we have $D_{\mathit{eff}}/D \sim \mathrm{Pe}_D^2$, whereas at $\mathrm{Pe}_D \ll 1$, we have $D_{\mathit{eff}}/D \sim 1$.

6.7 Diffusion Flux in a Natural Convection

One of the frequently used methods of solution mixing is natural convection. This term implies that the motion of the solution is influenced by forces arising in the course of heterogeneous reactions, which results in a change of solution density. Natural convection arises only when the change of density occurs in a gravitational field. It is also required that density increases with height (i.e. the angle between gravity and the density gradient can vary from 90° to 180°).

Consider a vertical plate in a gravitational field [3]. Assume that a fast chemical reaction takes place at the surface of this plate, so that the concentration of reacting substance at the plate is equal to zero. Far from the plate, this concentration is equal to C_0. Suppose that the density of the solution does not strongly depend on the concentration C_0, so we can write

$$\rho(C_0) \approx \rho(0) + \left(\frac{\partial \rho}{\partial C}\right)_{C=0} C_0. \tag{6.142}$$

Similarly, at any arbitrary point in the solution, we have

$$\rho(C) \approx \rho(C_0) + \left(\frac{\partial \rho}{\partial C}\right)_{C=C_0} (C - C_0). \tag{6.143}$$

Choose the system of coordinates such that the x-axis is vertical and the y-axis is perpendicular to the plate (and directed toward the solution). The bottom edge of the plate corresponds to $x = 0$. Assume that the change of concentration mostly occurs in the diffusion boundary layer. Since the flow of liquid is driven by the gradient of concentration, the flow occurs in this layer, too; that is, the viscous boundary layer coincides with the diffusion layer. The gravity force ρg acts on a unit solution volume. Since ρ changes with height, so does the force. At $\rho = \rho(C_0) = \text{const}$, this force does not cause the flow since the force is counterbalanced by the pressure gradient. The flow may be caused by density variations from $\rho(C_0)$. Since the difference $\Delta\rho = |\rho(C_0) - \rho(C)|$ is small, the equations of motion in the first approximation can be written in the form

$$u_x \frac{\partial u_x}{\partial x} + u_y \frac{\partial u_x}{\partial y} = \nu \frac{\partial^2 u_x}{\partial y^2} + g \frac{\rho(C_0) - \rho(C)}{\rho(C_0)}, \tag{6.144}$$

$$\frac{\partial u_x}{\partial x} + \frac{\partial u_x}{\partial y} = 0. \tag{6.145}$$

The concentration distribution is described by the equation of convective diffusion:

$$u_x \frac{\partial C}{\partial x} + u_y \frac{\partial C}{\partial y} = D \frac{\partial^2 C}{\partial y^2}. \tag{6.146}$$

Introduce the dimensionless concentration,

$$\varphi = (C_0 - C)/C_0.$$

Then Eqs. (6.144) and (6.146) (subject to the condition (6.143)) are transformed to:

$$u_x \frac{\partial u_x}{\partial x} + u_y \frac{\partial u_x}{\partial y} = \nu \frac{\partial^2 u_x}{\partial y^2} + g\alpha\varphi, \tag{6.147}$$

$$u_x \frac{\partial \varphi}{\partial x} + u_y \frac{\partial \varphi}{\partial y} = D \frac{\partial^2 \varphi}{\partial y^2}, \tag{6.148}$$

where $\alpha = \frac{C_0}{\rho(C_0)} \left(\frac{\partial \rho}{\partial C}\right)_{C=C_0}$.

The boundary conditions are:

$$\begin{aligned} &u_x = u_y = 0,\ \varphi = 1 \quad \text{at } y = 0, \\ &u_x = u_y = 0,\ \varphi = 0 \quad \text{at } y \to \infty. \end{aligned} \tag{6.149}$$

The system of equations (6.145), (6.147), and (6.148) with the boundary conditions (6.149) has a self-similar solution. Indeed, if we introduce the dimensionless variable

$$\eta = \left(\frac{g\alpha}{4\nu^2}\right)^{1/4} \frac{y}{x^{1/4}} \tag{6.150}$$

and seek for the stream function in the form

$$\psi = 4\nu \left(\frac{g\alpha}{4\nu^2}\right)^{1/4} x^{3/4} f(\eta), \tag{6.151}$$

we will obtain

$$\begin{aligned} u_x &= \frac{\partial \psi}{\partial y} = 4\nu \left(\frac{g\alpha x}{4\nu^2}\right)^{1/2} f'(\eta), \\ u_y &= -\frac{\partial \psi}{\partial x} = \nu \left(\frac{g\alpha}{4\nu^2}\right)^{1/4} \frac{\eta(f' - 3f)}{x^{1/4}}. \end{aligned} \tag{6.152}$$

Substitution of u_x and u_y into (6.147) leads to the equation

$$f''' + 3ff'' - 2(f')^2 + \varphi = 0. \tag{6.153}$$

Similarly, we can derive an equation for $\varphi = \varphi(\eta)$

$$\varphi'' + 3\text{Sc} f \varphi' = 0, \tag{6.154}$$

where $\text{Sc} = \nu/D$ is Schmidt's number.

Accordingly, the boundary conditions are transformed to

$$f = f' = 0,\ \varphi = 1 \quad \text{at } \eta = 0,$$
$$f' = 0,\ \varphi = 0 \quad \text{at } \eta \to \infty. \tag{6.155}$$

The solution of Eqs. (6.153) and (6.154) can be sought for in the form $\varphi = \varphi(f)$. After simple transformations we obtain:

$$\varphi = 1 - \left(\int_0^{\eta} e^{-3\mathrm{Sc}\int_0^{\eta} f\,d\eta}\,d\eta\right)\left(\int_0^{\infty} e^{-3\mathrm{Sc}\int_0^{\eta} f\,d\eta}\,d\eta\right)^{-1}. \tag{6.156}$$

As it has been indicated already, for extremely diluted solutions, $\mathrm{Sc} \gg 1$. Due to this inequality, the integrals converge rapidly. Therefore at small values of η, the integrals are mostly determined by the value of f, while at large η, the behavior of f does not noticeably influence the distribution of φ. Because of this, without any detriment to the accuracy, we can replace the boundary condition at the infinity by the condition at a finite distance η_0 from the wall, taking it to be equal to the thickness of the boundary layer δ_D, that is,

$$f'' = 0,\ j = 0 \quad \text{at } \eta = \eta_0 = \left(\frac{g\alpha}{4\nu^2}\right)^{1/4} \frac{\delta_D}{x^{1/4}}. \tag{6.157}$$

Outside the boundary layer, that is, at $\eta > \eta_0$, the velocity is taken to be zero. Inside the layer ($0 < \eta < \eta_0$), the function $f(\eta)$ is sought for as a power series in η. Taking into account the condition (6.155), we can write:

$$f = \frac{\beta}{2}\eta^2 + \frac{\gamma}{32}\eta^3 + \cdots \tag{6.158}$$

Retaining only the first term in the series (6.158) and putting this term into (6.156), we obtain

$$\varphi = 1 - \left(\int_0^{\eta} e^{-\beta \mathrm{Sc}\eta^{3/2}}\,d\eta\right)\left(\int_0^{\infty} e^{-\beta \mathrm{Sc}\eta^{3/2}}\,d\eta\right)^{-1}. \tag{6.159}$$

To calculate the integrals, one can make a change of variable: $t = (\beta \mathrm{Sc}/2)^{1/3}\eta$. Then (6.159) transforms to

$$\varphi = 1 - \int_0^{(\beta \mathrm{Sc}/2)^{1/3}\eta} e^{-t^3}\,dt \Big/ \left(\frac{1}{3}\Gamma\left(\frac{1}{3}\right)\right). \tag{6.160}$$

If $(\beta \mathrm{Sc}/2)^{1/3}\eta \gg 1$, the upper limit can be replaced by ∞. Then $\varphi \approx 0$. In the case when $(\beta \mathrm{Sc}/2)^{1/3}\eta \ll 1$, the integrand can be expanded as a power series in the vicinity of the point $t = 0$. After some simple algebra, we obtain

$$\varphi = 1 - \int_0^{(\beta\text{Sc}/2)^{1/3}\eta} (1 - t^3 + \cdots)\, dt \Big/ \left(\frac{1}{3}\Gamma\left(\frac{1}{3}\right)\right)$$
$$\approx 1 - \frac{(\beta\text{Sc}/2)^{1/3}\eta}{\frac{1}{3}\Gamma\left(\frac{1}{3}\right)} = 1 - \frac{(\beta\text{Sc}/2)^{1/3}\eta}{0.89}. \tag{6.161}$$

Now, use the second condition (6.154) to determine the thickness of the boundary layer:

$$\eta_0 = \frac{0.89}{(\beta\text{Sc}/2)^{1/3}\eta}. \tag{6.162}$$

Go back to Eq. (6.153) and plug the expression (6.161) for φ into this equation:

$$f''' + 3ff'' - 2(f')^2 + 1 - \frac{(\beta\text{Sc}/2)^{1/3}\eta}{0.89} = 0. \tag{6.163}$$

Taking the expression (6.158) for f and putting it into the obtained equation, we get the following result:

$$f = \frac{\beta}{2}\eta^2 - \frac{1}{6}\eta^3 + \frac{(\beta\text{Sc}/2)^{1/3}}{24 \cdot 0.89}\eta^4 + \cdots \tag{6.164}$$

The first boundary condition (6.157) enables us to find β:

$$\beta = 0.48/\text{Sc}^{1/4}. \tag{6.165}$$

It is now possible to find all parameters of the problem. The concentration distribution is

$$C = 0.7\text{Sc}^{1/4}\left(\frac{g\alpha}{4\nu^2}\right)^{1/4}\frac{yC_0}{x^{1/4}}. \tag{6.166}$$

The diffusion flux to the plate is

$$J = D\left(\frac{\partial C}{\partial y}\right)_{y=0} = 0.7D\text{Sc}^{1/4}\left(\frac{g\alpha}{4\nu^2}\right)^{1/4}\frac{yC_0}{x^{1/4}}$$
$$= 0.7D\text{Sc}^{1/4}\left(\frac{gC_0}{4\nu^2\rho(C_0)}\left(\frac{\partial\rho}{\partial C}\right)_{C=C_0}\right)^{1/4}\frac{C_0}{x^{1/4}}. \tag{6.167}$$

If we integrate (6.167) over the entire surface of a plate of length h and width b, we will obtain the total diffusion flux to the plate:

$$I = 0.9\mathrm{Sc}^{1/4}\left(\frac{g\alpha}{4\nu^2}\right)^{1/4} bh^{3/4}DC_0. \qquad (6.168)$$

The last expression can be presented in dimensionless form. Introduce a dimensionless parameter, known in problems on natural convection as Grashof's number,

$$\mathrm{Gr} = g\alpha h^3/4\nu^2 \qquad (6.169)$$

and Nusselt's number,

$$\mathrm{Nu} = hI/DSC_0, \quad S = hb. \qquad (6.170)$$

Then (6.165) can be rewritten as

$$\mathrm{Nu} = 0.9\mathrm{Sc}^{1/4}\mathrm{Gr}^{1/4}. \qquad (6.171)$$

6.8
Dynamics of the Bubble in a Solution

Dynamics of gas bubbles in liquids presents significant interest for many reasons. First, bubble motion research provides information about properties of the elementary boundary between liquid and gaseous phases, about the laws governing phase transitions (evaporation, condensation), and about chemical reactions at the surface. Second, this process is also of interest from a purely technical viewpoint. Such branches of industry as gas, petroleum, and chemical engineering commonly utilize processes and devices whose operation is directly interrelated with the laws of bubbles motion. Applications include separation of gas from liquid; barbotage of bubbles through a layer of mixture, which is thus enriched by various reagents contained in the bubbles; flotation, which is employed in treatment of polluted liquids, etc.

In the present section, the main attention will be given to the motion of the bubble's surface caused by the change of its volume, rather than the motion of bubbles relative to the liquid. This volume can change as a result of evaporation of the liquid phase or condensation of the gaseous phase. If both the liquid and the bubble are multi-component mixtures, then chemical reactions are possible at the bubble surface, which too can result in a change of the bubble's size.

Consider a spherical gas bubble of initial radius R, placed in a quiescent liquid [16]. Assume that the bubble center does not move relative to the liquid, but the bubble volume changes with time due to the difference of pressures inside the bubble and in the ambient liquid, and also as a result of dynamic and heat-mass-exchange interactions between the liquid and gaseous phases. In the majority of works devoted to the research of bubble dynamics, it is assumed that the parameters of gas inside the bubble, such as pressure, temperature, concentration, and others, are homogeneous over the bubble volume and change only in time. Suppose that the bubble consists of inert gas (i) and vapor of liquid (v). The liquid is a binary solution able to evaporate from the bubble surface.

Write down the basic equations describing the behavior of the liquid:

Continuity equation

$$\frac{D\rho_L}{Dt} + \rho_L \nabla \cdot \boldsymbol{u} = 0. \tag{6.172}$$

Since the liquid is incompressible and the problem is spherically symmetric, (6.172) reduces to

$$ur^2 = u_R R^2 = f(t). \tag{6.173}$$

where u, u_R are the radial components of velocity in the liquid and at the bubble surface; R – the radius of the bubble; r – the radial coordinate (where $r = 0$ corresponds to the bubble center).

The liquid flow velocity u and the velocity of motion of the bubble's boundary are not equal if phase transitions happen at the interface, since they cause the appearance of mass flux:

$$\dot{m} = 4\pi R^2 \rho_L (\dot{R} - u_R), \tag{6.174}$$

where ρ_L is the density of a fluid; $\dot{R} = dR/dt$.

The expression (6.174) allows us to write the balance-of-mass equation for gas in the bubble:

$$\frac{d}{dt}\left(\frac{4}{3}\pi R^3 \rho_G\right) = 4\pi R^2 \rho_L (\dot{R} - u_R). \tag{6.175}$$

In the course of bubble growth, its volume drastically changes, while the density of gas ρ_G doesn't change much. Therefore, taking in (6.175) $\rho_G = \text{const}$, we obtain:

$$u_R = \dot{R}\frac{(\rho_L - \rho_G)}{\rho_L} = \varepsilon \dot{R}. \tag{6.176}$$

Note that as a rule, we have $\rho_L \gg \rho_G$ and $\varepsilon = 1$. It means that $u_R = \dot{R}$. Substitution of (6.176) into (6.173) yields

$$ur^2 = \varepsilon \dot{R} R^2. \tag{6.177}$$

Motion equation

$$\rho_L \frac{D\boldsymbol{u}}{Dt} = -\nabla \cdot \boldsymbol{T} + \rho_L \boldsymbol{f}. \tag{6.178}$$

Consider the case of no volume forces ($\boldsymbol{f} = 0$). Then (6.178) reduces to the equation

$$\frac{\partial u}{\partial t} + u\frac{\partial u}{\partial r} = \frac{1}{\rho_L}\frac{\partial \tau_{rr}}{\partial r} - 2\nu\frac{\partial^2 u}{\partial r^2}. \tag{6.179}$$

Substituting into (6.179) the expression (6.177) for liquid velocity and integrating the resultant equation over r from R to ∞, we obtain:

$$\frac{\tau_{rr}(\infty) - \tau_{rr}(R)}{\varepsilon\rho_L} = R\ddot{R} + \frac{3}{2}\dot{R}^2 + 4\nu\frac{\dot{R}}{R}. \tag{6.180}$$

The radial component of stress tensor τ_{rr} that enters (6.180) is equal to

$$\tau_{rr}(\infty) = -p_\infty$$

$$\tau_{rr}(R) = -p_\mathrm{v} - p_\mathrm{i} + \frac{2\Sigma}{R}. \tag{6.181}$$

Here p is the pressure in the liquid far from the bubble, p_v and p_i are the partial pressures of liquid vapor and inert gas, Σ is the coefficient of surface tension.

The second condition (6.181) represents the balance of forces (from the gas and from the liquid) acting on the bubble's boundary.

Taking into account (6.181), it is possible to present (6.180) in the form

$$\frac{p_\mathrm{v} + p_\mathrm{i} - p_\infty - 2\Sigma/R}{\varepsilon\rho_L} = R\ddot{R} + \frac{3}{2}\dot{R}^2 + 4\nu\frac{\dot{R}}{R}. \tag{6.182}$$

Eq. (6.182) is known as Rayleigh's equation. This equation describes the time rate of change of the bubble's radius due to the change of pressure difference inside and outside the bubble.

To determine the partial pressure of vapor p_v, one has to use thermodynamic relations. Let us assume (this assumption is common in such problems) that there exists at the bubble surface a local thermodynamic balance between the liquid and gaseous phases. Then it follows from thermodynamic relations that

$$p_\mathrm{v} = \gamma_1 x_1 p_1 + \gamma_2(1 - x_1)p_2, \tag{6.183}$$

where γ is the activity coefficient; x is the molar fraction of a solution component; p is the pressure of saturated vapor; and indices 1 and 2 designate the solute and the solvent, respectively.

Since partial pressures depend on the temperature and on component concentrations, Eqs. (6.182) and (6.183) must be considered together with the equations of energy and mass transfer.

Energy equation

Neglect viscous dissipation and assume that there is no external influx of heat into the system, and the transfer of energy is caused only by heat conductivity. Then the energy equation reduces to

$$\frac{\partial T}{\partial t} + u\frac{\partial T}{\partial r} = \alpha\left(\frac{\partial^2 T}{\partial r^2} + \frac{2}{r}\frac{\partial T}{\partial r}\right) \tag{6.184}$$

or, taking into account (6.177),

$$\frac{\partial T}{\partial t} = \alpha\left(\frac{\partial^2 T}{\partial r^2} + \frac{2}{r}\frac{\partial T}{\partial r}\right) - \frac{\varepsilon R^2 \dot{R}}{r^2}\frac{\partial T}{\partial r}. \tag{6.185}$$

Mass transport equation

For a binary mixture, in the absence of chemical reactions in the volume, mass transport equation reduces to the equation of convective diffusion. Let C denote the concentration of dissolved substance. Then the mass transport equation is

$$\frac{\partial C}{\partial t} + u\frac{\partial C}{\partial r} = D\left(\frac{\partial^2 C}{\partial r^2} + \frac{2}{r}\frac{\partial C}{\partial r}\right). \tag{6.186}$$

Put the expression for the velocity u into (6.186):

$$\frac{\partial C}{\partial t} = D\left(\frac{\partial^2 C}{\partial r^2} + \frac{2}{r}\frac{\partial C}{\partial r}\right) - \frac{\varepsilon R^2 \dot{R}}{r^2}\frac{\partial C}{\partial r}. \tag{6.187}$$

Now, formulate the initial and boundary conditions. The initial conditions are as follows:

$$R(0) = R_0 = \frac{2\Sigma}{p_{i0} + p_{v0} - p_\infty}, \quad \dot{R}(0) = 0,\ T(r,0) = T_0,\ C(r,0) = C_0. \tag{6.188}$$

As the first initial condition, we have taken the condition saying that the initial radius R is equal to the equilibrium radius corresponding to the balance of initial pressure and surface tension (even though we could, admittedly, have chosen any value for the initial radius).

The boundary conditions far from the bubble are taken in the form

$$T(\infty,t) = T_0, \quad C(\infty,t) = C_0. \tag{6.189}$$

The conditions at the bubble surface are more intricate. Among them is the mass balance condition for component 1:

$$m\rho_G\dot{R} = C(R,t)(\dot{R} - u_R) + D\left(\frac{\partial C}{\partial r}\right)_{r=R}, \tag{6.190}$$

where m is the mass quota of component 1.

The left-hand side of (6.190) gives the rate of change of mass of component 1 in the bubble. The right-hand side is the sum of convective and diffusion fluxes in the liquid. Using the expression (6.176) for u_R, we get:

$$m\rho_G\dot{R} = C(R,t)(1-\varepsilon)\dot{R} + D\left(\frac{\partial C}{\partial r}\right)_{r=R}, \tag{6.191}$$

This is our first boundary condition. The second condition is that of energy balance at the bubble surface. It is obvious that the change of kinetic energy is small. Therefore the energy equation is:

$$\begin{aligned}
&\frac{d}{dt}\left(\frac{4}{3}\pi R^3\rho_G(me_1+(1-m)e_2)+4\pi R^2\Sigma\right)+4\pi R^2\dot{R}p_\infty \\
&\quad = 4\pi R^2\left(h_1\left((C(R,t))(1-\varepsilon)\dot{R}+D\left(\frac{\partial C}{\partial r}\right)_{r=R}\right)\right. \\
&\qquad \left. + h_2\left((\rho_L - C(R,t))(1-\varepsilon)\dot{R} - D\left(\frac{\partial C}{\partial r}\right)_{r=R}\right) + \alpha\left(\frac{\partial T}{\partial r}\right)_{r=R}\right),
\end{aligned} \tag{6.192}$$

where e is the specific internal energy of vapor, h is the specific enthalpy of a liquid component, indices 1 and 2 designate their respective components.

The condition (6.192) means that the rate of change of the total energy of gas in the bubble (the left-hand side) is the sum of the internal energy of vapor and the energy of surface tension, plus the work performed by the bubble against the pressure forces. This rate is equal to the rate of change of energy coming into the bubble from the outside. It is composed of the energy of evaporation of components and the heat due to heat conductivity. In its general form, the condition (6.192) is bulky and inconvenient to use. However, the terms in Eq. (6.192) may well be of different orders. It is quite natural to assume that during the growth of the bubble, the change in vapor density and internal energy is insignificant. Also, the surface energy term and the work against the pressure forces are insignificant as well. If we introduce the average internal energy of vapor, $e = me_1 + (1-m)e_2$ and in addition assume the enthalpies of liquid components to be equal $(h_1 = h_2)$, the condition (6.192) will simplify and take the form

$$\rho_G\dot{R}\bar{e} = \rho_L h(1-\varepsilon)\dot{R} + \alpha\left(\frac{\partial T}{\partial r}\right)_{r=R}. \tag{6.193}$$

The average internal energy of vapor is equal to the sum of the heat of vaporization l and the specific enthalpy of vapor:

$$\bar{e} = l + c_G(T(R,t) - T_0). \tag{6.194}$$

The specific enthalpy of liquid, in its turn, is equal to

$$h = c_L(T(R,t) - T_0). \tag{6.195}$$

Here c_L and c_G are specific heat capacities of the liquid and the gas.

Even with the series of assumptions made thus far, the system of equations (6.182), (6.183), and (6.187) with the initial condition (6.188) and the boundary conditions (6.189), (6.191), and (6.192) is still too complex and can be solved only numerically. Additional simplifications are needed. The first one is based on the assumption that the terms in the right-hand side of (6.182) (inertia and viscosity forces at the bubble surface) are small. Taking these terms into consideration is important only at the initial stage of bubble growth. If, in addition, we suppose that the bubble consists of vapor only, then $p_i = 0$ and (6.182) reduces to

$$p_V = p_\infty + \frac{2\Sigma}{R}. \tag{6.196}$$

If the bubble is sufficiently big, then $2\Sigma/R \ll p_\infty$, and the pressure inside the bubble may be considered as constant:

$$p_V = p_\infty. \tag{6.197}$$

When solving the problem of gas-bubble dynamics, we take into account the processes of heat and mass exchange between the bubble and the ambient liquid. Depending on the relative importance of these processes, it may be possible to consider them separately.

If growth or reduction of a bubble is basically caused by heat exchange, then we can speak of a heat exchange controlled process. Equations describing the change of temperature in the liquid and the change of the bubble's radius reduce to

$$\frac{\partial\theta}{\partial t} = \alpha\left(\frac{\partial^2\theta}{\partial r^2} + \frac{2}{r}\frac{\partial\theta}{\partial r}\right) - \frac{\varepsilon R^2\dot{R}}{r^2}\frac{\partial\theta}{\partial r}. \tag{6.198}$$

$$\theta(r,0) = \theta(\infty,t) = 0; \quad \theta(R,t) = -\frac{T_0 - T_{sat}}{T_0},$$

$$\dot{\theta}(R,t) = \frac{\dot{R}}{\alpha}\left(\xi + (1-\varepsilon)\omega\frac{T_0 - T_{sat}}{T_0}\right),$$

where $\theta = \dfrac{T - T_0}{T_0}$; $\xi = \dfrac{\rho_G l}{\rho_L c_L T_0}$; $\omega = \dfrac{c_L - c_G}{c_L}$; T_{sat} is the saturation temperature at a given pressure.

The reader will find more detailed information on the bubble's dynamics in reviews [17, 18].

6.9 Evaporation of a Multi-component Drop Into an Inert Gas

Evaporation of drops of multi-component solution into an inert gas, plays an important role in such processes as dehydration and humidifying of gas by the method of jet spraying, combustion of liquid fuel injected into the combustion chamber of engines and heating systems, etc.

Consider a spherical drop consisting of n components that can evaporate into the ambient inert (neutral) gas [19]. Neutral gas is defined as a gas that does not participate in mass exchange processes; in other words, it does not condense (though the term inert means chemically inactive we use it also for neutral gas) and does not evaporate at a drop's surface. The main problem is to study the drop-gas dynamics: changes of temperatures, drop size, and component concentrations in the drop and the gas. As demonstrated in the previous section, the general treatment that would lead to exact formulas for temperature and concentrations as functions of position and time presents an extremely complex problem. Therefore, we should make a few simplifying assumptions, some of which are similar to the assumptions made in the previous section.

First, we shall use a quasi-stationary approach already mentioned earlier, based on the assumption that characteristic times of heat and mass transfer in the gaseous phase are much shorter than in the liquid phase, since the coefficients of diffusion and thermal conductivity are much greater in the gas than in the liquid. Therefore the distribution of parameters in the gas may be considered as stationary, while they are non-stationary in the liquid. On the other hand, small volume of the drop allows us to assume that the temperature and concentration distributions are constant within the drop, while in the gas they depend on coordinates. Another assumption is that the drop's center does not move relative to the gas. Actually, this assumption is too strong, because in real processes, for example, when a liquid is sprayed in a combustion chamber, drops move relative to the gas due to inertia and the gravity force. However, if the size of drops is small (less than 1 μm) and the processes of heat and mass exchange are fast enough, then this assumption is permissible. As usual, we assume the existence of local thermodynamic equilibrium at the drop's surface, as well as equal pressures in both phases. The last condition was formulated at the end of Section 6.7.

Denote by x_i and y_i the molar fractions of components in the liquid and gaseous phases. In this case, $i = \overline{1..n}$ so that y_{n+1} is the molar fraction of neutral component in the gas. By virtue of the above assumptions and in view of spherical symmetry of the problem, we now have $x_i = x_i(t)$ and $y_i = y_i(r)$.

Now use the expressions (4.29) and (5.47) for molar fluxes of components and the energy flux:

$$j_i^* = J_i^* + y_i \sum_{j=1}^{n+1} j_i^* = -C_{im}\frac{dy_i}{dr} + y_i \sum_{j=1}^{n+1} j_j^*, \tag{6.199}$$

$$q = -\mathsf{k}\frac{dT_G}{dr} + \sum_{j=1}^{n+1} h_{Gj} j_j^*, \tag{6.200}$$

where C_G is the molar density of gas; D_{im} is the coefficient of binary diffusion in gas; h_{Gj} stands for molar enthalpies of gas components; k is the heat conduction coefficient of gas.

At the surface of a drop of radius R, we have the following conditions:

$$y_i = y_{iw},\ T_G = T_L \quad \text{at } r = R(t), \tag{6.201}$$

and far from the drop in the bulk of the solution,

$$y_i = y_{i\infty},\ T_G = T_{G\infty} \quad \text{at } r \to \infty, \tag{6.202}$$

where T_L is the drop temperature.

The values $y_{i\infty}$ and T_{Gi} are given and considered to be constant, and the value of $y_{i\infty}$ is to be determined. Note that if we consider not one drop, but an ensemble of drops, then the values $y_{i\infty}$ and $T_{G\infty}$ will change with time, and their rate of change will depend on the volume concentration of drops. In this case, the condition at the infinity should be replaced by the condition at a finite value of radius equal to the average distance between drop centers. The values $y_{i\infty}$ and $T_{G\infty}$ are calculated from the equations of mass and energy balance between gas mixture and the drops in a unit volume.

From the continuity equation there follows that $\nabla \cdot \boldsymbol{j} = 0$, or, in the spherical system of coordinates

$$\frac{1}{r^2}\frac{d}{dr}(r^2 \boldsymbol{j}^*) = 0, \tag{6.203}$$

which leads to

$$r^2 \boldsymbol{j}^* = R^2 \boldsymbol{j}_w^*. \tag{6.204}$$

Substitution of the expression (6.197) for $\boldsymbol{j}_i^*$ gives

$$j_{iw}^* = -\frac{C_G D_{im}}{R^2} r^2 \frac{dy_i}{dr} + y_i \sum_{j=1}^{n+1} j_{jw}^*, \tag{6.205}$$

Denote by $j_w^* = \sum_{j=1}^{n+1} j_{jw}^*$ the total molar flux at the drop surface and perform coordinate substitution: $\eta = 1/r$. Then (6.205) transforms to the following equation with respect to y_i:

$$\frac{dy_i}{d\eta} = \frac{R^2}{C_G D_{im}}(j_{iw}^* - y_i j_w^*). \tag{6.206}$$

The solution of this equation is

$$y_i = A \exp\left(-\frac{j_w^* R^2}{C_G D_{im}} \frac{1}{r}\right) + \frac{j_{iw}^*}{j_w^*}. \tag{6.207}$$

The condition (6.202) enables us to find A, and then to obtain from (6.201) the following equation:

$$y_{iw} = \left(y_{i\infty} - \frac{j_{iw}^*}{j_w^*}\right) \exp\left(-\frac{j_w^* R}{C_G D_{im}}\right) + \frac{j_{iw}^*}{j_w^*}. \tag{6.208}$$

Resolve (6.206) with respect to

$$j_{iw}^* = j_w^* \frac{y_{iw} \exp\left(\dfrac{j_w^* R}{C_G D_{im}}\right) - y_{i\infty}}{\exp\left(\dfrac{j_w^*}{C_G D_{im}}\right) - 1} \quad (i = \overline{1, n+1}). \tag{6.209}$$

Since the flux of neutral gas is absent at the drop surface, $j_{(n+1)w}^* = 0$, and from (6.209), it follows that

$$j_w^* \frac{y_{(n+1)w} \exp\left(\dfrac{j_w^* R}{C_G D_{(n+1)m}}\right) - y_{(n+1)\infty}}{\exp\left(\dfrac{j_w^*}{C_G D_{(n+1)m}}\right) - 1} = 0,$$

which leads to

$$\frac{y_{(n+1)\infty}}{y_{(n+1)w}} = \exp\left(\frac{j_w^* R}{C_G D_{(n+1)m}}\right). \tag{6.210}$$

Denote:

$$B = \frac{y_{(n+1)\infty}}{y_{(n+1)w}} = \frac{1 - \sum_1^n y_{i\infty}}{1 - \sum_1^n y_{iw}}.$$

Then from (6.210), we can derive the total flux at the drop surface:

$$j_w^* = \frac{C_G D_{(n+1)m}}{R} \ln B, \tag{6.211}$$

and from (6.209) – the molar fluxes of components:

$$j_{iw}^* = \frac{C_G D_{(n+1)m}}{R} \ln B\left(\frac{y_{iw} B^{D_{(n+1)m}/D_{im}} - y_{i\infty}}{B^{D_{(n+1)m}/D_{im}} - 1}\right). \tag{6.212}$$

Suppose that the liquid phase is an ideal solution, and the gas is a mixture of ideal gases. It follows from the condition of thermodynamic equilibrium at the drop surface that the chemical potentials of components in both phases are equal:

$$\mu_i^L = \mu_i^G. \tag{6.213}$$

Using expressions for chemical potentials of ideal systems [8], we obtain

$$\mu_i^0(T, p) + AT_L \ln x_i = \mu_i^+(T) + AT_G \ln p_i. \tag{6.214}$$

Here p_i is the partial pressure of i-th component, μ_i^0 is the chemical potential of pure liquid, μ_i^+ is the chemical potential of vapor at pressure 1, A is the gas constant.

Eq. (6.214) can be rewritten in the form

$$p_i = \mathsf{k}_i x_i, \tag{6.215}$$

where

$$\mathsf{k}_i = \exp\left(\frac{\mu_i^0 - \mu_i^+}{AT}\right).$$

Consider special cases. For pure liquid, $x_i = 1$ and $\mathsf{k}_i = p_i^0$, where p_i^0 is the vapor pressure of pure liquid. Thus (6.215) becomes

$$p_i = p_i^0 x_i. \tag{6.216}$$

Consequently, in the ideal solution, vapor pressure of each component is proportional to its molar fraction in the solution, and the proportionality coefficient is equal to the vapor pressure of pure liquid. This law is known as Raoult's law. If the gas is a mixture of ideal gases, then the full pressure is:

$$p = \sum_i p_i = \sum_i p_i^0 x_i. \tag{6.217}$$

Consider now the equilibrium of an ideal infinite diluted solution with its vapor. Then $x_1 \to 1$, $\sum_{i=2}^{n} x_i \to 0$, and it follows from (6.216) that for the solvent,

$$p_1 = p_1^0 x_1, \tag{6.218}$$

and for all other components,

$$p_i = \mathsf{k}_i x_i \quad (i = \overline{2, n}). \tag{6.219}$$

Consequently, in the ideal extremely diluted solution the pressures of vapors of dissolved substances are proportional to their molar fractions in the solution. This law is known as Henry's law, and k_i are Henry's constants.

For non-ideal solutions, the relation (6.215) converts to:

$$p_i = \mathsf{k}_i x_i \gamma_i. \tag{6.220}$$

Go back to our problem. The assumption of ideal solution and ideal gas allows us to determine the molar concentrations in the gas, in the bulk of the solution, and at the drop surface:

$$y_{i\infty} = p_{i\infty}/p, \quad y_{iw} = p_i/p. \tag{6.221}$$

Using Raoult's law (6.216), we obtain:

$$y_{iw} = \frac{x_i p_i^0(T_L)}{p}, \tag{6.222}$$

where $p = \sum_i p_i$ is the total vapor pressure.

The dependence of vapor pressure of a pure liquid on the liquid's temperature is:

$$p_i^0(T_L) = p_c \exp(u_i - v_i T_c/T_L). \tag{6.223}$$

The constants u_i and v_i depend on the liquid's characteristics.

Substitution of (6.222) into (6.212) produces the ultimate expression for the molar flux of i-th evaporating component:

$$j_{iw}^* = \frac{C_G D_{(n+1)m}}{R} \ln B \left(\frac{x_i p_i^0(T_L) B^{D_{(n+1)m}/D_{im}} - p y_{i\infty}}{B^{D_{(n+1)m}/D_{im}} - 1} \right). \tag{6.224}$$

$$B = \frac{p - \sum_1^n p_i}{p - \sum_1^n x_i p_i^0(T_L)}.$$

Determine now the flux of heat. In view of the assumptions made above, the equation for energy conservation in the gas reduces to

$$\nabla \cdot \boldsymbol{q} = 0 \tag{6.225}$$

or

$$\frac{1}{r^2}\frac{d}{dr}(r^2 q) = 0, \tag{6.226}$$

which leads to

$$r^2 q = R^2 q_w, \tag{6.227}$$

where q_w is the energy flux toward the drop surface.

Putting into (6.227) the relation (6.200) for q, we obtain:

$$q_w = -\frac{\mathsf{k}}{R^2} r^2 \frac{dT_G}{dr} + \sum_{j=1}^{n} h_{Gj} j^*_{jw}. \tag{6.228}$$

For an ideal gas, the molar enthalpy is equal to

$$h_{Gj} = h'_{Gj} + c_{Gpj}(T_G - T'_G), \tag{6.229}$$

where h'_{Gj} is the enthalpy value at some reference temperature.

Therefore,

$$q_w = -\frac{\mathsf{k}}{R^2} r^2 \frac{dT_G}{dr} + \sum_{j=1}^{n} j^*_{jw}(h'_{Gj} + c_{Gpj}(T_G - T'_G)). \tag{6.230}$$

Eq. (6.230) is solved in the same manner as (6.205). Its solution has the form:

$$T_G(r) = A \exp\left(-\frac{dr}{R}\right) - \frac{b}{d}, \tag{6.231}$$

$$d = \frac{R}{\mathsf{k}} \sum_{j=1}^{n} j^*_{jw} c_{Gpj}, \tag{6.232}$$

$$b = \frac{R}{\mathsf{k}} \left(\sum_{j=1}^{n} j^*_{jw} h'_{Gj} - T'_G \sum_{j=1}^{n} j^*_{jw} c_{Gpj} - q_w \right). \tag{6.233}$$

Substitution of boundary conditions (6.201) and (6.202) gives

$$T_L = \left(T_{G\infty} + \frac{b}{d}\right) e^{-d} - \frac{b}{d},$$

which leads to

$$b = d\frac{T_{G\infty} - T_L e^d}{e^d - 1}.$$

Substituting the found expressions into (6.230), we get:

$$q_w = \sum_{j=1}^{n} j^*_{jw}(h'_{Gj} - c_{Gpj}T'_G) - \frac{\mathsf{k}}{R}d\frac{T_{G\infty} - T_L e^d}{e^d - 1}. \tag{6.234}$$

Taking $T_{G\infty} = T_G$ on the grounds that the change of gas temperature is negligible, we can write the relation (6.234) as

$$q_w = \sum_{j=1}^{n} j^*_{jw}(h'_{Gj} - c_{Gpj}T'_G) + \frac{\mathsf{k}d}{R}\frac{(T_L e^d - T_G)}{e^d - 1}. \tag{6.235}$$

It follows from (6.235) that the total flux of energy at the drop surface is equal to the sum of enthalpy fluxes of evaporating components and the energy flux transferred by the substance. If mass exchange is absent, then $d \to 0$ and $d/(e^d - 1) \to 1$. If we decide to account for the mass exchange, then $d \neq 0$ and $d/(e^d - 1) < 1$. We conclude that mass exchange diminishes the transport of heat.

Now, let us switch our focus to the liquid phase. Consider the equations for conservation of volume, mass, and energy of the drop. Let $V_d = 4\pi R^3/3$ be the volume of the drop. Then the volume conservation equation for the drop will be written as

$$\frac{dV_d}{dt} = -4\pi R^2 \sum_{j=1}^{n} \mathsf{v}_j j^*_{jw} \tag{6.236}$$

or

$$\frac{dR}{dt} = -\sum_{j=1}^{n} \mathsf{v}_j j^*_{jw}. \tag{6.237}$$

Here the right-hand side is the total volume flux of the evaporating component.

Express the drop's mass in moles. Let N_d be the number of moles in the drop. An obvious equality takes place:

$$V_d = \sum_{i=1}^{n} x_i \mathsf{N}_d \mathsf{v}_i = N_d \sum_{i=1}^{n} x_i \mathsf{v}_i = N_d \mathsf{v}_m, \tag{6.238}$$

where v_m is molar volume of the solution, averaged over the molar composition.

The condition of conservation of the number of moles for i-th component in the drop is:

$$\frac{d}{dt}(x_i N_d) = -4\pi R^2 j^*_{iw} \tag{6.239}$$

or

$$N_d \frac{dx_i}{dt} + \frac{dN_d}{dt} x_i = -4\pi R^2 j^*_{iw}. \tag{6.240}$$

Similarly, we can write the condition of conservation of the number of moles for the drop mixture:

$$\frac{dN_d}{dt} = -4\pi R^2 \sum_{i=1}^{n} j^*_{iw}. \tag{6.241}$$

Combining (6.238) with (6.239) and resolving the result with respect to dx_i/dt, we obtain:

$$\frac{dx_i}{dt} = \frac{3}{R} \mathsf{v}_m \sum_{i=1}^{n} j^*_{jw} \left(x_i \frac{j^*_{iw}}{\sum_j j^*_{jw}} \right). \tag{6.242}$$

The change of the drop's energy is made up from the energy carried away by evaporating components and the work W performed by the drop as its volume changes:

$$\frac{dE_d}{dt} = -4\pi R^2 q_w - W. \tag{6.243}$$

The total energy of the drop is equal to the sum of partial energies of components:

$$E_d = \sum_{i=1}^{n} N_d x_i E_i = \sum_{i=1}^{n} N_d x_i (h_{Li} - p\mathsf{v}_i), \tag{6.244}$$

where h_{Li} is the molar enthalpy of i-th component of the solution, equal to

$$h_{Li} = h'_{Li} + c_{Lpi}(T_L - T'_L). \tag{6.245}$$

From (6.244), one can find:

$$\frac{dE_d}{dt} = \sum_{i=1}^{n} \frac{d}{dt}(N_d x_i)(h_{Li} - p\mathsf{v}_i) + N_d \sum_{i=1}^{n} x_i \frac{dh_{Li}}{dt}$$

$$= -4\pi R^2 \sum_{i=1}^{n} j_{iw}^*(\mathsf{h}_{Li} - \mathsf{v}_i) + N_d \sum_{i=1}^{n} x_i c_{Lpi} \frac{dT_L}{dt}$$

$$= -4\pi R^2 \sum_{i=1}^{n} j_{iw}^*(\mathsf{h}_{Li} - p\mathsf{v}_i) + N_d c_{Lpm} \frac{dT_L}{dt}, \tag{6.246}$$

where $c_{Lpm} = \sum_{i=1}^{n} \mathsf{x}_i c_{Lpi}$ is the mean specific heat of the solution.

The work performed by the drop as its volume changes, is equal (with the opposite sign) to the work on the drop performed by the gas:

$$W = p\frac{dV_D}{dt} = -4\pi R^2 p \sum_{j=1}^{n} \mathsf{v}_j j_{jw}^*. \tag{6.247}$$

Substitution of (6.235), (6.246), and (6.247) into Eq. (6.243) gives the equation for the drop energy. By resolving it with respect to dT_L/dt, we get:

$$\frac{dT_L}{dr} = \frac{4\pi R^2}{c_{Lpm} N_d}\left(\frac{\mathsf{k}d}{R}\frac{(\mathsf{T}_G - \mathsf{T}_L e^d)}{e^d - 1} = -\sum_{i=1}^{n} j_{iw}^*(\mathsf{h}_{Gi} - \mathsf{h}_{Li})\right). \tag{6.248}$$

Consider the difference of enthalpies in the last term. Since at the surface, $T_G = \mathsf{T}_L$, then, setting $\mathsf{T}'_G = \mathsf{T}'_L$, we obtain:

$$\mathsf{h}_{Gi} - \mathsf{h}_{Li} = \mathsf{h}'_{Gi} - \mathsf{c}_{Gpi}(\mathsf{T}_L - \mathsf{T}'_L) - \mathsf{h}'_{Li} - \mathsf{c}_{Lpi}(\mathsf{T}_L - \mathsf{T}'_L)$$

$$= l'_i + (\mathsf{c}_{Gpi} - \mathsf{c}_{Lpi})(\mathsf{T}_L - \mathsf{T}'_L). \tag{6.249}$$

Here $\mathsf{l}'_i = \mathsf{h}'_{Gi} - \mathsf{h}'_{Li}$ is the heat of vaporization of *i*-th component at reference temperature.

Substituting (6.249) and (6.238) into Eq. (6.248), we finally obtain:

$$\frac{dT_L}{dr} = \frac{3\mathsf{v}_m}{c_{Lpm} R}\left(\frac{\mathsf{k}d}{R}\frac{(\mathsf{T}_G - \mathsf{T}_L e^d)}{e^d - 1} - \sum_{i=1}^{n} j_{iw}^*(l'_i + (\mathsf{c}_{Gpi} - \mathsf{c}_{Lpi})(\mathsf{T}_L - \mathsf{T}'_L))\right). \tag{6.250}$$

Thus, Eqs. (6.237), (6.242), (6.250) with the additional relations (6.226) and (6.232) form a closed system of equations for the molar fractions of components x_i in the drop, drop radius R, and drop temperature T_L. This system is a system of ordinary differential equations of the kind

$$\frac{dx_i}{dt} = \frac{1}{R^2} f_1(x_j, T_L),$$

$$\frac{dR^2}{dt} = f_2(x_j, T_L),$$

$$\frac{dT_L}{dt} = \frac{1}{R^2} f_3(x_j, T_L) \qquad (6.251)$$

with the initial conditions

$$x_i(0) = x_{i0}, \quad R(0) = R_0, \quad T_L(0) = T_{L0}.$$

6.10 Chromatography

Chromatography is a physico-chemical method of separation and analysis of mixtures and solutions, based on distribution of their components between two phases – the motionless phase (sorbent) and the mobile phase (eluent) that flows through the motionless phase. Depending on the nature of interaction that determines the phase distribution of components, one can identify the following types of chromatography: adsorption, partition, ion exchange, exclusion, and precipitation.

Adsorption chromatography is based on the difference in sorption capabilities of substances to be separated. The sorbent is usually a solid substance with a developed surface. Partition chromatography is based on the difference in solubility of mixture components in the motionless phase – a solid macroporous substance treated by a high-boiling liquid – and in the eluent. The ion-exchange equilibrium constant between the motionless phase (ionite) and a mixture component is different for various mixture components that we are trying to separate; this difference lies in the basis of ion exchange chromatography. Exclusion chromatography, also known as gel chromatography, or molecular-sieve chromatography, is based on the difference in permeability of the motionless phase (a highly porous non-ionic gel) with respect to molecules of different components. This type of the chromatography is, in turn, subdivided into gel-penetration chromatography, where the eluent is a non-water solvent, and gel-filtration chromatography, where the eluent is plain water. Precipitation chromatography utilizes different capabilities of separating components to precipitate and form a sediment on a solid motionless phase.

According to phase state of the eluent, we make a distinction between gas and liquid chromatography.

Depending on the method used to separate the mixture flowing along the sorbent layer, we recognize the following variants of chromatography: frontal, elution, and displacement. In frontal chromatography, the mixture consisting of a carrying medium (gas–carrier) and components $1, 2, 3 \ldots, n$ to be separated is entered continuously into the sorbent layer. The least occlusive component 1 will be the

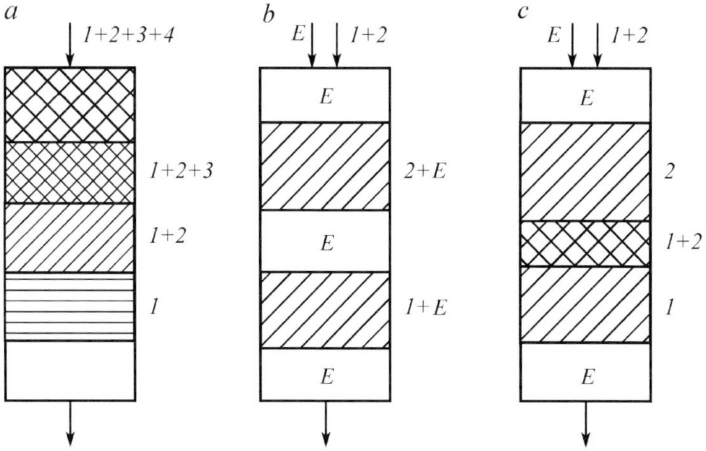

Fig. 6.10 Variants of chromatography: *a* – frontal; *b* – elution; *c* – displacement; *E* – eluent.

first one to leave the layer, forming a pure substance at the exit and leaving the other components behind. Then the other components will follow, in the order of their sorption capacities. At some moment, component 2 reaches the exit, and from this point on, the substance at the exit will be a mixture of components 1 and 2. After component 3 reaches the exit, the resultant mixture will contain three components, and so on. The consecutive mixture zones containing components 1, 1+2, 1+2+3, etc. are shown in (Fig. 6.10, a). In elution chromatography, the flow of eluent propagates continuously through a sorbent layer; periodically, a portion of the component mixture that we are trying to separate is entered (together with the eluent) into the sorbent layer. After a while, components propagating through the sorbent layer with different speeds will become spatially separated, forming pure substance zones separated from each other by layers of eluent (Fig. 6.10, b). In displacement chromatography, component mixture enters the sorbent, together with the gas-carrier that contains a displacement agent (eluent). During the motion of the eluent, the mixture forms pure substance zones separated from each other by zones containing component mixtures (Fig. 6.10, c).

Special devices – chromatographs – are installed at the exit to record the change of concentration with time as the mixture is passing by. The typical form of recordings (chromatograms) is depicted in Fig. 6.11.

We shall limit ourselves to consideration of exclusion (gel) chromatography only. The essence of this process is separating (fractionating) the solution components according to their molecular sizes. The sorbent, or the so-called "molecular sieve", consists of small spherical balls of porous polymeric material. During the passage of molecules through a sorbent layer, molecules whose size is smaller than the size of pores penetrate microscopic cavities of the balls, while molecules of larger size fail to penetrate the pores. Figure 6.12 shows the process of separation of two types of molecules that have different sizes [4].

The reservoir is filled with a layer of small porous balls made, for example, from resin; the solution is entered from the top. The substance dissolved in this

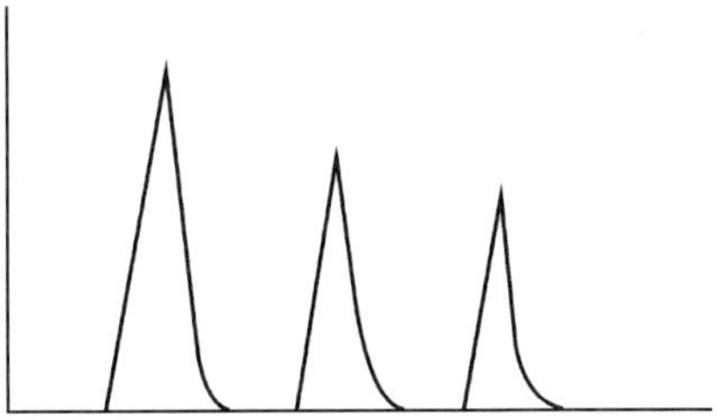

Fig. 6.11 A typical form of chromatograms.

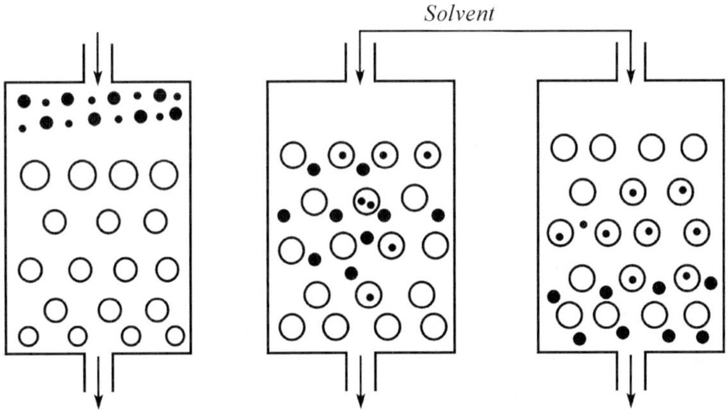

Fig. 6.12 Separation of a solution consisting of molecules of two types.

solution consists of molecules of two sizes – small and large as compared to the size of pores of resin balls. Big molecules pass through the sorbent layer unhindered, while small molecules penetrate the resin balls via the diffusion mechanism and leave the solution. The smaller the size of molecules, the greater their chance to penetrate the pores. Solvent passes freely from the top to the bottom of the layer with a small velocity that is sufficient for the establishment of diffusion equilibrium at the surface of sorbent particles. Thus, solution at the exit from the reservoir contains dissolved substance whose average molecular mass decreases with time.

The described process can be characterized by the separation coefficient σ, having the meaning of probability of capture of molecules by the layer of gel:

$$\sigma = \rho_{in}/\rho = m_{in}/\rho V_{in}, \tag{6.252}$$

where ρ is the mass concentration of dissolved substance in the solution entering the reservoir from the top; ρ_{in} and m_{in} are, respectively, the mass concentration and the mass of dissolved substance in the gel; V_{in} is the volume of pores of gel particles.

For an ideal gel, all particles are assumed to have pores of the same size. Therefore $\sigma = 1$ if all molecules have sizes smaller then the size of pores. It is obvious that $\sigma = 0$ for big molecules. If the sizes of pores are not equal, and are characterized by a continuous distribution, then σ is a function of the molecular mass M. A frequently used correlation between σ and M is:

$$\sigma = -a \ln M + b, \tag{6.253}$$

where a and b are empirical constants.

Let V be the volume of the reservoir, and ε – the volume fraction of interparticle space in the layer. Then the total volume of solution in the reservoir is equal to

$$V_e = \varepsilon V + \sigma V_{in}. \tag{6.254}$$

The first term in the right-hand side of (6.254) is the volume occupied by the solvent and by that part of the dissolved substance which consists of molecules with $\sigma = 0$, which are pushed away by the sorbent and also of molecules with $\sigma < 1$ that failed to penetrate the pores. The second term is the volume of that part of dissolved substance which consists of small molecules with $\sigma < 1$ that have succeeded in penetrating the pores of gel particles.

On the other hand, it is possible to write:

$$V_e = \alpha S L, \tag{6.255}$$

where S is the cross-sectional area of the reservoir; L is the height of the reservoir; α is the fraction of the cross section equal to

$$\alpha = \frac{1}{S}\left(\frac{\varepsilon V}{L} + \frac{\sigma V_{in}}{L}\right). \tag{6.256}$$

The first term in parenthesis is the net cross section of all spacings between particles, and the second term – the net cross section of all small cavities inside the sorbent particles, which are filled with dissolved substance.

The mean effective velocity of solution in the reservoir is

$$U_e = \frac{Q}{\alpha S} = \frac{L}{t}, \tag{6.257}$$

where Q is the constant volume flow rate of the solvent; t is the time needed for the solvent to reach the exit from the reservoir:

$$t = \frac{1}{Q}(\varepsilon V + \sigma V_{in}). \tag{6.258}$$

Consider what happens to a thin layer of solvent containing several types of dissolved component molecules. During this layer's downward motion, the fraction of molecules of different size penetrating the particles' pores, changes according to (6.253). Therefore the (initially thin) layer with the dissolved components at the reservoir entrance is expanding in course of its downward motion. The shape of the layer and the concentrations of different types of molecules in the layer influence the degree of separation of various types of molecules, and consequently, the shape of chromatograms. Distribution of component concentrations in the solvent at the reservoir exit can be found by solving the corresponding problem of convective diffusion of an impurity, delivered in the form of a thin layer at the entrance to the reservoir, analogously to the problem of admixture distribution in a channel (see Section 6.6).

The contribution from the input of a thin layer of dissolved substance can be considered as the initial concentration in the form of (6.134). Then the change of concentration over the length of the reservoir x at various moments of time is given by the expression (6.135), in which it is necessary to put $\pi R^2 = \alpha S$, $x - Ut = x - Qt/\alpha S$, $t = V/Q$, $x = L = V_e/\alpha S$, where V_e is the volume of the eluent, V is the total volume of the mixture flowing through the reservoir. Since concentration is measured with the help of the chromatograph at the exit from the reservoir, we may take $x = L$. Performing an appropriate replacement of variables in (6.135), we will obtain the following dependence of dissolved substance concentration in the solution on the volume flow rate of the flow:

$$\bar{C} = \frac{1}{2}\frac{N_0}{\alpha S}\left(\frac{Q}{\pi D_{eff} V}\right)^{1/2} \exp\left(-\frac{Q(V_e - V)^2}{4\alpha S^2 \pi D_{eff} V}\right), \tag{6.259}$$

The unknown coefficient of longitudinal diffusions D_{eff} should be determined experimentally.

It follows from (6.259) that reduction of D_{eff} means a reduction of the variance of the distribution $\bar{C}$ over the longitudinal coordinate x, which results in a more precise resolution of chromatograms. One of ways to decrease D_{eff} is to decrease permeability k of the sorbent layer, for example, by reducing particle sizes: it follows from that D_{eff} will diminish if we decrease the Peclet number $\mathrm{Pe}_D = RU/D$; here R is the radius of microcapillary, which decreases together with k. It should be kept in mind, however, that reduction of k necessitates an increase of the pressure drop in the reservoir in order to ensure a given volume flow rate of the mixture. High-resolution liquid chromatography requires the use of very small sorbent particles of radius 10 μm and a larger gradient of pressure. The obtained chromatograms are characterized by the presence of narrow and sharp peaks of the type shown in Fig. 6.11.

6.11
The Capillary Model of a Low-permeable Porous Medium

The flow of solution through a layer of gel that was considered in the previous section is an example of flow through a porous environment. At small Reynolds numbers, the distinguishing feature of such flows is that the pressure gradient is proportional to the flow (filtration) velocity U in accordance with Darcy's law:

$$\frac{dp}{dx} = -\frac{\mu}{\mathsf{k}}U. \tag{6.260}$$

The velocity U is defined as the ratio of the liquid's volume flow rate to the net cross section of all spacings between particles in the given layer of porous medium. It is obvious that $U < U_e$, since U_e also includes the volume flow rate of liquid through the pores of particles. The constant k is known as permeability (its dimensionality is m^2). In order to determine k, we must choose a certain model of porous medium. A low-permeable porous medium can be conceptualized as a medium consisting of a set of microchannels of diameter d_e (it is called hydraulic, or equivalent, diameter). This diameter is usually defined as

$$d_e = 4V_{\mathsf{V}}/S, \tag{6.261}$$

where V_{V} is the empty volume between particles. The coefficient 4 is introduced into this definition by analogy with the relation $\pi D^2 = S/4$, where D is the cross-sectional diameter of the reservoir.

Porosity is defined as the fraction of empty space in the gel layer:

$$\varepsilon = V_{\mathsf{V}}/V. \tag{6.262}$$

Now (6.262) will be rewritten as

$$d_e = 4\varepsilon V/S. \tag{6.263}$$

Introduce the parameter s – specific area which is equal to the ratio of the cross – sectional area of the reservoir to the volume fraction of gel particles in the reservoir:

$$s = S/(1-\varepsilon)V. \tag{6.264}$$

Then (6.263) can be rewritten again:

$$d_e = 4\varepsilon/(1-\varepsilon)s. \tag{6.265}$$

The velocity of the solution U_e is connected to the velocity U by the relation

$$U_e = U/\varepsilon. \tag{6.266}$$

Putting U_e instead of U into (6.260) and taking $\mathsf{k} \sim d_e^2$, we obtain:

$$\frac{dp}{dx} = -\mu U \frac{Ks^2(1-\varepsilon)^2}{\varepsilon^3}. \tag{6.267}$$

where K is the Kozeny constant.

Comparing (6.267) with (6.260), we get the following expression for permeability of a porous medium:

$$\mathsf{k} = \varepsilon^3 / Ks^2(1-\varepsilon)^2. \tag{6.268}$$

This formula for a low-permeable porous medium is known as the Kozeny-Carman formula. The Kozeny constant K is approximately equal to 5. In particular, if the porous medium consists of identical spherical particles of diameter d, then $s = 6/d$, and

$$\mathsf{k} = d^2\varepsilon^3/180(1-\varepsilon)^2. \tag{6.269}$$

7
Solutions of Electrolytes

7.1
Electrolytic Cell

Section 5.5 dwelled on the transport of charged mixtures and the derivation of the basic transport equations. Recall that for an infinite diluted mixture, the transport of ions takes place due to their migration in the electric field, diffusion and convection. As in the Section 5.5, we limit ourselves to the study of a binary electrolyte mixture, for which (in the case of electrically neutral mixture) the distribution of reduced ion concentration is described by a convective diffusion equation, with the effective diffusion coefficient given by (5.96). The solution of Eq. (5.94) allows us to find the distribution of electric potential. In Eq. (5.98), we can form scalar products of both parts with $\boldsymbol{dx}$, where $\boldsymbol{x}$ is the radius-vector, and then use the relation between diffusion coefficients of ions and their mobility: $D_{\pm} = AT\upsilon_{\pm}$. Integrating the resultant expression, we then find the potential difference $\Delta\varphi$ between two points of the mixture:

$$\Delta\phi = -\frac{AT}{F^2 z_{\pm}\nu_{+}(z_{+}D_{+} - z_{-}D_{-})}\int \frac{1}{C}\boldsymbol{i}\,\boldsymbol{dx} - \frac{AT(D_{+} - D_{-})}{F(z_{+}D_{+} - z_{-}D_{-})}\int \nabla(\ln C)\,\boldsymbol{dx}. \tag{7.1}$$

The expression for potential difference consists of two terms. The first term has the meaning of ohmic potential drop caused by the resistance of the medium to propagation of electric current of density $\boldsymbol{i}$. The second term, called the diffusion potential drop, is related to the gradient of concentration, that is, to the presence of regions of concentration polarization. This term is caused by the difference in diffusion rates of charged particles and the occurrence of diffusion flux (the second term in Eq. (5.98)).

In order to solve the equation for electric potential (5.94) and the diffusion equation (5.93), it is necessary to specify initial and boundary conditions. Usually the electrolyte is located in the region confined by the electrodes, which are metal plates (one example is copper plate electrodes immersed in a solution of copper vitriol).

The system consisting of two electrodes connected by a liquid conductor (electrolyte) is called an electrolytic cell. If electric current caused by an external EMF

Separation of Multiphase, Multicomponent Systems. E. G. Sinaiski and E. J. Lapiga

ISBN: 978-3-527-40612-8

passes through an electrolytic cell, we say that the process of electrolysis is taking place in the cell. In the case when the electric current is due to the EMF arising as a result of some chemical reaction at the electrode, such system is called a galvanic cell. Further on, our attention will be focused on the process of electrolysis.

Irrespective of whether or not there is an electric current, the electrolyte in the region between the electrodes can be regarded as electrically neutral, except for small areas near the electrodes. At the electrode, a thin layer is formed, which is known as the electric double layer, or the Debye layer. This is a flat layer with oppositely charged surfaces: on the external (from the viewpoint of an observer sitting inside the cell) side – the electrode surface bearing a certain charge (plus at the anode and minus at the cathode); on the internal side – the electrolyte surface with a charge of the opposite sign. The structure of the double layer will be discussed in more detail in Section 7.3. Note that when considering processes occurring in an electrolytic cell, the double layer can be neglected.

As an example, consider an electrolytic cell (Fig. 7.1) consisting of copper electrodes spaced by a layer of electrolyte – water solution of copper vitriol [3, 4].

The solution of copper vitriol contains Cu^{2+} and ${SO_4}^{2-}$ ions. When a potential difference is applied to the solution, the motion of cations Cu^{2+} to the cathode and of anions ${SO_4}^{2-}$ to the anode produces an electric current. At the anode, dissolution of copper takes place. This process can be written as a chemical reaction of the kind

$$Cu \to Cu^{2+} + 2e^- . \tag{7.2}$$

At the same time, precipitation of copper is observed at the cathode. Though the electrolytic cell contains both Cu^{2+} and ${SO_4}^{2-}$ ions, only Cu^{2+} ions enter into reactions with the electrodes, therefore each electrode can be visualized as a

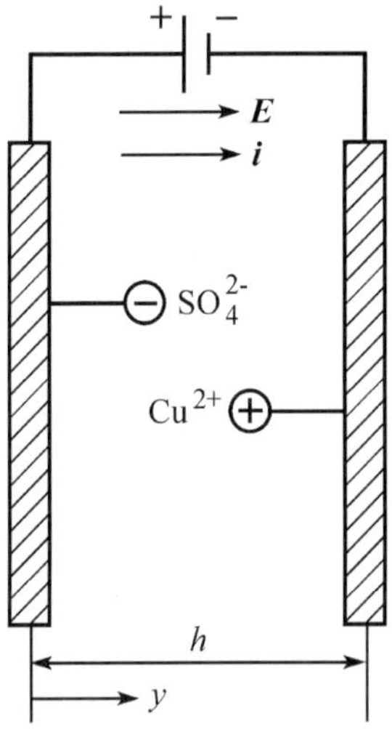

Fig. 7.1 Electrolytic cell.

semipermeable membrane which lets Cu^{2+} ions through and rejects $SO_4{}^{2-}$ ions. Anions gather near the anode and cations – near the cathode, which results in the emergence of an ion concentration gradient near the electrodes. Just as before, we call this effect polarization of concentration. According to (7.1), the emergence of concentration gradient near the electrode results in an additional potential drop. Since a thin double layer is formed near the electrode (there is no current flowing in this layer once the equilibrium is established), the potential drop in the double layer is caused only by the concentration gradient. The thickness of the double layer is very small, therefore we can regard the drop of potential at the electrode as being equal to the difference of potentials at the external and internal surfaces of the boundary dividing the solid and liquid phases.

All of this suggests that the electrolyte solution can be considered as electrically neutral. Only in the immediate vicinity of the electrode, at the metal-solution interface, one can observe a cluster of ions of the same sign. It is the so-called ionic part of the double layer.

Let us formulate the conditions for thermodynamic equilibrium at the metal – solution interface. To this end, consider a solution of the $Me^{z+}\,X^{z-}$ type, where Me^{z+} denotes ions of metal (cations) with charge z_+ and X^{z-} denotes ions of the solution (anions) with charge z_-. The metal-solution interface is impermeable for X^{z-} and permeable for Me^{z+}. On the cathode interface, the incoming Me^{z+} ions lose their charge, thansforming back into atoms. At the anode interface, the transition of Me^{z+} ions from the lattice into the solution takes place. By analogy with (7.2), one can write the last reaction as

$$\mathrm{Me} \rightarrow \mathrm{Me}^{z_+} + z_- e^- . \tag{7.3}$$

At stationary conditions, a partial equilibrium is established – equilibrium with respect to ions penetrating through the electrode boundary ("membrane"). The condition of thermodynamic equilibrium can written as the equality of chemical potentials:

$$\mu_{\mathrm{Me}} = \mu_{\mathrm{Me}^{z_+}} , \tag{7.4}$$

where μ_{Me} and $\mu_{\mathrm{Me}^{z_+}}$ are, respectively, the chemical potentials of particles in the metal, and of ions in the solution.

Since particles passing through the metal – solution boundary are charged, there is an electric field at this boundary, and

$$\mu_{\mathrm{Me}} = \mu_{\mathrm{Me}}^{(0)} + nF\phi_{\mathrm{Me}}; \quad \mu_{\mathrm{Me}^{z_+}} = \mu_{\mathrm{Me}^{z_+}}^{(0)} + nF\phi_{\mathrm{s}}, \tag{7.5}$$

where $\mu_{\mathrm{Me}}^{(0)}$, $\mu_{\mathrm{Me}^{z_+}}^{(0)}$ are the standard chemical potentials, i.e. the potentials that non-charged particles would possess; ϕ_{Me} and ϕ_{s} are the electric potentials in the corresponding phases; F is Faraday's constant; n is the valence of metal ions.

Denote the potential drop at the interface by $\Delta\phi = \phi_{Me} - \phi_s$. Then the condition of equilibrium (7.4), subject to the conditions (7.5), can be written as

$$\mu_{\mathrm{Me}}^{(0)} = \mu_{\mathrm{Me}^{z+}}^{(0)} - nF\Delta\phi. \tag{7.6}$$

Note that if the solution is not ideal, we should take the expression (5.53) for the standard potential:

$$\mu_{\mathrm{Me}^{z+}}^{(0)} = \psi(p,T) + AT\ln(a_{\mathrm{Me}^{z+}}), \tag{7.7}$$

where $a_{\mathrm{Me}^{z+}}$ is the activity of metal in the solution.

For an infinitely diluted solution, activity can be replaced by molar concentration of metal in the solution $C_{\mathrm{Me}^{z+}}$:

$$\mu_{\mathrm{Me}^{z+}}^{(0)} = \psi(p,T) + AT\ln(C_{\mathrm{Me}^{z+}}), \tag{7.8}$$

Substitution of (7.8) into (7.6) gives the potential drop at the interface:

$$\Delta\phi = \frac{AT}{nF}\ln(C_{\mathrm{Me}^{z+}}) + \frac{1}{nF}(\psi(p,T) - \mu_{\mathrm{Me}}^{(0)}) = \frac{AT}{nF}\ln(C_{\mathrm{Me}^{z+}}) + \phi_0^{(0)}. \tag{7.9}$$

The formula (7.9) describes the relation between the concentration of metal ions in the solution and the potential drop at the metal-solution interface. From this expression, there follows:

$$C_{\mathrm{Me}^{z+}} = \mathrm{const}\cdot\exp(nF\Delta\phi/AT). \tag{7.10}$$

The obtained expression is known as Nernst's formula. The sign of the potential drop is chosen such that $\Delta\phi < 0$ when deionization of ions of metal occurs at the electrode (cathode), and $\Delta\phi > 0$ when ions of metal pass into the solution at the electrode (anode).

In practical situations, the potential drop or the EMF between the two electrodes in the cell is either given or measured. Therefore it is necessary to write a relation similar to (7.9) for a system consisting of two identical electrodes. At each electrode, there is a potential drop. Hence, the potential drop between the electrodes (EMF) is equal to

$$V = \Delta\phi_1 - \Delta\phi_2 = \frac{AT}{nF}\ln(C_1) - \frac{AT}{nF}\ln(C_2) = \frac{AT}{nF}\ln\left(\frac{C_1}{C_2}\right), \tag{7.11}$$

where C_1 and C_2 are the concentrations of ions near the electrodes 1 and 2.

Let us now dwell on the mechanism of electrolysis. The passage of current through an electrolytic cell may be considered as a special case of heterogeneous chemical reaction. As was already mentioned (see Section 6.1), a heterogeneous reaction consists of three stages: 1) the transport stage, that is, the transport of ions from the bulk of the solution to the surface of the electrode; 2) the electrokinetic reaction accompanied by participation of ions and molecules; 3) formation of the final products of reaction and their precipitation on the electrode surface, or removal from the surface. The rate of a heterogeneous reaction is determined by the slowest stage. To maintain a finite rate of reaction, it is necessary to apply an external EMF. Most often, the first stage happens to be the slowest one. But in practical applications, there are also cases when the slow stage is either the second or the third one. In theoretical electrical engineering, the main attention is focused on the last two stages. We, however, limit ourselves to consideration of the first case only, i.e. we assume that the process is limited by the transport stage.

We have already ascertained that the passage of current through the electrolyte results in a change of concentration at one of the electrodes and, as a consequence, in the emergence of concentration overvoltage ϕ_{conc}. If the metal-solution boundary is in equilibrium state, there is the potential drop at the boundary, which is given by the formula (7.1). If we impose some other potential (distinct from the equilibrium one), the concentration C_c of ions discharging near the cathode becomes lower than the concentration C_0 of the same ions in the bulk of the solution. The reaction at the electrode goes very quickly, therefore it is possible to assume that C_c has equilibrium value and, applying Eq. (7.9) to the cathode, we have:

$$\phi_c = \phi_0 + \frac{AT}{nF} \ln C_c. \tag{7.12}$$

The corresponding equilibrium value in the solution is equal to

$$\phi_s = \phi_0 + \frac{AT}{nF} \ln C_0. \tag{7.13}$$

The drop of potential in the layer where the concentration of ions changes from C_0 to C_c is called concentration overvoltage and is equal to

$$\phi_{conc} = -(\phi_s - \phi_c) = -\frac{AT}{nF} \ln\left(\frac{C_0}{C_c}\right). \tag{7.14}$$

The change of potential has the minus sign, because at the equilibrium condition, this change tends to inhibit the further change of potential. A thermodynamic analogy of this assertion is known as the Le Chatelier – Braun principle [8]: "Any system, having been in the state of chemical equilibrium, as a result of change of one of the factors determining equilibrium, undergoes such change, which performed by itself, would cause the change of the considered factor in the opposite direction".

Thus, when the slowest stage of the process is the transport of ions, the drop of potential between the electrodes would consist of three parts: the ohmic drop $\Delta\phi_{om}$, the concentration overvoltage ϕ_{conc}, and the equilibrium drop $\Delta\phi_{eq}^{(0)}$ corresponding to the case of zero current. The last part, as a rule, is equal to zero. Accordingly, the EMF applied to the electrodes can be presented as

$$V = \Delta\phi_{om} + \frac{AT}{nF}\ln\left(\frac{C_c}{C_0}\right) + \Delta\phi_{eq}^{(0)}. \tag{7.15}$$

In this expression, the term describing concentration overvoltage at the anode is omitted because under real-life conditions, this term is usually much smaller than at the cathode [4].

The basic characteristic of an electrolytic cell is the volt-ampere diagram $i = f(V)$, known in electrochemistry as the polarization curve. For its determination, it is necessary to know C_c and $\Delta\phi_{om}$.

Assume the electrolyte to be quiescent, and electrodes to be infinite planes. Then the problem may be reduced to determination of the flux of metal ions (cations) to the cathode. The electric current density is $i = Fz_{\pm}j_{\pm}^{*}$. Using expressions (5.85) for molar fluxes $j_{\pm}^{*}$ under the condition $u = 0$, we get:

$$i_{+} = -D_{+}Fz_{+}\frac{dC_{+}}{dy} - \frac{F^2 z_{+}^2 D_{+}C_{+}}{AT}\frac{d\phi}{dy}, \tag{7.16}$$

$$i_{-} = -D_{-}Fz_{-}\frac{dC_{-}}{dy} - \frac{F^2 z_{-}^2 D_{-}C_{-}}{AT}\frac{d\phi}{dy}, \tag{7.17}$$

In the formula (5.85), mobilities are expressed via diffusion coefficients $\mathrm{v}_{\pm} = D_{\pm}/AT$.

From the condition of electric neutrality of the electrolyte, one can see that

$$z_{+}C_{+} + z_{-}C_{-} = 0. \tag{7.18}$$

Unlike the ions of metal, which are deposited freely at the cathode, the anions are halted by the anode, therefore the flux of anions is absent.

Consequently,

$$i_{-} = 0. \tag{7.19}$$

As in Section 5.5, we must introduce the reduced concentrations

$$C = C_{+}/\nu_{+} = C_{-}/\nu_{-}, \tag{7.20}$$

where ν_{+} and ν_{-} are the numbers of positive and negative ions forming in the process of dissociation of electrolyte molecules.

The expressions (7.16) and (7.17) for current densities may then be rewritten as

$$i_+ = -D_+ F z_+ v_+ \frac{dC}{dy} - \frac{F^2 z_+^2 D_+ C_+}{AT} \frac{d\phi}{dy}, \tag{7.21}$$

$$0 = -D_- F z_- v_- \frac{dC}{dy} - \frac{F^2 z_-^2 D_- C_-}{AT} \frac{d\phi}{dy}. \tag{7.22}$$

From (7.22), there follows:

$$\frac{d\phi}{dy} = -\frac{AT}{F z_- C} \frac{dC}{dy}. \tag{7.23}$$

It means that migration of anions in the electric field is counterbalanced by their diffusion under the influence of the concentration gradient.

Now $d\phi/dy$ can be eliminated from (7.21). Taking into account that $z_+ v_+ / z_- = -v_-$, we obtain:

$$i_+ = -D_+ F v_- (z_+ - z_-) \frac{dC}{dy}. \tag{7.24}$$

Since the electric current density in the electrolytic cell is constant, it follows from (7.24) that the distribution of cation concentration is a linear function of y. Further, it follows from the conditions $i_+ > 0$ and $z_+ - z_- > 0$ that $dC/dy < 0$, i.e. the concentration of cations decreases from the anode to the cathode. The drop of potential occurs in the same direction. Since the condition of electrical neutrality gives us $C_-/C_+ = \text{const}$, the distribution of anions behaves similarly to the distribution of cations.

Let us now obtain the distribution of concentration. Positive ions of metal are formed at the anode, therefore

$$C = C_a \quad \text{at } y = 0 \tag{7.25}$$

should be taken as the boundary condition.

Integration of (7.24) under condition (7.25) gives:

$$C = C_a - \frac{i_+ y}{D_+ F v_- (z_+ - z_-)}. \tag{7.26}$$

In particular, at the cathode $(y = h)$, the concentration of cations is equal to

$$C_c = C_a - \frac{i_+ y}{D_+ F v_- (z_+ - z_-)}. \tag{7.27}$$

Now the current density is

$$i_+ = (C_a - C_c)\frac{D_+ F v_-(z_+ - z_-)}{h}. \quad (7.28)$$

Eq. (7.28) implies that the current density achieves its maximum value i_m at $C_c = 0$. Denote by C_0 the initial value of reduced concentration and using the fact that for a binary electrolyte, the conservation condition $C_a + C_c = 2C_0$ should be satisfied, we obtain the following expression for the limiting electric current density:

$$i_m = 2C_0\frac{D_+ F v_-(z_+ - z_-)}{h}. \quad (7.29)$$

By implication, the limiting current density is achieved when cations are instantly absorbed by the cathode. This is possible at sufficiently high values of the applied potential drop.

Now, let us determine the distribution of potentials. To this end, substitute (7.26) into (7.21) and integrate the resultant expression over y from 0 to h. The outcome is:

$$\Delta\phi_{om} = \phi(0) - \phi(h) = \frac{AT(z_+ - z_-)}{z_+ F z_-}\ln\left(\frac{i_m - i_+}{i_m + i_+}\right) - \frac{AT}{z_+ F}\ln\left(\frac{C_a}{C_c}\right). \quad (7.30)$$

The formula (7.30) provides an expression for the ohmic potential drop. It consists of two parts. The first part is the common drop of potential in the solution. Electric conductivity of this solution is caused by the mobility of metal ions. Negative ions of the solution are motionless as a whole, and do not contribute to the current. Nevertheless, the presence of motionless anions results in an additional potential drop. It is expressed by the second term in (7.30).

In compliance with (7.15), the applied EMF is equal to

$$V = \Delta\phi_{om} + \phi_{conc} + \Delta\phi_{eq}^{(0)} = \frac{AT(z_+ - z_-)}{z_+ F z_-}\ln\left(\frac{i_m - i_+}{i_m + i_+}\right). \quad (7.31)$$

Resolving (7.31) with respect to i, one gets the volt-ampere characteristic:

$$\frac{i_+}{i_m} = \frac{1 - \exp(z_+ z_- F v/(z_+ - z_-)AT)}{1 + \exp(z_+ z_- F v/(z_+ - z_-)AT)}. \quad (7.32)$$

The dependence $i = f(V)$ is shown in Fig. 7.2. At small values of FV/AT, the dependence is close to linear. As FV/AT increases, the current density exponentially tends to i_m. Such a volt-ampere characteristic corresponds to the ideal electrolyte. Therefore no current greater than i_m can exist in an ideal electrolyte. This restriction is typical for quiescent electrolytes. If the electrolyte moves, for exam-

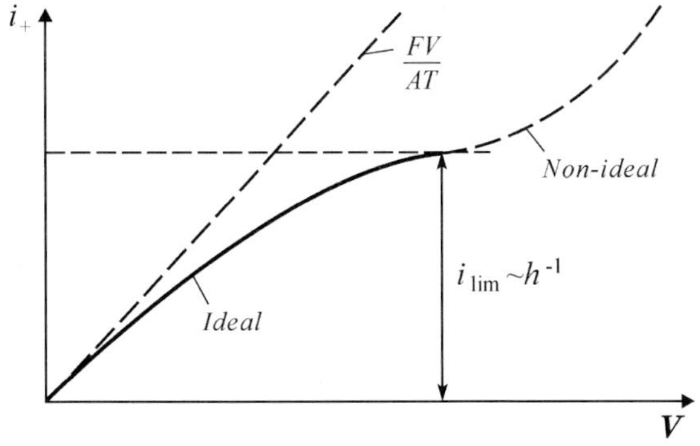

Fig. 7.2 The volt-ampere characteristic.

ple, being mixed, then polarization of ions in the electrolyte decreases, which is accompanied by an increase of the limiting flux and, consequently, i_m. For a non-ideal electrolyte, the values of i_+ greater than i are possible because of the dissociation of the solvent at high values of V.

7.2 Electrodialysis

Electrodialysis is a process in which ions contained in the solution are separated by membranes in the presence of an external electric field. The membranes employed in this process are called ion-exchange membranes. There are two types of membranes: cation-exchange membranes, which are penetrable only for cations, and anion-exchange membranes, penetrable for anions only.

The electrodialysis apparatus is a set of packets of flat membranes, one of which is shown in Fig. 7.3. Anion-exchange and cation-exchange membranes alternate inside the packets. The packet of membranes is limited from both sides by the electrodes. The solution containing ions (we assume for definiteness that these are ions of sodium chloride Na^+ and Cl^-) flows in flat channels between the membranes. Under the action of external electric field perpendicular to the membranes, Na^+ ions pass through cation-exchange membranes, while Cl^- ions pass through anion-exchange membranes. As a result, the concentration of salt decreases in the channel between the left pair of membranes, which is called the channel of dialyzate, and increases in the channel between the right pair of membranes, which is called the channel of concentrate. The salt solution flows through both channels, and during its flow, salt passes from the channel of dialyzate into the channel of concentrate. A section that consists of the dialyzate and concentrate channels together with the adjoining membranes is called the cellular pair. A typical electrodialysis packet may contain anywhere from 50 to 300 cellular pairs. The number of packets in the electrodialysis apparatus is chosen

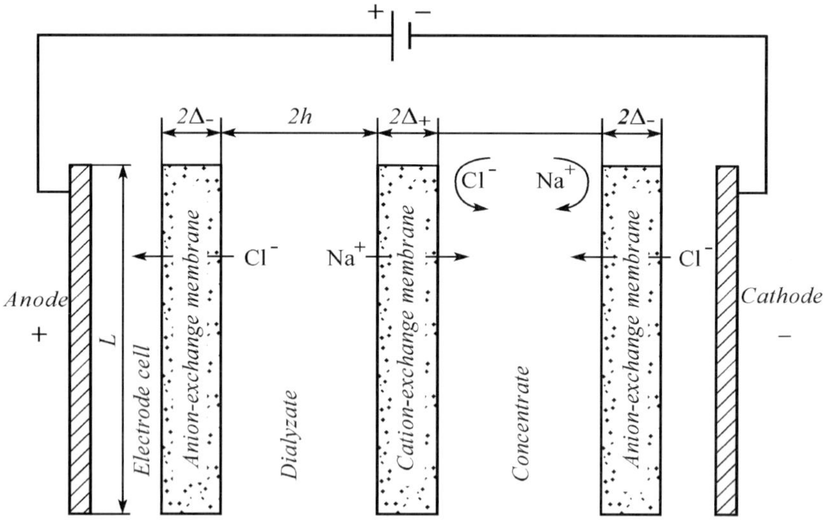

Fig. 7.3 Electrodialyzer packet.

based on the required degree of clearing of the solution from the salt. With the reduction of salt concentration in the channel of dialyzate, the conductivity of the solution falls. The characteristic width of the channels is equal to 1 mm; the channel length is equal to 0.25–1 m.

Ion-exchange membranes are made from organic polymeric materials; the thickness of a membrane is about 0.5 mm. Figure 7.4 shows the ion concentration distribution inside and outside a cation-exchange membrane for the case when the numbers of positive and negative ions are equal $(\nu_+ = \nu_-)$, so that $C_+ = C_-$.

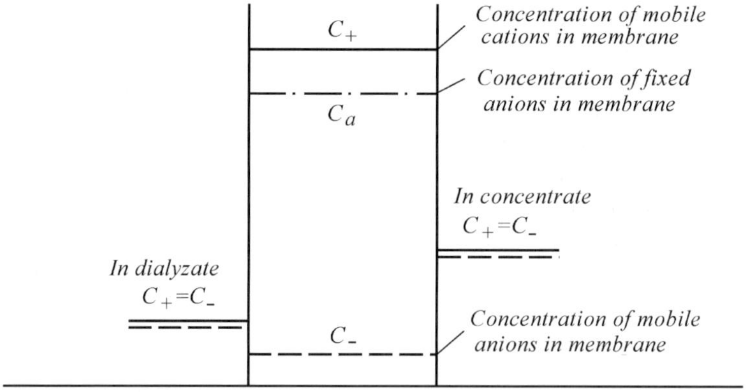

Fig. 7.4 Concentration distribution of ions inside and outside the membrane.

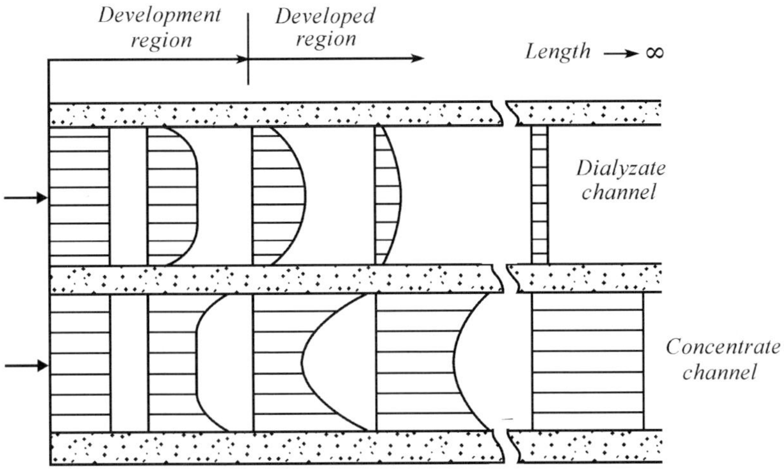

Fig. 7.5 Concentration profile development in dialyzate and concentrate channels.

Ionic motion leads to the appearance of an ion concentration gradient in the solution at the membrane surfaces. The resultant polarization of concentration is similar to polarization in the electrolytic cell. This polarization is responsible for the low concentration of salt and great strength of the electric field near the membrane in the dialyzate channel.

To illustrate the phenomenon of concentration polarization in the electrodialysis cell [20], consider the development of flow in the channels of concentrate and dialyzate (Fig. 7.5) under the condition of equal salt concentrations in the solution at the entrances to these channels. This condition implies that the solution has a constant electrical conductivity. At the entrance section, the velocity profile is considered as developed and the concentration profile is supposed to be uniform. Therefore the distribution of salt concentration near the entrance is close to uniform, and under the action of the electric field, the solution behaves as a medium whose electrical conductivity is constant over the cross section. In particular, in such a medium, as well as in membranes, the drop of potential is linear. Further downstream, the concentration near the dialyzate channel membranes decreases; however, it grows in the concentrate channel. A concentration boundary layer forms at the surface. Its thickness grows with the distance from the entrance. In the dialyzate channel, the potential drop caused by the concentration gradient at the membranes is greater than the potential drop in a solution with the same uniform mean electrical conduction. The sharp drop of potential near the membrane surface has the same nature as the potential drop near the electrode (see Section 7.1). After the concentration boundary layers reach the channel's axis, the ion concentration begins to change at the axes of dialyzate and concentrate channels as well. At a sufficiently large distance from the entrance, there emerges an established concentration distribution of ions over the channels' cross sections. Accordingly, an established potential drop emerges as well.

By analogy with the problem of convective diffusion in the channel with a soluble wall (see Section 6.3) and in the channel with membrane walls (see Section 6.4), we can introduce the concepts of the region of concentration development and the region of developed concentration (see Fig. 7.5).

Let the processes occurring in cells within the same package be identical. Consider a single cell containing dialyzate and concentrate channels with the adjoining membranes (see Fig. 7.5). If the potential drop across one electrode cell is negligible in comparison with the potential drop across the entire package of membranes, then the potential drops across all cells will have the same value, equal to the total potential drop across the package divided by the number of cells in the package. Since the distribution of parameters in each channel is symmetrical about its axis, the electrodialysis cell can be modeled by a half of the dialyzate channel and a half of concentrate channel, with a cation-exchange membrane (Fig. 7.6) dividing them.

Assume for simplicity that the ohmic resistance of membranes is small and can be ignored. Also, assume that the electrolyte consists of one positive component and one negative component with equal numbers of ions:

$$\nu_+ = \nu_- = \nu \tag{7.33}$$

and equal diffusion coefficients:

$$D_+ = D_- = D. \tag{7.34}$$

Note that these assumptions are not essential, they are needed only to simplify calculations.

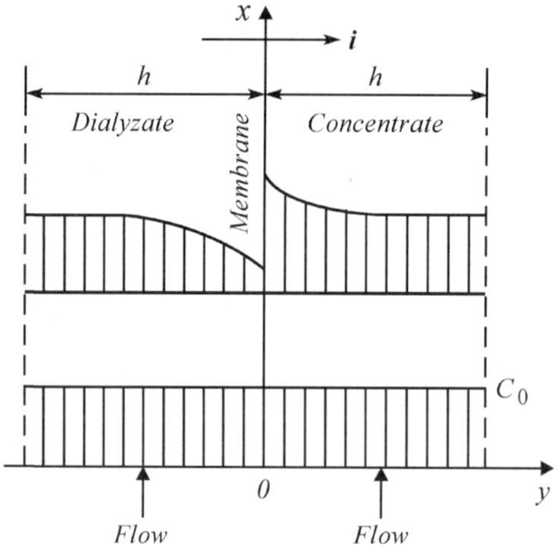

Fig. 7.6 Schematic distribution of concentration in the dialyzate and in the concentrate.

Suppose that there is a developed flow of solution in each channel, with the mean velocity U. The distribution of concentration is uniform at the entrance, i.e.

$$C = C_a \quad \text{at } x = 0. \tag{7.35}$$

Equations describing the distributions of ion concentration and electric potential have been derived in Section 5.5. Under the condition

$$C_+ = C_- = C \quad \text{and} \quad z_+ = -z_- = z \tag{7.36}$$

these equations become

$$u(y)\frac{\partial C}{\partial x} = D\frac{\partial^2 C}{\partial y^2}, \tag{7.37}$$

$$\frac{\partial \phi}{\partial y} = -\frac{i(y)AT}{2F^2z^2DC}. \tag{7.38}$$

Now, formulate the boundary conditions. At the cation-exchange membrane, the condition $i = 0$ is obeyed. In Section 7.1, this condition was applied to the cathode. In the present case, it follows from (7.17) that

$$D\frac{\partial C}{\partial y} = -\frac{i}{2z_+F} \quad \text{at } y = 0. \tag{7.39}$$

A similar condition is satisfied at the anion-exchange membrane, but the sign in the right-hand side of (7.39) changes to the opposite one.

The second boundary condition is the condition of symmetry at the axes of the channels:

$$\frac{\partial C}{\partial y} = 0 \quad \text{at } y = \pm h. \tag{7.40}$$

Eq. (7.37) with the conditions (7.35), (7.39), and (7.40) is the same as in the problem of concentration profile development in reverse osmosis (see Section 6.4). A comparison with this problem shows that the condition (7.39) takes place in the reverse osmosis problem if we replace $i/2z_+F$ by $v_w C_w$. The density of current is $i = Fz_+j_+$, where j_+ is the flux of ions. It is now possible to use the solution derived in Section 6.4. At small distances from the entrance $(x/h \to 0)$, diffusion layer thickness is insignificant, and it is possible to introduce the self-similar coordinate (6.76)

$$\eta = y\left(\frac{3U}{hDx}\right)^{1/3}. \tag{7.41}$$

As the longitudinal coordinate, we shall introduce

$$\xi = \gamma\left(\frac{z_+FV}{2AT}\right)^3\left(\frac{D}{3Uh}\right)\frac{x}{h}, \tag{7.42}$$

where V is the constant external potential drop applied to the considered system.

The Peclet number is

$$\mathrm{Pe}_D = \mathrm{v_w}h/D.$$

A comparison of (6.75) with (7.42) gives us

$$\mathrm{v_w} = Dz_+FV/2ATh.$$

Then Peclet number then becomes

$$\mathrm{Pe}_D = z_+FV/2AT = V^*/2.$$

Here we have introduced the dimensionless potential drop

$$V^* = z_+FV/AT. \tag{7.43}$$

The estimation of diffusion layer thickness can be made in accordance with (6.29):

$$\delta_D^3 \sim h^3\left(\frac{D}{3Uh}\right)\frac{x}{h}. \tag{7.44}$$

Now we are in a position to estimate the concentration of ions at the surface of a membrane, using (7.39):

$$\frac{C_0 - C_\mathrm{w}}{C_0} \sim \frac{i\delta_D}{2z_+FC_0D}. \tag{7.45}$$

The quantity i appearing in the right-hand side of (7.45) may be estimated from (7.38):

$$i \sim \frac{V}{2h}\frac{2F^2z_+^2DC_0}{AT}. \tag{7.46}$$

Now, there follows from (7.45):

$$\frac{C_0 - C_\mathrm{w}}{C_0} \sim \xi^{1/3}. \tag{7.47}$$

Note that the estimation (7.47) is applicable for the dialyzate channel, since in this channel, $C_0 > C_\mathrm{W}$.

The diffusion equation (7.37) is solved by the same method as in Section 6.4. First, use the relation (6.76) to determine the function $f(\eta)$. Then solve the ordinary differential equation (6.78) for $f(\eta)$. As a result, (6.77) will give the concentration of ions at the wall:

$$\frac{C_\mathrm{W}}{C_0} = 1 - 1.53\xi^{1/3}. \tag{7.48}$$

Once the distribution of ion concentration is found, we can obtain the distribution of potential from (7.38). But we should note first that the total potential drop is a sum of the drop in the dialyzate and concentrate channels and the drop at the membrane. The potential drop at the membrane is analogous to concentration overvoltage at the electrode caused by the difference of ion concentrations at membrane surfaces. For a cation-exchange membrane, the potential drop is equal to

$$\eta_\mathrm{over} = -\frac{AT}{z_+F}\ln\frac{(C_\mathrm{W})_\mathrm{conc}}{(C_\mathrm{W})_\mathrm{dial}}, \tag{7.49}$$

where $(C_\mathrm{W})_\mathrm{conc}$ and $(C_\mathrm{W})_\mathrm{dial}$ are the values of ion concentration at membrane surfaces facing, respectively, the channels of concentrate and dialyzate.

Thus, applied external potential is equal to

$$V = \Delta\phi - \frac{AT}{z_+F}\ln\frac{(C_\mathrm{W})_\mathrm{conc}}{(C_\mathrm{W})_\mathrm{dial}}, \tag{7.50}$$

where $\Delta\phi$ is the potential drop in the liquid, which can be found from (7.38):

$$\Delta\phi = -\frac{iAT}{2F^2z_+^2D}\left(\int_{-h}^{0}\frac{dy}{C} + \int_0^h\frac{dy}{C}\right). \tag{7.51}$$

Substituting here the expression for C, we obtain:

$$i^*\frac{2}{V^*} = 1 - 3.072\frac{\xi^{1/3}}{V^*}, \tag{7.52}$$

where

$$i^* = \frac{ih}{2z_+FC_0D}. \tag{7.53}$$

The obtained solutions (7.48) and (7.52) are valid under the condition that $\xi^{1/3}$ and $\xi^{1/3}/V^*$ are small. There are two limiting cases: $\xi \to 0$ and $V^* \to \infty$. At high values of V^*, the current i is close to its limiting value

$$i^*_{\text{lim}} = V^*/2. \tag{7.54}$$

Eq. (7.48) gives us the value of ξ at which the concentration at the membrane surface would be equal to zero:

$$\xi_{\text{lim}} = (1.536)^{-1} = 0.276. \tag{7.55}$$

7.3 Electric Double Layer

The previous sections dealt with processes of charge transport in electrically neutral electrolytes as applied to electrolytic and electrodialysis cells. Most substances, when at contact with a surrounding water (polar) medium, acquire a surface electric charge due to ionization, adsorption of ions, and dissociation [21, 22]. If the charged surface is placed into an electrolyte solution, then ions of the opposite sign (counter-ions) contained in the solution will be attracted to the surface, while ions of the same sign (co-ions) will be repelled from the surface (Fig. 7.7).

Thermal motion of ions is superposed on the motion due to attraction and repulsion of ions. It produces "zitterbevegung" (random component of motion), that accompanies ionic migration. As a result, an electric double layer is formed near the surface, in which the number of counter-ions exceeds the number of co-ions. The electric charge of the double layer compensates (shields) the surface charge, the distribution of ions having diffusional character. If we did not account for the thermal motion of ions, the number of counter-ions in the double layer would exactly compensate the surface charge. Such a state is called ideal shielding. The random thermal motion of ions enables some ions of the same sign as

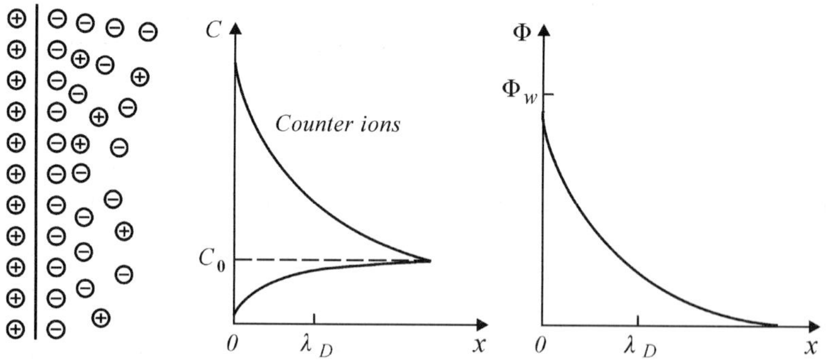

Fig. 7.7 Electric double layer on a charged surface in a solution.

the surface charge to overcome the potential barrier and approach the surface. On the other hand, ions of the opposite sign recede from the surface. As a result, the thickness of the double layer can be estimated by equating the potential energy to the thermal energy of counter-ions ($AT/2$ is the energy of one mole per degree of freedom, A is the gas constant, T is the absolute temperature).

Estimate the thickness of the double layer for the flat case (see Fig. 7.7), when the electric field is perpendicular to the charged plane $x = 0$. Consider a completely dissociated symmetric salt in a solution with equal positive and negative ion charges, that is,

$$z_+ = -z_- = z. \tag{7.56}$$

Suppose for definiteness that the surface has a positive charge. First, estimate the thickness of the double layer for the case of ideal shielding, that is, when there are no positive ions (co-ions) in the layer. Poisson's equation (5.88) says that the potential in the double layer satisfies the equation

$$\frac{d^2\phi}{dx^2} = -\frac{FzC}{\varepsilon}. \tag{7.57}$$

Here C is the average molar concentration of negative ions (counter-ions).

The molar electric energy of ions is equal to

$$W = -Fz\phi. \tag{7.58}$$

If we assume that the electric field vanishes on one side of the flat double layer, then, integrating (7.57) and imposing the condition $d\phi/dx = 0$ at $x = 0$, we can find the change W across the double layer:

$$\Delta W = -F^2Cz^2x^2/2\varepsilon. \tag{7.59}$$

The thickness of the double layer can be determined by demanding that W should be equal to the energy of thermal motion of ions (in the flat case, ions have two degrees of freedom):

$$\Delta W = AT. \tag{7.60}$$

Substituting the expression (7.59) into (7.60), we obtain the thickness of the double layer λ_D:

$$x = \lambda_D = \left(\frac{\varepsilon AT}{2F^2z^2C}\right)^{1/2}. \tag{7.61}$$

The electric double layer sometimes is called the Debye layer, and λ_D is known as the thickness (or radius) of the Debye layer. For water solution of symmetric electrolyte at $T = 25\ °\mathrm{C}$,

$$\lambda_D = \frac{9.61 \cdot 10^{-9}}{(z^2C)^{1/2}}. \tag{7.62}$$

The dimensionality of λ_D is meter; the dimensionality of C is mole/m^3. For a monovalent electrolyte, $z = 1$, and then we have $\lambda_D = 1$ mμ for $C = 100$ mole/m^3 and $\lambda_D = 10$ mμ for $C = 1$ mole/m^3.

We see from (7.62) that $\lambda_D \sim C^{-1/2}$. It means that the number of counter-ions per unit length increases with growth of C. The thickness of the layer also decreases with increase of the charge of ions, since the greater the charge, the smaller the number of ions required to shield the surface charge. The value of λ_D grows with AT. It means that in the absence of ionic thermal motion, the double layer thickness tends to zero. Now, the values obtained for λ_D enable us to refine the concept of electrically neutral solution. If the characteristic linear size of the system is $L \gg \lambda_D$, then any local charge that may appear in the solution will be shielded at the distance λ_D small in comparison with the size L of the electrolyte. Thus the bulk of the solution remains electrically neutral and the external potential in the bulk of the solution does not get distorted. Therefore, as long as $\lambda_D \sim 1$–10 μm, the solution can be considered electrically neutral in the majority of problems of practical interest. However, for systems with $L < \lambda_D$, for example, in microscopic capillaries or in microporous media, the solution cannot be considered electrically neutral, and it is necessary to take into account the structure of the double layer when calculating concentration and mass fluxes.

In the above-made estimation of double layer thickness, C was taken to be constant over the entire layer. Actually, the concentration of ions in the double layer is distributed according to Boltzmann's law:

$$C_\pm = C_0 \exp\left(\frac{\mp zF\phi}{AT}\right). \tag{7.63}$$

Here C_0 corresponds to the concentration of ions far from the charged surface ($C \to C_0$ at $\phi \to 0$). Thus the charge density in the double layer is

$$\rho_E = F\sum_i z_i C_i. \tag{7.64}$$

Taking the formula (7.63) for the distribution, we have:

$$\rho_E = F_z C_0 \left(\exp\left(-\frac{zF\phi}{AT}\right) - \exp\left(\frac{zF\phi}{AT}\right)\right) = -2FzC_0\,\mathrm{sh}\left(\frac{zF\phi}{AT}\right). \tag{7.65}$$

Now the Poisson equation may be written as

$$\frac{d^2\phi}{dx^2} = \frac{2FzC_0}{\varepsilon}\,\mathrm{sh}\left(\frac{zF\phi}{AT}\right). \tag{7.66}$$

For small values $zF\phi/AT \ll 1$, sh can be expanded as a power series. Retaining only the first term of this series, one obtains:

$$\frac{d^2\phi}{dx^2} = \frac{\phi}{\lambda_D^2}. \tag{7.67}$$

The approximation used in the derivation of Eq. (7.67) is known as the Debye-Huckel approximation. Integrating this equation and imposing the conditions $\phi = \phi_W$ at x = 0 and $\phi = d\phi/dx = 0$ at x $\to \infty$, we obtain:

$$\phi = \phi_W \exp\left(-\frac{x}{\lambda_D}\right). \tag{7.68}$$

Thus, in the considered case, the Debye thickness λ_D is the distance from the charged plane, at which the potential decreases by a factor of *e*. In fact, near the surface, the potential is not necessarily small, so the Debye-Huckel approximation can be violated, and ϕ can fall faster than suggested by (7.68). A double layer in which ϕ changes from ϕ_W to 0 as described by Eq. (7.66) or by Eq. (7.67) is called a diffusion layer.

A further improvement of the double layer model can be made if we take into consideration the finite size of ions (which were previously regarded as point charges). The boundary of the double layer adjoining the charged surface is then shifted by the distance of about one ionic radius (Fig. 7.8).

A proper consideration of this shift necessitates the introduction of the so-called Stern layer and a related notion – the Stern plane. Let $\zeta = \phi_\delta$ denote dzeta-potential. It is also known as the electrokinetic potential and characterizes the potential drop across the shear layer, i.e. the layer where the electrolyte can slip

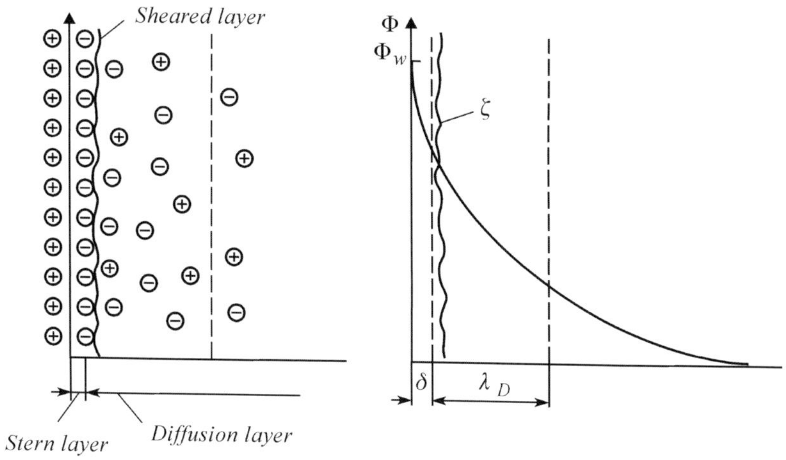

Fig. 7.8 Improved model of an electric double layer.

along the charged surface. A similar representation of the double layer structure leads to the concept of electrokinetic phenomenon, which is defined as a phenomenon that manifests itself in the interaction between the electrolyte moving along the charged surface, and the external electric field.

If the external electric field is parallel to the charged surface, then the electric force is acting on the ions in the diffusion electric double layer. This force is directed along the surface and causes migration of ions along the wall and consequently, motion of the solution as a whole.

If the electrolyte contains solid particles, then the double layer is formed on each particle. In this case the electrokinetic phenomenon is revealed in the possibility of particle motion in external electric field.

On a final note, keep in mind that the expression (7.61) for the thickness λ_D of the Debye layer is valid for an infinite diluted symmetric electrolyte. Paper [25] derives an expression for the electric double layer thickness for the case of an infinite diluted asymmetric electrolyte, and [26, 27] derive the following expression for λ_D for the case of a slightly diluted electrolyte:

$$\lambda_D^{-1} = \lambda_{D0}^{-1}(1 + f(C)),$$

where λ_{D0} is the Debye radius determined by the relation (7.61), and the term $f(C) \sim C^{1/2}$ is the correction term in the asymptotic expansion.

7.4
Electrokinetic Phenomena

For electrolyte solutions, the term "particle" will be applied to particles of sufficiently large size, which makes it possible to describe them as particles that form a separate phase. For example, this term may refer to colloidal particles of various nature and origin, particles of emulsions of water in oil (w/o) or oil in water (o/w) type, etc. These particles, as a rule, carry a negative charge.

At the boundary between the phases, one of which is an electrolyte solution, a region of charged solution is formed – an electric double layer whose presence causes peculiar kinds of electrohydrodynamic effects that manifest themselves in the motion of particles in the electrolyte solution and are called electrokinetic phenomena. All electrokinetic phenomena have common mechanism and arise from the relative motion of phases [3, 23].

If external electric field is applied to a solution containing dispersed particles, the particles will start moving. Such motion is called electrophoresis. An effect opposite to electrophoresis is also known: the emergence of an electric field due to particles' motion under non-electrical forces. For example, the so-called sedimentation potential arises in the case of particle sedimentation in a gravitational field. When particles move in a flow, there appears the electric potential of the flow. Generally, for any relative motion of phases, a corresponding potential drop arises in the mixture.

The same phenomenon takes place if the moving phase is the solution, rather than particles. Thus, if an electric field is applied to the interface between a solid phase and a solution (i.e. the solid phase is a motionless charged wall, rather than a disperse medium), the solution starts moving. This motion is known as electroosmosis.

It is obvious that only the relative motion of phases matters, therefore both phenomena should have the same physical nature even though they manifest themselves differently.

The flow of electrolyte solution along a wall gives rise to an effect known as the flow potential. This effect essentially does not differ from the appearance of sedimentation potential.

There are two limiting cases that are typical for electrokinetic phenomena. In the first case, the thickness of the electric double layer is much smaller then the characteristic linear size of the system. For an electroosmotic flow, this linear size could be the diameter of a capillary where the flow is taking place. The characteristic linear size for electrophoresis is the particle radius. In the given limiting case the solution can be considered electrically neutral everywhere, except for a very thin layer adjacent to the surface of the solid body. The characteristic feature of the second limiting case is that the thickness of the double layer exceeds the linear size of the system. Theoretical study of this case presents considerable difficulties because the solution cannot be considered electrically neutral. From the standpoint of applications, this case is of great importance in biology, because, as a rule, the size of biological particles is smaller than the Debye layer thickness. In the majority of practical cases one has to deal with particles whose size is much larger, therefore the subsequent discussion will be centered on the first limiting case.

7.5 Electroosmosis

It is known that particles of clay, sand, and other particles of mineral origin acquire a negative surface charge during their contact with water solutions (which usually contain dissolved salts). Therefore positively charged ions are predominant in electric double layers that form on particle surfaces. When an external electric field is applied, these ions tend to migrate to the anode. The forces of viscous friction cause the liquid to move together with the ions in the space between particles. If the solid phase is a porous medium, then the liquid flows through the interpore space. This phenomenon is known as electroosmosis.

Electroosmosis is used in many technical applications, for example, in removal of moisture from the ground during construction, in dehydration of industrial waste, in biological processes, etc.

Consider electroosmotic motion in a porous medium. We can model this medium by a system of parallel cylindrical microcapillaries. Consider one of such capillaries and assume that its wall carries a charge. The motion of liquid in the

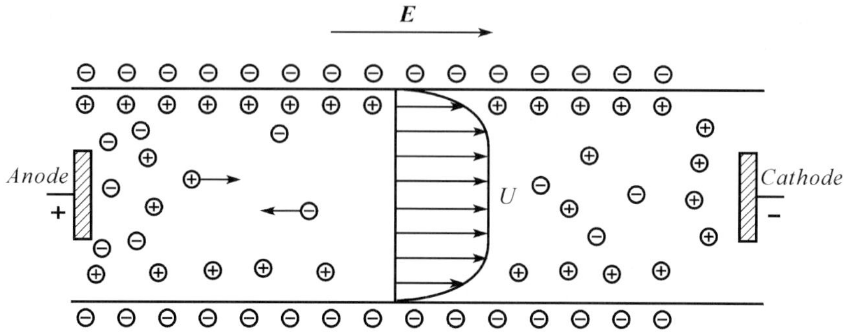

Fig. 7.9 Electroosmotic flow of liquid in a capillary.

capillary is taking place under the action of an external electric field which is taken to be parallel to the capillary axis (Fig. 7.9).

Let ρ_E be the volume density of electric charge in the liquid. Then the force acting on a unit volume of the liquid, is equal to $\rho \boldsymbol{f} = \rho_E \boldsymbol{E} + \boldsymbol{g}$, and the equation of motion of viscous liquid (5.19) is written as

$$\rho \frac{D\boldsymbol{u}}{Dt} = \nabla \cdot T + \rho \boldsymbol{g} + \rho_E \boldsymbol{E}. \tag{7.69}$$

In the current problem, it is safe to neglect both gravity and the liquid's inertia. In the case of zero gradient of pressure, Eq. (7.69) reduces to the equation balance of viscous and electric forces:

$$\mu \Delta \boldsymbol{u} = -\rho_E \boldsymbol{E}. \tag{7.70}$$

Direct the x-axis along the capillary axis (toward the cathode). Suppose the velocity has only one component $u(y)$. For a very narrow capillary, such an assumption is valid. Let the thickness of the Debye layer λ_D at the capillary wall be small in comparison with the capillary radius a, that is, $\lambda_D \ll a$. Then in the region adjacent to the wall, the curvature of the capillary wall can be neglected, and Eq. (7.70) reduces to

$$\mu \frac{d^2 u}{dy^2} = -\rho_E E_x = \varepsilon \frac{d^2 \phi}{dy^2} E_x. \tag{7.71}$$

The y-axis is directed perpendicularly to the wall inside the capillary; at the bottom wall, $y = 0$.

Integrating (7.71) and imposing the condition $du/dy = d\phi/dy = 0$ at $y \to \infty$, which corresponds to the external border of the double layer, one obtains:

$$\mu \frac{du}{dy} = \varepsilon E_x \frac{d\phi}{dy}. \tag{7.72}$$

After second integration over y, with y ranging from ∞ to δ (where δ is the thickness of the shear layer), with due account of the boundary conditions stating that far from the plane, the velocity is equal to the electroosmotic velocity ($u = U$) and that $\phi \to 0$ at $y \to \infty$, we get:

$$U = -\varepsilon\zeta E_x/\mu. \tag{7.73}$$

The formula (7.73) says that the electrolyte slips along the charged surface with the velocity U. This formula is known as the Helmholtz–Smoluchowski formula.

Note that in the small double layer thickness approximation, the character of motion of liquid in the capillary is that of plug flow with the velocity U. If the thickness of the double layer is small, but finite, the velocity profile looks like the one shown in Fig. 7.9. For the characteristic values $\zeta = 0.1$ V, $E_x = 10^3$ V/m, we have for water $U = 10^{-4}$ m/s. Thus, electroosmotic motion has a very low velocity.

Note that the liquid flow rate during electroosmotic motion is $Q_{el} = \pi a^2 U$, while the hydraulic flow rate of liquid due to a given pressure gradient be equal to $Q_{hydr} = \pi a^4 (dp/dx)/8\mu$. Hence, $Q_{el}/Q_{hydr} \sim 1/a^2$. It means that as the capillary radius gets smaller, the motion of the liquid becomes predominantly electroosmotic, rather than hydraulic. This conclusion is true only for $a \gg \lambda_D$.

The solution of a similar for the case of an arbitrary ratio λ_D/a is presented in [24]. The flow is assumed to be inertialess just as before, but this time, the gradient of pressure is taken into account. The electrolyte solution is considered as infinitely diluted and binary. The assumptions made allow us to write the equations of motion in the form:

$$-\nabla p + \mu\Delta \mathbf{u} - F(z_+C_+ + z_-C_-)\nabla\phi = 0. \tag{7.74}$$

Here the electric field strength $\mathbf{E}$ is replaced by $-\nabla\phi$ and the charge density is taken to be $\rho_E = F\sum_i z_i C_i$.

A further simplification can be achieved if we make the assumption that the solution contains a completely dissociated symmetric dissolved substance (salt), so that $z_+ = -z_- = z$. The current is axisymmetrical, and the velocity has only a single longitudinal component, therefore in the cylindrical coordinate system (x, r), Eq. (7.74) reduces to:

$$\frac{\mu}{r}\frac{\partial}{\partial r}\left(r\frac{\partial u}{\partial r}\right) = \frac{\partial p}{\partial x} + Fz(C_+ - C_-)\frac{\partial \phi}{\partial x}. \tag{7.75}$$

The electric field potential is described by Poisson's equation,

$$\Delta\phi = -\rho_E/\varepsilon. \tag{7.76}$$

The flow in a narrow capillary is characterized by the inequality $L \gg a$. Then it is possible to neglect the term $\partial^2\phi/\partial x^2$ in the Laplacian operator, and Eq. (7.76) can be written as

$$\frac{1}{r}\frac{\partial}{\partial r}\left(r\frac{\partial \phi}{\partial r}\right) = \frac{Fz}{\varepsilon}(C_+ - C_-). \tag{7.77}$$

We seek for the solution of Eq. (7.77) in the form

$$\phi(x, r) = \Phi(x) + \psi(x, r). \tag{7.78}$$

Since there is no the radial flux of ions at the surface of the capillary, we have $u = 0$. From the Nernst–Plank equation (5.85), there follows:

$$\frac{\partial C_\pm}{\partial r} = \mp\frac{zF}{AT}C_\pm\frac{\partial \psi}{\partial r}. \tag{7.79}$$

Integration of (7.79) under the condition that $C_\pm = C_0(x)$ at $\psi = 0$ (this condition corresponds to the condition of electric neutrality of the electrolyte solution) gives us a Boltzmann distribution (see (7.63)):

$$C_\pm(x, r) = C_0(x)\exp\left(\mp\frac{zF\psi}{AT}\right). \tag{7.80}$$

In dimensionless variables $r^* = r/a$, $\lambda^* = \lambda_D/a$, $\psi^* = zF\psi/AT$, Eq. (7.77), subject to (7.79), takes the form:

$$\lambda^{*2}\frac{\partial}{\partial r^*}\left(r^*\frac{\partial \psi^*}{\partial r^*}\right) = \sinh(\psi^*). \tag{7.81}$$

The boundary conditions are: the condition at the capillary surface,

$$\psi^* = \zeta^* \quad \text{at } r^* = 1 \tag{7.82}$$

and the symmetry condition at the capillary axis:

$$\frac{\partial \psi^*}{\partial r^*} = 0 \quad \text{at } r^* = 0. \tag{7.83}$$

The analytical solution of Eq. (7.81) can be obtained only at $\lambda_D/a \ll 1$. In the general case, when the value of λ_D/a is arbitrary, this equation needs to be solved numerically. Fig. 7.10 presents the results of numerical integration of Eq. (7.81) for various values of λ^*. For $\lambda^* < 0.1$, the potential ψ in the bulk of the solution is equal to zero. It is different from zero only near the charged wall of the capillary. At $\lambda^* > 10$, the distribution of ψ over the capillary section is constant.

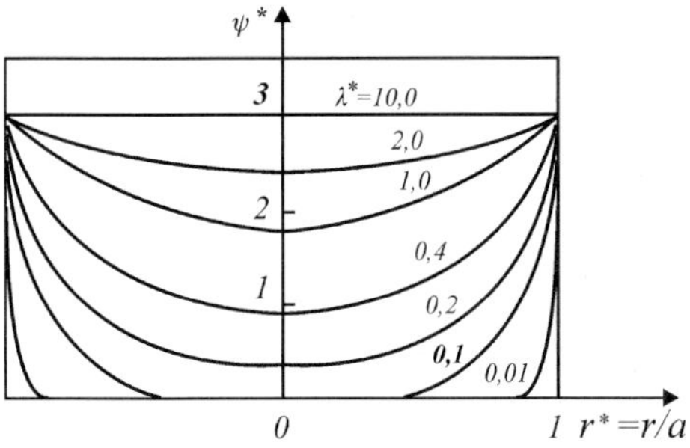

Fig. 7.10 Distribution of the dimensionless electric field potential in a capillary section for various values of $\lambda^* = \lambda_D/a$.

Let us now determine the velocity distribution and the flow rate of the liquid. From (7.75) and (7.77) there follows:

$$\frac{\mu}{r}\frac{\partial}{\partial r}\left(r\frac{\partial u}{\partial r}\right) = \frac{dp}{dx} - \frac{\varepsilon}{r}\frac{\partial}{\partial r}\left(r\frac{\partial \psi}{\partial r}\right)\frac{d\Phi}{dx}. \tag{7.84}$$

This equation should be integrated with the conditions

$$\frac{\partial u}{\partial r} = 0, \quad \frac{\partial \psi}{\partial r} = 0 \quad \text{at } r = 0;$$

$$u = 0, \quad \psi = \zeta \quad \text{at } r = a. \tag{7.85}$$

This gives us the following distribution of velocities:

$$u(r) = -\frac{\varepsilon(\psi - \zeta)}{\mu}\frac{d\Phi}{dx} + \frac{r^2 - a^2}{4\mu}\frac{dp}{dx}. \tag{7.86}$$

The volume flow rate is equal to

$$Q = \int_0^a u(r) 2\pi r\, dr = -\int_0^a \frac{\varepsilon(\psi - \zeta)}{\mu}\frac{d\Phi}{dx} 2\pi r\, dr - \frac{\pi a^4}{8\mu}\frac{dp}{dx}. \tag{7.87}$$

Consider the limiting case $\lambda_D/a \ll 1$, when it is possible to write the solution in the analytical form. Then $\psi = 0$ everywhere, except for a small region near the wall, where $r \sim a \gg \lambda_D$. In this region, it is possible to neglect the curvature of the wall in Eq. (7.81), so after introducing the dimensionless coordinate $y^* = y/\lambda_D$, where $y = a - r$, we obtain:

$$\frac{\partial^2 \psi^*}{\partial y^{*2}} = \sinh \psi^*. \tag{7.88}$$

The Debye-Huckel linearization (see (7.68)) leads to:

$$\psi = \zeta e^{-(a-r)/\lambda_D}. \tag{7.89}$$

Substitution of (7.89) into (7.86) and (7.87) results in:

$$u(r) = \frac{\varepsilon\zeta}{\mu}(1 - e^{-(a-r)/\lambda_D})\frac{d\Phi}{dx} + \frac{r^2 - a^2}{4\mu}\frac{dp}{dx}. \tag{7.90}$$

$$Q = \frac{\varepsilon\zeta}{\mu}\frac{d\Phi}{dx}\pi a^2\left(1 - 2\frac{\lambda_D}{a}\right) - \frac{\pi a^4}{8\mu}\frac{dp}{dx}. \tag{7.91}$$

At $dp/dx = 0$, $\lambda_D/a \to 0$ and $d\phi/dx = -E_x$, we get the expression (7.73) for velocity.

In the other limiting case, $\lambda_D/a \gg 1$, the whole capillary region is contained in the double layer, the potential is constant ($\phi = \zeta$), and the distribution of concentration is given by the expression

$$C_{\pm}(x, r) = C_0(x)\exp\left(\mp\frac{zF\zeta}{AT}\right). \tag{7.92}$$

In this case it is necessary to integrate Eq. (7.74) under the conditions $u = 0$ at $r = a$, $\partial u/\partial r = 0$ at $r = 0$. Note that there is no need to use the Poisson equation because $\phi = \zeta$, and the concentration is given by Eq. (7.92). As a result, one obtains the following:

$$u(r) = \frac{r^2 - a^2}{4\mu}\frac{d}{dx}\left(p - 2zFC_0\Phi\sinh\frac{zF\zeta}{AT}\right), \tag{7.93}$$

$$Q = -\frac{\pi a^4}{8\mu}\frac{d}{dx}\left(p - 2zFC_0\Phi\sinh\frac{zF\zeta}{AT}\right). \tag{7.94}$$

It follows from (7.94) that the ratio of flow rates with and without account taken of the pressure gradient does not depend on the capillary radius, even though it was shown earlier that it is proportional to $\sim a^2$. This outcome leads to the conclusion that at large values of λ_D/a, application of electric field is not a preferable choice as compared to application of pressure gradient, even in the case of very small capillaries.

References

1 Frank-Kamenetskiy D. F., Diffusion and heat transfer in chemical kinetics, Nauka, Moscow, 1987 (in Russian).

2 Schlichting H., Grenzschicht – theorie, G. Braun Verlag, Karlsruhe, 1964.

3 Levich V. G., Physicochemical hydrodynamics, Prentice Hall, Englewood Cliffs, N.J., 1962.

4 Probstein R. F., Physicochemical hydrodynamics, Butterworths, 1989.

5 Targ C. M., Main problems of laminar flow theory, GITTL, Moscow, 1959 (in Russian).
6 Carslow H. S., Jaeger J. C., Conduction of heat in solids, Clarendon Press, Oxford, 1959.
7 Dytnerskiy Yu. I., Reverse osmosis and ultrafiltration, Chemistry, Moscow, 1978 (in Russian).
8 Prigogine I., Defay R., Chemical thermodynamics, Longmans Green and Co, London – New York – Toronto, 1954.
9 Sherwood T. K., Brian P. L. T., Fisher R. E., Dresner L., Salt concentration at phase boundaries in desalination by reverse osmosis, Ind. & Eng. Chem. Fundamentals., 1965, Vol. 4, p. 113–118.
10 Solan A., Winograd Y., Boundary-layer analysis of polarization in electro-dialysis in a two-dimensional laminar flow, Phys. Fluids, 1969, Vol. 12, p. 1372–1377.
11 Berman A. S., Laminar flow in channels with porous walls, J. Appl. Phys., 1953, Vol. 24, p. 1232–1235.
12 Loitzyanskiy L. G., Mechanics of liquid and gas, Nauka, Moscow, 1970.
13 Gupalo Yu. P., Polyanin A. D., Ryasantzev Yu. S., Mass exchange of reacting particles with the flow, Nauka, Moscow, 1985 (in Russian).
14 Taylor G. I., Dispersion of soluble matter in solvent flowing slowly through a tube, Proc. Roy. Soc., 1953., A 219, p. 186–203.
15 Aris R., On the dispersion of a solute in a fluid flowing through a tube, Proc. Roy. Soc., 1956, A 235, p. 67–77.
16 Scriven L. E., On the dynamics of phase growths, Chem. Eng. Sci., 1959, Vol. 10(1), p. 71–80.
17 Plesset M. S., Prosperetti A., Bubble dynamics and cavitation, Ann. Rev. Fluid Mech., 1977, Vol. 9, p. 145–185.
18 Bankoff S. G., Diffusion-controlled bubble growth, Advan. Chem. Eng., 1966, Vol. 6, p. 160.
19 Newbold F. R., Amundson N. R., A model for evaporation of a multicomponent droplet, AICHE Journal, 1973, Vol. 19(1), p. 22–30.
20 Probstein R. F., Desalination: some fluid mechanical problems, Trans. ASME J. Basic Eng., 1972, Vol. 94, p. 286–313.
21 Kruyt H. P., Colloid science. Vol. 1: Irreversible systems, Elsevier, Amsterdam, 1952.
22 Duhin S. S., Deryaguin B. V., Electrophoresis, Nauka, Moscow, 1976 (in Russian).
23 Saville D. A., Electrokinetic effects with small particles, Ann. Rev. Fluid Mech., 1977, Vol. 9, p. 321–337.
24 Gross R. J., Osterle J. F., Membrane transport characteristics of ultrafine capillaries, J. Chem. Phys., 1968, p. 228–234.
25 Mitchel D., Ninham B., Phys. Rev., 1968, Vol. 174, p. 280.
26 Mitchel D., Ninham B., Chem. Phys. Lett., 1978, Vol. 53, p. 397.
27 Kekicheff P., Ninham B., Eurphys. Lett., 1990, Vol. 12, p. 471.

IV
Suspensions and Colloid Systems

Separation of Multiphase, Multicomponent Systems. E. G. Sinaiski and E. J. Lapiga

ISBN: 978-3-527-40612-8

8
Suspensions Containung Non-charged Particles

8.1
Microhydrodynamics of Particles

This chapter is essentially concerned with the hydrodynamics of mixtures containing non-charged particles and macromolecules, that is, molecules with molecular mass exceeding 10^5 and particles with sizes in the range 0.1–100 μm. For the branch of hydrodynamics concerned with the study of motion of liquids containing macromolecules and small particles, G.K. Batchelor [1] coined the term "microhydrodynamics". The distinctive features of microhydrodynamics may be summarized as follows:

1. As a rule, the force of inertia is small in comparison with the viscous force, and therefore the hydrodynamic equations may be reduced to Stokes equations appropriate to hydrodynamics at small Reynolds numbers.
2. Consideration of the Brownian motion of the particles in a liquid is essential.
3. The sedimentation rate of particles in a liquid in a gravitational field is small, and so the particles can be considered to be freely buoyant in the liquid. The volume concentration of particles affects the rheological properties of the liquid.
4. The force of surface tension plays a greater role than the volume force because the surface force is proportional to the particle surface area, that is, to the square of the linear particle size, whereas the volume force is proportional to the cube of the linear particle size.
5. It is also essential to consider electrokinetic phenomena, because solid and liquid particles in aqueous solutions have an electric (volume or surface) charge. In the present section, however, this matter is not addressed.

Consider first some basic problems of hydrodynamics at small Reynolds numbers (for a monograph that deals specifically with these issues, see Ref. [2]).

Separation of Multiphase, Multicomponent Systems. E. G. Sinaiski and E. J. Lapiga

ISBN: 978-3-527-40612-8

We begin with the behavior of isolated particles in a boundless volume of liquid.

If in the Navier-Stokes equations we neglect inertial terms and volume forces, then they reduce to Stokes equations:

$$\nabla p = \mu \Delta \boldsymbol{u}, \tag{8.1}$$

$$\nabla \cdot \boldsymbol{u} = 0, \tag{8.2}$$

The distinctive feature of these equations is their linearity, therefore a complex motion can be considered as a superposition of simpler motions. One consequence of this is that the force acting on a body of an arbitrary shape [3] moving with the translational velocity U is expressed as

$$F_i = 6\pi\mu R_{ij} U_j = f_{ij} U_j. \tag{8.3}$$

Sometimes it is more convenient to write this relation in the form

$$U_i = \mathrm{v}_{ij} F_j, \tag{8.4}$$

The tensor R_{ij} is called the translational tensor, or the resistance tensor. Its components depend on particle's size and shape and have the dimensionality of length. They can be interpreted as equivalent radii of the body. The tensor f_{ij} is called the friction tensor, and the values v_{ij} are known as mobilities. They are similar to mobilities introduced in Section 4.5. Therefore the tensor with components v_{ij} is called the mobility tensor.

The formulas (8.3) and (8.4) in the abbreviated form can be written as

$$\boldsymbol{F} = \boldsymbol{f} \cdot \boldsymbol{U}, \quad \boldsymbol{U} = \boldsymbol{V} \cdot \boldsymbol{F}, \tag{8.5}$$

where $\boldsymbol{f}$ and $\boldsymbol{V}$ are, respectively, the tensors of friction and mobility, which are connected by the obvious relation

$$\boldsymbol{f} = \boldsymbol{V}^{-1}, \tag{8.6}$$

If the body is a liquid spherical particle of radius a and internal viscosity μ_i, immersed in a liquid of some other viscosity μ_e, then

$$R_{ij} = \frac{2}{3}\delta_{ij} a \frac{(\mu_e + 3\mu_i/2)}{\mu_e + \mu_i}, \tag{8.7}$$

At $\mu_i/\mu_e \to \infty$ we have

$$R_{ij} = \frac{2}{3}\delta_{ij} a, \tag{8.8}$$

which corresponds to the motion of a bubble in a liquid.

At $\mu_e/\mu_i \to \infty$, we have the following expression for a solid particle:

$$R_{ij} = \delta_{ij} a. \tag{8.9}$$

Many particles have a non-spherical shape. Sometimes their shape can be approximated by an ellipsoid. Let a_1, a_2, a_3 be the semi-axes of this ellipsoid, and R_1, R_2, R_3 – the components of the translational tensor along these semi-axes. In the special case when $a_1 = a$, $a_2 = a_3 = b$ we have

$$R_1 = \frac{8}{3} \frac{a^2 - b^2}{(2a^2 - b^2)S - 2a},$$

$$R_2 = R_3 = \frac{16}{3} \frac{a^2 - b^2}{(2a^2 - b^2)S + 2a}, \tag{8.10}$$

whereas for the elongate ellipsoid $(a > b)$ we have

$$S = 2(a^2 - b^2)^{-1/2} \ln\left(\frac{a + (a^2 - b^2)^{1/2}}{b}\right),$$

and for oblate ellipsoid $(a < b)$

$$S = 2(a^2 - b^2)^{1/2} \tan^{-1}\left(\frac{(b^2 - a^2)^{1/2}}{a}\right).$$

In the case $a = b$, the value of R for the sphere ensues from (8.10). The Brownian motion of an ellipsoidal particle or a particle of an arbitrary shape has a stochastic character. Accordingly, the orientation of the particle in space is also stochastic. Therefore, for such a motion, the concepts of average coefficients of resistance, friction, and mobility are introduced:

$$\frac{1}{\bar{R}} = \frac{1}{3}\left(\frac{1}{R_1} + \frac{1}{R_2} + \frac{1}{R_3}\right),$$

$$\frac{1}{\bar{f}} = \frac{1}{3}\left(\frac{1}{f_1} + \frac{1}{f_2} + \frac{1}{f_3}\right), \tag{8.11}$$

$$\bar{\mathrm{v}} = \frac{1}{3}(\mathrm{v}_1 + \mathrm{v}_2 + \mathrm{v}_3).$$

Indices 1, 2, 3 designate the principal axes of the quadratic form describing the particle surface.

There are two formulas for the resistance coefficient of a very elongated $(a \gg b)$ spheroid. For the motion along the elongated semi-axis a,

$$R_a = \frac{2a/3}{\ln(2a/b) - 0.5}, \tag{8.12}$$

whereas for the motion along the semi-axis b,

$$R_b = \frac{4/3a}{\ln(2a/b) + 0.5}, \tag{8.13}$$

Approximate expression for resistance coefficient in translational motion of elongated cylinder with length $2a$ and with radius b along its axis is similar to (8.12), but has the term 0.72 instead of 0.5. In case when such a cylinder moves perpendicular to the axis, the resistance coefficient is approximately estimated by formula (8.13). At $a \gg b$, the ratio of resistance forces in these two cases is estimated as $F_b/F_a \sim 2$. It appears that similar estimation is true for an arbitrarily elongated axially-symmetric body, that is, the resistance force at motion perpendicular to the body's axis is almost twice the resistance force at motion along the body axis [4].

Another important parameter is the angular momentum of a particle, which is spinning with angular velocity

$$T_i = -6\mu\Omega_{ij}\omega_j = -f_{ij}^r\omega_j. \tag{8.14}$$

Here ω_j are components of angular velocity relative to coordinate axes, f_{ij}^r are components of rotational inertia tensor. The dimension of Ω_{ij} is cube of length, therefore the tensor is interpreted as equivalent volume. Note, that the relation (8.14) can be represented similar to (8.5), namely, as

$$\mathbf{T} = \Omega \cdot \omega, \quad \omega = \Omega^{-1} \cdot \mathbf{T}, \tag{8.15}$$

where Ω is the rotational tensor.

If the particle has the form of a sphere of radius a, then

$$\Omega_{ij} = \frac{4}{3}\pi a^3 \delta_{ij} = V_m \delta_{ij}. \tag{8.16}$$

For an ellipsoid of revolution we have exact formulae

$$\Omega_1 = \frac{16}{9}\pi\frac{(a^2 - b^2)b^2}{(2a - b^2 S)},$$

$$\Omega_2 = \Omega_3 = \frac{16}{9}\pi\frac{a^4 - b^4}{(2a - b^2)S - 2a},$$

In case of a thin elongated ellipsoid $a \gg b$, the coefficient of rotation around the axis b is equal to

$$\Omega_b = \frac{4\pi a^3/9}{\ln(2a/b) - 0.5}.$$

If very long cylinder of radius b rotates around axis b, then

$$\Omega_b = \frac{4\pi a^3/9}{\ln(2a/b) - 1.14}, \quad \Omega_b = 0.$$

Since the expressions (8.5) and (8.15) are identical, they can be combined, by introducing so-called global tensors of friction $\boldsymbol{f}$ and mobility $\boldsymbol{V}$, including translational and rotational components. In the Stokes flow, these tensors have some universal properties [2], of which the most important are dependence on instant configuration and independence of velocity, as well as symmetry and positive definiteness of matrixes $\{f_{ij}\}$ and $\{V_{ij}\}$.

Inasmuch as Stokes equations are linear, the velocity field at a given point is the sum of velocities caused by actions of several forces. This approach can be used to estimate the influence of ambient particles on motion of considered particle (test particle). For this purpose, it is necessary to use the expression for the velocity u of liquid at a point r due to a force F applied at the origin of coordinates [2]:

$$\boldsymbol{u} = \frac{1}{8\pi\mu r}\left(\boldsymbol{F} + \frac{(\boldsymbol{F}\cdot\boldsymbol{r})\boldsymbol{r}}{r^2}\right). \tag{8.17}$$

The presence in the liquid of particles of another phase results in change of rheological properties of the mixture. If we consider the motion of a test particle in such disperse medium, then such motion would be similar to motion in viscous liquid with effective coefficients of viscosity μ_{eff} depending on volume concentration of particles W. In particular, at $W \to 0$, we have $\mu_{eff} \to \mu_e$.

Of special interest is hydrodynamic influence of other particles on the considered solid particle, moving in a viscous liquid. If the volumetric content of solid particles is small, then particles are rather far apart and are scarcely affecting each other. Hence, as a first approximation one can assume that particles move independently and the given expressions for velocities and resistances of particles are true. However, at particles' approach down to distances comparable to their sizes, hydrodynamic interactions between them become essential and cannot be neglected.

The degree of hydrodynamic interaction of particles depends on the following parameters:

a) form and sizes of particles;
b) distances between them;
c) orientation of particles relative to each other;
d) orientation of particles relative to the gravity force;
e) velocity of translational and rotational motion of particles;
f) positions of particles with respect to boundaries of the flow (walls, interfaces).

Of basic interest is the influence of above-mentioned parameters on hydrodynamic forces and moments exerted on particles by liquid. If these forces and moments for a certain ensemble of particles are known, it is possible to solve an inverse problem – determining the particles' motion from known forces and moments (for example, from gravitational forces).

In such problems, the pressure and velocity fields are described by Stokes equations (8.1) and (8.2). In the most general case particles can perform both translational and rotational motion. Since the Stokes equations and the appropriate boundary conditions are linear, it is possible to consider both motions separately and then form a superposition of solutions. For small volume concentrations of particles, the analysis may be limited to pair interactions between them. For translational motion of particles, the boundary conditions at the particle's surface are:

$$\boldsymbol{u} = \boldsymbol{U}_a \quad \text{at } S_a, \qquad \boldsymbol{u} = \boldsymbol{U}_b \quad \text{at } S_b. \tag{8.18}$$

If, in addition, the motion of particles happens in quiescent liquid, then the following condition must be satisfied at the infinity:

$$\boldsymbol{u} \to 0 \quad \text{at } r \to \infty. \tag{8.19}$$

The solution of the problem just formulated presents significant mathematical difficulties. Well-developed analytical methods exist, however, for the case of spherical particles.

There are two ways to solve this problem. The first method, known as the method of reflections, is an approximate method. It is similar to the method of successive approximations, and its application was pioneered by Smoluchowski for a system of n solid particles located relatively far from each other. Let us set forth the general idea of this method for the case of two spherical particles undergoing translational motion in the xz plane with the velocities U_a and U_b (Fig. 8.1).

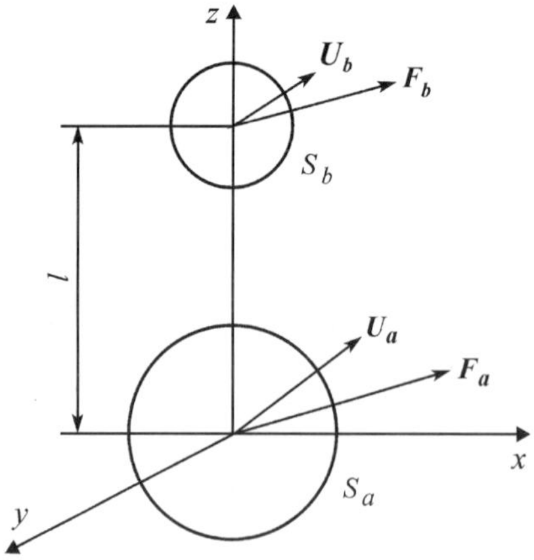

Fig. 8.1 Translational motion of two spherical particles.

Eqs. (8.1) and (8.2) should be solved together with the boundary conditions (8.18) and (8.19). For the zeroth approximation, we take the motion of an isolated particle S_a in an infinite liquid. The resistance force acting on particle S_a is equal to

$$\boldsymbol{F}_a^{(1)} = -\mu K_a \boldsymbol{U}_a = -\mu K_a (U_{ax}\boldsymbol{i} + U_{az}\mathbf{k}), \tag{8.20}$$

where $K_a = 6\pi a$ is the resistance coefficient of particle S_a.

The effect of particle S_b on particle S_a is considered to the first approximation as the effect of the force field (8.20) applied at the center of particle S_a. According to the data presented in [2], the fields of velocity $\boldsymbol{u}^{(1)}$, satisfying the condition $\boldsymbol{u}^{(1)} = \boldsymbol{U}_a$ at S_a, and of pressure $p^{(1)}$, caused by the force $\boldsymbol{F}_a^{(1)}$, are, respectively,

$$\begin{aligned} \boldsymbol{u}^{(1)} &= \frac{\boldsymbol{F}_a^{(1)}}{6\pi\mu r} - \frac{r^2}{24\pi\mu}\nabla(\boldsymbol{F}_a^{(1)}\cdot\nabla)\frac{1}{r}, \\ p^{(1)} &= \frac{1}{4\pi}(\boldsymbol{F}_a^{(1)}\cdot\nabla)\frac{1}{r}, \end{aligned} \tag{8.21}$$

where r is the distance from the center of the sphere S_a.

In the chosen system of coordinates (see Fig. 8.1),

$$\boldsymbol{u}^{(1)} = \frac{K_a U_{ax}}{8\pi r}\left(\boldsymbol{i} + \frac{x}{r^2}\boldsymbol{r}\right) + \frac{K_a U_{az}}{8\pi r}\left(\mathbf{k} + \frac{z}{r^2}\boldsymbol{r}\right). \tag{8.22}$$

The value $\boldsymbol{u}^{(1)}$ in the centre of particle S_b ($x = 0, y = 0, z = 1$) is equal to

$$\boldsymbol{u}_b^{(1)} = \frac{K_a}{8\pi l}(\boldsymbol{i}U_{ax} + 2\mathbf{k}U_{az}). \tag{8.23}$$

It is now possible to calculate the force acting on particle S_b

$$\boldsymbol{F}_b^{(2)} = -\mu K_b(\boldsymbol{U}_b - \boldsymbol{u}_b^{(1)}) = \mu K_b\left(U_{bx} - \frac{K_a U_{ax}}{8\pi l}\right)\boldsymbol{i} - \mu K_b\left(U_{bz} - \frac{K_a U_{az}}{4\pi l}\right)\mathbf{k}. \tag{8.24}$$

Using the same method that was applied to particle S_a, we may calculate the velocity field $\boldsymbol{u}^{(2)}$, caused by the force $\boldsymbol{F}_b^{(2)}$ and satisfying the condition $\boldsymbol{u}^{(2)} = -\boldsymbol{u}^{(1)} + \boldsymbol{U}_b$ at S_b. As a result, we obtain the second approximation of the velocity:

$$\boldsymbol{u}_a^{(2)} = \frac{K_b}{8\pi l}\left(U_{bx} - \frac{K_a U_{ax}}{8\pi l}\right)\boldsymbol{i} + \frac{K_b}{4\pi l}\left(U_{bz} - \frac{K_a U_{az}}{4\pi l}\right)\mathbf{k} \tag{8.25}$$

and the contribution to the force operating on the particle S_a

$$\boldsymbol{F}_a^{(3)} = \mu K_a \boldsymbol{u}_a^{(2)} = \frac{K_b U_a}{8\pi l}\left(U_{bx} - \frac{K_a U_{ax}}{8\pi l}\right)\boldsymbol{i} + \frac{K_b K_a}{4\pi l}\left(U_{bz} - \frac{K_a U_{az}}{4\pi l}\right)\mathbf{k}. \tag{8.26}$$

and so on.

Hence the force acting on particle S_a is

$$\begin{aligned}\boldsymbol{F}_a &= \boldsymbol{F}_a^{(1)} + \boldsymbol{F}_a^{(3)} + \boldsymbol{F}_a^{(5)} + \cdots \\ &= -\mu K_a\left(\frac{U_{ax} - K_b U_{bx}/8\pi l}{1 - K_a K_b/(8\pi l)^2}\boldsymbol{i} - \frac{U_{az} - K_b U_{bz}/4\pi l}{1 - K_a K_b/(4\pi l)^2}\mathbf{k}\right).\end{aligned} \tag{8.27}$$

To determine the force F_b, it is necessary to interchange positions of indices b and a in (8.27).

If particle velocities are known, then by using (8.27), one can obtain the forces acting on particles. In the case when forces are known, for example, when considering particle sedimentation in the gravitational field, it is possible to find the velocity of particle motion. In the specific case when spherical particles are identical ($a = b$, $\boldsymbol{U}_a = \boldsymbol{U}_b$), the expression (8.27) will become

$$\boldsymbol{F} = -6\pi\mu a\left(\frac{U_x}{1 + 3a/4l}\boldsymbol{i} + \frac{U_z}{1 + 3a/2l}\mathbf{k}\right). \tag{8.28}$$

Since the expression (8.28) is true at $a/l \ll 1$, the force $\boldsymbol{F}$ can be presented as a power series in a/l

$$\boldsymbol{F} = -6\pi\mu a\left(\boldsymbol{U} - \frac{a}{l}\left(\frac{3}{4}U_x\boldsymbol{i} + \frac{3}{4}U_z\mathbf{k}\right) + \cdots\right). \tag{8.29}$$

Another method consists in the exact solution of the problem of motion of two spherical particles at any value of a/l. The solution is derived in a special bipolar system of coordinates [5]. As an illustration of the method, consider the motion of two identical solid spheres with constant and equal velocities along the line of centers. Introduce the vorticity vector

$$\omega = \nabla \times \mathbf{u} = \left(\frac{\partial u_r}{\partial z} - \frac{\partial u_z}{\partial r}\right)\boldsymbol{i}_\varphi = \left(\frac{\partial}{\partial z}\left(\frac{1}{r}\frac{\partial\psi}{\partial z}\right) + \frac{\partial}{\partial r}\left(\frac{1}{r}\frac{\partial\psi}{\partial r}\right)\right)\boldsymbol{i}_\varphi = \frac{1}{r}E^2\psi\boldsymbol{i}_\varphi, \tag{8.30}$$

where $E^2 = r\dfrac{\partial}{\partial r}\left(\dfrac{1}{r}\dfrac{\partial}{\partial r}\right) + \dfrac{\partial^2}{\partial z^2}$ is the stream function.

Then we can bring Eqs. (8.1) and (8.2) to the form

$$E^4\psi = 0. \tag{8.31}$$

In the derivation of (8.30) and (8.31), we used the fact that the flow is axially-symmetric and $u_\varphi = 0$, while u_r and u_z depend only on z, r. Let us switch from coordinates z, r to coordinates ξ, η though conformal transformations

$$z + ir = ci \cot \frac{(\eta + i\xi)}{2}. \tag{8.32}$$

where c is a positive constant and

$$z = c\frac{\sinh \xi}{\cosh \xi - \cos \eta}; \quad r = c\frac{\sin \xi}{\cosh \xi - \cos \eta}. \tag{8.33}$$

Eliminating from (8.33), we shall obtain

$$(z - c \coth \xi)^2 + r^2 = c^2/\cosh^2 \xi. \tag{8.34}$$

Thus, $\xi = \xi_0$ corresponds to a family of spheres with the center on the z-axis. If $\xi_0 > 0$, the sphere is located above the plane $z = 0$, and vice versa: if $\xi_0 < 0$, the sphere is under this plane (Fig. 8.2).

So, two spheres external with respect to one another are defined by the equalities $\xi = \alpha$ and $\xi = \beta$, where $\alpha > 0$ and $\beta < 0$. If the radii of spheres r_1, r_2 and the distances of their centers from the origin, d_1 and d_2 are given, then

$$r_1 = \frac{c}{\sinh \alpha}; \quad r_2 = -\frac{c}{\sinh \beta}, \quad d_1 = c \coth \alpha, \quad d_2 = -c \coth \beta. \tag{8.35}$$

The solution of Eq. (8.31) in a bipolar system of coordinates is given in [2]. In the specific case of spheres with equal radii a, the resistance forces are equal

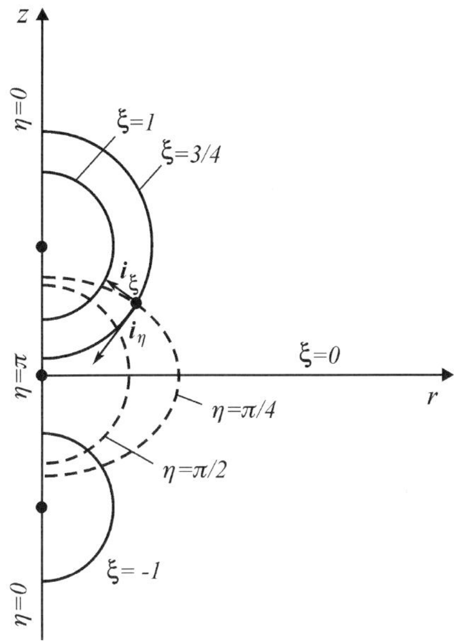

Fig. 8.2 A bipolar coordinate system.

$(\boldsymbol{U}_a = \boldsymbol{U}_b = \boldsymbol{U})$

$$F = -6\pi\mu a U\lambda,$$

$$\lambda = \frac{3}{4}\sinh\alpha\sum_{n=1}^{\infty}\frac{n(n+1)}{(2n-1)(2n+3)} \times\left(1-\left(\frac{2\sinh(2n+1)\alpha+(2n+1)\sinh 2\alpha}{4\sinh^2(n+1/2)\alpha-(2n+1)^2\sinh^2\alpha}\right)^{-1}\right). \tag{8.36}$$

Of special interest is the motion of two spheres with different radii and velocities. The approach used to solve this problem is the same as before. In a special case, when one of the spheres is much bigger then the other, the latter can be considered as moving near a solid plane. If the small sphere moves perpendicular to flat wall, the coefficient λ in the expression (8.36) for resistance force is determined by

$$\lambda = \frac{3}{4}\sinh\alpha\sum_{n=1}^{\infty}\frac{n(n+1)}{(2n-1)(2n+3)} \times\left(\frac{2\sinh(2n+1)\alpha+(2n+1)\sinh 2\alpha}{4\sinh^2(n+1/2)\alpha-(2n+1)^2\sinh^2\alpha}-1\right), \tag{8.37}$$

where $\alpha = \mathrm{arch}(h/a)$; h is the distance between sphere's center and the plane.

Note that at $h/a \to 1$, (that is, when the clearance between the sphere and the plane approaches zero), $\lambda \to \infty$ as $a/(h-a)$.

The motion of suspension in a pipe [6] is of interest in view of many technical applications (for example, the flow of liquid containing macromolecules, the motion of cellulose pulps, the flow of blood containing various particles and cells, the flow of polymer solutions etc.).

Assume, that the suspension represents Newtonian liquid, the volume content of particles in liquid is small, the suspension flow is laminar, the motion of particles relative to liquid is inertialless. Then the particles' motion is determined by two parameters:

$$\mathrm{Re}_t = U_m R/\nu, \quad \lambda = a/R, \tag{8.38}$$

where R is the tube's radius; a is the characteristic size of a particle; U_m is the velocity of flow on the tube's axis; Re_t is Reynolds number of the flow.

The flow of pure liquid in a cylindrical pipe (Fig. 8.3) is described by the Navier-Stokes equations

$$\mu\Delta\boldsymbol{u} - \nabla p = \rho(\boldsymbol{u}\cdot\nabla)\boldsymbol{u}, \tag{8.39}$$

$$\nabla\cdot\boldsymbol{u} = 0. \tag{8.40}$$

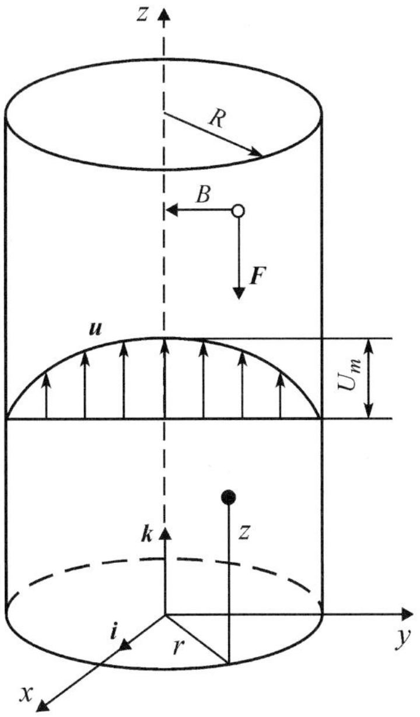

Fig. 8.3 Particle's motion in a tube.

If the velocity at pipe's wall is zero, the longitudinal velocity of developed current $\boldsymbol{u} = (u, 0)$ and the pressure gradient are

$$\boldsymbol{u} = U_m \left(1 - \frac{r^2}{R^2}\right)\mathbf{k}. \tag{8.41}$$

$$\frac{dp}{dz} = -4\mu \frac{U_m}{R^2}. \tag{8.42}$$

If $\mathrm{Re}_t \to 0$, then the Navier-Stokes equations reduce to the Stokes equations

$$\mu \Delta \boldsymbol{u} - \nabla p = 0, \tag{8.43}$$

$$\Delta \cdot \boldsymbol{u} = 0. \tag{8.44}$$

The linearity of equations (8.43) and (8.44) allows us to make some general conclusions on the character of particle's motion in liquid. The most important one is that if an external field $\boldsymbol{F}$ (for example, the gravity field) is parallel to the pipe's axis, then particle's established velocity is also parallel to this axis, that is, the particle does not move in radial direction.

Consider a particle moving in liquid with velocity $U\mathbf{k}$ and spinning with angular velocity $\Omega \boldsymbol{i}$. The force $\boldsymbol{F}$ and torque $\boldsymbol{G}$ acting on the particle have been were calculated in work [2] by method of reflections. In case $\lambda \ll 1$ they are given by:

$$F = -6\pi\mu a\left(\frac{U - U_m(1-\beta^2) + 2U_m\lambda^2/3}{1-\lambda f(\beta)} + U_m O(\lambda^3)\right)\mathbf{k}, \tag{8.45}$$

$$G = -8\pi\mu a^2 \boldsymbol{i}(a\Omega - U_m\beta\lambda) + (U - U_m(1-\beta^2))g(\beta)(1+\lambda f(\beta)) + O(\lambda^4)), \tag{8.46}$$

where $\beta = b/R$; $\lambda \ll 1 - \beta$ is the condition that the particle's distance from the pipe's wall is great in comparison with particle size; functions $f(\beta)$ and $g(\beta)$ are adduced in [2].

If $\boldsymbol{F}$ and $\boldsymbol{G}$ are given, then one can find U and Ω from (8.45) and (8.46). In particular, if the particle is buoyant free in liquid ($\boldsymbol{F} = \boldsymbol{G} = 0$), then

$$\begin{aligned} U &= U_m\left((1-\beta^2) - \frac{2}{3}\lambda^2 + O(\lambda^3)\right), \\ \Omega &= \frac{U_m}{a}(\beta\lambda + O(\lambda^4)). \end{aligned} \tag{8.47}$$

If the particle is balanced in liquid under action of gravity $-F\mathbf{k}$, then $\boldsymbol{G} = 0$ and

$$\begin{aligned} U &= -\frac{F}{6\pi\mu a}(1 - \lambda f(\beta) + O(\lambda^3)), \\ \Omega &= -\frac{F}{6\pi\mu a^2}(\lambda^2 g(\beta) + O(\lambda^4)). \end{aligned} \tag{8.48}$$

If a particle is non-spherical, the general expressions for components of force and torque, acting on it, are

$$\begin{aligned} F_i &= -\mu(A_{ij}U_j + B_{ij}\Omega_j), \\ G_i &= -\mu(C_{ij}U_j + D_{ij}\Omega_j), \end{aligned} \tag{8.49}$$

where A_{ij}, B_{ij}, C_{ij}, D_{ij} are tensors depending on particles' form and size.

Experimental studies of particles' motion in a pipe have shown that the particles suspended in the flow of liquid, move in radial direction. Thus, if the particles' density is greater then density of the liquid, then such particles migrate in ascending flow in radial directions to the pipe's wall. The same effect occurs in descending flow, when particles' density is less than density of liquid. In particular, the particles freely suspended in the flow gather at the pipe wall at distances from 0.5 R up to 0.6 R from the pipe axis. It has been found, that if the suspension containing solid particles moves so that particles migrate from the wall to the axis of the pipe, then a layer of liquid free from particles forms near the wall, and its thickness δ_∞ depends on volume concentration of particles W (Fig. 8.4).

Theoretical research has shown that the radial migration of particles is caused by the influence of nonlinear inertial terms in the equation of motion (8.39). For

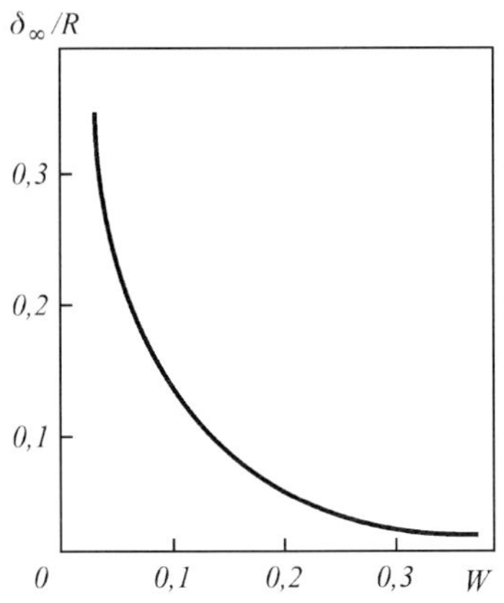

Fig. 8.4 Thickness dependence of clean layer of liquid at pipe wall on volume concentration of particles.

small values $\mathrm{Re}_U = aU/v$ the force $\boldsymbol{F}$ and torque $\boldsymbol{G}$ on a spherical particle moving with velocity $\boldsymbol{U}$ and spinning with angular velocity Ω are

$$\boldsymbol{F} = -6\pi\mu a\boldsymbol{U}\left(1 + \frac{3}{8}\mathrm{Re}_U\right) + \frac{\pi\mu a^3}{U}\mathrm{Re}_U\Omega \times \boldsymbol{U} + \mu a\boldsymbol{U}o(\mathrm{Re}_U),$$

$$\boldsymbol{G} = -8\pi\mu a^3\Omega(1 + o(\mathrm{Re}_U)). \tag{8.50}$$

The expression for the force contains two terms. The first term is the resistance due to translational motion of particles, and the second one represents a force, perpendicular to vectors Ω and $\boldsymbol{U}$. If $\boldsymbol{U} = \pm U\mathbf{k}$; this force is radial, which explains the lateral drift of particles at their motion with the flow in a pipe.

At present, there are numerous research works aimed at slowing down relative motion of two particles of different sizes. In these works practically all kinds of motion have been studied (see, for example, [7–9]).

The above-mentioned expressions for resistance force tested by a particle at slow motion in viscous liquid, are valid provided the particles are solid. In practice one deals not only with solid particles, but also with liquid and gaseous particles – drops and bubbles. Such particles in the flow of carrying liquid can be deformed under action of non-uniform velocity fields and pressures due to external and internal fluid flows. Deformation of comparatively large particles and the particles moving near the boundary of flow region (walls, interfaces), where the flow is changing significantly at distances comparable to particles' sizes, is especially noticeable. If liquid or gaseous particles are close to each other, then their relative motion produces strong hydrodynamic resistance force, depending on distance between their surfaces. In particular, at particles' approach along the centers' line, the resistance force at small clearance δ between surfaces grows as $1/\delta^\alpha$, where $\alpha = 1$ for solid particles and $\alpha = 0.5$ for liquid particles [8]. The degree of particles' deformation is determined by modified capillary number $\mathrm{Ca} = \mu_e Uab/(a+b)\Sigma$ [10], where μ_e is viscosity of carrying liquid; U is the approach velocity of drops with radii a and b; Σ is the coefficient of drops' surface tension. At Ca $\gg 1$ the drop deformation is small.

Slow motion of drops subject to deformations presents rather difficult mathematical problems because it involves solving the Stokes equation for external and internal liquids taking into account kinematic and dynamic conditions at originally unknown mobile interface of the drop. The method of solution of such problems is based on integral representation [11]

$$u_i(\mathbf{x}) = u_i^{\infty}(\mathbf{x}) - \frac{1}{8\pi\mu}\sum_{\alpha=1}^{N}\int_{S_\alpha} J_{ij}(\mathbf{x}-\mathbf{y}) f_j(\mathbf{y})\, dS_{\mathbf{y}}, \tag{8.51}$$

where $u_i(\mathbf{x})$ are velocity components of a liquid in point x exterior to particles; $u_i^{\infty}(\mathbf{x})$ are velocity components of liquid at a point x in the absence of particles; S_a is a surface of particle α; $\mathbf{y}$ is the coordinate of a point at the particle's surface; J_{ij} is the Green function of free (without particles) Stokes flow, equal to

$$J_{ij}(r) = \frac{\delta_{ij}}{r} + \frac{r_i r_j}{r^3};$$

$\boldsymbol{r} = \mathbf{x} - \boldsymbol{y}$, $r = |\mathbf{r}|$, $f_j(\boldsymbol{y})$ is the density of the i-th force component at a point $\boldsymbol{y}$ of particle surface. This force is expressed in terms of components of the stress tensor τ_{jk}:

$$f_j(\boldsymbol{y}) = \tau_{jk}(\boldsymbol{y}) n_k(\boldsymbol{y}),$$

where $n_k(\boldsymbol{y})$ are the components of normal vector, exterior to S_a at point $\boldsymbol{y}$. Summation in the right part is carried out over all N particles.

The total force $\boldsymbol{F}$ and angular momentum $\boldsymbol{T}$ acting on a particle α are determined by the equations

$$F_i^\alpha = -\int_{S_\alpha} f_i(\boldsymbol{y})\, dS_{\boldsymbol{y}}, \quad T_i^\alpha = -\int_{S_\alpha} \varepsilon_{ijk}(\boldsymbol{y}_j - \mathbf{x}_j^\alpha) f_k(\boldsymbol{y})\, dS,$$

where $\mathbf{x}_j^\alpha$ are coordinates of particles' centers, ε_{ijk} is the Levi-Civitta symbol, is zero if among indexes i, j, k there are at least two equal, is 1 if indexes form cyclic permutation, and is -1 in others cases. This method is known as method of boundary integral equations. A review of appropriate works is given in [12].

Up to now we were limited to considering the behavior of isolated particles or hydrodynamic interactions of two particles (pair interactions). However, in concentrated suspensions multi-particle interactions are possible, and one cannot neglect them. The problem of determining force and torque acting on a particle in suspension in the presence of other particles is very difficult and so far has not been completely solved. However, in the last years, a significant progress [13] in this direction has been made. One promising approach to the problem of Brownian diffusion in shear flow, which is being developed now, will be outlined at the end of the following section. Here we consider the contribution to tensors of friction $\boldsymbol{f}$ and mobility $\boldsymbol{V}$, obtained from the account taken of multi-particle (more than two) hydrodynamic interactions.

In the Ref. [14], the method of reflections was applied to calculations of three-particle and four-particle interactions. It was shown that, as compared to pair interactions, three- and four-particle interactions introduce corrections of the order $0(1/r^4)$ and $0(1/r^7)$ to the corresponding velocity perturbations, where r is the characteristic distance between particles. A generalization for the N-particle case was made in [15]. The velocity perturbation is found to be of the order $0(1/r^{-3N+5})$. In the same work, expressions for the mobility functions are derived up to the terms of order $0(1/r^7)$. It should be kept in mind that the corresponding expressions are power series in $1/r$, so to calculate the velocities at small clearances between particles (it is this case has presents the greatest interest), one has to take into account many terms in the series, or to repeat the procedure of reflection many times. In addition to analytical solutions, numerical solutions of a similar problem are available, for example, in [16]. At small clearances between particles, the application of numerical methods is complicated by the need to increase the number of elements into which particle surfaces are divided in order to achieve acceptable accuracy of the solution.

[17] uses two approaches to the problem of many-particle interactions in concentrated suspensions. Both are based on the assumption of additivity of velocities and forces in pair interactions. The first method focuses on determination of the mobility tensor, and the second one – of the tensor of friction. As was mentioned earlier, the proper account of interactions involving three or more particles in mobility calculations necessitates corrections up to the order of $0(1/r^4)$, and in the calculation of the friction tensor components – up to the order of $0(1/r^2)$. Therefore, for concentrated suspensions in which the clearance between particles is small, it is better to use the second approach, whereas for low-concentrated suspensions, the first approach is preferable. The review [13] contains additional information on the above-considered problems.

8.2 Brownian Motion

We begin with consideration of particles with low volume concentration buoyant in a quiescent liquid or a liquid undergoing translational motion with constant velocity.

Random thermal motion of small particles buoyant in the liquid is known as Brownian motion. In the absence of external forces acting on particles (which can have an arbitrary size), the particles have equal kinetic energy of thermal motion $3kT/2$, where k is Boltzmann's constant, T is the absolute temperature.

Therefore, for any particle, it is possible to write

$$\frac{1}{2}m\langle U^2\rangle = \frac{3}{2}\mathrm{k}T, \tag{8.52}$$

which leads to

$$\langle U^2\rangle = 3\mathrm{k}T/m. \tag{8.53}$$

Here m is the mass of the particle, $\langle U^2\rangle$ – the mean square of particle's velocity. However, the observed average velocity of the particle's motion is smaller than the velocity given by (8.53). In particular, for a particle of radius 1 μm and density equal to that of water, the value $\sqrt{\langle U^2\rangle}$ is equal to 1.7 mm/s at the room temperature. In his studies of Brownian motion, Einstein [18] concluded that, since in the process of thermal motion the particle collides with the molecules of the ambient liquid and changes its direction about a million times per second, the fine details of the particle's actual trajectory are impossible to observe visually. Einstein took the mean square of particle displacement per unit time as the observable parameter that could be determined from experiment and compared to its theoretical value. This parameter has the meaning of diffusion coefficient. Thus, we conclude that the Brownian motion of particles suspended in a liquid can be regarded as diffusion.

There are two ways to define the diffusion coefficient. Consider them in succession. For simplicity, begin with the one-dimensional case, that is, with the problem of one-dimensional random walk of a particle. The probability of the particle's displacement lying in the range $(x, x + dx)$ after n random displacements with step l, is given by the Gaussian distribution

$$P(n, x)\, dx = (2\pi n l^2)^{-1/2} e^{-x^2/2\pi l^2}. \tag{8.54}$$

Consider now the process of diffusion in a thin layer of liquid, into which a substance (particles) with concentration C_0 is introduced at the initial moment $t = 0$, at the point $x = 0$. The substance diffuses in the liquid. In view of the analogy between diffusion and random walk of particles, it is assumed that in time t, the particle makes n displacements, where the number n is proportional to t:

$$n = Kt. \tag{8.55}$$

The random motion of particles along the x-axis results in their scattering, so the concentration at the point x at the moment of time t will be equal to

$$C = C_0 P(x, t). \tag{8.56}$$

On the other hand, if the problem of diffusion is considered, then C satisfies the one-dimensional diffusion equation

$$\frac{dC}{dt} = D\frac{d^2C}{dx^2}, \tag{8.57}$$

which in the case of injection of the substance at the initial moment at the point $x = 0$ has the following solution:

$$C = \frac{C_0}{2\sqrt{\pi Dt}} e^{-x^2/4Dt}. \tag{8.58}$$

Comparing the expressions (8.54) to (8.56) with (8.58), we find that if

$$K = -2D/t^2. \tag{8.59}$$

then in a mixture containing non-interacting particles, the random thermal motion of these particles is analogous to the molecular diffusion in a binary infinitely dilute solution with the diffusion coefficient D.

For spatial Brownian motion, the formula (8.58) becomes

$$C = \frac{C_0}{8(\pi Dt)^{3/2}} e^{-r^2/4Dt}. \tag{8.60}$$

The probability to find the particle in the distance range $(r, r + dr)$ from the origin at the moment t is given by

$$P(r,t)\,dr = \frac{C}{C_0} 4\pi r^2\,dr = \frac{1}{2(\pi D^3 t^3)^{1/2}} e^{-r^2/4Dt} r^2\,dr. \tag{8.61}$$

It is obvious that the mean displacement is equal to zero, because the particle can be displaced in the positive and the negative direction with equal probability. Therefore $\langle r \rangle$ cannot characterize the motion of a particle. The appropriate characteristic is $\sqrt{\langle r^2 \rangle}$. It can be derived by using the distribution (8.61):

$$\langle r^2 \rangle = \int_0^\infty r^2 P(r,t)\,dr = 6Dt. \tag{8.62}$$

The coefficient D is known as the translational coefficient of diffusion.

So far, only random translational motions of particles were considered. However, particles can be involved in a random rotational motion simultaneously with the translational motion. Denote by s the angle of random turn between two consecutive positions of the particle. By reasonings similar to the ones carried out above, we arrive at the formula

$$\langle s^2 \rangle = 4D_{rot}t. \tag{8.63}$$

The coefficient D_{rot} is known as the rotational diffusion coefficient.

Proceed now to define the diffusion coefficient.

The system "liquid – particles" comes to a stationary state when the flux due to the translational diffusion is counterbalanced by the convective flux caused by the stationary hydrodynamic force acting on each particle. This force can be represented by the Stokes force

$$\boldsymbol{F} = -\bar{f}\boldsymbol{U} = -\boldsymbol{U}/\bar{\mathrm{v}}, \tag{8.64}$$

where $\bar{f}$ is the mean friction coefficient in translational motion; $\bar{\mathrm{v}}$ is the mean mobility.

Since particle sizes are small, it is possible to consider their motion as inertia-less, so the particles acquire their velocities very quickly, in spite of the frequent changes of the direction of motion.

Denote as n the particles number density. Then the condition of equilibrium is

$$-D\nabla n = n\boldsymbol{U} \tag{8.65}$$

or, in view of (8.64)

$$\bar{f}D\nabla(\ln n) = \boldsymbol{F}. \tag{8.66}$$

On the other hand, the system is in thermodynamic equilibrium. Einstein had assumed that a particle is in the thermodynamic equilibrium if the hydrodynamic force acting on it is balanced by thermodynamic force. The potential of this force is equal to Gibbs's free energy per particle.

As was already stressed earlier, a mixture consisting of liquid and colloidal particles or macromolecules of small volume concentration, could be considered as completely diluted solution because the particles' interactions are absent. To such solution the restricted thermodynamic relations for an ideal solution can be applied [19]. In particular, Gibbs's free energy per molecule of the dissolved substance, is equal to

$$\frac{G}{N_2} = A(p, T) - \mathrm{k}T \ln n. \tag{8.67}$$

Therefore the thermodynamic force acting on one particle is

$$\boldsymbol{F}_{therm} = -\nabla\left(\frac{G}{N_2}\right) = \mathrm{k}T\nabla \ln n. \tag{8.68}$$

Putting (8.68) into (8.66), we obtain the expression for the coefficient of diffusion transfer

$$D_{br} = \frac{\mathrm{k}T}{\bar{f}} = \bar{\mathrm{v}}\mathrm{k}T. \tag{8.69}$$

If the particle is a solid sphere of radius a, then

$$D_{br} = \mathrm{k}T/6\pi\mu a. \tag{8.70}$$

This expression is known as the Stokes–Einstein equation.
The appropriate expression for coefficient of rotational diffusion is

$$D_{rot} = \mathrm{k}T/\bar{f}_{rot}. \tag{8.71}$$

In particular, for a spherical particle

$$D_{rot} = \mathrm{k}T/8\pi\mu a^3. \tag{8.72}$$

Another approach [20] uses the description of particle's motion by the Langevin equation:

$$m\frac{d^2\mathbf{r}}{dt^2} = \mathbf{G}(t) - \boldsymbol{F}. \tag{8.73}$$

There are two terms in the right part of this equation. The first one, $\mathbf{G}(t)$, is the random force due to the molecules of the liquid. This force changes in magnitude and direction in a random manner, and the characteristic time of this change is very small ($\sim 10^{-13}$ s). The second term $\boldsymbol{F}$ is the force of hydrodynamic resistance to the particle's motion. Using the expression for $\boldsymbol{F}$ in Stokes approximation, we obtain

$$m\frac{d^2\mathbf{r}}{dt^2} = \mathbf{G}(t) - f\frac{d\mathbf{r}}{dt}. \tag{8.74}$$

It is pointless to average Eq. (8.74), since the average values of both velocity and acceleration are equal to zero. Let us therefore multiply both parts of the equation by $\boldsymbol{r}$, forming scalar products,

$$\frac{m}{2}\frac{d^2(\mathbf{r}^2)}{dt^2} - m\left(\frac{d\mathbf{r}}{dt}\right)^2 = \mathbf{r}\cdot\mathbf{G}(t) - \frac{\bar{f}}{2}\frac{d\mathbf{r}^2}{dt},$$

and then average the resulting expression:

$$\frac{m}{2}\frac{d^2\langle\mathbf{r}^2\rangle}{dt} + \frac{\bar{f}}{2}\frac{d\langle\mathbf{r}^2\rangle}{dt} = 3kT. \tag{8.75}$$

In derivation of (8.75), the term $\langle\mathbf{r}^2\cdot\mathbf{G}(t)\rangle$ was neglected because the characteristic time of fluctuation $\mathbf{G}(t)$ is small, and $m\langle(d\mathbf{r}/dt)^2\rangle = 3\mathrm{k}T$ is two times the kinetic energy of the particle's thermal motion.

Integration of Eq. (8.75), with the condition $\langle \mathbf{r}^2 \rangle = 0$ at $t = 0$, yields

$$\frac{d\langle \mathbf{r}^2 \rangle}{dt} = \frac{3kT}{\bar{f}}\left(1 - e^{-\bar{f}t/m}\right). \tag{8.76}$$

from which it follows that the characteristic time required for the establishment of $\langle \mathbf{r}^2 \rangle$ is equal to $t \sim m/\bar{f}$. After this time, $d\langle \mathbf{r}^2 \rangle/dt \sim 6\mathrm{k}T/\bar{f}$ and

$$\langle \mathbf{r}^2 \rangle \sim \frac{6\mathrm{k}T}{\bar{f}}t. \tag{8.77}$$

But according to (8.62), $\langle \mathbf{r}^2 \rangle \sim 6Dt$. Therefore

$$D = \mathrm{k}T/\bar{f}, \tag{8.78}$$

which coincides with the expression (8.69) obtained earlier.

Thus the coefficient of Brownian diffusion of particles with small volume concentration W, suspended in a liquid that is at rest or undergoing translational motion with a constant velocity, has a constant value and is identical in all directions.

Consider now the case when the liquid moves with a non-uniform profile of velocity, which can happen, for example, during the flow in a pipe (Poiseuille flow) or in a channel whose walls are moving with different velocities (Couette flow) [21]. Since the size of particles is very small, at distances of about several particle sizes, the non-uniform velocity profile can be regarded as linear, and the flow can be regarded as a shear flow. For simplicity, consider the flow to be two-dimensional. Inasmuch as $W \ll 1$, it is possible to be limited to the consideration of one particle, as before. Denote by $\boldsymbol{V} = (\alpha y, 0)$ the velocity of the liquid at the point $\boldsymbol{X} = (x(t), y(t))$, and by $\boldsymbol{U} = (u(t), \mathrm{v}(t))$ – the velocity of the particle.

Particle motion occurs under the action of hydrodynamic resistance force $f(\boldsymbol{U} - \mathbf{V})$ and the random Brownian force $\mathbf{G}(t)$, therefore the Langevin equation (8.73) can be written as

$$m\frac{d\boldsymbol{U}}{dt} = -f(\boldsymbol{U} - \mathbf{V}) + \mathbf{G}(t), \tag{8.79}$$

where f is the coefficient of hydrodynamic resistance for the particle; m is the particle mass.

As for the random force $\mathbf{G}(t)$, it is supposed to be Gaussian, therefore its average value and the mutual correlation function are equal to [22]

$$\langle \mathbf{G}_i(t) \rangle = 0, \quad \langle \mathbf{G}_i(t_1)\mathbf{G}_i(t_2) \rangle = 2\mathrm{k}Tf\delta_{ij}\delta(t_1 - t_2), \tag{8.80}$$

where G_i is i-th component of the force $\mathbf{G}(i, j = x, y)$; δ_{ij} is Kroneker's symbol; $\delta(t)$ is Dirac's delta function. Notice that the averaging is carried out over the ensemble of random forces acting on the particle.

Let the considered particle be initially located at the origin (0, 0). Then Eq. (8.79) gives:

$$x(t) = \frac{1}{\gamma}u(0)(1 - e^{-\gamma t}) + \frac{\alpha}{\gamma}\mathrm{v}(0)t(1 + e^{-\gamma t})$$

$$- \frac{2\alpha}{\gamma^2}\mathrm{v}(0)(1 - e^{-\gamma t}) + \frac{\alpha}{\gamma}\int_0^1 \mathrm{d}\tau \int_0^\tau (1 + e^{-\gamma(t-\sigma)})P_y(\sigma)\,\mathrm{d}\sigma$$

$$- \frac{2\alpha}{\gamma^2}\int_0^t (1 - e^{-\gamma(t-\tau)})P_y(\tau)\,\mathrm{d}\tau + \frac{1}{\gamma}\int_0^t (1 - e^{-\gamma(t-\tau)})P_x(\tau)\,\mathrm{d}\tau, \tag{8.81}$$

$$y(t) = \frac{1}{\gamma}u(0)(1 - e^{-\gamma t}) + \frac{1}{\gamma}\int_0^t (1 - e^{-\gamma(t-\tau)})P_y(\tau)\,\mathrm{d}\tau, \tag{8.82}$$

where $\gamma = f/m$, $\mathbf{P}(t) = \mathbf{G}(t)/m$. Now average $x(t)$ and $y(t)$ over the initial velocities. Taking into account the conditions (8.80), we obtain

$$\langle x(t)\rangle = \frac{1}{\gamma}\langle u(0)\rangle_0(1 - e^{-\gamma t}) + \frac{\alpha}{\gamma}\langle \mathrm{v}(0)\rangle_0 t(1 + e^{-\gamma t}) - \frac{2\alpha}{\gamma^2}\langle \mathrm{v}(0)\rangle_0(1 - e^{-\gamma t}),$$

$$\langle y(t)\rangle = \frac{1}{\gamma}\langle \mathrm{v}(0)\rangle_0(1 - e^{-\gamma t}). \tag{8.83}$$

Here $\langle\ \rangle$ denotes the averaging over the random forces and initial velocities, and $\langle\ \rangle_0$ denotes the averaging over the initial velocities only.

To determine the coefficient of Brownian diffusion, it is necessary to find the root-mean-square (instead of the mean) displacement of the particle (see (8.62)). These values are easy to derive with the help of the expressions (8.81) and (8.82). Suppose the distribution of particles' initial velocities is Maxwellian, as in the collisionless discharged rarefied gas, with the zero average value (this can be the case if the velocity of the liquid along the x-axis is zero). Then

$$\langle u(0)\rangle_0 = \langle \mathrm{v}(0)\rangle_0 = \langle u(0)\mathrm{v}(0)\rangle_0 = 0,$$

$$\langle u^2(0)\rangle_0 = \langle \mathrm{v}^2(0)\rangle_0 = \mathrm{k}T/m. \tag{8.84}$$

For times $t \gg 1/\gamma$, the expressions (8.81) and (8.82) become simplified. Omitting long calculations, we present the final expressions for the root-mean-square displacement of the particle at $t \gg 1/\gamma$ without their derivation:

$$\langle x^2(t)\rangle = 2Dt + \frac{2}{3}D\alpha^2 t^3,$$

$$\langle y^2(t)\rangle = 2Dt, \tag{8.85}$$

where $D = \mathrm{k}T/f$ is the coefficient of Brownian diffusion for a quiescent liquid.

It follows from (8.85) that in the absence of shear ($\alpha = 0$), we have $\langle x^2(t)\rangle = \langle y^2(t)\rangle = 2Dt$, that is, the diffusion coefficients in the directions of the coordinates axes are equal to D. On the other hand, the shear flow of liquid results in anisotropy of Brownian diffusion, with different values for the diffusion coefficient along and perpendicular to the direction of flow of the carrying liquid. The respective diffusion coefficients are

$$D_x = \frac{\langle x^2(t)\rangle}{2t} = D\left(1 + \frac{1}{3}\alpha^2 t^2\right),$$

$$D_y = \frac{\langle y^2(t)\rangle}{2t} = D. \tag{8.86}$$

The considered cases are limited by the absence of interaction between the particles, which is true only for suspensions with very low volume concentrations of particles. An increase in volume concentration of particles results in a reduction of the average distance between them, and consequently, in the necessity to take into account the hydrodynamic interactions between particles. Also, for the Brownian motion of particles of a suspension that is located in a bounded region, for example, in a porous medium, one has to make a proper account of the particles' interaction with the boundary. A discussion addressing these issues can be found in the review [23].

The hydrodynamic interaction of particles of the same radius a in their Brownian diffusion in a quiescent liquid was considered in [24–26]. These papers introduce a factor λ into the coefficient of Brownian diffusion (8.70), in order to make a correction for the deviation of resistance to particles' motion from Stokes law

$$D = D_0/\lambda(r/a), \tag{8.87}$$

Here D_0 is the coefficient of Brownian diffusion determined by the formula (8.70), and corresponding to the free Brownian motion of particles. The factor λ depends on the relative distance between approaching particles and can be determined from the resistance law $F = 6\pi\mu a U\lambda(r/a)$, which is applicable to the relative motion of particles along their line of centers with the velocity U (see expression (8.36)).

To properly account for the interaction of particles of different radii in a moving and sufficiently diluted suspension (so that it is possible consider only pair interactions), it is convenient to introduce a pair (two-particle) distribution function $P(\mathbf{r})$ [27], having meaning of probability to find the center of the particle of radius a_1 at the end of the radius vector $\mathbf{r}$, given that the center of the second particle of radius a_2 is coincident with the origin of the chosen coordinate system. This function satisfies the Fokker-Planck quasi-stationary equation:

$$\nabla \cdot (P(\mathbf{r})\mathbf{V}_{12}(\mathbf{r})) = 0, \tag{8.88}$$

where $\mathbf{V}_{12}(\mathbf{r}) = \mathbf{V}_1 - \mathbf{V}_2$ is the relative velocity of the considered pair of particles.

In order to find $V_{12}(r)$, it is necessary to solve the appropriate hydrodynamic problem of slow inertialess relative motion of two interacting spherical particles freely suspended in the external liquid. If two particles are suspended in a boundless liquid that moves relative to the particles in such a way that at a sufficiently large distance from the particles, its velocity is equal to the one expected for a shear flow, the expression for the relative velocity of the particle pair has the form [27]

$$\begin{aligned} \boldsymbol{V}_{12}(\boldsymbol{r}) &= (\nabla \boldsymbol{V}_{\infty}) \cdot \boldsymbol{r} - (A\boldsymbol{nn} + B(\boldsymbol{I} - \boldsymbol{nn})) \cdot \boldsymbol{E} \cdot \boldsymbol{r} - D_{12}^{(0)}(\boldsymbol{nn}G \\ &\quad + (\boldsymbol{I} - \boldsymbol{nn})H \cdot \nabla(\ln(P(\boldsymbol{r}))) - \frac{D_{12}^{(0)}}{\mathrm{k}T}(\boldsymbol{nn}G + (\boldsymbol{I} - \boldsymbol{nn})H) \cdot \nabla\phi_{12}(\boldsymbol{r}), \end{aligned} \tag{8.89}$$

where $\boldsymbol{V}_{\infty}$ is the velocity of the liquid at the infinity; $\boldsymbol{E}$ is the rate-of-strain tensor; $\boldsymbol{n} = \boldsymbol{r}/r$ is the unit vector directed along the line of centers of the two particles; $\boldsymbol{I}$ is the unit tensor; $D_{12}^{(0)}$ is the coefficient of relative Brownian diffusion of two particles whose interactions are disregarded; $\phi_{12}(\boldsymbol{r})$ is the potential of molecular interaction of particles; k is Boltzmann's constant; T is the absolute temperature; A, B, G, and H are mobility functions depending on the ratio of particle radii $\bar{\lambda} = a_2/a_1$, on the ratio of internal and external liquid viscosities $\bar{\mu} = \mu_i/\mu_e$, and on the relative gap between particle surfaces $s = 2r/(a_1 + a_2)$. Expressions for the mobility functions and a review of the pertinent works are given in [27].

It is necessary to add the following boundary conditions to the Eq. (8.88). The first one is the requirement that each particle collision results in coagulation for solid particles or in coalescence for drops. Then

$$P = 0 \quad \text{at } r = a_1 + a_2. \tag{8.90}$$

If the particles are far apart, then

$$P \to 1 \quad \text{at } r \to \infty. \tag{8.91}$$

Second, the frequency of pair collisions is determined by diffusion flux through the spherical surface of radius $a_1 + a_2$ (in more detail about it, see Section 10):

$$J_{12} = -n_1 n_2 \int_{r=a_1+a_2} P\boldsymbol{V}_{12} \cdot \boldsymbol{n}\, ds, \tag{8.92}$$

where n_1 and n_2 are numerical concentrations of particles of the first and second type, respectively.

Since the numerical concentration n_{12} of particles of both types is determined by the pair distribution function

$$n_{12}(r,t) = n_1 n_2 P(r,t), \tag{8.93}$$

the Eq. (8.88) reduces to

$$\nabla \cdot (n_{12}\mathbf{V}_{12}(r)) = 0,$$
$$n_{12}(r \to a_1 + a_2) = 0; \quad n_{12}(r \to \infty) \to n_1 n_2. \tag{8.94}$$

The Eq. (8.94) is similar to the equation of stationary diffusion. Relative contributions of convection and diffusion terms are characterized by the diffusion Peclet number

$$\mathrm{Pe}_D = \frac{|\nabla V_\infty|\bar{a}^2}{D_{12}^{(0)}}; \quad \bar{a} = \frac{a_1 + a_2}{2}. \tag{8.95}$$

The majority of publications on Brownian motion are limited to the cases $\mathrm{Pe}_D \ll 1$ (purely Brownian diffusion) or $\mathrm{Pe}_D \gg 1$ (particle interaction without taking into account Brownian diffusion). The case of arbitrary Peclet number is considered in [28].

Besides hydrodynamic interaction between particles, molecular and electrostatic interactions are also possible. The reader interested in this problem may turn to works [22–29].

Of the particular value is the case of concentrated suspensions for which the volume concentration of disperse phase is not small. The microstructure of such suspensions depends on relations between hydrodynamic forces of particle interactions and thermodynamic forces causing Brownian motion. In the last years the research of dynamics of concentrated suspensions (Stokes's dynamics [30]) was based on use of the Langevin equation for ensemble of N particles

$$m_i \frac{d\boldsymbol{U}_i}{dt} = \boldsymbol{F}_H^i + \boldsymbol{F}_B^i + \boldsymbol{F}_P^i \quad (i = \overline{1, N}), \tag{8.96}$$

where m_i is the mass of i-th particle; $\boldsymbol{F}_H^i$ is the hydrodynamic force acting on i-th particle; $\boldsymbol{F}_B^i$ is stochastic force from molecules of liquid surrounding the particle; $\boldsymbol{F}_P^i$ is the interaction force of non-hydrodynamic character, for example the force of molecular interaction of particles or external force.

The expression for the hydrodynamic force in the case of shear flow at small Reynolds numbers $\mathrm{Re} = \rho a^2\dot{\gamma}/\mu \ll 1$ has the form

$$\boldsymbol{F}_H^i = -\boldsymbol{f}_{FU}^i \cdot (\boldsymbol{U}^i - \boldsymbol{U}_0^i) + \boldsymbol{f}_{FE}^i : \boldsymbol{E}, \tag{8.97}$$

where $\boldsymbol{U}_0^i = \Gamma^i \cdot \boldsymbol{x}^i$ is the velocity of liquid at the point occupied by the centre of i-th particle; $|\Gamma^i| = \dot{\gamma}$ is the rate of shear; $\boldsymbol{E}$ is the rate-of-strain tensor of the liquid; $\boldsymbol{f}_{FU}$ and $\boldsymbol{f}_{FE}$ are, respectively, the tensors of friction for the particle motion relative to the liquid, and together with the liquid. The methods of determining these values and the resulting expressions for many-particle interaction in the limiting case of small gaps between particles can be found in works [13, 31].

The stochastic Brownian force satisfies the conditions

$$\langle \boldsymbol{F}_B \rangle = 0 \quad \langle \boldsymbol{F}_B(0)\boldsymbol{F}_B(t) \rangle = 2\mathrm{k}T\boldsymbol{f}_{FU}\delta(t) \tag{8.98}$$

where k is Boltzmann's constant; T – the absolute temperature; δ – the delta function. The first condition (8.98) is self-obvious, and the second one is the consequence of the theorem about dissipation of fluctuations in a system containing N particles. This condition was used earlier (see (8.80)).

From the Eqs. (8.96) there follow the basic dimensionless parameters influencing the system's evolution: the Peclet number $\mathrm{Pe} = \dot{\gamma}a^2/D_{\delta p} = 6\pi\mu a^3\dot{\gamma}/\mathrm{k}T$ that gives the ratio of the hydrodynamic force of shear flow to the thermodynamic Brownian force; dimensionless shear rate $\bar{\dot{\gamma}} = 6\pi\mu a^2\dot{\gamma}/|F_p|$, equal to the ratio of the hydrodynamic force of shear flow to the force of non-hydrodynamic interaction; and the volume concentration of particles W.

The symbol $\langle\ \rangle$ in Eq. (8.98) stands for the averaging of random forces over the ensemble. As a result of integration of Eqs. (8.96) in the range $\Delta t \gg \tau = m/\pi\mu a$, where τ is the characteristic time of establishment of $\langle \mathbf{r}^2 \rangle$ (see (8.76)), is possible to determine the displacement of particles Δx^i during the time Δt.

The macroscopic parameters can be determined by averaging over the particle ensemble, and the time-averaging.

In the same fashion as it was done earlier for a single particle in a quiescent liquid, it is possible to find the coefficient of Brownian diffusion

$$\boldsymbol{D}_{\delta p} = \mathrm{k}T\boldsymbol{f}_{FU}^{-1}. \tag{8.99}$$

However, in the present case, this coefficient is not a scalar, but a tensor of the second rank.

For concentrated suspensions and incompressible liquids, the phenomenological relation connecting the stress tensor and the rate-of-strain tensor is

$$\langle \boldsymbol{T} \rangle = -p\boldsymbol{I} + 2\mu\boldsymbol{E} + \langle \boldsymbol{E}_P \rangle \tag{8.100}$$

A comparison with Navier-Stokes law (4.13) shows that the presence of particles in a liquid results in additional stresses characterized by $\langle \boldsymbol{E}_P \rangle$. The corresponding term can be presented as

$$\begin{aligned}
\langle \boldsymbol{E}_P \rangle &= -n\mathrm{k}T\boldsymbol{I} + n\{\langle \boldsymbol{S}_H \rangle + \langle \boldsymbol{S}_P \rangle + \langle \boldsymbol{S}_B \rangle\},\\
\langle \boldsymbol{S}_H \rangle &= -\langle \boldsymbol{f}_{SU} \cdot \boldsymbol{f}_{FU}^{-1} \cdot \boldsymbol{f}_{FE} - \boldsymbol{f}_{SE} \rangle : \langle \boldsymbol{E} \rangle,\\
\langle \boldsymbol{S}_P \rangle &= -(\langle \boldsymbol{f}_{SU} \cdot \boldsymbol{f}_{FU}^{-1} + \boldsymbol{x}\boldsymbol{I}) \cdot \boldsymbol{F}_P \rangle,\\
\langle \boldsymbol{S}_B \rangle &= -\mathrm{k}T\langle \nabla \cdot (\boldsymbol{f}_{SU} \cdot \boldsymbol{f}_{FU}^{-1}) \rangle.
\end{aligned} \tag{8.101}$$

The first term in the right-hand side of (8.101) represents the isotropic stress similar to pressure and caused by the thermal (Brownian) motion of particles with number concentration n. The second term can be decomposed into three parts, each part characterizing a specific contribution to the total stress: $\langle \boldsymbol{S}_H \rangle$ is the hydrodynamic component due to the shear flow of the carrying liquid; $\langle \boldsymbol{S}_P \rangle$ is the contribution from non-hydrodynamic interactions of particles; $\langle \boldsymbol{S}_B \rangle$ is the deviatoric part of the stress due to the Brownian motion. Expressions for the friction tensors $\boldsymbol{f}_{SU}$, $\boldsymbol{f}_{FU}$, $\boldsymbol{f}_{FE}$ and tensor $\boldsymbol{f}_{SE}$ figuring in (8.101), are obtained in works [13, 31]. Using the solution of Eq. (8.96), it is possible to find from (8.101) the rheological properties of suspension. In particular, having set the initial configuration of suspension, one can calculate its further evolution and predict the emergence of aggregations and other complex structures [30].

8.3
Viscosity of Diluted Suspensions

Consider an extremely diluted suspension consisting of a viscous liquid and small spherical particles suspended in it. Small volume concentration of particles allows to neglect their interactions and to assume that each particle behaves as if it were surrounded by an infinite volume of pure liquid. It is obvious that with the increase of volume concentration of particles their mutual influence will become more and more prominent, and at a certain point it will become impossible to neglect it. Further on, for simplicity, the Brownian motion will be neglected. Besides, we assume that particles are small, so it is possible to neglect the influence of gravity and consider their motion as inertialess. It means that the velocity of particle motion is equal to the velocity of the flow, in other words, the particles are freely suspended in the liquid.

The presence of particles in the liquid causes a disturbance of the velocity field that would establish in the liquid in the absence of particles. It is known [32] that for Stokes translational motion of an isolated solid sphere in an unbounded volume of viscous liquid, hydrodynamic disturbances of the velocity field attenuate with the increase of r as $1/r$. It is a sufficiently slow attenuation, which causes mathematical complications when we want to find disturbances caused by the presence of a large number of particles in the liquid. In particular, it results in slowly convergent (and sometimes even divergent) integrals.

The addition of solid particles to a pure liquid results in an increase of viscous dissipation inasmuch as the total solid surface area increases. Therefore, if the suspension is to be considered as a Newtonian liquid, its viscosity should be greater than that of a pure liquid. Einstein [18] was the first to consider the problem of finding the viscosity of a suspension. He limited his analysis to the case of the Couette current of an infinitely dilute suspension.

We mentioned in Section 4.2 that the Couette flow (shear flow) is characterized by the linear structure of the velocity field (Fig. 8.5), which in the Cartesian system of coordinates looks like

$$u_0 = \dot{\gamma}_0 y, \quad v_0 = 0, \quad w_0 = 0, \tag{8.102}$$

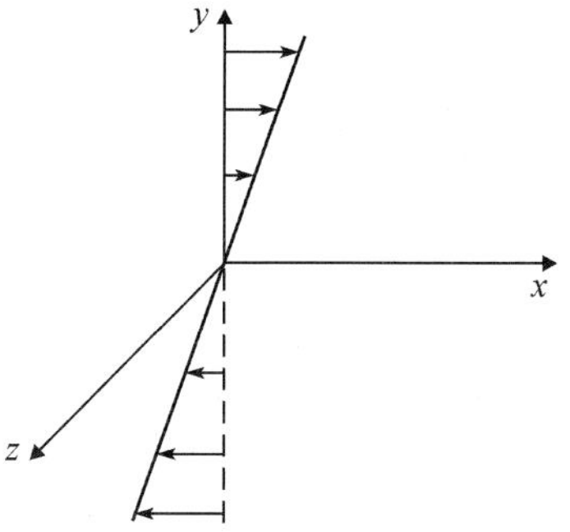

Fig. 8.5 The velocity profile in a shear flow.

where the shear rate is

$$\dot{\gamma}_0 = 2\varepsilon_{yx} = \frac{du_0}{dy} = \frac{U_0}{h}. \tag{8.103}$$

Here h is the distance between the walls, one of which moves parallel relative to the other with the velocity U_0, which is taken to be constant.

The corresponding shear stress τ_0 and viscous dissipation Φ_0 are equal to

$$\tau_0 = \tau_{yx} = \mu\frac{du_0}{dy} = \mu\frac{U_0}{h}, \tag{8.104}$$

$$\Phi_0 = \tau_{yx}\varepsilon_{yx} = \mu\left(\frac{du_0}{dy}\right)^2 = \mu\left(\frac{U_0}{h}\right)^2. \tag{8.105}$$

The presence of particles in the liquid complicates the hydrodynamic problem very considerably, because in order to determine the velocity field, it is necessary to solve the boundary value problem with the additional boundary "conditions of sticking" at the surface of each particle (i.e., zero relative velocity of liquid at the particle's surface). Physically, the problem reduces to the determination of disturbances that arise in the velocity field of a pure liquid due to the presence of particles in the liquid. The assumption of low volume concentration of particles allows us to begin with finding the disturbance provoked by one particle. Then, because in the considered approximation, particles do not interact with each other, the total disturbance will be the superposition of disturbances, assuming a uniform particle distribution in the liquid volume. The solution, which was obtained in [33], is presented below.

Consider a solid spherical particle suspended in the Couette flow of a viscous liquid. There are two boundary conditions: the conditions of sticking at the surface of the particle, and the equality of the flow velocity far from the particle to the velocity given by the expression (8.102). The translational velocity of the particle's center $\boldsymbol{u}_0$ is equal to the velocity of an unperturbed flow of liquid at the point of space occupied by the particle center. In the system of coordinates bound to the particle center and moving with the velocity $\mathbf{u}_0$, the velocity components of the fluid flow are given by (8.102) (see Fig. 8.5). In the chosen system of coordinates, the translational velocity of the particle is equal to zero, but the particle can rotate around the z-axis with the angular velocity

$$\omega_0 = -\frac{1}{2}|\nabla \times \mathbf{u}_0| = \frac{1}{2}\left(\frac{\partial \mathrm{v}_0}{\partial x} - \frac{\partial u_0}{\partial y}\right) = -\frac{\dot{\gamma}_0}{h}. \tag{8.106}$$

The minus sign is caused by the fact that, looking from the positive direction of the z-axis, the particle's rotation occurs clockwise.

It follows from the form of the angular velocity that the velocity at a point on the spherical surface is equal to

$$\mathbf{u}_r = \omega_r \times \mathbf{r}$$

and

$$u_r = \frac{1}{2}\dot{\gamma}_0 x, \quad \mathrm{v}_r = -\frac{1}{2}\dot{\gamma}_0 x, \quad w_r = 0. \tag{8.107}$$

The velocity field at the spherical surface corresponds to influence of an unperturbed external flow on the sphere. In its turn, the rotation of the sphere influences the distribution of velocities in the liquid. Let these perturbations $\mathbf{u}' = (u', \mathrm{v}', w')$ be small in comparison with the unperturbed velocity. The velocity of the liquid is now equal to

$$\mathbf{u}_0 + \mathbf{u}' = (u_0 + u', \mathrm{v}', w'), \tag{8.108}$$

In addition, we should have $\mathbf{u}' \cdot \mathbf{n} = 0$ on the sphere, while far from the sphere $(r \to \infty)$, $\mathbf{u}' \to 0$ should hold.

The solution of the Stokes equations with the given boundary conditions results in the following expressions for the components of perturbed velocity

$$\begin{aligned}
u' &= -\frac{5}{2}\frac{a^3\dot{\gamma}_0 x^2 y}{r^5} + \frac{1}{6}\dot{\gamma}_0 a^5\left(\frac{3y}{r^5} - \frac{15x^2 y}{r^7}\right),\\
\mathrm{v}' &= -\frac{5}{2}\frac{a^3\dot{\gamma}_0 x y^2}{r^5} + \frac{1}{6}\dot{\gamma}_0 a^5\left(\frac{3x}{r^5} - \frac{15xy^2}{r^7}\right),\\
w' &= -\frac{5}{2}\frac{a^3\dot{\gamma}_0 xyz}{r^5} + \frac{1}{6}\dot{\gamma}_0 a^5\left(\frac{15xyz}{r^7}\right),
\end{aligned} \tag{8.109}$$

where a is the particle radius and $r = (x^2 + y^2 + z^2)^{1/2}$.

If we choose to consider only the perturbations taking place far away from the particle, then at $a/r \ll 1$, we may leave only the first terms in the right-hand side

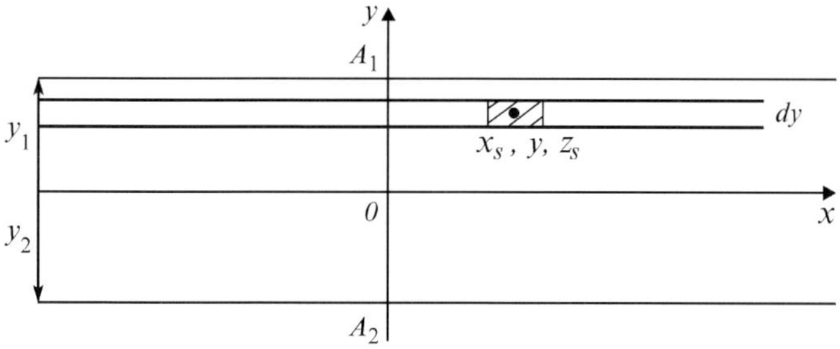

Fig. 8.6 To the calculation of viscosity of an infinite, diluted suspension.

of the expressions (8.109)

$$u' \approx -\frac{5}{2}\frac{a^3\dot{\gamma}_0 x^2 y}{r^5}, \quad v' \approx -\frac{5}{2}\frac{a^3\dot{\gamma}_0 x y^2}{r^5}, \quad w' \approx -\frac{5}{2}\frac{a^3\dot{\gamma}_0 xyz}{r^5}. \tag{8.110}$$

It follows from (8.110) that velocity perturbations in the liquid attenuate with the distance from the particle as a/r^2.

Consider now the planes $y = y_1$ and $y = -y_2$ (Fig. 8.6). The unperturbed velocities of the liquid at the points belonging to these planes are equal to $u_{01} = \dot{\gamma}_0 y_1$ and $u_{02} = -\dot{\gamma}_0 y_2$. Let solid spherical particles be present in the liquid and be homogeneously distributed in the volume with the number concentration n. At an arbitrary y, select a suspension layer of thickness dy parallel to the plane xz. Denote by x_s, y, z_s the particle center coordinates in this layer. The perturbation of the flow velocity due to a particle is given by the expression (8.110). In particular, at the point $A_1(0, y_1, 0)$

$$u'_s = -\frac{5}{2}\frac{a^3\dot{\gamma}_0 x_s^2 (y_1 - y)}{r_s^5}, \quad r_s^2 = x_s^2 + (y_1 - y)^2 + z_s^2. \tag{8.111}$$

In the small volume shaded in Fig. 8.6, there are $n\,dx_s\,dy\,dz_s$ particles. Therefore the x-th component of velocity perturbation due to all these particles is equal to $u' n\,dx_s\,dy\,dz_s$. Integrating this expression over all layers, we get the change of the longitudinal velocity at the point A_1 caused by presence of particles in the considered liquid layer

$$\Delta u_1 = -\frac{5}{2}\dot{\gamma}_0 a^3 (y_1 - y) n\,dy \int_{-\infty}^{\infty}\int_{-\infty}^{\infty} \frac{x_s^2}{r_s^5}\,dx_s\,dz_s. \tag{8.112}$$

The integral is calculated easily, with the final result

$$\Delta u_1 = -\frac{5}{2}\pi\dot{\gamma}_0 a^3 n\,dy. \tag{8.113}$$

The change in the longitudinal velocity is negative, in other words, the presence of solid particles reduces the velocity of the liquid in the x-direction. This change does not depend on layer's position, i.e., on y and on the choice of point A_1 on the $y = y_1$ plane.

The change of the flow velocity at point A_2 on the $y = -y_2$ plane is determined in the same manner

$$\Delta u_2 = \frac{5}{3}\pi\dot{\gamma}_0 a^3 n\,dy. \tag{8.114}$$

The velocity at the top plane relative to the velocity at the bottom plane in the unperturbed flow is equal to $u_1 - u_2 = \dot{\gamma}_0(y_1 + y_2)$. The presence of solid particles changes this velocity by the amount

$$\Delta u_{1,2} = \Delta u_1 + \Delta u_2 = -\frac{10}{3}\pi\dot{\gamma}_0 a^3 n\,dy. \tag{8.115}$$

Integrating (8.115) over all y from y_2 to y_1, we now obtain the expression for the relative velocity of two planes

$$u_{1,2} = \dot{\gamma}_0(y_1 + y_2)\left(1 - \frac{10}{3}\pi a^3 n\right). \tag{8.116}$$

The volume concentration of particles is equal to

$$W = \frac{4}{3}\pi a^3 n, \tag{8.117}$$

therefore (8.116) can be written in the form

$$u_{1,2} = \dot{\gamma}_0(y_1 + y_2)(1 - 2.5W). \tag{8.118}$$

Determine now the additional shear stress acting on the plane $y = y_1$ according to the formula

$$\tau_1' = \mu\frac{du'}{dy}. \tag{8.119}$$

Taking the expression (8.111) for u', we then obtain the shear stress due to the influence of one particle

$$\tau_1' = \frac{5}{2}\dot{\gamma}_0 a^3\left(\frac{x_s^2}{r_s^5} - \frac{5x_s^2(y_1 - y)^2}{r_s^7}\right). \tag{8.120}$$

Integrating (8.120) over all particles and over all layers, we get $\tau_1' = 0$. Similarly, we have $\tau_2' = 0$ at $y = -y_2$. Hence, addition of particles to the liquid reduces the velocity of the liquid, but does not affect the shear stress on the planes $y = y_1$ and $y = y_2$.

Now, go back to the Couette flow. Let the planes $y = y_1$ and $y = y_2$ be the walls of a channel, $y_1 + y_2 = h$, and let U be the relative velocity of motion of these walls. When a pure liquid is flowing in the channel, the undisturbed shear stress is

$$\tau_0 = \mu_0 \frac{U_0}{h}. \tag{8.121}$$

The presence of particles in the flow acts on the flow in the same way as a change in the relative velocity of channel walls, given by the expression (8.118)

$$U = \dot{\gamma}_0 h(1 - 2.5W). \tag{8.122}$$

Since the suspension is considered as a Newtonian liquid, then, by definition, the shear stress at the wall will be

$$\tau = \mu_* \frac{U}{h} = \mu_* \frac{U_0}{h}(1 - 2.5W). \tag{8.123}$$

where μ_* is the viscosity coefficient of the suspension.

It was shown above that the shear stress at the wall is the same for a pure liquid and for an infinite diluted suspension, i.e. $\tau = \tau_0$. Hence

$$\mu_0 \frac{U}{h} = \mu_* \frac{U_0}{h}(1 - 2.5W), \tag{8.124}$$

from which it is possible to find the viscosity coefficient for an infinite diluted suspension

$$\mu_* = \frac{\mu_0}{1 - 2.5W} \approx \mu_0(1 + 2.5W). \tag{8.125}$$

Here it is taken into account that $W \ll 1$ for an infinite diluted suspension.

The formula (8.125) for was first obtained by Einstein and is named after him.

Experiment has shown that Einstein's formula is suitable for describing the viscosity of suspensions with the volume concentration of solid particles $W < 0.02$, though sometimes this formula is used for the values up to $W = 0.1$.

If instead of solid particles the suspension contains drops of internal viscosity μ_i different from the viscosity μ_e of the ambient liquid (in this case we talk of emulsion rather than suspension), then the viscosity is determined by Taylor's formula [34]:

$$\mu_* = \mu_0 \left(1 + \frac{2.5\mu_i + \mu_e}{\mu_i + \mu_e} W \right). \tag{8.126}$$

For solid particles, $\mu_i/\mu_i \to \infty$, and Einstein's formula (8.125) follows as a limiting case of (8.126). the particles are bubbles of gas, then $\mu_i/\mu_i \to 0$ and (8.126) leads to

$$\mu_* = \mu_0(1 + W). \tag{8.127}$$

Let us summarize the obtained results. From the expression for the viscosity of an infinite diluted suspension, it follows that the viscosity factor does not depend on the size distribution of particles. The physical explanation of this fact is that in an infinite diluted suspension ($W \ll 1$), particles are spaced far apart (in comparison with the particle size), and the mutual influence of particles may be ignored. Besides, under the condition $a/h \ll 1$, we can neglect the interaction of particles with the walls. It is also possible to show that in an infinite diluted suspension containing spherical particles, Brownian motion of particles does not influence the viscosity of the suspension. However, if the shape of particles is not spherical, then Brownian motion can influence the viscosity of the suspension. It is explained by the primary orientation of non-spherical particles in the flow. For example, thin elongated cylinders in a shear flow have the preferential orientation parallel to the flow velocity, in spite of random fluctuations in their orientation caused by Brownian rotational motion.

At we increase the volume concentration of particles, the impact of particles on each other can no longer be neglected. The appropriate correction to Einstein's formula was obtained by Batchelor and has the order of W^2 [35]:

$$\mu = \mu_0(1 + 2.5W + 6.2W^2). \tag{8.128}$$

Some works have suggested empirical dependences of the kind

$$\mu = \mu_0(1 + 2.5W + kW^2). \tag{8.129}$$

The empirical constant k is of the order of 10, and the formula (8.129) is successfully used for suspensions with volume concentrations of particles up to $W = 0.4$. This value of W is much greater than those values for which the basic theoretical formulas are valid. Indeed, in the basis of all theories lies the assumption that the suspension is infinite and diluted. The value $W = 0.4$ is close to the values that apply for suspensions with dense packing of particles. Thus, for solid spherical particles, the limiting value of W, corresponding to dense packing in a motionless suspension, is equal to 0.74. For moving suspension, it is smaller. In particular, in a strongly sheared flow, $W_{max} = 0.62$.

8.4 Separation in the Gravitatonial Field

If the density of particles (solid or liquid) is greater than that of the ambient liquid, then these particles move in the liquid in the direction of gravity; otherwise particles will rise to the surface. The former process is known as sedimentation, and the latter – as flotation. Both processes play an important role in various branches of engineering.

The present section will be concerned with the process of sedimentation. Suppose that the transfer of particles due to gravity prevails over the diffusion flux in the opposite direction.

A particle moving freely in a liquid in a gravitational field is subject to the net force, which is just the difference between gravitational and Archimedean forces:

$$F_i = (\rho - \rho_f) g_i V = m\left(1 - \frac{\rho_f}{\rho}\right) g_i, \tag{8.130}$$

where ρ and ρ_f are densities of the particle and the external liquid; V is the volume of the particle; m is the particle mass.

At $\rho > \rho_f$, the particle moves downwards (sedimentation), and at $\rho < \rho_f$ – upwards (flotation). Note that the value F_i does not depend on the particle shape and its orientation in space.

On the other hand, the particle is subject to the force of resistance from the liquid. The velocity of particle sedimentation is established very quickly. The characteristic time of this process is estimated by the value a^2/ν equal to the relaxation time of the viscous force. Thus, for values a^2/ν m^2/s, $a \sim 100$ μm, this time is equal to 0.02 s. Of greatest interest is the case of small-sized particles, therefore it is possible to assume that their motion is inertialess, taking the expression (8.3)

$$(F_i)_V = -f_{ij} U_j. \tag{8.131}$$

as the force of viscous resistance.

In the inertialess approximation, the equation of particle's motion is

$$F_i + (F_i)_V = 0 \tag{8.132}$$

or

$$f_{ij} U_j = m\left(1 - \frac{\rho_f}{\rho}\right) g_i. \tag{8.133}$$

Components of the tensor f_{ij} depend on the shape and orientation of the particle. If the particle has spherical shape, then f_{ij} does not depend on the particle's orientation, and (8.133) can be rewritten as

$$6\pi\mu a U = \frac{4}{3}\pi a^3 (\rho - \rho_f) g, \tag{8.134}$$

leading to a well-known expression for the velocity of free sedimentation of a spherical solid particle in an infinite volume of viscous liquid:

$$U = \frac{2}{9}\frac{a^2}{\nu}\left(\frac{\rho}{\rho_f} - 1\right)g. \qquad (8.135)$$

Thus the velocity of particle sedimentation depends strongly on its radius. The Reynolds number found from the sedimentation velocity is

$$\mathrm{Re} = \frac{Ua}{\nu} = \frac{2}{9}\frac{a^3}{\nu^2}\left(\frac{\rho}{\rho_f} - 1\right)g. \qquad (8.136)$$

For the values $\nu \sim 10^{-6}\ \mathrm{m^2/s}$, $\rho/\rho_f \sim 2$, we have $\rho/\rho_f \sim 2(10^4 a)^3$. Therefore the number $\mathrm{Re} < 1$ for particles with $a < 75\ \mu\mathrm{m}$.

One can derive an analogous expression giving the sedimentation velocity of a spherical drop of viscosity μ_i that differs from the viscosity of the external liquid μ_e (the Hadamar- Rybczynski formula):

$$U_d = \frac{1}{3}\frac{a^2}{\nu_e}\frac{\mu_e + \mu_i}{\mu_e + 3\mu_i/2}\left(\frac{\rho}{\rho_f} - 1\right)g. \qquad (8.137)$$

In the limit $\mu_i/\mu_e \to \infty$, the formula (8.137) transforms to (8.135) – the sedimentation velocity of solid particles, but at $\mu_i/\mu_e \to 0$ and $\rho/\rho_f \to 0$, the formula (8.137) gives the flotation velocity of gas bubbles

$$U_b = \frac{1}{3}\frac{a^2}{\nu_e}g. \qquad (8.138)$$

If Reynolds number is $\mathrm{Re} > 1$, then the resistance force on the particle differs from Stokes force and is equal to

$$F = 0.5 f \pi a^2 \rho U^2 \qquad (8.139)$$

where f is the coefficient of resistance

$$f = \begin{cases} 24/\mathrm{Re} & \text{at } \mathrm{Re} < 2, \\ 18.5/\mathrm{Re}^{0.6} & \text{at } 2 < \mathrm{Re} < 500, \\ 0.44 & \text{at } \mathrm{Re} < 500. \end{cases} \qquad (8.140)$$

Equating the force of resistance to the Archimedean force, we get the equation for particle velocity. The approximate expression for U has the form

$$U = \frac{\mu \mathrm{Ar}}{2a\rho(18 + 0.575\mathrm{Ar}^{0.5})} \qquad (8.141)$$

where ρ and μ are the density and viscosity of the carrying liquid; $\mathrm{Ar} = \dfrac{8a^3\rho^2 g\Delta\rho}{\rho\mu^2}$ is the Archimedean number; $\Delta\rho$ is the density difference between the particles and the carrying medium.

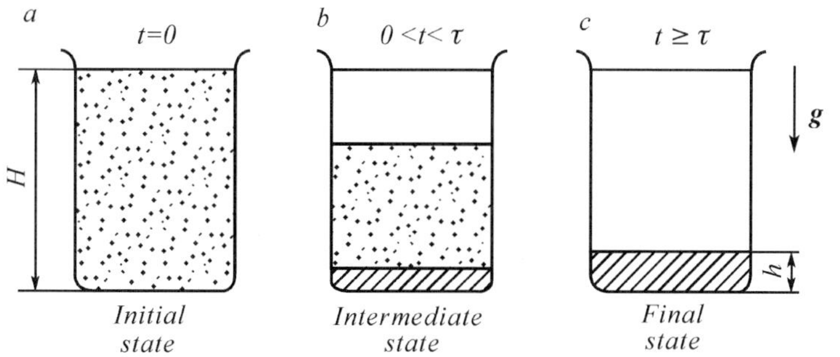

Fig. 8.7 Sedimentation in a container.

Consider now the process of sedimentation in a container filled with suspension, where particles are homogeneously distributed over the volume (Fig. 8.7, a) [36] at the initial moment.

Particles are precipitated at the bottom of the container and in some time three areas with precise borders (Fig. 8.7, b) can be distinguished in the volume. The pure liquid layer is located on the top, followed by the suspension layer (note that the top border of the second layer shifts downwards with time), and finally, the last layer consists of solid sediment. After a certain time τ all particles will precipitate from the liquid into the sediment, the suspension will be completely separated into the pure liquid, and the solid sediment layer and the process of sedimentation will be brought to completion by the establishment of sedimentation balance (Fig. 8.7, c). The boundaries between layers are characterized by jumps of density and known as contact discontinuities. Let us determine the velocities of motion of discontinuity surfaces. Consider the motion of the top border of the second layer in Fig. 8.7. Denote by u the velocity of the border's motion directed downwards. Following a common practice in hydrodynamics, choose the system of coordinates attached to the moving surface. In this system, the surface of discontinuity is motionless. Denote the values of parameters before the jump (above) by the index 1, and behind the jump (below) – by the index 2 (Fig. 8.8, a).

Write down the conservation condition for the solid phase at the discontinuity

$$\rho_1(U_1 - u) = \rho_2(U_2 - u), \tag{8.142}$$

where ρ is the solid phase density, that is, the density of the mass concentration of particles (not to confuse with the density of a particle).

Designate through $j = \rho U$ the solid phase flux relative to the motionless system of coordinates, for example relative to the walls of the container. Then (8.142) may be rewritten as

$$u = (j_2 - j_1)/(\rho_2 - \rho_1). \tag{8.143}$$

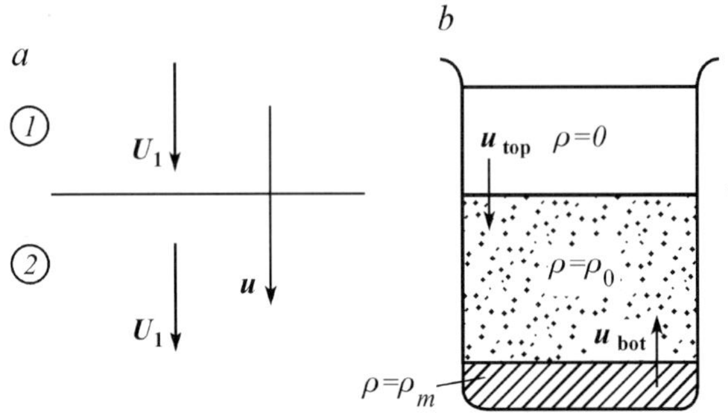

Fig. 8.8 Contact discontinuity.

For an infinite diluted suspension, it is possible to set $U_1 = U_2 = U_0$ where U_0 is the sedimentation velocity of an isolated particle determined by the formula (8.135). At the initial moment, the considered surface of discontinuity coincides with the free border of the suspension. Denote the velocity of the border at the initial moment of time by u_{top}. It is obvious, that at the same time, $\rho_1 = 0$ and $\rho_2 = \rho_0$ where ρ_0 is the mass concentration of the solid phase in the initial suspension. Then from (8.142) there follows

$$u_{top} = U_0. \tag{8.144}$$

This relation is self-obvious, since at the initial moment, particles in an infinite diluted suspension precipitate with Stokes velocity.

The initial velocity of the discontinuity surface of the bottom layer (solid deposit) is found in the same manner (this velocity is directed upwards) (Fig. 8.8, b). To this end, we should plug $\rho_1 = \rho_0$ and $\rho_2 = \rho_m$ into (8.143), where ρ_m is the maximum mass density of the solid phase in the sediment (it can be the density corresponding to stationary close packing of particles). Besides, at this surface, $j_2 = 0$, since there is no flux of particles behind the discontinuity surface. Thus

$$u_{bot} = \frac{-\rho_0 U_0}{\rho_m - \rho_0} = \frac{\rho_0}{\rho_m - \rho_0} u_{top}. \tag{8.145}$$

The minus sign means that the velocity is directed upwards. From now on, we shall agree to count the velocity as positive if its direction coincides with g.

The position discontinuity of surfaces can be represented schematically on the (x, t) plane, where x is the vertical coordinate counted from the free surface downwards (Fig. 8.9).

At the initial stage of sedimentation, the top and bottom discontinuity surface positions change linearly with time:

$$x_{top} = u_{top} t,$$

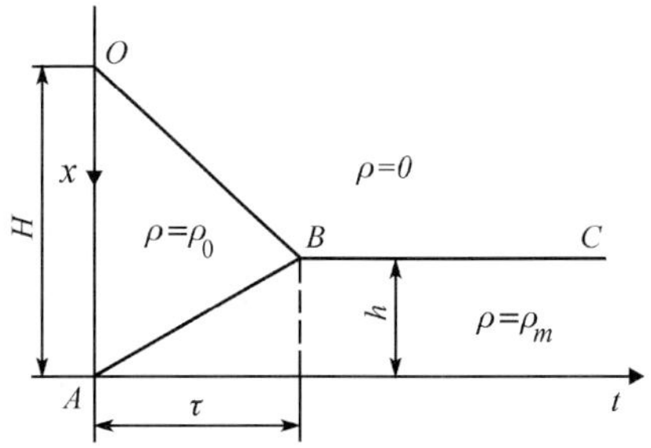

Fig. 8.9 The position of a discontinuity surface on the (x, t) plane for non-hindered sedimentation.

$$x_{bot} = H - u_{top}t. \tag{8.146}$$

The time τ needed to establish equilibrium can be estimated by equating $x_{top} = x_{bot}$, from which it follows that

$$\tau = \frac{H}{U_0}\left(1 - \frac{\rho_0}{\rho_m}\right). \tag{8.147}$$

Accordingly, it is possible to find the height of the established sediment layer

$$h = H - u_{top}\tau = H\frac{\rho_0}{\rho_m}. \tag{8.148}$$

Using the formulas (8.147) and (8.148), it is possible to determine experimentally the properties of infinite diluted suspensions containing same-sized particles (a monodisperse suspension), for example, the mass concentration and size of particles. If the suspension contains particles of different sizes (a polydisperse suspension), then dividing the entire spectrum of particle sizes from a_{min} to a_{max} into a finite number of fractions, it is possible to carry out the argumentation stated above for each fraction, and to determine the laws of motion for the corresponding discontinuity surfaces. Measuring the velocities of discontinuity surfaces in an experiment, it is possible to determine the characteristics of each fraction and thereby the size distribution of particles.

So far, the research was limited to an infinite diluted suspension. Proceed now to consider the case when particle concentration is not small, so the sedimentation velocity of a particle cannot be determined by the formula (8.135), but it is necessary to take into account the hindered character of particle movement. In this case, the velocity of motion should depend on the volume concentration of particles.

The particle motion in a liquid where the interactions with adjacent particles are important is known as hindered motion. For a monodisperse system the effect of constraint should be taken into account starting from the volume concentrations of about $W \approx 0.15$. In a general form, the sedimentation velocity of a particle is equal to

$$U = U_0 G(W), \tag{8.149}$$

where U_0 is the sedimentation velocity of a single particle given by the formula (8.135), and $0 < G(W) < 1$. The function $G(W)$ characterizes the deceleration of particle sedimentation due to the hindered character of motion; it depends only on the volume concentration of particles in the suspension.

There are two ways to determine the function $G(W)$ [37, 38]. In accordance with the first, the reduction of velocity is due to the increase of suspension viscosity with the increase of W. The second approach is based on modeling the suspension as a porous medium, and the resistance force on the particle is determined during its motion in microchannels of the porous medium.

The following empirical expression for $G(W)$ [39] has gained a broad acceptance:

$$G(W) = (1 - W)^n, \tag{8.150}$$

where $n \approx 4.7$.

Let us make a proper account of the dependence of particle sedimentation velocity on the volume concentration (or, which is the same thing, on the mass concentration ρ) of particles in the process of suspension sedimentation, and take a look at the results.

Sedimentation of particles leads to an increase in their mass concentration in the bottom part of the container. This process is extended upward, since particles entering the region of high values of ρ are sedimenting slower and slower. The propagation of the change of ρ to the top can be considered as the motion of the surface of perturbation ρ.

Going back to the scheme of sedimentation with discontinuity surfaces, it is possible to show that now, instead of the condition (8.143), the surface of discontinuity should satisfy the following condition

$$U(\rho) = -\frac{dj}{d\rho}, \quad j = \rho U(\rho). \tag{8.151}$$

This condition follows from (8.143) by going to the limit $j_2 \to j_1$, $\rho_2 \to \rho_1$ under the continuity condition for the particle flux $j(\rho)$ that is directed downwards.

It should be emphasized that instead of the motion of discontinuity surfaces, we are interested in the motion of perturbations of mass concentration of particles ρ through the liquid. These perturbations are known as concentration (or kinematic) waves. Since $j(\rho)$ is, as a rule, a nonlinear function of ρ, by analogy to the theory of a compressed, nonviscous liquid flow, kinematic waves are similar to Riemann's waves. It is known [40] that the propagation of such waves results in the formation of breaks.

The change of particle mass concentration $\rho(x, t)$ is described by the continuity equation:

$$\frac{\partial \rho}{\partial t} + \frac{\partial j}{\partial x} = 0, \tag{8.152}$$

from which it follows that the surfaces of constant density (characteristics of the Eq. (8.152)) propagate upwards with the velocity

$$\frac{dx}{dt} = -u(\rho). \tag{8.153}$$

Since along these characteristics $\rho = \text{const}$, and $u(\rho) = \text{const}$, we conclude that the characteristics are straight lines on the x, t plane. The equation of vertical (upward) motion of the condensation wave is

$$x = H - u(\rho)t. \tag{8.154}$$

In Fig. 8.10, these characteristics are shown by dashed lines. The diagram on the x, t plane differs somewhat from the diagram for the case of an infinite diluted suspension (see Fig. 8.9).

Accordingly, the depression wave, extending downwards, is given by the expression

$$\frac{dx}{dt} = U(\rho). \tag{8.155}$$

The research of sedimentation process subject to constraints has been described in [41], therefore details of the analysis are not presented here. Some features, however, are worth mentioning here. One of them is that the flux $j(\rho) = \rho U(\rho)$ is a non-monotonous function ρ (Fig. 8.11).

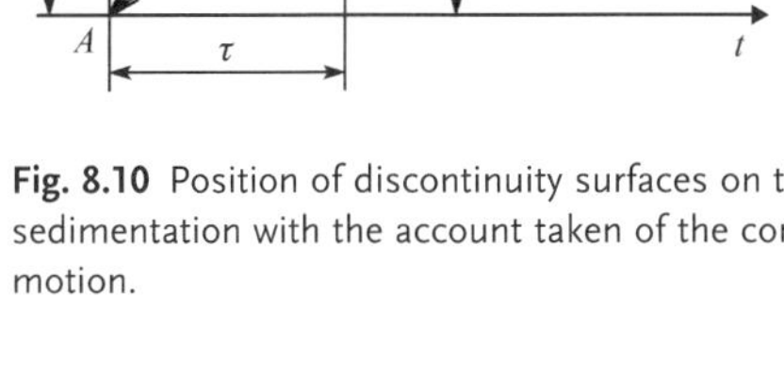

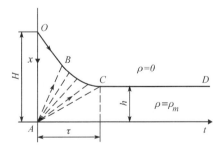

Fig. 8.10 Position of discontinuity surfaces on the (x, t) plane at sedimentation with the account taken of the constraints in particle's motion.

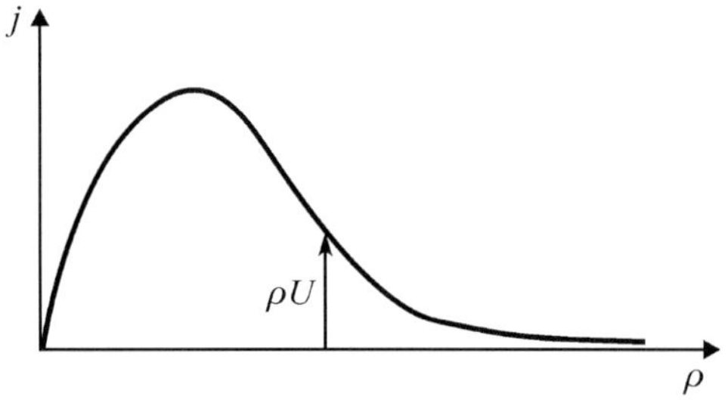

Fig. 8.11 Mass flux at sedimentation with constraint motion of particles.

At small values of ρ (extremely diluted suspension) the sedimentation occurs quickly, then ρ increases at distances close to the bottom of the container, and j decreases almost down to zero near the sediment layer. The other feature is the upward propagation of the compression wave (increase in ρ), which slows down the process of sedimentation. In order to prevent this process in practice, it is necessary to remove the solid phase formed at the bottom of the container. It is possible to select such removal rate that the upward velocity of the compression wave be equal to zero. In this case the flux j consists of two contributions – the sedimentation and convection fluxes. The dependence of the total flux on ρ in this case is shown in Fig. 8.12 [42].

At present an effective method of separation of liquid from particles in an inclined channel is being used. The increase of the sedimentation velocity of particles was first shown in [43] in the case of particles' sedimentation in blood. This effect is known as Boycott's effect. The picture of particle sedimentation is shown in Fig. 8.13.

At sedimentation in the inclined channel particles are deposited not only at the bottom of the channel, as in the vertical channel, but also on the lateral wall. The sedimentation velocity is determined by the rate of decrease of height H of the suspension layer. At sedimentation, a layer of pure liquid is formed above the surface of the suspension. The thickness of this layer is far less than the

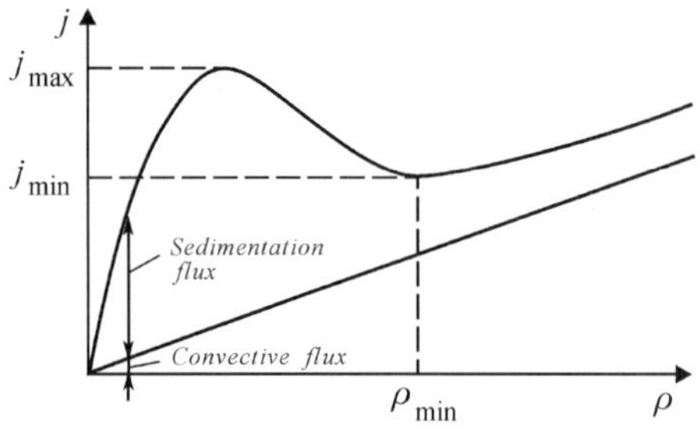

Fig. 8.12 Mass flux at sedimentation with removal of dispersed phase from the bottom of container.

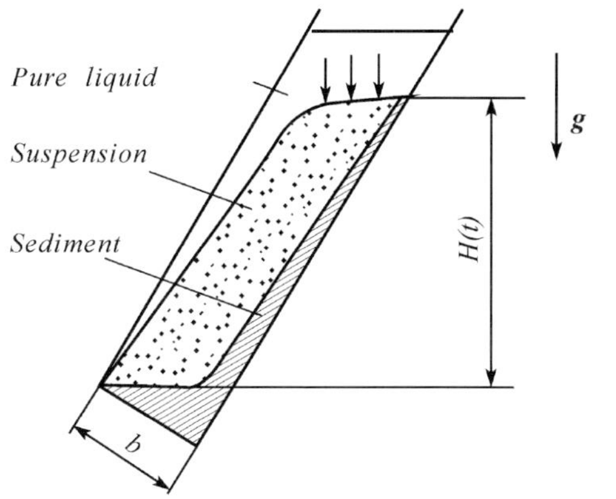

Fig. 8.13 Sedimentation in an inclined channel.

thickness of the channel b. The most of pure liquid collects in the top part. The kinematic "jump" formed at the top moves with greater vertical velocity than the velocity U of constrained motion of sedimenting particles in the vertical channel. Hydrodynamic analysis of gravitational sedimentation of particles in inclined channel has been carried out in [44]. The rate of change of the interface, that is the surface separating the layer of pure liquid from suspension layer, is given by

$$\frac{dH}{dt} = -U\left(1 + \frac{H}{b}\cos\theta\right). \tag{8.156}$$

The second term in the right-hand part of (8.156) characterizes the increase of sedimentation rate due to inclination of the channel. The smaller is the slope and cross-sectional dimension of the channel b, the greater is this increase.

The separation of the particles from carrying medium has important practical applications in petroleum and gas industry. Before delivery of oil and natural gas into the oil and gas main pipelines, it is necessary first to separate water from oil and mechanical admixtures, as well as gas condensate and water from gas. These processes are performed in special devices – the settling tanks, separators, and multiphase dividers, in which separation of phases occurs under action of gravitational, centrifugal and other forces. The methods used in modeling separation processes of hydrocarbon systems are described in work [45].

8.5 Separation in the Field of Centrifugal Forces

In the previous section, the method of gravitational sedimentation has been considered as applied to separation of relatively large particles from the liquid. A peculiarity of this process is the negligible diffusion flux and the presence of the expensive dividing borders between pure liquid, suspension, and solid sediment.

However, if the particles' size is very small, diffusion can dominate over gravitational sedimentation. The diffusion rate is inversely proportional to particle's mass, while gravitational sedimentation is directly proportional to mass. The diffusion mass transfer does not give rise to discontinuities in mass density distribution. It follows that for very small particles the gravitational separation is very slow and thus, inefficient. More effective separation method of small particles from liquid is centrifugation [46]. This method uses rapid rotation of suspension, which results in the emergence of large centrifugal forces acting on the particles and equal to $m\omega^2 r$, so that the particles are forced to move relative to liquid. Since the centrifugal force increases with radial distance r from the rotational axis, particles under the action of this force will not reach constant velocity as they do in a gravitational field near the ground level. The radial component of the particle's velocity is called the drift- or migration velocity. Due to rotation of the liquid there arises also the radial gradient of pressure, which serves as driving force of particles' diffusion.

Consider rotation of a liquid with constant angular velocity. Relative the co-rotating coordinate system associated with the rotating liquid, the liquid is at rest. If particles move together with liquid with the same angular velocity, they are under the centrifugal force, the gravitational force being neglected. Then the particles are being forced to the periphery, if $\rho > \rho_f$ or, on the contrary, to the axis of rotation, if $\rho < \rho_f$. Equating the inertial (centrifugal) force to the resistance force, yields the velocity of radial drift of the particle

$$U_r = \frac{m\omega^2 r}{\bar{f}}\left(1 - \frac{\rho_f}{\rho}\right), \tag{8.157}$$

where f is mean coefficient of viscous friction.

The centrifugation method is widely used in biology for separation of macromolecules in physiological solutions and is known as ultra-centrifugation. Frequently, instead of particles' (e.g., macromolecules) density ρ, the specific volume of macromolecules is used. Then instead of the velocity U, we can introduce parameter

$$s = \frac{U_r}{\omega^2 r} = \frac{m(1 - \bar{\mathbf{v}}\rho_f)}{\bar{f}}, \tag{8.158}$$

called the sedimentation coefficient. Typical values of s in the ultra-centrifugation of solutions lay within the interval from 1S up to 100S, where $1S = 10^{13}$ s is the corresponding unit of measurement called the "sverdberg". Modern ultracentrifuges are characterized by high values of angular velocity (up to 70 000 revolutions/minute), at which very great accelerations are attained.

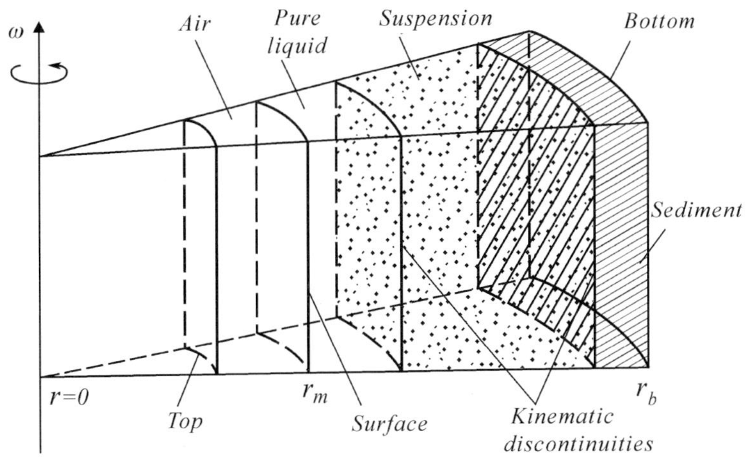

Fig. 8.14 Separation of mixture in a centrifuge.

In principle, the concentration distributions of macromolecules or colloidal particles over r are about the same, as are distributions over height in a container at gravitational sedimentation.

Consider centrifugation process for a mixture in a centrifuge depicted in Fig. 8.14 [36, 46, 47]. We can identify the following surfaces: 1) the free surface $r = r_m$ dividing separated liquid (solvent) and air; 2) the kinematic break $r = r^*$, a surface dividing solvent and mixture; 3) the kinematic break $r = r_b$, a surface dividing mixture and sediment layer at the outer wall of centrifuge. The exhibited cylindrical sector filled with binary mixture, rotates as a whole with angular velocity ω. Introduce the cylindrical system of coordinates attached to rotating mixture. The motion of dissolved substance relative to this coordinate system occurs in radial direction and all parameters of the mixture depend only on r and t. Denote, again, as ρ the mass concentration of dissolved substance. The continuity equation in moving coordinate system is

$$\frac{\partial \rho}{\partial t} = \frac{1}{r}\frac{\partial}{\partial r}\left(r\left(D\frac{\partial \rho}{\partial r} - \rho s \omega^2 r\right)\right). \tag{8.159}$$

In the literature devoted to centrifugation processes this equation is known as the Lemm equation.

If the mixture represents extremely diluted solution, then D and s do not depend on ρ and are constants. In this case, the Eq. (8.159) may be reduced to

$$\frac{\partial \rho}{\partial t} = \frac{D}{r}\frac{\partial}{\partial r}\left(r\frac{\partial \rho}{\partial r}\right) - \frac{s\omega^2}{r}\frac{\partial}{\partial r}(\rho r^2). \tag{8.160}$$

Let at the initial moment the mixture represents homogeneous solution, so that the initial condition is

$$\rho = \rho_0 (r_m < r < r_b) \quad \text{at } t = 0. \tag{8.161}$$

As the boundary conditions, we should take the absence of flux of dissolved substance at the free surface and at the lateral wall of the centrifuge

$$D\left(\frac{\partial \rho}{\partial r}\right)_{r=r_m} - s\omega^2 r_m \rho_m = 0, \quad D\left(\frac{\partial \rho}{\partial r}\right)_{r=r_b} - s\omega^2 r_b \rho_b = 0. \tag{8.162}$$

If the diffusion were neglected, then the distribution $\rho(r)$ would be similar to distribution $\rho(h)$ over the variable altitude h in gravitational sedimentation. In particular, the waves of sharp change of ρ (the kinematic shocks) would appear: one wave separating the layer of pure solvent from mixture would move to the centrifuge's wall, while another wave front separating the mixture from the sediment layer would converge toward the centrifuge axis. But even in this case, there is an essential difference from gravitational sedimentation. Namely, the concentration of particles of equal size in the mixture-layer does not remain constant, as is the case in gravitational sedimentation; instead, it falls with time, because increase of r causes the corresponding increase of the cross-sectional area, and consequently, of the volume. This effect is called radial dilution of mixture.

Since the mixture may contain particles of various sizes, consider first the particles that are large enough so that their diffusion transfer could be ignored. Also, take into account the constrained motion of particles. Then $s = s(\rho)$, and the Eq. (8.159) will take the form

$$\frac{\partial \rho}{\partial t} + r\omega^2 \frac{d(s\rho)}{d\rho}\frac{\partial \rho}{\partial r} = -2\rho s\omega^2. \tag{8.163}$$

The two conditions (8.162) contain more information than was required. Therefore we should take as boundary condition only the first of these conditions, and use (8.161) as the initial condition.

The Eq. (8.163) is solved by the method of characteristics.

For an extremely diluted suspension under the condition $D = 0$, the general solution is [46]

$$\begin{aligned} &\rho = 0 \quad \text{at } r_m \le r < r^*,\ t > 0, \\ &\rho = \rho_0 e^{-2s_0\omega^2 t} \quad \text{at } r^* \le r < r_b,\ t > 0, \end{aligned} \tag{8.164}$$

The surface $r = r^*$ is the surface of kinematic jump. The law of motion for this surface is determined by the equation $\dfrac{dr^*}{dt} = r^*\omega^2 s_0$, whose solution is

$$r^*(t) = r_m e^{s_0\omega^2 t}. \tag{8.165}$$

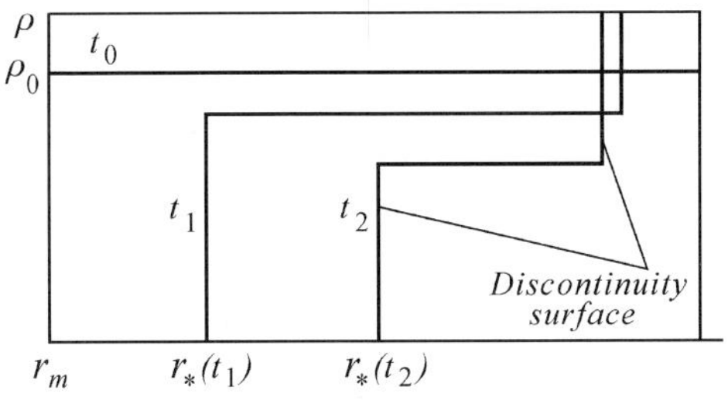

Fig. 8.15 Distribution of particle concentration over the radial variable *r*.

The distribution of concentration over r in the absence of diffusion is shown in Fig. 8.15.

The expression (8.165) can be rewritten as

$$\ln r^* = \ln r_m + \omega^2 t s_0. \tag{8.166}$$

The linear dependence (8.166) is convenient for determining the sedimentation coefficient s_0 from experimental observations of discontinuity surface motion, i.e. $r^*(t)$.

It follows from (8.164) and (8.165) that

$$\frac{\rho}{\rho_0} = \frac{r_m^2}{r^{*2}} \quad \text{or} \quad r^{*2} = \text{const.} \tag{8.167}$$

Turn now to the next case, when diffusion is taken into account. Restrict ourselves to the case $D = \text{const}$ and $s = \text{const}$. Introduce new variables into (8.160)

$$\pi = \frac{\rho}{\rho_0} e^{2s\omega^2 t}, \quad \tau = 2s\omega^2 t, \quad \xi = \ln\left(\frac{r}{r_m}\right)^2. \tag{8.168}$$

Then Eq. (8.160) transforms to

$$\frac{\partial \pi}{\partial \tau} + \frac{\partial \pi}{\partial \xi} = \frac{e^{-\xi}}{\mathrm{Pe_D}} \frac{\partial^2 \pi}{\partial \xi^2}. \tag{8.169}$$

Here the Peclet diffusion number is introduced

$$\mathrm{Pe_D} = \frac{s\omega^2 r_m^2}{D} = \frac{U_r r_m}{D}. \tag{8.170}$$

The solution of Eq. (8.169) presents significant mathematical difficulties. Therefore we should proceeded as we did in Section 6.4 when deriving the solution of the diffusion problem of reverse osmosis. Consider the solution at the values of r close to r_m, i.e. in the area adjacent to the free surface. It means that the initial stage of process is considered. Then it is possible to set $e^{-\xi} \approx 1$ and $r_b/r_m = \infty$, and the Eq. (8.169) reduces to

$$\frac{\partial \pi}{\partial \tau} + \frac{\partial \pi}{\partial \xi} = \frac{1}{\mathrm{Pe}_D} \frac{\partial^2 \pi}{\partial \xi^2}. \tag{8.171}$$

The initial condition (8.161) transforms into

$$\pi = 1 \quad \text{at } \tau = 0,\ 0 < \xi < \infty. \tag{8.172}$$

For the boundary conditions, we take the condition at the free surface

$$\pi = \frac{1}{\mathrm{Pe}_D} \frac{\partial \pi}{\partial \xi} \quad \text{at } \xi = 0,\ \tau > 0, \tag{8.173}$$

and the condition that π is finite at $\xi \to \infty$.

The formulated boundary value problem has an analytical solution [46], but it is unwieldy and is not presented here. It should be noted that the necessary condition for the considered approximation is $\mathrm{Pe}_D \gg 1$, otherwise it would be impossible to formulate the condition at the infinity. At centrifugation, the characteristic values of the Peclet number are 10^2–10^3, therefore this assumption is justified. Another assumption refers to the short time intervals. Time enters the expression for the Strouhal number

$$\mathrm{St} = \frac{1}{\omega^2 st} = \frac{r_m}{U_r t}, \tag{8.174}$$

therefore using the condition $\mathrm{St} \gg 1$ is mandatory. For the characteristic values $s \approx 10^{-13}$ s, $\omega \approx 5000$ s, this inequality holds for the times $t \gg 4 \cdot 10^3$ s ≈ 1 hour. Inasmuch as the centrifugation lasts several hours, this assumption is reasonable.

The use of centrifugal forces has found wide application in separation of gas-condensate hydrocarbon mixtures. In the previous section, the process of separation in a gravitational field has been considered. However, it has low efficiency, especially at high gas flow rates. In order to increase the efficiency of separation, separators are equipped with special devices capable of trapping drops which had not been separated in the gravitational settling section [45]. Centrifugal branch pipes (cyclones), i.e. vertical cylinders in which the flow of gas containing fine drops is whirled at the entrance, may be used in this capacity. Depending on the method of whirling, cyclones are classified as axial cyclones, in which the flow is curled as it goes though a swirler placed at the entrance; and tangential cyclones, in which the flow is delivered into the cyclon through tangential slits in the walls. In cyclones of the first type, the local rotational velocity of the flow is a function of the radial distance r, and can be approximated by the law $u_\varphi = Cr^{\mathrm{k}}$. At $\mathrm{k} = -1$, the flow is curled according to the law of constant circulation (potential rotation); at $k = 0$, the flow exhibits independence of the angle of rotation from radius; at $k = 1$ the rotational velocity follows the law of a solid-body rotation (a quasi-solid rotation).

The influence of the circulation mechanism (determined by k) on separation efficiency of gas-condensate mixtures has been studied theoretically in [45]. The methods of calculations of separators equipped by cyclones are considered in [45] as well. Section 19.2 contains a detailed analysis of these processes.

9
Suspensions Containing Charged Particles

9.1
Electric Charge of Particles

Most particles, as they come into contact with a polar liquid, acquire an electric surface charge. The resulting potential jump in the thin surface layer is known as ζ-potential. Let us determine the relation between the electric charge q of a particle and its ζ-potential. For this purpose, we use conservation of the electric charge, the Poisson equation, and the condition that q should be equal and opposite in sign to the total electric charge of the electric double layer.

Consider an example of such calculation [36] for a spherically-symmetric diffusion double layer, formed on a non-conducting spherical particle of radius a. A spherical double layer of thickness dr contains the charge

$$dq = 4\pi r^2 \rho_E \, dr, \tag{9.1}$$

where ρ_E is the volume charge density; r is the distance from the particle center to the considered layer.

Substituting ρ_E in (9.1) by Poisson's equation

$$\rho_E = -\varepsilon \frac{1}{r^2} \frac{\partial}{\partial r}\left(r^2 \frac{\partial \phi}{\partial r}\right) \tag{9.2}$$

and integrating over r, one obtains the total charge of the double layer

$$q = -\int_{\alpha}^{\infty} 4\pi\varepsilon \frac{\partial}{\partial r}\left(r^2 \frac{\partial \phi}{\partial r}\right) dr. \tag{9.3}$$

Application of the condition $d\varphi/dr \to 0$ at $r \to \infty$ to (9.3) yields

$$q = 4\pi\varepsilon a^2 \left(\frac{\partial \phi}{\partial r}\right)_{r=a}. \tag{9.4}$$

Separation of Multiphase, Multicomponent Systems. E. G. Sinaiski and E. J. Lapiga

ISBN: 978-3-527-40612-8

Let $-q$ be the electric charge of the particle. For conducting particle the total charge of the particle is distributed over the surface. Consider now the surface charge density q_s determined by the relation

$$-q = 4\pi a^2 q_s. \tag{9.5}$$

Then it follows from (9.4) and (9.5) that

$$q_s = -\varepsilon \left(\frac{\partial \phi}{\partial r} \right)_{r=a}. \tag{9.6}$$

Thus, q is associated with the distribution of ϕ over the double layer. In the case of a very thin layer, r should be understood as a direction normal to the surface. In the Debye-Huckel approximation, we have:

$$\frac{1}{r^2} \frac{\partial}{\partial r} \left(r^2 \frac{\partial \phi}{\partial r} \right) = \frac{\phi}{\lambda_D^2}. \tag{9.7}$$

Denote $\xi = r\phi$. Then (9.7) transforms into the equation

$$\frac{d^2 \xi}{dr^2} = \frac{\xi}{\lambda_D^2}. \tag{9.8}$$

This equation coincides with (7.67). Its solution, satisfying the condition $\phi \to 0$ at $r \to \infty$, is

$$\phi = \frac{A}{r} e^{-r/\lambda_D}. \tag{9.9}$$

The constant A can be found from the condition that the potential at the particle's surface $r = a$ (practically at the Shtern plane) is $\phi = \zeta$:

$$A = a\zeta e^{a/\lambda_D}. \tag{9.10}$$

Substituting (9.10) into (9.9), we obtain

$$\left(\frac{\partial \phi}{\partial r} \right)_{r=a} = -\zeta \left(\frac{1}{a} + \frac{1}{\lambda_D} \right). \tag{9.11}$$

Now one gets from (9.6)

$$q_s = \varepsilon \zeta \left(\frac{1}{a} + \frac{1}{\lambda_D} \right), \tag{9.12}$$

and from (9.5)

$$\zeta = \frac{q}{4\pi\varepsilon a(1 + a/\lambda_D)} \tag{9.13}$$

or

$$\zeta = \frac{q}{4\pi\varepsilon a} - \frac{q}{4\pi\varepsilon a(a + \lambda_D)}. \tag{9.14}$$

Thus, the ζ-potential represents the sum of two potentials: the first potential is caused by the charge q at the surface of a sphere of radius a, and the second potential is caused by the charge $-q$ at the concentric sphere of radius $a + \lambda_D$. In other words, the ζ-potential is equal to the potential difference between two concentric spheres with equal but opposite charges, spaced by λ_D.

It follows from (9.12) that for a thin Debye's layer $\lambda_D \ll a$ we have

$$q_s = \varepsilon\zeta/\lambda_D. \tag{9.15}$$

In various problems, the approximation of constancy of either the surface potential, or the surface charge of the particle is used. For a charged electrolytic mixture the condition $q_s = \text{const}$ is usually used, whereas in colloidal chemistry problems, where the surface charge of colloids can change, one sets $\zeta = \text{const}$.

9.2 Electrophoresis

Consider the electrophoretic motion of a charged spherical particle of radius a in a solution of electrolyte in the presence of applied electric field (Fig. 9.1).

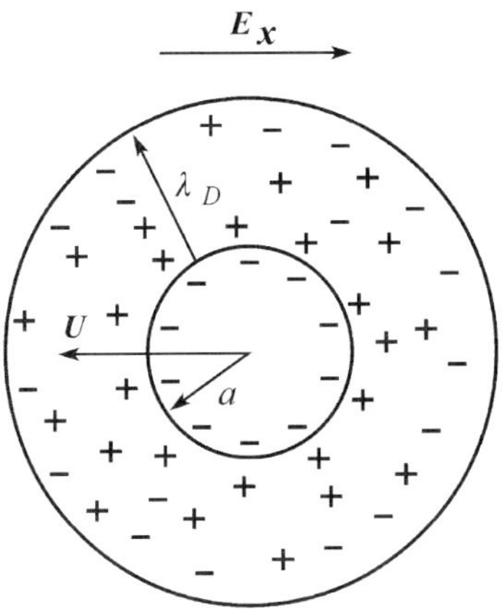

Fig. 9.1 Electrophoretic motion of a particle.

Consider first the limiting case $a \ll \lambda$. In this case, the particle can be considered as a point electric charge, placed in a non-perturbed electric field $\boldsymbol{E}_x$. Equating the electric force of the point charge to the viscous resistance force of the particle gives

$$qE_x = 6\pi\mu a U. \tag{9.16}$$

Substituting here instead of q the expression (9.13), one obtains

$$U = \frac{2}{3}\frac{\zeta\varepsilon(1 + a/\lambda_D)E_x}{\mu} \approx \frac{2}{3}\frac{\zeta\varepsilon E_x}{\mu}. \tag{9.17}$$

The expression for electrophoresis velocity of a particle, whose size is small in comparison with the thickness of the Debye layer, is called the Huckel equation.

Sometimes instead of particle's electrophoresis velocity, the mobility is introduced

$$\mathsf{V} = \frac{U}{E_x} \approx \frac{2}{3}\frac{\zeta\varepsilon}{\mu}. \tag{9.18}$$

Another limiting case, $\lambda_D \ll a$, is mostly realized in disperse systems. In particular, for colloidal systems a ranges between 0.1 and 1 μm, and at $\lambda_D \sim 10$ μm we have $\lambda_D/a \sim 10^{-1}$–10^{-2}. In this case the curvature of the particle's surface can be neglected, and the diffusion layer may be considered as locally flat. In this approximation, the electric field in the double layer is parallel to the surface. The motion of charged ions in the double layer may be considered as an electroosmotic motion along the surface just as in paragraph 7.5. It follows from (7.72) that

$$\eta\phi E_x = \mu u + \text{const}, \tag{9.19}$$

where u is the velocity of liquid flow along the particle's surface; $\boldsymbol{E}_x$ is electric field strength parallel to the surface.

From boundary conditions $\phi = 0$, $u = 0$ on the outer face of the double layer and $\phi = \zeta$, $u = U$ at the particle's surface, and using (9.19), one obtains

$$U = \zeta\varepsilon E_x/\mu. \tag{9.20}$$

This equality is known as the Helmholtz–Smoluchowski equation. Hence, at $\lambda_D \ll a$ the velocity of electrophoresis coincides with the velocity of electroosmosis, – precisely what one would expect, because electrophoresis is a phenomenon opposite to electroosmosis.

The comparison of (9.20) with (9.17) shows that the electrophoresis velocities for two limiting cases differ from each other by the factor 2/3.

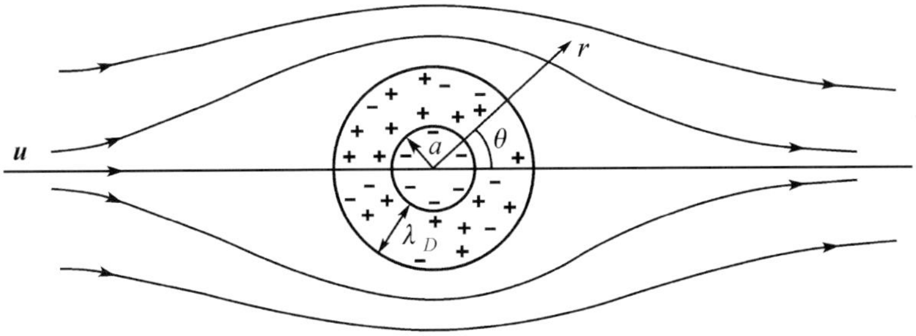

Fig. 9.2 Electrophoresis at finite values λ_D/a.

Consider now the case of a double layer with the finite thickness λ_D. Here we should distinguish between three effects [48–51], due to which the velocity of electrophoresis will differ from the Huckel (9.17) and the Helmholtz–Smoluchowski (9.20) equations: electrophoresis retardation; surface conductivity; relaxation. Consider them successively.

The electrophoresis retardation is motion of ions in the double layer in direction opposite to particles' motion. Due to forces of viscous friction, the ions cause the electroosmotic motion of liquid, which retards the particle's motion. Following the approach presented in [48], consider the electrophoresis motion of a particle, assuming that the double layer remains spherical during the motion, and the potential of the particle's surface is small enough, so Debye-Huckel approximation is valid. The motion is supposed to be inertialess. Introduce a coordinate system moving with the particle's velocity U so that in the chosen system of coordinates the particle is motionless, and the flow velocity at infinity is equal to $-U$ (Fig. 9.2).

The electric field is superposition of external field φ and a field ψ, created by a non-conducting charged sphere, with $\boldsymbol{E} = -\nabla\varphi$. The external field satisfies the equation

$$\nabla\varphi = 0 \tag{9.21}$$

with boundary conditions

$$\varphi = -E_x r \cos\theta \quad \text{at } r \to \infty,$$
$$\frac{\partial\varphi}{\partial r} = 0 \quad \text{at } r = a. \tag{9.22}$$

The second condition (9.22) is that the normal component of the current density at the surface of dielectric is zero.

The solution of the equation (9.21) with conditions (9.22) is

$$\varphi = -E_x\left(r + \frac{1}{2}\frac{a^3}{r^2}\right)\cos\theta. \tag{9.23}$$

The potential in the double layer satisfies the Poisson equation; therefore the perturbation of electric field ψ is given by

$$\Delta\psi = -\rho_E/\varepsilon. \tag{9.24}$$

In the Debye-Huckel's approximation the function (is equal to (see (9.9) and (9.10))

$$\psi = \zeta\left(\frac{a}{r}\right)e^{-(r-a)/\lambda_D}. \tag{9.25}$$

In order to determine velocity field one uses the Stokes equations

$$\nabla\cdot\boldsymbol{u} = 0; \quad -\mu\Delta\boldsymbol{u} + \nabla p = -\rho_E\nabla(\varphi+\psi) \tag{9.26}$$

in which, in accord with (9.24), we should set $\rho_E = -\varepsilon\Delta\psi$.

Boundary conditions are

$$\begin{aligned} &u_r = -U\cos\theta, \quad u_\theta = -U\sin\theta, \quad \psi = 0 \quad \text{at } r\to\infty,\\ &u_r = u_\theta = 0, \quad \psi = 0 \quad \text{at } r = a. \end{aligned} \tag{9.27}$$

The solution of Eq. (9.26) with conditions (9.27) can be obtained by the same method, as the solution of the problem on the Stokes flow over a sphere [51]. As a result, we will find the pressure and velocity distributions at the surface of the sphere, and the tangential stress τ_{rx} at the sphere. Therefore the hydrodynamic force acting on the sphere, is equal to

$$2\pi a^2\int_0^\pi \tau_{rx}\sin\theta\, d\theta.$$

On the other hand, there is an electric force

$$-qE_x = -4\pi\varepsilon a^2\left(\frac{\partial\psi}{\partial r}\right)_{r=a}E_x,$$

exerted on the sphere due to the surface charge q (see (9.4)).

Here the condition $(\partial\varphi/\partial r)_{r=a} = 0$ has been used (see (9.92)).

Since in the chosen coordinate system the particle is at rest, the net force must be equal to zero. Using this condition and omitting intermediate calculations, we write down the final result

$$-6\pi\mu a U + 6\pi\zeta\varepsilon E_x a\left(1 + 5a^2\int_\infty^a \frac{\psi}{\zeta r^6}dr - 2a^3\int_\infty^a \frac{\psi}{\zeta r^4}dr\right) = 0. \tag{9.28}$$

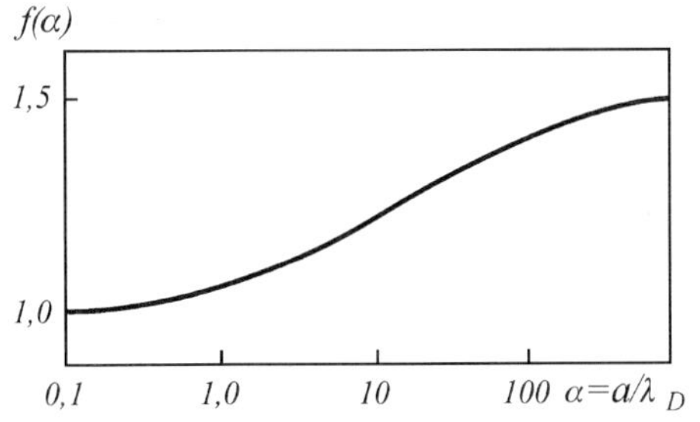

Fig. 9.3 Correction factor f(α) in the expression for electrophoresis velocity.

Substituting here the value ψ from (9.25) and performing the integration, one obtains

$$U = \frac{2}{3}\frac{\zeta \varepsilon E_x}{\mu} f(\alpha), \tag{9.29}$$

where $\alpha = a/\lambda_D$ is reverse Debye radius, and

$$f(\alpha) = 1 - \frac{1}{16}\alpha^2 - \frac{5}{48}\alpha^3 - \frac{1}{96}\alpha^4 + \frac{1}{96}\alpha^5 + \frac{1}{8}\alpha^4 e^{\alpha}\left(1 - \frac{\alpha^2}{12}\right)\int_{\infty}^{\alpha} \frac{e^{-t}}{t}\,dt.$$

This equation is known as the Henry's equation. In cases $\alpha \to \infty$ and $\alpha \to 0$, it is possible to get from this equation the Hemlholz–Smoluchowski solution ($\lambda_D \ll a$) and the Huckel solution ($\lambda_D \gg a$). The function $f(\alpha)$ is monotonously increasing with α from 1 up to $3/2$ c (Fig. 9.3).

Consider now the second effect known as surface conductivity [50]. Since we consider the double-layer with finite thickness, there is an area near the particle's surface, where the solution is not neutral, and there is an excess of gegenions in comparison with the bulk of solution. High concentration of gegenions results in formation of a layer with increased electric conductivity causing the reduction of electric field strength. We estimate this effect under condition $\lambda_D/a \ll 1$, and to this end we accept the following model.

Let there be a thin spherical mantle with conductivity σ_s' around the surface of a spherical particle. Outside of the double layer, the conductivity σ_b is equal to conductivity of the electrolyte. The electric potential satisfies the Laplace equation, whose solution in the layered medium is

$$\phi = \begin{cases} -E_x\left(r + \dfrac{Aa^3}{r^2}\right)\cos\theta & \text{in a liquid,} \\ -E_x\left(Br + \dfrac{Ca^3}{r^2}\right)\cos\theta & \text{in a double layer.} \end{cases} \tag{9.30}$$

The constants A, B, C may be determined from boundary conditions of continuity ϕ and $\sigma\nabla\phi$ at the particle's surface ($r = a$) and at the external boundary of the double layer ($r = a + \delta$), where δ is the thickness of the mantle. As a result, we get

$$\frac{Aa^3}{(a+\delta)^3} = \frac{2(a+\delta)^3(\sigma_b - 2\sigma_s') + a^3(\sigma_b + 2\sigma_s')}{2(a+\delta)^3(2\sigma_b + \sigma_s') + 2a^3(\sigma_b - 2\sigma_s')}. \tag{9.31}$$

Denote the surface conductivity in (9.31) as $\sigma_s'\delta = \sigma_s$, and assume $\delta/a \ll 1$. Then (9.31) may be simplified to

$$A = \frac{\sigma_b - 2\sigma_s/a}{2(\sigma_b - 2\sigma_s/a)}. \tag{9.32}$$

In the $\delta/a \to 0$ limit, (9.30) changes to

$$\phi = -(1 + A)E_x a \cos\theta. \tag{9.33}$$

as we take $r = a$ at the surface.

Neglecting surface conductivity, we write $A = 1/2$ and

$$\phi = -\frac{3}{2}E_x a \cos\theta, \tag{9.34}$$

which coincides with (9.23) for electrophoretic retardation. Comparing (9.33) with (9.34), we see that if E_x is replaced by the new effective field $E_{eff} = 2(1 + A)E_x/3$, then the electrophoresis velocity will coincide with (9.29), therefore

$$U = \frac{2}{3}\frac{\zeta\varepsilon E_x}{\mu(1 + \sigma_s/\sigma_b a)} f(\alpha). \tag{9.35}$$

This effect is called relaxation [51] and is related to the deformation of the double layer due to the motion of ions, whose direction is opposite to the direction of particle motion (Fig. 9.4).

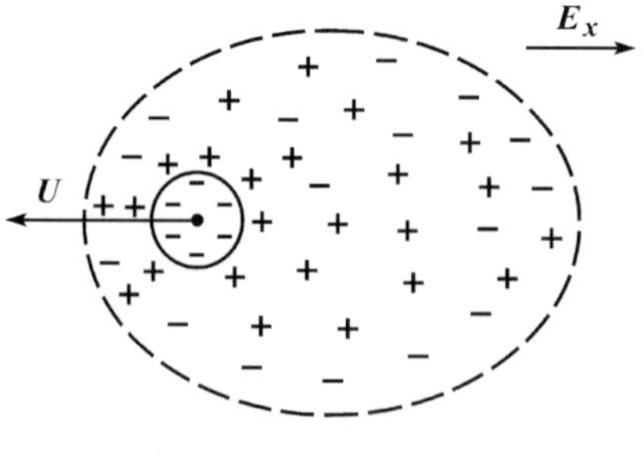

Fig. 9.4 Effect of relaxation.

The motion of ions can be pictured as attempting to tear off the double layer from the particle, so that the centre of the double layer lies behind the centre of the particle. The word relaxation means that a certain finite time is needed for the restoration of the double layer via migration and diffusion of ions.

For the limiting cases of large and small values of λ_D/a, the influence of this phenomenon can be neglected.

9.3
The Motion of a Drop in an Electric Field

Consider the motion of a drop in an electrolyte solution under the action of applied electric field [52]. Suppose that the thickness of the electric double layer is small in comparison with the drop radius ($\lambda_D \ll a$), and that the drop is ideally polarizable, i.e. no discharge or formation of ions occurs at the drop surface and no current flow through the drop. It is also assumed that the drop has spherical shape. It will be shown in Part V that the drop may deform under the action of external electric field, extending parallel to the field strength vector and assuming the shape of on ellipsoid. The spherical shape assumption is valid if the strength of external electric field does not surpass some critical value. Electric potential is described by the Poisson equation (9.24)

$$\Delta\phi = -\rho_E/\varepsilon. \tag{9.36}$$

Far from the drop, $\phi = Er\cos\theta$ coincides with the potential of the applied field, and at the surface $\phi = \phi_w$ (for convenience, from now on, it will be taken as the zero potential, that is, $\phi = 0$ at $r = a$).

We are primarily interested in the case when drop conductivity is much greater than the solution conductivity, which is true, for example, for a liquid metal drop or for a drop of brine water in oil. Denote by $\Delta\phi = \phi_0 - \phi_p$ the potential difference between the drop and the solution. The state of the surface layer and therefore the surface tension Σ at the drop–solution boundary is uniquely determined by the value $\Delta\phi$. When the drop moves, ions in the double layer are displaced to the rear part of the drop, and thus $\Delta\phi$ changes along the drop surface. The value of Σ changes as well. At the drop surface, there appears an additional tangential force equal to $\boldsymbol{F}_t = \nabla_\theta\Sigma$, that is, to the gradient of surface tension. The surface tension Σ is related to the potential jump $\Delta\phi$ and the surface charge of the drop q_s (i.e. the unit surface area charge of the mobile part of the double layer), through the Lippmann-Helmholtz equation

$$\frac{\partial\Sigma}{\partial(\Delta\phi)} = -q_s. \tag{9.37}$$

Since $\Sigma = \Sigma(\theta)$ and $d\Sigma = (d\Sigma/d\theta)\,d\theta = (\partial\Sigma/\partial(\Delta\phi))(\partial(\Delta\phi)/\partial\theta)\,d\theta$, we have

$$\Sigma(\theta) = \Sigma_{\pi/2} + \int_{\pi/2}^{\theta} \frac{\partial\Sigma}{\partial(\Delta\phi)} \frac{\partial(\Delta\phi)}{\partial\theta}\,d\theta = \Sigma_{\pi/2} - \int_{\pi/2}^{\theta} \frac{\partial(\Delta\phi)}{\partial\theta}\,d\theta,$$

where $\Sigma_{\pi/2}$ is the value of Σ at the equator of the drop.

If the applied field is small, it has a weak influence on q_s, so q_s can be considered independent of θ and

$$\Sigma(\theta) = \Sigma_{\pi/2} - q_s\Delta\phi. \tag{9.38}$$

The presence of surface tension varying with θ results in the emergence of normal $\tau_{rr}^{(\Sigma)} = F_n$ and tangential $\tau_{r\theta}^{(\Sigma)} = -F_t$ stresses at the drop surface, where

$$F_n = \frac{2\Sigma}{a} = \frac{2\Sigma_{\pi/2}}{a} - \frac{2q_s\Delta\phi}{a}, \tag{9.39}$$

$$F_t = \nabla_\theta\Sigma = -q_s\nabla_\theta(\Delta\phi) = -q_s\frac{1}{a}\frac{\partial(\Delta\phi)}{\partial\theta}. \tag{9.40}$$

Outside of the double layer, the solution is electrically neutral, so the potential there must satisfy the Laplace equation

$$\Delta\phi = 0 \tag{9.41}$$

The boundary conditions for (9.41) are: the condition that holds on an infinite distance from the drop

$$\phi \to Er\cos\theta \quad \text{at } r \to \infty, \tag{9.42}$$

and the condition at the solution-double layer boundary ($r = a + \lambda_D$). Under the action of the external field, ions in the double layer move with the velocity u in the tangential direction to the drop surface. This motion results in the occurrence of surface convective current whose density is $\boldsymbol{i}_s = q_s\boldsymbol{u}_\theta$. Since u_θ changes along the drop surface, $\nabla_\theta \cdot (q_s\boldsymbol{u}_\theta) \neq 0$. From the law of charge conservation at the solution-double layer boundary, there follows

$$\sigma_b\frac{\partial\phi}{\partial r} = \nabla_\theta \cdot (q_s\boldsymbol{u}_\theta), \tag{9.43}$$

where σ_b is the electric conductivity of the solution.

The physical meaning of the condition (9.43) consists in the equality of the convective flux of ions in the double layer to the flux of ions through the external boundary of the double layer. It should be noted that the condition (9.43) is written in the assumption of an ideally polarizable drop. If the drop is not ideally polarizable, then ions can be discharged or formed at the drop surface, in other words, there may occur an exchange of ions between the drop and the medium. In this case ions can be exchanged, or, to use another term, current can flow through the drop surface. The density of this current should be added to the right-hand side of (9.43). Thus the condition (9.43) serves as the second boundary condition for Eq. (9.41).

Proceed now to the formulation of the hydrodynamical problem. The motion of the drop is supposed to be inertialess. Introduce a coordinates system moving with the drop. Then, by virtue of spherical symmetry, the problem would be similar to the problem of Stokesian flow around a liquid drop. In the spherical system of coordinates, the equations describing the flow inside and outside the drop are

$$\mu\left(\frac{\partial^2 u_r}{\partial r^2}+\frac{1}{r^2}\frac{\partial^2 u_r}{\partial \theta^2}+\frac{2}{r}\frac{\partial u_r}{\partial r}+\frac{\cot\theta}{r^2}\frac{\partial u_r}{\partial \theta}-\frac{2}{r^2}\frac{\partial u_\theta}{\partial \theta}-\frac{2u_r}{r^2}-\frac{2\cot\theta}{r^2}u_\theta\right)=\frac{\partial p}{\partial r}, \tag{9.44}$$

$$\mu\left(\frac{\partial^2 u_\theta}{\partial r^2}+\frac{1}{r^2}\frac{\partial^2 u_\theta}{\partial \theta^2}+\frac{2}{r}\frac{\partial u_\theta}{\partial r}+\frac{\cot\theta}{r^2}\frac{\partial u_\theta}{\partial \theta}+\frac{2}{r^2}\frac{\partial u_r}{\partial \theta}+\frac{2}{r^2}\frac{\partial u_r}{\partial \theta}-\frac{u_\theta}{r^2\sin^2\theta}\right)$$
$$=\frac{1}{r}\frac{\partial p}{\partial \theta}, \tag{9.45}$$

$$\frac{\partial u_r}{\partial r}+\frac{1}{r}\frac{\partial u_\theta}{\partial \theta}+\frac{2u_r}{r}+\frac{u_\theta\cot\theta}{r}=0. \tag{9.46}$$

Here (9.44) and (9.45) are the equations of motion and (9.46) is the continuity equation for both the external and internal flows. Parameters of the current inside the drop will be denoted below by primed symbols.

The boundary conditions for equations (9.44)–(9.46) are the conditions that apply far from the drop

$$u_r \to U\cos\theta, \quad u_\theta \to -U\sin\theta \quad \text{at } r\to\infty, \tag{9.47}$$

the conditions of finiteness of u_r' and u_θ at the center of the drop, and, finally, the equality of velocities and stresses at the drop surface

$$u_r = u_r' = 0, \quad u_\theta = u_\theta' \quad \text{at } r = a, \tag{9.48}$$

$$\tau_{rr}^{(\mathrm{V})}+\tau_{rr}^{(\Sigma)}=\tau_{rr}', \quad \tau_{r\vartheta}^{(\mathrm{V})}+\tau_{r\theta}^{(\Sigma)}=\tau_{r\theta}'. \tag{9.49}$$

Here $\tau^{(w)}$ and $\tau^{(\Sigma)}$ denote viscous and capillary stresses, and

$$\tau_{rr}^{(\mathrm{V})}=-p+2\mu\frac{\partial u_r}{\partial r}, \quad \tau_{r\theta}^{(\mathrm{V})}=\mu\left(\frac{1}{r}\frac{\partial u_r}{\partial \theta}+\frac{\partial u_\theta}{\partial r}-\frac{u_\theta}{r}\right). \tag{9.50}$$

Thus, the task reduces to solving the hydrodynamic and electrodynamic equations with the appropriate boundary conditions. Omitting the mathematics, we present the final results for the electric field potential

$$\phi = E\cos\theta\left(r + \left(\frac{1}{2} - \frac{3}{2}\frac{q_s U}{\sigma_b Ea}\right)\frac{a^3}{r^2}\right) \tag{9.51}$$

and for the velocity of drop motion

$$U = \frac{q_s Ea}{2\mu + 3\mu' + q_s^2/\sigma_b}. \tag{9.52}$$

It follows from (9.51) that the maximum potential difference at the surface of a motionless drop is equal to $\Delta\phi_0 = 3Ea$. If we introduce the mobility b of a drop of radius $a = 1$, then

$$b = \frac{U(a=1)}{E}\frac{q_s Ea}{2\mu + 3\mu' + q_s^2\sigma_b}$$

and (9.52) will be written as

$$U = \frac{1}{3}b\Delta\phi_0. \tag{9.53}$$

Consider now the case of motion of emulsion drops in an electric field (this case is important for many applications). In contrast to the case considered above, electric conductivity of emulsion drops can be smaller or equal to that of the external liquid. For such a system, an electric field can exist inside the drop, and the condition (9.41) holds not only for the external, but also for the internal liquid.

The appropriate expression for the drop velocity is

$$U = \frac{q_s Ea}{2\mu + 3\mu' + q_s^2(1/\sigma_b + 2/\sigma_b')}, \tag{9.54}$$

where σ_b' is the electric conductivity of the internal liquid.

At $\sigma_b' \gg \sigma_b$, the expression (9.54) transforms into (9.52). For the water-in-oil type of the emulsion (for example, water drops in oil), this inequality is obeyed, and the motion velocity of water drops is governed by the law (9.52). For emulsion of the oil-in-water type, $\sigma_b' \ll \sigma_b$ and the formula (9.54) transforms into

$$U = \frac{q_s Ea}{2\mu + 3\mu' + q_s^2/\sigma_b'} \approx \frac{\sigma_b' Ea}{2q_s}. \tag{9.55}$$

Since σ_b' is small, the velocity of oil drops in the water is also small.

The presence of a charge at the drop surface is of essential importance not only for drop motion, but also for sedimentation in the gravitational field. The problem is formulated in a similar manner, but in the hydrodynamic equations for the external liquid, it is necessary to take into account the Archimedian force

$$\nabla p = \mu \Delta u + (\rho - \rho')g. \tag{9.56}$$

Introducing pressure $\pi = (\rho - \rho')gz$, where the z-axis is directed we arrive at the following expression

$$\nabla (p - \pi) = \mu \Delta \mathbf{u}. \tag{9.57}$$

whose form is identical to that of Eqs. (9.44) and (9.45) if we replace p with $\mathrm{p} - \pi$. Therefore the solution may be carried out in a similar way. For a drop whose conductivity exceeds that of the external liquid, the velocity of sedimentation in the gravitational field is

$$U = \frac{2(\rho - \rho')ga^2}{3\mu} \left(\frac{\mu + \mu' + q_s^2/3\sigma_b}{2\mu + 3\mu' + q_s^2/\sigma_b} \right). \tag{9.58}$$

At $2\mu + 3\mu' \gg q_s^2/\sigma_b$, the expression (9.58) transforms into the Hadamar-Rybczynski formula. If $2\mu + 3\mu' \ll q_s^2/\sigma_b$, then the drop velocity coincides with the velocity of a solid sphere (the Stokes formula).

The distribution of the electric field potential over the drop surface is

$$\phi = \frac{q_s(\rho - \rho')ga}{3(2\mu + 3\mu' + q_s^2/\sigma_b)\sigma_b} \frac{a^3}{r^2} \cos\theta, \tag{9.59}$$

from which it follows that the maximum value of potential on the drop surface is

$$\Delta\phi = \frac{2q_s(\rho - \rho')ga}{3(2\mu + 3\mu' + q_s^2/\sigma_b)\sigma_b}. \tag{9.60}$$

9.4 Sedimentation Potential

In the previous section, it was shown that in the process of sedimentation of a charged drop in an electrolyte solution, a potential difference is formed at the drop ends, which is determined from the formula (9.60). If the solution contains a series of drops arranged in a line, one after another, then along the column of the mixture, a potential drop is established. This potential drop is known as sedimentation potential [52]. The purpose of this section is to determine the sedimentation potential.

Let the volume concentration of drops be small, so the drops are spaced far apart, the sedimentation of each drop occurs independently from the others, and the electric fields created by the drops are additive. Consider first the case of a single drop descending in a column of liquid. The linear size of the column is much greater than the drop radius. Let us select an imaginary plane above the drop in such a way that the distance between the drop and the plane is small in comparison with the radius of the column of drops, and try to find the average value of potential $\bar{\phi}_d$ in some region on this plane. Place the origin of coordinates in the drop center. The area of the region that lies on the plane between the angles θ and $\theta + d\theta$ is equal to $2\pi r^2 \tan\theta\, d\theta$. Then the average potential value $\bar{\phi}_d$ is derived by using the formula (9.59):

$$\bar{\phi}_d = \frac{1}{S}\int 2\pi r^2 \phi \tan\theta\, d\theta = \frac{2\pi q_s(\rho-\rho')ga^4}{S3\sigma_b(2\mu+3\mu'+q_s^2/\sigma_b)}\int_{\pi/2}^{\pi}\sin\theta\, d\theta$$

$$= \frac{2\pi}{3}\frac{q_s(\rho-\rho')ga^4}{S\sigma_b(2\mu+3\mu'+q_s^2/\sigma_b)}. \tag{9.61}$$

Analogously, for a region of the similar plane below the drop,

$$\bar{\phi}_d' = \frac{2\pi}{3}\frac{q_s(\rho-\rho')ga^4}{S\sigma_b(2\mu+3\mu'+q_s^2/\sigma_b)}. \tag{9.62}$$

It follows that the sedimentation potential caused by the downward motion of one drop is

$$\phi_{sed} = \phi_d' - \phi_d = \frac{4\pi}{S}\frac{q_s(\rho-\rho')ga^4}{3\sigma_b(2\mu+3\mu'+q_s^2/\sigma_b)}. \tag{9.63}$$

Let n be number of drops in a unit volume. Then the sedimentation potential in a column of the mixture is equal to

$$\phi_{sed} = nS(\phi_d' - \phi_d) = \frac{4\pi q_s(\rho-\rho')ga^4}{3\sigma_b(2\mu+3\mu'+q_s^2/\sigma_b)}. \tag{9.64}$$

If we close the electrical circuit formed by the column of liquid of length L and cross section S through the external resistance W, then equalization of potential will occur in the external circuit, and there will appear an electric current:

$$I = \frac{\phi_{sed} L}{L/S\sigma_b + W}. \tag{9.65}$$

10
Stability of Suspensions, Coagulation of Particles, and Deposition of Particles on Obstacles

10.1
Stability of Colloid Systems

In the absence of external forces (gravitational, centrifugal, electrical), uncharged particles dispersed in a quiescent liquid should be distributed homogeneously. Actually, there is always an interaction between particles: electrostatic repulsion (for charged particles surrounded by electric double layers), molecular attraction (Van der Waals forces), hydrodynamic forces (forces arising due to the mutual influence of the velocity fields of liquid and particles).

Electrostatic repulsion forces between equally charged particles support a homogeneous distribution. The ability to maintain a homogeneous distribution of particles in a liquid for a long time is referred to in the colloid chemistry as stability of a colloid system. This concept only applies to liophobic colloids, i.e. colloids insoluble in liquid.

In practice, in the majority of biphase colloid systems, the number of particles decreases with time, and their size is simultaneously increased. In an emulsion, collision of two drops results in their coalescence with the formation of one drop of greater size. Solid particles undergoing collision do not coalesce, but form aggregates. Aggregation of particles occurs due to attractive Van der Waals forces between molecules.

Systems in which particle aggregation occurs are referred to as unstable.

In the absence of external hydrodynamic forces, the stability of a colloid depends on particles' interaction caused by surface forces: electrostatic repulsion and molecular attraction [52]. In order for the particles to interact with each other under influence of these forces, they need to be sufficiently close to one another. The particles' approach in a liquid occurs under the action of Brownian motion, due to the influence of external forces, for example, gravity, or due to hydrodynamic forces. Studies of stability of the colloid systems should be carried out with due consideration of all the factors listed. Generally, this problem is very difficult, and therefore we consider first the interaction of particles under the action only of electrostatic and molecular forces. The theory of stability of a colloid system subject to such interactions is called DLFO theory as an acronym of its founders – Derjaguin, Landau, Ferwey, and Overbeck [53].

Separation of Multiphase, Multicomponent Systems. E. G. Sinaiski and E. J. Lapiga

ISBN: 978-3-527-40612-8

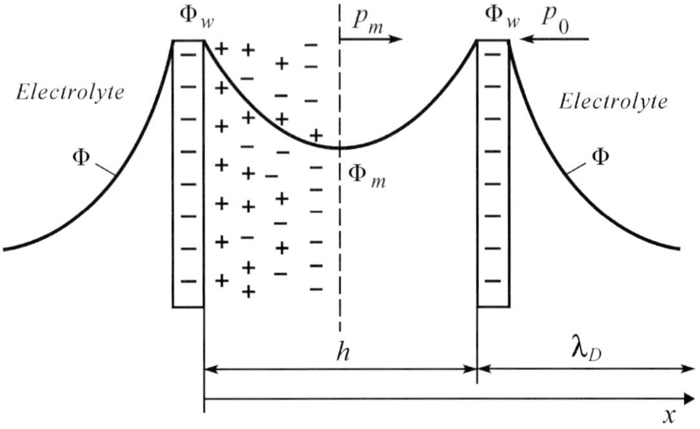

Fig. 10.1 Electrostatic interaction of two like-charged planes in an electrolyte solution.

The energy of inter-particle repulsion can be determined using the theory of electrical double layers (see Section 7.3). For simplicity, consider first the case of electrostatic repulsion of two parallel planes carrying surface charge of the same sign (negative) and having identical potentials f_w. These planes are placed in an infinite volume of electrolyte with a concentration C_0 of dissolved ionized substances far from the plane (Fig. 10.1).

Near each plane an electrical double layer of thickness l_D is formed. If the distance between planes is such that the double layers are superimposed, there is interaction between the planes. Due to symmetry, the minimum value of the potential distribution between the planes is achieved at the symmetry axis of the planes ($x = h/2$), where $d\phi/dx = 0$, that is, the electric field is absent. At the axis, there is an excess of ions of the same sign as the charge on the planes, which results in increase of osmotic pressure trying to force the planes apart.

The balance condition for such planes in the case of a quiescent liquid reduces to a balance of the electric force and the pressure gradient

$$-\nabla p + \rho_E \mathbf{E} = 0. \tag{10.1}$$

Since $\mathbf{E} = -\nabla\phi$, Eq. (10.1) reduces to

$$\frac{dp}{dx} + \rho_E \frac{d\phi}{dx} = 0. \tag{10.2}$$

Substituting ρ_E with the help of the Poisson equation, Eq. (5.88), one obtains:

$$\frac{dp}{dx} - \varepsilon \frac{d^2\phi}{dx}\frac{d\phi}{dx} = 0. \tag{10.3}$$

From this equation, the first integral follows:

$$p - \frac{\varepsilon}{2}\left(\frac{d\phi}{dx}\right)^2 = const. \tag{10.4}$$

A consequence of this equation is the constancy of the difference between hydrostatic and electrostatic pressures. Taking $d\phi/dx = 0$ and $p = p_m$ at $x = h/2$, one can derive the constant. It is obtained as:

$$p - \frac{\varepsilon}{2}\left(\frac{d\phi}{dx}\right)^2 = p_m. \tag{10.5}$$

Note that the value p_m is not yet known. One takes advantage now of expression Eq. (7.65) for rE, which is substituted into Eq. (10.2). One then obtains:

$$dp = 2FzC_0 \sinh\left(\frac{zF\phi}{AT}\right) d\phi. \tag{10.6}$$

Use the conditions $p = p_0$ at $\phi = 0$ and $p = p_m$ at $\phi = \phi_m$. Then from Eq. (10.6), there follows

$$\pi = p_m - p_0 = 2ATC_0\left(\cosh\left(\frac{zF\phi_m}{AT}\right) - 1\right). \tag{10.7}$$

Pressure drop p has the meaning of force acting on a unit area of the boundary plane and aspiring to pull the planes apart. In a sense, p is excessive osmotic pressure at the symmetry axis of the planes in comparison with the pressure in the bulk of the solution.

Expression (10.7) for small values of $zF\phi_m/AT$ can be simplified to

$$\pi \approx ATC_0\left(\frac{zF\phi_m}{AT}\right)^2. \tag{10.8}$$

The Debye–Huckel approximation gives the potential distribution in the electrical double layer: $\phi_w \exp(-x/\lambda_D)$. Then we have at the symmetry axis

$$\phi_m = \phi_1 + \phi_2 = 2\phi_w \exp(-x/\lambda_D). \tag{10.9}$$

Substitution of (10.9) into (10.8), gives us the force acting on a unit plane area

$$\pi \approx \frac{2\varepsilon\phi_w^2}{\lambda_D^2} \exp\left(-\frac{h}{\lambda_D}\right). \tag{10.10}$$

Here the expression (7.61) for λ_D has been used. Note that ϕ_w usually means the ξ-potential.

In the literature on colloid chemistry, interaction between particles is commonly defined in terms of potential energy of interactions with a unit surface area. For determination of this energy, it is necessary to integrate π over the range of h (from ∞ to h), keeping in mind that energy tends to 0 at $x \to \infty$:

$$V_R^{plane} = -\int_{\infty}^{h} \pi \, dh = \frac{2\varepsilon\phi_w^2}{\lambda_D} \exp\left(-\frac{h}{\lambda_D}\right). \tag{10.11}$$

From Eq. (10.11), it follows that the potential of repulsive forces decreases with the increase of concentration of counter-ions of electrolyte C_0, since, according to Eq. (7.61), $\lambda_D \sim C_0^{-1/2}$ and

$$V_R^{plane} \sim C_0^{-1/2} \exp(-constC_0^{-1/2}). \tag{10.12}$$

Note that expression (10.11) is valid for small values of ϕ_m, and $\phi_w = \zeta$. Skipping derivation, we adduce V_R^{plane} at small ϕ_m and large ϕ_w:

$$V_R^{plane} = \frac{2\varepsilon}{\lambda_D}\left(\frac{4AT\gamma}{zF}\right)^2 \exp\left(-\frac{h}{\lambda_D}\right); \quad \gamma = \tanh\left(\frac{zF\phi_w}{4AT}\right). \tag{10.13}$$

A similar calculation can be carried out for the potential energy of repulsive forces between two identical spherical particles. In the case when the thickness of the electrical double layer around these particles is small compared to their radii, interaction of electrical double layers of these spheres may, according to Derjaguin, be considered as a superposition of interactions of infinitely narrow parallel rings (Fig. 10.2) [53].

The full energy of repulsion of spheres is equal to

$$V_R^{sphere} = -\int_{\infty}^{0} V_R^{plane} 2\pi H \, dH. \tag{10.14}$$

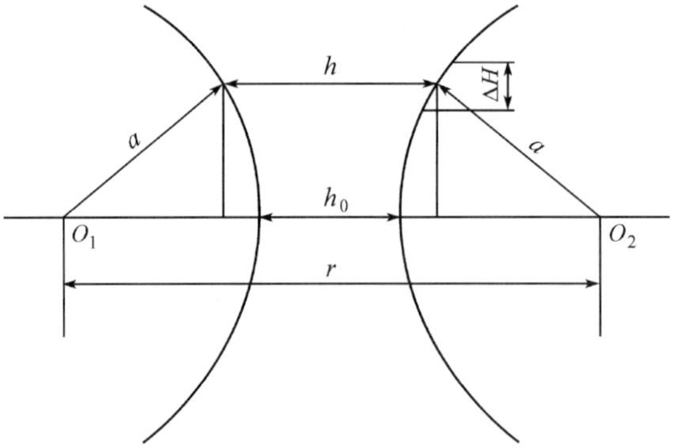

Fig. 10.2 On calculation of repulsion forces between two identical spherical particles.

The main contribution to the interaction is provided by the integral values in the vicinity of the minimum distance h_0 between the surfaces. In this region, $(h - h_0)/2 \ll a$, $2H\,dH = a\,dh$, and (10.14) will become

$$V_R^{sphere} = -\pi a \int_{\infty}^{h_0} V_R^{plane}\,dh. \tag{10.15}$$

Plug in V_R^{plane} from (10.11) for small values of ϕ_m and ϕ_w. Then

$$V_R^{sphere} = 2\pi\varepsilon a\phi_w^2 \exp(-h_0/\lambda_D). \tag{10.16}$$

For small ϕ_m and big ϕ_w, we obtain from (10.13)

$$V_R^{sphere} = 2\pi\varepsilon a\left(\frac{4aT\gamma}{zF}\right)^2 \exp\left(-\frac{h_0}{\lambda_D}\right). \tag{10.17}$$

The electrostatic potential for two spheres of radii a and b $(a < b)$ with the distance r between their centers, is equal to [55]

$$V_R^{sphere} = \frac{\varepsilon b\phi_b^2}{4}\frac{a}{(a+b)}\left(-2\frac{\phi_a}{\phi_b}\ln\frac{1-e^{-\chi h_0}}{1+e^{-\chi h_0}} + \left(1+\frac{\phi_a^2}{\phi_b^2}\right)\ln(1-e^{-\chi h_0})\right), \tag{10.18}$$

where ϕ_a and ϕ_b are the potentials on particle surfaces; χ is the inverse radius of the electrical double layer.

The Van der Waals attractive forces are forces of molecular origin, though at their basis lie electrical interactions. By their nature, these forces are caused by molecular polarization under the influence of fluctuations of charge distribution in the neighboring molecule and vice versa. These forces are also known as London's dispersive forces. The potential energy of molecular interaction (the London attraction energy) is equal to

$$V_A = b_{12}/r^6, \tag{10.19}$$

where r is the distance between molecules; b_{12} is a factor dependent on substance properties.

Consider two particles with finite volumes and denote the molecular concentrations (number of molecules per unit volume) in each particle as n_1 and n_2, respectively. Then the attraction energy of these particles is

$$V_A = -b_{12}n_1n_2\int\frac{dV_1\,dV_2}{r^6} = \frac{\Gamma}{\pi^2}\int\frac{dV_1\,dV_2}{r^6}. \tag{10.20}$$

The constant Γ is referred to as the Hamaker constant; its characteristic values are range within $(10^{-20}–10^{-19})$ J. The forces of molecular attraction between two parallel planes, and between two spherical particles have been calculated by Hamaker [56]. He has shown that the attraction force between the particles falls off with increase of distance between them slower than it does according to the London's law for interaction between molecules. In particular, the attraction energy per unit area for two parallel planes is given by

$$V_R^{plane} = -\Gamma/12\pi h^2, \tag{10.21}$$

where h is the distance between planes. And the interaction energy for two identical spherical particles of radius r under condition $h_0 \ll a$, where h_0 is the minimum distance between their surfaces, is equal to

$$V_R^{sphere} = -a\Gamma/12h_0. \tag{10.22}$$

For identical spheres at any distance r between their centers, this gives the expression

$$V_R^{sphere} = -\frac{\Gamma}{6}\left(\frac{2a^2}{r^2-4a^2} + \frac{2a^2}{r^2} + \ln\left(\frac{r^2-4a^2}{a^2}\right)\right). \tag{10.23}$$

In case of two different spheres with radii a and b we have:

$$V_R^{sphere} = -\frac{\Gamma}{6}\left(-\frac{8(a/b)}{(s^2-4)(1+a/b)^2} + \frac{8(a/b)}{s^2(a/b+1)^2 - 4(1-a/b)^2} + \ln\left(\frac{(s^2-4)(1+a/b)^2}{s^2(1+a/b)^2 - 4(1-a/b)^2}\right)\right). \tag{10.24}$$

where $s = 2r/(a+b)$.

The total potential energy of interaction between two spherical particles is equal to the sum of attraction and repulsion energies:

$$V = V_A + V_R. \tag{10.25}$$

Shown in Fig. 10.3, is the characteristic dependence $V(r)$ between the two identical spherical particles, for three different repulsion energies $V_R^{(1)}$, $V_R^{(2)}$, $V_R^{(3)}$, and the same attraction energy V_A.

The repulsion energy exponentially decreases with increase of h_0, with the characteristic distance λ_D. Energy of attraction decreases as $1/h_0$. Therefore the Van der Waals attraction force prevails at rather small and great distances between particles while at intermediate distances force of repulsion dominates. The curve $V^{(1)}$ corresponds to a stable system, since the repulsion force between particles at intermediate distances prevents their coagulation. The curve $V^{(3)}$ corresponds to unstable state of the system. Such a system is said to be subject to a fast coagulation. The curve $V^{(2)}$ corresponds to an intermediate region between stable and unstable states.

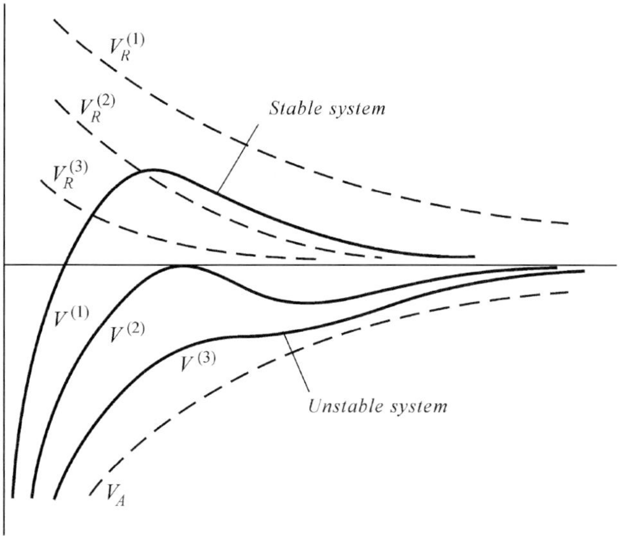

Fig. 10.3 Energy of particles' interaction.

If the maximum value of V by far exceeds the particles' thermal energy AT, the system is stable, otherwise it is capable of coagulation, that is, unstable. The energy barrier depends on value of the ζ-potential of the particles and radius of the electric double layer. Another distinctive feature of colloid liophobic systems is the sensitivity of particle' coagulation to concentration of the electrolyte. According to (7.61), the increase of electrolyte concentration results in decrease of the double-layer's thickness. This means that particles can be brought closer together without interaction of double layers, and thus begin to coagulate.

As is known in the colloid chemistry, the electrolyte concentration C_0 necessary for fast coagulation of colloids, strongly depends on charge of the "anti-ions", that is, the ions with the charge opposite to that of the particles in question. On the other hand, stability of colloid system practically does not depend on the ions' charge and on concentration of colloid particles. This is consistent with the Schulze–Hardy law, according to which, it is the valence of the "anti-ions" that exerts the basic influence on stability of colloidal systems.

This conclusion follows directly from the DLFO-theory [36]. Indeed, the first maximum on potential curve $V^{(2)}$ in Fig. 10.3 separates the stable and unstable regions for the system [57]. At this point, the two conditions are satisfied:

$$V = V_A + V_R = 0. \tag{10.26}$$

$$\frac{dV}{dh} = \frac{dV_A}{dh} + \frac{dV_R}{dh} = 0. \tag{10.27}$$

Consider the case when the values f_w are large. Substitution of (10.17) and (10.22) into (10.26) yields

$$2\pi\varepsilon a\left(\frac{4aT\gamma}{zF}\right)^2 \exp\left(-\frac{h_0}{\lambda_D}\right) - \frac{a\Gamma}{12h_0} = 0. \tag{10.28}$$

The condition (10.27) takes the form

$$-\frac{V_R}{\lambda_D} - \frac{V_A}{h_0} = 0. \tag{10.29}$$

It follows from this equation that the first maximum of the curve $V^{(2)}$ is reached at $h_0 = \lambda_D$. Substituting this value into (10.28), one obtains the approximate value of the Debye critical radius:

$$(\lambda_D)_{crit} \sim \Gamma z^2/\gamma^2. \tag{10.30}$$

According to (7.61), we have $\lambda_D \sim (Cz^2)^{-1/2}$. Therefore it follows from (10.30) that the critical value of electrolyte concentration, at which the colloid system loses stability, is equal to

$$C_{crit} \sim \gamma^4/\Gamma^2 z^6. \tag{10.31}$$

At high values of the surface potential of colloid particles ϕ_w the parameter $\gamma \sim 1$, therefore it follows from (10.31) that

$$C_{crit} \sim 1/z^6. \tag{10.32}$$

Let the electrolyte solution contain "anti-ions" Na^+, Ca^{2+} and Al^{3+}. Then the ratio of charges is $z_1 : z_2 : z_3 = 1 : 2 : 3$. According to (10.32), for a spontaneous coagulation of colloids, the critical concentration of counter-ions should satisfy the proportions $1 : 2^{-6} : 3^{-6}$ or 100 : 1.56 : 0.137.

Note that the critical concentration of electrolyte at low particle surface potentials ϕ_w strongly depends on ζ-potential, while it is practically independent of ζ at high values of ϕ_w. Besides, the critical concentration does not depend on particles' size at a given value of ϕ_w.

10.2
Brownian, Gradient (Shear) and Turbulent Coagulation

In the previous section, the possibility of the particles' coagulation was considered for the case when the particles were pulled close enough together. For the coagulation to be possible under this condition, the suspension should be unstable. The coagulation rate in the system is determined by the particles' collision frequency as a result of their relative motion in the ambient liquid.

There are some mechanisms, leading to the particle's mutual approach. The first one is the Brownian motion. The corresponding coagulation type is referred to in this case as perikinetic. The Brownian coagulation is the basic coagulation mechanism for particles, whose size is less than one micrometer. The basis of the second mechanism is the relative motion of particles in the field of velocity gradient of carrying liquid. This type of coagulation is called gradient, shift, and also ortokinetic coagulation. It is characteristic of particles, whose size exceeds one micrometer. Also possible is the coagulation of particles due to different velocities of their motion in quiescent liquid under gravity (at sedimentation). Such coagulation is called the gravitational coagulation.

Listed mechanisms, as a rule, exist in a quiescent liquid or in liquid in a state of laminar flow.

The third mechanism, named turbulent coagulation, is characteristic of coagulation of particles suspended in turbulent flow, for example in a pipe or in some special mixing devices – mixers, agitators, etc. In some aspects, the turbulent coagulation is similar to the Brownian one, since in the first case the particles' approach is due to random turbulent pulsations, and in the second case it is due to random thermal motion of particles.

Regarding the coagulation rate, all kinds of coagulation are divided into two classes: the fast and slow coagulation. If the system is completely unstable, that is, the repulsion forces are small enough to be neglected, then each collision of particles results in their coagulation. Such coagulation is called fast. The presence of a stabilizer in liquid – the electrolyte – results in the emergence of repulsive forces due to the electric double layers at particle surfaces. This means that coagulation is slowed down. Therefore such coagulation is called slow.

For a quantitative description of slow coagulation, Smoluchowski has suggested to formally introduce into the expression for the particle collision frequency a factor $\alpha \leq 1$, describing the share of collisions resulting in formation of aggregates. Introduction of this factor is equivalent to increase of the characteristic coagulation time by a factor $1/\alpha$. The coagulation rate is characterized by stability factor W. It is the ratio of the particles' collision rates without and with the force of electrostatic repulsion [58]

$$W = I/I_R. \tag{10.33}$$

At fast coagulation $I_R = I$ and $W = 1$, while at slow coagulation $I_R < I$ and $W > 1$.

Introduction of the stability factor gives the physical meaning to factor α. Since $\alpha = 1/W$, the coagulation rate slows down by a factor of W.

In particular, for Brownian coagulation of identical spherical particles of radius [53],

$$W = 2a \int_{2\alpha}^{\infty} e^{V(r)} \frac{dr}{r^2}, \tag{10.34}$$

where $V(r)$ is the particles interaction potential; r is the distance between the particle centers.

Further on we will discuss the fast coagulation, paying main attention to determining the particles collision frequency. Slow coagulation will be considered in section V.

10.2.1
Brownian Coagulation

Consider a problem on definition of collision frequency of small spherical particles executing Brownian motion in a quiescent liquid. In Section 8.2, Brownian motion was considered as diffusion with a effective diffusion factor. It was supposed that suspension is sufficiently diluted, so it is possible to consider only the pair interactions of particles. To simplify the problem, consider a bi-disperse system of particles, that is, a suspension consisting of particles of two types: particles of radius a_1 and particles of radius a_2. In this formulation, the problem was first considered by Smolukhowski [59].

In the course of Brownian motion particles are in state of chaotic motion and randomly collide with each other. Choose one particle of radius a_1 called the test particle with a system of coordinates connected to this particle, with the origin at the particle's center (Fig. 10.4).

The adopted diffusion model of Brownian motion allows to us consider the collision frequency of particles of radius a_2 with the test particle of radius a_1 as a diffusion flux of particles a_2 toward the particle a_1. Assume the surface of the particle a_1 to be ideally absorbing. It means that as soon as the particle a_2 will come into contact with the particle a_1, it will be absorbed by this particle. In other words, absorption occurs as soon as the center of the particle a_2 reaches the surface of a sphere of radius $R_c = a_1 + a_2$. The quantity R_c is called the coagulation radius. Hence, the concentration of particles $a2$ should be equal to zero at $r = a_1 + a_2$.

The characteristic time of Brownian diffusion is estimated by the expression

$$t_{brown} \sim \frac{a^2}{D_{br}} = \frac{6\pi\mu a^3}{kT}. \tag{10.35}$$

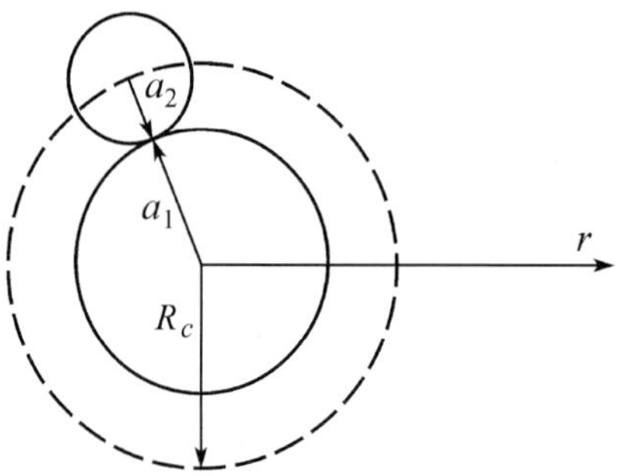

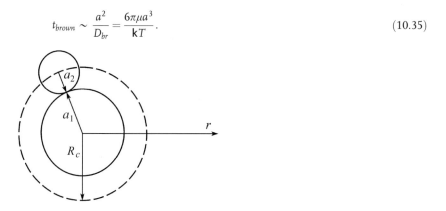

Fig. 10.4 The model of Brownian coagulation of particles according to Smoluchowski.

For particles of radius $a \sim 0.1$ mm in a water solution under normal conditions, this time is $t_{brown} \sim 5 \cdot 10^{-3}$ s. Hence, if we consider times greater than t_{brown}, the process of diffusion of particles a_2 can be regarded as stationary. The number concentration of these particles n is described by the diffusion equation

$$D_{12} \frac{1}{r^2} \frac{d}{dr} \left(r^2 \frac{dn}{dr} \right) = 0. \tag{10.36}$$

with the diffusion factor D_{12} known as the factor of Brownian diffusion and describing the relative motion of two particles. Since the motion of particles is considered independently, $D_{12} = D_1 + D_2$, where D_1 and D_2 are the factors of Brownian diffusion of particles a_1 and a_2, given by the formula (8.70).

The boundary conditions are: the condition of constant concentration of particles $a2$ in the bulk of the liquid, and the condition of ideal absorption at the surface of the particle a_1

$$n \to n_2 \quad \text{at } r \to \infty; \qquad n = 0 \quad \text{at } r = a_1 + a_2. \tag{10.37}$$

The solution of Eq. (10.36) with conditions (10.37) is

$$n = n_2 \left(1 - \frac{a_1 + a_2}{r} \right). \tag{10.38}$$

By its definition, the collision frequency of particles $a2$ with the test particle a_1 is equal to

$$\omega_{12} = D_{12} \left(4\pi r^2 \frac{dn}{dr} \right)_{r=a_1+a_2}. \tag{10.39}$$

If ω_{12} is multiplied by the number of particles a_1, the product will give us the collision frequency ($1/\text{m}^3\cdot\text{s}$) of particles a_2 with particles a_1 under the condition $a_1 \neq a_2$:

$$\beta_{12} = 4\pi n_1 n_2 D_{12}(a_1 + a_2). \tag{10.40}$$

In the case of a monodisperse system ($a_1 = a_2$), the expression for collision frequency needs to be divided by 2 because the same collisions are counted twice: the particle is first considered as the test particle and then as the diffused particle.

$$\beta_{ii} = 8\pi n_i^2 a_i D_i = \frac{4\mathrm{k}T n_i^2}{3\mu}. \tag{10.41}$$

In a simple model, when the considered volume is filled with identical particles engaged in the process of Brownian coagulation, and the suspension is assumed to remain monodisperse, the change of number concentration of particles with time is described by the balance equation for the particles:

$$\frac{dn}{dt} = \frac{4}{3}\frac{\mathrm{k}T}{\mu}n^2. \tag{10.42}$$

If the initial concentration of particles is $n = n_0$, then

$$n = \frac{n_0}{1 + t/\tau}, \quad \tau = \left(\frac{4}{3}\frac{\mathrm{k}T}{\mu}n_0\right)^{-1}. \tag{10.43}$$

where t is the characteristic time of coagulation. After time $t = \tau$, the initial number of particles will be reduced by the factor of 2.

Note that one can properly account for the forces of surface interaction of particles by introducing into the diffusion equation (10.36) the convective flux caused by the interaction of particles via central force:

$$\frac{1}{r^2}\frac{d}{dr}\left(D_{12}r^2\frac{dn}{dr} + \frac{Fn}{6\pi\mu a}\right) = 0. \tag{10.44}$$

Recalling that $F = -dV/dr$, using the expressions for $V(r)$ given above, and integrating (10.44), one can find the diffusion flux and thus the system's stability factor.

10.2.2 Gradient (Shear) Coagulation

The second mechanism of coagulation is coagulation of particles suspended in a laminar flow of liquid whose velocity is a regular function of position. As the characteristic distance of particle interaction is small, the structure of the velocity field on such linear scales can be considered as linear. In other words, the flow may be regarded as a shear flow, hence the name of the coagulation mechanism. This problem was also studied by Smolukhowski [59].

In the inertialess approximation, particles move in the field of a shear translational flow along the straight lines if we neglect their rotation. Note that rotation of particles results in their transverse drift. If one particle moves with a high velocity, and the other – with a parallel but lower velocity, and the distance between the particles along the velocity gradient does not exceed $a_1 + a_2$, the particles will collide. Such kind of collision has a pure geometrical character, since Brownian motion and hydrodynamical interaction of particles are ignored.

Consider a test particle of radius a_1 whose center coincides with the origin of coordinates, and a particle of radius a_2 moving relative to the test particle in a linear velocity field (Fig. 10.5).

One can see from Fig. 10.5 that collision of particle a_2 with the test particle a_1 occurs when the distance over x-axis between them is less or equal to $(a_1 + a_2)\sin\theta$. The velocity of particle a_2 relative to a_1 is

$$u = \frac{du}{dy}y = \dot{\gamma}y. \tag{10.45}$$

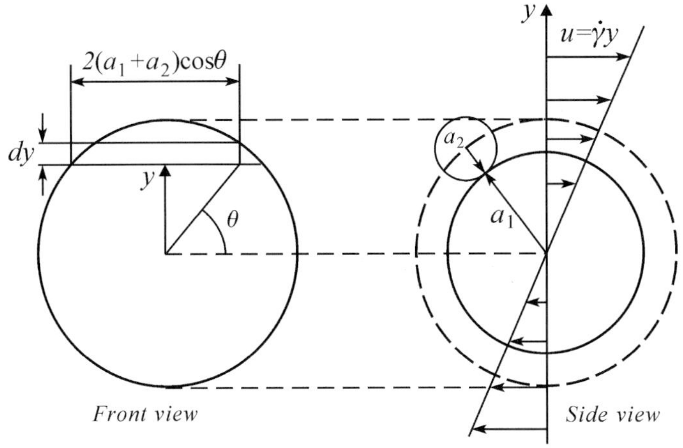

Fig. 10.5 The model of gradient (shear) coagulation of particles according to Smoluchowski.

Here $\dot{\gamma}$ denotes the shear rate.

Consider a strip of width dy. The number of particles a_2 crossing the strip dy in a unit time is

$$d\omega_{12} = un_2 2(a_1 + a_2) \cos\theta\, dy. \tag{10.46}$$

The total collision frequency is equal to

$$\omega_{12} = 2\int_0^{a_1+a_2} 2n_2 u(a_1 + a_2) \cos\theta\, dy. \tag{10.47}$$

Substituting into (10.47) the expression (10.45) for u and recalling that $y = (a_1 + a_2)\sin\theta$, one obtains

$$\omega_{12} = 4n_2\dot{\gamma}(a_1 + a_2)^3 \int_0^{\pi/2} \cos^2\theta \sin\theta\, d\theta. \tag{10.48}$$

Multiplying (10.48) by the number of test particles n_1 and integrating, we obtain the collision frequency of particles a_2 with particles a_1:

$$\beta_{12} = \frac{4}{3} n_1 n_2 \dot{\gamma}(a_1 + a_2)^3. \tag{10.49}$$

If the particles have equal size ($a_1 = a_2$), it is necessary to divide the resulting collision frequency by two, which results in

$$\beta_{ii} = \frac{16}{3}\dot{\gamma} n_i^2 a_i^3. \tag{10.50}$$

Introduce the volume concentration of particles

$$W = \frac{4}{3}\pi a^3 n. \tag{10.51}$$

Then the equation describing the change in time of the number concentration of particles in a monodisperse suspension has the form

$$\frac{dn}{dt} = \frac{4}{\pi}\dot{\gamma} W n. \tag{10.52}$$

Comparing Eq. (10.42) with Eq. (10.52), one can draw the conclusion that Brownian coagulation (10.42) corresponds to a chemical reaction of the second order, and gradient coagulation (10.52) – to a reaction of the first order.

Integrating (10.52) with the initial condition $n = n_0$ and taking into account that for the coagulation process, $W = const$, we get

$$n = n_0 e^{-t/\tau}, \quad \tau = \pi/4\dot{\gamma} W. \tag{10.53}$$

Compare the frequencies of shear and Brownian coagulation:

$$\frac{\omega_{shear}}{\omega_{brown}} = \frac{3}{\pi}\frac{\dot{\gamma} W}{(\mathrm{k}T/\mu)n} = \frac{4a^3\dot{\gamma}}{\mathrm{k}T/\mu}. \tag{10.54}$$

From (10.54), it follows that increase of the particle size *and* the shear rate makes the frequency of shear coagulation larger in comparison with that of Brownian one. The particle radius has an especially strong influence ($\sim a^3$) on the ratio (10.54). It is worth to note that the parameter (10.54) is just the Peclet number, equal to the ratio of the characteristic time of Brownian coagulation to the characteristic time of gradient coagulation.

10.2.3 Turbulent Coagulation

The mechanism of coagulation in a laminar flow has limited practical application since in the majority of applications the motion of liquid has a turbulent character. At turbulent motion, the collision frequency of particles increases very considerably in comparison with a quiescent environment or with laminar motion. Now consider the mechanism of particle coagulation in a turbulent flow, following the work [52].

Let particles be suspended in turbulent flow, with the average particle concentration n. Turbulent pulsations are characterized by the velocity v_λ, and by the length λ over which the pulsation velocity undergoes a noticeable change. In a turbulent flow, there are large-scale pulsations (with the upper cap on them imposed by the characteristic linear size of the flow region l, for example, the diameter of the pipe), and small-scale pulsations. Most of the kinetic energy of motion is stored in large-scale pulsations. To each pulsation corresponds its own Reynolds number $\mathrm{Re}_\lambda = v_\lambda \lambda / \nu$, where n is the coefficient of kinematical viscosity of the liquid. For large-scale pulsations, $\mathrm{Re}_\lambda \gg 1$, so these pulsations have nonviscous character. At some $\lambda = \lambda_0$, we have $\mathrm{Re}_{\lambda_0} = 1$. It means that small-scale pulsations with $\lambda < \lambda_0$ have viscous character. The value $\lambda_0 = l/\mathrm{Re}^{3/4}$, where Re is Reynolds number of the flow, is called the inner scale of turbulence (Kolmogorov's scale). One of the characteristic parameters of turbulent motion is specific dissipation of energy ε_0, having the order of U^3/l, where U is the average velocity of the flow. Then

$$\lambda_0 = \left(\nu^3/\varepsilon_0\right)^{1/4}. \tag{10.55}$$

When there is intensive turbulent mixing of the liquid, $\lambda_0 \sim 10^{-4}$ m.

Consider particles of radius $a \ll \lambda_0$ and assume that in the course of their motion in the liquid, they are completely entrained by turbulent pulsations that play the basic role in the mechanism of mutual approach of suspended particles. Then it can be assumed that particle transport is performed via isotropic turbulence. Since particles move chaotically in the liquid volume, their motion is similar to Brownian one and can be considered as diffusion with some effective factor of turbulent diffusion D_T. In the same manner as in the case of Brownian coagulation, it is possible to consider the diffusion flux of particles of radius a_2 toward the test particle of radius a_1. The distribution of particles a_2 is characterized by the stationary diffusion equation

$$\frac{1}{r^2}\frac{d}{dr}\left(D_T r^2 \frac{dn}{dr}\right) = 0. \tag{10.56}$$

The expression for D_T depends on the scale of turbulent pulsations λ and is estimated by the expression [51]

$$D_T \sim \mathrm{v}_\lambda \lambda \sim \begin{cases} \left(\varepsilon_0/\lambda\right)^{1/3}\lambda & \text{at } \lambda > \lambda_0, \\ \sqrt{\varepsilon_0/\nu}\,\lambda^2 & \text{at } \lambda > \lambda_0. \end{cases} \tag{10.57}$$

Just as we did in the case of Brownian coagulation, introduce the radius of coagulation $R_c = a_1 + a_2$. Then the boundary conditions for Eq. (10.56) become

$$n = 0 \quad \text{at } r = R_c = a_1 + a_2, \qquad n = n_2. \tag{10.58}$$

As shown in [52], the relatively large-scale pulsations ($\lambda > \lambda_0 \gg R_c$) stir the liquid vigorously enough to ensure a uniform distribution of particles in the liquid volume. Then the main diffusion resistance lies in the region $r < \lambda_0$. Omitting intermediate calculations, we give the final expression derived in [52] for the coagulation frequency of particles a_2 with the test particle a_1

$$\omega_{12} \approx 12\pi n_2 \left(\frac{\varepsilon_0}{\nu}\right)^{1/2} (a_1 + a_2)^3. \tag{10.59}$$

Multiplying w_{12} by the number of test particles, one finds the collision frequency of particles a_2 with particles a_1:

$$\beta_{12} = 12\pi n_1 n_2 \left(\frac{\varepsilon_0}{\nu}\right)^{1/2} (a_1 + a_2)^3. \tag{10.60}$$

Expressing ε_0 in terms of flow velocity U and the characteristic linear size l, one gets

$$\beta_{12} = 12\pi n_1 n_2 \frac{U^{3/2}}{(\nu l)^{1/2}} (a_1 + a_2)^3 = 12\pi (a_1 + a_2)^3 \frac{\mathrm{Re}^{3/2}}{l^2} \nu n_1 n_2. \tag{10.61}$$

In the case of a monodisperse suspension ($a_1 = a_2$), the expression for the coagulation frequency must be divided by 2. The result is

$$\beta_{ii} = 48\pi \nu a_i^3 n_i^2 \frac{\mathrm{Re}^{3/2}}{l^2}. \tag{10.62}$$

Thus, in the case of coagulation in a turbulent flow, coagulation frequency corresponds to a reaction of the second order and is proportional to a_i^3, as in the laminar flow.

The change of the number of particles is described by the equation

$$\frac{dn}{dt} = -36 \frac{\nu \mathrm{Re}^{3/2}}{l^2} n W, \tag{10.63}$$

where W is the volume concentration of particles, which remains constant during coagulation.

The solution of Eq. (10.63) with the initial condition $n = n_0$ is

$$n = n_0 e^{-t/\tau}, \quad \tau = l^2 / 36 \nu \mathrm{Re}^{3/2} W. \tag{10.64}$$

Compare the frequencies of Brownian and turbulent coagulations:

$$\frac{(\beta_{12})_T}{(\beta_{12})_{br}} \sim \left(\frac{\varepsilon_0}{\nu}\right)^{1/2} \frac{R_c^2}{D_{br}}. \tag{10.65}$$

For particles whose size exceeds 0.1 μm, we get $(\beta_{12})_T > (\beta_{12})_{br}$.

The expressions given in this section for collision frequencies in the processes of Brownian, shear, and turbulent coagulation are derived with no account taken of hydrodynamic, molecular, and electrostatic interactions of particles. Taking them into account considerably complicates the problem. In particular, in the factors of Brownian and turbulent diffusion one must take into account the hydrodynamic resistance acting on the particle due to distortion of the velocity field in the presence of neighboring particles. One should also correct the diffusion equation to account for the convective flux due to the force of molecular interaction acting on particles. In the case of gradient coagulation in a laminar flow, one must consider trajectories of relative motion of particles, making necessary corrections to account for hydrodynamic and molecular interaction forces.

The role of hydrodynamic interaction in Brownian diffusion was discussed in Section 8.2. Consider now its effect on turbulent coagulation. Formally, it can be taken into account in the same manner as in Brownian motion, by introducing a correction multiplier into the factor of turbulent diffusion (10.57). Another, more correct way (see Section 11.3) is to use the Langevin equation that helped us determine the factor of Brownian diffusion in Section 8.2. As was demonstrated in [60], the factor of turbulent diffusion is inversely proportional to the second power of the hydrodynamic resistance factor:

$$D_T = D_T^{(0)} \frac{1}{\beta^2(r/a)} . \tag{10.66}$$

A further discussion of these issues and the review of the pertinent works in this field is offered in [61–63] (see also part V).

10.3 Particles' Deposition on the Obstacles

Consider now some technical applications of the models of particle interaction (with no hydrodynamic interaction forces) that have been presented earlier.

A common way to clear a liquid from particles suspended in it is to use particle deposition as the liquid flows past various obstacles (collectors). In this process, larger particles, filters, porous media, grids, and other obstacles can serve as collectors. Particles deposited on an obstacle form a layer of solid sediment. It should be noted that, as a rule, the particles' size does not exceed the linear size of obstacles. Therefore the capture of particles by an obstacle is not determined entirely by its geometry, but also depends on the character of the liquid's flow past obstacles, and on the forces of molecular and electrostatic interaction of particles with the collector. These forces operate when particles are close enough to the collector surface, therefore it is important to know the particle trajectories in the flow of carrying liquid. Following [61], we shall limit the discussion to the case of slow suspension flow around the collector, in the assumption that particle sizes are much smaller than the linear size of collector elements. The present section will consider two basic mechanisms of particle capture by an obstacle: Brownian diffusion of very small particles ($a \leq 1$ μm), and capture of relatively large particles ($a > 1$ μm). The latter process should be distinguished from diffusion, as it has a different character. Due to the small size of particles, it can be considered inertialess and viewed as a geometrical collision with an obstacle that occurs when the particle trajectories, which are coincident with streamlines, cross the obstacle. Note that this picture applies only to particles whose density does not differ much from that of the liquid. For a similar problem involving gas flow with suspended solid particles, the large density difference between particles and the gas means that particles can move relative to the gas, particles' inertia must be taken into account. This is especially true in the vicinity of an obstacle, since in this region particles are slowed down, change their direction of motion, and acquire significant negative accelerations. This mechanism of particle collisions with obstacles or among themselves is called inertial in work [52].

In the following discussion we neglect the inertia of particles. We also ignore surface (molecular and electrostatic) interactions. The proper account of surface forces will be made in the following section, and of inertia – in Section 10.5. Problems on deposition of small particles on obstacles are presented in works [36, 52, 58, 61].

10.3.1
Brownian Diffusion

Consider deposition of particles on an obstacle due to Brownian diffusion in a slow flow whose velocity far from the obstacle is U [36]. As an obstacle, we take a solid sphere of radius a.

We assume that deposition on the sphere is ideal, that is, each collision of a particle with the sphere results in the particle being captured. The factor of Brownian diffusion $D_{br} = \mathrm{k}T/6\pi\mu a_p$, where a_p is the particle's radius, is much smaller than the factor of molecular diffusion, therefore the Peclet diffusion number is $\mathrm{Pe}_D = Ua/D_{br} \gg 1$. By virtue of this inequality (see Section 6.5), the diffusion flux of particles toward the sphere can be found by solving the stationary equation of convective diffusion with a condition corresponding to a thick or thin diffusion-boundary layer. Particles may then be considered as point-like, and the diffusion equation will become:

$$u_r \frac{\partial \rho}{\partial r} + \frac{u_\theta}{r}\frac{\partial \rho}{\partial \theta} = D_{br}\frac{\partial^2 \rho}{\partial r^2}, \qquad (10.67)$$

where r is the mass concentration of particles, and $(2/r)\partial\rho/\partial r \ll \partial^2\rho/\partial r^2$.

The boundary conditions are

$$\rho = 0 \quad \text{at } r = a, \qquad \rho \to \rho_0 \quad \text{at } r \to \infty. \qquad (10.68)$$

The first condition corresponds to the ideal absorption on the sphere, and the second one – to a constant concentration far away from the sphere.

Recall that Eq. (10.67) is written in the boundary-layer approximation, and u_r and u_θ are the components of the Stokes flow velocity near the sphere. Since the Schmidt number is $\mathrm{Sc} \gg 1$, the thickness of the viscous boundary layer is much greater than that of the diffusion boundary layer, therefore it is possible to find the solution in the same manner as in the problem on diffusion toward a rigid particle moving in the solution.

The components of velocity at a slow flow near the sphere are:

$$u_r = -U\cos\theta\left(1-\frac{3}{2}\frac{a}{r}+\frac{1}{2}\frac{a^3}{r^3}\right),\quad u_\theta = U\sin\theta\left(1-\frac{3}{4}\frac{a}{r}-\frac{1}{4}\frac{a^3}{r^3}\right). \qquad (10.69)$$

Of the main interest is the velocity distribution near the sphere in the diffusion boundary layer. Introduce local system of coordinates (y, θ), where the y-axis is perpendicular and θ is tangential to the corresponding area element of the surface. Then $r/a = 1 + y/a$. Considering the case $y/a \ll 1$, expand (10.69) as a power series in y/a. In a result, we obtain

$$u_r \approx -\frac{3}{2}\left(\frac{y}{a}\right)^2 U\cos\theta,\quad u_\theta \approx \frac{3}{2}\left(\frac{y}{a}\right) U\sin\theta. \qquad (10.70)$$

Thus, the problem reduces to solution of equation (10.67) with the expressions (10.70) for velocity components, and the boundary conditions (10.68). The solution of this problem has been presented in Section 6.5. In particular, the diffusion flux to the sphere is

$$I_s = 7.98\rho_0 a D_{br}\left(\frac{Ua}{D_{br}}\right)^{1/3}. \qquad (10.71)$$

The efficiency of particle's capture can be characterized by dimensionless parameter

$$E_s = \frac{I_s}{\pi a^2 U\rho_0} = \frac{2.5}{\mathrm{Pe}_D^{2/3}}. \qquad (10.72)$$

This parameter is the ratio of number of particles, captured by the sphere per unit time, to number of particles per unit time that would pass through cross-sectional area equal to the sphere's projection, far from the sphere. Since $D_{br} \sim a_p^{-1}$ and $\mathrm{Pe}_D \sim D_{br}^{-1}$, $E_{sph} \sim a_p^{-2/3}$, that is, the efficiency of capture decreases with increase of particle's size. When the collector is a grid, the deposition on it can be modeled by deposition on cylinders in a cross flow. In work [61] the expression for efficiency of particle's capture by a cylinder has been obtained:

$$E_{cyl} = \frac{I_{cyl}}{2aU\rho_0} = \frac{4.63\rho_0 (UaD_{br}^2)^{1/3}}{2aU\rho_0} = \frac{2.32}{\mathrm{Pe}_D^{2/3}}. \qquad (10.73)$$

Note the similarity of expressions for the efficiency of particles' capture by a sphere and a cylinder.

10.3.2 Particles' Collisions with an Obstacle

When size of the particles is $a_p > 1$ μm, the Brownian diffusion does not play a noticeable role, and deposition of the particles suspended in the flow of liquid can be studied by a method of trajectory analysis of the flow-streamlines in the carrying liquid near the obstacle. The particles' collisions with an obstacle occur at contact with its surface. The case of the particles' collision with a cylinder at cross flow of suspension is illustrated in Fig. 10.6.

Consider collision of a particle with an obstacle, neglecting the surface forces and the viscous resistance force arising from squeezing out the layer of liquid film between the particle's surface and the obstacle.

Consider first collision of particles with a sphere. We assume the motion of particles to be inertialess. Then the particles' trajectories coincide with the streamlines. The stream function of the Stokes flow past the sphere is given by

$$\psi = \frac{1}{2} U r^2 \sin^2\theta \left(1 - \frac{3}{2}\frac{a}{r} - \frac{1}{2}\frac{a^3}{r^3}\right). \tag{10.74}$$

Introduce a local system of coordinates (y, θ), where $r/a = 1 + y/a$. Then near the surface we have $y/a \ll 1$ and

$$\psi \approx \frac{3}{4} U y^2 \sin^2\theta = \frac{3}{4} U (r - a)^2 \sin^2\theta. \tag{10.75}$$

By its physical meaning, the difference $2\pi(\psi_1 - \psi_2)$ is equal to the volume flow rate of liquid between two stream surfaces $\psi = \psi_1$ and $\psi = \psi_2$. If the particles' distribution far from the sphere is homogeneous, with mass concentration ρ_0, the mass flux of the particles to the sphere is equal to

$$I_{sph} = 2\pi\rho_0\Delta\psi, \tag{10.76}$$

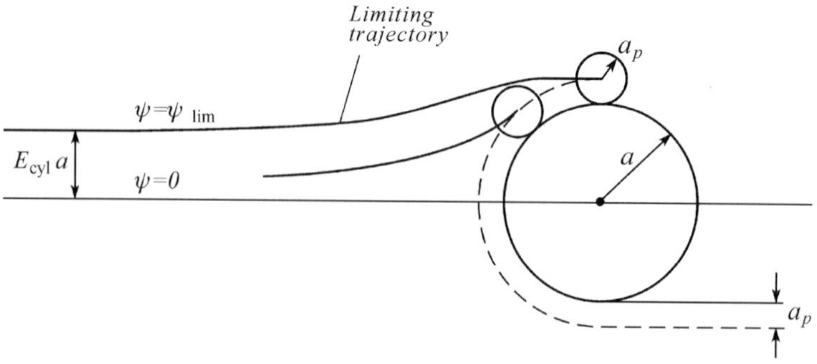

Fig. 10.6 Particles' collision with a cylinder.

where $\Delta\psi = \psi_{lim}$ (see Fig. 10.6). Here ψ_{lim} is the value of stream function at the limiting trajectory.

To determine ψ_{lim}, put $r = a + a_p$ and $\theta = \pi/2r$ into Eq. (10.75). This yields:

$$\psi_{lim} = \frac{3}{4} U a_p^2. \tag{10.77}$$

Thus,

$$I_{sph} = \frac{3}{2} \pi U a_p^2 \rho_0. \tag{10.78}$$

Using (10.72), one finds the efficiency of particles capture by the sphere

$$E_{sph} = \frac{3}{2} \left(\frac{a_p}{a} \right)^2. \tag{10.79}$$

In a similar way one can obtain the expression for the flux to a cylindrical obstacle at a cross flow. The difficulty here is that the Stokes flow past the cylinder does not exist. However, it is possible to use the Oseen solution, in which the inertial terms in the Navier-Stokes equations are partially taken into account by replacing the term $(\boldsymbol{u} \cdot \nabla)\boldsymbol{u}$ with $(\boldsymbol{U} \cdot \nabla)\boldsymbol{u}$. This approximation is valid at $\mathrm{Re} \leq 1$. We can use the solution [64] of this problem, from which it follows, that near the surface of the cylinder

$$\psi \approx \frac{U A_{cyl}}{a} (r - a)^2 \sin\theta,$$

$$A_{cyl} = \left(2 - \ln\left(\frac{2aU}{\nu} \right) \right)^{-1}. \tag{10.80}$$

Setting in (10.80) $r = a + a_p$ and $\theta = \pi/2$, one obtains $\psi_{lim} = U A_{cyl} a_p^2 / a$. Now find the mass flux of particles to the cylinder

$$I_{cyl} = 2\rho_0 \psi_{lim} = 2\rho_0 U A_{cyl} a_p^2 / a, \tag{10.81}$$

and the capture section

$$E_{cyl} = \frac{I_{cyl}}{2aU\rho_0} = A_{cyl} \left(\frac{a_p}{a} \right)^2. \tag{10.82}$$

Shown in Fig. 10.6, is the distance $E_{cyl} a$ of limiting trajectory from the axis $\theta = 0$ at $r \to \infty$, so that $\psi_{lim} = a E_{cyl} U$. Note that for a sphere the distance of the limiting trajectory from the axis $\theta = 0$ at $r \to \infty$ is equal to $E_{sph}^{1/2} a$.

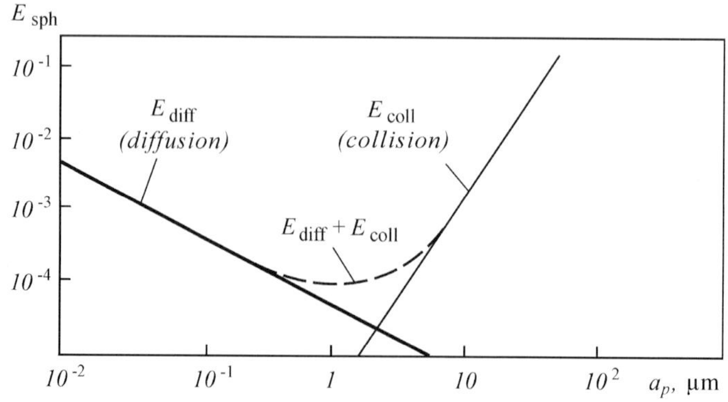

Fig. 10.7 The capture efficiency of a particles with radius a_p by a sphere.

The comparison of the capture efficiency of particles by a sphere (10.79) and by a cylinder (10.82) shows, that they differ only by a numerical factor. It can be explained by similarity of velocity fields near their surfaces. The basic difference of capture of large particles from the capture of small particles is in dependence on particles' radius, due to diffusion mechanism of deposition. For the diffusion mechanism of capture we have $E_{diff} \sim a_p^{-2/3}$, while for large particles $E_{coll} \sim a_p^2$.

In Fig. 10.7 the capture efficiency of particles in the water-suspension flow at normal temperature over spherical collector is shown [36]. The capture vs. particle-size dependencies for the two above-considered mechanisms of capture are shown, and in the intermediate area the dependence obtained by adding the efficiencies $E_{diff} + E_{coll}$ due to both mechanisms of capture is depicted.

10.4
The Capture of Particles Due to Surface and Hydrodynamic Forces

The discussion of the particles capture in the previous section was based on the assumption, that the particles are not subject to the forces from the system liquid – collector. Actually a particle near the surface of collector, is subject to the surface forces of (molecular and electrostatic) interactions, as well as the hydrodynamic force of viscous resistance from the liquid film between the particle and collector. The account of these forces considerably complicates the problem of determining the particles' capture efficiency for the given obstacle.

Consider the collision of particles of size more than 1 μm with a collector, taking into account only forces of molecular attraction and the hydrodynamic resistance force. The particles are assumed to be uncharged, and their deposition due to gravity is neglected. Consider deposition of particles on the cylinder, whose radius is much greater than radius of particles. The last condition allows us to assume that the particles practically do not disturb the velocity field far from the

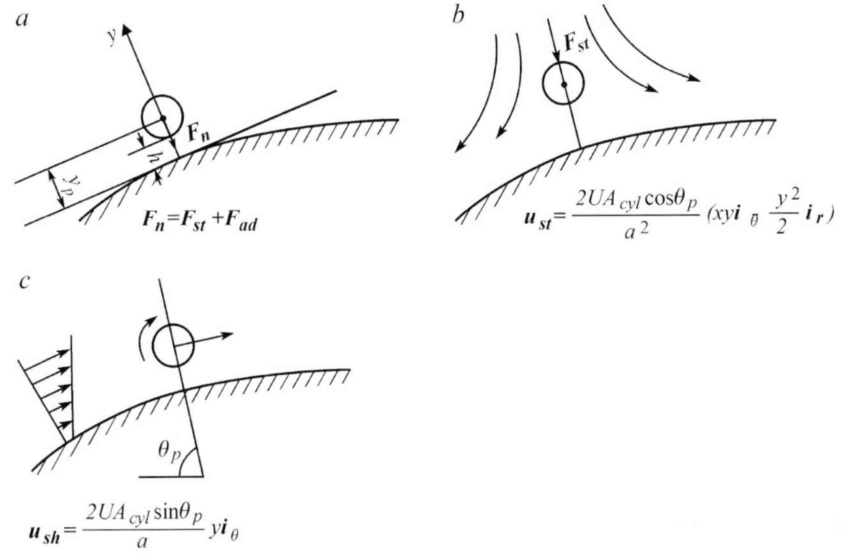

Fig. 10.8 Motion of a particle near the obstacle's surface:
a – motion under action of applied force; b – motionless particle;
c – free motion of the particle.

cylinder. Particles can change the velocity field only in the immediate proximity of the cylinder surface. Besides, in considering the particle's motion near the cylinder, the cylinder's surface can be considered as flat.

We also assume that the undisturbed velocity field corresponds to the Oseen solution, which is, near the cylinder's surface, described by the Eq. (10.80). The motion of particle in the vicinity of the cylinder's surface is schematically shown in Fig. 10.8 [65].

The radial component of the hydrodynamic force is denoted as F_{st}. The net hydrodynamic force is a sum of the external force acting on the particle from the liquid flowing around the obstacle, and the force of viscous resistance of the liquid film dividing surfaces of particle and cylinder. The external force can push the particle closer to or pull it away from the obstacle's surface. Note that the force of viscous resistance is negative. Next, denote as F_{ad} the molecular force of the Van der Waals attraction. This force is directed along the perpendicular line from the particle to the symmetry axis of the cylinder. Since the Navier–Stokes equations in the Oseen's approximation are linear, the forces and velocity fields induced by them are additive.

Consider undisturbed velocity field (10.80) near the cylinder's surface. It can be represented as the sum of two flows. The first is a flat flow such as the flow in the vicinity of stagnation point (Fig. 10.8, *b*). The velocity in the vicinity of this point is associated with the velocity component directed along the centerline. The other flow is a shear flow, perpendicular to centerline and shown in Fig. (10.8, *c*). Under condition $a \gg (x^2 + y^2)^{1/2} \gg a_p$, the velocities corresponding to these two flows, are [65]

$$\boldsymbol{u}_{st} = \frac{U}{a^2} A_{cyl} \cos\theta_p (2xy\boldsymbol{i}_\theta - y^2\boldsymbol{i}_r), \tag{10.83}$$

$$\boldsymbol{u}_{sh} = \frac{2U}{a} A_{cyl} \sin\theta_p \, y\boldsymbol{i}_\theta. \tag{10.84}$$

Here $y = r - a$, $x = a(\theta - \theta_p)$, θ_p is the angle between the position vectors of the particle and the forward critical point. As is seen from Fig. 10.8, the coordinates of the particle's center are $x = 0$, $y_p = a_p + h$, where h is the clearance between the surfaces of particle and cylinder.

As mentioned above, by virtue of linearity of the problem, the forces acting on the particle are additive, and the total force is equal to

$$\boldsymbol{F}_n = \boldsymbol{F}_{st} + \boldsymbol{F}_{ad}. \tag{10.85}$$

Let the particle under the action of force $\boldsymbol{F}_n$ approach the cylinder. Denote as u_r the radial component of its velocity. It is possible to show, that from the condition of inertialess motion of the particle near the collector's surface and the requirement that its velocity be zero at this surface, it follows

$$F_n = 6\pi\mu a_p u_r f_1(h/a_p). \tag{10.86}$$

It is obvious, that

$$u_r = \frac{dy_p}{dt} = \frac{dh}{dt}. \tag{10.87}$$

If the particle is far from the cylinder, that is, $h \gg a_p$, the force F_n is close to the Stokes force and

$$f_1(h/a_p) \approx 1 \quad \text{at } h/a_p \gg 1. \tag{10.88}$$

If the particle is near the collector surface, then the force F_n is essentially different from the Stokes force. The solution of the corresponding problem can be found in [2] (see Section 8.1). It follows from this solution that

$$f_1(h/a_p) \approx a_p/h \quad \text{at } h/a_p \ll 1. \tag{10.89}$$

Note that the dependence (10.89) corresponds to the case of approach of rigid spherical particle to a rigid plane.

The potential of molecular force of attraction between the two spherical particles of radiuses a_1 and a_2 is determined by formula (10.24)

$$V_{ad} = -\frac{\Gamma}{6}\left(-\frac{8\mathsf{k}}{(s^2-4)(1+\mathsf{k})^2} + \frac{8\mathsf{k}}{s^2(\mathsf{k}+1)^2 - 4(1-\mathsf{k})^2} + \ln\left(\frac{(s^2-4)(1+\mathsf{k})^2}{s^2(\mathsf{k}+1)^2 - 4(1-\mathsf{k})^2}\right)\right). \quad (10.90)$$

where G is the Hamaker constant, $\mathsf{k} = a_2/a_1$, $s = 2r/(a_1+a_2) = 2h/(a_1+a_2) + 2$, r is the distance between centers of particles.

The attraction force is determined by the expression

$$F_{ad} = -\frac{dV_A}{dr} = -\frac{2\Gamma}{3a_1(1+\mathsf{k})}\left(\frac{s}{(s^2-4)(1+\mathsf{k})^2}(8\mathsf{k} - (1+\mathsf{k})^2(s^2-4)) + \frac{s(1+\mathsf{k})^2}{(s^2(\mathsf{k}+1)^2 - 4(1-\mathsf{k})^2)^2}(8\mathsf{k} + s^2(1+\mathsf{k})^2 - 4(1-\mathsf{k})^2)\right). \quad (10.91)$$

When the gap between the particles is small, we can use the approximation

$$F_{ad} \approx -\frac{2\Gamma\mathsf{k}}{3a_1(1+\mathsf{k})^3(s-2)^2}. \quad (10.92)$$

If we consider the attraction force between a plane and a sphere of radius a_p, then, assuming in (10.91) $a_2 = a_p \ll a_1$, one obtains:

$$F_{ad} \approx -\frac{2}{3}\frac{\Gamma}{a_p}\frac{1}{(2+h/a_p)^2} = -\frac{2}{3}\frac{\Gamma}{a_p} f_{ad}\left(\frac{h}{a_p}\right). \quad (10.93)$$

When the particle is far from the plane, we have $h/a_p \gg 1$ and

$$f_{ad}\left(\frac{h}{a_p}\right) \approx \left(\frac{a_p}{h}\right)^4 \quad \text{at } h/a_p \gg 1. \quad (10.94)$$

In case of a small gap between the particle and the plane,

$$f_{ad}\left(\frac{h}{a_p}\right) \approx \left(\frac{a_p}{2h}\right)^2 \quad \text{at } h/a_p \ll 1. \quad (10.95)$$

Consider now the component of hydrodynamic force acting on the particle from the external flow. Since we are focused on the approach to the surface, that is, on the motion along the normal to the cylinder's surface, we can take as the hydrodynamic force F_{st}, the force depicted in Fig. 10.8, *b*. In this case it is necessary to solve a hydrodynamic problem of Stokesian motion of the particle a_p in the vicinity of the plane in the velocity field of carrying liquid (10.80), under condition of the liquids' velocity being zero at the cylinder's or particle's surface. The solution of this problem gives

$$F_{st} = \frac{6\pi\mu a_p^3}{a^2} UA_{cyl} \cos\theta_p f_2\left(\frac{h}{a_p}\right). \tag{10.96}$$

Far from the plane ($h \gg a_p$) the resistance force is close to Stokesian, and $F_{ad} \to 0$. Then the equation of particle's motion (10.85) reduces to

$$F_n = F_{st} = 6\pi\mu a_p u_r, \tag{10.97}$$

where u_r is the velocity of the carrying liquid, which can be found from (10.80) assuming $u_r = -(1/r)\, d\psi/d\theta$:

$$u_r = \frac{UA_{cyl}}{a^2} \cos\theta_p (h + a_p)^2 \approx UA_{cyl} \cos\theta_p \left(\frac{h}{a}\right)^2. \tag{10.98}$$

Comparing (10.96) and (10.97) with (10.98), we find that at $h \gg a_p$,

$$f_2(h/a_p) \approx (h/a_p)^2. \tag{10.99}$$

For a particle near the surface, we have $h \ll a_p$. In this case, $f_2 \to const$ [66]

$$f_2(h/a_p) \approx 3.2 \quad \text{at } h \ll a_p. \tag{10.100}$$

A comparison of the molecular attraction force F_{ad} (10.93) with the hydrodynamic force component F_{st} (10.96) by the order of magnitude shows that

$$\frac{F_{ad}}{F_{st}} = N_{ad} f\left(\frac{h}{a_p}\right) \frac{1}{\cos\theta_p},$$
$$N_{ad} = \frac{\Gamma a^2}{9\pi\mu a_p^4 UA_{cyl}}, \tag{10.101}$$

where N_{ad} is the dimensionless parameter of adhesion. In the case $N_{ad} \gg 1$ the molecular force exceeds the component of hydrodynamic force F_{st}.

Generally, to solve the problem of collisions of particles a_p with the surface of a cylinder, it is necessary to solve the equation of motion of particle a_p under the action of molecular attractive forces (10.93) and forces of hydrodynamic resistance (10.86) and (10.96). This problem can be solved numerically; however, in the limiting case $N_{ad} \gg 1$ and $h \gg a_p$, the efficiency of capture of particles from the oncoming flow by the cylinder can be determined analytically [65]. Recall that the capture efficiency determined in Section 10.3 (see the formula (10.82)

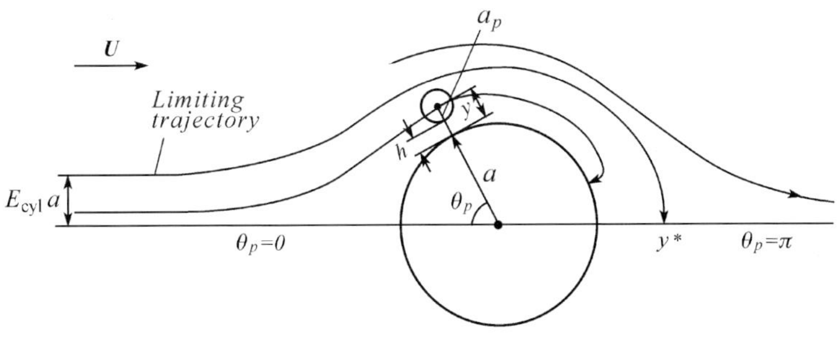

Fig. 10.9 Collision of particles with the cylinder, with due consideration of surface and hydrodynamic forces.

does not take into account the forces of molecular attraction and hydrodynamic resistance.

In the considered case, the deposition of particles on the surface of the cylinder is represented schematically in Fig. 10.9.

The basic difference between particle capture in the case of non-zero molecular and hydrodynamic forces (Fig. 10.9) from particle capture in the case when these forces are ignored (Fig. 10.7) is in the limiting trajectory dividing the family of trajectories into those that lead the particles to collide with the cylinder, and those that lead the particles away from the cylinder. In the first case the limiting trajectory ends at the back critical point of the cylinder $\theta_p = \pi$, and in the second case – at the point of contact between particle a_p and the cylinder surface at $\theta_p = \pi/2$.

In the case $N_{ad} \gg 1$, it is possible to assume that the limiting trajectory passes through the region $h \gg a_p$. When these inequalities are satisfied, the equation of inertialess particle motion normally to the cylinder surface, with due consideration for the expressions (10.86), (10.93), (10.94), (10.96), and (10.99), will become

$$6\pi\mu a_p \frac{dh}{dt} = -\frac{2}{3}\frac{\Gamma}{a_p}\left(\frac{a_p}{h}\right)^4 - \frac{6\pi\mu a_p^3}{a^2} U A_{cyl} \cos\theta_p \left(\frac{h}{a_p}\right)^2. \tag{10.102}$$

This equation can be transformed to

$$\frac{dh}{dt} = -\frac{U A_{cyl} a_p^2}{a^2}\left(N_{ad}\left(\frac{a_p}{h}\right)^4 + \cos\theta_p\left(\frac{h}{a_p}\right)^2\right). \tag{10.103}$$

To determine the projection of the equation of motion on a tangent to the cylinder's surface, one needs to find the tangential component of velocity $u_\theta = a(d\theta_p/dt) = u_{sh}$ from (10.84) (see Fig. 10.8, *c*). Then in the approximation $h \gg a_p$, we obtain

$$\frac{d\theta_p}{dt} = \frac{2 U A_{cyl} a_p}{a^2} \sin\theta_p \left(\frac{h}{a_p}\right). \tag{10.104}$$

Thus, Eqs. (10.103) and (10.104) describe trajectories of particles a_p flowing around the cylinder surface, including the limiting trajectory, at the conditions $N_{ad} \gg 1$ and $h/a_p \gg 1$.

Let us show how one can find the limiting trajectory and collision frequency of particles with the cylinder without solving these equations. The limiting trajectory (see Fig. 10.9) ends at the back point $\theta_p = \pi$, $y = y^*$. To determine y^*, one should proceed as follows. When the particle is moving along the critical trajectory near the back stagnation point, the velocity is $u_r = u_\theta = 0$ and since the particle motion in the vicinity of this point is parallel to the cylinder surface, $dh/dt = 0$. Then it follows from (10.103) that

$$N_{ad}\left(\frac{a_p}{h}\right)^4 + \cos\theta_p\left(\frac{h}{a_p}\right)^2 = 0. \tag{10.105}$$

Setting in this equation $\theta_p = \pi$, $h = y^* - a_p \approx h^*$ and using the inequality $h \gg a_p$, one gets

$$\frac{y^*}{a_p} \approx \frac{h^*}{a_p} = N_{ad}^{1/6} \gg 1. \tag{10.106}$$

It follows from the Eqs. (10.103) and (10.104):

$$\frac{d(h/a_p)^4}{d\theta} + 3\cot\theta_p\left(\frac{h}{a_p}\right)^6 = -3\frac{N_{ad}}{\sin\theta_p}. \tag{10.107}$$

The solution to this equation can be easily obtained. In particular, along the limiting trajectory,

$$\left(\frac{h}{a_p}\right)^2 \sin\theta_p = \left(3N_{ad}\int_{\theta_p}^{\pi} \sin^2\theta\, d\theta\right)^{1/3}. \tag{10.108}$$

Using now the definitions (10.81) and (10.82), it is possible to find the cross section of capture of particles a_p by the cylinder

$$E_{cyl} = \psi_{lim}/aU. \tag{10.109}$$

To determine y_{lim}, we should use the expression (10.80) at $h \gg a_p$. The resulting expression is

$$\frac{\psi_{lim}}{aU} = \frac{A_{cyl}y^2}{a^2}\sin\theta_p = A_{cyl}\left(\frac{a_p}{a}\right)^2\left(\frac{h}{a_p}\right)^2\sin\theta_p. \tag{10.110}$$

At $\psi \to \psi_{lim}$, we find that

$$E_{cyl} = \lim_{\theta_p \to 0}\left(A_{cyl}\left(\frac{a_p}{a}\right)^2\left(\frac{h}{a_p}\right)^2\sin\theta_p\right) = A_{cyl}\left(\frac{a_p}{a}\right)^2\left(\frac{3\pi}{2}N_{ad}\right)^{1/3}. \tag{10.111}$$

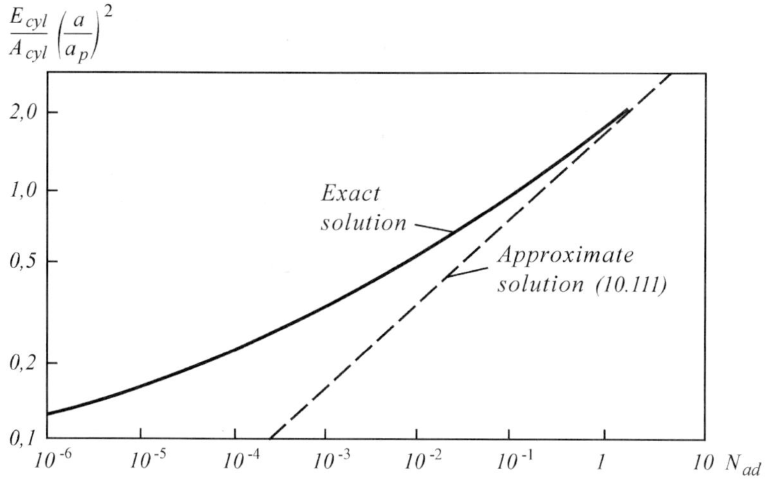

Fig. 10.10 Cross section of capture of particles by the cylinder.

Since $N_{ad} \sim 1/a_p^4$ and $E_{cyl} \sim a_p^{2/3}$, the dependence of E_{cyl} on the particle radius a_p is weaker than in the case of negligible molecular and hydrodynamic forces.

For the case of capture of particles by a sphere, with the same assumptions as above, the capture efficiency is written as

$$E_{sph} = \frac{3}{2}\left(\frac{a_p}{a}\right)^2\left(\frac{4}{3}\left(\frac{9}{5}N_{ad}^{sph}\right)^{1/3}\right), \qquad (10.112)$$

where N_{ad}^{sph} is determined by the expression (10.101), in which we should set $A_{cyl} = 1$.

Figure 10.10 compares the dependence (10.111) with the dependence obtained by numerical solution of equations of particle motion near the cylinder [67].

Finally, we should mention that if the effect of gravity is significant, there will be one more force acting on the particles:

$$F_{gr} = -\frac{4}{3}\pi a_p^3(\rho - \rho_f)g\cos\theta_p. \qquad (10.113)$$

In this case an additional dimensionless parameter

$$N_{gr} = \frac{F_{gr}}{F_{st}} = \frac{2(\rho - \rho_f)ga^2}{9\mu U A_{cyl}}, \qquad (10.114)$$

will appear in the equations of motion, influencing the efficiency of capture of particles by the cylinder.

A review of works devoted to the considered subject is given in [61].

10.5
Inertial Deposition of Particles on the Obstacles

Consider now the effect of inertial deposition, which plays a noticeable role in deposition on obstacles of rather large particles whose density differs strongly from that of the ambient liquid.

Let the particles be suspended in a laminar flow that is going around the body, and let them have sufficiently large sizes so that Brownian diffusion can be neglected. On the other hand, the particles' sizes remain negligible in comparison with the size of the body. Besides, it is assumed that a particle's mass density is much greater than that of the liquid, so when considering particles' motion relative to the body, their inertia cannot be ignored. If calculation of particle's motion does not take into account hydrodynamic and molecular interactions with the body, then without inertia on the one hand, and finiteness of the particle size on the other, they could not reach the body's surface. Inertia of the particles results in deviation of their trajectories from streamlines of the liquid's flow round the body, which is especially noticeable at points of streamlines bending near the surface of the body. If the particles' inertia were not taken into account, their trajectories would coincide with streamlines. Thus the particles of the zero size would flow around the body without touching it. Particles of finite size can collide with the body if they move along streamlines whose minimal distance from the body surface does not exceed the particle radius. If the flow around a sphere of radius a is considered, then the collision cross section is

$$E_0^{sph} = \pi h_\infty^2 / \pi a^2, \tag{10.115}$$

where h_∞ is the distance between the limiting streamline and a straight line passing through the center of the sphere and parallel to the velocity of the flow running toward the sphere.

If inertia of particles is neglected, the limiting trajectory passes at a distance a_p from the body. In Ref. [58] this effect is called the effect of hooking. It has been shown in the same work that, for a potential flow around the sphere we have

$$E_0^{sph} = \left(1 + \frac{a_p}{a}\right)^2 - \frac{1}{1 + a/a_p} \approx \frac{3a_p}{a}. \tag{10.116}$$

In another limiting case of highly inertial particles, it is possible to assume that particles move along straight trajectories. Then the cross section of collision can be found from simple geometrical arguments:

$$E_0^{sph} = \left(\frac{a_p + a}{a}\right)^2 \approx 1 + \frac{2a_p}{a}. \tag{10.117}$$

In the general case, trajectories of particles entrained by the flow are determined from the equation of motion, which can be written as

$$\frac{4}{3}\pi a_p^3 \rho \frac{d^2\boldsymbol{r}}{dt^2} = 6\pi\mu a\left(\boldsymbol{U} - \frac{d\boldsymbol{r}}{dt}\right). \tag{10.118}$$

Here it is supposed that particles' motion obeys the Stokes law, and $\boldsymbol{U}$ is the flow velocity of the carrying liquid.

Introducing dimensionless variables

$$\boldsymbol{r}' = \frac{\boldsymbol{r}}{a}, \quad \tau = \frac{U_\infty t}{a}, \quad \boldsymbol{V} = \frac{\boldsymbol{U}}{U_\infty}, \quad S = \frac{4U_\infty \rho a_p^2}{18\mu a},$$

we can transform the equation (10.116) to

$$S\frac{d^2\boldsymbol{r}'}{d\tau^2} + \frac{d\boldsymbol{r}'}{d\tau} = \boldsymbol{V}(\boldsymbol{r}). \tag{10.119}$$

The dimensionless parameter S is referred to as the Stokes number. It characterizes the measure of particle's inertia. At $S \ll 1$, the particle's inertia is small, and it follows from (10.119) that the trajectory of the particle is close to the streamline. At $S \gg 1$, one finds from (10.119) that $d^2\boldsymbol{r}/dt^2 \approx 0$, i.e. trajectories are the straight lines.

Having found the law describing the flow around the body, we can obtain from (10.119) the family of particle trajectories, and then determine the limiting trajectory and cross section of particles' collisions with the body.

A notable feature of inertial deposition of particles on the body is the existence of the Stokes critical number S_{cr} such that at $S < S_{cr}$ the capture of particles by the body is impossible; in other words, a particle's inertia is not sufficient to overcome particle's entrainment by the flow of carrying liquid [68]. The values of S_{cr} for the cylinder are as follows: 0.0625 for potential flow, approximate solution [58]; 0.1 for potential flow, numerical solution. If we take into account formation of the boundary layer at the surface, the resulting value will be $S_{cr} \approx 0.25$ [68].

The existence of the critical Stokes number means that there exists a minimal radius for a particle that can be captured by an obstacle:

$$(a_p)_{\min} = 1.5\left(\frac{2\mu a}{U_\infty \rho} S_{cr}\right)^{1/2}. \tag{10.120}$$

10.6 The Kinetics of Coagulation

In Section 10.2 the basic mechanisms of particle coagulation have been presented, and a few simple problems considered on the time dependence of particle's concentration, in the assumption, that at the initial moment and at subsequent moments of time suspension is mono-disperse. Such assumption is of course an idealization of real process. Actually, even if all the particles are initially of the same size, and the system may be considered as mono-disperse at the initial stage of coagulation, the particles merge together in the course of the coagulation process, forming new particles of different mass and sizes. Hence, in the course of time, the initial mono-disperse distribution becomes poly-disperse. A state's change in time and (in case of non-uniform particles' distribution) also in space, refers to kinetics of coagulation. Further we will be limited to the case when the particles' concentration is constant over the suspension volume.

We will consider the Brownian coagulation [53], but the approximation to be used can be generalized to the case of any other coagulation mechanism.

Let at a moment t there is a poly-disperse distribution of particles in a given volume. Divide particles into groups of identical size (i). Number these groups according to their particles' size, so that the group 1 includes initial (single) particles, group 2 – double particles, etc. The collision frequency (number of collisions per unit time) between the initial particles, that is, particles with the smallest radius, is defined by expression (10.41)

$$\beta_{11} = \frac{1}{2}(4\pi D_{11} R_{11} n_1^2), \tag{10.121}$$

where D_{11} is the factor of mutual diffusion of single particles; $R_{11} = 2a_1$ is the coagulation radius; n_1 is numerical concentration of single particles.

It the same way one can obtain the collision frequency between particles of types i and j (see (10.40)):

$$\beta_{ij} = 4\pi D_{ij} R_{ij} n_i n_j, \tag{10.122}$$

where $D_{ij} = D_i + D_j$; n_i and n_j are the numerical concentrations of particles of type i and j, respectively, at a moment t; $R_{ij} = a_i + a_j$.

Consider the balance of particles of the sort k. Their number increases at collisions with particles of sort i and $\mathrm{k} - i$, and decreases at collisions with particles of sorts k and i. The equation describing coagulation kinetics, has the form

$$\frac{dn_\mathrm{k}}{dt} = \frac{1}{2} \sum_{\substack{i<\mathrm{k} \\ j=\mathrm{k}-i}}^{j=\mathrm{k}-1} 4\pi D_{ij} R_{ij} n_i n_j - n_\mathrm{k} \sum_{i=1}^{i=\infty} 4\pi D_{i\mathrm{k}} R_{i\mathrm{k}} n_i. \tag{10.123}$$

At coagulation of rigid particles, they stick together, therefore approximately it is possible to take $R_{ij} \approx R_i + R_j$.

The coefficient of Brownian diffusion is $D_i \sim 1/a_i$ (see (8.70)), therefore

$$D_{ij} R_{ij} = (D_i + D_j)(a_i + a_j) = D_1 a_1 \left(\frac{1}{a_i} + \frac{1}{a_j}\right)(a_i + a_j). \tag{10.124}$$

If a_i and a_j are slightly different, which can be the case at the initial stage of coagulation in an initially mono-disperse mixture, $(a_i + a_j)(a_i^{-1} + a_j^{-1}) \approx 4$. Therefore $D_{ij} R_{ij} \approx 4 D_1 a_1 = 2 D_1 R$, where $R = 2a_1$.

The simplifications made allow us to transform the equation (10.123) to

$$\frac{dn_k}{dt} = 4\pi R D_1 \left(\sum_{\substack{i<k \\ j=k-i}}^{j=k-1} n_i n_j - 2 n_k \sum_{i=1}^{i=\infty} n_i \right). \tag{10.125}$$

Carrying out the summation over k and denoting $\sum_k n_k = n$, where n is the particles' net concentration, one obtains

$$\begin{aligned} \frac{dn}{dt} &= 4\pi R D_1 \left(\sum_{k=1}^{k=\infty} \sum_{j=1}^{j=\infty} n_i n_j - 2 \sum_{k=1}^{k=\infty} \sum_{i=1}^{i=\infty} n_k n_i \right) \\ &= 4\pi R D_1 \sum_{i=1}^{i=\infty} \sum_{j=1}^{j=\infty} n_i n_j = -4\pi R D_1 \left(\sum_{i=1}^{i=\infty} n_k \right)^2 = -4\pi R D_1 n^2. \end{aligned} \tag{10.126}$$

If $n = n_0$ at $t = 0$, then from (10.126) it follows, that

$$n = \frac{n_0}{1 + t/\tau}, \qquad \tau = 1/4\pi R D_1 n_0. \tag{10.127}$$

The obtained expression for the particles' net concentration as a function of time is identical with expression (10.43).

Using now (10.127), it is possible to find from (10.125) the expressions for concentration of single n_1, secondary n_2 and other particles:

$$n_1 = \frac{n_0}{(1 + t/\tau)^2}, \; n_2 = \frac{n_0 t/\tau}{(1 + t/\tau)^3}, \ldots, n_k = \frac{n_0 (t/\tau)^{k-1}}{(1 + t/\tau)^{k+1}}. \tag{10.128}$$

In Fig. 10.11 the dependencies $n_k(t)$ and total concentration $n(t)$ of particles are shown.

Note, that characteristic coagulation time is $\tau = 3\mu/4kTn_0$. If the liquid phase is water at $T = 298$ °K, then $\tau \sim 2 \cdot 10^{11}/n_0$, that is, the characteristic time depends only on initial concentration of particles. The initial particles' volume content is $W_0 = 4\pi a_0^3 n_0/3$. If $n_0 = 10^{14}$ m^{-3}, then $a_0 = 1$ μm, $W_0 = 4 \cdot 10^{-14}$, $\tau = 2 \cdot 10^{-3}$.

Considering coagulation in poly-disperse systems, it is convenient to use the continuous particles' distribution over volumes, rather than discrete distribution. If we assume that a distribution is homogeneous in suspension volume, then $n = n(V, t)$, where V is the particle volume. In this case $n\,dV$ is number of particles with volumes within the range $(V, V + dV)$ in a unit volume of suspension.

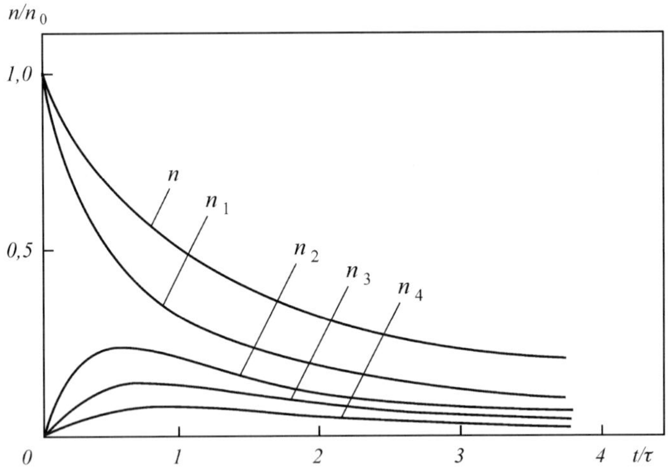

Fig. 10.11 Kinetics of Brownian coagulation.

The total number of particles in a unit volume of suspension (numerical concentration N) and volumetric concentration of these particles W are determined by

$$N = \int_0^\infty n(V,t)\,dV, \quad W = \int_0^\infty Vn(V,t)\,dt. \tag{10.129}$$

Knowing N and W, it is possible to determine average volume of particles

$$V_{av} = W/N. \tag{10.130}$$

The equation describing $n(V,t)$ as a function of time, is similar to (10.123), but one should replace the summation with integration (see (5.84))

$$\frac{\partial n(V,t)}{\partial t} = \frac{1}{2}\int_0^V K(V-u,u)n(V-u,t)n(u,t)\,du - n(V,t)\int_0^\infty K(u,V)n(u,t)\,du. \tag{10.131}$$

Here $K(u,V)$ is a function called coagulation kernel or constant of coagulation. It characterizes the collision frequency for particles of volumes $V-u$ and u and satisfies the symmetry condition $K(u,V) = K(V,u)$.

We seek the solution of the integro-differential kinetic equation (10.131), which would satisfy the initial condition $n(V,0) = n_0(V)$ and the boundary conditions $n(V,t) \to 0$ at $V \to 0$ and $V \to \infty$. The solution of this equation presents significant mathematical difficulties. A number of analytical and approximate solutions for the certain kinds of the coagulation constants is known. Detailed analysis of the equation (10.131) can be found in Ref. [69].

10.7
The Filtering and a Model of a Highly Permeable Porous Medium with Resistance

Filtering is a term for a process of separation of particles from the carrying liquid as the suspension passes through the permeable material called filter. The basic difference from process of filtration is in high permeability of the porous material (filter). A porous granulated material or fiber material can serve as a filter. The separated substance can form a solid deposit at a filter's surface or in the internal pores of material. Particles with sizes smaller than filter's pores, deposit on the surface of granules of which the filter consists, due to forces of hydrodynamic, molecular, and electrostatic interaction. Other kinds of interactions are also possible.

In many cases filters represent porous media of high permeability. For example, a rather widespread way of gas purification (clearing of liquid or rigid impurities) is the use of mesh nozzles. The impurity particles are deposited on the surface of the grid as the gas passes through it. It is obvious, that such a filter has a high permeability and the capillary model is unsuitable.

In Section 6.11 the question on slow liquid flow through high-permeable porous medium has been discussed in connection with chromatography. The corresponding velocity was determined by Darcy's law. Calculated permeability k of the medium depends on the accepted model of porous medium. In particular, if the medium consists of identical spherical particles, then the Kozeny–Karman formula (6.266) for k is valid. This formula has been obtained in the assumption, that motion of liquid can be considered as motion through system of micro-capillaries, whose diameters are determined by the formula (6.263), therefore such model is a capillary model. It is valid for a medium of rather small permeability. It follows from the Kozeny–Karman formula that permeability k exhibits a sharp increase at $\varepsilon \to 1$, where ε is the porosity of medium equal to the ratio of the void volume V_v to the net volume of medium V. At $\varepsilon \ll 1$ the representation of a porous medium as a system of capillaries is reasonable. However, at $\varepsilon \to 1$ the volume occupied by a solid phase of the medium is small, and the flow through porous medium represents a fluid flow through a system of solid particles placed rather far from each other and forming the stuffing of a filter. Hence, transition from the case $\varepsilon \ll 1$ to the case $\varepsilon \to 1$ results in a basic change of flow structure. The capillary model is not suitable any more, and it is necessary to consider the flow of one motionless particle, taking into account the influence of neighboring particles, that is, the constrained motion of particles. Such model of high-permeable porous medium is close to model with resistance. The solution of this problem has been obtained in work [2].

Consider a porous medium consisting of ensemble of randomly located solid spherical particles of identical radius. The account of presence of neighboring

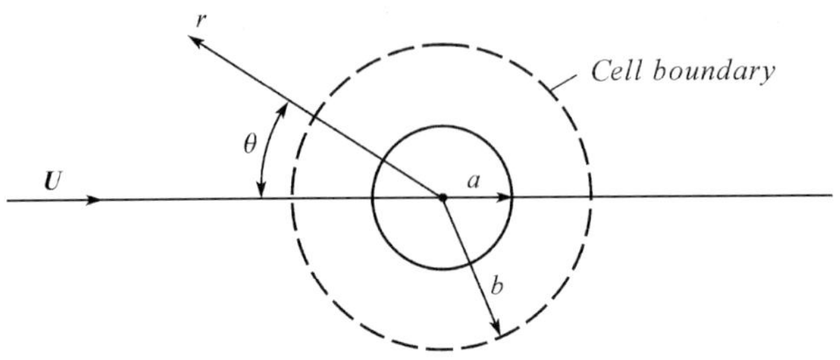

Fig. 10.12 Illustration with the account taken of the constrained motion of the particle.

particles (the account of constraints) can be taken on the basis of cellular model whose essence is as follows. Introduce a volume fraction $W = 1 - \varepsilon$ of rigid particles in considered volume of porous medium and denote as a radius of particles. Surround each particle by a liquid spherical layer with radius (Fig. 10.12)

$$b = a/W^{1/3}. \tag{10.132}$$

If the particles are equidistant from each other, the spheres surrounding particles fill the volume of the filter and do not intersect. There are obvious relations

$$V = V_\mathrm{v} + V_p, \quad V_p = VW, \tag{10.133}$$

where V_p is the volume occupied by particles.

Since $V_p = 4\pi a^3 n/3$ and all the particles are identical, we have

$$\frac{4\pi a^3 n}{3W} \approx V_\mathrm{v} + \frac{4\pi a^3 n}{3} = \frac{4\pi(b^3 - a^3)n}{3} + \frac{4\pi a^3 n}{3},$$

and it follows, that

$$a^3 = Wb^3.$$

Suppose, that each cell remains spherical and the tension on its external surface is zero. The flow of liquid through the cell is assumed to be slow, so that $\mathrm{Re} \ll 1$. Then it is described by the Stokes equations

$$\nabla \cdot \boldsymbol{u} = 0, \quad \mu\Delta\boldsymbol{u} = \nabla p \tag{10.134}$$

with boundary conditions

$$u_r = u_\theta = 0, \quad \text{at } r = a; \quad u_r = U\cos\theta, \quad \tau_{r\theta} = 0 \quad \text{at } r = b. \tag{10.135}$$

Note that the flow is, as usual, considered in the system of coordinates attached to the particle. In case of a flow around a single spherical particle in unbounded liquid, the stream function is determined by (10.74). The account of constraints is necessary due to the boundary condition (10.135) at $r = b$ and results in the following expression for the stream function

$$\psi = \frac{3}{4} U A_{sph} (r - a)^2 \sin^2 \theta,$$

$$A_{sph} = \frac{2(1 - W^{5/3})}{2 - 3W^{1/3} + 3W^{5/3} - 2W^2}. \tag{10.136}$$

The resistance force per one particle is

$$F = 6\pi\mu a U \frac{3 + 2W^{5/3}}{3 - 4.5W^{1/3} + 4.5W^{5/3} - 3W^2}. \tag{10.137}$$

According to the model of porous medium with resistance, the pressure drop per unit length of the medium is equal to the force F divided by the cell volume $4\pi b^3/3$, that is

$$-\frac{dp}{dx} = \frac{3F}{4\pi b^3}. \tag{10.138}$$

Since $dp/dx = -\mu U/\mathrm{k}$, we obtain from (10.137) and (10.138) the permeability of filter

$$k = \frac{3 - 4.5W^{1/3} + 4.5W^{5/3} - 3W^2}{18W(3 + 2W^{5/3})} (2a)^2. \tag{10.139}$$

Using $W = 1 - \varepsilon$ and taking the limit $\varepsilon \to 0$, one obtains from (10.139)

$$\mathrm{k} \approx \varepsilon^3 d^2/162. \tag{10.140}$$

A similar expansion in the Kozeny-Karman formula yields

$$\mathrm{k} \approx \varepsilon^3 d^2/180. \tag{10.141}$$

Expressions (10.140) and (10.141) for permeability differ by a numerical factor. This difference is explained by the fact that the formula (10.140) has been obtained from (10.139), which is valid at $W \to 0$ or $\varepsilon \to 1$ (highly-permeable medium), while the formula (10.141) is used at $\varepsilon \to 0$.

The amount of solid phase in the suspension flowing through the filter and captured by the filter's particles, can be found by considering the suspension flow over spherical or cylindrical obstacles constituting the filter. Methods of solution of such problems have been described in the previous sections of this chapter.

10.8
The Phenomenon of Hydrodynamic Diffusion

It is now time to discuss yet another mechanism of particle motion in a suspension in view of influence of ambient particles. Up till now, slow motion of a test particle was usually considered with the proper account taken of interactions with the neighboring particles, but with the assumption of sufficiently small volume concentration of particles W. In spite of this restriction, the formula (8.150) can be used to find the velocity of constrained motion of a particle at rather high values of j, i.e. in highly concentrated suspensions. In such suspensions, particle motion (gravitational sedimentation or motion in a shear flow) has a random character, which is different, however, from the character of Brownian motion, because particles sizes are much greater. This type of motion is caused by hydrodynamic interactions with the neighboring particles.

So, in concentrated suspensions subject to shear flow, particles do not obligatory move along the streamline, but may perform random motion among the ambient particles (Fig. 10.13, *a*) in the average direction opposite to gradients of number of particles and suspension velocity. The reason for this is that the collision frequency of the test particle with ambient particles is different at the opposite sides of the test particle's surface. Another example is gravitational sedimentation of particles in a concentrated suspension (Fig. 10.13, *c*). Particles do not descend with a constant velocity or with the velocity given by the formula for hindered sedimentation. Instead, they perform random motion along the direction of gravitational acceleration **g**, as well as in the cross direction which is superimposed on the motion given by the formula. This type of particle motion has all characteristics of diffusion and is therefore referred to as hydrodynamic diffusion [70].

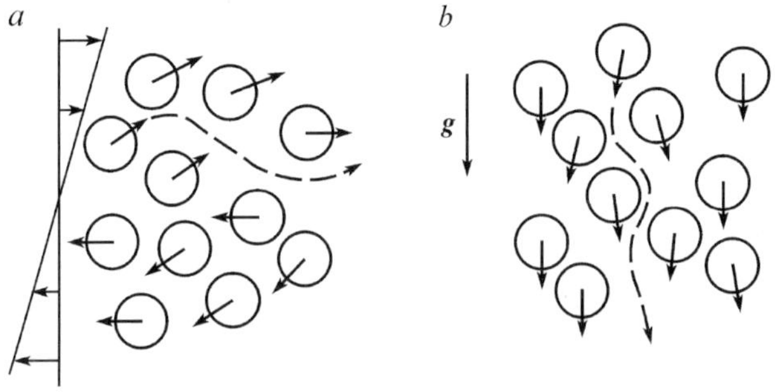

Fig. 10.13 Schematic representation of particle velocities and the trajectory of a test particle (dotted line): a – motion of suspension in a shear flow; b – gravitational sedimentation.

The motion of particles during gravitational sedimentation can be considered as a self-diffusion phenomenon if particle distribution in the suspension is homogeneous. Heterogeneity in the distribution of particles results in the phenomenon of gradient or ordinary diffusion. Experiments [71] have shown that fluctuations of particle velocity can be as high as the average velocity of particles, and sometimes even higher, so that particles can even move against gravity. Strong anisotropy of hydrodynamic diffusion results in the fact that the factor of self-diffusion in the direction $\boldsymbol{g}$ is equal to $D_{\parallel}^{s} = 8aU$ and in the cross direction – to $D_{\perp}^{s} = 2aU$, where a is the radius of particles, U – the average velocity of hindered sedimentation of particles. It has also been observed that the effect of self-diffusion decreases appreciably when the concentration of particles becomes higher than 30%. Self-diffusion has also been observed in experiments on sedimentation of heavy particles in a suspension composed of light particles. If we take into account only pair hydrodynamic interactions of particles in a Stokesian flow, the horizontal component of hydrodynamic self-diffusion appears to be equal to zero [72]. This fact indicates that the cross-component of self-diffusion in a suspension is caused, apparently, by multi-particle (rather than paired) hydrodynamic interactions.

Gradient diffusion appears as a drift of particles against the particle concentration gradient. [72] derives the following expression for the factor of cross diffusion: $D_{\perp}^{s} = 2aU$.

The phenomenon of hydrodynamic diffusion in a shear flow, and also its possible applications are discussed in [70].

References

1 Batchelor G. K., Developments in microhydrodynamics in theoretical and applied mechanics; Ed. W. T. Koiter, North Holland, Amsterdam, 1976, p. 33–55.

2 Happel J., Brenner H., Low Reynolds number hydrodynamics, Prentice-Hall, 1965.

3 Brenner H., Hydrodynamic resistance of particles at small Reynolds numbers, Advan. Chem. Eng., 1966, Vol. 6, p. 287–438.

4 Taylor G. I., Motion of axisymmetric bodies in a viscous fluid in problems of hydromechanics and continuum mechanics, SIAM, Philadelphia, 1969, p. 718–724.

5 Morse P. M., Feshbach H., Methods of Theoretical physics (in 2 vols.), McGraw-Hill, New York, 1953.

6 Cox R. G., Mason S. G., Suspended particles in fluid flow through tubes, Ann. Rev. Fluid Mech., 1971. Vol. 3, p. 291–316.

7 Haber S., Hetsroni G., Solan A., On the low Reynolds number motion of two drops, Int. J. Multiphase Flow, 1973. Vol. 1(1), p. 57–71.

8 Zinchenko A. Z., To calculation of hydrodynamical interaction of drops at small Reynolds numbers, Appl. Math. Mech., 1978, Vol. 42. No. 5. p. 955–959 (in Russian).

9 Davis R. H., Schonberg J. A., Rallison J. M., The lubrication force between two viscous drops, Phys. Fluids A, 1989, Vol. 1, p. 77–81.

10 Yiantsios S. G., Davis R. H., Close approach and deformation of two viscous drops due to gravity and Van

der Waals forces, J. Colloid Interface Sci., 1991, Vol. 144, p. 412–433.

11 Ladyszenskaya O. A., Mathematical problems in viscous incompressible fluid dynamics, Nauka, Moscow, 1970 (in Russian).

12 Rallison J. M., The deformation of small viscous drops and bubbles in shear flows, Ann. Rev. Fluid Mech., 1984, Vol. 16, p. 45–66.

13 Brady J. F., Bossis G., Stokesean dynamics, Ann. Rev. Fluid Mech., 1988, Vol. 20, p. 111.

14 Kynch G. J., The slow motion of two or more spheres through a viscous fluid, J. Fluid Mech, 1959, Vol. 5, p. 193–208.

15 Mazur P., van Saarloos W., Many-sphere hydrodynamic interaction and mobilities in a suspension, Physica, 1982. Vol. 115 A, p. 21–57.

16 Ganatos P., Pfeffer R., Weinbaum S., A numerical-solution technique for three-dimensional Stokes flows, with application to the motion of strongly interacting spheres in a plane, J. Fluid Mech., 1978, Vol. 84, p. 79–111.

17 Bossis G., Brady J. F., Dynamic simulation of shear suspensions. I. General Method, J. Chem. Phys., 1984, Vol. 80, p. 5141–5154.

18 Einstein A., Smoluhowski M., Brownian motion, ONTI, Moscow, 1936 (in Russian).

19 Prigogin I., Defay R., Chemical thermodynamics, Longmans Green and Co., London, New York, Toronto. 1954.

20 Leontovich M. L., Introduction to thermodynamics. Statistical physics, Nauka, Moscow, 1983 (in Russian).

21 Katayama Y., Terauti R., Brownian motion of a single particle under shear flow, Eur. J. Phys., 1966, Vol. 17, p. 136–140.

22 Van Kampen N. G., Stochastic processes in physics and chemistry, North Holland, Amsterdam, 1992.

23 Russel W. B., Brownian motion of small particles suspended in liquids, Ann. Rev. Fluid Mech., 1981, Vol. 13, p. 425–455.

24 Deryagin B. V., Muller V. M., On slow coagulation of hydrophobic colloids, Papers Acad. Science, 1967, Vol. 176, p. 869–872 (in Russian).

25 Spielman L. A., Viscous interaction in Brownian coagulation, J. Colloid Interface Sci., 1970, Vol. 33, p. 562–571.

26 Honig E. P., Roebersen G., Wiersema P. H., Effect of hydrodynamic interaction on the coagulation rate of hydrophobic colloids, J. Colloid Interface Sci., 1971, Vol. 36, p. 91–109.

27 Batchelor G. K., Brownian diffusion of particles with hydrodynamic interaction, J. Fluid Mech., 1976, Vol. 74, p. 1–29.

28 Zinchenko A., Davis R. H., Collision rates of spherical drops or particles in a shear flow at arbitrary Peclet numbers, Phys. Fluids A, 1995, Vol. 7, p. 2310–2327.

29 Sinaiski E. G., Brownian and turbulent coagulation of drops in a viscous liquid and stability of emulsions, Colloid. J., 1994, Vol. 56, No. 1, p. 105–112.

30 Phung T. N., Brady J. F., Bossis G., Stokesean dynamics simulation of Brownian syspensions, J. Fluid Mech., 1996, Vol. 313, p. 181–207.

31 Durlofsky L. J., Brady J. F., Bossis G., Dynamic simulation of hydrodynamically interacting particles, J. Fluid Mech., 1987, Vol. 180, p. 21.

32 Lamb H., Hydrodynamics, Dover, New York, 1945.

33 Sadron Ch., Dilute solutions of impenetrable rigid particles in flow properties of disperse systems (Ed. J. J. Hermans), North-Holland, Amsterdam, 1953, p. 131–198.

34 Taylor G. I., The viscosity of a fluid containing small drops of another fluid, Proc. Roy. Soc., 1932, A 138, p. 41–48.

35 Batchelor G. K., The effect of Brownian motion on the bulk stress in a suspension of spherical particles, J. Fluid Mech, 1977, Vol. 831, p. 97–117.

36 Probstein R. F., Physicochemical hydrodynamics, Butterworths, 1989, p. 318.

37 Mandersloot W. G. B., Scott K. J., Geyer C. P., Sedimentation in the

hindered settling regime: Advances in solid-liquid separation (Ed. H. S. Muralidhara), Columbus: Battelle, 1986, p. 63–77.
38 Davis R. H., Acrivos A., Sedimentation of non-colloidal particles at low Reynolds numbers, Ann. Rev. Fluid Mech., 1985. Vol. 17, p. 91–118.
39 Wallis G. B., One-dimensional two-phase flows, McGraw-Hill, New York, 1969.
40 Mises R., The mathematical theory of compressible fluid flow, Academic Press, New York, 1958.
41 Kynch G. J., A theory of sedimentation, Trans. Faraday Soc., 1952, Vol. 48, p. 166–176.
42 Petty C. A., Continuous sedimentation of suspension with a nonconvex flux law, Chem. Eng. Sci., 1975, Vol. 30, p. 1451–1458.
43 Boycott A. E., Sedimentation of blood corpuscles, Nature, 1920, Vol. 104, p. 532.
44 Activos A., Herbolzheimer E., Enhanced sedimentation in settling tanks with inclined walls, J. Fluid Mech., 1979, Vol. 92, p. 435–457.
45 Sinaiski E. G., Separation of two-phase multicomponent mixtures in oil-gas field equipment, Nedra, Moscow, 1990.
46 Fujita H., Foundation of ultracentrifugal analysis, Wiley, New York, 1975.
47 Greenspan H. P., On centrifugal separation of a mixture, J. Fluid Mech., 1983, Vol. 127, p. 91–101.
48 Duhin S. S., Deryaguin B. V., Elektroforesis, Nauka, Moscow, 1976 (in Russian).
49 Henry D. C., The catophoresis of suspended particles. Part I. The equation of catophoresis, Proc. Roy. Soc. 1931, A 133, p. 106–129.
50 Henry D. C., The catophoresis of suspended particles. Part IV. The surface conductivity effect, Trans. Faraday Soc., 1948, Vol. 44, p. 1021–1026.
51 Wiersema P. H., Loeb A. L., Overbeek J. Th. G., Calculation of the electrophoretic mobility of spherical colloid particle, J. Colloid Interface Sci., 1966, Vol. 22, p. 78–99.
52 Levich V. G., Physico-chemical hydrodynamics, Prentice-Hall, Englwood Cliffs, N.J., 1962.
53 Kruyt H. R. (Ed.), Colloid science, V. I. Irreversible system, Elsevier, Amsterdam, 1952.
54 Hiemenz P. C., Principles of colloid and surface chemistry, Marcel Dekker, New York, 1986.
55 Hogg R., Healy T. W., Fuerstenau D. W., Mutual coagulation of colloidal dispersions, Trans. Faraday Soc., 1966, Vol. 62, p. 1638–1651.
56 Hamaker H. C., The London-Van der Waals attraction between spherical particles, Physica, 1937, Vol. 4, p. 1058–1072.
57 Muller V. M., Deryagin B. V., On dependence of threshold concentration of electrolyte on the size of colloid particles, Papers. Acad. Sci., 1967, Vol. 176, p. 1111–1113.
58 Fuks N. A., Mechanics of aerosols, Acad. Sci., 1955, McGraw-Hill, New York (in Russian).
59 Smoluchowski M., Colloid Coagulation, ONTI, Moscow, 1936.
60 Entov V. M., Kaminski V. A., Lapiga E. J., On calculation of the coalescence rate of an emulsion in a turbulent stream, Fluid Dyn., 1976, No. 3. p. 47–55 (in Russian)
61 Spielman L. A., Particle capture from low-speed laminar flows, Ann. Rev. Fluid Mech., 1977, Vol. 9, p. 297–319.
62 Schowalter W. R., Stability and coagulation of colloids in shear fields. Ann. Rev. Fluid Mech., 1984, Vol. 16, p. 214–261.
63 Sinaiski E. G., Coagulation of drops in a turbulent flow of a viscous liquid, Colloid. J., 1993, Vol. 55, No. 4, p. 91–103.
64 Batchelor G. K., An introduction to fluid dynamics, Univ. Press, Cambridge, 1973.
65 Spielman L. A., Goren S. L., Capture of small particles by London forces from low-speed liquid flows, Environ. Sci. Technol., 1970, Vol. 42, p. 135–140.

66 Goren S. L., The normal force exerted by creeping flow on a small sphere touching a plane, J. Fluid Mech., 1970, Vol. 41. p. 619–625.
67 Spielman L. A., Fizpatrick J. A., Theory for particle collection under London and gravity forces, J. Colloid Interface Sci., 1973, Vol. 42, p. 607–623.
68 Sinaiski E. G., Tolstov V. A., Gorbatkin A. T., Nikiforov A. N., A method of calculation of separators equipped with mesh droplet-capture sections. Proceedings of high school. Oil and gas, 1987, No. 11, p. 51–54 (in Russian).
69 Voloshuk V. M., Kinetic theory of coagulation. Hydrometizdat, Moscow, 1984 (in Russian).
70 Davis R. H., Hydrodynamic diffusion of suspended particles: a symposium, J. Fluid Mech., 1996, Vol. 310, p. 325–335.
71 Nicolai H., Herzhaft B., Hinch E. J., Oger L., Guazzelli E., Particle velocity fluctuations and hydrodynamic self-diffusion of sedimenting non-Brownian spheres, Phys. Fluids, 1995, Vol. 7, p. 12–23.
72 Davis R. H., Hassen M. A., Spreading of the interface at the top of a slightly polydisperse sedimentating suspension, J. Fluid Mech, 1988, Vol. 196, p. 107–134.

V
Emulsions

Emulsion is defined as a disperse system consisting of two insoluble liquids. One example is water-oil emulsion, which is a coarsely dispersed system with the disperse phase consisting of drops of diameter 0.1 μm or higher. Water-oil emulsions can be of two kinds. If the volume concentration of water is small in comparison with that of oil, then the oil forms the continuous phase, and water is present as the disperse phase – in the form of drops suspended in oil. Such kind of emulsion is known as a reverse, or the water-in-oil (w/o) emulsion. If, on the other hand, the volume concentration of oil is small in comparison with that of water, then such an emulsion is called a direct, or oil-in-water (o/w) emulsion, where the water serves as the continuous phase, and oil – as the disperse phase.

Emulsions are quite common in both nature and engineering. In addition to water-oil emulsions, there exist other kinds of emulsions: milk, bitumens, plastic lubricants, water-emulsion paints etc. The type of emulsions that is most widespread in nature is an emulsion formed by water and some organic liquid.

Emulsion is formed when two insoluble liquids mix together. Its dispersiveness depends on stability of the emulsion. If we mix two liquids with different volumes, the outcome will be the formation of drops of one liquid suspended in the other liquid. In the absence of electrolytes in the system (electrolytes are known to stabilize emulsions), the emulsion is thermodynamically unstable. When the emulsion is at rest, droplets tend to integrate due to the process of coalescence. If it is moving in a non-uniform velocity field, droplets can be distorted and broken, or, alternatively, come within short distance of each other and coalesce. The result is the development of two competing processes – breakage and coalescence of droplets. Depending on the characteristic times of these processes, the emulsion can become more finely dispersed (the rate of breakage surpasses the rate of coalescence) or more coarsely dispersed in the opposite case. If the rates of breakage and coalescence are equal, the emulsion is in the state of dynamic equilibrium and the disperse state does not change.

Processes of breakage and coalescence can be controlled. If we add demulsifier to the emulsion, it will be adsorbed at the drop surface, reducing the surface tension S and thus increasing Weber's number. At a certain demulsifier concentration, Weber's number exceeds the critical value and the drop splits. On the other hand, we can increase the collision frequency and coalescence rate of drops by

Separation of Multiphase, Multicomponent Systems. E. G. Sinaiski and E. J. Lapiga

ISBN: 978-3-527-40612-8

mixing the emulsion. However, an overly intensive mixing can promote breakage of drops in addition to coalescence.

Natural liquids, for example oil, contain a plenty of insoluble finely dispersed solid substances which, being adsorbed at the surface of drops, form a strong protective envelope, reducing mobility of the drop surface and interfering with coalescence. Thus, the presence of pollutants promotes emulsion stabilization. Therefore, after a certain time, emulsions become stabilized, or, to use another term, "grow old".

Further on, we shall be studying reverse water-oil emulsions of the w/o type. The continuous phase – the oil – is a substance with very low conductivity (10^{-13}–10^{-6} 1/ohm·м). The disperse phase (water) contained in the oil output has many soluble mineral salts that causes its high conductivity (10^{-4}–10^{-1} 1/ohm·м). Therefore a reverse water-oil emulsion can be considered as a disperse system, in which the disperse phase (water droplets) is conductive, and the continuous phase (oil) is dielectric. It means that we can always act selectively on the disperse phase of a w/o emulsion with external electric field. Under the action of electric field, water drops become polarized, get drawn to each other, collide and coalesce. Thus the external electric field promotes integration of the emulsion. Later on, it will be shown that a high intensity of the electric field may also cause droplets to break.

One technological process that is important on preparation of oil for transportation is dewatering (dehydration). This process is performed in special capacities (settlers), in which water drops are separated from oil by gravitational sedimentation. The size of the settler should provide for the separation of sufficiently fine droplets from oil. The size of drops is usually small, so the velocity of their sedimentation obeys Stokes law $U = 2\Delta\rho g R^2/9\mu_e$, where $\Delta\rho$ is the difference of phase densities, μ_e – the dynamic viscosity of the continuous phase. For the characteristic values $\Delta\rho = 200$ kg/m^3, $\mu_e = 10^{-2}$ Па·s, $R = 10$ μm, we have $U = 0.5 \cdot 10^{-5}$ m/c. It means that all drops of radius above 10 μm will settle from a layer of w/o emulsion of height 1 m in time $t \sim 2 \cdot 10^5$ s $= 50$ h. For $R = 100$ μm, this time will be $t \sim 0.5$ h. Thus, if we can increase the radius of water drops in the emulsion by a factor of 10 (for example, from 10 μm to 100 μm), the time needed for separation of the emulsion will decrease by two orders of magnitude, and consequently, the volume (length) of the settler can be reduced by the same factor. Such a large increase of the drop size in a rather short time can be achieved by putting the emulsion into a uniform external electric field. In order to determine the time necessary to enlarge the water drop by a required factor, it is necessary to find the rate of drop coalescence, in other words, we have to study the dynamics of drop integration in the emulsion.

11
Behavior of Drops in an Emulsion

11.1
The Dynamics of Drop Enlargement

Consider a polydisperse emulsion, assuming a spatially homogeneous case, with low volume concentrations of the disperse phase. Assume further that it is possible to limit ourselves to consideration of pair interactions of drops. The dynamics of enlargement (integration) of drops due to their collision and coalescence is then described by the following kinetic equation

$$\frac{\partial n(V,t)}{\partial t} = \frac{1}{2}\int_0^V K(V-\omega,\omega)n(V-\omega,t)n(\omega,t)\,d\omega - n(V,\omega)\int_0^\infty K(V,\omega)n(\omega,t)\,d\omega, \tag{11.1}$$

where V is the drop volume, $n(V,t)$ is the volume distribution of drops at the moment of time t, $K(V,\omega)$ is the kernel of coagulation, whose physical meaning is the collision frequency for the drops of volumes V and w in a unit volume of disperse phase, for a unit concentration of these drops.

The first term in the right-hand side of equation (11.1) corresponds to the formation rate of drops of volume V due to the coalescence of drops with volumes $V-\omega$ and ω, and the second term – to the rate of population decrease of drops of volume V at their coalescence with other drops.

The solution of the equation (11.1) for a given initial distribution $n_0(V)$ allows to trace the evolution of volume distribution of drops and to determine the following key parameters of distribution: numerical concentration of drops (the number of drops per unit volume of emulsion)

$$N(t) = \int_0^\infty n(V,t)\,dV, \tag{11.2}$$

volume concentration (content) of drops (volume of drops per unit volume of emulsion),

Separation of Multiphase, Multicomponent Systems. E. G. Sinaiski and E. J. Lapiga

ISBN: 978-3-527-40612-8

$$W(t) = \int_0^\infty V n(V,t)\, dV, \tag{11.3}$$

and the average volume of drops

$$V_{av}(t) = W(t)/N(t). \tag{11.4}$$

The kernel $K(V,\omega)$ defines the mechanism of drop interactions, therefore the study of its general properties along with its concrete forms for various processes is an independent problem of physics of disperse medium.

The important property of the kernel of kinetic equation is symmetry with respect to sizes of colliding drops, that is $K(V,\omega) = K(\omega, V)$. Multiply both parts of the equation (11.1) by V and integrate the result over V from 0 up to ∞. Taking into account the expression (11.3), one obtains

$$\begin{aligned}\frac{dW}{dt} &= \frac{1}{2}\int_0^\infty \int_0^V VK(V-\omega,\omega)n(V-\omega,t)n(\omega,t)\, dV\, d\omega \\ &\quad - \int_0^\infty \int_0^V VK(V,\omega)n(V,t)n(\omega,t)\, dV\, d\omega.\end{aligned} \tag{11.5}$$

The change of variables $z = V - \omega$, $\omega = \omega$ in the first integral (11.5) allows us to transform the integration area $0 \le V < \infty$, $0 \le \omega < V$ into $0 \le z < \infty$, $0 \le \omega < \infty$. Thus the equation (11.5) will take the form

$$\begin{aligned}\frac{dW}{dt} &= \frac{1}{2}\int_0^\infty \int_0^\infty (z+\omega)K(z,\omega)n(z,t)n(\omega,t)\, dz\, d\omega \\ &\quad - \int_0^\infty \int_0^\infty K(V,\omega)n(V,t)n(\omega,t)\, dV\, d\omega.\end{aligned} \tag{11.6}$$

Re-denote now z as V. Then (11.6) can be rewritten as

$$\frac{dW}{dt} = \frac{1}{2}\int_0^\infty \int_0^\infty (\omega K(V,\omega) - VK(V,\omega))n(V,t)n(\omega,t)\, dV\, d\omega. \tag{11.7}$$

Since only coalescence of drops is considered, the total volume of drops, that is the volume content of drops W, remains constant. It means, that

$$\int_0^\infty \int_0^\infty (\omega K(V,\omega) - VK(V,\omega))n(V,t)n(\omega,t)\, dV\, d\omega = 0. \tag{11.8}$$

Exchanging V and w in the first term of the integrand (11.8), will not affect the value of the integral, therefore one obtains

$$\int_0^\infty \int_0^\infty V(K(\omega, V) - K(V, \omega))n(V, t)n(\omega, t)\, dV\, d\omega = 0. \tag{11.9}$$

It follows from (11.9) that the symmetry of the kernel $K(V, \omega) = K(\omega, V)$ is sufficient condition for constancy of volume concentration of drops W. Let us show the necessity of this condition. The relation (11.9) as a consequence of the kinetic equation (11.1), must hold for any physically possible drop distribution over volumes $n(V, t)$, including a bi-disperse distribution, that is a system containing drops of two volumes V_1 and V_2. For such a system,

$$n(V, t) = n_0(\xi\delta(V - V_1) + (1 - \xi)\delta(V - V_2)). \tag{11.10}$$

Here ξn_0 is number of drops with volume V_1, $(1 - \xi)n$ – number of drops with volume V_2, and $\delta(x)$ – delta – function.

Putting (11.10) into (11.9) and using property of the delta function $\int_0^\infty F(x)\delta(x - x_0)\, dx = F(x_0)$, one obtains:

$$n_0^2(V_1 - V_2)\xi(1 - \xi)(K(V_2, V_2) - K(V_1, V_2)) = 0. \tag{11.11}$$

For a bi-disperse system we have $n_0 \neq 0$, $V_1 \neq V_2$, $\xi \neq 0$ and $\xi \neq 1$, therefore

$$K(V_2, V_1) = K(V_1, V_2). \tag{11.12}$$

Since the choice of V_1 and V_2 is arbitrary, the equality (11.12) must hold for drops of any volumes within the range of their definition.

Violation of symmetry condition of the coagulation kernel means, that volumetric concentration of drops does not remain constant, which is equivalent to implicit introduction of sources and drains into the system, whose intensity would depend on degree and form of kernel's asymmetry. The symmetry of the kernel $K(V, \omega)$ has another consequence, namely, that a function $K(V, \omega)$ at a fixed net volume of drops $V + \omega = z = const$, has an extreme value at $V = \omega$.

The kernel of the kinetic equation characterizes the frequency of drop collisions, and in most cases is determined by numerical solution of the problem on drop interaction, and subsequent approximation of results. Therefore the condition of symmetry, being necessary, imposes restrictions on the choice of approximating expressions. The effective method of kernel symmetrization on available numerical data is described in work [1].

The kinetic equation (11.1) is a nonlinear integro-differential equation, general theory of which does not exist. Its known exact solutions are based on the use of operational methods with reference to a case of linear dependence $K(V, \omega)$ on each of drop volumes [2]. To solve the equation (11.1) with more general kernel, the approximate methods are used – parametric methods and method of moments, and also numerical methods. Parametric methods and method of moments are based on transforming the kinetic equation into a system of equations for the moments of drop distribution over volumes. However, the resulting system of equations is, as a rule, incomplete, since, apart from the integer moments, there appear fractional moments in it as well, whose presence is caused by power dependence of the kernel $K(V,\omega)$ on drop volumes. Additional relations or constraints on the form of drop distribution are necessary for closure of system of equations. So, the use of parametric method is based on the assumption, that distribution belongs to a certain class, for example, to the class of logarithmically normal distributions, whose parameters change in time and are to be determined. The method of interpolation of fractional moments is not confined to the form of distribution, but demands additional relations. Further on, both methods will be used for solving specific problems. Therefore let us consider them in more detail.

In most practical cases, it is not the volume distribution of drop concentration $n(V,t)$ that is of greatest interest, but rather some of its moments or their combinations, which have clear physical meaning and can rather easily be found experimentally. Experimental results can give the answer to a question whether these moments agree with an accepted model of drop interactions, since the choice of the mechanism of drop coalescence is, as a rule, based on a number of assumptions.

Moments of distribution of drops or other particles over volumes are determined by the following expressions:

$$m_{\mathsf{k}} = \int_0^\infty V^{\mathsf{k}} n(V,t)\,dV; \quad (\mathsf{k} = 0, 1, 2, \ldots). \tag{11.13}$$

The value $\mathsf{k} = 0$ corresponds to the zero moment m_0, which is just the number of drops (numerical concentration) per unit volume of emulsion; the value $\mathsf{k} = 1$ corresponds to the first-order moment, equal to the net drop volume per unit volume of emulsion (volume concentration or volume contents of drops). Combinations of moments give: average radius of drops $\left(\frac{3}{4\pi}\right)^{1/3}\frac{m_{1/3}}{m_0}$, average volume of drops $\frac{m_1}{m_0}$, area of all drop surfaces (the interface area) $(36\pi)^{1/3} m_{2/3}$ in a unit volume of emulsion.

Besides the first five moments m_i $(i = \overline{0,4})$, it is possible to define, using the Pearson's diagram [3], the class of distribution $n(V,t)$ (Fig. 11.1). On the diagram, the parameters β_1 and β_2, which are referred to as a square of asymmetry and an excess of distribution, are plotted along the coordinate axes; they are expressed in terms of distribution moments as follows:

$$\beta_1 = \left(\frac{\mu_3}{\mu_2^{3/2}}\right)^2, \quad \beta_1 = \frac{\mu_4}{\mu_2^2},$$

$$\mu_2 = \frac{m_2}{m_0} - \left(\frac{m_1}{m_0}\right)^2, \quad \mu_3 = \frac{m_3}{m_0} - 3\frac{m_2 m_1}{m_0^2} + 2\left(\frac{m_1}{m_0}\right)^3,$$

$$\mu_4 = \frac{m_4}{m_0} - 4\frac{m_3 m_1}{m_0^2} + 6\frac{m_2 m_1^2}{m_0^3} - 3\left(\frac{m_1}{m_0}\right)^4.$$

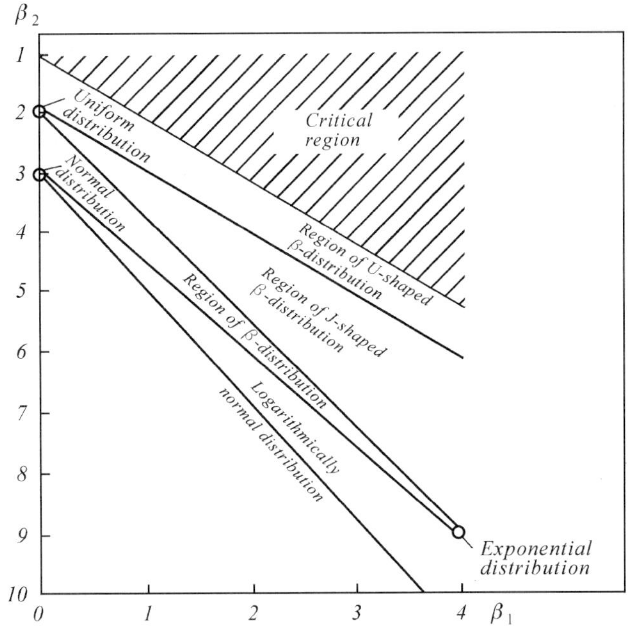

Fig. 11.1 The Pearson diagram.

In order to transform the kinetic equation (11.1) into the equations for moments m_k, one should multiply both parts of this equation by V^k and integrate the obtained expression over V from 0 up to ∞. As a result one obtains:

$$\frac{d}{dt}\left(\int_0^\infty V^k n(V,t)\,dV\right) = \frac{1}{2}\int_0^\infty \int_0^V V^k K(V-\omega,\omega)n(V-\omega,t)n(\omega,t)\,d\omega\,dV - \int_0^\infty \int_0^\infty V^k K(V,\omega)n(V,t)n(\omega,t)\,d\omega\,dV. \tag{11.14}$$

Acting in the same way as the one used in deriving the equation (11.6), we now change the variables $z = V - \omega$, $\omega = \omega$ in (11.14). Then (11.14) will transform to

$$\frac{dm_k}{dt} = \frac{1}{2}\int_0^\infty \int_0^\infty (z+\omega)^k K(z,\omega)n(z,t)n(\omega,t)\,d\omega\,dz - \int_0^\infty \int_0^\infty V^k K(V,\omega)n(V,t)n(\omega,t)\,d\omega\,dV.$$

Replacing now in the first integral z by V, rewrite this equation in the form, commonly used in the method of moments:

$$\frac{dm_k}{dt} = \int_0^\infty \int_0^\infty \left(\frac{1}{2}(V+\omega)^k - V^k\right) K(V,\omega) n(V,t) n(\omega,t)\, d\omega\, dV. \tag{11.15}$$

The majority of known kernels of kinetic equation have the form of power functions of particle volumes. Taking into account, that $K(V,\omega)$ should obey symmetry condition with respect to V and w, it generally can be presented as

$$K(V,\omega) = \sum_{j=0}^{l} G_j (V^{\alpha_j}\omega^{\beta_j} + \omega^{\alpha_j} V^{\beta_j}). \tag{11.16}$$

Substitute now expression (11.6) for the kernel in the equation (11.15) and remove binomial in the integrand. It follows then

$$\frac{dm_k}{dt} = \sum_{j=0}^{l} G_j \left(\frac{1}{2}\sum_{i=0}^{k} \binom{i}{k} (m_{i+\alpha_j} m_{k-i+\beta_j} + m_{i+\beta_j} m_{k-i+\alpha_j}) - (m_{\beta_j} m_{k+\beta_j} + m_{\alpha_j} m_{k+\alpha_j})\right); \quad (k = 0,1,2,\ldots). \tag{11.17}$$

Expression (11.17) represents an infinite system of ordinary differential equations with respect to the moments m_k. If α_j and β_j are distinct from zero or one, the number of equations of any finite subsystem is less than the number of unknowns and all finite subsets of system of equations are incomplete. Some expressions for the kernels contain fractional powers of the volume, therefore the fractional moments appear in the right part of (11.17). In this case the system of equations (11.17) can be supplemented by interpolations of fractional moments with integer moments using the method proposed in [1], or by making an additional assumption about the form of distribution (parametric method). Let us start with considering the first method.

Suppose that at the initial moment of time $t = 0$, the moments $m_k(0)$ are given. Introduce dimensionless moments $\hat{m}_k(t) = m_k(t)/m_k(0)$, where k can be integer or fractional number. Integer moments appear only in the left-hand side of equations (11.17), while the right-hand side can contain both integer and fractional moments. The problem is to express fractional moments via the integer ones. Consider a fractional moment of the order $j+\xi$, where j is the nearest integer number and $0 < \xi < 1$, and integer moments $m_s, m_{s+1}, \ldots, m_{s+r}$, where $s < j+\xi < s+r$. We shall seek for the logarithm of the fractional moment as an interpolation polynomial of the form

$$\ln(\hat{m}_{j+\xi}) = \sum_{q=s}^{s+r} L_q^{(r)}(j+\xi) \ln(\hat{m}(t)); \quad (j = 1,2,\ldots), \tag{11.18}$$

where $L_q^{(r)}(j+\xi)$ are coefficients of the interpolation (Lagrange) polynomial, which is interpolated over $r+1$ points that are determined from the equation

$$L_q^{(r)}(j+\xi) = \prod_{\substack{p=s \\ p\neq q}}^{s+r} \frac{j+\xi-r}{q-p}. \tag{11.19}$$

Equation (11.19) allows us to present fractional moments as

$$\frac{m_{j+\xi}(t)}{m_{j+\xi}(0)} = \prod_{q=s}^{s+r} \left(\frac{m_q(t)}{m_q(0)}\right)^{L_q^{(r)}(j+\xi)}; \quad s \le j \le s+r,\ 0 < \xi < 1 \tag{11.20}$$

and thereby to supplement the system of equations (11.17).

Let us apply this method to the kinetic equation with the kernel of the form

$$K(V,\omega) = GV^{\alpha}\omega^{\alpha}; \quad 0 \le \alpha \le 1. \tag{11.21}$$

Take the first four moments of the system of equations (11.17), and for the fractional moments entering the right-hand side, use two-point interpolation, expressing moments of the order $m_{j+\alpha}$ via integer moments m_j and m_{j+1}. The outcome is a closed system of equations with respect to the first four moments

$$\begin{aligned}
\frac{dm_0}{dt} &= -\frac{1}{2} G m_\alpha^2(0) \left(\frac{m_0(t)}{m_0(0)}\right)^{2-2\alpha} \left(\frac{m_1(t)}{m_1(0)}\right)^{2\alpha}, \\
\frac{dm_1}{dt} &= 0, \\
\frac{dm_2}{dt} &= G m_{1+\alpha}^2(0) \left(\frac{m_1(t)}{m_1(0)}\right)^{2-2\alpha} \left(\frac{m_2(t)}{m_2(0)}\right)^{2\alpha}, \\
\frac{dm_3}{dt} &= 3G m_{2+\alpha}^2(0) m_{1+\alpha}(0) \left(\frac{m_1(t)}{m_1(0)}\right)^{1-\alpha} \left(\frac{m_2(t)}{m_2(0)}\right)^{\alpha}.
\end{aligned} \tag{11.22}$$

For $\alpha \neq 0.5$, the solution of this system of equations is

$$\begin{aligned}
m_0(\tau) &= m_0(0)\left(1+\frac{B_0\tau}{\delta}\right)^{-\delta}, \quad m_1(\tau) = m_1(0), \\
m_2(\tau) &= m_2(0)\left(1+\frac{B_0\tau}{\delta}\right)^{-\delta}, \\
m_3(\tau) &= m_3(0)\left(1+\frac{B_3}{2B_2}\left(1+\frac{B_2\tau}{\delta}\right)^{1+\delta}\right)^{1/(1-\alpha)},
\end{aligned} \tag{11.23}$$

When $\alpha = 0.5$, the solution is

$$\begin{aligned}
m_0(\tau) &= m_0(0)e^{B_0\tau}; \quad m_1(\tau) = m_1(0), \\
m_2(\tau) &= m_2(0)e^{B_2\tau}; \quad m_3(\tau) = m_3(0)\left(\frac{B_3}{B_2}e^{B_2\tau}+1\right)^2,
\end{aligned} \tag{11.24}$$

Here

$$\tau = Gm_1^{2\alpha}t; \quad B_0 = \frac{m_\alpha^2(0)}{2m_0(0)m_1^{2\alpha}}; \quad \delta = \frac{1}{1-2\alpha};$$

$$B_2 = \frac{m_{1+\alpha}^2(0)}{m_2(0)m_1^{2\alpha}}; \quad B_3 = \frac{m_{1+\alpha}(0)m_{2+\alpha}(0)}{m_3(0)m_1^{2\alpha}}.$$

Now, consider the parametric method. As was already mentioned, the underlying assumption is that the distribution belongs to a certain class. The choice of the distribution class can be made on the basis of general considerations about the kind of distribution that is likely to be formed in a specific physical process. For example, for a turbulent flow of emulsion in a pipe, the distribution can be described with sufficient accuracy by a logarithmic normal distribution or a gamma-distribution. Consider these two cases successively.

Let the volume distribution of drops belong to the class of lognormal distributions [4]

$$n(V,t) = \frac{N}{3\sqrt{2\pi}\sigma V}\exp\left(-\frac{\ln^2(V/V_0)}{18\ln^2\sigma}\right), \tag{11.25}$$

where N is the numerical concentration of drops; σ^2 – the variance of the distribution; V_0 – a parameter connected to the average volume V_{av} of drops by the relation $V_{av} = V_0\exp(1.5\ln^2 s)$.

The shape of the distribution does not change with time; only its parameters $N(t)$, $V_0(t)$ and $\sigma(t)$ change.

Now substitute (11.25) into the equations (11.15) for moments and take successively $\mathsf{k} = 0, 1, 2$. In the case of power dependence of the kernel $K(V,\omega)$ on V and w, fractional and integer moments (which are distinct from the first three moments) appear in the right-hand side of the equations. To the obtained system of equations we should add the property of distributions of the kind (11.25), namely

$$m_\mathsf{k} = m_1 V_0^{\mathsf{k}-1}\exp\left(-\frac{9}{2}(\mathsf{k}^2-1)\ln^2\sigma\right). \tag{11.26}$$

This results in a closed system of equations with respect to m_1, V_0 and σ. Actually, there are only two unknowns, since $m_1 = const$.

Let the sought-for distribution belong to the class of gamma distributions [4]

$$n(V,t) = \frac{W}{V_0^2}\frac{(i+1)}{i!}\left(\frac{V}{V_0}\right)^i\exp\left(-\frac{V}{V_0}\right), \tag{11.27}$$

where W is the volume concentration of drops, which remains constant in the process of coalescence, and V_0 and i are parameters of the distribution connected to the average volume and variance by the relations $V_0 = V_{av}/(i+1)$,

$\sigma = V_{av}/\sqrt{i+1}$. The shape of the distribution does not change with time, only N and V_0 change. The distribution (11.27) allows us to look for the solution of equations (11.15) for moments in the form of expansion of $n(V, t)$ over a system of orthogonal Laguerre associated polynomials [5] L_{ki}:

$$n(V,t) = \left(\frac{V}{V_0}\right)^i \exp\left(-\frac{V}{V_0}\right) \sum_{k=0}^{\infty} a_k L_{ki}\left(\frac{V}{V_0}\right). \tag{11.28}$$

The first two polynomials are

$$L_{0i} = 1; \quad L_{1i} = i + 1 - \frac{V}{V_0},$$

and the others are given by the recurrent relation

$$(n+1)L_{n+1,i} - \left(\frac{V}{V_0} - 2n + i + 1\right)L_{ni} + (n+i)L_{n-1,i} = 0.$$

The orthogonality condition for polynomials allows us to find the factors of the expansion (11.28)

$$a_k = \frac{k!}{\Gamma(k+i+1)} \int_0^{\infty} L_{ki}\left(\frac{V}{V_0}\right) n(V,t) d\left(\frac{V}{V_0}\right), \tag{11.29}$$

where $\Gamma(x)$ is a gamma-function.

In particular, the first two factors are

$$a_0 = \frac{k!}{\Gamma(k+1)} \int_0^{\infty} n(V,t) d\left(\frac{V}{V_0}\right); \quad a_1 = 0.$$

If we use only the first two moments, there follows from (11.15) and (11.28)

$$\begin{aligned} n(V,t) &= \frac{N}{V_0 i!}\left(\frac{V}{V_0}\right)^i \exp\left(-\frac{V}{V_0}\right), \\ \frac{d(N/V_0)}{dt} &= -\frac{1}{2}\left(\frac{N}{V_0}\right)^2 \int_0^{\infty}\int_0^{\infty} \exp\left(-\frac{\omega+V}{V_0}\right) K(V,\omega) \\ &\quad \times \left(\frac{V}{V_0}\right)^i \left(\frac{\omega}{V_0}\right)^i d\omega\, dV, \\ (i+1)NV_0 &= W. \end{aligned} \tag{11.30}$$

The system of equations (11.30) in the two-moment approximation makes it possible to find the time rate of change of the volume distribution of drops, and the parameters of this distribution.

The practical application of these results shows that the method of moments provides an accurate description of the kinetics of coalescence only at its initial stage. With some restrictions imposed on the form of the kernel of the kinetic equation and on the initial distribution [6], a self-similar solution of the kinetic equation for a longer time range can be obtained. In the general case, the solution can be obtained by numerical methods.

11.2
The Basic Mechanisms of Drop Coalescence

The small difference between densities of drops and the ambient liquid, as well as small size of drops in the emulsion, result in a low sedimentation rate of drops in gravitational field. Thus the main challenge in the process of emulsion separation is to increase the drop size. This problem can be addressed by intensifying the coalescence of drops. The factors utilized to enhance the rate of drop integration may include the application of electric field and "turbulization" of the flow. Before we proceed to describe these effects, consider in general the process of drop coalescence in the emulsion.

On a conceptual level, the process of drop coalescence can be divided into two stages. At the first stage, called the transport stage, drops approach each other and come into contact, and at the second stage known as the kinetic stage they merge. Experimental research of drop coalescence on a flat interface between two liquids gives reason to believe [7] that merging of drops takes a much shorter time than their mutual approach. This means that coalescence time depends primarily on the duration of the transport stage. It follows that the kernel of the kinetic equation can be calculated by performing a detailed analysis of the mutual approach of drops up to their point of contact, with the assumption that every collision of drops results in their coalescence.

The characteristics of mutual approach of drops, and consequently, of their collisions, depend on the hydrodynamic regime of emulsion motion. Consider first the coalescence of drops settling under gravity in a quiescent liquid. Such kind of coalescence is called the gravitational coalescence.

Drops of different sizes are settling under gravity with different velocities. As a result, larger drops overtake smaller ones, and their collision becomes possible. Each drop has its own trajectory, whose determination should be our primary goal if we want to calculate the collision frequency of drops. In most cases, the volume concentration of drops is assumed to be low, so it is possible to consider only the relative motion of two drops. The analysis will be carried out in a spherical system of coordinates attached to the center of the larger drop (Fig. 11.2).

In this coordinate system, the flow of the ambient liquid moves relative to the larger drop. On a large distance from the drop, the velocity can be assumed constant and equal to the sedimentation velocity of the drop. Another drop of smaller size moves relative to the larger drop together with the flow, goes around it, and either touches it or passes by. Due to their small sizes, the motion of drops can be

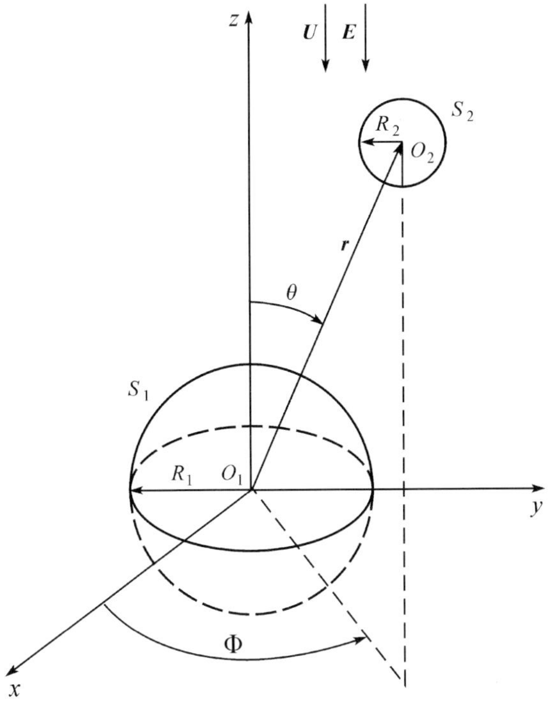

Fig. 11.2 Spherical system of coordinate (r, θ, Φ), with the origin in the center of the larger drop.

considered inertialess. Then, far away from a larger drop (at distances substantially greater than its radius), the trajectory of the smaller drop relative to the larger one coincides with the stream line of the external liquid, while at small distances it noticeably deviates from the stream line, because of the interaction between the drop and the ambient liquid and the mutual interaction of drops. Interaction forces can be hydrodynamic, molecular, or electrostatic. Hydrodynamic forces of resistance to the drop motion increase with decrease of clearance between the drop surfaces. Molecular forces are the Van der Waals–London attractive forces acting at small distances. Electrostatic forces are the forces of repulsion caused by double electric layers at the drop surfaces and/or the forces of interaction between conducting drops (electrically charged or neutral), placed in an external electric field. In order to determine the relative trajectories of drops, it is necessary to know all of these forces. Determining them involves solving of the appropriate hydrodynamic and physical problems. Later on, these questions will be discussed in more detail.

All sets of trajectories of smaller drops relative to the larger ones can be divided into two classes: trajectories which do not end at the surface of the large drop, and trajectories resulting in a collision of drops (Fig. 11.3).

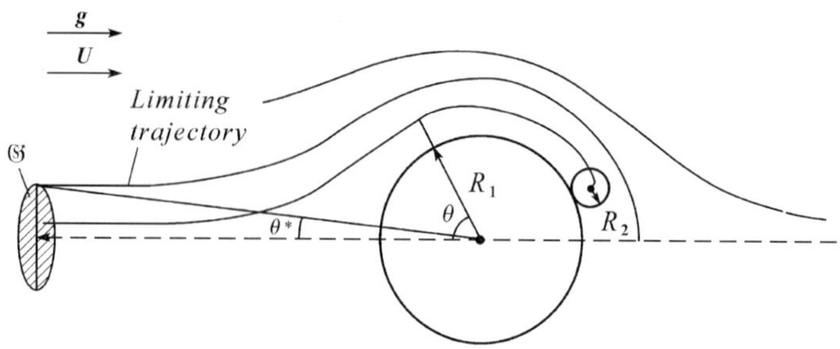

Fig. 11.3 Motion trajectories of particle 2 relative to particle 1.

The trajectory dividing these two classes is called the limiting trajectory. Then, looking at the limiting trajectories in the oncoming flow on a large distance from the big drop, we see that they form a tube of current. The area of the normal cross-section $G(V, \omega)$ of this tube has the meaning of collision cross-section of two drops of given volumes V and ω $(V > \omega)$. Then the collision frequency of drops with volumes V and ω can be determined as

$$K(V, \omega) = G(V, \omega)|U_V - U_\omega|, \tag{11.31}$$

where U_V and U_ω stand for the sedimentation velocities of drops with volumes V and ω located far away from each other. They can be taken to be equal to the corresponding Stokesian velocities.

The drop trajectories are determined from the equations

$$m_i \frac{d\boldsymbol{u}_i}{dt} = \sum_k \boldsymbol{F}_k^{(i)}; \quad J_i \frac{d\boldsymbol{\Omega}_i}{dt} = \sum_k \boldsymbol{T}_k^{(i)}; \quad \boldsymbol{u}_i = \frac{d\boldsymbol{r}_i}{dt}, \tag{11.32}$$

where m_i is the mass of the drop; $\boldsymbol{F}_k^{(i)}$, $\boldsymbol{T}_k^{(i)}$ are the forces and moments acting on the drops; J_i is the moment of inertia, $i = 1, 2$.

If drop sizes are small enough, and the difference in densities of the internal and external liquid is sufficiently small, the flow of the liquid can be considered as slow, and the motion of drops – as inertialess. This is the case when we separate emulsions of the w/o (water-oil) type. Then equations (11.32) are reduced to the equations of inertialess motion of drops in the quasi-stationary approximation

$$\sum_k \boldsymbol{F}_k^{(i)} = 0; \quad \sum_k \boldsymbol{T}_k^{(i)} = 0; \quad \boldsymbol{u}_i = \frac{d\boldsymbol{r}_i}{dt}. \tag{11.33}$$

Supposing the initial positions of drop 2 to be given, and integrating (11.33), we find the family of trajectories of drop 2 relative to drop 1. In a spherical coordinate system, the initial position of drop 2 is specified by the initial coordinates of the particle's center r_0, θ_0, Φ_0. To determine the capture cross-section, we must consider a stream tube whose surface contains the limiting trajectories (see Fig. 11.3). Denote by $\theta^*(\Phi)$ the angle between the vector $\boldsymbol{r}_0$ and the straight line connecting the particle center with the point on the critical (limiting) trajectory at the fixed normal cross-section of this stream tube $z = r_0 \cos\theta^*$. In the general case, the normal cross-section of this current tube at a large distance from drop 2 (i.e. $r_0 \to \infty$ and $\theta^* \to 0$) is non-circular, and its area can be determined from simple geometrical considerations:

$$G = \lim_{r_0\to\infty}\left(\frac{1}{2}r_0^2\int_0^{2\pi}\sin^2(\theta^*(\Phi))\,d\Phi\right). \tag{11.34}$$

If the liquid flow and the relative motion of drops are axially symmetric, then θ^* does not depend on Φ, the shape of the collision cross-section is circular, and

$$G = \lim_{\substack{r_0\to\infty\\ \theta^*\to 0}}(\pi r_0^2\sin^2(\theta^*)). \tag{11.35}$$

The whole bulk of works devoted to the calculation of collision cross-section of particles can be divided into three groups depending on the extent to which interaction forces between particles are taken into account. Some of the works are worth mentioning. The first works were published by Smoluchowski [8], who developed the theory of colloid coagulation neglecting hydrodynamic interactions of particles. The majority of subsequent papers dealt with the motion of particles in low-viscous liquids in the context of problems having to do with drop and particle coagulation, and with particles in the atmosphere [9, 10]. Works [11–13] took into account the hydrodynamic interaction of two slowly moving spherical particles in a viscous liquid. The approximate expressions used in these works are obtained by the method of reflection and are correct only for particles placed relatively far from each other. In particular, the collision cross-section for two spherical particles of different radii settling under the action of gravitational field was determined in [11]. The results of this work were used in [12] to calculate the collision cross-section of particles of comparable sizes in an electric field. The calculation of collision cross-section for two charged particles, one of which is much smaller than another was performed in [14]. A more accurate accounting of hydrodynamic forces was done in [15, 16]. We shall mention that in [15], the collision cross-section of drops of various sizes in an external electric field was determined for conducting drops, and in [16] – for charged drops. The last two works take into account both hydrodynamic and electric forces, deriving them from the exact solution of the corresponding hydrodynamic and electrostatic problems. All works mentioned so far analyze interaction of particles without taking into account internal viscosity. Paper [17] derives the collision cross-section of two spherical drops whose internal viscosity is different from the viscosity of the ambient liquid. This paper also considers the force of molecular interaction of drops, which is responsible for their coalescence.

The use of turbulent emulsion flow regime to facilitate integration of drops is justified by the substantial increase of collision frequency that is achieved in a turbulent flow as compared to the collision frequency during the sedimentation of drops in a quiescent liquid or in a laminar flow. Particles suspended in the liquid are entrained by turbulent pulsations and move chaotically inside the volume in a pattern similar to Brownian motion. Therefore this pulsation motion of particles can be characterized by the effective factor of turbulent diffusion D_T, and the problem reduces to the determination of collision frequency of particles in the framework of the diffusion problem, as it was first done by Smoluchowski for Brownian motion [18]. A similar approach was first proposed and realized in [19] for the problem of coagulation of non-interacting particles. The result was that the obtained frequency of collisions turned out to be much greater than the frequency found in experiments on turbulent flow of emulsion in pipes and agitators [20, 21].

The effect of hydrodynamic interaction of particles on the factor of mutual diffusion of particles has been studied in paper [22]. It formulated a theoretical basis for the determination of collision frequency of particles in a turbulent flow. The effect of internal viscosity of drops on their collision frequency was studied in [23–26]. It was shown that the correct accounting for hydrodynamic interaction ensures a good agreement between the theory and the experiment.

The number of collisions of particles with radii R_1 and R_2 in a unit time when they approach each other via the diffusion mechanism is equal to the flux of particles of radius R_2 towards the particle of radius R_1. If we assume that diffusion equilibrium is established much faster than concentration equilibrium, the problem reduces to solving the stationary diffusion equation in a force field [27]

$$\nabla \cdot \left(D_T \nabla C - \frac{\boldsymbol{F}}{h} C \right) = 0, \tag{11.36}$$

where C is the number concentration of particles 2; $\boldsymbol{F}$ – the interaction force between particles 1 and 2 due to the presence of fields (electric, gravitation, molecular, electrostatic); h – the hydrodynamic resistance factor for the particles approaching each other along the line connecting their centers. If $\boldsymbol{u}$ denotes the relative velocity of the approaching particles (again, it is directed along the line connecting the centers), and $\boldsymbol{F}_h$ – the hydrodynamic resistance force experienced by particle 2, then

$$\boldsymbol{F}_h = h\boldsymbol{u}. \tag{11.37}$$

Sometimes instead of the hydrodynamic resistance factor, we use mobility $b = h^{-1}$, thus expressing velocity in terms of force:

$$\boldsymbol{u} = b\boldsymbol{F}. \tag{11.38}$$

If every collision of particles results in their coalescence, the boundary conditions for equation (11.36) are

$$C = 0 \quad \text{at } r = R_1 + R_2; \quad C = C_0 \quad \text{at } r \to \infty. \tag{11.39}$$

Here C_0 is the concentration of particles 2 far away from the test particle 1.

Having found the distribution C from (11.36) and (11.39), it is possible to find the flux of particles 2 towards a fixed particle 1

$$J = \iint_s \left(D_T \nabla C - \frac{\boldsymbol{F}}{h} C \right) \cdot \boldsymbol{n}\, ds. \tag{11.40}$$

There is a number of works attempting to derive the collision frequency of particles approaching each other via the diffusion mechanism. However, only a few of then take into account the hydrodynamic interaction of particles. For Brownian diffusions, it is taken into account in works [28–31], and for particle motion in a turbulent flow – in [22–26, 32].

Thus, to determine the frequency of collision of particles or drops, it is necessary to determine the forces of particle interaction first, and then to find the trajectories of their motion and the collision cross-section or the diffusion flux. In the latter case, it is necessary to find the turbulent diffusion factor. As a result, the kernel of the kinetic equation is determined. If the kernel thus derived appears to be asymmetric, it should be symmetrized. After that, one can proceed to study the kinetics of coalescence for the considered process, including the time rate of change of size distribution of particles and the parameters of this distribution.

11.3 Motion of Drops in a Turbulent Flow of Liquid

Small particles and drop transfer in turbulent flow can be considered as diffusion with an effective diffusion factor [19]. Before defining it, let us consider basic laws of motion of liquid in turbulent flow. In case of developed turbulence, these laws are well studied and described in works [19, 33–35], therefore we will limit here to the aspects concerning definition of particle and drop collisions.

In turbulent flow of liquid, random pulsation motions characterized by a set of pulsation velocities u_λ are imposed on average movement with velocity U in a certain direction. Turbulent pulsations are characterized not only by velocities, but also by distances at which these velocities undergo noticeable change. These distances are referred to as pulsation scales and are denoted as λ. The set of values λ represents a spectrum of turbulent pulsations varying from 0 up to a maximal value, having the order of linear scale of cross-sectional area of current flow. So, at motion in a pipe of diameter L the greatest value λ is equal to L. Every pulsation movement is characterized by its Reynolds number $\mathrm{Re}_\lambda = \lambda u_\lambda / \nu$, where

ν is kinematics viscosity of carrying liquid. Pulsations with $\lambda \sim L$ are referred to as large-scale pulsations. For them $\mathrm{Re}_\lambda \gg 1$, therefore the flow of liquid caused by these pulsations, has a non-viscous character. Reduction of pulsation scale results in decrease of Reynolds number. At some value $\lambda = \lambda_0$, called the inner scale of turbulence, Reynolds number

$$\mathrm{Re} = \mathrm{Re}_{\lambda_0} = \lambda u_{\lambda_0}/\nu \sim 1. \tag{11.41}$$

It means that pulsations with $\lambda < \lambda_0$, have viscous character and pulsation motion at such scale is accompanied by energy dissipation. Pulsations with $\lambda \ll L$ are referred to as small-scale pulsation. They are generated by large-scale pulsations, whose energy is transferred to small-scale movements, and then dissipates into heat. Thus, turbulent flow of liquid is accompanied by significant dissipation of energy, which losses per unit mass and unit time is characterized by ε_0, called specific dissipation of energy. Since the energy is taken from large-scale pulsations, ε_0 depends on U and L. The value ε_0 can be estimated from dimension reasons

$$\varepsilon_0 \sim U^3/L. \tag{11.42}$$

Consider small-scale non-viscous pulsations with $\lambda_0 \ll \lambda \ll L$. The velocity of such pulsations can depend only on ε_0 and λ, therefore

$$u_\lambda \sim (\varepsilon_0 \lambda)^{1/3}. \tag{11.43}$$

Since at $\lambda \sim \lambda_0$, the condition $\mathrm{Re}_{\lambda_0} \sim 1$ should be satisfied, and the inner scale of turbulence can be obtained from (11.41) and (11.43)

$$\lambda_0 \sim \frac{\nu}{u_{\lambda_0}} \sim (\nu^3/\varepsilon_0)^{1/4}. \tag{11.44}$$

As mentioned above, motion of liquid with pulsations with $\lambda < \lambda_0$ has viscous character. According to the Landau–Levich hypothesis, such pulsations do not disappear suddenly but attenuate gradually. Motion of liquid can be represented as a combination of independent periodic motions, whose periods T are constant for all $\lambda < \lambda_0$. The value of T can be estimated by $T \sim \sqrt{\nu^3/\varepsilon_0}$. In this case, the velocities of pulsations of scales $\lambda < \lambda_0$ are estimated as

$$u_\lambda \sim \frac{\lambda}{T} \sim \lambda\sqrt{\varepsilon_0/\nu}. \tag{11.45}$$

Combining (11.43) and (11.45), one receives general expression for velocity of turbulent pulsations of scale l

$$u_\lambda \sim \begin{cases} (\varepsilon_0 \lambda)^{1/3} & \text{at } \lambda > \lambda_0, \\ \lambda(\varepsilon_0/\nu)^{1/2} & \text{at } \lambda < \lambda_0. \end{cases} \tag{11.46}$$

Consider now motion of small particles in turbulent flow of liquid. Assume that the volume concentration of particles is small enough, so it is possible to neglect their influence on the flow of liquid. The large-scale pulsations transfer a particle together with layers of liquid adjoining to it. Small-scale pulsations with $\lambda \ll R$, where R is particle radius, cannot involve the particle in their motions – the particle behaves in this respect as a stationary body. Pulsations of intermediate scales do not completely involve the particle in their motion. Consider the case most interesting for applications, when respective densities of particle ρ_i and external liquid ρ_e are only slightly different from one another, and radius of the particle is much less than inner scale of turbulence, that is $R \ll \lambda_0$. Thus, for water-oil emulsion $\rho_i/\rho_e \sim 1.1$–1.5. Let $\boldsymbol{u}_0$ be the velocity of liquid at the particle's location, and $\boldsymbol{u}_1$ the velocity of particle relative to liquid. At full entrainment of particle by the liquid, the same force would act on the particle as on the liquid enclosed in the same volume $\frac{4}{3}\pi R^3 \rho_e \frac{d\boldsymbol{u}_0}{dt}$. However, due to incomplete entrainment of the particle by liquid the particle exhibits resistance $\boldsymbol{F}$, which at $R \ll \lambda_0$ and under condition of $u_1 < u_{\lambda_0}$ is determined by the Stokes formula $\boldsymbol{F} = -6\pi\rho\nu R\boldsymbol{u}$. The equation of particle's motion will be [19, 36]:

$$m\frac{d\boldsymbol{u}_1}{dt} + \frac{4}{3}\pi R^3 \rho_e \left(\frac{d\boldsymbol{u}_0}{dt} + \frac{d\boldsymbol{u}_1}{dt} \right) = \frac{4}{3}\pi R^3 \rho_e \frac{d\boldsymbol{u}_0}{dt} - 6\pi\rho\nu R\boldsymbol{u}_1, \tag{11.47}$$

where $m = 2\pi R^3 \rho_e/3$ is the effective mass of the particle.

Estimate the accelerations figuring in (11.47). Since for motions of the scale $\lambda < \lambda_0$ the period T is constant, we have

$$\frac{du_1}{dt} \sim \frac{u_1}{T} \sim u_1 \sqrt{\varepsilon_0/\nu}. \tag{11.48}$$

To estimate the acceleration of liquid at the particle's location, one should take the maximal acceleration of pulsation movements within considered range of scales, that is at $\lambda = \lambda_0$

$$\frac{du_0}{dt} \sim \frac{u_{\lambda_0}}{T} \sim \varepsilon_0 \frac{\lambda_0}{\lambda} \sim \left(\frac{\varepsilon_0^3}{\nu} \right)^{1/4}. \tag{11.49}$$

Substituting (11.48) in (11.47), one obtains the result, that vectors $\boldsymbol{u}_1$ and $d\boldsymbol{u}_0/dt$ are approximately collinear, and in view of relation (11.49), the solution of equation (11.47) is found as

$$u_1 \approx \frac{2R^2 |\rho_e - \rho_i| \lambda_0 \varepsilon_0 \nu^{-1}}{2R^2(\rho_i + 0.5\rho_e)\varepsilon_0^{1/2} + 9\rho\nu}. \tag{11.50}$$

The degree of particle entrainment by pulsation of scale λ can be estimated by the value of ratio u_1/u_{λ_0}. So if $u_1/u_{\lambda_0} \ll 1$, we have the particle completely entrained, and at $u_1/u_{\lambda_0} \gg 1$ the entrainment of particle does not occur. For pulsations with $\lambda < \lambda_0$ from (11.46) and (11.50) it follows, that

$$\frac{u_1}{u_\lambda} \approx \frac{2R^2|\rho_e - \rho_i|\varepsilon_0^{1/2}\nu^{-1/2}}{R^2(\rho_e + 2\rho_i)\varepsilon_0^{1/2}\nu^{-1/2} + 9\rho\nu}. \tag{11.51}$$

For water-oil emulsion of the type w/o we have $\rho_e \approx 800$ kg/m^3, $\rho_i \approx 1200$ kg/m^3, $\nu \approx 10^{-5}$ m^2/s, $\lambda_0 \approx 10^{-4}$ m, $\varepsilon_0 \approx 10$ J/kg·s. Then it follows from (11.51)

$$\frac{u_1}{u_\lambda} \approx \frac{8R^2}{28R^2 + 72 \cdot 10^{-3}} \ll 1.$$

Hence, water drops of radius $R < \lambda < \lambda_0$ are practically completely entrained by pulsations of scale λ.

Transfer of drops with radius $R \ll \lambda_0$ is determined by two parameters: velocity u_λ and pulsation scale λ. They can be combined into a parameter having dimension of diffusion

$$D_T \sim \lambda u_\lambda. \tag{11.52}$$

Substituting in (11.52) the velocity of pulsations from (11.46), one obtains expression for diffusion factor of particles suspended in turbulent flow of liquid

$$D_T \cong \begin{cases} (\varepsilon_0\lambda^4)^{1/3} & \text{at } \lambda > \lambda_0, \\ \lambda^2(\varepsilon_0/\nu)^{1/2} & \text{at } \lambda < \lambda_0. \end{cases} \tag{11.53}$$

Consider now application of the obtained relation to diffusion interaction of particles. Denote through r the distance between centers of two particles. As was shown in [19], main diffusion resistance to the particles' approach lies within the region $r < \lambda_0$, that is the basic resistance that particles exhibit, occurs when the clearance between their surfaces is $\delta < \lambda_0 - (R_1 + R_2)$, where R_1 and R_2 are radii of particles. The factor of diffusion according to (11.53) is equal to

$$D_T = \frac{\varepsilon_0^{1/2}\lambda^2}{\nu^{1/2}} \sim \nu\frac{\lambda^2}{\lambda_0^2}. \tag{11.54}$$

The question about pulsation scale at which the particles are drawn together, is left without answer, since large pulsations will carry two drops together with ambient liquid as a single whole and are unable to noticeably draw particles toward each other. It is obvious, that in order to draw together particles of about identical radiuses $R_1 \sim R_2$, the pulsation scales should be, by the order of magnitude, equal to the distance between the particles' centers, that is $\lambda \sim r$. Particles of significantly different sizes ($R_1 \gg R_2$) can be drawn together by pulsations of the scale equal to distance between the centre of small particle and the surface of the big one, that is, pulsations with $\lambda \sim r - R_1$, since in this case he big particle can be considered as a flat wall, at whose vicinity the pulsations are damped. To determine scales of pulsations able to bring the two particles into intermediate area of their sizes, one should use an approximation, which satisfies the two above indicated limiting cases, and also an obvious condition of symmetry of diffusion factor with respect to particles' radii $D_T(R_1, R_2) = D_T(R_2, R_1)$,

$$\lambda \sim (R_1 + R_2)\left(\frac{r - R_1 - R_2}{R_1 + R_2} + \frac{R_1 R_2}{R_1^2 + R_2^2 - R_1 R_2}\right). \tag{11.55}$$

Returning to (11.54), one obtains the following expression for the factor of mutual diffusion of particles with radiuses R_1 and R_2:

$$D_T = \beta \frac{\nu}{\lambda_0^2}(R_1 + R_2)^2 \left(\frac{r - R_1 - R_2}{R_1 + R_2} + \frac{R_1 R_2}{R_1^2 + R_2^2 - R_1 R_2}\right)^2. \tag{11.56}$$

Here a correction factor β of the order 1 has been introduced, since expressions (11.54) and (11.55) are estimated by the order of magnitude.

Expression (11.56) is obtained under the assumption, that particles are involved completely in relative motion by pulsations of scale λ. Therefore the formula (11.56) can be used only if particles are relatively far apart. However, when the particles approach each other so that the clearance δ between them is about radius of a smaller particle, then the velocity of their approach will be affected by the force of hydrodynamic resistance, which, as was emphasized earlier in section 8.1, grows unboundedly at $\delta \to 0$. To account for this force, one should apply the approach, which is used in statistical physics for description of the Brownian motion of a particle under action of random external force and based on Langevin equation [37, 38] (see also section 8.2).

The particle under action of random turbulent pulsations many times changes direction of motion during a small time interval. Therefore it is difficult to trace visually the particle's actual trajectory. So the particle's displacement cannot serve as characteristic of its motion, since the average value of displacements in finite time interval may be zero. For such time interval, the average square of displacement can be used as the main characteristic of particle's motion. At this point one should bear in mind, that "average" is understood here not as average over time, but as average over set of particles. The average square of displacement has meaning of diffusion factor of particle under action of random external force.

Consider two moments of time: t_1 and $t > t_1$ and accept, that at these times the particle many times changes direction of motion, and displacements of particle in two non-overlapped time intervals are independent. Restrict ourselves for simplicity to the case of one-dimensional motion. Denote s the way traveled by a particle in time interval $(0, t)$, s_1 – the way traveled within the interval $(0, t_1)$, and s_2 – the distance traveled during the interval (t_1, t). From statistical independence of s_1 and s_2 and equal probability of positive and negative displacements it follows, that average square of displacement is

$$\langle s^2 \rangle = \langle (s_1 + s_2)^2 \rangle = \langle s_1^2 \rangle + \langle s_2^2 \rangle. \tag{11.57}$$

Denoted $\langle s^2 \rangle = \psi(t)$, rewrite the equation (11.57) as

$$\psi(t) = \psi(t_1) + \psi(t - t_1). \tag{11.58}$$

The solution of this equation is of the form

$$\psi(t) = \langle s^2 \rangle = 2D_T t, \tag{11.59}$$

where D_T is the factor of turbulent diffusion to be determined.

Consider now the equation of particle motion under action of a force from ambient liquid. This force consists of regular part, friction force, and random force F, whose average value is equal to zero. Apart from these forces, there can be forces external to the system particles-liquid, for example, the gravity force. These external forces will be neglected further on. In inertia-neglecting approximation the equation of particle motion along x-axis (the Langevin equation), is of the form

$$h\frac{dx}{dt} + F = 0, \tag{11.60}$$

where h is the factor of hydrodynamic resistance.

It follows from (11.60) that displacement of particle from initial position x_0 at a time t is given by

$$x - x_0 = \frac{1}{h}\int_0^t F(\tau)\,d\tau. \tag{11.61}$$

The integral in the right part (11.61) has the meaning of impulse of random force F during time t, and for any $0 < t_1 < t$ it can be presented as

$$\int_0^t F(\tau)\,d\tau = \int_0^{t_1} F(\tau)\,d\tau + \int_{t_1}^t F(\tau)\,d\tau. \tag{11.62}$$

Taking the square of both parts (11.62), averaging the result and taking into account independence of random impulses for two non-overlapping time intervals, one obtains

$$\left\langle \left(\int_0^t F(\tau)\,d\tau \right)^2 \right\rangle = \left\langle \left(\int_0^{t_1} F(\tau)\,d\tau \right)^2 \right\rangle + \left\langle \left(\int_{t_1}^t F(\tau)\,d\tau \right)^2 \right\rangle. \tag{11.63}$$

The equation (11.63) is similar to the equation (11.57) and its solution has the same form, as (11.59)

$$\left\langle \left(\int_0^t F(\tau)\, d\tau \right)^2 \right\rangle = 2Bt. \tag{11.64}$$

Having determined from (11.61) the average square of displacement of the particle

$$\langle (x - x_0)^2 \rangle = \frac{1}{h^2} \left\langle \left(\int_0^t F(\tau)\, d\tau \right)^2 \right\rangle. \tag{11.65}$$

and substituting in the obtained expression relations (11.59) and (11.64), one gets

$$D_T = \frac{B}{h^2}; \quad B = \frac{1}{2t} \left\langle \left(\int_0^t F(\tau)\, d\tau \right)^2 \right\rangle. \tag{11.66}$$

Since similar approach was used in [37] for Brownian diffusions, it should be noted the principal difference of turbulent diffusion from Brownian one. In the process of Brownian diffusion, the particles perform random thermal motion due to collisions with molecules of ambient liquid. In [37] the appropriate force acting on the considered particle, is taken into account as quasi-elastic force proportional to the particle's displacement $F_{contr} = -\alpha x$. As a result, the form of the equation (11.60) changes, there appears a term proportional to x, and from the condition of thermodynamic equilibrium of the system it follows that

$$B = h\mathrm{k}T, \quad D_{br} = \mathrm{k}T/h. \tag{11.67}$$

Hence, the factor of Brownian diffusions is inversely proportional to the first power of factor of hydrodynamic resistance of particle.

In case of turbulent diffusion, the situation is somewhat different. Motion of particles under action of turbulent pulsations is not connected to thermal fluctuations. Therefore $B = const$ and the factor of turbulent diffusion is inversely proportional to the second power of factor of hydrodynamic resistance.

If the change of hydrodynamic resistance factor with distance is great as compared with particle's displacement in pulsation, then the quantity h figuring in (11.66) depends on displacement x. The same takes place in motion of particles near the wall or at mutual approach of particles.

Consider approach of two spherical particles of radiuses R_1 and R_2 ($R_1 \geq R_2$) along the line of centers. In the "non-inert" approximation, the equation of particle's motion relative to another particle has the form

$$\boldsymbol{F}_{1r} + \boldsymbol{F}_r = 0, \quad \boldsymbol{F}_{1r} = h_r^0 \boldsymbol{u}_r^0, \quad \boldsymbol{F}_r = -h_r u_r. \tag{11.68}$$

Here $\boldsymbol{F}_{1r}$ is the force exerted on the "reference particle" by the liquid flowing around it, $\boldsymbol{F}_r$ the force of resistance to the motion of particle in question, $\boldsymbol{u}_r^0$ is the velocity of the liquid at the particle's location, $\boldsymbol{u}_r$ is the particle's velocity.

At the distances between particles big in comparison with their sizes, each of them is completely entrained by the liquid and $h_r^0 = h_r$, $\boldsymbol{u}_r^0 = \boldsymbol{u}_r$. Decrease of the distances between particles results in change of factors of hydrodynamic resistance of particles h_r^0 and h_r. The first factor changes slightly, whereas the second one grows and becomes infinite at the particles' contact.

It follows from (11.68) that the velocity of particle's motion in the vicinity of another particle is constrained and is determined by

$$\boldsymbol{u}_r = \frac{h_r^0}{h_r}\boldsymbol{u}_r^0. \tag{11.69}$$

The equation (11.69) describes, for small displacements, the motion similar to motion of a non-constrained particle with similarity factor, equal to h_r^0/h_r. This result allows one to present the factor of turbulent diffusion (11.66) in a similar form

$$D_T = D_{T0}(r)(h_r^0/h_r)^2, \tag{11.70}$$

where $D_{T0}(r)$ is the factor of turbulent diffusion under conditions of non-constrained motion of particles, determined by the formula (11.56).

Expression (11.70) for the factor of turbulent diffusion does not take into account motion of the second particle. To take proper account of mutual influence of particles on the velocity of their approach, it should proceed as follows. Let $\boldsymbol{u}$ be the velocity of one particle relative to another, and $\boldsymbol{u}_1$ and $\boldsymbol{u}_2$ – the particles' velocities relative to a reference frame whose origin lies between the particles on the line connecting their centers. Then $\boldsymbol{u} = \boldsymbol{u}_1 - \boldsymbol{u}_2$. Forces $\boldsymbol{F}$ acting on particles, are equal in magnitude and opposite in directions. Then the factor of hydrodynamic resistance to particles' approach can be written as

$$h_r = \frac{F}{u_1 - u_2} = \frac{F}{F/h_1 + F/h_2} = \frac{h_1 h_2}{h_1 + h_2}. \tag{11.71}$$

Here h_i denote the factors of hydrodynamic resistance to motion of each particle $(i = 1, 2)$.

At great distances between particles their interaction is insignificant, $h_i = 6\pi\rho\nu R_i$, and from it follows from (11.71)

$$h_r = h_r^0 = 6\pi\rho\nu\frac{R_1 R_2}{(R_1 + R_2)}; \quad r \gg R_1, R_2. \tag{11.72}$$

At small distances $\delta = r - R_1 - R_2$ between particles' surfaces, the resistance factor behaves as $1/\delta$ for rigid particles [13], and as $1/\sqrt{\delta}$ for drops [39] having mobile surface. We will still limit ourselves to the case of rigid particles. For them, at small clearance between particles, the factor of resistance is

$$h_r = h_r^0 \frac{R_1 R_2}{(R_1 + R_2)^2} \frac{(R_1 + R_2)}{r - R_1 - R_2}. \tag{11.73}$$

Combining the "far" (11.72) and "near" (11.73) asymptotic equations, one obtains the following approximate expression

$$h_r = 6\pi\rho v \frac{R_1 R_2}{(R_1 + R_2)} \left(1 + \frac{R_1 R_2}{(R_1 + R_2)^2} \frac{(R_1 + R_2)}{(r - R_1 - R_2)}\right). \tag{11.74}$$

Expression for hydrodynamic resistance factor of drops with mobile interface will be given in section 13.7, in which the coalescence of drops in emulsion will be considered.

Substitution of relations (11.56), (11.72), and (11.74) into (11.70) results in the following expression for the mutual turbulent diffusion factor of spherical particles in the presence of hydrodynamic interaction

$$D_T = \beta \frac{v}{\lambda_0^2} \frac{(R_1 + R_2)^2 (s + \gamma)^2 s^2}{(s + \chi)^2}. \tag{11.75}$$

Here the following dimensionless parameters have been introduced:

$$s = (r - R_1 - R_2)/(R_1 + R_2), \quad \gamma = R_1 R_2/(R_1^2 + R_2^2 - R_1 R_2),$$
$$\chi = R_1 R_2/(R_1 + R_2)^2.$$

11.4 Forces of Hydrodynamic Interaction of Drops

In order to determine the collision frequency of drops, it is necessary to consider the relative motion of drops subject to interaction forces – mutual interaction of drops and interaction of drops with the ambient liquid.

Hydrodynamic forces acting on drops reveal themselves as drops move relative to the ambient liquid and relative to each other. These forces, generally speaking, can deform drops, especially when they approach each other and the clearance between them becomes smaller than the drop radius. However, if the size of drops is small enough and their surface is covered with protective envelope, drop deformation can be neglected. Therefore, further one, drops will be considered spherical and non-deformable. Hydrodynamic forces are influenced by the velocity of the surrounding liquid and by the presence of neighboring drops. For drops with sizes up to 100 μm, we can assume that the motion occurs at small Reynolds numbers. Therefore, if the translation velocity and the angular velocity of rotation of the drop are given at each moment, the hydrodynamic force acting on the drop can be determined from the Stokes equations. In such problems, it is common to assume that the volume concentration of the disperse phase is small.

The assumption makes it possible to consider only interactions between two drops (pair interactions). In the coordinate system attached to the centre of the large drop S_1 (see Fig. 11.2), the Stokes equations and boundary conditions are:

$$\nabla \cdot \boldsymbol{u} = 0; \quad \mu \nabla \boldsymbol{u} = \Delta p, \tag{11.76}$$

$$\begin{aligned} &\boldsymbol{u}|_{S_i} = \mathbf{v}_i + \Omega_i \times (\boldsymbol{r} - \boldsymbol{r}_i), \quad |\boldsymbol{r} - \boldsymbol{r}_i| = R_i; \quad (i = 1, 2), \\ &\boldsymbol{u} \to \boldsymbol{U} = (-U \cos\theta, U \sin\theta) \quad \text{at } \boldsymbol{r} \to \infty, \end{aligned} \tag{11.77}$$

where $\mathbf{v}_i$ is the translational velocity of the drop center S_i; $\boldsymbol{U}$ – the velocity of the carrying liquid far away from the drops (further on, we shall limit ourselves to the case $U = const$); Ω_i – the angular velocity of the drop S_i; $\boldsymbol{r}_i$ – the radius-vector of the drop center; R_i – particle radii.

Force and torque acting on the drop from the external liquid are given by the expressions [13]

$$\boldsymbol{F}_{ih}^0 = \iint_{S_i} d\boldsymbol{s} \cdot (p\boldsymbol{I} + 2\mu\boldsymbol{E}); \quad \boldsymbol{T}_{ih}^0 = \iint_{S_i} (\boldsymbol{r} - \boldsymbol{r}_i) \times (-p\boldsymbol{I} + 2\mu\boldsymbol{E}) \cdot d\boldsymbol{s}, \tag{11.78}$$

where $\boldsymbol{E}$ is the strain rate tensor; $\boldsymbol{I}$ – the unit tensor; and $d\boldsymbol{s} = ds\boldsymbol{n}$ is the surface element, which is directed toward the external liquid.

So, the solution of the boundary value problem (11.76), (11.77) gives us $\boldsymbol{F}_{ih}^0$ and $\boldsymbol{T}_{ih}^0$ as functions of sizes and velocities of particles, distances between them, and the viscosity μ_e of the external liquid. If the problem of hydrodynamic interaction of drops with a mobile surface is considered, then, in addition to the mentioned parameters, there is yet another one – the viscosity of the internal liquid μ_i. Then

$$\begin{aligned} \boldsymbol{F}_{ih} &= F_i(R_1, R_2, \boldsymbol{v}_1, \boldsymbol{v}_2, \Omega_1, \Omega_2, \boldsymbol{r}_2 - \boldsymbol{r}_1, \mu_e, \mu_i), \\ \boldsymbol{T}_{ih} &= T_i(R_1, R_2, \boldsymbol{v}_1, \boldsymbol{v}_2, \Omega_1, \Omega_2, \boldsymbol{r}_2 - \boldsymbol{r}_1, \mu_e, \mu_i). \end{aligned} \tag{11.79}$$

Because of the linearity of Stokes equations, the required expressions for $\boldsymbol{F}_{ih}$ and $\boldsymbol{T}_{ih}$ can be found by superposition and approximation of the known particular solutions. We can take the particular solutions for the motion of drop S_2 relative to drop S_1 in a quiescent liquid, and in a flow with the velocity $\boldsymbol{U}$ at the infinity that goes around the two motionless drops *S2* and *S1* separated by a fixed distance. Based on the above, the force $\boldsymbol{F}_h$ and the torque $\boldsymbol{T}_h$ can be presented as

$$\boldsymbol{F}_h = \boldsymbol{F}_s + \boldsymbol{F}_e; \quad \boldsymbol{T}_h = \boldsymbol{T}_s + \boldsymbol{T}_e. \tag{11.80}$$

The terms $\boldsymbol{F}_s$ and $\boldsymbol{T}_s$ are the Stokesian components, corresponding to a fixed sphere S_1 under the condition $\boldsymbol{v}_1 = \Omega_1 = 0$ and $\boldsymbol{U} \neq 0$, while the terms $\boldsymbol{F}_e$ and $\boldsymbol{T}_e$ are the proper components, contributing to the hydrodynamic force and proper angular momentum of drop S_2 at $\boldsymbol{v}_2 \neq 0$, $\Omega_2 \neq 0$, the liquid being at rest at the infinity ($\boldsymbol{U} = 0$). The expressions for $\boldsymbol{F}_s$ and $\boldsymbol{T}_s$ can be determined as follows. Denote by $\boldsymbol{u}_0$ the velocity of the Stokesian flow around a single drop S_1. Then the projections of this velocity onto the meridian plane ($\varphi = const$) are equal to [19]

$$u_{or} = U\cos\theta\left(\frac{3R}{2r} - \frac{R^3}{2r^3} - 1\right); \quad u_{0\theta} = U\sin\theta\left(\frac{3R}{4r} - \frac{R^3}{4r^3} + 1\right). \tag{11.81}$$

The characteristic scale of velocity variation (11.81) is equal to R. It means that for any region whose size is much smaller than R, the flow can be approximated as a quasi-planar flow in the meridian plane, formed by superposition of uniform flow and simple shear flow. The uniform flow induces the force $\boldsymbol{F}_s$ on the drop S_2, while the shear flow produces torque $\boldsymbol{T}_s$. At a sufficient distance from the drop S_2 we have

$$\boldsymbol{F}_s = 6\pi\mu_e R_2 \mathbf{u}_0; \quad \boldsymbol{T}_s = 4\pi\mu_e R_2^3 \nabla \times \mathbf{u}_0. \tag{11.82}$$

A decrease of the clearance d between drops S_1 and S_2 results in the velocity figuring in the dependences (11.82). To take a proper account of the influence of distortion of the velocity field at small δ, one has to include the resistance factors into (11.82). It should be kept in mind that resistance factors have different forms for the motion of drop S_2 parallel and perpendicular to the surface S_1, therefore

$$F_{sr} = 6\pi\mu_e R_2 f_{sr} u_{0r}; \quad F_{s\theta} = 6\pi\mu_e R_2 f_{s\theta} u_{0\theta}; \quad T_{s\varphi} = 8\pi\mu_e R_2^3 t_{s\varphi} G, \tag{11.83}$$

where f_{sr} and $f_{s\theta}$ are translational resistance factors for the motion perpendicular and parallel the surface S_1; $t_{s\varphi}$ – rotational resistance factors; $G = 0.5|\nabla \times \mathbf{u}_0|$. Note that these factors depend on the relative clearance between drops $\Delta = (r - R_1 - R_2)/(R_1 + R_2)$ and the ratio of drop radii $\mathsf{k} = R_2/R_1$.

A review of early works on hydrodynamic interaction between two solid spherical particles is contained in [13]. For the most part, these works focus on derivation of forces and torques acting on particles placed relatively far apart, forces and torques on a particle moving perpendicular or parallel to a flat surface, and forces and torques on a particle moving relative to another particle along, or perpendicular to, the line connecting the particle centers. The more general case of translation and rotation of two rigid particles was considered in [40, 41]. Hydrodynamic interaction of two drops with a proper account of the mobility of their surfaces and the internal circulation is analyzed in [39, 42–50].

The analysis of the solution of appropriate hydrodynamic problems shows that the factors f_{sr}, $f_{s\theta}$ and $t_{s\varphi}$ are close to unity at distances from S_1 less or equal to several times the radius of the smaller particle S_2. At small clearances between particles, these factors remain finite. It will be shown later that the analogous factors in the force $\boldsymbol{F}_e$ and the torque $\boldsymbol{T}_e$ grow unbounded when the clearance between drops tends to zero. So we can take f_{sr}, $f_{s\theta}$ and $t_{s\varphi} \approx 1$ without any detriment to the accuracy of calculations.

Consider now the hydrodynamic force and torque caused by the proper motion of drop S_2. This problem has been studied extensively. The available solutions enable us to cover all three types of proper motion of the particle S_2: motion along the line of centers; motion perpendicular to the line of centers in the meridian plane; and rotation around the axis perpendicular to the meridian plane [41]. Numerical results obtained in these works are supplemented by asymptotic expressions [39, 41, 50] for the case of motion of a sphere in the vicinity of another sphere (or in the vicinity of a plane, if the clearance between spheres is small in comparison with the radius of the smaller sphere). The three specified types of proper motion result in the following expressions for the torque $\boldsymbol{T}_e$ and the projection of force $\boldsymbol{F}_e$ onto the meridian plane, acting on a rigid particle

$$\begin{aligned} F_{er} &= -6\pi\mu_e R_2 f_{er} v_r, \quad F_{e\theta} = -6\pi\mu_e R_2 f_{e\theta} v_\theta + 6\pi\mu_e R_2^2 f_{e\theta_1}\Omega, \\ T_{e\varphi} &= 8\pi\mu_e R_2^2 t_{e\varphi} v_\varphi - 8\pi\mu_e R_2^3 t_{e\varphi_1}\Omega. \end{aligned} \tag{11.84}$$

Let us adduce the asymptotic expressions for the resistance factors of rigid particles. Introduce a dimensionless distance from the center of the large particle S_1 to the surface of the small particle S_2 along the line of centers, $x_1 = (r - R_2)/R_1$. For small values of the clearance ($1 < x_1 < 1.5$), it is possible to use the near-asymptotic relations

$$\begin{aligned} f_{e\theta} &\approx -\frac{2(2+k+2k^2)}{15(1+k)^3}\ln(x_1-1) + 0.959, \\ f_{e\theta_1} &\approx -\frac{2(1+4k)}{15(1+k)^2}\ln(x_1-1) - 0.2526, \\ f_{e\varphi} &\approx -\frac{1+4k}{10(1+k)^2}\ln(x_1-1) + 0.29, \\ f_{e\varphi_1} &\approx -\frac{2}{5(1+k)^2}\ln(x_1-1) + 0.3817, \\ f_{er} &\approx -\frac{1}{(1+k)^2(x_1-1)} - \frac{1+7k+k^2}{5(1+k)}\ln(x_1-1) + 0.97. \end{aligned} \tag{11.85}$$

In the region $x_1 \gg 1/\mathsf{k}$ for $\mathsf{k} \ll 1$, the following asymptotic relations are valid

$$\begin{aligned} & f_{e\theta} \approx \left(1 - \frac{9}{15x_1} + \frac{1}{8x_1^3}\right)^{-1}; \quad f_{e\theta_1} \approx \frac{1}{8x_1^4}, \\ & f_{e\varphi} \approx \frac{1}{32x_1^4}; \quad f_{e\varphi_1} \approx 1 + \frac{5}{16x_1}, \\ & f_{er} \approx 1 + \frac{1.125}{15x_1} + \frac{1.266}{x_1^2}, \end{aligned} \tag{11.86}$$

Finally, for any $\mathsf{k} \leq 1$,

$$f_{er} \approx \frac{1}{1 - 2.25(kx_1 + 1)^{-2}}; \quad f_{e\theta} \approx \frac{1}{1 - 0.56k(kx_1 + 1)^{-2}};$$
$$t_{e\varphi} \approx \frac{0.56k^2(kx_1 + 1)^{-3}}{1 - 0.56k(kx_1 + 1)^{-2}}. \tag{11.87}$$

The factors of resistance in the intermediate region are presented in the above-mentioned works in the form of infinite series, and their numerical values are given in tables.

A detailed analysis of relative motion of drops along the line of centers for small clearances between drops is performed in [49]. The approximate expression for the resistance factor of drop S_2 looks like

$$h = 6\pi\mu_e R_2 f_{er} \approx 6\pi\mu_e \left(\frac{R_1 R_2}{R_1 + R_2}\right)^2 \frac{f(m)}{\delta}, \tag{11.88}$$

where $\delta = r - R_1 - R_2$ is the clearance between drops;

$$f(m) = \frac{1 + 0.402m}{1 + 1.711m + 0.461m^2}; \quad m = \frac{1}{\bar{\mu}}\left(\frac{R_1 R_2}{R_1 + R_2}\right)^{1/2}; \quad \bar{\mu} = \frac{\mu_i}{\mu_e}.$$

The parameter m characterizes the mobility of the drop surface. So, at $m \ll 1$, the drop surface is almost completely retarded and from the hydrodynamic standpoint, it behaves like a solid particle. The resistance factor is then

$$h \approx 6\pi\mu_e \left(\frac{R_1 R_2}{R_1 + R_2}\right)^2 \frac{1}{\delta}. \tag{11.89}$$

The expression (11.89) coincides with the first term of asymptotic relations (11.85) for f_{er}. From the form of the resistance factor, it follows that it has a non-integrable singularity at $\delta \to 0$, so drops with fully retarded surface cannot come into contact in a finite time under the action of a finite force.

Another limiting case $m \gg 1$ corresponds to an almost fully mobile surface. Then the drop behaves as a gas bubble and the resistance factor has the form

$$h \approx 6\pi\mu_e \left(\frac{R_1 R_2}{R_1 + R_2}\right)^2 \frac{0.872\bar{\mu}}{\delta^{1/2}} \left(\frac{R_1 + R_2}{R_1 R_2}\right)^{1/2}. \tag{11.90}$$

At $\delta \to 0$, the expression for h of a drop with mobile surface has an integrable singularity. So, in a laminar flow, particularly, in the process of gravitational sedimentation, a contact between drops is possible even in the absence of molecular forces. In a turbulent flow, the turbulent diffusion factor is $D_T \sim 1/h^2$, therefore in the absence of molecular forces, any contact between drops with fully mobile surfaces is also impossible.

To summarize, note that asymptotic expressions were found in paper [39] for the resistance factors of relative motion of two spherical drops with different internal viscosities $\mu_i^{(1)}$ and $\mu_i^{(2)}$ placed into a liquid of viscosity μ_e for $\delta \to 0$

$$h \approx \frac{\pi^2}{16} \frac{\mathsf{k}^{1/2}(\mu_i^{(1)} + \mu_i^{(2)})}{(1+\mathsf{k})^2 (s-2)^{1/2}}, \tag{11.91}$$

where $s = 2r/(R_1 + R_2)$; $\bar{\mu}_i^{(1,2)} = \mu_i^{(1,2)}/\mu_e$.

If one of the drops has a fully retarded surface, which corresponds to the case of relative motion between a drop and a rigid particle, the asymptotic expression for the resistance factor at $\delta \to 0$ looks like

$$h \approx \frac{\mathsf{k}}{2(1+\mathsf{k})^3(s-2)} + \frac{9\pi^2 \mathsf{k} \bar{\mu}_i}{64(1+\mathsf{k})^2(s-2)^{1/2}}. \tag{11.92}$$

11.5
Molecular and Electrostatic Interaction Forces Acting on Drops

Forces of molecular and electrostatic interactions between particles were discussed in great detail in Section 10.1, so we only need to cover some additional aspects that will be useful later on.

At the absence of external forces (gravitational, centrifugal, electric), particles with zero charge and drops suspended in a quiescent liquid should be distributed uniformly. There is always interaction between particles: electrostatic repulsion (for charged particles and particles surrounded by electric double layers), molecular attraction (the Van der Waals–London forces), hydrodynamic forces (forces arising due to the mutual influence of velocity fields of the liquid and the particles).

Electrostatic forces between particles with equal charges help to ensure a uniform distribution of particles. The ability of the system to maintain a uniform distribution of particles in the liquid over a long time characterizes the system's stability.

In practice, in the majority two-phase disperse systems, the number of particles decreases with time, while their size simultaneously increases. For emulsions, the collision of two drops results in their coalescence with the formation of one drop of larger size. Colliding particles can form aggregates. Aggregation of particles occurs due to the forces of molecular attraction.

The systems in which aggregation, coagulation, or coalescence of particles and drops is taking place are called unstable systems.

In the absence of external and hydrodynamic forces, stability of disperse systems depends on particle interaction caused by surface forces: forces of electric repulsion and molecular attraction [51].

Electrostatic forces are repulsive forces caused by electric double layers at the surfaces of particles or drops. The electrostatic potential for two various spheres of radii R_1 and R_2 ($R_1 > R_2$) with the distance r between its centers is equal to [52]

$$V_R^{sphere} = \frac{\varepsilon R_1 \phi_1^2 R_2}{4(R_1 + R_2)} \left(-2\frac{\phi_2}{\phi_1} \ln\left(\frac{1 - e^{-\chi h_0}}{1 + e^{-\chi h_0}}\right) + \left(1 + \frac{\phi_2^2}{\phi_1^2}\right) \ln(1 - e^{-2\chi h_0}) \right), \tag{11.93}$$

where ϕ_1 and ϕ_2 are the particle surface potentials; χ is the inverse Debye radius.

The force of electrostatic repulsion between two spherical particles is found from the relation

$$F_R = -\frac{dV_R}{dr} = \frac{\varepsilon R_1 \phi_1^2 \chi}{2} f_R,$$

$$f_R = \frac{\mathsf{k}}{\mathsf{k}+1} \frac{e^{-a}\varepsilon}{1 - e^{-2a}} \left(2\frac{\phi_2}{\phi_1} - \left(1 + \frac{\phi_2^2}{\phi_1^2}\right) e^{-a} \right), \tag{11.94}$$

$$a = 0.5 R_1 \chi (1 + \mathsf{k}) \left(\frac{2r}{R_1 + R_2} \right).$$

There follows from (11.94) a frequently used formula for electrostatic interaction of identical particles of radius R with equal surface potentials ϕ_0

$$F_R = \varepsilon R \phi_0^2 \chi / 2(1 + e^a). \tag{11.95}$$

The Van–der Waals–London attractive forces have molecular origin, although on the basic level, they are still caused by electric interactions. By their nature, these forces are caused by the polarization of a molecule under the action of charge distribution fluctuations in the neighboring molecule and vice versa. They are also known as London dispersive forces.

The force of molecular attraction between two parallel planes and two spherical particles was derived in [53]. The corresponding energy of attraction is

$$V_A^{plane} = -\Gamma / 12\pi h^2, \tag{11.96}$$

where h is the distance between planes and Γ is the Hamaker constant, whose characteristic values lie in the range 10^{-20}–10^{-19} J.

For two identical spherical particles of radius R with the condition $h_0 \ll R$ where h_0 is the minimum distance between their surfaces, the energy is equal to

$$V_A^{plane} = -R\Gamma / 12 h_0. \tag{11.97}$$

For identical spheres,

$$V_A^{plane} = -\frac{\Gamma}{6} \left(\frac{2R^2}{r^2 - 4R^2} + \frac{2R^2}{r^2} + \ln\left(\frac{r^2 - 4R^2}{r^2}\right) \right). \tag{11.98}$$

where the distance r between their centers can be arbitrary.

In the case of two spheres with different radii R_1 and R_2

$$V_A^{plane} = -\frac{\Gamma}{6}\left(\frac{8\mathsf{k}}{(s^2-4)(1+\mathsf{k})^2} + \frac{8\mathsf{k}}{s^2(\mathsf{k}+1)^2 - 4(1-\mathsf{k})^2} + \ln\left(\frac{(s^2-4)(1+\mathsf{k})^2}{s^2(\mathsf{k}+1)^2-4(1-\mathsf{k})^2}\right)\right), \tag{11.99}$$

where $s = 2r/(R_1 + R_2)$; $\mathsf{k} = R_2/R_1$.

The corresponding molecular attraction force between two various spherical particles of different radii is

$$F_A = -\frac{dV_A}{dr} = -\frac{2\Gamma}{3R_1} f_A,$$

$$F_A = -\frac{1}{(1+\mathsf{k})}\left(\frac{s}{(s^2-4)(1+\mathsf{k})^2}(8\mathsf{k} - (1+\mathsf{k})^2(s^2-4)) + \frac{s(1+\mathsf{k})^2}{(s^2(1+\mathsf{k})^2 - 4(1-\mathsf{k})^2)^2}(8\mathsf{k} + s^2(1+\mathsf{k})^2 - 4(1-\mathsf{k})^2\right). \tag{11.100}$$

The expression (11.100) does not take into account the effect of electromagnetic retardation that results in a reduction of the Van der Waals force. The approximate expression for the molecular force in view of this effect is given in [54] and has the form

$$F_A = \frac{2\Gamma}{3R_1} f_A \Phi(p), \tag{11.101}$$

where

$$\Phi(p) = \begin{cases} \dfrac{1}{1+1.77p} & \text{at } p \le 0.57, \\[2ex] -\dfrac{2.45}{60p} + \dfrac{2.17}{120p^2} - \dfrac{0.59}{420p^3} & \text{at } p > 0.57, \end{cases}$$

here $p = 2\pi(r - R_1 - R_2)\lambda_L$; λ_L is London's wavelength having the order 103.

The total potential energy of two spherical particles is equal to the sum of electrostatic potential energy and the potential energy of molecular interaction between the particles:

$$V = V_A + V_R. \tag{11.102}$$

The energy of repulsion decreases exponentially with the growth of h_0, with characteristic linear size λ_D. The energy of attraction decreases as $1/h_0$. Therefore the attractive force prevails on very small and very large distances between the particles, while the repulsive force dominates in the intermediate range.

Coagulation can be classified as fast or slow. Fast coagulation is coagulation that takes place under the influence of molecular attractive forces. If in addition to attractive forces, we also take into account the forces of electrostatic repulsion, the resulting coagulation is called slow. It is known that the electrolyte concentration C_0 necessary for fast coagulation depends strongly on the charge of counter-ions, i.e. ions whose charge sign is opposite to that of particles. On the other hand, stability of the system practically does not depend on the charge of ions and on the concentration of particles. This corresponds to the Shulze-Hardi rule, according to which the system stability mostly depends on the valence of counter-ions. Molecular and electrostatic potentials at the point separating stable and unstable conditions for the system can be found from the following system of equations:

$$V = V_A + V_R = 0. \tag{11.103}$$

$$\frac{dV}{dr} = \frac{dV_A}{dr} + \frac{dV_R}{dr} = 0. \tag{11.104}$$

If the system is unstable, that is, repulsive forces are not taken into account, then each collision of particles results in their coagulation. The presence of a stabilizer (electrolyte) in the liquid gives rise to additional forces due to the formation of electric double layers at particle surfaces. It means that coagulation will slow down. Therefore such coagulation is referred to as slow coagulation.

The rate of coagulation is characterized by the stability factor W, which is equal to the ratio of the number of particle collisions without electrostatic repulsion and the number of collisions with the repulsion present [9]:

$$W = I/I_R. \tag{11.105}$$

For fast coagulation, $I_R = I$ and $W = 1$, while for slow coagulation, $I_R < I$ and $W > 1$.

11.6 The Conducting Drops in an Electric Field

This section will focus on the general behavior of drops in an external electric field, assuming a low volume concentration of drops. When drops are relatively far apart, the influence of neighboring drops is insignificant and is possible to consider just a single drop in an infinite liquid. Smaller distances between drops result in the distortion of external electric field near the drop surfaces, which has a significant effect on the shape of drops. Interaction of two conducting drops in an external electric field will be discussed in detail in the following chapter. For now, we shall only consider the behavior of a single drop, and also discuss the problem of drop's stability in an electric field.

If both internal and external liquids happen to be ideal dielectrics and there are no free charges at the interface, or if liquid inside the drop is highly conductive and the external liquid is an isolator, the external electric field leads to the appearance of a force distributed over the surface of drop. This phenomenon is caused by the discontinuity of the electric field at the interface [55]. The force is perpendicular to the interface and is directed from the liquid with higher dielectric permittivity (or from the conducting liquid) toward the liquid with lower dielectric permittivity (or toward the isolator). For the equilibrium shape of a motionless drop in a quiescent liquid to be conserved, the condition of equality of the surface electric force and the surface tension force must be satisfied. As a result, at static conditions, the drop assumes the shape of an ellipsoid extended along the direction of the external electric field.

The theory of static equilibrium of a drop in an electric field (electrohydrostatics) is advanced further for ideal media – dielectrics and conductors – in works [56–62]. However, real liquids are liquids with a finite conductivity and dielectrics with a finite dielectric permittivity. The exception is a superconducting liquid at a very low temperature, for example, liquid helium. The proper account of finite conductivity considerably complicates the problem mathematically as well as physically, because the possible drop shapes are different from the shapes of ideally conducting drops. Thus, a drop can assume the shape of an ellipsoid extended along the direction of the electric field or along the direction perpendicular to the field, and even the spherical shape that was observed in experiments [63]. The theoretical explanation of these phenomena is given in [64]. This paper shows that for a drop of finite conductivity, the electric charge is concentrated in the surface layer, producing a non-uniform, tangential surface electric stress. This stress, in turn, induces a tangential hydrodynamic stress that influences the drop's deformation. The magnitude of the stress depends on the properties of the liquid and on the intensity of the external electric field. Therefore the drop can assume one of the above-mentioned shapes depending on the relationship between the electric and hydrodynamic surface stresses. The solution of the problem with a proper account of the internal circulation of the liquid has been obtained in the paper [64], which assumed a small deformation of the drop surface and a slow Stokesian flow, thus making it possible to obtain an approximate asymptotic solution.

Experiments whose outcomes are listed in [65] have shown only a qualitative agreement with the theory [64], while quantitative differences, in particular, in the extent of drop deformation, appeared to be significant. An attempt to take into account the higher-order terms in the asymptotic expansion was made in [66], but it did not eliminate the discrepancy with experimental results. It should be noted that performing a successful experiment is a difficult task because one has to impose a strict requirement that drops should remain motionless (no sedimentation or lifting), for which it is necessary to make sure that the drops are very small, and the densities of the internal and external liquids should not differ by a lot.

A more general case is considered in [67], which studies the behavior of a drop suspended in a liquid, where the conductivities of both the drop and the liquid are finite. The equation of motion takes into account nonlinear inertial terms. The problem is solved by the numerical method of finite elements. It is shown that a proper account of finite conductivity and nonlinear effects results in a good agreement with experimental data.

All of the above-mentioned works neglect the convection of a surface charge and assume that the transport of charge is caused only by ohmic current through the interface. Therefore the boundary condition for an electrically polarized drop plays a subsidiary role and serves only to determine the surface charge density. The solution of the problem with the account of this effect can be found in [68].

It is worth mentioning some works, in which the effect of electric field on other processes has been studied. In [69–76] it has been shown that the electric field intensifies the heat-mass-transfer between drops and the ambient liquid. In [77], the breakage of drops of finite conductivity in the electric field has been discussed.

We shall limit ourselves to simple estimations. To this end, consider an ideally conducting drop suspended in an ideal dielectric. The motion of the liquid inside and outside of the drop is neglected. In this case, conducting drops suspended in a dielectric liquid are polarized and deformed under the action of the external electric field $\mathbf{E}_0$, assuming the shape of an ellipsoid whose greater semi-axis is parallel to the field.

Consider the behavior of a single conducting spherical drop of radius R suspended in a quiescent dielectric liquid of a constant dielectric permittivity e in the presence of a uniform external electric field of strength $\mathbf{E}_0$ (Fig. 11.4).

The charge outside of the sphere is zero, so the electric potential ϕ satisfies Laplace's equation

$$\Delta\phi = 0. \tag{11.106}$$

The surface of a conducting drop is an equipotential, therefore one can assume that

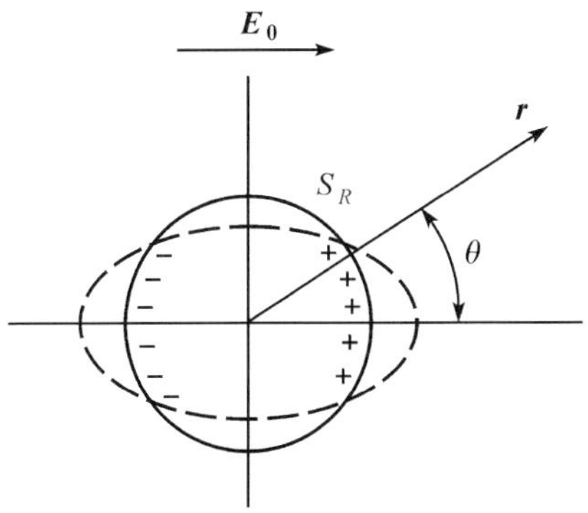

Fig. 11.4 Deformation of a conducting drop in an electric field.

$$\phi = 0 \quad \text{on } S_R. \tag{11.107}$$

At a large distance from the drop, the potential field balances the external electric field:

$$\Delta\phi \to -\boldsymbol{E}_0 \quad \text{at } \boldsymbol{r} \to \infty. \tag{11.108}$$

Inside the drop, the electric field is absent, so the charge induced by the external field is distributed over the surface. Its surface density σ is given by

$$2\pi\sigma = -\partial\phi/\partial n, \tag{11.109}$$

where $\boldsymbol{n}$ is the normal external to S_R.

The total charge of the drop is equal to zero, so we must impose the condition

$$\int_{S_R} \frac{\partial\phi}{\partial n} ds = 0. \tag{11.110}$$

Suppose that the drop is deformed slightly. Then the shape of the drop can be approximated by a sphere. Laplace's equation in a spherical coordinate system under the spherical symmetry condition is written as

$$\sin\theta \frac{\partial}{\partial r}\left(r^2 \frac{\partial\phi}{\partial r}\right) + \frac{\partial}{\partial\theta}\left(\sin\theta \frac{\partial\phi}{\partial\theta}\right) = 0. \tag{11.111}$$

The boundary conditions are:

$$\phi = 0 \quad \text{at } r = R; \qquad -\frac{\partial\phi}{\partial r} \to E_0 \cos\theta \quad \text{at } r \to \infty. \tag{11.112}$$

The solution is

$$\phi = E_0\left(r - \frac{R^3}{r^2}\right)\cos\theta. \tag{11.113}$$

From (11.113), we can determine the electric field strength at the surface of the drop

$$E_R = \left(-\frac{\partial\phi}{\partial r}\right)_{r=R} = 3E_0\cos\theta. \tag{11.114}$$

Now, find the force acting on the conducting drop. The momentum flux density in an electric field is defined by Maxwell's stress tensor [77]

$$t_{ik} = -\frac{1}{4\pi}\left(\frac{E^2}{2}\delta_{ik} - E_i E_k\right). \tag{11.115}$$

The force acting on a directed surface element $\boldsymbol{ds}$ is equal to

$$t_{ik}\, ds_k = t_{ik} n_k\, ds.$$

The quantity $t_{ik}n_k$ is the force acting on a unit surface. The electric field at the surface of the drop is directed along the normal to the surface, so $E_i E_k = E_R^2 \delta_{ik}$. The force acting from the inside on a unit surface area is equivalent to pressure

$$\Delta p = \frac{E_R^2}{8\pi} - \frac{9E_0^2}{8\pi}\cos^2\theta. \tag{11.116}$$

This pressure reaches the greatest value at $\theta = 0$ and $\theta = \pi$, that is, at the drop poles lying on a straight line parallel to $\boldsymbol{E}_0$. Therefore the deformation of the drop becomes is maximized in the vicinity of these points. As a result, the drop assumes the shape of an ellipsoid with the greater semi-axis directed along $\boldsymbol{E}_0$.

The equilibrium condition for the drop is the equality of the electric pressure and surface tension. If the force of electric pressure surpasses the force of surface tension, the drop will continue to deform until the decreasing main radiuses of curvature at the drop poles causes an increase of the surface tension that is sufficient to balance the internal pressure. A significant deformation of the drop, however, may destabilize the drop and cause it to break. Therefore the critical value of external electric field strength E_{cr} can be estimated from the equality

$$\frac{9E_{cr}^2}{8\pi} \approx \frac{2\Sigma}{R}. \tag{11.117}$$

from which it follows that the stability of a drop is characterized by the dimensionless parameter

$$\chi^2 = \frac{\text{electric force}}{\text{force of surface tension}} = E_0^2 \frac{R}{\Sigma}. \tag{11.118}$$

The simple estimation discussed above gives us the critical parameter value $\chi_{cr} \sim 4\sqrt{\pi}/3$. A more accurate calculation that takes into account the deformation of the drop [58, 79] gives $\chi_{cr} = 1.625$. For water in oil, $\Sigma = 3 \cdot 10^{-2}$ H/m and for water droplet of 1 cm radius, the critical strength of the electric field $E_{cr} = 2.67$ KV/cm. Therefore an electric field with strength $\boldsymbol{E}_0 < 3$ KV/cm will not cause any noticeable deformation of a drop with radius $R \ll 1$ cm placed far away from the other drops.

The situation is different when drops are in close proximity. In this case, the smaller distance between drops causes a substantial increase of local strength of the electric field in the gap between drop surfaces. It is shown in [80] that the

Table 11.1

δ/R	∞	10	1	0.1	0.01	0.001
χ_{cr}	1.625	1.555	0.9889	0.0789	$3.91 \cdot 10^{-3}$	$1.9 \cdot 10^{-4}$

stability of a system of two conducting drops is also determined by the parameter χ, but, as opposed to the single drop case, this parameter depends not only on $\boldsymbol{E}_0$, R, and Σ, but also on the relative clearance δ/R between the drop surfaces. Table 11.1 gives the dependence of χ_{cr} on δ/R.

From the values of χ_{cr} given in the table, it follows that even small drops, when placed close enough to each other, can lose stability in the presence of rather small external electric fields. However, the distances between drops at which such destabilization can occur are so small that the forces of molecular interaction will promote capture and coalescence of fine drops formed as the result of breakage.

11.7 Breakup of Drops

In the previous section, the problem of deformation of conducting drops and the possibility of their breakage was examined. Consider now the problem of breakage of non-conducting drops.

Examining the motion of small drops and the ambient liquid in the Stokesian approximation framework, one sees that it does not result in a large deformation of the drop surface, and consequently, in their breakage. Breakage of a drop is preceded with a significant deformation of its surface, which is possible if in the liquid layers adjoining the drop surface from both sides, there are significant gradients of velocity and pressure, capable of overcoming the surface tension of the interface. Therefore, in order to describe the deformation of a drop, one has to take into account the joint influence of inertial and viscous effects and the surface tension force. An exhaustive review of problems on deformation and breakage of drops in a viscous liquid at small Reynolds numbers is contained in [81]. Some data on drop breakage at high Reynolds numbers can be found in [82].

The field water-oil emulsion contains drops of water with sizes from fractions of a micron and higher. The main difficulties arising in water-oil emulsion separation are caused by the finely-dispersed component, that is, drops with sizes of up to 100 μm. The deformation of such drops during gravitational sedimentation or in a laminar flow is insignificant. Therefore, the breakage of drops that small mostly occurs when the emulsion moves in a turbulent regime under the action of turbulent pulsations, generating in the vicinity of the drops an average local shear flow with the rate of shear $\dot{\gamma} = (4\varepsilon_0/15\pi\nu_e)^{1/2}$, where ε_0 is the specific dissipation of energy.

In Section 11.1, we obtained the kinetic equation describing the dynamics of the volume distribution of drops $n(V, \omega)$ in the coarse of coalescence. The behavior of water-oil emulsion during production, preparation, and transportation of oil is characterized by the simultaneous occurrence of two processes – coalescence of drops in the disperse phase, and breakage of drops whose size exceeds some critical value. As a rule, the motion of emulsions inside the elements of technological equipment and in pipelines occurs in a turbulent flow regime. Therefore breakage of drops should be considered as a random process characterized by the following parameters: the breakage frequency $f(V)$ of drops whose volumes lie in the interval $(V, V + dV)$; the probability $P(V, \omega)$ of formation of a drop with volume in the interval $(V, V + dV)$ when a larger drop which volume in the interval $(\omega, \omega + d\omega)$ splits; the minimum radius R_{min} of drops subject to breakage. The kinetic equation that takes into account the coalescence and breakage of drops is

$$\begin{aligned}\frac{\partial n(V,t)}{dt} = \frac{1}{2}\int_0^V & K(V-\omega,\omega)n(V-\omega,t)n(\omega,t)\,d\omega \\ & - n(V,t)\int_0^\infty K(V,\omega)n(\omega,t)\,d\omega \\ & + \int_0^V f(\omega)P(V,\omega)n(\omega,t)\,d\omega - f(V)n(V,t). \qquad (11.119)\end{aligned}$$

Start with the estimation of the minimum radius of drops that are subject to breakage. For this purpose, it is necessary to estimate the forces acting on the drop in a turbulent flow that are capable of deforming it.

A drop placed in a field of uniform and isotropic turbulence experiences the following forces from the external liquid: dynamic pressure $Q = \mathsf{k}_f\rho_e u^2/2$ (where k_f is a factor of the order 0.5; ρ_e – density of the external liquid; u – velocity of the external liquid relative to the drop); viscous friction force $F_v \sim \mu_e\dot{\gamma}$ (where μ_e is the viscosity coefficient of the external liquid; $\dot{\gamma} = (4\varepsilon_0/15\pi\nu_e)^{1/2}$ – average shear rate; ε_0 – specific dissipation of energy; $\nu_e = \mu_e/\rho_e$ – coefficient of kinematic viscosity). Besides, acting on the drop surface is the force of surface tension $F_{cap} = 2\Sigma/R$, where Σ is the surface tension factor; R is the radius of the drop. Depending on which one of the external forces acting on the drop surface dominates, two mechanisms of drop breakage are possible.

Suppose the predominant influence is exerted by the dynamic pressure [19]. Then drop deformation is caused by the difference of dynamic pressures acting at the opposite sides of the drop:

$$Q = k_f\frac{\rho_e(u_1^2 - u_2^2)}{2}, \qquad (11.120)$$

where u_1 and u_2 are velocities at the opposite sides of the drop (separated from each other by the distance $2R$).

First, consider the drops of size $R > \lambda_0$, where λ_0 is the inner scale of turbulence. Then large-scale pulsations ($\lambda_0 \ll \lambda \ll L$), which don't vary too much on distances of the order of the drop size, do not exert a noticeable influence on these drops. Hence, the deformation and breakage of such drops can be caused only by small-scale pulsations. For such pulsations, the change of pulsation velocity u_1 on the distance equal to the drop size $2R$ is

$$u_\lambda \approx (\varepsilon_0 \lambda)^{1/3} \approx (\varepsilon_0 2R)^{1/3}. \tag{11.121}$$

as follows from (11.43).

From (11.120) and (11.121), it follows that

$$Q \approx \frac{\mathsf{k}_f \rho_e}{2} \varepsilon_0^{2/3} (2R)^{2/3}. \tag{11.122}$$

If the dynamic pressure difference (11.122) exceeds the surface tension of the drop, if can affect a significant deformation of the drop and the drop can break. Therefore the drop breakage condition will be

$$\frac{\mathsf{k}_f \rho_e}{2} \varepsilon_0^{2/3} (2R)^{2/3} \approx \frac{2\Sigma}{R}. \tag{11.123}$$

Plugging in the expression (11.42) for ε_0, we obtain the following expression for the minimum radius of a drop that is subject to breakage

$$R_{\min} \sim \sqrt{2} \frac{L^{2/5} \Sigma^{3/5}}{\mathsf{k}_f^{3/5} \rho_e^{3/5} U^{6/5}}. \tag{11.124}$$

Here L is the characteristic linear scale of the flow region, U is the average velocity of the flow. Notice that the derivation of (11.124) hinges on the assumption that the difference between densities of internal and external liquid is small. Otherwise it would be necessary to take into account the dynamic pressure from the internal liquid.

Introduce Weber's number

$$\mathrm{We} = \frac{\text{dynamic pressure}}{\text{force of surface tension}} = \frac{2R\rho_e U^2}{\Sigma}. \tag{11.125}$$

Then the expression (11.124) can be rewritten as

$$\frac{R_{\min}}{L} = \frac{2^{11/4}}{\mathsf{k}_f^{3/2}} \mathrm{We}^{-3/2} \sim C\mathrm{We}^{-3/2}. \tag{11.126}$$

Consider now the drops whose size is smaller than the inner scale of turbulence ($R \ll \lambda_0$). It is obvious that the breakage of such drops can be caused only by pulsations with scale $\lambda < \lambda_0$, i.e. pulsations whose motion is accompanied by large forces of viscous friction. Therefore only the force of viscous friction at the drop surface can function as the main mechanism causing drop deformation. The criterion of strong deformation of a drop is the equality of forces of viscous friction and surface tension

$$\mu_e \left(\frac{4\varepsilon_0}{15\pi\nu_e} \right)^{1/2} \approx \frac{2\Sigma}{R}. \tag{11.127}$$

From this relation one obtains the minimum drop radius

$$R_{\min} \sim \sqrt{15\pi} \frac{\Sigma}{\rho_e (\nu_e \varepsilon_0)^{1/2}} \sim \frac{\Sigma}{\rho_e (\nu_e \varepsilon_0)^{1/2}}. \tag{11.128}$$

This is the form in which the expression for the minimum radius of a drop is given in [83].

Introduce a dimensionless parameter called the Ohnesorge number

$$\mathrm{Oh} = \mu_e / (2R\rho_e\Sigma)^{1/2}. \tag{11.129}$$

Then the relation (11.128) can be rewritten as

$$R_{\min} = \left(\frac{15\pi}{4} \right)^{1/4} \mathrm{Re}^{-3/4} \mathrm{Oh}^{-1}, \tag{11.130}$$

where $\mathrm{Re} = UL/\nu$ is Reynolds number of the flow.

Proceed now to determine the breakage frequency of drops $f(V)$, following [84], in which a model of drop breakage in a locally isotropic, developed turbulent flow is offered. In the basis of model lies the assumption that the breakage of a single drop is completely defined by fluctuations of energy dissipation in its vicinity. Breakage occurs when the value of specific energy dissipation, averaged over the volume of the order of V (i.e. the drop volume), exceeds the critical value $\varepsilon_{cr}(V)$. For a given size of the drop, the critical value of specific energy dissipation is equal to the corresponding value of ε_0, entering the formula for the minimum drop radius. Assume that viscous friction is the main force causing the deformation and breakage of the drop. Then it is possible to utilize the formula (11.128), from which it follows that

$$\varepsilon_{cr}(V) \sim \left(\left(\frac{4\pi}{3} \right)^{1/3} C \frac{\Sigma}{\rho_e} \right)^2 \frac{1}{\nu_e V^{2/3}}. \tag{11.131}$$

Assume that energy dissipation in the vicinity of the drop has a uniform distribution, with the average value $\varepsilon(t)$. Then breakage frequency, that is, the number of drops splitting in a unit time, can be treated as the probability that $\varepsilon(t)$ in a random process exceeds the critical level $\varepsilon_{cr}(V)$, which remains constant. Adopting this definition, we can then use the approach that was advanced in [84].

Consider two moments of time t and $t+\Delta t$ separated by a short interval Δt. At the moment t, the energy is $\varepsilon(t) < \varepsilon_{cr}(V)$, and at the moment $t+\Delta t$ it is $\varepsilon(t+\Delta t) > \varepsilon_{cr}(V)$. Then breakage frequency can be presented as

$$f(V) = \lim_{\Delta t \to 0}\left(\frac{1}{\Delta t}\,\frac{p(s(t) < \varepsilon_{cr}(V); \varepsilon(t+\Delta t) > \varepsilon_{cr}(V))}{p(s(t) < \varepsilon_{cr}(V))}\right). \qquad (11.132)$$

In the right-hand side there are probabilities of the average specific energy dissipation being smaller than the critical value at the moment t, and larger than the critical value at the moment $t+\Delta t$. In the denominator, there is the probability that at the initial moment t, the value of specific energy dissipation is smaller than the critical value.

Expanding $\varepsilon(t+\Delta t)$ in power series in Δt and assuming the random process to be stationary, we can transform the relation (11.132) into

$$f(V) = \int_0^\infty \dot{\varepsilon}\,p(\varepsilon_{cr}, \dot{\varepsilon})\,d\dot{\varepsilon}\Big/\int_0^{\varepsilon_{cr}} p(\varepsilon)\,d\varepsilon, \qquad (11.133)$$

where $p(\varepsilon, \dot{\varepsilon})$ is the joint distribution density of random variables $\varepsilon(t)$ and $\dot{\varepsilon}(t)$ at one and the same moment of time t; $p(\varepsilon)$ is a one-dimensional distribution of the random variable $\varepsilon(t)$.

According to [85], the joint distribution density $p(\varepsilon_{cr}, \dot{\varepsilon})$ is expressed in terms of joint distribution densities of the random variable $\varepsilon(t)$ at different moments of time:

$$p(\varepsilon_{cr}, \dot{\varepsilon}) = \lim_{\Delta t \to 0} \Delta t g\left(\varepsilon + \frac{\Delta t}{2}\dot{\varepsilon}, \varepsilon - \frac{\Delta t}{2}\dot{\varepsilon}\right). \qquad (11.134)$$

To determine the one-dimensional distribution $p(\varepsilon)$, one should use the conclusion made in [86] that $p(\varepsilon)$ is well approximated by the lognormal distribution law

$$p(\varepsilon) = \frac{1}{\sqrt{2\pi}\alpha\varepsilon}\exp\left(-\frac{1}{2\alpha^2}(\ln \chi\varepsilon)^2\right), \qquad (11.135)$$

where $\alpha^2 = \ln(\sigma^2/\bar{\varepsilon}^2 + 1)$; $\chi = \exp(\alpha^2/2)/\bar{\varepsilon}$; $\bar{\varepsilon}$ and σ^2 are the mean and the variance of specific energy dissipation distribution. [87] uses Millionshikov's hypothesis that the relation between the fourth and the second moments of velocity gradients is the same as suggested by the normal distribution, and shows that the assumption of its validity leads to

$$\sigma^2 = 0.4\bar{\varepsilon}^2. \qquad (11.136)$$

The joint distribution density is determined in [84]:

$$p(\varepsilon_{cr}, \dot{\varepsilon}) = \frac{T_0}{2\pi\alpha c\varepsilon^2} \exp\left(-\frac{T_0^2}{2c^2}\left(\frac{\dot{\varepsilon}}{\varepsilon}\right)^2 - \frac{1}{2\alpha^2}(\ln \chi\varepsilon)^2 \right),$$
$$T_0 = \sqrt{\nu/\bar{\varepsilon}}, \quad c = 1 - \exp(-\alpha^2). \tag{11.137}$$

The substitution of (11.135) and (11.137) into (11.133) gives the expression for breakage frequency

$$f(y) = \frac{c}{\sqrt{2\pi}T_0\alpha}\frac{d}{dy}(\ln \Phi(y)),$$
$$\Phi(y) = \frac{1}{\sqrt{2\pi}}\int_{-\infty}^{\infty} e^{-y^2/2}\,dy, \quad y = \frac{1}{\alpha}\ln(\chi\varepsilon_{cr}(V)). \tag{11.138}$$

Replacing ε_{cr} with (11.131) and taking into account the relation (11.136), one finally obtains

$$f(x) = \frac{1}{2.03\sqrt{2\pi}T_0}\frac{d}{dx}(\ln \Phi(x)),$$
$$x = -1.1\ln\left(1.3\frac{V}{V_{\min}}\right), \quad V_{\min} = \frac{4\pi}{3}R_{\min}^3. \tag{11.139}$$

Paper [84] offers an expression which approximates the expression (11.139) to a sufficient accuracy and is convenient in calculations

$$2.03\sqrt{2\pi}T_0 f(x) = \begin{cases} -x + \dfrac{0.798}{1 - 0.65x} & \text{at } x \le 0, \\ \dfrac{1}{\sqrt{2\pi}}\exp\left(-\dfrac{x^2}{2}\right)(1 + \exp(-1.65x)) & \text{at } x \ge 0. \end{cases}$$

Now, let us determine the probability of drop formation $P(V, \omega)$. In drop breakage experiments [83], various types of drop disintegration were observed, depending on the hydrodynamical conditions. Breakup of a single drop most often resulted in the formation of two identical "daughter drops" and several drops of much smaller size (satellites). No correlation was observed between the type of breakage and the sizes of breaking drops. This allows us to assume that with properties of the internal and external liquids being fixed, the breakup probability depends only on the ratio of volumes ω and V of breaking and newly-formed drops and is written as

$$P(V, \omega) = \mathsf{k}\frac{1}{\omega}g\left(\frac{V}{\omega}\right). \tag{11.140}$$

Impose the condition of conservation of the total volume of drops that are formed after the breakage

$$\int_0^{\infty} VP(V,\omega)\,dV = \omega \tag{11.141}$$

From this and (11.140), it follows that the mean of the density distribution $g(y)$ is equal to

$$\langle y\rangle = \int_0^1 yg(y)\,dy = \frac{1}{\mathsf{k}}.$$

From what we said about the character of drop breakage, it follows that the function $g(y)$ is bimodal with two distinct maxima, one of which is in the interval of satellite sizes, and the other – in the interval of daughter drop sizes. Then it can be presented as a sum of two weighed unimodal distribution densities $g_1(y)$ and $g_2(y)$, defined on the interval $(0,1)$ and having the average values $\bar{g}_1 = V_1/\omega$ and $\bar{g}_2 = V_2/\omega$:

$$P(V,\omega) = \mathsf{k}_1 \frac{1}{\omega} g_1\left(\frac{V}{\omega}\right) + \mathsf{k}_2 \frac{1}{\omega} g_2\left(\frac{V}{\omega}\right). \tag{11.142}$$

To satisfy the conservation condition (11.141), we must require that

$$\mathsf{k}_1 y_1 + \mathsf{k}_2 y_2 = 1. \tag{11.143}$$

In the limiting case, when the drop splits into daughter drops of identical sizes and satellites whose sizes are also identical, but different from the size of daughter drops, the variances of distribution densities $g_1(y)$ and $g_2(y)$ are equal to zero and (11.142) takes the form

$$P(V,\omega) = \mathsf{k}_1\delta(V - y_1\omega) + \mathsf{k}_2\delta(V - y_2\omega), \tag{11.144}$$

where $\delta(x)$ is the delta – function.

In the case when the probability of breakage is a multimodal, rather than a bimodal, function, it can be presented as

$$P(V,\omega) = \sum_{i=1}^{n} \mathsf{k}_i\delta(V - y_i\omega); \quad \sum_{i=1}^{n} \mathsf{k}_i y_i = 1. \tag{11.145}$$

The majority of theoretical works that take into account the breakage of drops assume that two identical drops are formed after the breakage. The probability of breakage in this elementary model is

$$P(V,\omega) = 2\delta(V - 0.5\omega). \tag{11.146}$$

If we use the breakage probability (11.146) and substitute it into equation (11.119) disregarding the first two integrals in the right-hand side (this corresponds to drop breakage without coalescence), then, using well-known properties of the delta-function [89], we obtain

$$\frac{\partial n(V,t)}{\partial t} = 2f(2V)n(2V,t) - f(V)n(V,t). \tag{11.147}$$

Paper [84] discusses the solution of this equation in the form of the sum of independent solutions with discrete spectra. Special cases of monodisperse and uniform over some interval of initial distributions are examined. On the basis of the obtained solutions, the first four moments of the volume distribution of drops are found, and a Pearson diagram (see Fig. 11.1) is used to show that the solution converges to a lognormal distribution. It is consistent with the result of [86], where it was supposed that breakage frequency $f(V)$ is constant and does not depend on the size of drops.

12
Interaction of Two Conducting Drops in a Uniform External Electric Field

12.1
Potential of an Electric Field in the Space Around Drops

Consider two conducting spherical particles with radii R_1 and R_2, carrying charges q_1 and q_2 and placed in a uniform external electric field of strength $\boldsymbol{E}_0$ (Fig. 12.1). The angle between the center line of the particles and vector $\boldsymbol{E}_0$ is denoted by θ. The space between the particles is occupied by a quiescent homogeneous isotropic dielectric with dielectric permittivity ε; the particles do not move relative to this dielectric medium. There are no free charges outside of the spheres, and the potential of the electric field ϕ in this area satisfies the Laplace equation:

$$\Delta\phi = 0. \tag{12.1}$$

We now formulate the boundary conditions. Far from the particles, the strength of the electric field tends towards that of the external field:

$$\boldsymbol{E} = -\nabla\phi \rightarrow \boldsymbol{E}_0 = -\nabla\phi_0. \tag{12.2}$$

Since the system of coordinates is chosen such that the vector $\boldsymbol{E}_0$ is in the plane $x0z$, it is evident that ϕ_0 does not depend on y and is equal to:

$$\phi_0 = E_0(z\cos\theta + x\sin\theta). \tag{12.3}$$

From the condition of the equipotentiality of the surfaces of the particles, it follows that:

$$\phi_{s_1} = V_1, \quad \phi_{s_2} = V_2, \tag{12.4}$$

where V_1 and V_2 are the constant potentials of the surfaces. To determine these potentials, it is necessary to take advantage of the condition that the charge of a conducting particle is distributed over its surface [78]:

Separation of Multiphase, Multicomponent Systems. E. G. Sinaiski and E. J. Lapiga

ISBN: 978-3-527-40612-8

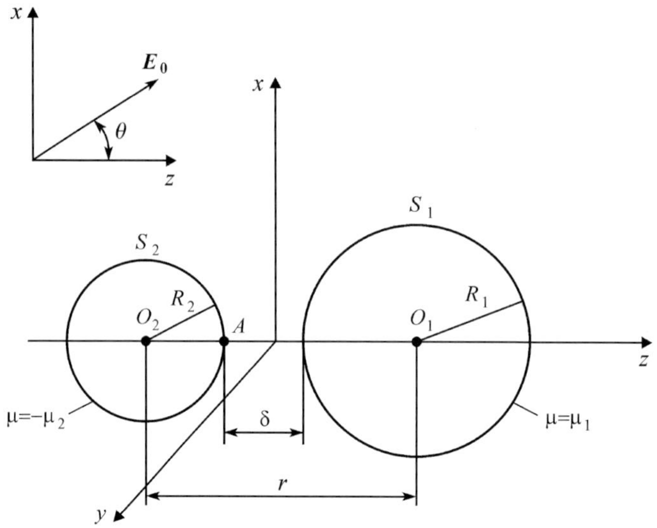

Fig. 12.1 Two conducting spherical particles in an external uniform electric field.

$$-\frac{\varepsilon}{4\pi}\int_{S_i}\frac{\partial\phi}{\partial n}ds = q_i \quad (i=1,2). \tag{12.5}$$

Here, $\boldsymbol{n}$ is the external normal to surface of the particle.

The solution of the boundary value problem, Eqs. (12.1)–(12.5), should be sought in a bispherical system of coordinates (η, μ, Φ) in conjunction with a Cartesian system of coordinates (x, y, z) (see Fig. 12.1) according to ref. [88]:

$$x = \frac{a \sin\eta \cos\Phi}{\cosh\mu - \cos\eta}, \quad y = \frac{a \sin\eta \sin\Phi}{\cosh\mu - \cos\eta}, \quad z = \frac{a \sinh\eta}{\cosh\mu - \cos\eta} \tag{12.6}$$

$$(0 \le \eta < \pi, -\infty < \mu < \infty, 0 \le \Phi \le 2\pi).$$

A bispherical system of coordinates is characterized by coordinate surfaces as depicted in Fig. 12.2. In particular, the surfaces of spheres S_1 and S_2 are characterized by $\mu = \mu_1$ and $\mu = -\mu_2$, where μ_1 and μ_2 are positive constants. The parameter a that appears in Eq. (12.6) is expressed through the radii of the particles by the relationship:

$$a = R_1 \sinh\mu_1 = R_2 \sinh\mu_2. \tag{12.7}$$

In the bispherical system of coordinates, the Laplace equation, Eq. (12.1), transforms to:

$$\frac{\partial}{\partial\eta}\left(\frac{\sin\eta}{\cosh\mu - \cos\eta}\frac{\partial\phi}{\partial\eta}\right) + \frac{\partial}{\partial\mu}\left(\frac{\sin\eta}{\cosh\mu - \cos\eta}\frac{\partial\phi}{\partial\mu}\right)$$
$$+ \frac{1}{\sin\eta(\cosh\mu - \cos\eta)}\frac{\partial^2\phi}{\partial\Phi} = 0. \tag{12.8}$$

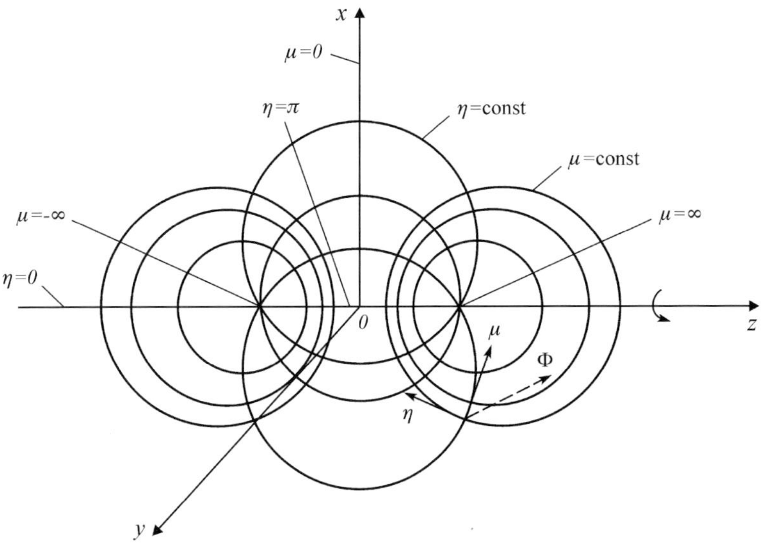

Fig. 12.2 Bispherical system of coordinates (η, μ, Φ).

Consider now the boundary conditions of Eqs. (12.4) and (12.5). The first conditions are obvious:

$$\phi_{\mu=\mu_1} = V_1, \quad \phi_{\mu=\mu_2} = V_2. \tag{12.9}$$

To transform the second conditions, one should take advantage of expressions for the derivative with respect to the external normal to the surface of a sphere:

$$\frac{\partial}{\partial n} = -\frac{1}{a}(\cosh \mu - \cos \eta)^{1/2}\frac{\partial \phi}{\partial \mu} \tag{12.10}$$

and for an element of the surface of a sphere $\mu = \text{const}$.

$$ds = \frac{a^2 \sin \eta}{(\cosh \mu - \cos \eta)^2} d\eta \, d\Phi.$$

The conditions of Eq. (12.5) then become:

$$\frac{\varepsilon a}{4\pi}\int_{\substack{S_i \\ (\mu=\mu_i)}} \frac{\sin \eta}{(\cosh \mu - \cos \eta)^{3/2}}\frac{\partial \phi}{\partial \mu} d\eta \, d\Phi = q_i \quad (i = 1, 2). \tag{12.11}$$

The condition of Eq. (12.2) far from the particles is easily transformed, using Eq. (12.3) for ϕ_0 and Eq. (12.6) for Cartesian coordinates.

The general solution of Eq. (12.8), limited at $\eta = 0$, takes the form of a series of functions [88]:

$$\phi_{m,n} = (\cosh\mu - \cos\eta)^{1/2}\begin{pmatrix}\exp((n+1/2)\mu)\\ \exp(-(n+1/2)\mu)\end{pmatrix} P_n^m(\cos\eta)\begin{pmatrix}\cos m\Phi\\ \sin m\Phi\end{pmatrix}. \qquad (12.12)$$

Here, P_n^m are the associated Legendre polynomials [5].

We now use the condition of Eq. (12.2), having first expanded φ_0 in terms of a series of Legendre polynomials:

$$\phi_0 = -\sqrt{2}aE_0(\cosh\mu - \cos\eta)^{1/2}$$
$$\times\left(\sum_{n=0}^{\infty}(2n+1)\exp\left(-\left(n+\frac{1}{2}\right)|\mu|\right)P_n(\cos\eta)\frac{\mu}{|\mu|}\cos\theta\right.$$
$$\left.+2\sum_{n=1}^{\infty}\exp\left(-\left(n+\frac{1}{2}\right)|\mu|\right)\frac{\mu}{|\mu|}P_n^1(\cos\eta)\sin\theta\cos\Phi\right). \qquad (12.13)$$

As a solution of Eq. (12.8) in a form similar to Eq. (12.13), we obtain:

$$\phi = (\cosh\mu - \cos\eta)^{1/2}\left(\sum_{n=0}^{\infty}\left(A_n\exp\left(\left(n+\frac{1}{2}\right)\mu\right) + B_n\exp\left(-\left(n+\frac{1}{2}\right)\mu\right)\right.\right.$$
$$\left.-\sqrt{2}aE_0(2n+1)\exp\left(-\left(n+\frac{1}{2}\right)|\mu|\right)\frac{\mu}{|\mu|}\cos\theta\right)P_n(\cos\eta)$$
$$+\cos\Phi\sum_{n=1}^{\infty}\left(G_n\exp\left(\left(n+\frac{1}{2}\right)\mu\right) + H_n\exp\left(-\left(n+\frac{1}{2}\right)\mu\right)\right.$$
$$\left.\left.-2\sqrt{2}aE_0\sin\theta\exp\left(-\left(n+\frac{1}{2}\right)|\mu|\right)\right)P_n^1(\cos\eta)\right). \qquad (12.14)$$

To determine the unknown factors A_n, B_n, G_n, and H_n, we divide both parts of Eq. (12.4) by $(\cosh\mu - \cos\eta)^{1/2}$, and we take advantage of the boundary conditions of Eq. (12.9) and the linear independence of Legendre polynomials for any value of m. Expanding $(\cosh\mu - \cos\eta)^{-1/2}$ in a series of Legendre polynomials:

$$(\cosh\mu - \cos\eta)^{-1/2} = \sum_{n=0}^{\infty}P_n(\cos\eta)\exp\left(-\left(n+\frac{1}{2}\right)|\mu|\right), \qquad (12.15)$$

one obtains an infinite system of linear equations with respect to factors A_n, B_n, G_n, and H_n:

$$\left(\sqrt{2}V_1 \exp\left(-\left(n+\frac{1}{2}\right)\mu_1\right) - A_n \exp\left(\left(n+\frac{1}{2}\right)\mu_1\right) - B_n \exp\left(-\left(n+\frac{1}{2}\right)\mu_1\right)\right.$$
$$\left. + \sqrt{2}(2n+1)aE_0 \exp\left(-\left(n+\frac{1}{2}\right)\mu_1\right) \cos\theta\right) P_n(\cos\eta) = 0.$$
$$\left(G_n \exp\left(\left(n+\frac{1}{2}\right)\mu_1\right) + H_n \exp\left(-\left(n+\frac{1}{2}\right)\mu_1\right)\right.$$
$$\left. - 2\sqrt{2}aE_0 \exp\left(-\left(n+\frac{1}{2}\right)\mu_1\right) \sin\theta\right) \cos\Phi P_n^1(\cos\eta) = 0,$$
$$\left(\sqrt{2}V_2 \exp\left(-\left(n+\frac{1}{2}\right)\mu_2\right) - A_n \exp\left(-\left(n+\frac{1}{2}\right)\mu_2\right) - B_n \exp\left(\left(n+\frac{1}{2}\right)\mu_2\right)\right.$$
$$\left. - \sqrt{2}(2n+1)aE_0 \exp\left(-\left(n+\frac{1}{2}\right)\mu_2\right) \cos\theta\right) P_n(\cos\eta) = 0,$$
$$\left(G_n \exp\left(-\left(n+\frac{1}{2}\right)\mu_2\right) + H_n \exp\left(\left(n+\frac{1}{2}\right)\mu_2\right)\right.$$
$$\left. - \sqrt{2}aE_0 \exp\left(-\left(n+\frac{1}{2}\right)\mu_2\right) \sin\theta\right) \cos\Phi P_n^1(\cos\eta) = 0. \quad (12.16)$$

For any n ($n = 0, 1, 2, \ldots$), the system of equations, Eqs. (12.16), has the solution:

$$A_n = \sqrt{2}\frac{(2n+1)aE_0(\exp((2n+1)\mu_2) + 1)\cos\theta + V_1 \exp((2n+1)\mu_2) - V_2}{\exp((2n+1)(\mu_1+\mu_2)) - 1},$$
$$B_n = \sqrt{2}\frac{(2n+1)aE_0(\exp((2n+1)\mu_1) + 1)\cos\theta + V_1 - V_2 \exp((2n+1)\mu_1)}{\exp((2n+1)(\mu_1+\mu_2)) - 1},$$
$$G_n = 2\sqrt{2}\frac{aE_0(\exp((2n+1)\mu_2) - 1)\sin\theta}{\exp((2n+1)(\mu_1+\mu_2)) - 1},$$
$$H_n = 2\sqrt{2}\frac{aE_0(\exp((2n+1)\mu_1) - 1)\sin\theta}{\exp((2n+1)(\mu_1+\mu_2)) - 1}. \quad (12.17)$$

Factors A_n, B_n, G_n, and H_n are expressed through unknown potentials of the surfaces of the particles, V_1 and V_2. To determine these potentials, one should use the conditions of Eq. (12.10) and relationships that follow from Eq. (12.14):

$$\left(\frac{\partial\phi}{\partial n}\right)_{\mu=\mu_1-0} = -\frac{1}{a}(\cosh\mu_1 - \cos\eta)^{3/2}$$
$$\times\left(\sum_{n=0}^{\infty}(2n+1)\exp\left(\left(n+\frac{1}{2}\right)\mu_1\right)A_n P_n(\cos\eta)\right.$$
$$\left. + \cos\Phi\sum_{n=0}^{\infty}(2n+1)\exp\left(\left(n+\frac{1}{2}\right)\mu_1\right)C_n P_n^1(\cos\eta)\right), \quad (12.18)$$

$$\left(\frac{\partial\phi}{\partial n}\right)_{\mu=\mu_2+0} = -\frac{1}{a}(\cosh\mu_2 - \cos\eta)^{3/2}$$
$$\times\left(\sum_{n=0}^{\infty}(2n+1)\exp\left(\left(n+\frac{1}{2}\right)\mu_2\right)B_n P_n(\cos\eta)\right.$$
$$\left.+\cos\Phi\sum_{n=1}^{\infty}(2n+1)\exp\left(\left(n+\frac{1}{2}\right)\mu_2\right)H_n P_n^1(\cos\eta)\right).$$

As a result, we obtain from Eq. (12.11):

$$\int_{-1}^{1}\frac{a}{(\cosh\mu_1 - t)^{1/2}}\sum_{n=0}^{\infty}(2n+1)\exp\left(\left(n+\frac{1}{2}\right)\mu_1\right)A_n P_n(t)\,dt = \frac{2}{\varepsilon}q_1,$$
$$\int_{-1}^{1}\frac{a}{(\cosh\mu_1 - t)^{1/2}}\sum_{n=0}^{\infty}(2n+1)\exp\left(\left(n+\frac{1}{2}\right)\mu_2\right)B_n P_n(t)\,dt = \frac{2}{\varepsilon}q_2. \tag{12.19}$$

Here, t denotes $\cos\eta$.

Having taken advantage of Eq. (12.15) and the condition of orthogonality of Legendre polynomials in the interval $(-1, 1)$:

$$\int_{-1}^{1}P_n(t)P_m(t)\,dt = \frac{2}{n+1}\delta_{mn},$$

we obtain:

$$\sum_{n=0}^{\infty}A_n = \frac{1}{\sqrt{2}a\varepsilon}q_1; \quad \sum_{n=0}^{\infty}B_n = \frac{1}{\sqrt{2}a\varepsilon}q_2. \tag{12.20}$$

Substitution of Eqs. (12.17) for A_n and B_n into Eqs. (12.20) gives the potentials of the particles as:

$$V_1 = \alpha_{11}\frac{q_1}{a\varepsilon} + \alpha_{12}\frac{q_2}{a\varepsilon} + \alpha_{13}aE_0\cos\theta,$$
$$V_2 = \alpha_{21}\frac{q_1}{a\varepsilon} + \alpha_{22}\frac{q_2}{a\varepsilon} + \alpha_{23}aE_0\cos\theta, \tag{12.21}$$

where

$$\alpha_{11} = \frac{s_0^1}{2d}; \quad \alpha_{12} = \alpha_{21} = \frac{s_0^0}{2d}; \quad \alpha_{22} = \frac{s_0^2}{2d};$$
$$\alpha_{13} = \frac{s_1^1 s_0^0 - s_1^2 s_0^1 + s_1^0 s_0^0 - s_0^1 s_1^0}{d}; \quad \alpha_{23} = \frac{s_1^1 s_0^2 - s_1^2 s_0^0 + s_0^2 s_1^0 - s_0^0 s_1^0}{d};$$

$$d = s_0^2 s_0^1 - s_0^0 s_0^0; \quad s_m^1 = s_m(\mu_1); \quad s_m^2 = s_m(\mu_2); \quad m = 0, 1, 2,$$

$$s_m(\xi) = \sum_{n=0}^{\infty} \frac{(2n+1)^m \exp((2n+1)\xi)}{\exp((2n+1)(\mu_1+\mu_2)) - 1}, \quad \xi < \mu_1 + \mu_2.$$

Substituting the factors obtained into Eq. (12.14), one obtains:

$$\phi = E_0 R_2 \varphi_1 \cos\theta + E_0 R_2 \varphi_2 \sin\theta + \frac{q_1}{\varepsilon R_2}\varphi_3 + \frac{q_2}{\varepsilon R_2}\varphi_4. \tag{12.22}$$

Here, the following dimensionless parameters are introduced:

$$\varphi_1 = \sinh\mu_2(\cosh\mu - \cos\eta)^{1/2} \sum_{n=0}^{\infty}\left(d_{1n}\exp\left(\left(n+\frac{1}{2}\right)\mu\right)\right.$$
$$\left. + d_{2n}\exp\left(-\left(n+\frac{1}{2}\right)\mu\right) + d_{3n}\exp\left(-\left(n+\frac{1}{2}\right)|\mu|\right)\frac{\mu}{|\mu|}\right)P_n(\cos\eta),$$

$$\varphi_2 = \sinh\mu_2(\cosh\mu - \cos\eta)^{1/2} \sum_{n=1}^{\infty}\left(d_{4n}\exp\left(\left(n+\frac{1}{2}\right)\mu\right)\right.$$
$$+ d_{5n}\exp\left(-\left(n+\frac{1}{2}\right)\mu\right)$$
$$\left. - 2\sqrt{2}d_{3n}\exp\left(-\left(n+\frac{1}{2}\right)|\mu|\right)\right)\cos\Phi P_n^1(\cos\eta),$$

$$\varphi_3 = \cosh\mu_2(\cosh\mu - \cos\eta)^{1/2} \sum_{n=1}^{\infty} d_{6n}\exp\left(\left(n+\frac{1}{2}\right)\mu\right)$$
$$+ d_{7n}\exp\left(-\left(n+\frac{1}{2}\right)\mu\right)P_n^1(\cos\eta),$$

$$\varphi_4 = \cosh\mu_2(\cosh\mu - \cos\eta)^{1/2} \sum_{n=1}^{\infty} d_{8n}\exp\left(\left(n+\frac{1}{2}\right)\mu\right)$$
$$+ d_{9n}\exp\left(-\left(n+\frac{1}{2}\right)\mu\right)P_n^1(\cos\eta),$$

where

$$d_{1n} = a_{1n} + a_{2n}\alpha_{13} + a_{3n}\alpha_{23}; \quad d_{2n} = b_{1n} + b_{2n}\alpha_{13} + b_{3n}\alpha_{23};$$
$$d_{3n} = \sqrt{2}(2n+1); \quad d_{4n} = 2(\exp((2n+1)\mu_2) - 1)L_n;$$
$$d_{5n} = 2(\exp((2n+1)\mu_1) - 1)L_n;$$

$$d_{6n} = a_{2n}\alpha_{11} + a_{3n}\alpha_{21}; \quad d_{7n} = b_{2n}\alpha_{11} + b_{3n}\alpha_{21};$$

$$d_{8n} = a_{2n}\alpha_{12} + a_{3n}\alpha_{22}; \quad d_{9n} = b_{2n}\alpha_{12} + b_{3n}\alpha_{22};$$

$$a_{1n} = (2n+1)(\exp((2n+1)\mu_2) + 1)L_n; \quad a_{2n} = (\exp((2n+1)\,\mu_2))L_n;$$

$$b_{1n} = -(2n+1)(\exp((2n+1)\mu_1) + 1)L_n; \quad b_{2n} = a_{3n} = -L_n;$$

$$b_{3n} = 2(\exp((2n+1)\mu_1) - 1)L_n; \quad L_n = \frac{1}{\exp((2n+1)(\mu_1 + \mu_2)) - 1}.$$

Thus, Eq. (12.22) gives the solution of the stated problem and represents the distribution of electric field potential in the space around the spherical particles.

To calculate the potential, it is necessary to specify values of the particle radii R_1, R_2, the distance between the centers of the particles r, and the coordinates x, y, z of the point at which the potential is sought. Through these values, one may determine the parameters μ_1, μ_2, μ, η, Φ. For this purpose, it is necessary to take advantage of Eqs. (12.6) and (12.7), which connect Cartesian coordinates to bi-spherical ones. As was noted earlier, spheres of radii $a|\mathrm{csh}\,\mu|$ correspond to surfaces with $\mu = \text{const.}$:

$$x^2 + y^2 + (z - \coth\mu)^2 = \frac{a^2}{\sinh^2\mu}; \quad (-\infty < \mu < \infty). \tag{12.23}$$

From Eqs. (12.7) and (12.23), it follows that, having set R_1, R_2, and r, it is possible to find μ_1, μ_2, and a from the system of equations:

$$R_1 = a\cosh\mu_1, \quad R_2 = a\cosh\mu_2, \quad r = a(\coth\mu_1 + \coth\mu_2). \tag{12.24}$$

Solving Eqs. (12.24), one obtains:

$$\begin{aligned} &\mu_1 = \ln\left(b_1 + \sqrt{b_1^2 - 1}\right); \quad \mu_2 = \ln\left(b_2 + \sqrt{b_2^2 - 1}\right); \\ &a = R_1\sqrt{b_1^2 - 1} = R_2\sqrt{b_2^2 - 1}, \end{aligned} \tag{12.25}$$

where

$$b_1 = 1 + \frac{u\mathrm{k}^2(1 + u/2)}{1 + \mathrm{k}(1 + u)}; \quad b_2 = \frac{u(1 + u\mathrm{k}/2)}{1 + \mathrm{k}(1 + u)};$$

$$u = (r - R_1 - R_2)/R_2; \quad \mathrm{k} = R_2/R_1; \quad \mathrm{k} \le 1.$$

To determine the bispherical coordinates of a point μ, η, Φ from specified values of Cartesian coordinates x, y, z, it is necessary to invoke formulae of reverse transformation [89]:

$$\begin{aligned} &\eta = \frac{i}{2}\ln\frac{((x^2 + y^2)^{1/2} - ia)^2 + z^2}{((x^2 + y^2)^{1/2} + ia)^2 + z^2}, \quad (i^2 = -1), \\ &\mu = \frac{i}{2}\ln\frac{x^2 + y^2 + (z + a)^2}{x^2 + y^2 + (z - a)^2}, \quad \Phi = \arctan\frac{y}{x} \end{aligned} \tag{12.26}$$

12.2
Strength of an Electric Field in the Gap Between Drops

We consider now the definition of the strength of an electric field in the space around particles. It is known [90] that the approach of conducting, non-charged spherical particles in an electric field is accompanied by an increase in the strength of the electric field in the gap between the particles. At a small gap compared to the particle sizes, the strength of the electric field can be tens or hundreds of times increased, which leads to destruction of the dielectric properties of the continuum. The authors of ref. [90] even observed spark discharge between closely located particles. As will be shown later, redistribution of charges between particles as a result of their collision significantly influences interaction forces between the particles. In view of the above, there is increased interest in calculating the strengths of electric fields in gaps between particles, especially for small gap distances.

For any point in space, the strength of the electric field is determined by the relationship:

$$\boldsymbol{E} = -\nabla\phi. \tag{12.27}$$

The potential of an electric field was determined in Section 12.1 in the form of an infinite series, the convergence rate of which decreases with decreasing clearance between the particles, that is, in the area of greatest interest. For a small gap between the surfaces of spherical particles, the electric field strength vector deviates little from the direction of the surface normal. This allows us to carry out asymptotic analysis in calculating the strength of the electric field near the surface of one of the particles.

Consider an area in the gap between two particles near the smaller one, that is, in the vicinity of a point with bispherical coordinates $\eta = \pi$, $\mu = -\mu_2$, $\Phi = 0$ (see Fig. 12.1, point A). At this point, the conditions $t = \cos\eta = -1$, $P_n(-1) = (-1)^n$, and $P_n^1(-1) = 0$ are satisfied. Then, from Eqs. (12.10) and (12.18) it is found that:

$$\begin{aligned} E_A &= -\left(\frac{\partial\varphi}{\partial n}\right)_{\substack{\mu=-\mu_2+0 \\ \eta=\pi}} \\ &= \frac{1}{a}(\cosh\mu_2 + 1)^{3/2}\sum_{n=0}^{\infty}(-1)^n(2n+1)\exp\left(\left(n+\frac{1}{2}\right)\mu_2\right)B_n. \end{aligned} \tag{12.28}$$

Substituting in Eq. (12.28) the expressions Eq. (12.17) for B_n and Eq. (12.21) for V_1 and V_2, one obtains:

$$E_A = \frac{1}{\varepsilon R_2^2}(E_1 q_1 + E_2 q_2) + E_3 E_0 \cos\theta, \tag{12.29}$$

where

$$E_2 = \frac{R_2^2}{a^2}\sqrt{2}(\cosh \mu_2 + 1)^{3/2}\left(\hat{s}_1\left(\mu_1 + \frac{\mu_2}{2}\right)a_{21} - \hat{s}_1\left(\frac{\mu_2}{2}\right)a_{11}\right);$$

$$E_2 = \frac{R_2^2}{a^2}\sqrt{2}(\cosh \mu_2 + 1)^{3/2}\left(\hat{s}_1\left(\mu_1 + \frac{\mu_2}{2}\right)a_{22} - \hat{s}_1\left(\frac{\mu_2}{2}\right)a_{12}\right);$$

$$E_3 = \sqrt{2}(\cosh \mu_2 + 1)^{3/2} \tag{12.30}$$

$$\times\left(\hat{s}_1\left(\mu_1 + \frac{\mu_2}{2}\right)a_{23} - \hat{s}_1\left(\frac{\mu_2}{2}\right)a_{13} - \hat{s}_2\left(\mu_1 + \frac{\mu_2}{2}\right) - \hat{s}_2\left(\frac{\mu_2}{2}\right)\right);$$

$$\hat{s}_m(\xi) = \sum_{n=0}^{\infty}(-1)^n \frac{(2n+1)^m \exp((2n+1)\xi)}{\exp((2n+1)(\mu_1+\mu_2)) - 1}, \quad \xi < \mu_1 + \mu_2,\ m = 0, 1, 2.$$

The factors E_1, E_2, and E_3 depend on the relative clearance between particles, $\Delta = (r - R_1 - R_2)/R_1$, and on the ratio of the particle radii $\mathsf{k} = R_2/R_1$. In Fig. 12.3, (a), (b) show the dependence $E_i(\Delta, \mathsf{k})$, from which it follows that at small

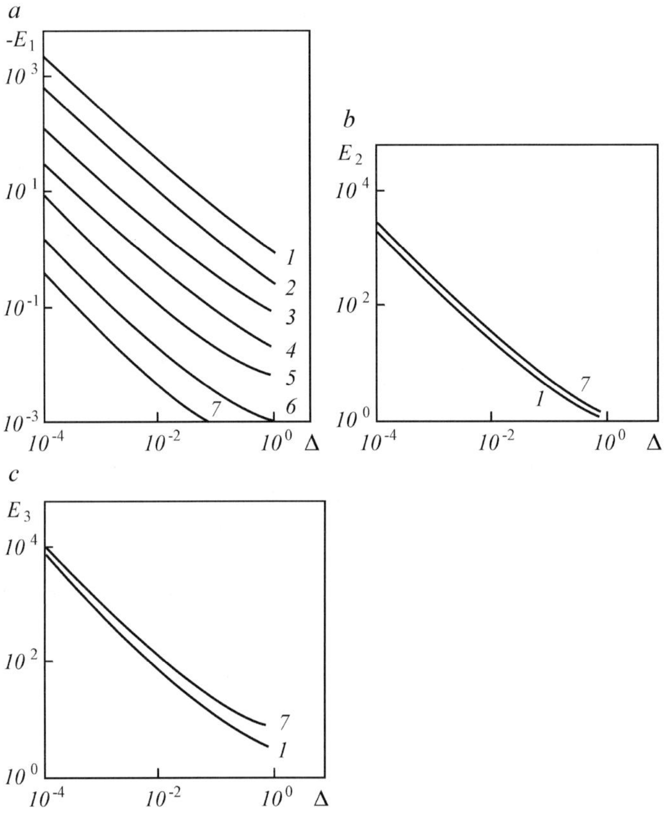

Fig. 12.3 Dependence of the coefficients E_i upon the relative clearance between drops Δ and the ratio of drop radii k: *1*: $k = 1$; *2*: 0.5; *3*: 0.2; *4*: 0.1; *5*: 0.05; *6*: 0.02; *7*: 0.01.

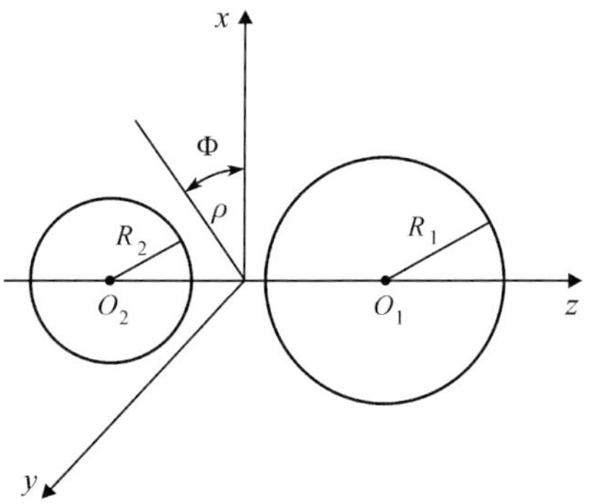

Fig. 12.4 Cylindrical system of coordinates (ρ, Φ, z).

clearance, $\Delta \ll 1$, the dependence of E_i on Δ is close to a power of one for all values of $\boldsymbol{k}$. Thereby, sufficiently accurate values of E_i may be obtained by interpolation of numerical data.

Another method for determining dependences $E_i(\Delta, \mathbf{k})$ at small gaps between particles involves analysis of asymptotic behavior of factors of E_i as $\Delta \to 0$. It can be shown that if the size of the gap between particles is much less than their radii, the electric field in the clearance near to the line between particle centers is close to uniform. Here, we introduce a cylindrical system of coordinates (ρ, Φ, z), as shown in Fig. 12.4.

Since particles are conducting near their surfaces, the strength vector deviates slightly from the normal to the surface; at $\rho \ll R_j$ $(j = 1, 2)$ projections of the strength vector onto the axes r and z are estimated as:

$$E_{j\rho} = E_{jn}\frac{\rho}{R_j} + 0\left(\left(\frac{\rho}{R_j}\right)^2\right); \quad E_{jz} = E_{jn}\left(1 - \frac{\rho^2}{R_j^2}\right) + 0\left(\left(\frac{\rho}{R_j}\right)^3\right). \tag{12.31}$$

Here, the index j corresponds to the number of the sphere. Since the spheres have different radii $(R_2 \leq R_1)$, the greatest deviation of the distribution of electric field strength will be near the surface of the smaller sphere; therefore, consideration may be restricted to the area near sphere 2.

Substituting Eqs. (12.31) in the equation $\nabla \cdot \boldsymbol{E} = 0$, one obtains the main terms of an asymptotic expansion:

$$\frac{1}{\rho}\frac{\partial}{\partial \rho}\left(\rho E_{2n}\frac{\rho}{R_2}\right) + \frac{\partial E_{2n}}{\partial z} = 0. \tag{12.32}$$

Close to sphere 2, $E_{2n} \sim E_{2z}$ and as $\rho \to 0$:

$$\frac{1}{\rho}\frac{\partial}{\partial \rho}\left(\rho E_{2n}\frac{\rho}{R_2}\right) = \frac{\rho}{R_2}\frac{\partial E_{2z}}{\partial \rho} + 2\frac{E_{2z}}{R_2} \sim 2\frac{E_{2z}}{R_2}.$$

As a result, Eq. (12.32) becomes:

$$2\frac{E_{2z}}{R_2} + \frac{\partial E_{2z}}{\partial z} = 0. \tag{12.33}$$

From Eq. (12.33), it follows that at small clearance between the spheres, $\Delta z \sim r - R_1 - R_2$:

$$\left|\frac{\Delta E_{2z}}{E_{2z}}\right| \sim 2\frac{\Delta z}{R_2} \sim 2\Delta \ll 1.$$

Hence, for $|\Delta E_{2z}| \ll E_{2z}$, the component of strength along the line joining the centers as $\Delta \to 0$ can be considered uniform and equal to:

$$E = \frac{V_2 - V_1}{r - R_1 - R_2} = \frac{V_2 - V_1}{\Delta R_2}. \tag{12.34}$$

Here, values of the potentials of spheres V_1 and V_2 should be taken as $\Delta \to 0$, for which it is necessary to find asymptotic μ_1, μ_2, a, and $s_m(\xi)$. The first three parameters are easily obtained from Eq. (12.25), expanding them in power series in terms of Δ:

$$\begin{aligned}&\mu_1 = \mathrm{k}\sqrt{2\Delta/(\mathrm{k}+1)} + 0(\Delta); \quad \mu_2 = \sqrt{2\Delta/(\mathrm{k}+1)} + 0(\Delta);\\ &a = R_2\sqrt{2\Delta/(\mathrm{k}+1)} + 0(\Delta).\end{aligned} \tag{12.35}$$

We now turn to the determination of asymptotic values $s_m(\xi)$ at μ_1, μ_2, and $\xi \ll 1$. Consider first the case $m = 0$. We have:

$$s_0(\xi) = \sum_{n=0}^{\infty} \frac{\exp((2n+1)\xi)}{\exp((2n+1)\beta) - 1}; \quad \xi < \beta = \mu_1 + \mu_2. \tag{12.36}$$

We may represent $s_0(\xi)$ as:

$$\begin{aligned}s_0(\xi) &= \sum_{n=1}^{\infty} \frac{\exp((2n-1)\xi)}{\exp((2n-1)\beta) - 1} + \sum_{n=N+1}^{\infty} \frac{\exp((2n-1)\xi)}{\exp((2n-1)\beta) - 1}\\ &= s_0^N(\xi) + s_0^*(\xi),\end{aligned} \tag{12.37}$$

where $N \gg 1$, and choose N to obey the condition $\sqrt{1/\beta} \ll N \ll 1/2\beta$. So, if $N \sim \beta^{-2/3}$, these conditions are satisfied. It is then possible to find $s_0^N(\xi)$, expanding the exponent in series at small values of $N\beta$, and calculating the remainder term $s_0^*(\xi)$ having replaced the sum by an integral. As a result, as $\beta \to 0$ we obtain:

$$\sum_{n=1}^{\infty}\frac{\exp((2n-1)\xi)}{\exp((2n-1)\beta)-1}=\frac{1}{\beta}\sum_{n=1}^{\infty}\frac{1}{2n-1}+\left(a-\frac{1}{2}\right)N$$
$$+\beta\left(\frac{\alpha^2}{2}-\frac{\alpha}{2}+\frac{1}{12}\right)\sum_{n=1}^{\infty}(2n-1)+0(\beta N),$$

where $\alpha=\xi/\beta$.

Having determined the sums included in this expression, one obtains:

$$s_0^N(\xi)=\frac{1}{\beta}\left(\frac{C}{2}+\frac{\ln N}{2}+\ln 2+\frac{1}{48N^2}\right)+\left(a-\frac{1}{2}\right)N$$
$$+\beta N^2\left(\frac{\alpha^2}{2}-\frac{\alpha}{2}+\frac{1}{12}\right)+0(\beta N), \tag{12.38}$$

where $C=0.5772\ldots$ is Euler's constant.

In order to calculate s_0^*, the sum is replaced by an integral, in much the same way as in numerical integration using the trapezium formula:

$$s_0^*(\xi)=\frac{1}{2(2N+1)\beta}+\frac{1}{2}\left(a-\frac{1}{2}\right)+\frac{1}{2\beta}\int_{(2N+1)\beta}^{\infty}\frac{\exp(\alpha z)}{\exp(z)-1}dz+R(N), \tag{12.39}$$

where $R(N)$ is the remainder term estimated as $R(N)\le 1/2N^2$. If we choose $N\sim\beta^{-2/3}$, then $R(N)\sim\beta^{-4/3}$ and at $\beta\ll 1$ we obtain $R(N)\sim 0(\beta)$. The integral in Eq. (12.39) may be transformed to:

$$\int_{(2N+1)\beta}^{\infty}\frac{\exp(\alpha z)}{\exp(z)-1}dz=\int_0^{\infty}\frac{\exp(\alpha z)-1}{\exp(z)-1}dz+\int_{(2N+1)\beta}^{\infty}\frac{1}{\exp(z)-1}dz$$
$$-\int_0^{(2N+1)\beta}\frac{\exp(\alpha z)-1}{\exp(z)-1}dz.$$

The last integral can be determined if we take into account that $(2N+1)\beta\ll 1$ and expand the integrand series in z. As a result, we obtain:

$$\int_{(2N+1)\beta}^{\infty}\frac{\exp(\alpha z)}{\exp(z)-1}dz=-\frac{\ln((2N+1)\beta)}{2\beta}-\frac{C+\psi(1-\alpha)}{2\beta}+\frac{2N+1}{4}(1-2\alpha)$$
$$+\frac{(2N+1)^2}{48}\beta(6\alpha-6\alpha^2-1)+0(\beta N). \tag{12.40}$$

Substituting Eqs. (12.38) and (12.39) in Eq. (12.37) and taking into account Eq. (12.40), one obtains:

$$s_0(\xi)=-\frac{\ln\beta}{2\beta}-\frac{1}{2\beta}\psi\left(1-\frac{\xi}{\beta}\right)+\frac{\ln 2}{2\beta}0(\beta N). \tag{12.41}$$

Here, $\psi(x)$ is the Euler psi-function [5].

To determine $s_m(\xi)$ at $m = 1, 2, 3, \ldots$ one should take advantage of the obvious property:

$$s_m(\xi) = \frac{\partial^m}{\partial \xi^m}(s_0(\xi)),$$

use of which gives factors that appear in Eq. (12.21):

$$s_0^0 = -\frac{1}{2\beta}\left(\ln\frac{\beta}{2} + \psi(1)\right), \quad s_0^1 = -\frac{1}{2\beta}\left(\ln\frac{\beta}{2} + \psi\left(\frac{\mu_2}{\beta}\right)\right),$$

$$s_0^2 = -\frac{1}{2\beta}\left(\ln\frac{\beta}{2} + \psi\left(\frac{\mu_1}{\beta}\right)\right),$$

$$s_1^0 = \frac{1}{2\beta^2}\psi'(1), \quad s_1^1 = -\frac{1}{2\beta^2}\psi'\left(\frac{\mu_2}{\beta}\right), \quad s_1^2 = -\frac{1}{2\beta^2}\psi'\left(\frac{\mu_1}{\beta}\right),$$

$$a_{11} = \frac{\beta}{s}\left(\ln\frac{\beta}{2} + \psi\left(\frac{\mu_2}{\beta}\right)\right) + 0(\beta); \quad a_{12} = a_{21} = \frac{\beta}{s}\left(\ln\frac{\beta}{2} + \psi(1)\right) + 0(\beta),$$

$$a_{13} = \frac{1}{\beta s}\left(\ln\frac{\beta}{2}\left(\psi'\left(\frac{\mu_1}{\beta}\right) - \psi'\left(\frac{\mu_2}{\beta}\right)\right) + \psi\left(\frac{\mu_2}{\beta}\right)\left(\psi'(1) + \psi'\left(\frac{\mu_1}{\beta}\right)\right)\right.$$
$$\left. - \psi(1)\left(\psi'(1) + \psi'\left(\frac{\mu_2}{\beta}\right)\right)\right) + 0(\beta),$$

$$a_{22} = \frac{\beta}{s}\left(\ln\frac{\beta}{2} + \psi\left(\frac{\mu_1}{\beta}\right)\right) + 0(\beta); \quad a_{23} = \frac{1}{\beta s}\left(\ln\frac{\beta}{2} + \left(\psi'\left(\frac{\mu_1}{\beta}\right) - \psi'\left(\frac{\mu_2}{\beta}\right)\right)\right.$$
$$\left. - \psi\left(\frac{\mu_1}{\beta}\right)\left(\psi'(1) + \psi'\left(\frac{\mu_2}{\beta}\right)\right) + \psi(1)\left(\psi'(1) + \psi'\left(\frac{\mu_1}{\beta}\right)\right)\right) + 0(\beta),$$

$$s = \ln\frac{\beta}{2}\left(\psi\left(\frac{\mu_1}{\beta}\right) + \psi\left(\frac{\mu_2}{\beta}\right) - 2\psi(1)\right) + \psi\left(\frac{\mu_1}{\beta}\right)\psi\left(\frac{\mu_2}{\beta}\right) - (\psi(1))^2.$$

Substituting the derived factors a_{ij} into Eq. (12.21) and taking into account that at $\Delta \ll 1$ the equalities $\mu_1/\beta = \mathrm{k}/(\mathrm{k}+1) + 0(\Delta)$ and $\mu_2/\beta = 1/(\mathrm{k}+1) + 0(\Delta)$ are valid, one can determine the potentials of the particles, V_1 and V_2, and from Eq. (12.34) the average strength of the electric field along the line joining the particle centers under the condition $\beta \ll 1$:

$$E = \frac{q_1}{\varepsilon R_2^2}E_1^* + \frac{q_2}{\varepsilon R_2^2}E_2^* + E_2^* E_0 \cos\theta, \tag{12.42}$$

where

$$E_i^* = \frac{c_i(\mathsf{k})}{\Delta(\ln\Delta + c_0(\mathsf{k}))} + 0(\Delta); \quad i = 1, 2, 3; \ \Delta \ll 1$$

$$c_0(\mathsf{k}) = \frac{2\left(\psi(1)\psi(1) - \psi\left(\frac{\mathsf{k}}{\mathsf{k}+1}\right)\psi\left(\frac{1}{\mathsf{k}+1}\right)\right)}{d} + \ln\left(\frac{\mathsf{k}+1}{2}\right);$$

$$c_1(\mathsf{k}) = \frac{2(\mathsf{k}+1)\left(\psi\left(\frac{1}{\mathsf{k}+1}\right) - \psi(1)\right)}{d};$$

$$c_2(\mathsf{k}) = \frac{2(\mathsf{k}+1)\left(\psi(1) - \psi\left(\frac{\mathsf{k}}{\mathsf{k}+1}\right)\right)}{d};$$

$$c_3(\mathsf{k}) = 2\left(\psi'(1)d + \psi(1)\left(\psi'\left(\frac{\mathsf{k}}{\mathsf{k}+1}\right) + \psi'\left(\frac{1}{\mathsf{k}+1}\right)\right)\right.$$

$$\left. - \psi\left(\frac{\mathsf{k}}{\mathsf{k}+1}\right)\psi'\left(\frac{1}{\mathsf{k}+1}\right) - \psi'\left(\frac{\mathsf{k}}{\mathsf{k}+1}\right)\psi'\left(\frac{1}{\mathsf{k}+1}\right)\right) \Big/ d(\mathsf{k}+1)$$

$$d = 2\psi(1) - \psi\left(\frac{\mathsf{k}}{\mathsf{k}+1}\right) - \psi\left(\frac{1}{\mathsf{k}+1}\right).$$

Comparison with the results of numerical calculations of E_i by means of exact formulae, Eq. (12.30), has shown that discrepancies between the values of E_i^* obtained and the respective values of E_i do not exceed 2% in the region $\Delta < 0.1$, and as $\Delta \to 0$ $E_i^* \to E_i$.

12.3 Interaction Forces of Two Conducting Spherical Drops

Forces $\boldsymbol{F}_1$ and $\boldsymbol{F}_2$, acting on the first and second particles bearing charges q_1 and q_2, are connected by the obvious relationship:

$$F_1 + F_2 = E_0(q_1 + q_2) \tag{12.43}$$

where $\boldsymbol{E}_0$ is the strength of a uniform external field.

Therefore, determination of one of the forces, for example $\boldsymbol{F}_2$, is sufficient for our purposes.

From the definition of Maxwellian stress tension [91], it follows that projection of the force acting on particle 2 in the direction of a unit vector $\boldsymbol{p}$ is equal to:

$$\boldsymbol{F} \cdot \boldsymbol{p} = \frac{\varepsilon}{8\pi} \cdot \int_{S_2} \left(\frac{\partial\phi}{\partial n}\right)^2 \boldsymbol{n} \cdot \boldsymbol{p}\, ds \tag{12.44}$$

Taking for $\boldsymbol{p}$ the unit vectors $\mathbf{e}_1$ and $\mathbf{e}_2$ in the direction of axes OZ and OH and using the properties $e_1 \cdot \boldsymbol{n} = n_z$, $e_2 \cdot \boldsymbol{n} = n_x$ and expressions for projections of the external normal $\boldsymbol{n}$ in a bispherical system of coordinates:

$$n_z = \frac{1 - \cosh \mu \cos \eta}{\cosh \mu - \cos \eta}; \quad n_x = \frac{\sinh \mu \sin \eta}{\cosh \mu - \cos \eta} \cos \Phi,$$

from Eq. (12.44) and taking into account the relationships of Eqs. (12.10) and (12.18) at $\mu = -\mu_2$, we obtain:

$$F_{2z} = \frac{\varepsilon}{8\pi} \int_0^{2\pi} d\Phi \int_{-1}^{1} \left(\sum_{n=0}^{\infty} (2n+1) \exp\left(\left(n + \frac{1}{2} \right) \mu_2 \right) B_n P_n(t) \right.$$
$$\left. + \cos \Phi \sum_{n=0}^{\infty} (2n+1) \exp\left(\left(n + \frac{1}{2} \right) \mu_2 \right) H_n P_n^1(t) \right)^2 (1 - \cosh \mu_2 t)\, dt,$$
$$F_{2x} = \frac{\varepsilon}{8\pi} \int_0^{2\pi} d\Phi \int_{-1}^{1} \sinh \mu_2 \cos \Phi \left(\sum_{n=0}^{\infty} (2n+1) \exp\left(\left(n + \frac{1}{2} \right) \mu_2 \right) B_n P_n(t) \right.$$
$$\left. + \cos \Phi \sum_{n=0}^{\infty} (2n+1) \exp\left(\left(n + \frac{1}{2} \right) \mu_2 \right) H_n P_n^1(t) \right) dt. \tag{12.45}$$

Integrating Eqs. (12.45) with respect to Φ from 0 to 2π and t from -1 to 1 and using the properties of Legendre polynomials:

$$(2n+1)tP_n(t) = (n+1)P_{n+1}(t) + nP_{n-1}(t),$$
$$(2n+1)tP_n^1(t) = (n+1)P_{n-1}^1(t) + nP_{n+1}^1(t),$$
$$(1-t^2)^{1/2}(2n+1)tP_n^1(t) = n(n+1)(P_{n-1}(t) - P_{n+1}(t)),$$
$$\int_{-1}^{1} P_n(t)P_m(t)\, dt = \frac{2}{2n+1}\delta_{mn}; \quad \int_{-1}^{1} P_n^1(t)P_m^1(t)\, dt = \frac{2n(n+1)}{2n+1}\delta_{mn},$$

gives

$$F_{2z} = \frac{\varepsilon}{2} \left(\sum_{n=0}^{\infty} (2n+1) \exp((2n+1)\mu_2) B_n \left(B_n - \frac{n+1}{2n+1}(1 + \exp(2\mu_2)) B_{n+1} \right) \right.$$
$$+ \frac{1}{2} \sum_{n=1}^{\infty} n(n+1)(2n+1) \exp((2n+1)\mu_2) H_n$$
$$\left. \times \left(H_n - \frac{n+2}{2n+1}(1 + \exp(2\mu_2)) H_{n+1} \right) \right),$$
$$F_{2z} = \frac{\varepsilon}{2} \sinh \mu_2 \left(\sum_{n=0}^{\infty} (n+1) \exp((2n+1)\mu_2)((n+2)B_n H_{n+1} - nB_{n+1}H_n) \right).$$

Substituting the factors B_n and H_n from Eq. (12.17) and V_1, V_2 from Eq. (12.21) into the obtained expressions, we finally obtain:

$$E_{2z} = \varepsilon R_2^2 E_0^2 (f_1 \cos^2\theta + f_2 \cos^2\theta) + E_0 \cos\theta (f_3 q_1 + f_4 q_2) + \frac{1}{\varepsilon R_2^2}(f_5 q_1^2 + f_6 q_1 q_2 + f_7 q_2^2) + E_0 q_2 \cos\theta, \qquad (12.46)$$

$$E_{2x} = \varepsilon R_2^2 E_0^2 f_8 \sin 2\theta + E_0 \sin\theta (f_9 q_1 + f_{10} q_2) + E_0 q_2 \sin\theta, \qquad (12.47)$$

where

$$f_1 = \frac{a^2}{R_2^2}(s_1 + s_2\alpha_{13}^2 + s_3\alpha_{23}^2 + s_4\alpha_{11} + s_5\alpha_{23} + s_6\alpha_{13}\alpha_{23}); \quad f_2 = \frac{2a^2}{R_2^2}s_4;$$

$$f_3 = s_4\alpha_{13} + s_5\alpha_{23} + 2s_2\alpha_{11}\alpha_{13} + 2s_3\alpha_{21}\alpha_{23} + s_6(\alpha_{13}\alpha_{21} + \alpha_{11}\alpha_{23});$$

$$f_4 = s_4\alpha_{12} + s_5\alpha_{22} + 2s_2\alpha_{12}\alpha_{13} + 2s_3\alpha_{22}\alpha_{23} + s_6(\alpha_{13}\alpha_{22} + \alpha_{12}\alpha_{23});$$

$$f_5 = \frac{R_2^2}{a^2}(\alpha_{11}^2 s_2 + \alpha_{21}^2 s_1); \quad f_6 = \frac{R_2^2}{a^2}(\alpha_{11}\alpha_{12} + \alpha_{22}\alpha_{21})s_6;$$

$$f_7 = \frac{R_2^2}{a^2}(\alpha_{12}^2 s_2 + \alpha_{22}^2 s_1); \quad f_8 = \frac{a^2}{R_2^2}(s_8 + s_9\alpha_{12} + s_{10}\alpha_{13});$$

$$f_9 = 2(s_9\alpha_{21} + s_{10}\alpha_{11}); \quad f_{10} = 2(s_9\alpha_{22} + s_{10}\alpha_{12});$$

$$s_1 = \sum_{n=0}^{\infty} \frac{a_{2,n}b_n}{a_{3,n}}\left((2n+1)\frac{b_n}{a_{3,n}} - (n+1)c_1\frac{b_{n+1}}{a_{3,n+1}}\right);$$

$$s_1 = \sum_{n=0}^{\infty} \frac{a_{1,n}a_{2,n}}{a_{3,n}}\left(\frac{2n+1}{a_{3,n}} - c_1\frac{n+1}{a_{3,n+1}}\right);$$

$$s_3 = \sum_{n=0}^{\infty} \frac{a_{1,n}a_{2,n}}{a_{3,n}}\left((2n+1)\frac{a_{1,n}}{a_{3,n}} - (n+1)c_1\frac{a_{1,n+1}}{a_{3,n+1}}\right);$$

$$s_4 = \sum_{n=0}^{\infty} \frac{a_{2,n}}{a_{3,n}}\left(2(2n+1)\frac{b_n}{a_{3,n}} - (n+1)c_1\frac{b_n + b_{n+1}}{a_{3,n+1}}\right);$$

$$s_5 = -\sum_{n=0}^{\infty} \frac{a_{2,n}}{a_{3,n}}\left(2(2n+1)\frac{a_{1,n}b_n}{a_{3,n}} - (n+1)c_1\frac{b_n a_{1,n+1} + b_{n+1}a_{1,n}}{a_{3,n+1}}\right);$$

$$s_6 = -\sum_{n=0}^{\infty} \frac{a_{2,n}}{a_{3,n}}\left(2(2n+1)\frac{a_{1,n}}{a_{3,n}} - (n+1)c_1\frac{a_{1,n} + a_{1,n+1}}{a_{3,n+1}}\right);$$

$$s_7 = \sum_{n=0}^{\infty} n(n+1)\frac{a_{2,n}}{a_{3,n}}d_n\left((2n+1)\frac{d_n}{a_{3,n}} - (n+2)c_1\frac{d_{n+1}}{a_{3,n+1}}\right);$$

$$s_8 = \sum_{n=1}^{\infty} c_2(n+1)\frac{a_{2,n}}{a_{3,n}}\left(n\frac{d_n b_{n+1}}{a_{3,n+1}} - (n+2)\frac{d_{n+1}b_n}{a_{3,n+1}}\right);$$

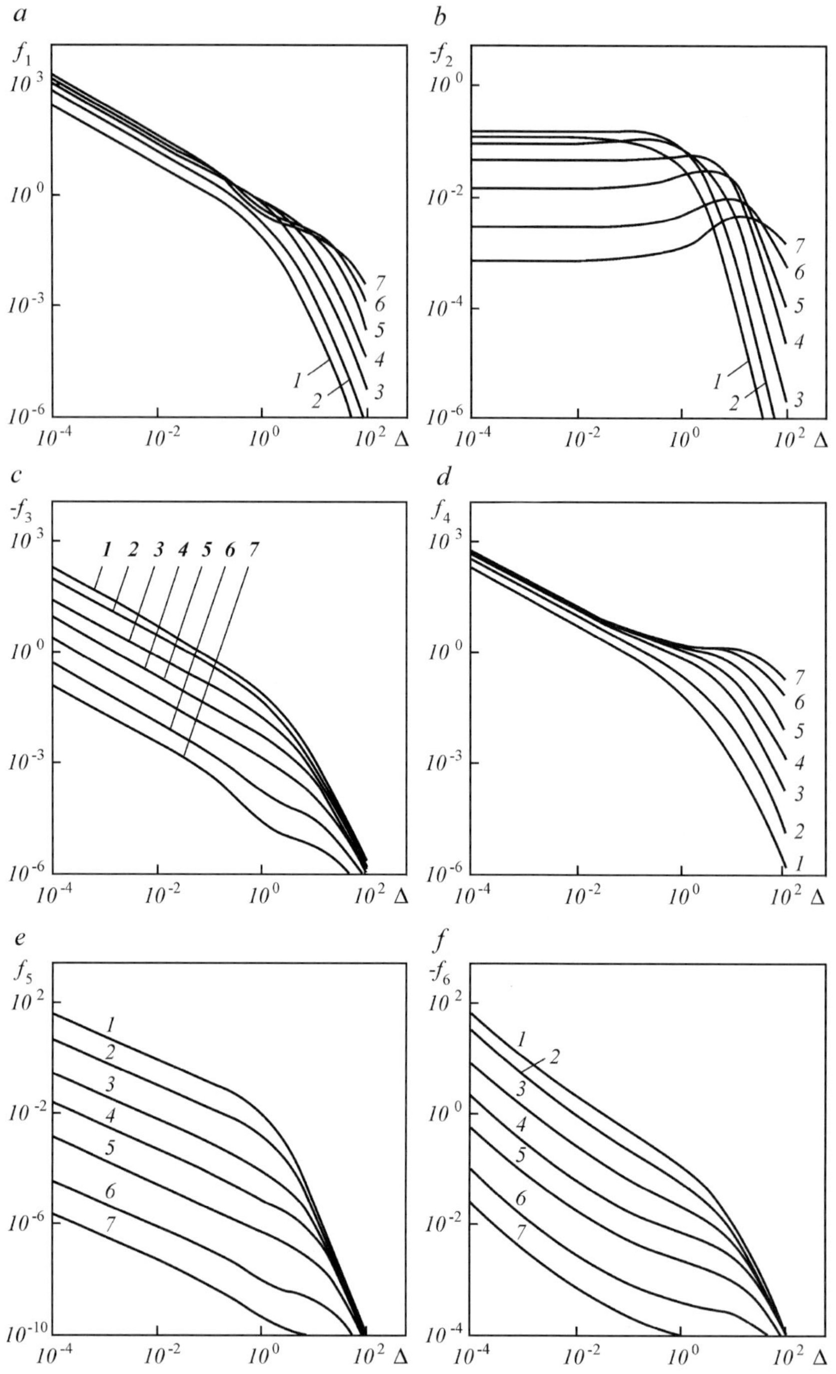

Fig. 12.5 Dependence of factors f_i upon the relative distance Δ between particles at various ratios of the particle radii *k*. *1*: $k = 1$, *2*: 0.5, *3*: 0.2, *4*: 0.1, *5*: 0.05, *6*: 0.02, *7*: 0.01.

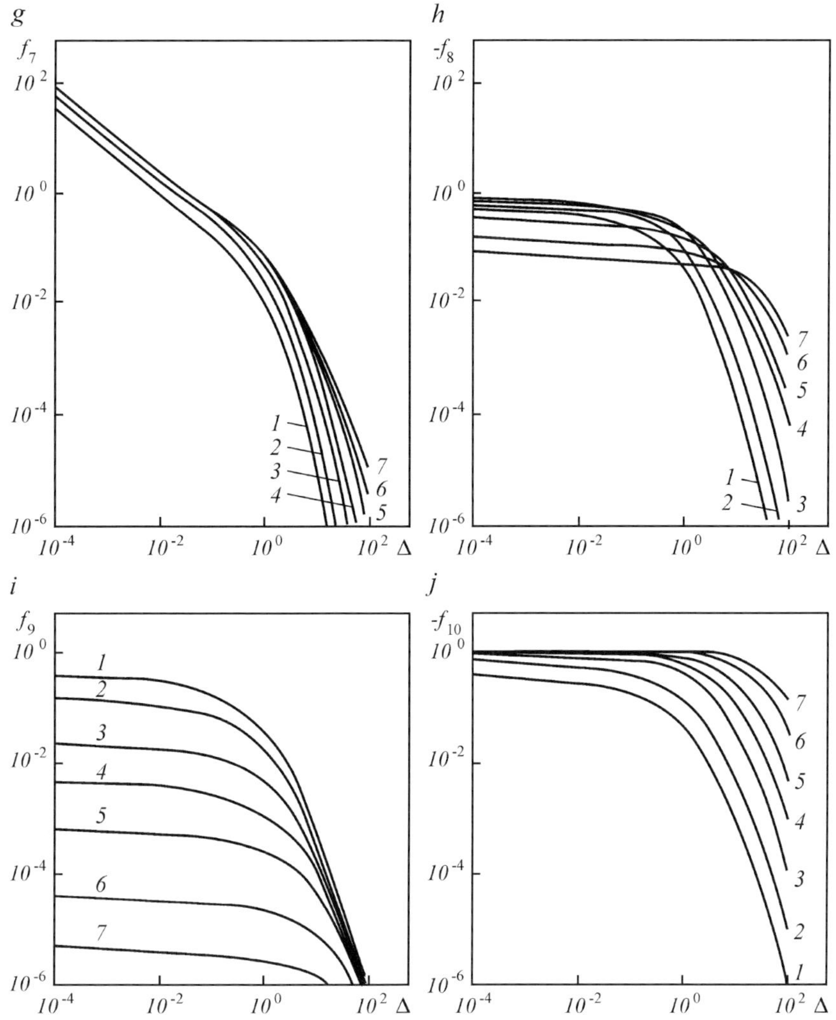

Fig. 12.5 *(continued)*

$$s_9 = \sum_{n=1}^{\infty} c_2(n+1)\frac{a_{2,n}}{a_{3,n}}\left((n+2)\frac{d_{n+1}a_{1,n}}{a_{3,n+1}} - n\frac{d_n a_{1,n+1}}{a_{3,n+1}}\right);$$

$$s_{10} = \sum_{n=1}^{\infty} c_2(n+1)\frac{a_{2,n}}{a_{3,n}}\left((n+2)\frac{d_{n+1}}{a_{3,n+1}} - n\frac{d_n}{a_{3,n+1}}\right);$$

$$a_{1,n} = \exp((2n+1)\mu_1); \quad a_{2,n} = \exp((2n+1)\mu_2);$$

$$a_{3,n} = a_{1,n}a_{2,n} - 1; \quad b_n = (2n+1)(a_{1,n}+1);$$

$$c_1 = 1 + \exp(2\mu_2); \quad c_2 = \frac{a}{R_2} = \sinh\mu_2; \quad d_n = a_{1,n} - 1.$$

Thus, Eqs. (12.46) and (12.47), supplemented by Eqs. (12.25) for μ_1 and μ_2, permit the determination of the interaction forces of two conducting spherical particles of radii R_1 and R_2 separated by a distance between their centers of $r > R_1 + R_2$. Factors f_i are dimensionless quantities dependent on the relative gap between the particle surfaces $\Delta = (r - R_1 - R_2)/R_1$ and the ratio of their radii $k = R_2/R_1$. These dependences are shown graphically in Fig. 12.5. It should be noted that the rate of series convergence quickly decreases with decreasing Δ, and therefore to maintain calculation accuracy at small values of Δ it is necessary to retain an increasing number of series terms.

From the depicted graphs, it can be seen that in the dependence of f_i upon the relative gap between the particles, three areas may be distinguished: the area of small gaps $\Delta < 0.1$; the area of large gaps $\Delta > 3/k$, and the intermediate area $0.1 < \Delta < 3/k$. In the first and third areas, the dependences of f_i on Δ are close to power law. In particular, as $\Delta \to 0$, the following power approximation of factors f_1, f_2, and f_8 applies, these being of the greatest interest since they correspond to interaction forces of non-charged particles:

$$f_i(k) \approx \alpha_i(k)\frac{1}{\Delta\beta_i(k)} \quad (i = 1, 2, 8). \tag{12.48}$$

The values of $\alpha_i(k)$ and $\beta_i(k)$ are listed in Table 12.1.

Along with the general solution of the problem of the definition of interaction forces of two conducting particles, of great interest is the asymptotic behavior of forces at great and small distances between particle surfaces. Interaction of particles at great distances in approximations of dipole–dipole, dipole–coulomb, and coulomb interactions have been investigated in depth [92]. Therefore, the basic question in this area concerns the accuracy of the cited approximations. In the case of small distances between particles, the force of electrostatic interaction has been investigated less thoroughly. In the following sections, two limiting cases are considered in more detail.

Table 12.1

k	α_1	β_1	α_2	β_2	α_8	β_8
0.01	−0.980	0.824	0.0010	−0.0120	−0.048	0.059
0.05	−0.975	0.823	0.0035	−0.0111	−0.091	0.061
0.1	−0.949	0.810	0.0491	−0.0063	−0.300	0.071
0.2	−0.831	0.805	0.1037	−0.0015	−0.381	0.082
0.5	−0.492	0.805	0.1400	0.0111	−0.298	0.133
1.0	−0.214	0.815	0.0915	0.0141	−0.170	0.122

12.4 Interaction Forces Between Two Far-spaced Drops

When solving problems involving the calculation of forces of electrostatic interaction between two conducting particles, approximations of dipole, dipole–coulomb, and coulomb interactions are frequently used. Within the framework of these approximations, particle interactions are considered as interactions of two electric dipoles, a dipole and a point charge, or two point charges. Here, the sizes of the particles define only the values of the dipole moments. Therefore, such approximations may only be used if the distance between the particles is much greater than the sum of their radii. However, in practice, these approximations are frequently used when the gap between the surfaces of the particles is comparable to or less than the size of the particles. Consequently, this raises doubts concerning the accuracy of cited approximations for various distances between particles.

The forces acting on two distant conducting spherical particles are set out in ref. [92]. They can be presented in a form corresponding to Eqs. (12.46) and (12.47) if we write:

$$
\begin{aligned}
f_1^0 &= 6\frac{R_1^3 R_2}{r^4}, \quad f_2^0 = 3\frac{R_1^3 R_2}{r^4}, \quad f_3^0 = 2\frac{R_2^3}{r^3}, \quad f_4^0 = 2\frac{R_1^3}{r^3},\\
f_5^0 &= \frac{R_2^5(2r^2 - R_2^2)}{r^3(r^2 - R_2^2)^2}, \quad f_7^0 = \frac{R_2^5(2r^2 - R_2^2)}{r^3(r^2 - R_2^2)^2},\\
f_6^0 &= \frac{R_2^2}{r^2} + R_1 R_2^3\left(\frac{1}{r^4} + \frac{1}{(r^2 - R_1^2 - R_2^2)^2} - \frac{1}{(r^2 - R_2^2)^2} - \frac{1}{(r^2 - R_1^2)^2}\right),\\
f_8^0 &= 3\frac{R_1^3 R_2}{r^4}, \quad f_9^0 = \frac{R_2^3}{r^3}, \quad f_{10}^0 = \frac{R_1^3}{r^3}.
\end{aligned}
\tag{12.49}
$$

where r is the distance between the centers of the particles.

Compare now the exact values of the factors f_i, given by Eqs. (12.46) and (12.47), with the approximate ones given by Eq. (12.49). For this, we introduce relative deviations:

$$
\sigma_j = \left|\frac{f_j - f_j^0}{f_j}\right|.
$$

Figure 12.6 shows the dependences of σ_j (full line) upon the relative distance between the surfaces Δ and the ratio of the radii of the particles, $k = R_2/R_1$.

Analysis of the results of this comparison shows that for the same values of Δ and k, the accuracy of the approximation f_j^0 to f_j is variable. The most accurate approximations are $f_1^0 - f_3^0$ and $f_8^0 - f_{10}^0$, that is, those factors basically responsible for interaction between non-charged particles in an electric field. For distances

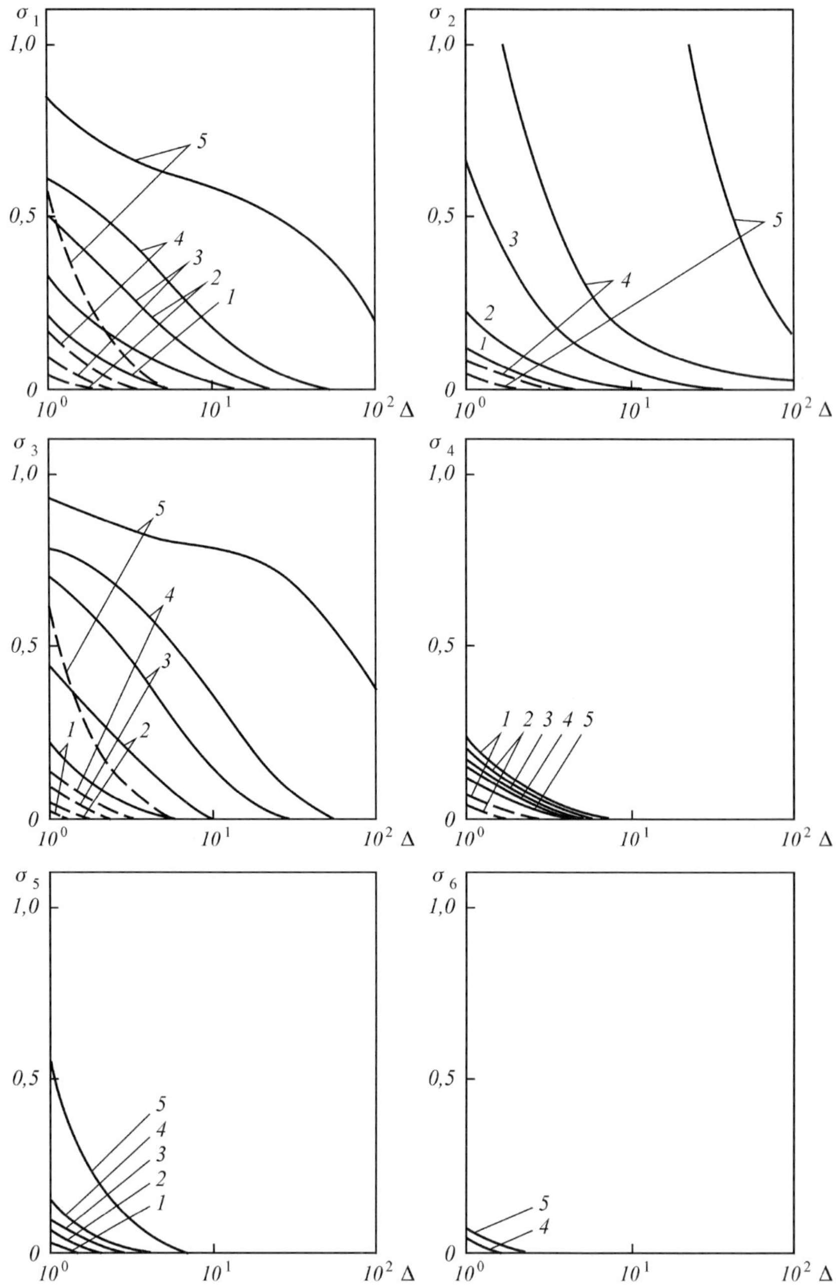

Fig. 12.6 Relative deviations σ_i of factors f_i from f_j^0 (full line) and from f_j^1 (dashed line): *1*: $k = 1$, *2*: 0.5, *3*: 0.2, *4*: 0.1, *5*: 0.05.

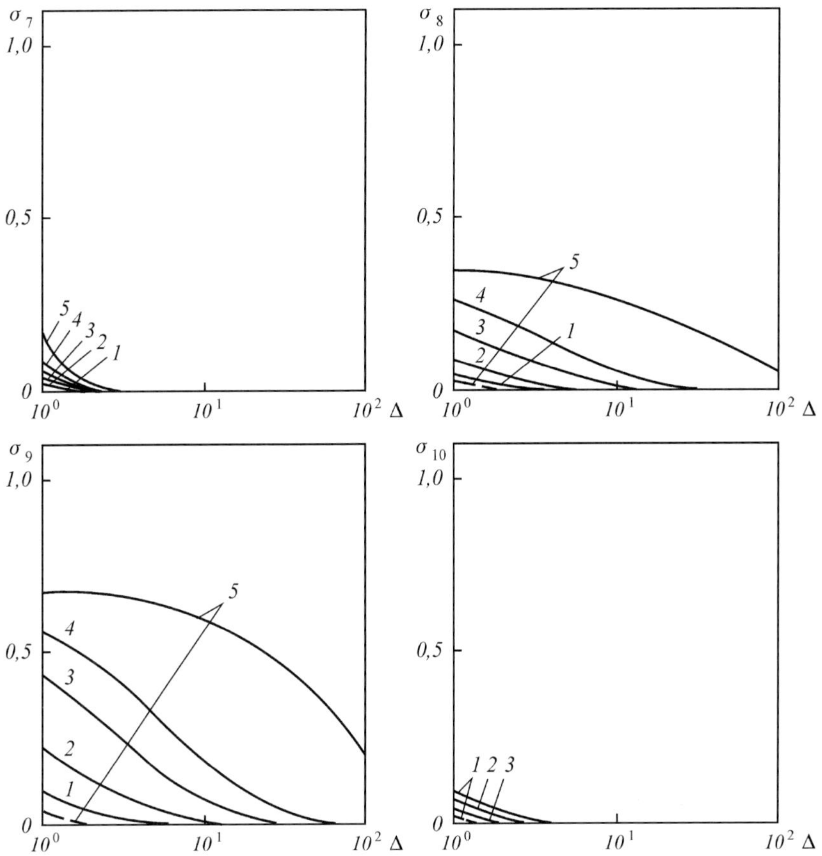

Fig. 12.6 *(continued)*

$\Delta > 3(1/\mathsf{k}+1)$, the error incurred in the definition of factors f_i by the approximate formulae, Eqs. (12.49), is no more than 2%.

To increase the accuracy of the definition of factors f_i in cases when the particles are located relatively far apart, one should take advantage of the method of reflections, which allows one to obtain an approximate solution in the form of a power series, r^{-k}. In this way, terms of higher order are taken into account in Eqs. (12.49). Omitting simple but bulky calculations, accurate to terms of r^{-3}, we obtain:

$$f_1^1 = 6\frac{R_1^3 R_2}{r^4}\left(1+\frac{2R_2^2 r^3}{(r^2-R_1^2)^3}+\frac{2R_1^3 r^3}{(r^2-R_2^2)^3}\right),$$

$$f_2^1 = 3\frac{R_1^3 R_2}{r^4}\left(1-\frac{R_2^3 r^3}{(r^2-R_1^2)^3}-\frac{R_1^3 r^3}{(r^2-R_2^2)^3}\right),$$

$$f_3^1 = 2\frac{R_2^3}{r^3}\left(1+5\frac{R_1^3 r^3}{(r^2-R_2^2)^3}\right), \quad f_4^1 = 2\frac{R_1^3}{r^3}\left(1+5\frac{R_2^3 r^3}{(r^2-R_1^2)^3}\right),$$

$$f_5^1 = f_5^0, \quad f_6^1 = f_6^0, \quad f_7^1 = f_7^0,$$

$$f_8^1 = 3\frac{R_1^3 R_2}{r^4}\left(1+\frac{1}{2}\frac{R_1^3 r^3}{(r^2-R_2^2)^3}+\frac{1}{2}\frac{R_2^3 r^3}{(r^2-R_1^2)^3}\right),$$

$$f_9^1 = \frac{R_2^3}{r^3}\left(1+\frac{2R_1^3 r^3}{(r^2-R_2^2)^3}\right), \quad f_{10}^1 = \frac{R_1^3}{r^3}\left(1+\frac{2R_2^3 r^3}{(r^2-R_1^2)^3}\right). \tag{12.50}$$

Comparison of factors f_j^1 with the exact values f_j shows that for $\Delta \geq 5$ for all k the difference between them does not exceed 2% and for $\Delta \geq 3$ does not exceed 3–5%. In Fig. 12.6, the dashed lines show the dependences of the errors σ_j upon Δ when f_j^1 are taken instead of f_j^0.

12.5 Interaction of Two Touching Drops

Before proceeding to the situation when particles are located close to each other ($\Delta \ll 1$), we consider the interaction of particles in the limiting case of touching particles ($\Delta = 0$).

Consider two touching spherical particles with radii R_1 and R_2 placed in a uniform external electric field of intensity $\boldsymbol{E}_0$ making an angle θ with the line joining the centers of the particles (Fig. 12.7).

The charges on the particles are unknown, but the total charge of both particles is considered to be specified and equal to Q. The particles are conducting, while the space around them is occupied by a homogeneous isotropic dielectric with dielectric permittivity ε. The primary goal here is definition of the forces of interaction between the particles and of their charges.

Equations and boundary conditions for the electric field potential in the area around the particles have been formulated in Section 12.1, and look like:

$$\Delta\phi = 0, \tag{12.51}$$

$$\phi_{S_1} = \phi_{S_2} = V, \quad \nabla\phi \to -\boldsymbol{E}_0 \quad \text{at } \boldsymbol{r} \to \infty, \tag{12.52}$$

$$-\frac{\varepsilon}{4\pi}\int_{S_1+S_2}\frac{\partial\phi}{\partial n}ds = Q, \tag{12.53}$$

where S_i are the surface areas of the particles and V is the potential of the particles to be determined. The difference between the problem represented by Eqs. (12.51)–(12.53) and the similar problem considered in Section 12.1 lies in the fact that here the potentials of the two particles are equal and are given not as charges on the particles, but as their total charge since the particles are touching each

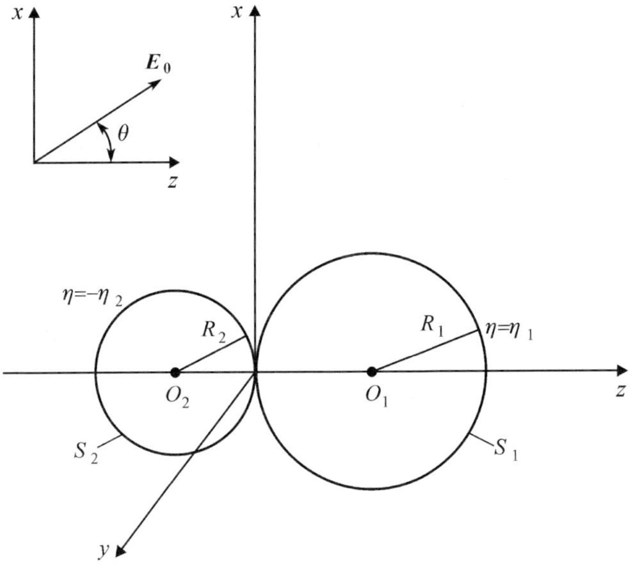

Fig. 12.7 Two touching spherical particles.

other. The latter condition means that we cannot use a bispherical system of coordinates, Eq. (12.6), but rather we need a confluent bispherical system of coordinates (ξ, η, Φ) (Fig. 12.8), derived from the bispherical system (ξ', η', Φ) by the limiting transition $\xi' \to 0$ and $\eta' \to 0$ [93].

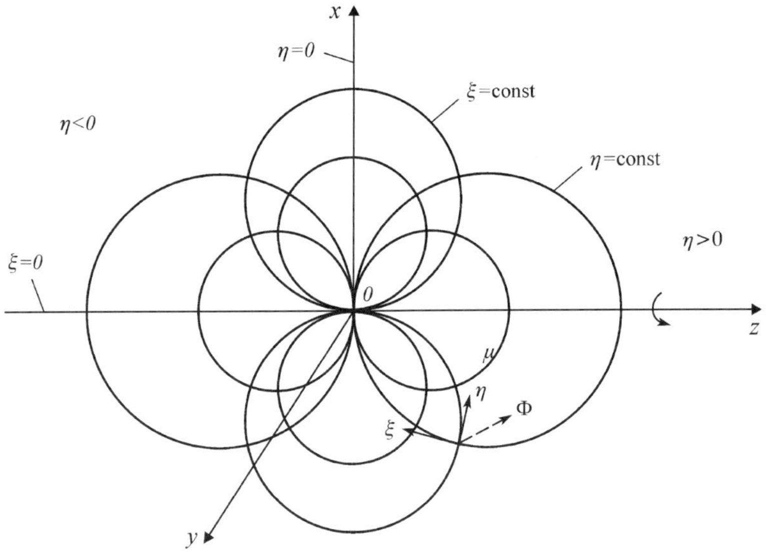

Fig. 12.8 Confluent bispherical system of coordinates (ξ, η, Φ).

These confluent bispherical coordinates are related to Cartesian coordinates as follows:

$$x = \frac{a\xi \cos \Phi}{\eta^2 + \xi^2}; \quad y = \frac{a\xi \sin \Phi}{\eta^2 + \xi^2}; \quad z = \frac{a\eta}{\eta^2 + \xi^2}; \tag{12.54}$$

where $0 \leq \xi < \infty$, $-\infty < \eta < \infty$, $0 \leq \Phi \leq 2\pi$, and the parameter specifying linear size is:

$$a = 2R_1\eta_1 = 2R_2\eta_2. \tag{12.55}$$

Equations for touching spheres in the confluent bispherical system of coordinates are $\eta = \eta_1$ and $\eta = \eta_2$. We also need expressions for the surface element ds, and projections of the external normal to this element on axes of Cartesian coordinates:

$$ds = \frac{a^2\xi \, d\xi \, d\Phi}{\eta^2 + \xi^2}, \tag{12.56}$$

$$n_x = \frac{2\xi\eta \cos \Phi}{\eta^2 + \xi^2}, \quad n_y = \frac{2\xi\eta \sin \Phi}{\eta^2 + \xi^2}, \quad n_z = \frac{\xi^2 - \eta^2}{\eta^2 + \xi^2}, \tag{12.57}$$

and the derivative with respect to the external normal to the sphere:

$$\frac{\partial}{\partial n} = -\frac{1}{a}(\xi^2 + \eta^2)\frac{\partial}{\partial \eta}. \tag{12.58}$$

The general solution of the Laplace equation, Eq. (12.51), in the degenerate bispherical system of coordinates, limited at $\xi \to 0$, takes the form:

$$\phi_{m,t} = (\xi^2 + \eta^2)^{1/2} \exp(\pm t\eta) J_m(t\xi) \begin{bmatrix} \cos m\Phi \\ \sin n\Phi \end{bmatrix}, \tag{12.59}$$

where $J_m(t\xi)$ is a Bessel function of the mth order.

Since the Cartesian system of coordinates is chosen so that the vector $\boldsymbol{E}_0$ is in the plane XOZ, the potential of the uniform external field is given by:

$$\phi_0 = -E_0(x \sin \theta + z \cos \theta). \tag{12.60}$$

In presenting φ_0 as an expansion in terms of Bessel functions, one should take advantage of the known equality [5]:

$$(\xi^2 + \eta^2)^{-1/2} = \int_0^\infty \exp(-t|\eta|) J_0(t\xi) \, dt. \tag{12.61}$$

Differentiating Eq. (12.61) with respect to η and separately with respect to ξ and comparing the obtained expressions with Eq. (12.54), we obtain:

$$z = a(\xi^2+\eta^2)^{1/2}\frac{\eta}{|\eta|}\int_0^\infty t\exp(-t|\eta|)J_0(t\xi)\,dt.$$

$$x = a(\xi^2+\eta^2)^{1/2}\cos\Phi\int_0^\infty t\exp(-t|\eta|)J_1(t\xi)\,dt.$$

The obtained expressions us allow to write a general solution of the Laplace equation limited at $\xi \to \infty$ and satisfying the second condition, Eq. (12.52):

$$\phi(\xi,\eta,\Phi) = (\xi^2+\eta^2)^{1/2}\int_0^\infty \Bigg(\Bigg(M(t)\exp(t\eta)+N(t)\exp(-t\eta) - aE_0 t\exp(-t|\eta|)\cos\theta\frac{\eta}{|\eta|}\Bigg)J_0(t\xi)+(K(t)\exp(t\eta) + L(t)\exp(-t\eta) - aE_0 t\exp(-t|\eta|)\sin\theta)J_1(t\xi)\cos\Phi\Bigg)\,dt. \quad (12.62)$$

To determine the unknown coefficients M, N, K, and L, one should substitute Eq. (12.62) into the first condition, Eq. (12.52), and then use Eq. (12.61), taking into account the fact that V depends only on η. As a result, a system of equations is obtained, having the solution:

$$M(t) = \frac{\exp(-t\eta_1)(V\sinh(t\eta_2)+aE_0\cosh(t\eta_2)\cos\theta)}{\sinh(t(\eta_1+\eta_2))},$$

$$N(t) = \frac{\exp(-t\eta_1)(V\sinh(t\eta_1)-aE_0\cosh(t\eta_1)\cos\theta)}{\sinh(t(\eta_1+\eta_2))}, \quad (12.63)$$

$$K(t) = \frac{aE_0 t\exp(-t\eta_1)\sinh(t\eta_2)\sin\theta}{\sinh(t(\eta_1+\eta_2))},\quad L(t) = \frac{aE_0 t\exp(-t\eta_1)\sinh(t\eta_2)\sin\theta}{\sinh(t(\eta_1+\eta_2))}.$$

The coefficients thus obtained allow us to determine the potential V of the particles. From the expression of the general solution, Eq. (12.62), and the first condition, Eq. (12.52), it follows that:

$$\frac{V}{(\xi^2+\eta^2)^{1/2}} = \int_0^\infty \Bigg(\Bigg(M(t)\exp(t\eta)+N(t)\exp(-t\eta) - aE_0 t\exp(-t|\eta|)\cos\theta\frac{\eta}{|\eta|}\Bigg)J_0(t\xi)+(K(t)\exp(t\eta) + L(t)\exp(-t\eta) - aE_0 t\exp(-t|\eta|)\sin\theta)J_1(t\xi)\cos\Phi\Bigg)\,dt. \quad (12.64)$$

We may then express the left-hand side of Eq. (12.64) as an integral of Bessel functions with the aid of Eq. (12.61). One thereby obtains:

$$\int_0^\infty \Bigg(V \exp(-t|\eta|) J_0(t\xi) - \Bigg(M(t) \exp(t\eta) + N(t) \exp(-t\eta)$$

$$- aE_0 t \exp(-t|\eta|) \cos\theta \frac{\eta}{|\eta|} \Bigg) J_0(t\xi) + (K(t) \exp(t\eta) + L(t) \exp(-t\eta)$$

$$- aE_0 t \exp(-t|\eta|) \sin\theta) J_1(t\xi) \cos\Phi \Bigg) dt = 0. \tag{12.65}$$

Equation (12.65) is obeyed for all ξ and Φ, and therefore the integrand should be equal to zero. Multiplying the integrand in Eq. (12.65) by t, and then integrating the first term containing V, we obtain:

$$\frac{V}{(\xi^2 + \eta^2)^{1/2}} = \int_0^\infty t \Bigg(\Bigg(M(t) \exp(t\eta) + N(t) \exp(-t\eta)$$

$$- aE_0 t \exp(-t|\eta|) \cos\theta \frac{\eta}{|\eta|} \Bigg) J_0(t\xi) + (K(t) \exp(t\eta)$$

$$+ L(t) \exp(-t\eta) - aE_0 t \exp(-t|\eta|) \sin\theta) J_1(t\xi) \cos\Phi \Bigg) dt. \tag{12.66}$$

Taking from Eq. (12.62) the derivative with respect to the normal according to Eq. (12.58), and taking into account the relationship of Eq. (12.66), we obtain:

$$\left(\frac{\partial\phi}{\partial n}\right)_{\eta=\eta_1} = -2\frac{(\xi^2+\eta^2)^{3/2}}{a} \int_0^\infty t \exp(t\eta_1)(J_0(t\xi)(M(t) + J_1(t\xi)K(t) \cos\Phi)\, dt,$$

$$\left(\frac{\partial\phi}{\partial n}\right)_{\eta=-\eta_2} = -2\frac{(\xi^2+\eta^2)^{3/2}}{a} \int_0^\infty t \exp(t\eta_2)(J_0(t\xi)(N(t) + J_1(t\xi)L(t) \cos\Phi)\, dt. \tag{12.67}$$

Substitution of Eq. (12.67) and Eq. (12.56) into the condition according to Eq. (12.53) and integration of the obtained expression with respect to Φ between limits of 0 and 2π and with respect to ξ between limits of 0 and ∞ gives:

$$\int_0^\infty (M(t) + N(t))\, dt = \frac{Q}{a\varepsilon}. \tag{12.68}$$

Substituting Eqs. (12.63) for $M(t)$ and $N(t)$ into Eq. (12.68) and integrating the resulting expression, one obtains:

$$V = E_0 a_1 R \cos\theta + \frac{Qa_2}{\varepsilon R_2}, \tag{12.69}$$

where

$$a_1 = \frac{\psi'(\alpha) - \psi'(1-\alpha)}{2\psi(1) - \psi(\alpha) - \psi(1-\alpha)}; \quad a_1 = \frac{1}{2\psi(1) - \psi(\alpha) - \psi(1-\alpha)};$$

$$R = \frac{R_1 R_2}{R_1 + R_2}; \quad \alpha = \frac{\eta_2}{\eta_1 + \eta_2} = \frac{R_1}{R_1 + R_2}.$$

Here, $\psi(x)$ is the Euler psi-function.

Thus, Eqs. (12.62), (12.63), and (12.69) give the distribution of the electric field potential in the space around the touching particles. Next, we turn to the definition of the charges q_1 and q_2 and the interaction force between the particles.

The total charge on the particles, Q, is considered to be known:

$$q_1 + q_2 = Q. \tag{12.70}$$

The forces acting on the first and second particles are connected by:

$$\tilde{\boldsymbol{F}}_1 + \tilde{\boldsymbol{F}}_2 = \boldsymbol{E}_0 Q. \tag{12.71}$$

Equations (12.70) and (12.71) connecting the charges and forces permit us to limit the definition of these quantities to just one of the particles, for example, particle 2.

Thus, we express the charge on particle 2 through its surface density:

$$q_2 = -\frac{\varepsilon}{4\pi} \int_{S_2} \frac{\partial \phi}{\partial n} ds. \tag{12.72}$$

Substituting in this expression the normal derivative from Eq. (12.67) and Eq. (12.56) for ds, we obtain:

$$q_2 = a\varepsilon \int_0^{\infty} N(t)\, dt. \tag{12.73}$$

Taking advantage of the expression for $N(t)$ in Eq. (12.63) and integrating the obtained expression, one obtains:

$$q_2 = -\varepsilon E_0 R^2 A_1 \cos\theta + A_2 Q, \tag{12.74}$$

where

$$A_1 = \frac{(\psi(1-\alpha) + C)\psi'(\alpha) + (\psi(\alpha) + C)\psi'(1-\alpha)}{2C + \psi(\alpha) + \psi(1-\alpha)} + \frac{\pi^2}{6};$$

$$A_2 = \frac{\psi(\alpha) + C}{2C + \psi(\alpha) + \psi(1-\alpha)}; \quad \alpha = \frac{R_1}{R_1 + R_2};$$

where $C = 0.5772\ldots$ is the Euler constant.

If the external electric field is directed along the center line, that is $\theta = 0$, and if moreover the particles are of identical size ($R_1 = R_2$) and their total charge is equal to zero ($Q = 0$), which corresponds to a particle charging at a flat condenser, from Eq. (12.74) it follows that:

$$A_1 = \frac{2}{3}\pi^2; \quad q_2 = -\varepsilon E_0 R_2^2 \frac{\pi^2}{6}. \tag{12.75}$$

At $\theta = 0$, $Q = 0$, and $R_2 \ll R_1$, we have:

$$A_1 = \frac{\pi^2}{2}; \quad q_2 = -\varepsilon E_0 R_2^2 \frac{\pi^2}{2}. \tag{12.76}$$

From Eqs. (12.75) and (12.76), it follows that the charge acquired by a small particle on contact with a large one is three times greater than the charge acquired on contact with a particle of identical size. This may be rationalized by the fact that the maximum strength of the electric field near the smaller conducting particle is three times higher than the strength of the uniform external field. For particles of identical size, $R_1 = R_2$, the charge is distributed evenly between them. Note that in the case of two far-spaced particles having equal surface potentials, the charge on particle 2 is given by:

$$\frac{q_2}{Q} = \frac{R_2}{R_1 + R_2} \tag{12.77}$$

and in the case $R_2 \ll R_1$ we have $q_2 \sim QR_2/R_1 \ll q_1$.

Factors A_1 and A_2 are functions of $\mathsf{k} = R_2/R_1$ and have properties:

$$A_1(1/\mathsf{k}) = A_1(\mathsf{k}); \quad A_2(1/\mathsf{k}) = 1 - A_2(\mathsf{k}).$$

This means that if A_1 and A_2 are known for values $\mathsf{k} \le 1$, then it is possible to find these factors for $\mathsf{k} > 1$.

Figure 12.9 shows the dependences of factors A_1 and A_2 on k for $\mathsf{k} \le 1$. The dashed lines show asymptotes of the dependences $A_1(\mathsf{k})$ and $A_2(\mathsf{k})$ as $\mathsf{k} \to 0$. The plotted points show the relevant experimental values obtained in ref. [94]. Experiments were carried out on steel balls and water drops in petroleum. In both cases, good agreement between theory and experiment was obtained.

To estimate particle charges, the following approximation for numerical results at $\mathsf{k} \le 1$ may be used:

$$A_1 = -1.64(1-k)^2(3 - 3.5k(1 - 0.24k)^2);$$

$$A_2 = 1.6k^2\left(1 - \frac{0.96k}{1 + 0.4k^2}\right).$$

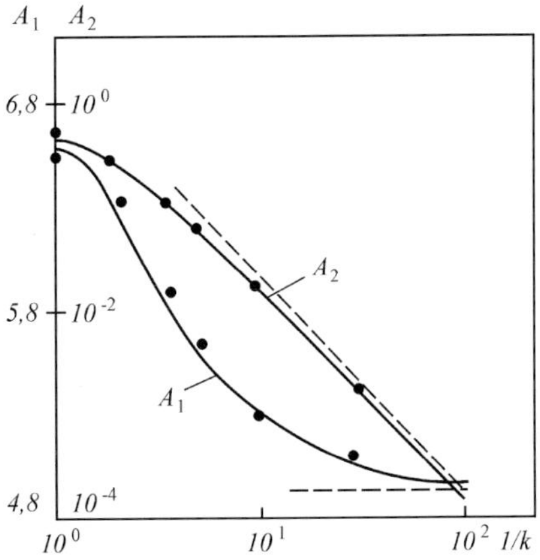

Fig. 12.9 Dependence of factors A_i upon k: dashed line: asymptotic values as $k \to 0$; points: experimental values.

To determine the force $\tilde{F}_2$ acting on particle 2, we utilize Eq. (12.44). Using Eqs. (12.56), (12.57), and (12.67), one obtains:

$$\tilde{F}_{2z} = \varepsilon \int_0^\infty \xi(\xi^2 - \eta_2^2)\left(\left(\int_0^\infty tN_1(t)J_0(t\xi)\,dt\right)^2 + \frac{1}{2}\left(\int_0^\infty t(L_1(t)J_1(t\xi)\,dt\right)^2\right)d\xi;$$

$$\tilde{F}_{2x} = -2\varepsilon \int_0^\infty \xi^2\eta_2\left(\int_0^\infty tN_1(t)J_0(t\xi)\,dt\right)\left(\int_0^\infty t(L_1(t)J_1(t\xi)\,dt\right)d\xi, \qquad (12.78)$$

where

$$N_1(t) = N(t)\exp(t\eta_2); \quad L_1(t) = L(t)\exp(t\eta_2);$$

To calculate the integrals in Eq. (12.78), it is necessary to take advantage of an equation [95] that stems from the Parcevale theorem for Henkel integral transformation:

$$\int_0^\infty \xi\varphi(t)J_{n-1}(t\xi)\,d\xi = \varphi(t)J_n(t\xi)|_0^\infty - \int_0^\infty t^n J_n(t\xi)d\left(\frac{\varphi(t)}{t^n}\right). \qquad (12.79)$$

Integrating Eq. (12.78) with respect to ξ taking into account Eq. (12.79), one obtains:

$$\tilde{F}_{2z} = -\varepsilon\eta_2^2\left(\int_0^\infty t\left(N_1^2(t) + \frac{1}{2}L_1^2(t)\right)dt\right)$$
$$+\varepsilon\left(\int_0^\infty t\left(\left(\frac{dN_1}{dt}\right)^2 + \frac{1}{2}t\left(\frac{d}{dt}\left(\frac{L_1}{t}\right)\right)^2\right)dt\right); \qquad (12.80)$$
$$\tilde{F}_{2x} = -2\varepsilon\eta_2 \int_0^\infty tL_1(t)\left(\frac{dN_1}{dt}\right)^2 dt.$$

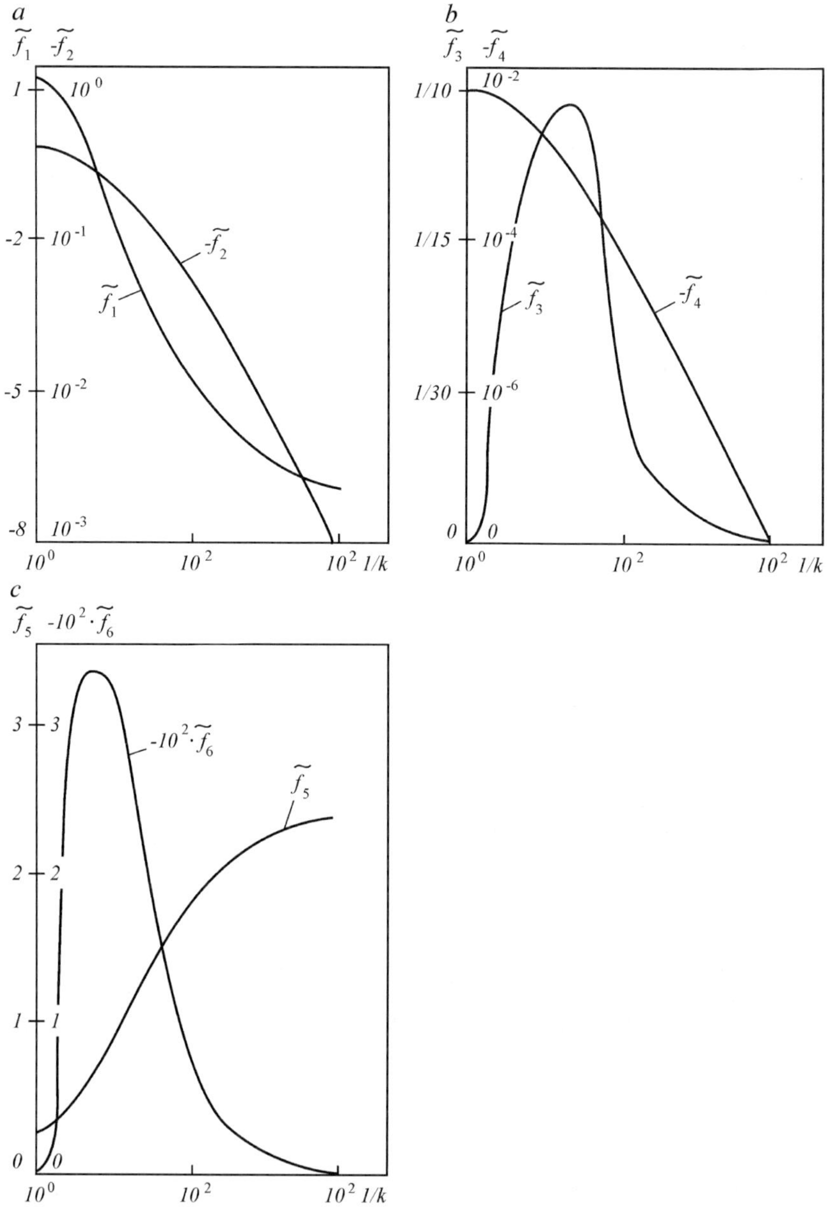

Fig. 12.10 Dependences of coefficients $\tilde{f}_i$ upon k.

Substituting $N_1(t)$ and $L_1(t)$ in Eq. (12.80), one eventually obtains:

$$\tilde{F}_{2z} = \varepsilon E_0 R^2 (\tilde{f}_1 \cos^2\theta + \tilde{f}_2 \sin^2\theta) + QE_0 \tilde{f}_3 \cos\theta + \frac{Q^2}{\varepsilon R^2} \tilde{f}_4$$
$$+ q_2 E_0 \cos\theta, \tag{12.81}$$
$$\tilde{F}_{2x} = \varepsilon E_0 R^2 \tilde{f}_5 \sin(2\theta) + QE_0 \tilde{f}_6 \sin\theta + q_2 E_0 \sin\theta,$$

which contains the following dimensionless factors, $\tilde{f}_j$, dependent on the ratio of the radii of particles ***k***:

$$\tilde{f}_1 = j_1 + A_1; \quad \tilde{f}_2 = j_2; \quad \tilde{f}_3 = j_3 + A_2;$$
$$\tilde{f}_4 = j_4; \quad \tilde{f}_5 = j_5 + A_1/2; \quad \tilde{f}_6 = j_6 + A_2;$$
$$j_1 = \int_0^\infty (c_3^2 - c_0^2) t\, dt; \quad j_2 = \int_0^\infty (c_1^2 - b_1^2) t^3\, dt;$$
$$j_3 = 2a_2 \int_0^\infty (c_1 c_3 - b_1 c_0) t\, dt; \quad j_4 = \int_0^\infty (c_1^2 - b_1^2) t\, dt;$$
$$j_5 = -2 \int_0^\infty b_1 c_3 t^2\, dt; \quad j_6 = -4a_2 \int_0^\infty b_1 c_1 t^2\, dt;$$
$$c_0 = \alpha(a_1 b_1 - 2t b_2); \quad c_1 = \beta b_1 - b_2 \cosh(t);$$
$$c_2 = b_2 + \beta t b_1 - t b_2 \coth(t); \quad c_3 = a_1 c_1 - 2c_2;$$
$$b_1 = \frac{\sinh(\beta t)}{\sinh(t)}; \quad b_2 = \frac{\cosh(\beta t)}{\cosh(t)}; \quad \alpha = \frac{R_1}{R_1 + R_2}; \quad \beta = 1 - \alpha.$$

Parameters a_1, a_2, A_1, and A_2 are determined according to Eqs. (12.69) and (12.74).

The dependences of these factors on the ratio of the radii of the particles are shown in Fig. 12.10. As one would expect, touching conducting particles are repelled from each other irrespective of the orientation of the pair relative to the external field.

12.6 Interaction Forces Between Two Closely Spaced Drops

Consider now the case when the gap between particles is small in comparison with particle size, that is $\Delta \ll 1$. Since the forces acting on the particles are connected by Eq. (12.43), the problem may be limited to the definition of force $\boldsymbol{F}_2$. Taking advantage of Eq. (12.44), and dividing the surface of the particle, $S2$, into two parts, S^* and $S_2 - S^*$, where S^* is that part of the surface located close to the line joining the centers of the particles, we obtain:

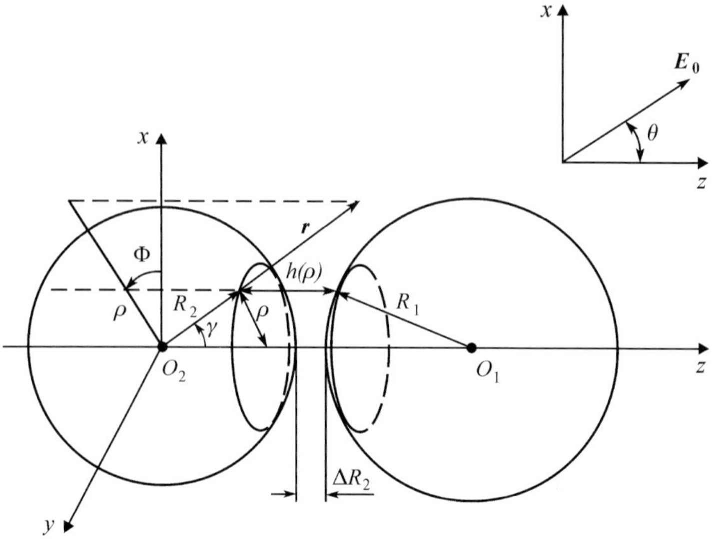

Fig. 12.11 Cartesian (x, y, z), cylindrical (ρ, Φ, z), and spherical (r, γ, Φ) systems of coordinates relating to a smaller particle.

$$\boldsymbol{F}_2 \cdot \boldsymbol{p} = \frac{\varepsilon}{8\pi}\int_{S^*} E^2 \boldsymbol{n} \cdot \boldsymbol{p}\, ds + \frac{\varepsilon}{8\pi}\int_{S_2 - S^*} E^2 \boldsymbol{n} \cdot \boldsymbol{p}\, ds. \tag{12.82}$$

The rationale for such a division of particle area lies in the fact that the strength of the electric field in the gap at $\Delta \to 0$ increases indefinitely along the line joining the centers, but remains finite in the other part of the gap area.

We now estimate the electric field strength in the gap near the line joining the centers. For this, we introduce a cylindrical system of coordinates (ρ, Φ, z) (Fig. 12.11), in which the equation $\nabla \cdot \boldsymbol{E} = 0$ becomes:

$$\frac{1}{\rho}\frac{\partial}{\partial \rho}(\rho E_\rho) + \frac{1}{\rho}\frac{\partial E_\Phi}{\partial \Phi} + \frac{\partial E_z}{\partial z} = 0. \tag{12.83}$$

The area of greatest distortion of the electric field is located close to the surface of the smaller particle 2, the vector $\boldsymbol{E}$ in this area deviating slightly from the external normal $\boldsymbol{n}$ to the surface S_2. Therefore, in the considered area, relationships similar to Eq. (12.31) are valid:

$$\begin{aligned} &E_\rho = E_n \frac{\rho}{R_2} + 0\left(\frac{\rho^2}{R_2^2}\right); \quad E_z = E_n\left(1 - \frac{\rho^2}{2R_2^2}\right) + 0\left(\frac{\rho^2}{R_2^2}\right); \\ &E_\Phi = 0\left(\frac{\rho^2}{R_2^2}\right); \quad \frac{\rho}{R^2} \ll 1. \end{aligned} \tag{12.84}$$

From Eq. (12.84), it follows that:

$$E_\rho = E_z \frac{\rho}{R_2} + \left(1 + 0\left(\frac{\rho}{R_2}\right)\right);$$
$$E_\Phi = E_z 0\left(\frac{\rho^2}{R_2^2}\right); \quad \frac{\partial E_\Phi}{\partial \Phi} \approx E_z 0\left(\frac{\rho^2}{R_2^2}\right). \tag{12.85}$$

Substituting Eq. (12.85) in Eq. (12.83), we obtain:

$$\frac{\partial E_z}{\partial z} = -\frac{1}{\rho}\frac{\partial}{\partial \rho}\left(E_z \rho\left(\frac{\rho}{R_2} + 0\left(\frac{\rho}{R_2}\right)\right)\right). \tag{12.86}$$

Integrating Eq. (12.86) with respect to z, one obtains:

$$E_z(\rho,\Phi,z) - E_z(\rho,\Phi,z_2)$$
$$= \frac{1}{\rho}\frac{\partial}{\partial \rho}\left((\phi(\rho,\Phi,z) - \phi(\rho,\Phi,z_2))\frac{\rho^2}{R_2^2}\left(1 + 0\left(\frac{\rho}{R_2}\right)\right)\right).$$

Differentiating the right-hand side of the obtained equation and taking into account that:

$$\frac{\partial \phi}{\partial \rho} = -E_\rho = -E_z \frac{\rho}{R_2}\left(1 + 0\left(\frac{\rho}{R_2}\right)\right),$$

one arrives at the expression:

$$E_z(\rho,\Phi,z) - E_z(\rho,\Phi,z_2) = -(E_z(\rho,\Phi,z) - E_z(\rho,\Phi,z_2))\frac{\rho}{R_2}\left(1 + 0\left(\frac{\rho}{R_2}\right)\right)$$
$$+ \frac{2}{R_2}(\phi(\rho,\Phi,z) - \phi(\rho,\Phi,z_2))\left(1 + 0\left(\frac{\rho}{R_2}\right)\right).$$

This equation allows the estimation that:

$$|E_z(\rho,\Phi,z) - E_z(\rho,\Phi,z_2)| \le \frac{2}{R_2}(\phi(\rho,\Phi,z) - \phi(\rho,\Phi,z_2))\left(1 + 0\left(\frac{\rho}{R_2}\right)\right)$$
$$\le \frac{2}{R_2}|V_1 - V_2|; \quad \Delta \ll 1, \; \frac{\rho}{R_2} \le 1. \tag{12.87}$$

Here, it is taken into account that the electric field potential in the gap between the particles reaches maximum and minimum values at the surfaces of the particles, and that $V_2 \ge \phi(\rho,\Phi,z) \ge V_1$ if $V_2 > V_1$, and $V_1 \ge \phi(\rho,\Phi,z) \ge V_2$ if $V_1 > V_2$.

Upon the approach of particles, that is, as $\Delta \to 0$, $|V_1 > V_2| \to 0$ according to $1/\ln \Delta$. Therefore, the deviation of the electric field strength from the average value in the gap between the particles $\langle E_z(\rho, \Phi)\rangle$ at $\rho = \text{const.}$ also tends to zero as $\Delta \to 0$ with $\rho/R_2 \ll 1$. Hence, in order to estimate the electric field strength close to S_2, it is possible to utilize its average value over z:

$$E_z(\rho, \Phi, z_2) = \langle E_z(\rho, \Phi)\rangle \left(1 + 0\left(\frac{\rho}{R_2}\right)\right)$$
$$= \frac{1}{h(\rho)} \int_{z_2}^{z_1} E_z(\rho, \Phi, z)\left(1 + 0\left(\frac{\rho}{R_2}\right)\right) dz = \frac{V_2 - V_1}{h(\rho)}\left(1 + 0\left(\frac{\rho}{R_2}\right)\right),$$
$$h(\rho) = \left(\Delta + \left(\frac{\rho}{R_2}\right)^2 \frac{\mathrm{k}+1}{2}\right) R_2 + R_2 0\frac{\rho^2}{R_2^2}. \tag{12.88}$$

Consider next the definition of the force acting on particle 2. For this purpose, we introduce a spherical system of coordinates (r, γ, Φ) (see Fig. 12.11). In this system of coordinates, Eq. (12.44) for the projection of the force in the direction of a unit vector $\boldsymbol{p}$ in view of the relationship represented by Eq. (12.88) at $\Delta \ll \gamma^* \ll 1$ can be rewritten as:

$$\boldsymbol{F}\cdot\boldsymbol{p} = \iint_{\substack{\gamma<\gamma^* \\ 0\le\Phi<2\pi}} \left(\frac{V_2 - V_1}{h(\rho)}\right)^2 \boldsymbol{n}\cdot\boldsymbol{p}\, ds + \iint_{\substack{\gamma>\gamma^* \\ 0\le\Phi<2\pi}} E^2 \boldsymbol{n}\cdot\boldsymbol{p}\, ds + 0\left(\frac{\rho}{R_2}\right). \tag{12.89}$$

Choosing the unit individual vector first in the direction of the OZ axis ($\mathbf{p} = \mathbf{e}_1$), and then in the direction of the OX axis ($\mathbf{p} = \mathbf{e}_2$), going from $h(p)$ to $h(\gamma, R_2)$, and taking into account that at $\gamma < \gamma^* \ll 1$:

$$n_z = \cos\gamma = 1 + 0(\gamma^2), \quad n_x = \sin\gamma\cos\Phi = \gamma\cos\Phi + 0(\gamma^3),$$
$$ds = \frac{\rho^2}{R_2^2}\sin\gamma\, d\gamma\, d\Phi = \frac{\rho^2}{R_2^2}\gamma\, d\gamma\, d\Phi + 0(\gamma^2),$$

from Eq. (12.89) we obtain:

$$F_{2z} = \frac{\varepsilon}{8\pi}\int_0^{2\pi} d\Phi \int_0^{\gamma^*} \frac{(V_1 - V_2)^2}{(\Delta + \gamma^2(\mathrm{k}+1)/2)^2}\gamma^2\, d\gamma + \iint_{\substack{\gamma>\gamma^* \\ 0\le\Phi<2\pi}} E^2 n_z\, ds + 0(\gamma^*),$$
$$F_{2x} = \frac{\varepsilon}{8\pi}\int_0^{2\pi} \cos\Phi\, d\Phi \int_0^{\gamma^*} \frac{(V_1 - V_2)^2}{(\Delta + \gamma^2(\mathrm{k}+1)/2)^2}\gamma^2\, d\gamma$$
$$+ \iint_{\substack{\gamma>\gamma^* \\ 0\le\Phi<2\pi}} E^2 n_x\, ds + 0(\gamma^*). \tag{12.90}$$

Calculation of the first integrals in Eq. (12.90) gives:

$$F_{2z} = \frac{\varepsilon(V_1 - V_2)^2}{4\Delta(\mathsf{k}+1)} - \frac{\varepsilon(V_1 - V_2)^2}{2(\mathsf{k}+1)(2\Delta + (\gamma^*)^2(\mathsf{k}+1))} + \iint_{\substack{\gamma > \gamma^* \\ 0 \le \Phi < 2\pi}} E^2 n_z \, ds + 0(\gamma^*), \tag{12.91}$$

$$F_{2x} = \iint_{\substack{\gamma > \gamma^* \\ 0 \le \Phi < 2\pi}} E^2 n_x \, ds + 0(\gamma^*).$$

The first term in Eq. (12.91) for F_{2z} is the principal member of an asymptotic expansion as $\Delta \to 0$ and $V_1 - V_2 \neq 0$; the second term tends to zero as $1/(\ln \Delta)^2$ as $\Delta \to 0$, and the third term tends to a constant. In the limit as $\Delta \to 0$, the spheres become touching, as considered in the previous section. Since for touching spheres $V_1 = V_2$ and $F_{iz} = \tilde{F}_{iz}$, in the limit as $\Delta \to 0$ we have:

$$\iint_{\substack{\gamma > \gamma^* \\ 0 \le \Phi < 2\pi}} E^2 n_z \, ds + 0(\gamma^*) = \tilde{F}_{2z} + 0(\Delta).$$

Returning now to Eq. (12.91), one obtains:

$$F_{2z} = \frac{\varepsilon(V_1 - V_2)^2}{4\Delta(\mathsf{k}+1)} + \tilde{F}_{2z} + 0(\Delta); \quad \Delta \ll 1. \tag{12.92}$$

Similarly, for F_{2x} the following asymptotic relationship can be obtained:

$$F_{2x} = \tilde{F}_{2x} + 0(\Delta); \quad \Delta \ll 1. \tag{12.93}$$

Expressing V_1 and V_2 at $\Delta \ll 1$ through E_0, q_1, q_2, R_1, R_2, and Δ by way of Eq. (12.21) and substituting the obtained expressions into Eq. (12.92), one obtains:

$$F_{2z} = \tilde{F}_{2z} + \varepsilon E_0^2 R_2^2 \cos^2\theta f_1^* + E_0 \cos\theta (f_3^* q_1 + f_4^* q_2) + \frac{1}{\varepsilon R_2^2}(f_5^* q_1^2 + f_6^* q_1 q_2 + f_7^* q_2^2) + 0(\Delta), \tag{12.94}$$

where

$$f_j^* = \frac{b_j(\mathsf{k})}{\Delta(\ln\Delta + c_0)^2}; \quad j = 1, 3, 4, 5, 6, 7;$$

$$b_1(\mathsf{k}) = \frac{c_3^2}{\mathsf{k}+1}; \quad b_3(\mathsf{k}) = 2c_1c_3; \quad b_4(\mathsf{k}) = 2c_2c_3;$$

$$b_5(\mathsf{k}) = c_1^2(\mathsf{k}+1); \quad b_6(\mathsf{k}) = 2c_1c_3(\mathsf{k}+1); \quad b_7(\mathsf{k}) = 2c_2^2(\mathsf{k}+1);$$

$$c_0 = \frac{2\left(\psi\left(\frac{k}{k+1}\right)\psi\left(\frac{1}{k+1}\right) - \psi(1)\psi(1)\right)}{d} + \ln\left(\frac{k+1}{2}\right);$$

$$c_1 = \frac{\psi\left(\frac{1}{k+1}\right) - \psi(1)}{d}; \quad c_2 = \frac{\psi(1) - \psi\left(\frac{k}{k+1}\right)}{d};$$

$$c_3 = \frac{\psi(1)\left(\psi'\left(\frac{k}{k+1}\right) + \psi'\left(\frac{1}{k+1}\right)\right) - \psi\left(\frac{k}{k+1}\right)\psi'\left(\frac{1}{k+1}\right) - \psi'\left(\frac{k}{k+1}\right)\psi'\left(\frac{1}{k+1}\right)}{d}$$

$$+ \psi(1);$$

$$d = 2\psi(1) - \psi\left(\frac{k}{k+1}\right) - \psi\left(\frac{1}{k+1}\right); \quad k = \frac{R_2}{R_1} \le 1.$$

Here, $\psi(x)$ is the Euler psi-function.

Comparing Eq. (12.46) for the electrostatic force at an arbitrary distance Δ between particles with Eq. (12.94) at $\Delta \ll 1$, it is possible to conclude that factors f_j^* are the principal members of an asymptotic expansion of electrostatic force along the line joining the centers of the particles at small values of the gap between them. From the form of Eq. (12.92), it follows that the force acting on particle 2 along the line joining the centers consists of two terms, the first of which is positive and defines the attraction force of particles, while the second relates to the force of touching particles, which, as was shown in the previous section, is negative, that is, a repulsive force. To answer the question as to whether particles at a small distance from each other are attracted or repelled, one needs to estimate the ratio of these terms. If the particles have equal potentials, $V_1 = V_2$, they are repelled. If $V_1 \neq V_2$, there is not a unique answer. Nevertheless, it is possible to show that in this case attractive forces can act between particles even when they bear charges of the same sign.

For simplicity, consider the case of two charged particles in the absence of an external electric field. According to Eq. (12.46), the electrostatic force acting on particle 2 along the line joining the centers of the particles is equal to:

$$F_{2z} = \frac{1}{\varepsilon R_2^2}(f_5 q_1^2 + f_6 q_1 q_2 + f_7 q_2^2). \tag{12.95}$$

On the other hand, if the gap between the particles is small, then according to Eq. (12.92) and Eq. (12.94) we have:

$$F_{2z} = \tilde{F}_{2z} + \frac{\varepsilon(V_1 - V_2)^2}{4\Delta(k+1)} + 0(\Delta)$$

$$= \tilde{F}_{2z} + \frac{1}{\varepsilon R_2^2}(f_5^* q_1^2 + f_6^* q_1 q_2 + f_7^* q_2^2) + 0(\Delta); \quad \Delta \ll 1. \tag{12.96}$$

Since $\tilde{F}_{2z}$ determines the repulsive force pushing the particles apart, and the second term in Eq. (12.96) determines the attractive force, the repulsive force at $\Delta \ll 1$ will be maximal under the condition $V_1 = V_2$ or when the following relationship between the charges of the particles is satisfied:

$$f_5^* q_1^2 + f_6^* q_1 q_2 + f_7^* q_2^2 = 0 \tag{12.97}$$

Substitution of the above cited expressions for f_j^* into Eq. (12.97) gives a relationship connecting the charges and sizes of the particles:

$$-\left(\psi(1) - \psi\left(\frac{1}{\mathsf{k}+1}\right)\right) q_1 + \left(\psi(1) - \psi\left(\frac{\mathsf{k}}{\mathsf{k}+1}\right)\right) q_1 = 0;$$

from which it follows that at $\Delta \ll 1$ the repulsive force will be maximal under the condition:

$$\chi_0 = \frac{q_1}{q_2} = \frac{\psi(1) - \psi(\mathsf{k}/(\mathsf{k}+1))}{\psi(1) - \psi(1/(\mathsf{k}+1))}. \tag{12.98}$$

From the theorem of impossibility of stable equilibrium of charges under the action of electrostatic forces alone [78], it follows that if at $\Delta = \Delta_0$ the particles are repelled, they will be repelled at all $\Delta > \Delta_0$, and conversely if at $\Delta = \Delta_0$ the particles are attracted, they will be attracted at all $\Delta < \Delta_0$. This theorem allows us to divide the plane XOZ into areas of attraction and repulsion of particles on the basis of k for any ratio of particle radii. We denote through $\chi(\Delta_0, \mathsf{k})$ the ratio q_1/q_2, such that at $\Delta = \Delta_0$ and a given value of χ the condition $F_{2z} = 0$ is satisfied. Then, at $\Delta > \Delta_0$ the particles are repelled, whereas at $\Delta < \Delta_0$ they are attracted. This permits us to find an equation for the curve dividing the plane XOZ into areas of attraction and repulsion:

$$F_{2z} = (\chi, \Delta, \mathsf{k}) = 0$$

or

$$f_5(\Delta, \mathsf{k})\chi^2 + f_6(\Delta, \mathsf{k})\chi + f_7(\Delta, \mathsf{k}) = 0. \tag{12.99}$$

Solving the quadratic equation, Eq. (12.99), one obtains

$$\chi_{1,2} = \frac{-f_6 \pm \sqrt{f_6^2 - 4 f_5 f_7}}{2 f_5}. \tag{12.100}$$

Since $f_5 > 0$, $f_7 > 0$, and $f_6 < 0$ for all k and Δ, both roots are positive and $\chi_1 > \chi_2$. Then, for any Δ at $\chi_2 < \chi < \chi_1$ particles are repelled, and at other values of χ they are attracted. With increase of the relative distance between the surfaces of the particles, Δ, the area of repulsion is expanded. In fact, from Eq. (12.49), it

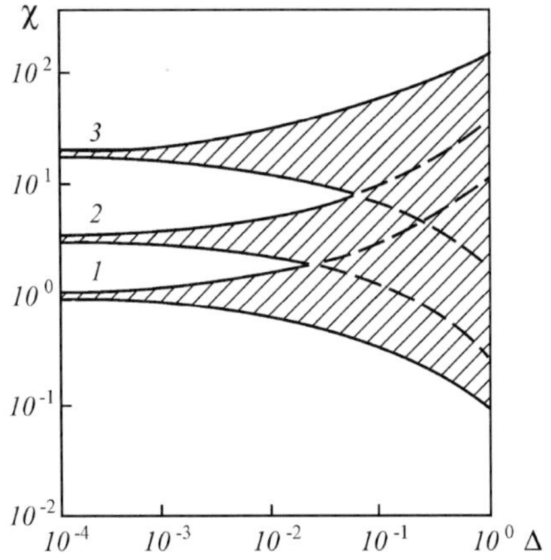

Fig. 12.12 Regions of attraction and repulsion (dashed region) of charged particles of equal sign in the plane (χ, Δ): *1*: $k = 1$, *2*: 0.5, *3*: 0.2.

follows that at large distances as $\Delta \to \infty$, factors f_j behave as $f_5 \sim \Delta^{-3}$, $f_7 \sim \Delta^{-3}$, and $f_6 \sim \Delta^{-2}$. Thus, $\chi_2 \to 0$ and $\chi_1 \to \infty$. On the other hand, as $\Delta \to 0$, areas corresponding to repulsion of the particles are narrowed. In fact, $f_6 \sim 2\sqrt{f_5 f_7}$ and $f_5, f_6, f_7 \to \infty$ as $(\Delta \ln \Delta)^{-1}$, and from Eq. (12.94) it follows that Eq. (12.99) can be transformed into:

$$\xi^2\chi^2 - 2\xi\chi + 1 + \alpha = 0, \tag{12.101}$$

where $\xi = \sqrt{f_5 f_7^{-1}}$ and $\alpha \to 0$ at $\Delta \to 0$ as $\Delta \ln \Delta$. The latter means that to within terms of the order Δ, the roots of Eq. (12.101) coincide, that is, $\chi_1 \sim \chi_2 \to 1/\xi$.

Figure 12.12 shows areas of attraction and repulsion for charged particles of the same sign and with ratios of radii of $k = 1, 0.5$, and 0.2 in the plane (χ, Δ). For particles of identical size ($k = 1$) at $\Delta = 1$, $\chi_1 = 11.3$, which is in good agreement with the value of $\chi_1 \approx 12$ reported in ref. [96].

The narrowing of the area of particle repulsion (dashed area in Fig. 12.12) as $\Delta \to 0$ means that even a seemingly insignificant deviation of the ratio of the charges from the equilibrium value χ_0, as determined by Eq. (12.98), will lead to attraction of the particles.

Figure 12.13 shows the dependences of the dimensionless electrostatic force

$$\hat{F}_{2z} = \frac{\varepsilon R_2^2 F_{2z}}{q_2^2} = \chi^2 f_5 + \chi f_6 + f_7, \tag{12.102}$$

acting on particle 2 along the line joining the centers of two charged particles of the same sign upon Δ for various values of ratios of charges q_1/q_2 and radii

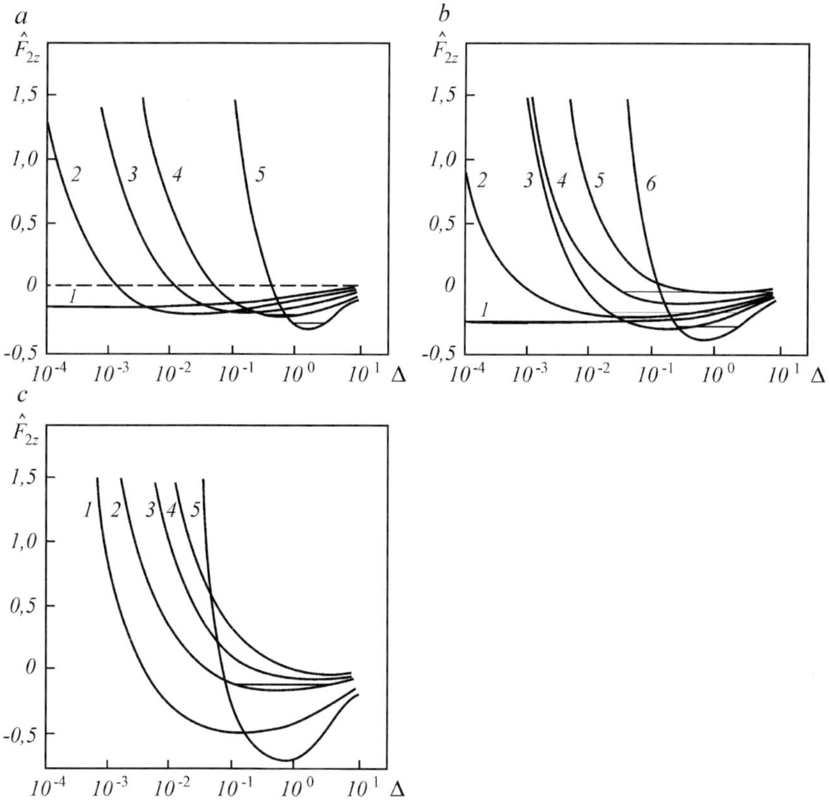

Fig. 12.13 Dependences of the dimensionless electrostatic force acting on particle 2 along the center line upon Δ, $\chi = q_1/q_2$, and $k = R_1/R_2$: *a*) $k = 1$; *1*: $\chi = 1$, *2*: 1.2, *3*: 1.5, *4*: 2, *5*: 5; *b*) $k = 0.5$; *1*: $\chi = 3.45$, *2*: 3, *3*: 5, *4*: 2; *5*: 1, *6*: 10; *c*) $k = 0.2$; *1*: $\chi = 25$, *2*: 10, *3*: 5, *4*: 2, *5*: 50.

$k = R_2/R_1$. At $\hat{F}_{2z} > 0$, the particles are attracted, while at $\hat{F}_{2z} < 0$ they are repelled.

In qualitative analysis of the force of electrostatic interaction between charged particles of the same sign, one needs to define the character of this force, that is, whether it is an attractive or repulsive force for various values of the ratio of the radii of the particles k. In fact, for various values of k, areas of attraction and repulsion in the plane (χ, Δ) can be brought into coincidence by a transformation from χ to a new parameter χ^*:

$$\chi^* = \frac{\chi}{\chi_0} = \chi \frac{\psi(1) - \psi(1/(k+1))}{\psi(1) - \psi(k/(k+1))}. \tag{12.103}$$

Thus, in the plane (χ^*, Δ), areas of attraction and repulsion for particles with various values of k coincide (Fig. 12.14).

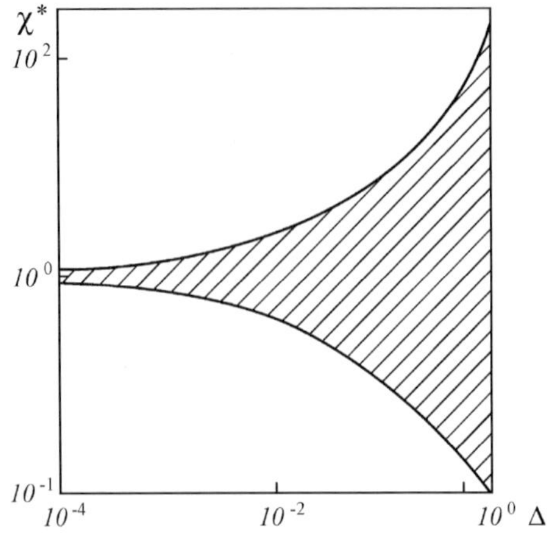

Fig. 12.14 Regions of attraction and repulsion (dashed area) of particles with different values of k in the plane (χ^*, Δ).

12.7
Redistribution of Charges

Research into the behavior of disperse media is connected with the problem of the definition of charges on particles. Initially neutral conducting particles placed in an external electric field can, in time, acquire a charge through contact with each other, through breakdown of the dielectric layer dividing them, as a result of approach of particles, or upon their breakage, for example, the breakage of emulsion drops. Besides, particles can be charged due to selection of the charges of the ions surrounding them, and also upon contact with surfaces [97].

The charging process of a conducting spherical particle on contact with a flat electrode is described in ref. [98]. In this work, the charge acquired by the particle was determined, and the force of particle interaction with the flat electrode was calculated.

Particles can acquire a charge not necessarily as a result of contact with each other, but also as a result of an electric discharge in the clearance between them. In this case, it is necessary to know the electric field strength in the clearance.

The natural condition of particles in direct contact is equality of the potentials of their surfaces. If the charge is acquired as a result of a breakdown, this condition is not always obeyed.

Thus, the process of charge redistribution between particles should begin even before their contact, as a result of significant growth of the strength of the electric field in the clearance between them (see Section 12.2). However, because of incomplete equalization of the potentials of the particles and the presence of molecular forces of interaction, it can proceed unnoticed. In coarsely disperse systems, in which the size of particles, R, is around 1–200 μm, the process of charge redistribution can be the main barrier to integration of the dispersed phase. Thus, in the processing of water/oil emulsions in an electric field of low strength ($E_0 < 2$ kV cm^{-1}) the rate of water drop integration increases, while in processing in fields of high strength ($E_0 > 2$ kV cm^{-1}) this rate decreases. Such a change in the behavior of water/oil emulsions is commonly associated [99] with drop breakage. However, the process of breakage of even relatively large drops ($R \sim 100$ μm) begins at a field strength of the order of 10 kV cm^{-1}. Redistribution of charges and the subsequent interaction of particles offers a satisfactory explanation for this effect.

Consider the process of redistribution of charges between two conducting spherical particles of radii R_1 and R_2 upon their collision in a uniform external electric field of strength E_0. The charges on the particles before collision are specified and are equal to q_1^0 and q_2^0. Note that upon redistribution of the charges the total charge on the particles $Q = q_1^0 + q_2^0$ is maintained. The problem of definition of the charges and of the forces of electrostatic interaction of the particles after their contact reduces to a problem of the interaction of two touching particles, as considered in Section 12.5. According to Eq. (12.74), the charges on the particles after collision are equal to:

$$q_2 = -\varepsilon E_0 R^2 A_1 \cos\theta + A_2(q_1^0 + q_2^0); \quad q_1 = q_1^0 + q_2^0 - q_2, \tag{12.104}$$

where

$$A_1 = \frac{(\psi(1-\alpha) + C)\psi'(\alpha) + (\psi(\alpha) + C)\psi'(1-\alpha)}{2C + \psi(\alpha) + \psi(1-\alpha)} + \frac{\pi^2}{6};$$

$$A_2 = \frac{\psi(\alpha) + C}{2C + \psi(\alpha) + \psi(1-\alpha)}; \quad \alpha = \frac{R_1}{R_1 + R_2}; \quad R = \frac{R_1 R_2}{R_1 + R_2};$$

where $C = 0.5772$ is the Euler constant.

The forces of electrostatic interaction between particles after contact are determined by Eqs. (12.81) if the particles continue to be in contact, and by Eqs. (12.46) and (12.47) if redistribution of charges results in separation of the particles.

Consider now the case when redistribution of charges occurs due to electric discharge in the clearance between closely approached particles without their direct contact. Assume that the electric field strength E_m at which the electric discharge in the dielectric is possible is rather greater than the strength of the external electric field E_0 and that the electric discharge results in full equalization of the potentials of the particles without affecting the total charge. Since by virtue of the first assumption $E_m/E_0 \gg 1$, from Eq. (12.29) and Fig. 12.3 for the strength of the electric field in the gap between the particles, it follows that this is only possible at $\Delta \ll 1$, that is, when the gap between particles is sufficiently small. By virtue of the second assumption, after redistribution of the charges of the particles, the condition $V_1 = V_2$ is satisfied. Thus, due to Eq. (12.34), the value of the average strength along the line joining the centers in a narrow clearance between particles is equal to zero. Taking into account the connection of the average strength of the electric field in the gap with the charges on the particles (see Eq. (12.42)), the following equations are obtained for definition of the charges on the particles after redistribution:

$$q_1 E_1^* + q_2 E_2^* + E_3^* E_0 \cos\theta = 0,$$
$$q_1 + q_2 = q_1^0 + q_2^0. \qquad (12.105)$$

The solution of the system of equations, Eq. (12.105), gives the same values for q_1 and q_2 as Eq. (12.104) in the case of contact of the particles. Besides, since $\Delta \ll 1$, from Eqs. (12.92) and (12.93), it follows that at $V_1 = V_2$, to within terms of the order of Δ, the electrostatic interaction force of particles after redistribution of charges is the same as for touching particles.

Clearly, the condition of full equalization of the potentials of the particles due to electric discharge in the clearance is an idealization of the real process. We now determine conditions under which this is possible. Consider the asymptotic expression Eq. (12.92) for the electrostatic force along the line joining the centers of two particles. In order for the electrostatic force after redistribution of the charges upon discharge to be the same as that upon contact of particles, it would be necessary that at $\Delta \ll 1$ the following condition be satisfied:

$$\tilde{F}_{2z} \gg \frac{\varepsilon}{4(\mathrm{k}+1)} \frac{(V_1 - V_2)^2}{\Delta}. \qquad (12.106)$$

At $\Delta \ll 1$, from Eq. (12.34) it follows that $V_2 - V_1 = R_2 \Delta E_s$, and from Eq. (12.81) we obtain:

$$\tilde{F}_{2z} = \varepsilon E_0^2 R^2 + Q^2/\varepsilon R_2^2. \qquad (12.107)$$

Then, the condition of Eq. (12.106) is fulfilled if even just one of the following conditions is satisfied:

$$\frac{E_s}{E_0} \ll \frac{1}{\sqrt{\Delta}} \quad \text{or} \quad \Delta \ll \left(\frac{E_0}{E_s}\right)^2, \quad Q = 0;$$
$$\frac{E_s R_2^2}{Q} \ll \frac{1}{\sqrt{\Delta}} \quad \text{or} \quad \Delta \ll \left(\frac{Q}{E_s R_2^2}\right)^2, \quad E_0 = 0, \qquad (12.108)$$

where E_s is the electric field strength in the gap at which the process of redistribution of the charges stops.

Next, we estimate the value of the gap δ between particles, at which their potentials become equal. We set $E_s/E_0 \sim 10^2$ as typical for a water/oil emulsion at $E_0 \sim 1$–5 KV cm^{-1} and $R_2 \sim 10$ µm. Then, from Eq. (12.108), we find:

$$\delta \ll R_2 (E_0/E_s)^2 \sim 10^{-3}\ \mu\text{m}.$$

Such a small value of δ testifies that under the given conditions the potentials of the particles are equalized practically upon their contact. Note that the obtained value of δ is less than the radius of action of attractive molecular forces ($\sim 5 \times 10^{-2}$ μm). This fact explains why the process of redistribution of charges between fine particles can proceed unnoticed. Moreover, from the cited estimations, it follows that studies of the interaction of conducting particles in a dispersed medium in the presence of an external electric field should be carried out with due regard for both electrostatic and molecular forces between particles, especially in the final stages of the approach of the particles.

In summary, it should be noted that when the following inequality is satisfied:

$$\tilde{F}_{2z} > \frac{\varepsilon}{4(\mathsf{k}+1)} \frac{(V_1 - V_2)^2}{\Delta}. \tag{12.109}$$

the electrostatic force of the particle interactions will be one of repulsion.

13
Coalescence of Drops

The process of drop coalescence consists of two stages. The first one – the transport stage – involves the mutual approach of drops until their surfaces come into contact. The second – kinetic stage – involves coalescence itself, that is, merging of drops into a single drop. The transport stage is assumed to take a predominant share of the total duration of the process. We also assume that each collision of drops results in their coalescence. Then the primary goal is to determine the collision frequency for drops of various sizes.

Consider the mutual approach and subsequent collision of two spherical conducting drops of different radii suspended in a dielectric liquid in the presence of a uniform external electric field of strength $\boldsymbol{E}_0$. The assumption that drops maintain their spherical shape until they come into contact is not absolutely correct: as was remarked earlier, when the clearance between the approaching drops becomes small, electric and hydrodynamic forces grow boundlessly, which can lead to a significant deformation of drop surfaces and promote their breakage. However, if drops are small, the strength of the external electric field does not surpass E_{cr}, and drop surfaces are retarded by secure envelopes, they can be considered as slightly deformed. All this allows us to make the additional assumption that drops move as rigid particles.

13.1
Coalescence of Drops During Gravitational Settling

Consider a slow motion of conducting, charged drops under the action of gravity in a quiescent liquid in the presence of a uniform external electric field.

Sedimentation of drops occurs under the influence of gravitational and buoyancy forces. The net force acting on i-th drop is

$$F_i^A = \frac{4}{3}\pi R_i^3 \Delta\rho g, \tag{13.1}$$

where $\Delta\rho = |\rho_e - \rho_i|$ is the density difference between the continuous and the disperse phase.

Separation of Multiphase, Multicomponent Systems. E. G. Sinaiski and E. J. Lapiga

ISBN: 978-3-527-40612-8

Consider first the case when the external electric field $\boldsymbol{E}_0$ is parallel to the direction of the gravity force, that is, to the vector $\boldsymbol{g}$. Then the motion of drop S_2 relative to drop S_1 can be considered as planar motion in a meridian plane $\Phi = const$ of spherical system of coordinates (r, θ, Φ) (see Fig. 11.2). In this case the system is axially-symmetric with respect to axis O_1Z, and the angle between vector $\boldsymbol{E}_0$ and the line of centers of drops coincides with the angle θ of spherical system of coordinates. The expressions for electric forces acting on two conducting charged drops have been obtained received in Section 12.3. The components of these forces along axes r and θ for each drop are

$$\begin{aligned} F_{2r}^{e1} &= F_{2z}; \quad F_{1r}^{e1} = E_0(q_1 + q_2)\cos\theta - F_{2r}^{e1}; \\ F_{2\theta}^{e1} &= F_{2x}; \quad F_{1\theta}^{e1} = E_0(q_1 + q_2)\cos\theta - F_{2\theta}^{e1}, \end{aligned} \tag{13.2}$$

where F_{2x} and F_{2z} are determined by expressions (12.46) and (12.47).

At the final stage of drops approach it is necessary to take into account molecular forces of attraction (11.100), which at small clearance between the drops are given by

$$F_{2r}^{mol} = -\frac{\Gamma R_1}{6R_2(R_1 + R_2)}\frac{1}{\Delta^2}, \tag{13.3}$$

where Γ is Hamaker constant and $\Delta = (r - R_1 - R_2)/R_1$ – is the dimensionless clearance between surfaces of drops.

The type of dominating interaction forces between drops can vary, depending on particles' size, difference in phase densities, external electric field strength, and drop charges. Thus, in case of approaching colloidal particles ($R \le 1\ \mu m$), the basic role is played by forces of molecular and electrostatic interactions (due to charges). For coarse-disperse systems ($R > 1\ \mu m$) it is necessary to take into account all the forces involved.

Hydrodynamic forces and torgues acting on drops in the external liquid were considered in Section 11.4. We express the forces and torgues acting on drop 2 in terms of drops' linear and angular velocities [15]:

$$\begin{aligned} \boldsymbol{F}_2^{hyd} &= \boldsymbol{F}_{2s}^{hyd} + \boldsymbol{F}_{2e}^{hyd}, \quad \boldsymbol{T}_2^{hyd} = \boldsymbol{T}_{2s}^{hyd} + \boldsymbol{T}_{2e}^{hyd}, \\ F_{2r}^{hyd} &= 6\pi\mu R_2(u_{0r}f_{sr} - v_{2r}f_{er}), \\ F_{2\theta}^{hyd} &= 6\pi\mu R_2(u_{0\theta}f_{s\theta} - v_{2\theta}f_{e\theta} + R_2\Omega f_{e\theta 1}), \\ T_{2s\varphi} &= 8\pi\mu R_2^2(R_2 G t_{s\varphi} + v_\theta f_{e\theta} - R_2\Omega f_{e\varphi 1}). \end{aligned} \tag{13.4}$$

Here F_{ir} and $F_{i\theta}$ are the longitudinal and transverse components (with reference to the line of the drops' centers) of hydrodynamic force acting on i-th drop, u_{0r} and $u_{0\theta}$ are projections of the Stokes's velocity vector of the flow around the drop 2 (see (11.81)), v_{2r} and $v_{2\theta}$ are the velocity components of drop 2 relative drop 1, $G = 0.5|\nabla \times \mathbf{u}_0|$. Factors f_{sr}, $f_{s\theta}$, f_{er}, $f_{e\theta}$, $f_{r\theta 1}$, $f_{s\varphi}$, $f_{e\varphi}$, $f_{e\varphi 1}$ have been introduced in Section 11.4.

In the zero-mass approximation the equations of relative motion for drops 1 2 and are reduced to:

$$\begin{aligned}
&\boldsymbol{F}_2^A + \boldsymbol{F}_2^{mol} + \boldsymbol{F}_2^{e1} + \boldsymbol{F}_2^{hyd} = 0,\\
&\boldsymbol{T}_{2s}^{hyd} + \boldsymbol{T}_{2e}^{hyd} = 0,\\
&\frac{dr}{dt} = \mathsf{v}_{2r}, \quad r\frac{d\theta}{dt}\mathsf{v}_{2\theta}.
\end{aligned} \tag{13.5}$$

The form of the second equation (13.5) is reflects the fact that the torques of molecular and electric forces are equal to zero.

Projecting (13.5) on axes r and θ and taking into account expressions (13.1)–(13.4) for interaction forces between the drops, one obtains a system of equations for components of linear velocity of drop 2 relative drop 1 and angular velocity of drop 2. Solving this system of equations with initial conditions $r(0) = r_0$, $\theta(0) = \theta_0$, one obtains the family of trajectories of drop with radius R_2 relative to drop with radius R_1.

Now introduce the following dimensionless variables

$$\Delta = \frac{r - R_1 - R_2}{R_2}, \quad \tau = \frac{Ut}{R_2}, \quad \mathsf{v}_r = \frac{\mathsf{v}_{2r}}{\mathsf{U}}, \quad \mathsf{v}_\theta = \frac{\mathsf{v}_{2\theta}}{\mathsf{U}}, \quad \omega = \frac{\Omega R_2}{U}, \quad \mathsf{k} = \frac{R_2}{R_1},$$

$$h = \frac{GR_2}{U}, \quad S_1 = \frac{2\Delta\rho g R_2^2}{9\mu U}, \quad S_2 = \left(\frac{\varepsilon R_2 E_0^2}{6\pi\mu U}\right)^{1/2}, \quad S_2 = \frac{\Gamma}{36\pi\mu R_2^2 U},$$

$$N_1 = \frac{q_1}{(6\pi\mu\varepsilon R_2^3 U)^{1/2}}, \quad N_2 = \frac{q_2}{(6\pi\mu\varepsilon R_2^3 U)^{1/2}}, \quad U = \frac{2R_1^2\Delta\rho g}{9\mu} - \frac{q_1 E_0}{6\pi\mu R_2^2}.$$

In these variables, the system of equations (13.5) will become

$$\begin{aligned}
&f_{sr}\mathsf{v}_r - f_{er}\frac{d\Delta}{d\tau} + S_1 \cos\theta - S_2(f_1 \cos^2\theta + f_2 \sin^2\theta)\\
&\qquad - S_2(N_1 f_3 + (f_4 + 1)N_2)\cos\theta - N_1^2 f_5 - N_1 N_2 f_6 - N_1^2 f_7 - S_3 f_{11} = 0,\\
&f_{s\theta}\mathsf{v}_\theta - f_{e\theta}\left(\Delta + 1 + \frac{1}{\mathsf{k}}\right)\frac{d\theta}{d\tau} + S_2^2 f_8 \sin 2\theta - S_1 \sin\theta\\
&\qquad + S_2(f_9 N_1 + (f_{10} + 1)N_2) + \omega f_{e\theta 1} = 0,\\
&f_{e\theta}\left(\Delta + 1 + \frac{1}{\mathsf{k}}\right)\frac{d\theta}{d\tau} - \omega t_{e\varphi 1} + h t_{s\varphi} = 0.
\end{aligned} \tag{13.6}$$

Factors f_i at $i = 1, 2, \ldots, 10$ are determined from (12.46) and (12.47), and $f_{11} = (1 + \mathsf{k})^{-1}\Delta^{-2}$.

Eliminating ω from (13.6), one obtains:

$$\frac{d\Delta}{dt} = \frac{1}{f_{er}}(\xi_1 \cos^2\theta + \xi_2 \cos\theta + \xi_3),$$
$$\frac{d\theta}{d\tau} = \alpha \frac{\sin\theta}{(t_{e\varphi} f_{e\theta} - f_{e\theta 1} t_{e\varphi 1})}(\xi_4 \cos\theta + \xi_5), \tag{13.7}$$

where

$$\xi_1 = -S_2^2(f_1 - f_2), \quad \xi_2 = f_{sr}\mathsf{v}_r + S_1 - S_2(N_1 f_3 + (f_4 + 1)N_2),$$
$$\xi_3 = -N_1^2 f_5 - N_1 N_2 f_6 - N_2^2 f_7 - S_3 f_{11} - S_2^2 f_2, \quad \xi_4 = 2S_2^2 f_8,$$
$$\xi_5 = -\overline{\mathsf{v}}_\theta f_{s\theta} - S_1 + S_2(N_1 f_9 + (f_{10} + 1)N_2) + \frac{3}{4}\alpha \frac{t_{s\varphi} f_{e\theta 1}}{t_{e\varphi}},$$
$$\alpha = \frac{\mathsf{k}}{(\mathsf{k} + 1 + \mathsf{k}\Delta)^2}, \quad \overline{\mathsf{v}}_r = \frac{3}{2}\frac{1}{(\mathsf{k} + 1 + \mathsf{k}\Delta)} - \frac{1}{2}\frac{1}{(\mathsf{k} + 1 + \mathsf{k}\Delta)^3},$$
$$\overline{\mathsf{v}}_\theta = 1 - \frac{3}{4}\frac{1}{(\mathsf{k} + 1 + \mathsf{k}\Delta)} - \frac{1}{4}\frac{1}{(\mathsf{k} + 1 + \mathsf{k}\Delta)^3}.$$

Solve the system of equation (13.7) with the initial conditions

$$\Delta(0) = \Delta_0, \quad \theta(0) = \theta_0. \tag{13.8}$$

Taking various values Δ_0 and θ_0, it is possible to determine trajectories of motion of drop S_2 relative to S_1. During the motion, the drop S_2 either collides with S_1 or passes by. Trajectories corresponding to these cases, form two families: trajectories terminating at $r = R_2 + R_1$ from the center of sphere S_1 (collision), and trajectories passing by S_1 and extended to infinity. These two families are separated by the limiting trajectory.

Select a cross section perpendicular to the vector $\boldsymbol{g}$ at a large distance from the sphere S_1, in the oncoming flow containing drops S_2. A cross section placed far away from the drop and selected in such a way that all limiting trajectories pass through its contour is called the collision cross section. If the angle θ_0 between vectors $\boldsymbol{E}_0$ and $\boldsymbol{g}$ is equal to zero, this contour will be a circle of radius d. It is obvious that drops S_2 that are remote from S_1 and move along trajectories that cross the circle confined by this contour, will eventually collide with drop S_1. The area of collision cross section G is called capture cross section of drops S_2 by drop S_1. The frequency of collisions of drops S_2 with drop S_1 is proportional to G. Therefore the primary goal is to determine the collision cross section.

The system of equations (13.7) with conditions (13.8) has been solved by numerical methods. But before presenting the results, we should discuss the system of equations (13.7) qualitatively.

The system of the equations (13.7) is autonomous, therefore the problem should be studied in the phase space (x, θ) where x denotes Δ. To find singular points of the velocity field of the smaller drop 2, we equate the right-hand sides of equations (13.7) to zero. This gives us the following system of equations:

$$\xi_1(x)\cos^2\theta + \xi_2(x)\cos\theta + \xi_3(x) = 0,$$
$$(\xi_4(x)\cos\theta + \xi_5(x))\sin\theta = 0. \tag{13.9}$$

The first of these two is the equation of zero-isoclinic line, and the second – infinite-isoclinic line. Singular points of the velocity field of drops 2 are the points of intersection of these isoclinic lines.

The solution of the first equation with respect to θ has the form

$$\theta(x) = \arccos\left(\frac{-\xi_2(x) \pm \sqrt{\xi_2^2(x) - 4\xi_1(x)\xi_3(x)}}{2\xi_1(x)}\right) \tag{13.10}$$

at

$$\xi_1(x) \neq 0, \quad \xi_2^2(x) \geq 4\xi_1(x)\xi_3(x), \quad \left|\frac{-\xi_2(x) \pm \sqrt{\xi_2^2(x) - 4\xi_1(x)\xi_3(x)}}{2\xi_1(x)}\right| \leq 1$$

and

$$\theta(x) = \arccos\left(\frac{\xi_3(x)}{\xi_2(x)}\right) \quad \text{at } \xi_1(x) = 0, \quad \left|\frac{\xi_3(x)}{\xi_2(x)}\right| \leq 1. \tag{13.11}$$

Thus, for each value of x, the first equation (13.9) allows for two solutions at most, and the second equation can have two or three solutions

$$\theta(x) = 0, \quad \theta(x) = \pi, \quad \theta(x) = \arccos\left(-\frac{\xi_5(x)}{\xi_4(x)}\right) \quad \text{at } \left|\frac{\xi_5(x)}{\xi_4(x)}\right| \leq 1. \tag{13.12}$$

Consider the behavior of the solution of system of equations (13.7) for the case of charged particles at $N_1 = N_2 = 0$. If, in addition to this, there is no external electric field ($S_2 = 0$), then $\xi_1(x) \equiv 0$, $\xi_4(x) \equiv 0$. Thus, it follows from (13.10)–(13.12) that in the phase space, there is one zero-isoclinic line $\theta(x) = \arccos\left(\frac{\xi_3(x)}{\xi_2(x)}\right)$ and two infinite-isoclinic lines $\theta(x) = 0$ and $\theta(x) = \pi$ (Fig. 13.1, a). The intersection of the zero-isoclinic line with the infinite-isoclinic line $\theta = \pi$ gives us the unique singular point A (saddle). The separatrix of point A divides the phase plane into two regions. All trajectories in the first region reach the boundary (drops collide), while all trajectories in the second region do not reach the boundary (collision of drops does not occur). Hence, the separatrix of a singular point A represents a limiting trajectory of a small drop relative to the large one. If there is an external electric field, then $S_2 \neq 0$; $\xi_1(x), \xi_2(x), \xi_4(x) < 0$,

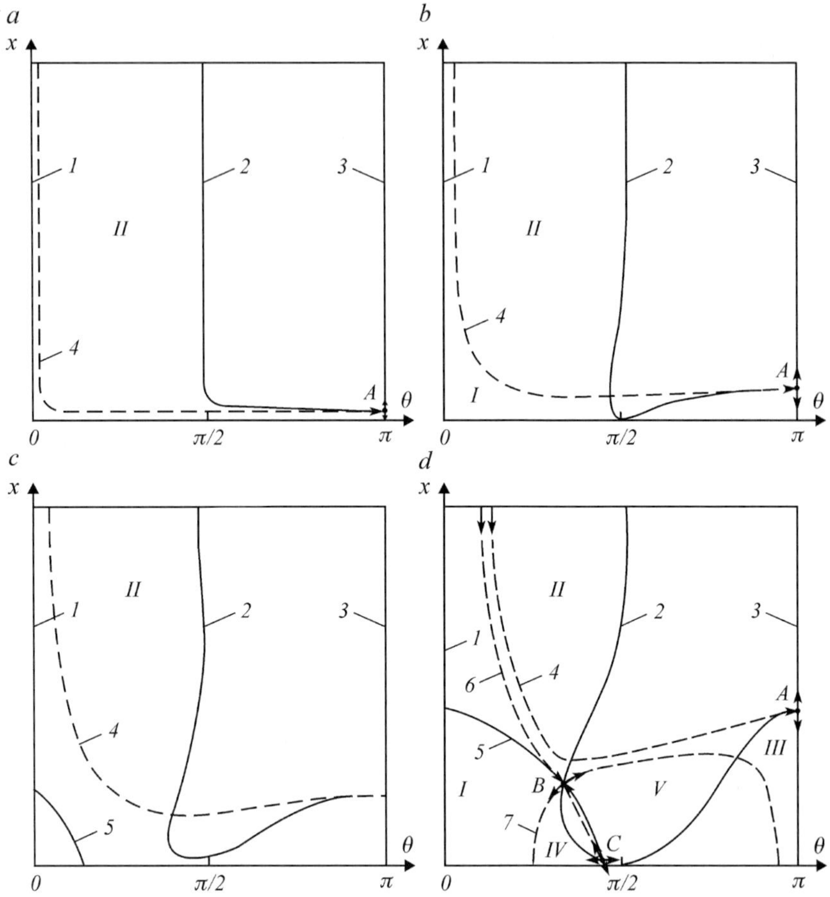

Fig. 13.1 Singular points of velocity field of small particle on phase plane (x, θ). Case of non-charged particles $(N_1 = N_2 = 0)$: $a - S_2^2 = 0$; $b - S_2^2 \sim 0.1$; $c - S_2^2 \leq 1$; $d - S_2^2 \geq 4$; 1, 3, 5 – infinite isoclinic lines; 2 – zero isoclinic line; 4 – separatrix of singular point A – saddle; 6, 7 – separatrix of singular point B – saddle; C – singular point – unstable node.

and $\xi_5(x) > 0$ for all x; $\xi_3(x) < 0$ only at $x \gg 1$, and for all other x we have $\xi_3(x) \geq 0$. At $x \to \infty$, we have $\xi_3(x) \to 0$, $\xi_4(x) \to 0$, $\xi_2(x) \sim -1$, $\xi_5(x) \sim 1$.

At sufficiently low strength of the external electric field $(|S_2| \leq 1.5)$ a set of phase trajectories is topologic equivalent to the set of trajectories at $S_2 = 0$ (Fig. 13.1, *a*, *b*, *c*). Considered cases are marked by presence of only one singular special point A (saddle), whose separatrix (curve A) represents critical trajectory of motion of small drop relative to the big one.

At $S_2^2 \geq 0.3$ we have $|\xi_4(0)| \geq |\xi_5(0)|$, and there appears one more infinite-isoclinic line (curve 5 in Fig. 13.1, *c*), which moves towards zero-isoclinic line (curve 2) with increase of S_2^2. Starting from $S_2^2 \sim 4$, the curves 5 and 4 intersect, and the topology of phase space essentially changes. There appears a pair of singular points: B (saddle) and C (unstable node), therefore the phase space is divided by separatrixes of points A and B into five separate areas (Fig. 13.1, *d*).

As to calculation of collision cross sections of drops, areas IV and V do not represent special interest, since in view of enclosure of these areas, the trajectories beginning far from the straight line $x = 0$, cannot pass through them. The phase trajectories of areas I and III reach the boundary $x = 0$, that is the drops collide. Trajectories entering the area II, do not reach the boundary, that is, the drops do not collide. Hence, as well as in the case considered above, the separatrix of

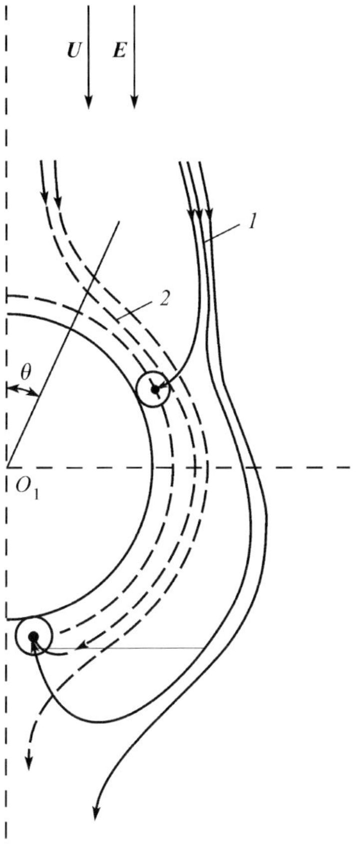

Fig. 13.2 Trajectories of motion of small drop relative big one at $N_1 = N_2 = 0$; $k = 0.1$: full lines $S_2^2 = 5$; dashed lines $S_2^2 = 0.1$; 1, 2 – trajectories close to critical.

singular point A is the limiting trajectory defining the cross section of drop collision.

The trajectories of motion of a small drop relative to the big one were determined by numerical integration of equations of motion (13.7) at parameter values $N_1 = N_2 = 0$, corresponding to the case of uncharged drops [16].

Figure 13.2 shows the characteristic trajectories of motion of a small drop relative to the big one ($k = 0.1$) for two values of parameter of electro-hydrodynamic interaction $S_2^2 = 5$ (full lines) and $S_2^2 = 0.1$ (dashed lines). The trajectories close to critical are denoted as 1 and 2. The corresponding phase trajectories for $S_2^2 = 5$ are shown in Fig. 13.3.

Since in the case considered the direction $\mathbf{E}_0$ is parallel to $\mathbf{g}(\theta_0 = 0)$, the cross section of collision represents a circle of radius d. Shown in Fig. 13.4 is the dependence of dimensionless radius of collision cross section d/R_1 on parameter of electro-hydrodynamic interaction S_2 and the ratio of drop radiuses $k = R_2/R_1$. The collision cross section grows with k and S_2.

Influence of viscous resistance of the liquid layer separating approaching drops can be traced by neglecting deviation of resistance factors from Stokesean ones, that is by setting $f_{er} = f_{sr} = f_{s\theta} = f_{e\theta} = 1$, $f_{e\theta 1} = 0$. For simplicity we also ignore

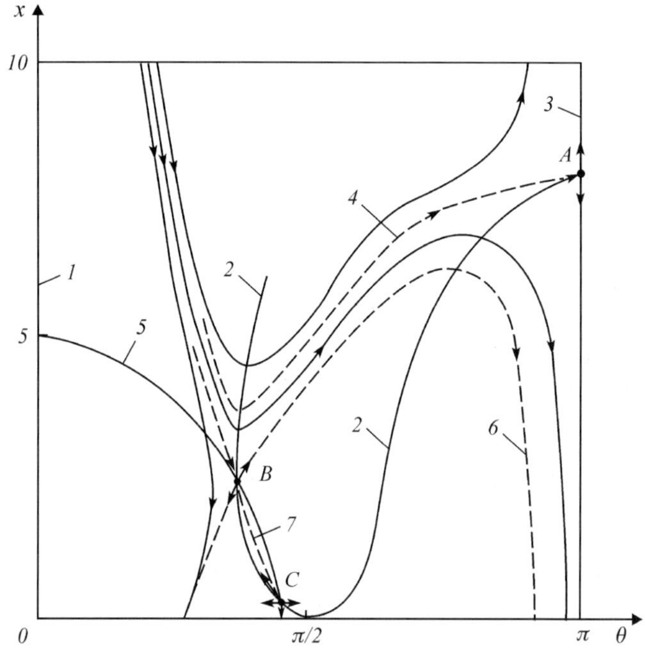

Fig. 13.3 Phase trajectories of motion of small drop relative big one ($N_1 = N_2 = 0$; $k = 0.1$; $S_2^2 = 5$): 1, 3, 5 – infinity isoclinic lines; 2 – zero isoclinic lines; 4 – separatrix of singular point A-saddle; 6, 7 – separatrix of singular point B – saddle; C – singular point – unstable node.

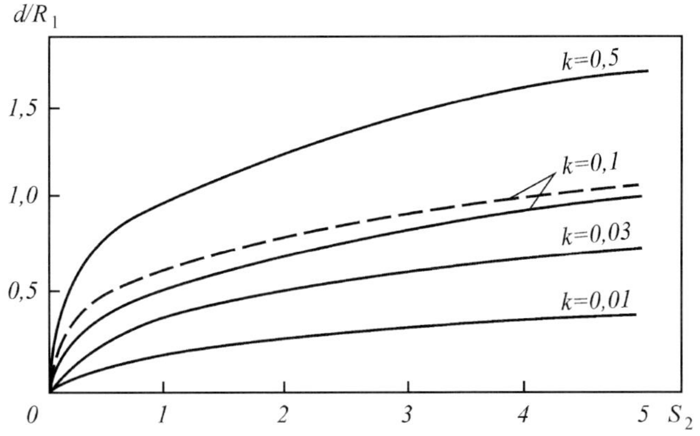

Fig. 13.4 Dependence of radius of collision sector upon S_2 and k.

the molecular interaction, setting $S_3 = 0$. The simplified system of equations takes the form

$$\frac{d\Delta}{d\tau} = u_r - S_2^2(f_1 \cos^2\theta + f_2 \sin^2\theta),$$

$$\frac{d\theta}{d\tau} = \frac{u_\theta + S_2^2 f_8 \sin 2\theta}{(x_1 + 1/\mathbf{k})}, \tag{13.13}$$

where u_r and u_θ are dimensionless components of Stokesian velocities (11.81).

The dashed line in Fig. 13.4 represents the case $k = 0.1$. It is seen that at $S_2^2 \geq 1$ the viscous resistance does not appreciably influence the collision cross section of drops. The reason for this is that at small values S_2^2 the drop 2 moves a long time near the surface of drop 1 experiencing significant viscous resistance, while at $S_2^2 > 1$ it quickly passes through dividing layer almost at a right angle to the surface of drop 1.

The situation is more complicated when $0 < \theta_0 < \pi/2$, that is, when the electric field $\mathbf{E}_0$ makes an acute angle with vector $\mathbf{U}$ or $\mathbf{g}$ (Fig. 13.5). In this case the motion of drop 2 is not planar, its trajectories do not lie in a meridian plane, and it is necessary to consider 3-dimensional motion. To simplify calculations, we neglect the viscous resistance, taking $S_2^2 \geq 1$. In this case the components of electric force are

$$\mathbf{F}_r = -(f_1 \cos^2\psi + f_2 \sin^2\psi)\mathbf{e}_r; \quad \mathbf{F}_\psi = f_8 \sin 2\psi \mathbf{e}_\psi, \tag{13.14}$$

where e_r and e_ψ are the unit vectors parallel and perpendicular to $\mathbf{r}$ in the plane $(\mathbf{r}, \mathbf{E}_0)$. In spherical system of coordinates (r, θ, Φ) (see Fig. 13.5), the equations of motion of drop 2 relative to drop 1 are

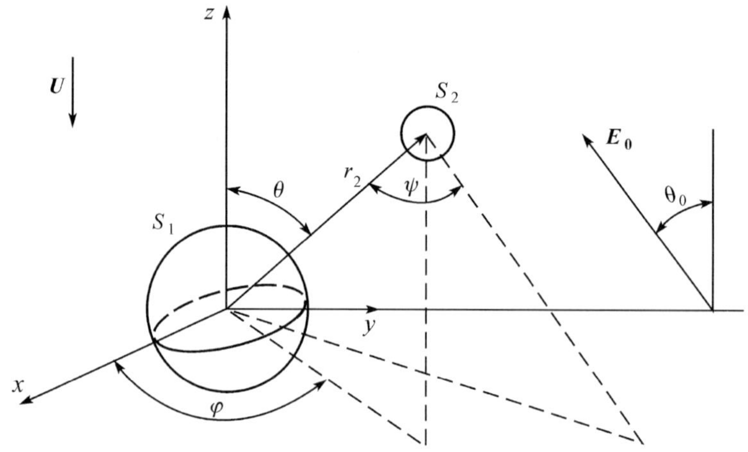

Fig. 13.5 Spatial configuration of interacting drops.

$$\frac{d\Delta}{d\tau} = u_r - S_2^2(f_1 \cos^2(\theta + \theta_0) + f_2 \sin^2(\theta + \theta_0)),$$

$$\frac{d\theta}{d\tau} = \frac{u_\theta}{x_1 + 1/\mathrm{k}} + \frac{S_2^2 f_8}{(x_1 + 1/\mathrm{k})\sqrt{\alpha^2 + \beta^2 + \gamma^2}} \times (\alpha \cos\theta \cos\Phi + \beta \cos\theta \sin\Phi - \gamma \sin\Phi), \tag{13.15}$$

$$\frac{d\Phi}{d\tau} = \frac{S_2^2 f_8 \sin\psi(\beta \cos\Phi - \alpha \sin\Phi)}{(x_1 + 1/\mathrm{k}) \sin\theta \sqrt{\alpha^2 + \beta^2 + \gamma^2}},$$

where $\cos\psi = \cos\theta_0 \cos\theta - \sin\theta_0 \sin\theta \sin\Phi$; $\alpha = xz \cos\theta_0 - xy \sin\theta_0$; $\beta = x^2 \sin\theta_0 + yz \cos\theta_0 + x^2 \sin\theta_0$; $\gamma = -x^2 \cos\theta_0 - y^2 \cos\theta_0 - yz \sin\theta_0$; $x = (x_1 + 1/\mathrm{k}) \sin\theta \cos\Phi$; $y = (x_1 + 1/\mathrm{k}) \sin\theta \sin\Phi$; $z = (x_1 + 1/\mathrm{k}) \cos\Phi$.

The collision cross sections obtained by numerical integration of equations (13.15) for various angles θ_0 are shown in Fig. 13.6.

The collision cross sections for various angles q0 are found to be almost equal. This means that drop collision frequency practically does not depend on orientation of electric field with respect to the direction of gravity. The collision section G is approximated by the following expression:

$$G = 0.57\pi R_1^2 S_2^2(1 - \exp(-5.75k)). \tag{13.16}$$

At gravitational sedimentation drops of the various size move with velocities $U_1 = 2\Delta\rho g R_1^2/9\mu_e$ and $U_2 = 2\Delta\rho g R_2^2/9\mu_e$. In this case the collision frequency of drops is determined by the expression

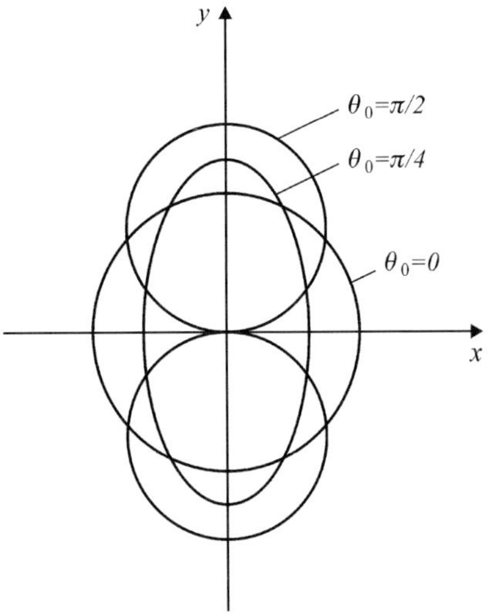

Fig. 13.6 Shapes of collision cross sections of drops for various values of angle θ_0 between E_0 and g.

$$\beta_{12} = \mathrm{G}|U_1 - U_2|n_2 n_1 = \frac{2\Delta\rho g R_1^2}{9\mu_e}(1 - \mathrm{k}^2)\mathrm{G}n_2 n_1. \tag{13.17}$$

The coagulation constant K figuring in the kinetic equation of coagulation (11.1), is, by its physical meaning, equal to the collision frequency of drops with unit concentration, therefore

$$K = \frac{2\Delta\rho g R_1^2}{9\mu_e}(1 - \mathrm{k}^2)\mathrm{G}. \tag{13.18}$$

As it was stressed in Section 11.1, the constant of coagulation is symmetric function of interacting drop volumes V and ω. Therefore condition $K(V, \omega) = K(\omega, V)$ should be satisfied. Since the dependence (13.18), in view of (13.16), does not provide such property, the approximate expression [100]

$$K(V, \omega) = 0.131\frac{\varepsilon E_0^2}{\mu_e}\frac{(V\omega)^{2/3}}{(V^{1/3} + \omega^{1/3})}\left(1 - 2.4\frac{(V\omega)^{1/3}}{(V^{1/3} + \omega^{1/3})^2}\right) \tag{13.19}$$

can be used as symmetric function K in this case.

Consider now the qualitative behavior of the solution of system of equations (13.7) when drops carry free charges, that is, when $N_1 \neq 0$ and $N_2 \neq 0$. If the external electric field is absent ($S_2 = 0$), then $\xi_1(x) = \xi_4(x) = 0$, $\xi_2(x) < 0$, and there are two infinite-isoclinic lines $\theta(x) = 0$, $\theta(x) = \pi$ and one zero-isoclinic line

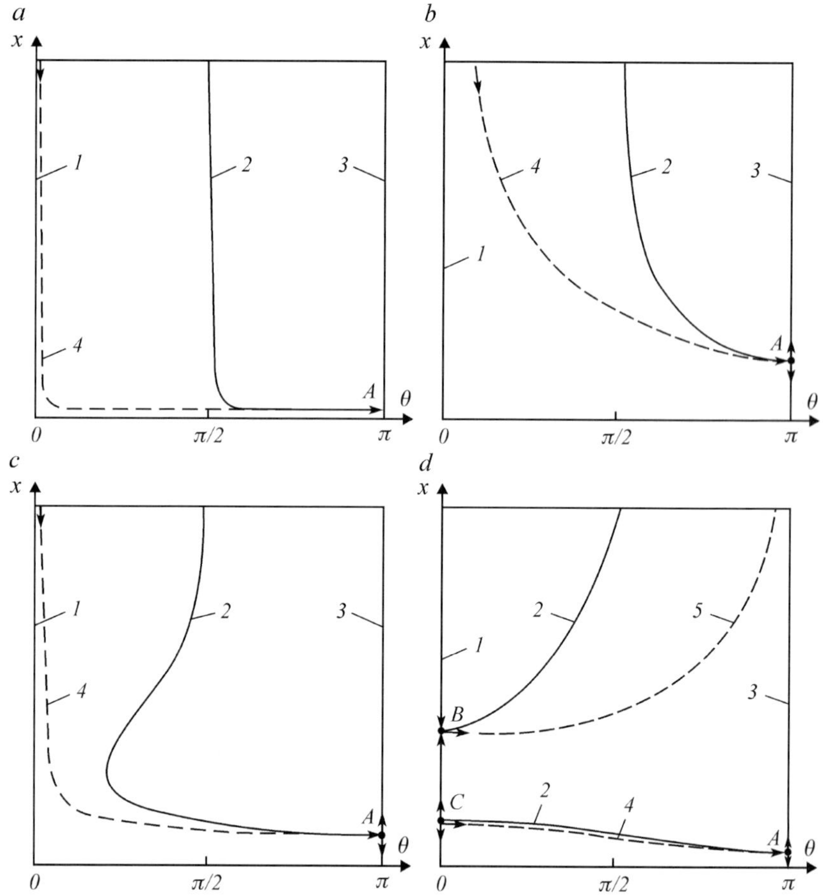

Fig. 13.7 Zero- and infinite-isoclinic lines and the singular point of the velocity of a small drop. The external electric field is absent ($S_2 = 0$): a – $N_1 = N_2 = 0$; b – $N_1 N_2 < 0$; c – $0 < N_1 N_2 < 1$; d – $N_1 N_2 \geq 1.5$; 1, 3 – infinite-isoclinic lines; 2 – zero-isoclinic lines; 4 – separatix of singular point A-saddle; 5 – separatix of singular point B-saddle; C – singular point – unstable node.

$\theta(x) = \arccos(-\xi_3(x)/\xi_2(x))$ in the phase space (x, θ). When the electric forces are attractive for all x (only one drop is charged ($N_1 N_2 = 0$) or the drops have opposite charges ($N_1 N_2 < 0$)), we have $-\xi_3(x)/\xi_2(x) < 0$ for all x, and the zero-isoclinic line lies wholly in the region $\theta(x) \geq \pi/2$ (Fig. 13.7, a, b). Intersection of the zero-isoclinic line and the infinite-isoclinic line $\theta(x) = \pi$ gives a unique singular point A (saddle). As well as in the case of non-charged drops, the separatrix of point A divides the phase space into two regions and represents the critical motion trajectory of the small drop relative to the large one.

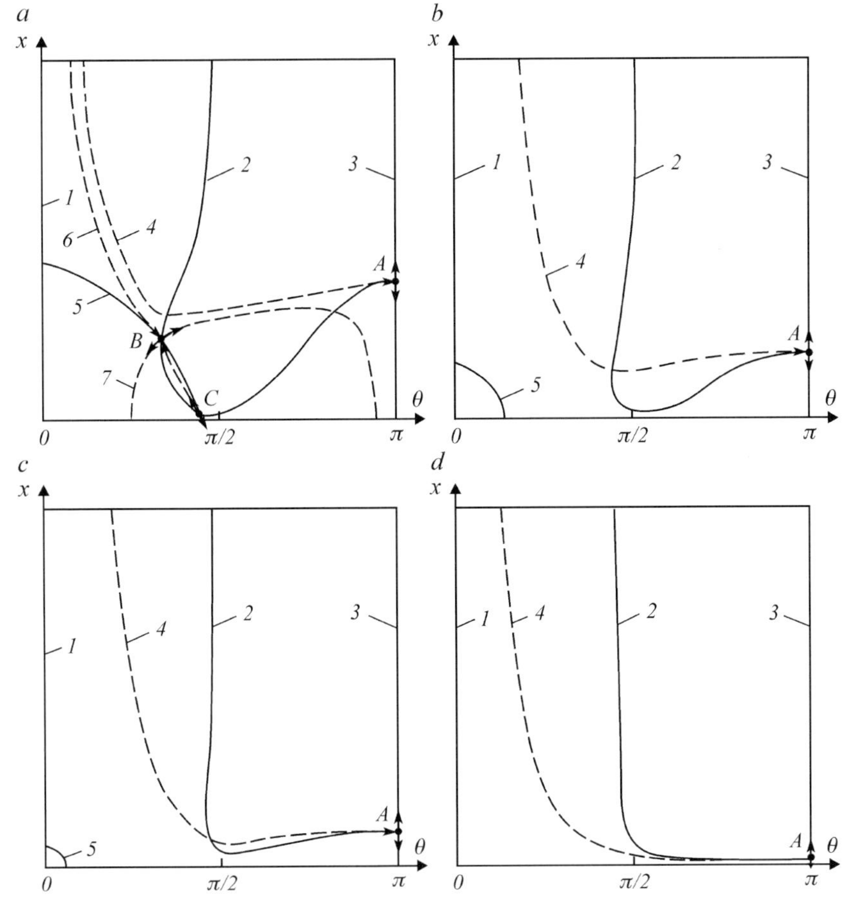

Fig. 13.8 Zero- and infinite-isoclinic lines and singular points of velocities of a small drop. The large drop has zero charge ($N_1 = 0$); $N_2S_2 > 0$: $a - S_2^3 \sim 4$; $S_2N_2 \ll 1$; $b - S_2^2 \leq 3$; $S_2N_2 \sim 0.5$; $c - 0.5 \leq S_2N_2 \leq 1$; $d - S_2N_2 \geq 1$; 1, 3, 5 – infinite-isoclinic lines; 2 – zero-isoclinic lines; 4 – separatix of singular point A-saddle; 6, 7 – separatix of singular point B-saddle; C – singular point – unstable node.

The situation changes when the drops have charges of the same sign, $N_1N_2 > 0$, because, as was shown in Section 12.6, depending on the ratio of the radii and charges of the drops (unclear: (ratio of radii) AND charges, or ratio of (radii AND charges) –VF) and on the distance between them, they can be attracted to as well as repelled from each other. For those values of x at which the electric force appears as a repulsive force, we have $\xi_3(x) > 0$, and the zero-isoclinic line $\theta(x) = \arccos(-\xi_3(x)/\xi_2(x))$ lies in the region $\theta(x) < \pi/2$. For all other x, this line lies in the region $\theta(x) \geq \pi/2$. If $N_1N_2 < 1.5$, then $\xi_3(x) < \xi_2(x)$

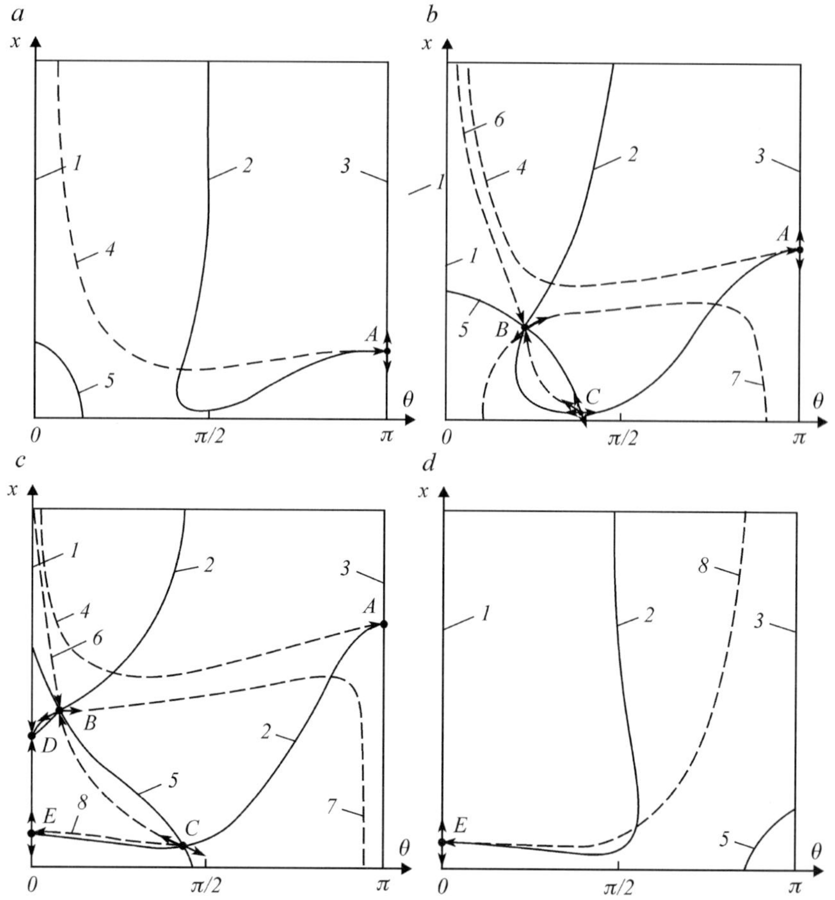

Fig. 13.9 Zero- and infinite-isoclinic lines of the velocity field of a small drop. The large drop is not charged ($N_1 = 0$); $N_2S_2 < 0$: a – $0 > S_2N_2 \geq -0.2$; b – $-0.2 > S_2N_2 \geq -0.5$; c – $-0.5 > S_2N_2 \geq -1$; d – $S_2N_2 < -1$; 1, 3, 5 – infinite-isoclinic lines; 2 – zero-isoclinic lines; 4 – separatix of singular point A (saddle); 6, 7 – separatix of singular point B (saddle); 8 – separatix of singular point E (saddle); C – singular point (unstable node); D – singular point (unstable node).

for all x. In this case, intersection of the zero-isoclinic line and the infinite-isoclinic line $\theta(x) = \pi$ gives the infinity point A (saddle). The collision cross section of the drops is determined by the separatrix of point A (Fig. 13.7, c). At $N_1N_2 \geq 1.5$, intersection of the zero-isoclinic line and the infinite-isoclinic line $\theta(x) = 0$ gives two singular points: B (saddle) and C (unstable node) (Fig. 13.7, d). The separatrix of point B (curve 5) does not intersect the boundary $x = 0$, therefore none of the trajectories from the region $x \gg 1$ reaches the boundary $x = 0$, and no collision of the drops occurs in this case.

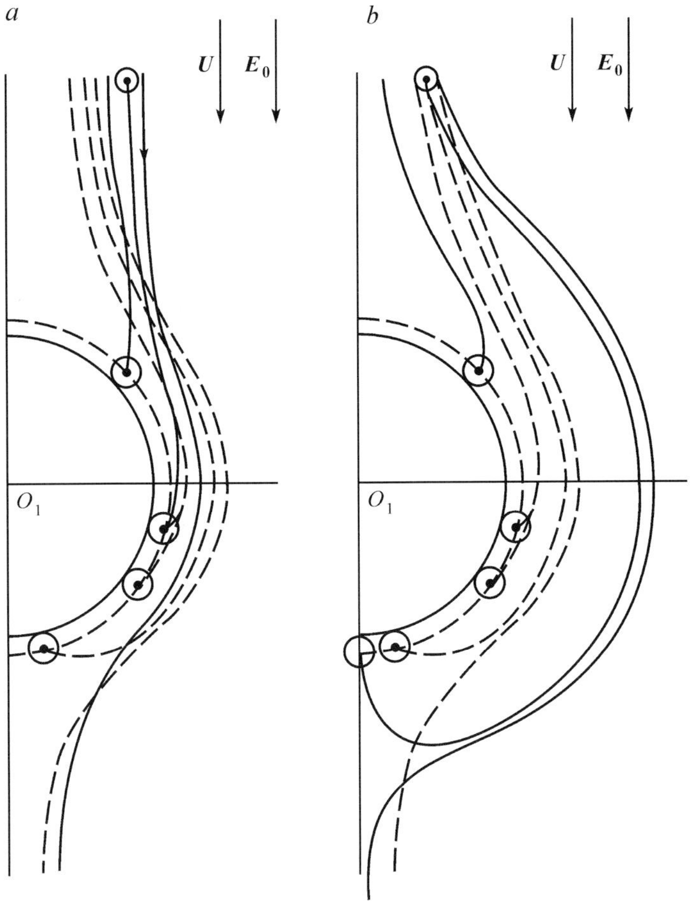

Fig. 13.10 Trajectories of motion of a small charged drop relative to a big drop with zero charge at $k = 0.1$; $N_1 = 0$: a – $N_2 > 0$; b – $N_2 < 0$; full lines – $S_2^2 = 1$; $|N_2| = 0.2$; dashed lines – $S_2^2 = 1$; $N_2 = 0$.

The behavior of the solution of equations (13.7) becomes considerably more complicated when the charged drops are placed in an external electric field ($S_2 \neq 0$). If $N_2 S_2 > 0$, but $N_1 S_2 < 0$, then $\xi_3(x)/\xi_1(x) > -\sqrt{2}/2$ and the zero-isoclinic line lies in the region $\theta(x) > \pi/4$. In this case the velocity field has either three or one singular point. These situations are shown in Fig. 13.8 for the special case $N_1 = 0$.

Consider them in succession. If $S_2^2 \geq 4$ and $S_2 N_2 \ll 1$, there are three singular points: A (saddle), B (saddle) and C (unstable node) (Fig. 13.8, *a*). If $S_2^2 < 3$ or $S_2 N_2 \geq 1$, there is only one singular point A (saddle) (Fig. 13.8, *b*, *c*, *d*). The critical trajectory at $N_2 S_2 > 0$ and $N_1 S_2 \leq 0$ is always defined by the separatrix of singular point A irrespective of the values S_2, N_1 and N_2.

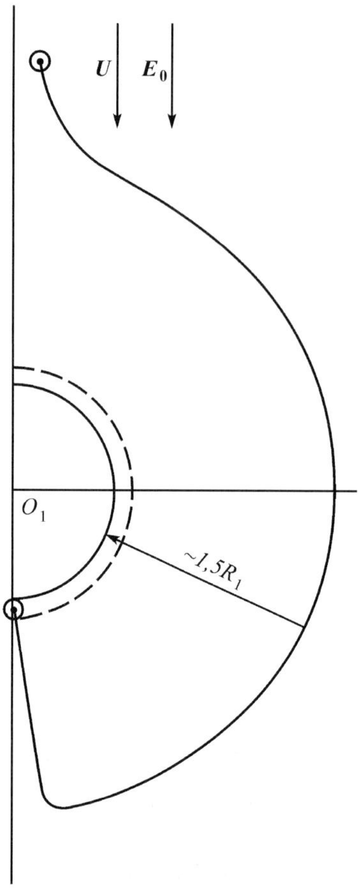

Fig. 13.11 Near-critical trajectories of a small charged drop relative to big drop with no charge for $k = 0.1$; $S_2 = 1$; $N_1 = 0$; $N_2 = -0.5$.

If $N_2 S_2 < 0$, then, depending on the values of parameters S_2, N_1 and N_2, the ratio $\xi_3(x)/\xi_1(x)$ can take various values from $-\infty$ up to $+\infty$. Then, for the value $N_1 = 0$, there are four various possible cases shown in Fig. 13.9.

1. $|S_2 N_2| \ll 1$, $S_2 N_2 < 0$. In this case there is only one singular point A (saddle), whose separatrix defines the collision cross section of drops (Fig. 13.9, a).

2. $-0.5 < S_2 N_2 \leq 0.2$. There are three singular points in this case: point A (saddle), point B (saddle), and point C (unstable node) (Fig. 13.9, b). The collision cross section is defined by the separatrix of point A.

3. $S_2 N_2 \geq 1$. There are five singular points: A (saddle), B (saddle), C (unstable node), D (stable node), E (saddle) (Fig. 13.9, *c*). The collision cross section is defined by the separatrix of points A and B.

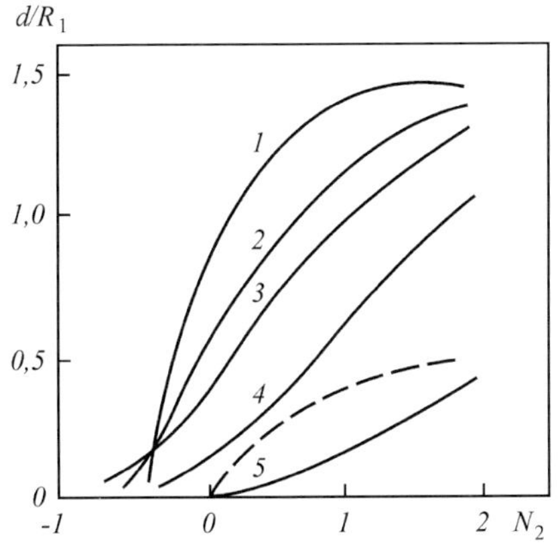

Fig. 13.12 Dependence of the radius of cross section of collision of a small charged drop with a big uncharged drop on N_2 for $k = 0.1$; $N_1 = 0$: 1 – 5 – $S_2^2 = 3; 1; 0.5; 0.1; 0$. Dashed line – dependence (13.20) at $k = 0.1$.

4. $S_2N_2 < -1$. In this case the relative motion of a small drop at a large distance from the big drop changes its direction and the collision cross section is defined in the same manner as it was done for $S_2N_2 > 0$ (Fig. 13.9, *d*).

The motion trajectories of small charged drop relative to big drop with zero charge, that were obtained by numerical integration of equations (13.7), [16] are shown in Fig. 13.10, a for $N_1 = 0$; $N_2 = 0.2$; $k = 0.1$, $S_2 = 1$ and in Fig. 13.10, *b* for $N_1 = 0$; $k = 0.1$; $S_2 = 1$, and $N_2 = -0.2$. For comparison purposes, Fig. 13.10 also provides trajectories of non-charged particles for the same values of parameters k and S_2; they are shown by dashed lines. Note that the near-critical trajectory in Fig. 13.11 corresponds to the case 2. The dependence of collision cross section radius on the small particle's charge for $k = 0.1$ and $N_1 = 0$ is given in Fig. 13.12.

Curves numbered from 1 up to 5 correspond to the values of the parameter of electrohydrodynamical interaction $S_2^2 = 3; 1; 0.5; 0.1$ and 0. At $N_2 > 0$, the smaller drop carries a positive charge and the collision cross section increases with N_2. At $N_2 < 0$, the drop has a negative charge and the collision cross section sharply decreases with the growth of $|N_2|$, vanishing at $N_2 = -1/S_2$. A similar problem was solved in [14] on the basis of approximated expressions for the interaction forces between drops. In this paper, the following expression was obtained for the collision cross section radius:

$$d/R_2 = (2\mathbf{k}N_2)^{1/2} \tag{13.20}$$

It is assumed that the external electric field is absent and $2\mathbf{k} \ll N_2$. The dependence (13.20) is shown in Fig. 13.12 by a dashed line. A comparison of this curve with the corresponding curve 5 obtained by using the exact expressions for the interaction forces between drops, shows that the use of approximate expressions for electric and hydrodynamic interaction forces in calculations results in a significant overestimation of the collision cross section, especially noticeable at small and intermediate values of the parameter N_2. As N_2 increases, the main part of the limiting trajectory recedes from the surface of the big drop to a distance at which interaction forces are accurately described by the approximate expressions. In addition to the increased collision cross section of drops, the growth of $|N_2|$ also results in the growth of their approach velocity, which causes an additional increase of the collision frequency. Therefore the rate of capture of fine charged drops by large uncharged drops depends not only on the drop charge but also on the direction of the strength vector of external electric field.

13.2
The Kinetics of Drop Coalescence During Gravitational Separation of an Emulsion in an Electric Field

To describe the dynamics of drop integration when the emulsion undergoes separation in a gravitational field, it is necessary to turn to the kinetic equation (11.1) describing the volume distribution of drops $n(V, t)$ as a function of time. The kernel of coalescence $K(V, \omega)$ entering this equation is determined by the expression (13.19). One remark should be made concerning this expression. It has the meaning of drop collision frequency for drop volumes V and ω. However, not every collision of conducting drops in an electric field results in their coalescence. Indeed, as was shown in Section 12.5, after the contact, the drop experiences repulsion, and if the molecular force of attraction does not compensate for the repulsion force, the coalescence of drops will not take place. The electric force of repulsion between the touching, initially uncharged drops can be estimated by the formula (12.81)

$$F_{e1} = \varepsilon E_0^2 \varphi(\mathbf{k}) \frac{R_1^2 R_2^2}{(R_1 + R_2)^2}, \quad \varphi(\mathbf{k}) = \tilde{f_1} - A_1. \tag{13.21}$$

The force of molecular attraction between two touching spherical drops is determined by Derjaguin's formula [101]

$$F_{mol} = \frac{2\pi \Sigma_0 R_1 R_2}{R_1 + R_2}, \tag{13.22}$$

where Σ_0 is the surface tension factor of the drop.

The condition of coalescence of drops after the contact now can be written as

$$\frac{2\pi \Sigma_0 R_1 R_2}{R_1 + R_2} > \varepsilon E_0^2 \varphi(\mathbf{k}) \frac{R_1^2 R_2^2}{(R_1 + R_2)^2}. \tag{13.23}$$

To estimate the size of drops, we shall set $k = 1$, in other words, we will be considering drops of equal size. In this case, $\varphi(1) = 5.46$, and (13.23) transforms to

$$R < 2.3\Sigma_0 \varepsilon^{-1} E_0^{-2}.$$

A more accurate estimate can be made using experimental data as discussed in [94, 101]:

$$R < 0.75\Sigma_0 \varepsilon^{-1} E_0^{-2} = R_{cr}. \tag{13.24}$$

The inequality (13.24) can be rewritten in terms of drop volumes as

$$V < V_{cr} = 1.75 \frac{\Sigma_0^3}{\varepsilon^3 E_0^6}. \tag{13.25}$$

Hence, for the drops of volumes $V < V_{cr}$, each collision results in coalescence, and the coalescence kernel of such drops is equal to their collision frequency.

For further calculations, it will be convenient to present $K(V, \omega)$ as a power function of volumes, using the approximation [1]

$$\frac{(V\omega)^{2/3}}{V^{1/3} + \omega^{1/3}} \approx 0.5 V^{1/2} \omega^{1/2},$$

which makes it possible to write the coalescence kernel as

$$K(V, \omega) = \begin{cases} 0.026 \dfrac{\varepsilon E_0^2}{\mu_e} V^{0.5} \omega^{0.5} & \text{at } V, \omega < V_{cr}, \\ 0 & \text{at } V, \omega > V_{cr}. \end{cases} \tag{13.26}$$

If the average volume of drops in the emulsion $V_{av} \ll V_{cr}$, then the fraction of drops with volumes $V \geq V_{cr}$ is small and it is safe to neglect these drops when doing calculations. Then we can say that

$$K(V, \omega) = 0.026 \frac{\varepsilon E_0^2}{\mu_e} V^{0.5} \omega^{0.5} \tag{13.27}$$

holds in the entire range of drop volumes.

On the other hand, if $V_{av} \gg V_{cr}$, then for all practical purposes, no integration of drops takes place.

For the characteristic values $E_0 \sim 2$ kV/cm, $\Sigma_0 \sim 10^{-2}$ N/m, and $\varepsilon \sim 2\varepsilon_0$, we have $R_{cr} \sim 10$ m. Because the size of drops that present practical interest is much smaller ($R \leq 10^{-4}$), further on we shall be using the expression (13.27) for the coalescence kernel.

Consider now the dynamics of drop integration. Start with the elementary case of a monodisperse emulsion. Assume that the emulsion remains monodisperse in the course of coalescence. Then drop collision frequency for drops of equal volume can be found from (13.27) by multiplying K by the number concentration of drops N and then dividing it by 2, because at $V = \omega$, every collision gets counted twice:

$$\beta_{11} = \frac{1}{2} K(V, V) N = 0.013 \frac{\varepsilon E_0^2}{\mu_e} VN = 0.013 \frac{\varepsilon E_0^2}{\mu_e} W, \tag{13.28}$$

where $W = VN$ is the volume concentration of drops in the emulsion. W remains constant in course of coalescence if the disperse phase is not removed from the considered volume.

The change of number concentration of drops is described by the drop-number-balance equation

$$\frac{dN}{dt} = -\beta_{11} N; \quad N(0) = N_0. \tag{13.29}$$

Substituting into (13.29) the expression (13.28) for β_{11}, we get

$$N(t) = N_0 \exp\left(-0.013 \frac{\varepsilon E_0^2}{\mu_e} Wt\right). \tag{13.30}$$

Since the drop volume $V = W/N$ changes with time, the radius of drops should be written as

$$R(t) = R_0 \left(\frac{V}{V_0}\right)^{1/3} = R_0 \exp\left(-0.004 \frac{\varepsilon E_0^2}{\mu_e} Wt\right), \tag{13.31}$$

where R_0 is the initial drop radius.

It follows from (13.31) that the characteristic time of drop integration, that is, the time during which the drop radius will increase by the factor of e is

$$T_{mono} = 250 \mu_e / \varepsilon E_0^2 W. \tag{13.32}$$

The rate of drop integration grows with the reduction of viscosity of the liquid surrounding the drop, and with the increase of electric field strength and volume concentration of drops. Note that within the framework of the adopted model of pair collisions it is necessary that $W \ll 1$. Thus the average distance between drops of radius R is estimated as $r_{av} \sim R/W^{1/3} \gg R$. This inequality enables us to limit ourselves to pair collisions only. For the characteristic parameter values $\mu_e = 10^{-2}$ Pa·s, $\varepsilon \sim 2\varepsilon_0$, $E_0 = 1$ kV/cm, $W = 0.01$, the characteristic time is $T_{mono} = 12.5$ s. So, to increase the drop radius from 10 to 100 μm, it is necessary to separate the emulsion for 25 s.

The simple model described above lies in the basis of the estimation of characteristic integration time for drops in the emulsion. Real drops have various sizes, and they can be characterized by a continuous volume distribution $n(V,t)$, where V is the drop volume.

Consider now drop coalescence in a polydisperse emulsion in the process of sedimentation in a gravitational field in the presence of a uniform external electric field. Since the capture cross section of drops does not depend much on the direction of the electric field, the direction of $\boldsymbol{E}_0$ is taken to be arbitrary.

We shall consider only the homogeneous case, i.e. the size distribution of drops depends only on V and t. Physically, it may correspond to the process of drop coalescence under the weak mixing condition and in the absence of drop breakage. In the current case the kinetic equation (11.1) will become

$$\frac{\partial n}{\partial t} = I_{coll}; \quad n(V,0) = n_0(V). \tag{13.33}$$

Note that the kernel of coagulation (13.27) does not obey the necessary conditions of existence of a self-similar solution. Therefore we shall consider some results of the numerical solution and obtain an approximate solution of equation (13.33) by the method of moments.

Suppose the initial distribution is a logarithmic normal distribution

$$n_0(V) = \frac{n^* V_0}{s_0 V} \exp\left(\frac{\ln^2(V/V_0)}{2s_0^2}\right). \tag{13.34}$$

which is a typical case for emulsions moving inside pipes. Here s_0^2 is the distribution variance, $V_0 = V_{av}^{(0)} \exp(-s_0^2/2)$, $V_{av}^{(0)}$ is the average drop volume in the initial distribution (13.34), and n^* is the parameter which is obtained from the condition that the volume concentration of drops is initially given by

$$W = \int_0^\infty V n_0 \, dn. \tag{13.35}$$

Fig. 13.13 shows the dynamics of radius distribution of drops for two initial distributions (13.34) with equal average drop volumes $V_{av}^{(0)}$ but different variances: $s_0^2 = 0.09$ (Fig. 13.13, *a*) and $s_0^2 = 1$ (Fig. 13.13, *b*). For convenience, the dimensionless drop radius $r_1 = (V/V_0)^{1/3}$ is plotted on the horizontal axis and a dimensionless time $\tau = 0.024\varepsilon E_0^2 Wt/\pi\mu_e$ is introduced. Here $\omega = VV_0 n/W$.

The coalescence rate decreases with time, because the number of drops in a unit volume of the emulsion $N(t)$ (Fig. 13.14) diminishes considerably. The characteristic coalescence time is estimated from the equality $\tau \sim 1$, which gives us

$$T_{poly} \approx 130\mu_e/\varepsilon E_0^2 W. \tag{13.36}$$

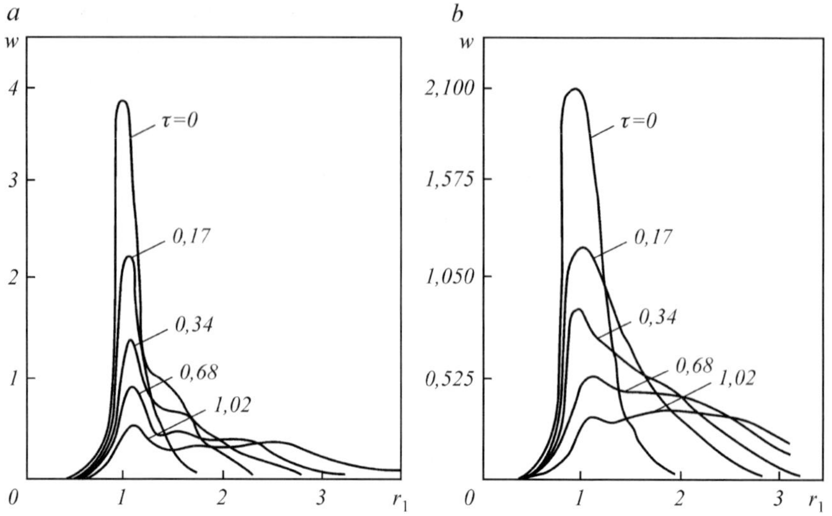

Fig. 13.13 Dynamics of the radius distribution of drops: $a - s_0^2 = 0.09$; $b - s_0^2 = 1$.

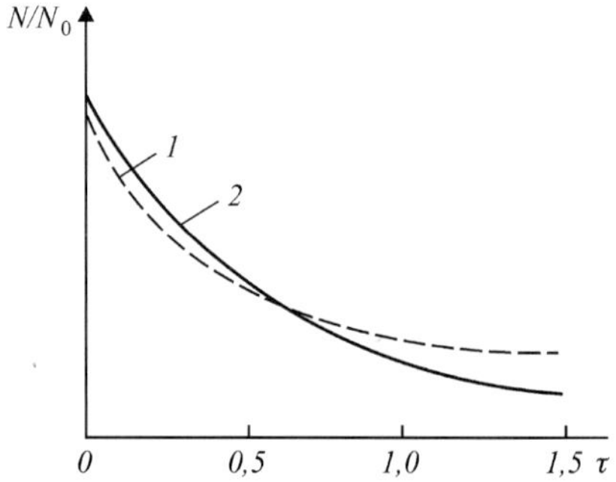

Fig. 13.14 Number concentration of drops as a function of time: 1 – numerical solution, 2 – approximate solution (3.41) at $k = 2$.

Now it is possible to compare the characteristic coalescence times in mono- and polydisperse emulsions. From (13.32) and (13.36), it follows that

$$T_{mono}/T_{poly} \approx 2. \tag{13.37}$$

Thus, the coalescence rate in a polydisperse emulsion is almost two times smaller than in a monodisperse one. It is explained by the fact that in an electric field, the collision frequency is higher for drops of commensurable sizes than for drops whose sizes differ by a lot (see Fig. 13.4).

We now obtain an approximate solution by the method of moments (see Section 11.1), using the parametric method and assuming that the initial distribution is a gamma – distribution,

$$n_0(V) = \frac{W(\mathsf{k}+1)}{V_0^{(0)^2}\mathsf{k}!}\left(\frac{V}{V_0^{(0)}}\right)^{\mathsf{k}} \exp\left(\frac{V}{V_0^{(0)}}\right). \tag{13.38}$$

The parameters V_0 and k that appear in this distribution are connected to the average volume $V_{av}^{(0)}$ and the variance s_0^2 by the relations

$$V_{av}^{(0)} = (\mathsf{k}+1)V_0; \quad \frac{s_0}{V_{av}^{(0)}} = (\mathsf{k}+1)^{-1/2}. \tag{13.39}$$

Suppose that in the course of coalescence, the distribution $n(V,t)$ remains in the class of gamma – distributions

$$n(V,t) = \frac{W(\mathsf{k}+1)}{V_0^2\mathsf{k}!}\left(\frac{V}{V_0}\right)^{\mathsf{k}} \exp\left(\frac{V}{V_0}\right), \tag{13.40}$$

and only the average drop volume $V_{av} = (\mathsf{k}+1)V_0$ and $s = V_{av}(\mathsf{k}+1)^{-1/2}$ change with time. The volume concentration of drops W remains constant, while the number concentration of drops $N = W/V_{av}$ varies.

As a result, the number of drops in a unit volume of the emulsion is obtained as a function of time

$$N = N_0 \exp(-\alpha_{\mathsf{k}}\tau). \tag{13.41}$$

Here τ is the dimensionless time introduced earlier and

$$\alpha_k = \frac{1.2(2k+2)!\Phi(1;2k+3;k+2.45;0.5)}{(k+1.45)(k+1)2^{2k+3}(k!)^2},$$

where $\Phi(\alpha;\beta;\gamma;z)$ is a hypergeometric function [5].

In Fig. 13.14, the dependence (13.41) is shown for $\mathsf{k}=2$. A comparison with the numerical solution shows that the approximate solution agrees with the exact one to a sufficient accuracy within the range of τ up to $\tau \sim 1$. At $\tau > 1$, the exact solution predicts a more rapid slowdown of the coalescence process than the approximate one.

Finally, estimate the time it takes for the process of drop coalescence in an electric field to be competed. This will happen when that the average drop radius becomes equal to R_{cr}, which is determined by the formula (13.24). Use the expression (13.31) the time dependence of the drop radius in a monodisperse emulsion. Then

$$t_{coal} \sim T_{mono} \ln(0.75\Sigma_0 \varepsilon^{-1} E_0^{-2} R_0^{-1}), \tag{13.42}$$

where R_0 is the initial drop radius before the electric field is turned on.

For the characteristic values $E_0 \sim 1$ kV/cm, $\Sigma_0 \sim 10^{-2}$ N/m and $\varepsilon \sim 2\varepsilon_0$, $\mu_e = 10^{-2}$ Pa·s we have

$$t_{coal} \sim 8.2 T_{mono} \sim 100 \text{ s}. \tag{13.43}$$

13.3 Gravitational Sedimentation of a Bidisperse Emulsion in an Electric Field

Consider one application – sedimentation of a bidisperse emulsion. Suppose the disperse phase of the emulsion at the initial moment consists of conducting uncharged drops of two kinds, with volumes V_1 and V_2, $(V_1 > V_2)$, and volume concentrations W_{10} and W_{20}. The emulsion is placed into an external uniform electric field of strength E_0 parallel to the direction of gravity. Sedimentation of drops takes place at $t > 0$. The volume of small drops V_2 and the number concentration of big drops $n1$ are assumed to remain constant in the course of sedimentation. Physically, this means that big drops do not interact among themselves, and are integrated by collisions with small drops of volume V_2, which, in their turn, do not interact among themselves. In view of these assumptions, we write the following equations describing the volume change of drops of kind 1:

$$\frac{dV_1}{dt} = \frac{4}{3}\pi R_2^3 G n_1 (V_1 - V_2); \quad V_2(n_{20} - n_{10}) = (V_2 - V_{10}) n_1; \\ V_1(0) = V_{10}. \tag{13.44}$$

Substituting into (13.44) the expression (13.16) for collision cross section G and introducing dimensionless variables $\tau = 0.024\varepsilon E_0^2 W t/\pi\mu_e$, $\mathsf{k} = R_2/R_1$, we obtain

$$\frac{dV_1}{dt} = -k^2(1 - \exp(-5.75k))\left(1 - \frac{W_{10}}{W}\frac{k_0^3}{k^3}\right); \\ \frac{W_2}{W} = 1 - \frac{W_{10}}{W}\left(\frac{k_0}{k}\right); \quad k(0) = k_0, \tag{13.45}$$

where W is the net volume concentration (concentration of the disperse phase as a whole).

In Fig. 13.15, the dependence of W_2/W on τ is shown at $W_{10}/W = 0.5$, for various initial ratios $\boldsymbol{k_0}$ of the drop radiuses.

At $\tau \to \infty$, the ratio $W_2/W \to 0$, and the characteristic time τ_{cr} during which large drops pick out small ones, decreases with the increase of $\boldsymbol{k_0}$. It means that the smaller the size difference between the two kinds of drops, the faster the drops of kind 1 will pick out drops of kind 2.

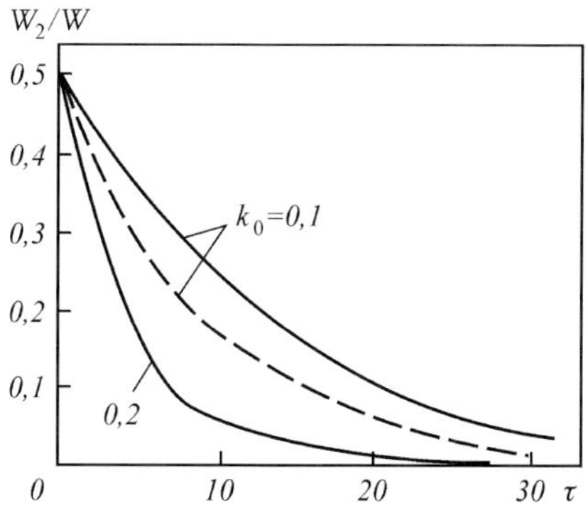

Fig. 13.15 Relative concentration of small drops as a function of time. Dashed line – dependence (13.49).

In the case $W_2/W \ll 1$, (13.45) gives us the following approximate solution

$$\frac{W_2}{W} \approx \begin{cases} \dfrac{W_{20}}{W} \exp(-17.25B^3\tau) & \text{at } k_0 \ll 0.15, \\ \dfrac{W_{20}}{W} \exp(-3B^2\tau) & \text{at } k_0 \geq 0.15, \end{cases} \tag{13.46}$$

where the parameter B equals $B = (W_{10}/W)^{1/3}\mathrm{k}_0$.

It follows from (13.46) that the characteristic sedimentation time is $\tau_{cr} \sim \mathrm{k}_0^{-2}$ for drops with $\mathrm{k}_0 \ll 1$ and $\tau_{cr} \sim \mathrm{k}_0^{-2}$ for drops of commensurable sizes. The dependence (13.46) is shown in Fig. 13.15 by the dashed line ($k_0 = 0.1$; $W_{20}/W = 0.5$).

Consider now a bidisperse emulsion, distributed uniformly in a layer between parallel horizontal planes $z = 0$ and $z = H$ at the initial moment. The vertical z-axis is directed parallel to gravity. The initial volume concentrations of drops are small ($W_{10} \ll 1$ and $W_{20} \ll 1$), so the "hindered" character of sedimentation can be neglected. After a certain time, all large drops of kind 1 will settle. We can select a column of a unit cross section in the emulsion volume and ask how many small drops will remain inside this column. A similar problem is of interest in laboratory studies on how the electric field affects the emulsion sedimentation rate.

Denote by $C_2(z)$ the volume concentration of drops of kind 2 that will remain in the cross-sectional layer at height z after the large drops that were initially in the layer above z pass through the layer z. The sedimentation rate of small drops is much lower than that of large ones, therefore for simplicity's sake, we can assume that there is no settling of small drops. Then after the sedimentation of large drops in the layer $0 < z < H$ is completed, the ratio of the volume of remaining drops to the total initial volume of drops will be equal to

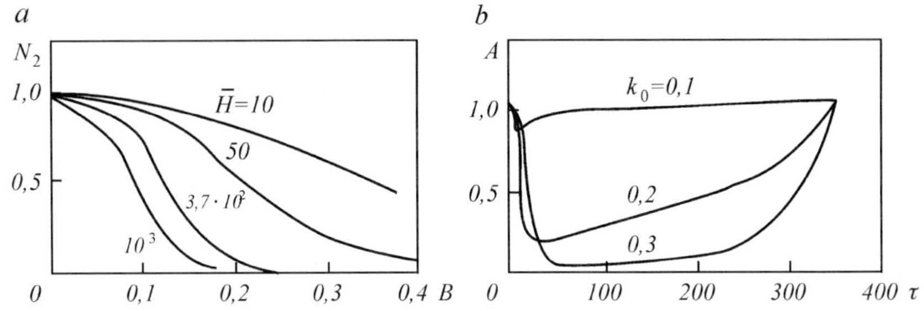

Fig. 13.16 Bidisperse emulsion: *a* – relative volume concentration of small drops; *b* – the effect of the electric field on the sedimentation of a bidisperse emulsion.

$$N_2 = \frac{1}{WH}\int_0^H W_2(z)\,dz. \tag{13.47}$$

Let us use Eq. (13.45) that was obtained earlier, and the approximate solution (13.46), making the substitution $dW_2/d\tau = \mathrm{k}_0^{-2}\,dW_2/dz$. For $W_2/W \ll 1$, we get

$$\frac{W_2}{W} \approx \begin{cases} \dfrac{W_{20}}{W}\exp(-17.25B^3\xi) & \text{at } k_0 \ll 0.15, \\[2ex] \dfrac{W_{20}}{W}\exp(-3B^2\xi) & \text{at } k_0 \geq 0.15, \end{cases} \tag{13.48}$$

where $\xi = W\varepsilon E_0^2 z/29.1\Delta\rho g R_2^2$.

The dependence of N_2 on B is shown in Fig. 13.16, *a* for various values of the dimensionless height of the layer $\overline{H} = W\varepsilon E_0^2 H/29.1\Delta\rho g R_2^2$.

Consider now the case of sedimentation of drops of both kinds. During the time t, small drops settling with the velocity $U_2 = 2\Delta\rho g R_2^2/9\mu_e$ will travel the distance $h_2 = U_2 t$, while the big drops will travel the distance $h_1 = \int_0^t U_1\,dt$. Using the relations (13.48), we obtain:

$$N_2(\tau) = W_{20}(1 - \exp(-(\overline{H}_1 - \tau)))\frac{1}{\alpha\overline{H}W} + 1 - \frac{\overline{H}}{H}, \tag{13.49}$$

$$\overline{H}_1 = \frac{1}{\alpha B^2}\left(\ln\left(\frac{e^{\alpha\tau} + W_{20}/3W}{1 + W_{20}/3W}\right) + \frac{W_{20}}{3W}\left(\frac{1}{e^{\alpha\tau} + W_{20}/3W} - \frac{1}{1 + W_{20}/3W}\right)\right),$$

$$\alpha = \begin{cases} 17.25B^3 & \text{at } k_0 \ll 0.15, \\ 3B^2 & \text{at } k_0 \geq 0.15, \end{cases}$$

The effect of the electric field on the process of sedimentation can be characterized by the parameter A, equal to the ratio of volume concentrations of small drops that remain in the layer after time t for sedimentation with and without the electric field:

$$A = \frac{WHN_2(\tau)}{W_{20}(H - h_2) + W_{10}(H - h_1)}. \tag{13.50}$$

The time dependence $A(\tau)$ is shown in Fig. 13.16, *b* for $W_{20}/W = 0.3$. Apparently, there are three dependencies distinct from each other. The first one occurs in the interval $0 < \tau < 40$. A sharp decrease of A is explained by the fact that a significant integration of big drops occurs as time goes on, and at $\tau \sim 40$ all of them are settled out from the layer. The second dependence has a small slope and holds in the interval $40 < \tau < 300$. This corresponds to the slow sedimentation of small drops. The increase of A at $300 < \tau < 370$ is caused by the fact that by this time the top emulsion layer, in which the concentration of small drops is maximal, reaches the bottom boundary of the selected layer.

13.4 The Effect of Electric Field on Emulsion Separation in a Gravitational Settler

An important part of oil processing is separation of water from oil. Water is present in oil in the form of drops whose radius can reach hundreds of microns. This process is carried out in big horizontal or vertical reservoirs – gravitational settlers. The separation of water from oil occurs inside them due to gravitational sedimentation of drops. The density difference between water and oil is insignificant ($\Delta\rho \sim 100$ kg/m^3) and drop sizes are small, therefore sedimentation rate is low, so a high-quality separation of water from oil takes a great amount of time and requires large overall dimensions of the settler. The size of devices can be reduced very substantially if the average size of water drops is increased during preliminary processing. One efficient way of preliminary treatment is applying electric field to the emulsion. The electric field can be created inside a settler called electrodehydrator or in special device (electrocoalescentor) placed before the settler.

Suppose the emulsion was first integrated in an electrocoalescentor before entering the settler. Let t denote the time spent by the emulsion in the electrocoalescentor, $n_0(V)$ – the volume distribution of drops in the emulsion at the entrance to, and $n(V, t)$ – at the exit from the electrocoalescentor.

After exiting the electrocoalescentor, the emulsion enters the settler, therefore $n(V, t)$ is also the volume distribution of drops in the emulsion at the settler entrance. Let $n_1(V, t)$ denote the distribution at the exit from the settler. The efficiency of emulsion separation inside the settler is characterized by the ablation factor, which is equal to the ratio of water volume per unit volume of the emulsion at the exit from the settler, to the same quantity at the entrance

$$\lambda = \frac{W_1}{W} = \int_0^\infty V n_1(V, t)\, dV \Big/ \int_0^\infty V n(V, t)\, dV. \tag{13.51}$$

It is evident that the smaller the factor λ, the higher the efficiency of emulsion separation in the settler.

The ratio below is known as the transfer function of the settler:

$$\Phi(r) = n_1(V,t)/n(V,t). \tag{13.52}$$

Transfer function depends on the hydrodynamical mode of drop motion in the settling section of the settler. In the elementary case of drop sedimentation, when drop interactions are not taken into account, the expressions for $\Phi(r)$ are easy to obtain when the emulsion motion is perpendicular or parallel to gravity [4]. In the former case the term "settler" refers to a horizontal one and

$$\Phi_h(r) = \begin{cases} 1-(r/r_{cr})^2 & \text{at } r < r_{cr}, \\ 0 & \text{at } r \geq r_{cr}, \end{cases} \tag{13.53}$$

In the latter case, the term "settler" refers to a vertical one and

$$\Phi_v(r) = \begin{cases} 1 & \text{at } r < r_{cr}, \\ 0 & \text{at } r \geq r_{cr}, \end{cases} \tag{13.54}$$

The quantity $r_{cr} = (9HQ\mu_e/2\Delta\rho V)^{1/2}$ corresponds to the maximum radius of drops at the exit from the settler, H and V are the height and volume of the settler, Q is the volume flow rate of the emulsion, μ_e is the viscosity of the continuous phase, $\Delta\rho$ is the density difference between the disperse and the continuous phase.

Substituting $\Phi(r)$ into the formula (13.51), we obtain expressions for the ablation factor of vertical and horizontal settlers, respectively

$$\lambda_v = \int_0^{V_{cr}} Vn_1(V,t)\,dV \Big/ \int_0^{\infty} Vn(V,t)\,dV, \tag{13.55}$$

$$\lambda_h = \lambda_v - \int_0^{V_{cr}} Vn_1(V,t)\,dV \Big/ \int_0^{\infty} Vn(V,t)\,dV, \tag{13.56}$$

where $V_{cr} = 4\pi r_{cr}^3/3$ is the critical drop volume.

It follows from (13.55) and (13.56) that $\lambda_h < \lambda_v$ if r_{cr} is the same for horizontal and vertical settlers. The higher efficiency of emulsion separation in a horizontal separator as compared to a vertical one is explained by of the shape of drop distributions at the settler exit (Fig. 13.17). In a vertical settler, the part of the initial distribution with $V > V_{cr}$ gets cut off at the entrance, since drops of such volume cannot move against gravity. In a horizontal settler, drops of all sizes settle continuously along the whole length, and not only drops with $V > V_{cr}$, but also a part of drops with $V < V_{cr}$ get settled out, which does not happen in a vertical settler.

To determine the ablation factor, one should first find $n(V,t)$ – the distribution of drops established in the emulsion after the emulsion was acted upon by elec-

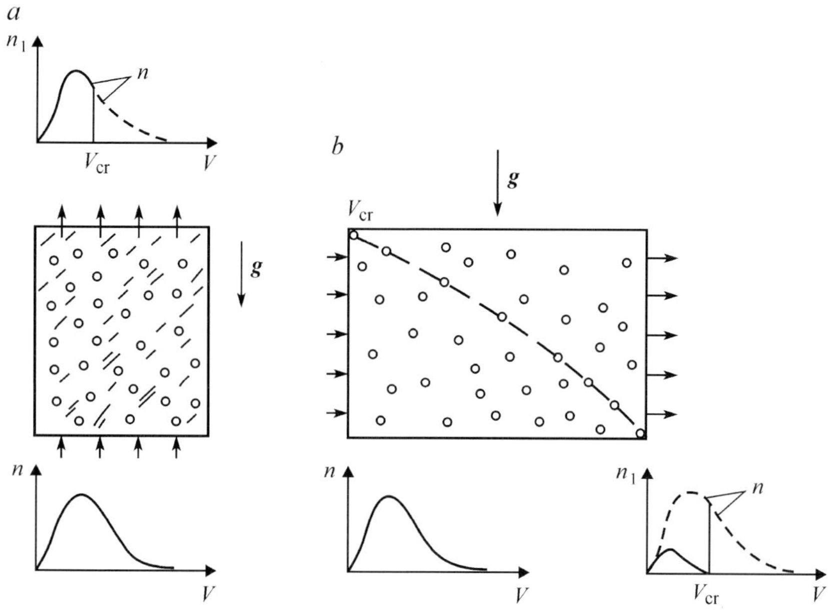

Fig. 13.17 Schematic sketch of a vertical (a) and a horizontal (b) settler.

tric field for the time t. Using the approximate solution (11.30) that was derived earlier, we write

$$n(V,t) = \frac{N^2(t)(k+1)}{k!W}\left(\frac{V}{V_0}\right)^4 \exp\left(-\frac{V}{V_0}\right). \tag{13.57}$$

Here $N = N_0 \exp(-\alpha k\tau)$, $V_0 = V_{av}^0 \exp(\alpha k\tau)/(k+1)$, $W = W_0$, $\tau = 0.024\varepsilon E_0^2 W_0 t/\pi\mu_e$, N_0, W_0, and V_{av}^0 are the initial values, that is, initial number and volume concentrations and initial average volume of drops, α_k is determined by the relation (13.41).

Substituting (13.57) into (13.55) and (13.56), we obtain

$$\lambda_v = \frac{\gamma(k+2,x)}{(k+1)!}; \quad \lambda_h = \lambda_v - \frac{\gamma(k+8/3,x)}{(k+1)!x^{2/3}}, \tag{13.58}$$

where $x = V_{cr}/V_0$; $\gamma(\alpha,x)$ is an incomplete gamma-function [5].

The expressions (13.58) describe the dependence of the ablation factor of vertical and horizontal settlers on the parameters of the initial drop distribution k and $V_{av}^0 = W_0/N_0$, the dimensionless time τ of emulsion treatment in the electric field, and also on the parameter V_{cr} describing the geometric and hydrodynamic parameters of the settler.

The dependences of λ_h (full lines) and λ_v (dashed lines) on the parameter $x_0 = V_{cr}/V_{av}^0$ for various values of τ are shown in Fig. 13.18. The curves at $\tau = 0$

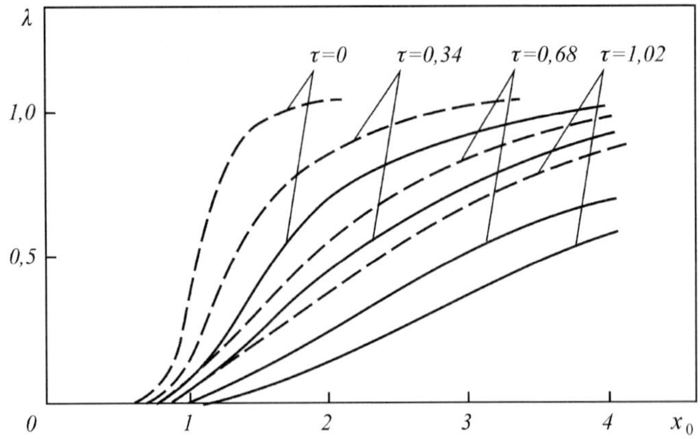

Fig. 13.18 The ablation factors of horizontal (full lines) and vertical (dashed lines) settlers as functions of duration τ of emulsion treatment in the presence of electric field.

correspond to the values $\lambda_h(0, x_0)$ and $\lambda_v(0, x_0)$ with no preliminary integration of the emulsion. Other curves correspond to longer emulsion treatment times: $\tau = 0.34; 0.68; 1.02$. As one should expect, the increase of τ results in a substantially lower ablation of water from the settler.

Now, compare the ablation factors of horizontal and vertical settlers. To this end, we shall introduce the parameter

$$\psi = \lambda_h(\tau, x_0)/\lambda_v(\tau, x_0) \tag{13.59}$$

At equal volume concentrations of water at the entrance of the two compared settlers, the parameter ψ is equal to the ratio of volume concentrations at the exit from the settler. It is obvious that $\psi < 1$ for all values of x_0 and that $\tau > 0$. Using the asymptotic behavior of the incomplete gamma – function, we find that at small values of x_0

$$\psi = \frac{2}{(3k+8)}\left(1 - x_0 \frac{k+2}{(k+3)(k+11/3)}\right), \tag{13.60}$$

while at $x_0 \gg 1$

$$\psi = 1 - x_0^{-2/3} \frac{\Gamma(k+8/3)}{(k+1)!}. \tag{13.61}$$

In particular, at $x_0 \to 0$ we have $\psi \to 2/(3k+8)$, and at $x_0 \to \infty$, $\psi \to 1$. Since ψ decreases at small values of x_0 and grows when x_0 is large, ψ has a minimum. The ablation factor of a horizontal settler is more sensitive to the duration of emulsion treatment in the electric field than that of a vertical settler.

The obtained dependences can be used to determine the duration of emulsion treatment in the electric field necessary to achieve a required volume concentration of the disperse phase at the settler exit.

13.5 Emulsion Flow Through an Electric Filter

The existing technology of removal of water from oil products enables us to produce fuel with residual water concentration of about 0.01%. However, even such small water contents may have a detrimental effect on serviceability of fuel in engines, especially when the engine is employed at low temperatures. An electric field can be used for deep dewatering of oil products, while removal of very low concentrations of water from oil emulsions can be carried out by using a package of mesh nozzles – electrodes arranged across the oncoming emulsion flow (Fig. 13.19).

The main principle of electric filter operation consists in the following. If we charge conducting water droplets and put them into the interelectrode space, the external electric field will push them toward one of electrodes, depending on the charge sign. When it touches the electrode, the water drop will be recharged and pushed away from the electrode – in the opposite direction. Thus, the space between the electrodes will be filled with two countercurrents of drops with opposite charges, which increases their frequency. As a result, drops will coalesce, increase in size and eventually settle. If electrodes form a mesh, then some of the drops can pass through the mesh without colliding. If the device consists of a series of meshes (a "mesh package"), drops will be successively captured by each mesh.

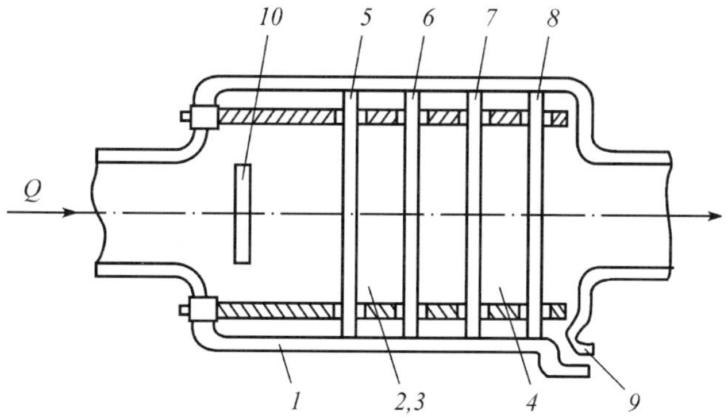

Fig. 13.19 Schematic sketch of an electric filter: 1 – body; 2 to 4 – filtering section; 5 to 8 – mesh electrodes; 9 – drainage system; 10 – damper.

The process can be described in terms of filtering of the disperse phase by a system of mesh electrodes. The primary goal is to determine the amount of residual water in the oil product at the filter exit. The volume concentration of water W_{out} at the exit depends on the volume concentration at the entrance W_{in}, disperse structure of the emulsion, electric field strength, construction parameters of the device, and physico-chemical properties of the emulsion.

Define the ablation factor of one filter section volume concentration of water in the emulsion at the exit divided by the concentration at the entrance:

$$\mathsf{K}_i = W_{out,\,i}/W_{in,\,i}. \tag{13.62}$$

Then the ablation factor of the filter is

$$\mathsf{K} = \prod_i \mathsf{K}_i. \tag{13.63}$$

Thus, to determine the volume concentration of water at the exit from the filter, $W_{out} = \mathsf{K} W_{in}$, we must know the ablation factor of each filter section. Take all sections of the filter to be identical and consider the process occurring in one section.

Suppose the emulsion consists of a non-conducting continuous phase and a conducting disperse phases in the form of identical small spherical drops of radius R carrying a constant charge q. The drops move between two parallel flat mesh electrodes. On a large distance from the electrodes, drops move perpendicularly to the mesh plane. Assume that drops acquire the charge as a result of collisions with electrodes. Then, according to [98],

$$q = \frac{2}{3}\pi^3 \varepsilon_0 \varepsilon E_0 R^3. \tag{13.64}$$

Note that the formula (13.64) was derived for a flat continuous electrode. If the drop size is small in comparison with the wire diameter of a mesh electrode, then in the vicinity of a cylindrical wire, drops behave as if the wire were a flat wall, therefore the use of the relation (13.64) is permissible.

Small volume concentrations of water in the emulsion make it possible to limit the analysis to the motion of isolated drops, ignoring their influence on each other and on the continuous phase. Since trajectories of drops are curved only near electrode, it suffices to consider the behavior of drops in the immediate vicinity of the electrode elements, that is, near cylindrical wires. The motion of charged particles flowing around a single cylindrical electrode has been considered in [102].

Choose the system of coordinates XYZ attached to the plane of one of the electrodes (Fig. 3.20).

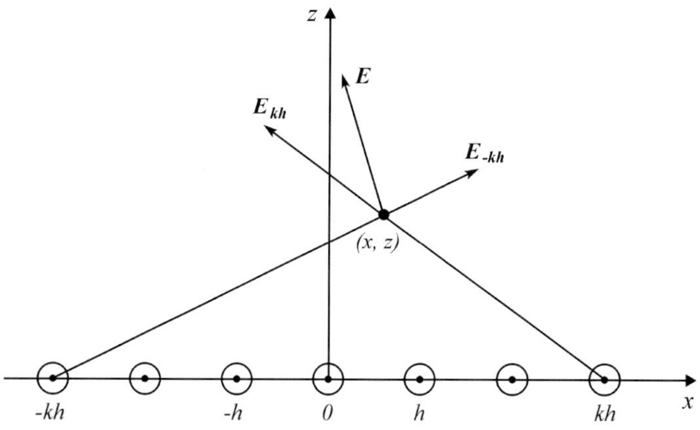

Fig. 13.20 System of coordinates attached to the electrode plane.

Taking the inertialess approximation and neglecting gravity, we can write the equation of motion for a drop as

$$\boldsymbol{F}_h + \boldsymbol{F}_{el} = 0, \tag{13.65}$$

where force of hydrodynamic resistance acting on the drop is

$$\boldsymbol{F}_h = -6\pi\mu_e Rf(\boldsymbol{u} - \boldsymbol{U}), \tag{13.66}$$

and the electric force is

$$\boldsymbol{F}_{el} = q\boldsymbol{E}. \tag{13.67}$$

Here $\boldsymbol{u}$ is the drop velocity, R – the drop radius, $\boldsymbol{U}$ – the velocity of the liquid, $\boldsymbol{E}$ – strength of the electric field. We shall assume from now on that the electric field is strong enough, so the resistance factor f in the expression for the hydrodynamic force can be put equal to unity when considering the capture of a drop by a cylindrical wire.

If the distance between two electrodes is much larger than the characteristic size of the cell of a mesh electrode, then any inhomogeneity in the distribution of fluid velocity or electric field strength is noticeable only near the electrode – up to distances of about the size of cell. Outside this region, the distributions of $\boldsymbol{E}$ and $\boldsymbol{U}$ can be considered as undisturbed. Model the mesh electrode by two mutually perpendicular systems of infinite parallel cylinders of radius R_c spaced at equal distances h from each other and all lying in one plane. Then the electric field will be given by the superposition of fields created by each electrode, and the velocity field can be determined from the problem of cross-flow around an infinite cylinder, provided that the deformation of the velocity field caused by the other cylinders is negligible. This picture will be distorted in the vicinity of a mesh node and near the cylinder edges. But if the size of the mesh is much greater than h, and the mesh period is much greater than R_c, this assumption is permissible. With this arrangement, the problem statement will be identical, whether we look for the distribution of hydrodynamic and electric parameters in the plane containing the vectors $\boldsymbol{E}$ and $\boldsymbol{U}$, or in the plane perpendicular to the electrode plane.

The electric field created by one electrode is

$$\boldsymbol{E} = 2\lambda \frac{\boldsymbol{r}}{r^2}, \tag{13.68}$$

where r is the distance from the center of the wire cross section to the given point; λ – the surface density of charge at the wire.

The total electric field is the superposition of fields produced by each cylinder, therefore

$$\boldsymbol{E} = \left(E_{0x} + \sum_{\mathrm{k}} E_{\mathrm{k}x} \right) \mathbf{e}_x + \left(E_{0z} + \sum_{\mathrm{k}} E_{\mathrm{k}z} \right) \mathbf{e}_z, \tag{13.69}$$

where E_{0x} and E_{0z} are components of the electric field produced by the chosen cylinder, and E_{kx} and E_{kz} – by the cylinders placed at the distances $h\mathbf{k}$ and $-h\mathbf{k}$ from this cylinder. From Eqs. (13.68) and (13.69), there follows:

$$\begin{aligned} E_x &= 2\lambda \frac{x}{r^2} + 4\lambda x \sum_{\mathrm{k}} \frac{(r^2 - h^2\mathrm{k}^2)}{(h\mathrm{k})^4 + 2(h\mathrm{k})^2(z^2 - x^2) + r^4}, \\ E_z &= 2\lambda \frac{z}{r^2} + 4\lambda z \sum_{\mathrm{k}} \frac{(r^2 + h^2\mathrm{k}^2)}{(h\mathrm{k})^4 + 2(h\mathrm{k})^2(z^2 - x^2) + r^4}. \end{aligned} \tag{13.70}$$

Figure 13.21 shows the distribution of the dimensionless electric field strength components $E'_{x,z} = hE_{x,z}/\lambda$ over the dimensionless coordinates $x' = x/h$ and $z' = z/h$. One can see that at distances of about the size of the mesh cell h from the plane of the electrode, the electric field becomes practically homogeneous and equal to the field of a flat capacitor. Hence, the influence of specific details of the mesh structure is noticeable only at distances $z \leq h$.

Now, look at the velocity field. For a slow flow of a viscous fluid around an infinite cylinder (Re < 1), where V_∞ is the flow velocity on a large distance from the cylinder, the flow can be considered as practically undisturbed at distances $r > R_c/\mathrm{Re}$, while in the region $r < R_c/\mathrm{Re}$, the velocity components expressed in cylindrical coordinates r, θ, (the coordinate system is attached to the cylinder), are equal to [103]

$$\begin{aligned} \frac{V_r}{V_\infty} &= \frac{2}{\mathrm{Re}^2 - 1 - 2\ln(\mathrm{Re})} \left(\ln \frac{r}{R_c} - \frac{1}{2} + \frac{R_c^2}{2r^2} \right) \sin\theta, \\ \frac{V_\theta}{V_\infty} &= \frac{2}{\mathrm{Re}^2 - 1 - 2\ln(\mathrm{Re})} \left(\ln \frac{r}{R_c} - \frac{1}{2} + \frac{R_c^2}{2r^2} \right) \cos\theta. \end{aligned} \tag{13.71}$$

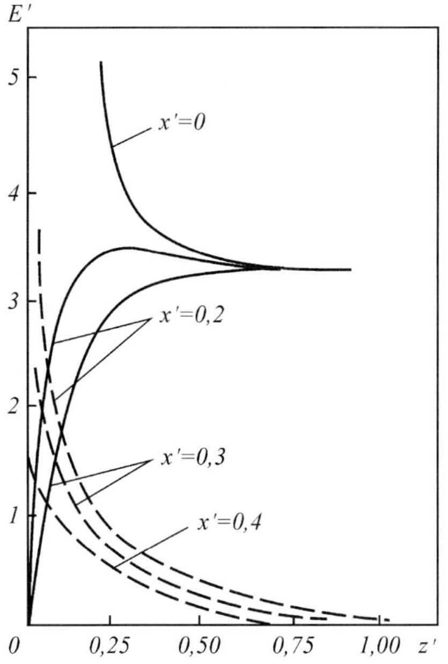

Fig. 13.21 Distribution of the electric field strength near the electrode: full lines E'_z, dashed lines E'_x.

Components of the hydrodynamic force are determined by relation (13.66)

$$F_r = -6\pi\mu R\left(-U_r + \frac{dr}{dt}\right); \quad F_\theta = -6\pi\mu R\left(-U_\theta + r\frac{d\theta}{dt}\right). \tag{13.72}$$

Introduce dimensionless variables

$$\xi = r/h, \quad \tau = V_\infty t/h, \quad U'_r = U_r/V_\infty, \quad U'_\theta = U_\theta/V_\infty, \quad b = 2q\lambda/6\pi\mu V_\infty R_c,$$
$$d = R_c/h, \quad f_r = hF_r/\lambda q, \quad f_\theta = hF_\theta/\lambda q.$$

In new variables, Eq. (13.72) will become

$$\frac{d\xi}{d\tau} = U'_r - bf_r; \quad \xi\frac{d\theta}{d\tau} = U'_\theta - bf_\theta. \tag{13.73}$$

The initial conditions are

$$\xi = \xi_0; \quad \theta = \theta_0 \quad \text{at } \tau = 0. \tag{13.74}$$

Solving equations (13.73) for various initial values ξ_0 and θ_0, we find a family of drop trajectories. Some of the drops will collide with the cylinder, recharge, reflect from the cylinder, and leave (moving either upstream or downstream depending on the charge). The trajectory dividing two specified families of trajectories is called the limiting trajectory. For any mesh cell, we can identify a cross section (far enough from the mesh), through which the trajectories of all drops leaving the mesh cell to move downstream must pass. Such a cross section is called the passage cross section of a mesh cell. If $x_k^{(0)}$ is the initial coordinate of a drop that moves along the limiting trajectory, then the passage cross section is

$$G = h^2 \left(1 - 2\frac{x_k^{(0)}}{h}\right)^2. \tag{13.75}$$

It is now possible to find the ablation factor of a mesh electrode

$$K_i = \left(1 - 2\frac{x_k^{(0)}}{h}\right)^2. \tag{13.76}$$

Thus, the problem of finding K_i is reduced to the problem of finding the limiting trajectory of drop motion. Consider the capture and reflection of drops by a cylinder (Fig. 13.22). Full lines show the trajectories of oncoming drops. Drops that are far from the cylinder move rectilinearly, because at distances $z > h$ the electric field and the fluid flow are practically uniform. At distances $z < h$, there appears a force component parallel to the plane of the electrode. Therefore, at distances $r < h/2$ from the mesh, the trajectories deviate noticeably from straight lines. At $r < R_c/\mathrm{Re}$, drops enter the region of disturbance produced by the mesh, and the fluid velocity decreases from the undisturbed flow velocity V_∞ to zero at the surface of mesh. Streamlines become distorted at the boundary of the disturbance region, but the absolute value of velocity is still close to V_∞.

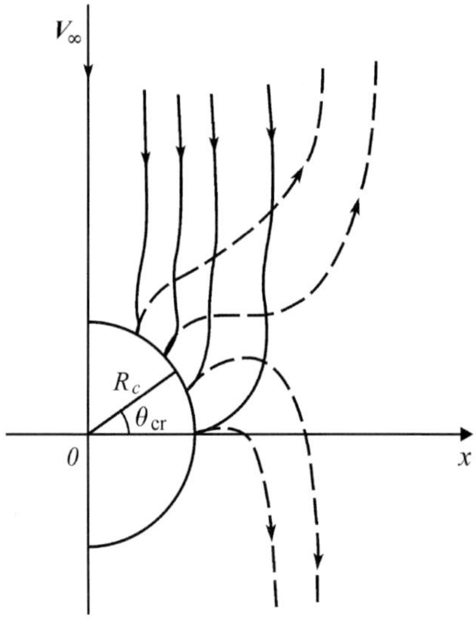

Fig. 13.22 Drop trajectories near the electrode.

The result is that the motion of a drop just changes its direction, and the drop gets displaced a little farther downstream, getting closer to the cylinder. Near the cylinder, however, the velocity falls, and the drop, acted upon by the electric force, is deposited at the cylinder. Dashed lines indicate the trajectories of reflected drops. There is a critical angle θ_{cr} such that for any $\varepsilon > 0$, after being recharged at the point $(R_c, \theta_{cr} + \varepsilon)$, the drop remains in the filtering zone, receding from the cylinder in the upstream direction, while if it recharges at the point $(R_c, \theta_{cr} - \varepsilon)$, it leaves the filtering zone and goes downstream. Trajectories of reflected drops at $\theta > \theta_{cr}$ are have a significant curvature. Thus, near the mesh electrode, there are two countercurrents of drops with opposite charges, which are characterized by the increased volume concentration of drops. These drops can interact with each other quite intensively, which results in a higher collision frequency and faster integration of drops. Taking this effect into account can be difficult, as we need to solve the kinetic equation that involves the charge distribution of drops in addition to their size distribution. But if this effect is ignored, the resulting ablation factor (ideal factor) will be overstated.

It is possible to determine the ideal ablation factor by using the results of numerical solution of equations (13.73) and (13.74). It depends on three dimensionless parameters: b, which describes the relation between electric and hydrodynamic forces acting on the drop; Re, which characterizes the flow structure; and d, which includes the mesh electrode parameters. The dependences K_i on these three parameters are shown in Fig. 13.23.

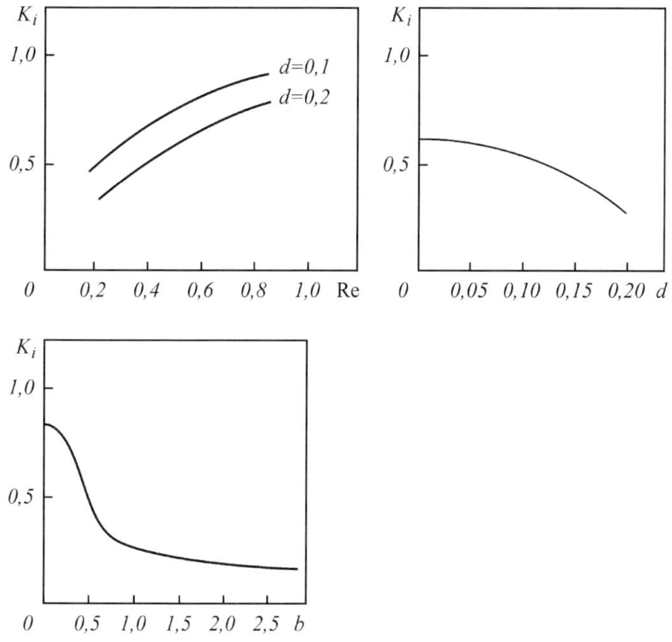

Fig. 13.23 Dependence of the ablation factor K_i of a mesh electrode on the parameters Re, d, and b.

To increase forces that promote emulsion separation, mesh electrodes can be installed at some angle to the flow. In this case, at the charge of drops exceeding some critical value, there appears an electric force component parallel to the force of gravity. Calculations have shown that a change of the inclination angle of the mesh electrodes from 0 to 45° does not cause a significant change of the passage factor. Hence, the arrangement of electrodes at an angle to the flow increases drop sedimentation rate without having a noticeable influence on the filtering characteristics of the mesh electrode.

13.6
Coalescence of Drops with Fully Retarded Surfaces in a Turbulent Emulsion Flow

Consider the coalescence of drops with fully retarded (delayed) surfaces (which means they behave as rigid particles) in a developed turbulent flow of a low-concentrated emulsion. We make the assumption that the size of drops is much smaller than the inner scale of turbulence ($R \ll \lambda_0$), and that drops are non-deformed, and thus incapable of breakage. Under these conditions, and taking into account the hydrodynamic interaction of drops, the factor of mutual diffusion of drops is given by the expression (11.70). To determine the collision frequency of drops with radii R_1 and R_2 ($R_2 < R_1$), it is necessary to solve the diffusion equation (11.36) with boundary conditions (11.39). Place the origin of a spherical system of coordinates (r, θ, Φ) into the center of the larger particle of radius R_1. If interaction forces between drops are spherically symmetrical, Eq. (11.36) with boundary conditions (11.39) assumes the form

$$\frac{1}{r^2}\frac{d}{dr}\left(r^2\left(D_r\frac{dn}{dr}-\frac{F}{h}n\right)\right)=0, \qquad (13.77)$$

$$n=0 \quad \text{at } r=R_1+R_2; \quad n=n_0 \quad \text{at } r\to\infty, \qquad (13.78)$$

where n is the number concentration of drops with radius R_2; D_T is the factor of mutual turbulent diffusion; F is the interaction force (molecular, electrostatic, electric) between the drops; h is the factor of hydrodynamic resistance to the drop's motion.

The general solution Eq. (13.77) has the form

$$n(r)=e^{g(r)}\left(C_1-\int_r^{\infty}C_2\frac{e^{-g(r)}}{\rho D_T^2(\rho)}d\rho\right), \quad g(\rho)=\int_{\rho}^{\infty}C_2\frac{F(r)}{h(r)D_T(r)}dr. \qquad (13.79)$$

The boundary conditions (13.78) give us constants C_1 and C_2:

$$C_1=n_0; \quad C_2=n_0\left(\int_{R_1+R_2}^{\infty}\frac{e^{-g(r)}}{z^2D_T(r)}dz\right)^{-1}.$$

The final result is:

$$n(r) = n_0 e^{g(r)} \left(1 - \int_r^\infty \frac{e^{-g(z)}}{z^2 D_T(z)} dz \Big/ \int_{R_1+R_2}^\infty \frac{e^{-g(z)}}{z^2 D_T(z)} dz \right). \tag{13.80}$$

The diffusion flux of drops 2 toward the given drop 1 is determined by the relation (11.40), which in spherically symmetrical case is written as

$$J = 4\pi (R_1 + R_2)^2 \left(D_T \frac{dn}{dr} - \frac{F}{h} n \right). \tag{13.81}$$

Substituting the expression (13.80) into (13.81), we get

$$J(R_1, R_2) = 4\pi n_0 \left(\int_{R_1+R_2}^\infty \frac{dr}{r^2 D_T(r)} \exp\left(\int_r^\infty \frac{F(z)}{h(z) D_T(z)} \right) \right)^{-1}. \tag{13.82}$$

Let us write the expression for the flux as

$$J(R_1, R_2) = C \frac{W}{t}. \tag{13.83}$$

Here $t = (\nu_e/\varepsilon_0)^{1/2}$ is the characteristic time scale in a turbulent flow with specific energy dissipation ε_0, $W = 4\pi\mu R_2^3 n_0/3$ is the volume concentration of drops of type 2 in the flow, C is the parameter whose value depends on the chosen model of drop coagulation in a turbulent flow.

The first theoretical analysis of particle coagulation in a turbulent flow was offered in paper [104], in which the turbulent flow in the vicinity of the assigned particle (target) was considered as a shear flow with the shear rate $\dot{\gamma} = (\varepsilon_0/\nu_e)^{1/2}$. The rate of coagulation was determined by Smolukchowski's method [105]. As a result, the value $C = 1.27$ was obtained. This approach was advanced further in [106] with more a detailed consideration of the flow field around the particle. The expression for the flux looks like (13.83) with $C = 1.23$. The correction for the velocity field distortion due to the influence of particle 2 (i.e. the small particle) was made in [107], resulting the same formula (13.83), but with a correction multiplier that depends on the ratio of particle radii R_2/R_1. As pointed out in [108], expressions for the flux J derived in the adduced works are valid for turbulent flows with a moderate specific energy dissipation $\varepsilon_0 < 0$ and large values of the inner scale of turbulence. For example, for water, $\lambda_0 > 5 \cdot 10^{-4}$ m and for air, $\lambda_0 > 5 \cdot 10^{-5}$ m. Since $\varepsilon_0 \sim U^3$ and $\lambda_0 \sim U^{-3/4}$, these conditions will hold only for a low-velocity flow. A more intense turbulence will result in larger values of ε_0 and smaller values of λ_0. In this case, when considering the relative motion of particles, we must take their inertia into account. Such an analysis is carried out in [108], and the minimum size of drops for which the suggested model is valid is estimated. A model of turbulent coagulation of particles based on the diffusion mechanism of particle collision, was proposed in [19]. For particles whose size is smaller than the inner scale of turbulence ($R \ll \lambda_0$), the diffusion flux looks like (13.83) with $C = 9.24$. We should mention another paper [20], which proposes a model of particle coagulation based on the concept of mean free path by analogy with the theory of molecular collisions in a rarefied gas. The resulting expression for the flux has the form (13.83), with $C = 0.77$.

The experimental data on particle coagulation in a turbulent flow in pipes and agitators is presented in works [20, 108]. These works show that among the suggested theoretical models of particle coagulation in a turbulent flow in the pipe, the best one is the model proposed in [20], while the diffusion model offered in [19] gives strongly overestimated values for the rate of coagulation in a turbulent flow in the pipe as well as in the agitator.

All of the above-mentioned works offer simple models of particle interaction, neglecting the hydrodynamic forces of resistance, which arise when particles approach each other and become especially noticeable at small clearance between particles. Also neglected are the forces of molecular interaction responsible for the coupling of particles at the collision stage. This is why the discrepancy between theoretical and experimental results for hydrosols is much greater than for aerosols.

To understand the principal cause of the discrepancy between the experiment and the theory of turbulent coagulation based on the diffusion mechanism of collision, we consider successively the interaction of drops with the proper account of hydrodynamic, molecular, electrostatic, and electric forces.

We start with the case when the drop surface is completely retarded, in other words, drops can be considered as undeformed particles. We also assume that coalescence occurs only due to the joint action of turbulent pulsations and molecular attractive forces. The force of molecular attraction between two spherical particles is given by the formula (11.100), which implies that this force is determined by the distance between particle surfaces and does not depend on their mutual orientation, i.e. is spherically-symmetrical with respect to the center of the particle of radius R_1. Since the force of molecular attraction manifests itself only at small clearances Δ between particles, we shall take its asymptotic expression at $\Delta \to 0$

$$F_A = \frac{\Gamma R_1 R_2}{6(R_1 + R_2)^3} \frac{1}{\Delta^2}, \tag{13.84}$$

where $\Delta = (r - R_1 - R_2)/(R_1 + R_2)$ is the dimensionless clearance between the drop surfaces; Γ is the Hamaker constant.

We take the following approximate expression for the hydrodynamic resistance factor

$$h = 6\pi \nu_e \rho_e \frac{R_1 R_2}{R_1 + R_2} \left(\frac{R_1 R_2}{(R_1 + R_2)^2 \Delta} + 1 \right), \tag{13.85}$$

This expression was obtained by combining the far and near asymptotics of the force of hydrodynamic interaction between approaching drops that move along the line of centers (see Section 11.4).

The factor of mutual turbulent diffusion was obtained earlier (see (11.70)):

$$D_T = \frac{\nu_e}{\lambda_0^2}\rho_e(R_1+R_2)^2\left(\frac{r-R_1-R_2}{R_1+R_2}+\frac{R_1R_2}{R_1^2+R_2^2-R_1R_2}\right)^2\left(\frac{h^{(0)}}{h}\right)^2. \tag{13.86}$$

Substituting the relations (13.84)–(13.86) into (13.82), we get

$$J(R_1,R_2) = \frac{4\pi n_0\nu_e(R_1+R_2)^3}{\lambda_0^2}\varphi_1, \tag{13.87}$$

$$\varphi_1^{-1} = \int_0^\infty \frac{(\Delta+\Delta_1)^2}{\Delta^2(1+\Delta)^2(\Delta+\Delta_2)^2}\exp\left(-S_A\int_A^\infty \frac{(\Delta_1+z)\,dz}{z^3(z+\Delta_2)^3}\right)d\Delta,$$

where $\Delta_1 = \dfrac{R_1R_2}{(R_1+R_2)^2}$; $\Delta_2 = \dfrac{R_1R_2}{R_1^2+R_2^2-R_1R_2}$ is the molecular interaction parameter $S_A = \Gamma\lambda_0^2/36\pi\rho_e\nu_e^2(R_1+R_2)^3$.

For an emulsion of the w/o type, this parameter is small. So, at $\Gamma\sim10^{-20}$ J, $\lambda_0\sim10^{-3}$ m, $\nu_e\sim5\cdot10^{-5}$ m^2/s, $\rho_e\sim10^3$ kg/m^3, we have $S_A\sim4\cdot10^{-23}(R_1+R_2)^{-3}$. The smallness of the parameter S_A allows us to find an asymptotic expression for φ_1. For drops of comparable sizes, the main contribution to the integral entering the expression for $\bar{\varphi}_1^{-1}$ is made by the quantities $\Delta\ll1,\Delta_1,\Delta_2$. Expanding the integrand in power series in Δ and keeping the principal terms of the expansion, we write

$$\varphi_1^{-1} = \int_0^\infty \frac{\Delta_1^2}{\Delta^2\Delta_2^2}\exp\left(-S_A\int_0^\infty\frac{\Delta_1\,dz}{z^3\Delta_2^3}\right)d\Delta. \tag{13.88}$$

Taking these integrals, we get

$$\varphi_1^{-1} = \frac{\Delta_1^{3/2}}{\Delta_2}\left(\frac{\pi}{2S_A}\right)^{1/2}. \tag{13.89}$$

Substituting (13.89) into (13.87), we find an asymptotic expression for the diffusion flux at $S_A\ll1$

$$J_1(R_1,R_2) = \frac{2\sqrt{2}\Gamma^{1/2}(R_1+R_2)^{9/2}}{\lambda_0\sqrt{\rho_e}R_1R_2(R_1^2+R_2^2-R_1R_2)}n_0. \tag{13.90}$$

Turbulent coalescence of drops was studied in works [22, 32]. In [22], an expression for the flux $J(R_1,R_2)$ was derived in the assumption that particle 1 (i.e. the large particle) is fixed. The obtained flux does not obey the condition of symmetry with respect to particle sizes.

With particle 1 fixed, the expression for the flux obtained in [22] has the form

$$J_1(R_1, R_2) = \frac{2\sqrt{2}\Gamma^{1/2}(R_1 + R_2)^{11/2}}{\lambda_0\sqrt{\rho_e}R_1^3 R_1^{1/2} R_2^{1/2}} n_0. \tag{13.91}$$

The ratio of fluxes for the cases of fixed and free particle can be estimated as $J_1/J \sim (R_1^3 + R_2^3)R_1^{-3}$ at $R_1 \geq R_2$, $J_1/J \sim 1$ at $R_1 \gg R_2$, and $J_1/J \sim 2$ at $R_1 \sim R_2$. Thus, the "fixed particle approximation" results in an increase of the flux and thereby of the collision frequency by a factor of 2 at most.

Another consequence of the asymmetry of the flux is the discontinuity of the derivative $\partial J/\partial R_2$ at the point $R_1 = R_2 = R$:

$$\left(\frac{\partial J}{\partial R_2}\right)_{R_2=R_1-0} = 9f_0 2^{7/2} R^{3/2}; \quad \left(\frac{\partial J}{\partial R_2}\right)_{R_2=R_1+0} = -3f_0 2^{7/2} R^{3/2},$$

where f_0 is a constant dependent on the particles' sizes.

[32] considers the case when both particles are moving, but the chosen scale of pulsations that are able to pull the particles together does not possess symmetry. Asymmetry of the flux with respect to particle sizes results in the asymmetry of the coalescence kernel, which violates the necessary condition that must hold for the kernel of the kinetic equation (11.1).

The principal shortcoming of the turbulent coagulation model offered by Levich [19] and rejected by many researchers is that it seriously overestimates the collision frequency of drops. Therefore the shear coagulation model [110] of particle coagulation in a turbulent flow has emerged as by far the most popular one. Since Levich's model does not take into account the hydrodynamic interaction of particles, let us estimate the effect of this interaction on the collision frequency.

Denote by J_0 the diffusion flux of non-hindered (free) motion of particles at $h = h_0 = 6\pi\rho_e\nu_e R$. Since coalescence is possible in this case regardless of the presence or absence of the molecular force, we can attempt to simplify the task by taking $F_A = 0$ and $S_A = 0$. Then (13.87) gives us

$$J_0 = \frac{4\pi\nu_e(R_1, R_2)^3}{\lambda_0^2}\left(\frac{1}{\Delta_2 - 1} + \frac{2}{(\Delta_2 - 1)^2} - \frac{2}{(\Delta_2 - 1)^3}\ln\Delta_2\right)^{-1} n_0. \tag{13.92}$$

At $R_1 = R_2$, the flux (13.92) coincides with the flux obtained in [19].

To estimate the influence of hydrodynamic forces, compose the ratio J_1/J, where for J_0 we take the expression (13.92), and for J – the expression (13.90) for identical particles with $R_1 = R_2 = R$

$$\frac{J_0}{J} = \frac{3\sqrt{2}\pi\nu_e}{\Gamma^{3/2}\lambda_0} R^{3/2}. \tag{13.93}$$

Let us take the parameter values $\lambda_0 \sim 10^{-3}$–10^{-4} m, $R \sim 10^{-5}$ m, $\nu_e \sim 10^{-5}$ m^2/s, $\rho_e \sim 10^{-3}$ kg/m^3, $\Gamma \sim 10^{-20}$ J, which are typical for water-oil emulsions. Then $J_0/J \sim 10$–10^2. It means that hydrodynamic resistance reduces the frequency of particle collisions by 1–2 orders of maginitude, which is consistent with the experimental data cited in publications, and the main contradiction in the turbulent coagulation model is thus eliminated. The influence of hydrodynamic interaction on the diffusion flux decreases with the reduction of particle size. At $R \sim 10^{-8}$ m, the ratio becomes $J_0/J \sim 1$, so for $R \leq 10^{-8}$ m, any hydrodynamic interaction of particles can be neglected. Note that for particles that small, the range of action of molecular forces becomes greater than the size of particles.

For Brownian diffusion of small particles, the influence of hydrodynamic interaction on the collision frequency was studied in works [28, 29], which also mention the decrease in the collision frequency by a factor of 1.5–2. This decrease is not as large as in the case of turbulent coagulation. There are two reasons why the effect of hydrodynamic interaction on the collision frequency of particles differs so substantially in the cases of turbulent flow and Brownian motion. First, the particle size is different in these two cases (the characteristic size of particles participating in Brownian motion is smaller than that of particles in a turbulent emulsion flow). Second, the hydrodynamic force behaves differently (the factor of Brownian diffusion is inversely proportional to the first power of the hydrodynamic resistance factor h, and the factor of turbulent diffusion – to the second power of h).

The asymptotic expression (13.90) for the diffusion flux is suitable for drops of roughly the same size, since at $\mathsf{k} = R_2/R_1 \to 1$, even at small values of S_A it is impossible to neglect Δ in comparison with Δ_1 and Δ_2 in the integrand expressions (13.87). At $\mathsf{k} \to 1$, in integrand l, the quantities Δ_1 and Δ_2 can be neglected in comparison with Δ. We get as a result:

$$\varphi_1^{-1} = \int_0^\infty \frac{1}{\Delta^2} \exp\left(-S_A \int_0^\infty \frac{dz}{z^4}\right) d\Delta. \tag{13.94}$$

Integration gives us

$$\varphi_1^{-1} = \frac{3^{1/3}\Gamma(4/3)\lambda_0^2}{S_A^{1/3} \nu_e (R_1 + R_2)^3}, \tag{13.95}$$

where $\Gamma(x)$ is the gamma-function.

Now, going back to (13.87), we obtain the following asymptotic expression for the diffusion flux of small particles toward the big particle ($R_2 \ll R_1$)

$$J(R_1, R_2) = \frac{2^{4/3}\pi}{3\Gamma(4/3)} (R_1 + R_2)^2 \left(\frac{\Gamma \nu_e}{\pi \rho_e \lambda_0^4}\right)^{1/3} n_0. \tag{13.96}$$

The dependence of φ_1 on the ratio of particle radii $\mathsf{k} = R_2/R_1$ for various values of molecular interaction parameter S_A obtained by numerical integration of the expression (13.87) is shown in Fig. 13.24. Dashed lines indicate the asymptotics (13.95) for $R_2 \ll R_1$.

We now proceed to determine the coalescence kernel $K(V, \omega)$. For the diffusion mechanism of drop collision, the coalescence kernel is equal to the diffusion flux

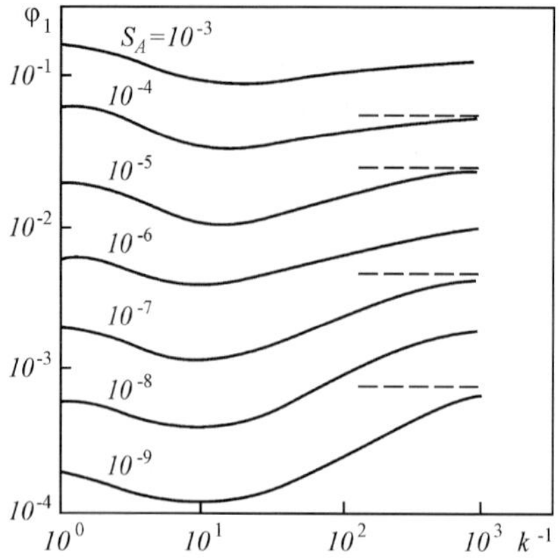

Fig. 13.24 Dependence of φ_1 on the ratio of drop radii k and the parameter of molecular interaction S_A.

of particles of radius R_2 at their individual concentration toward the particle of radius R1. Using expressions (13.90) and (13.96), respectively, for the flux of particles with a small and large size difference, we get

$$K_1^0(V,\omega) = \frac{2\sqrt{2}\Gamma^{1/2}(V^{1/3}+\omega^{1/3})^{9/2}(V\omega)^{-1/6}}{\lambda_0\sqrt{3\rho_e\pi}(V^{2/3}+\omega^{2/3}-V^{1/3}\omega^{1/3})} \quad \text{at } V\sim\omega. \tag{13.97}$$

$$K_2^0(V,\omega) = \frac{1}{3^{1/3}\Gamma(4/3)}(V^{1/3}+\omega^{1/3})^2\left(\frac{\Gamma\nu_e}{\pi\rho_e\lambda_0^4}\right)^{1/3} \quad \text{at } V\gg\omega. \tag{13.98}$$

The matching of received asymptotics thus obtained can be made by using the weight function $\alpha = 4V\omega/(V+\omega)^2$ [1]:

$$K^0(V,\omega) = \alpha K_1^0(V,\omega) + (1-\alpha)K_2^0(V,\omega). \tag{13.99}$$

The kernel (13.99) is inconvenient for use in calculations, therefore, it should be replaced by the approximating expression

$$K^0(V,\omega) = 16[\Gamma/3\pi\rho_e\lambda_0]^{1/2}(V^{1/3}\omega^{1/6}+V^{1/6}\omega^{1/3}). \tag{13.100}$$

13.7 Coalescence of Drops with a Mobile Surface in a Turbulent Flow of the Emulsion

Consider an emulsion that consists of a continuous phase – a viscous liquid of density ρ_e and viscosity μ_e, and a disperse phase – various spherical drops of another liquid of density ρ_i and viscosity μ_i. Assume that drops are undeformed and remain spherical. It should be noted that the mutual approach of drops at small clearances δ between their surfaces leads to an increase in the force of viscous friction, and when the thin layer of external liquid is pushed away, it may cause the drops to deform. However, if the modified capillary number is

$$\mathrm{Ca} = \frac{\nu_e u R_1 R_2}{\Sigma(R_1 + R_2)\delta} \ll 1, \tag{13.101}$$

where Σ is the drop's coefficient of surface tension and u is the relative velocity of approaching drops, the deformation of drops is small and can be ignored.

At $\delta \to 0$, the condition (13.101) can be violated. However, at such distances, the force of molecular attraction becomes essential, and it is this force that is responsible for the final stage of drop coalescence. Therefore the possible deformation of the drop can be noticeable only at final stage of drops' mutual approach, and it can only slow coalescence by a little, but it will have no significant effect on the collision frequency of drops.

The assumptions made allow us to consider the coalescence of drops with a mobile surface in the same manner as that of drops with a fully retarded surface. The main difference from the case considered in Section 13.6 is in the form of the hydrodynamic resistance factor. If the drops are placed far apart, the factor of hydrodynamic resistance $h^{(0)}$ for the relative motion of drops is determined by the formula (11.71), where each of the factors h_1 and h_2 is determined according to Hadamar-Rubczynski's formula

$$h^{(0)} = \frac{h_1^{(0)} h_2^{(0)}}{h_1^{(0)} + h_2^{(0)}} = 6\pi\mu_e \frac{R_1 R_2}{R_1 + R_2} \frac{2 + 3\bar{\mu}}{3 + 3\bar{\mu}}. \tag{13.102}$$

At small values of the gap δ between the drops, the following asymptotic expression is valid [see (11.88)]:

$$\begin{aligned} h_\delta &= 6\pi\mu_e \left(\frac{R_1 R_2}{R_1 + R_2}\right)^2 \frac{f(m)}{\delta} = \frac{1 + 0.402m}{1 + 1.711m + 0.461m^2}; \\ m &= \frac{1}{\bar{\mu}} \left(\frac{R_1 R_2}{R_1 + R_2}\right)^{1/2}; \quad \delta = r - R_1 - R_2, \end{aligned} \tag{13.103}$$

where the dimensionless parameter $\bar{\mu} = \mu_i/\mu_e$ has been introduced. The case of $\bar{\mu} \to 0$ corresponds to gas bubbles.

The factor of hydrodynamic resistance is given by the expression

$$h = h^{(0)} \left[1 + \frac{R_1 R_2}{R_1 + R_2} \frac{3 + 3\bar{\mu}}{2 + 3\bar{\mu}} \frac{f(m)}{\delta}\right], \tag{13.104}$$

which accurately reflects the behavior of h at large and small values of the clearance between the drops.

The factor of turbulent diffusion is given by (13.86); it was derived using the assumption that the size of drops is small in comparison with the inner scale of turbulence $R \ll \lambda_0$.

Now take into account forces of molecular and electrostatic interactions of drops (see Section 11.5). Total force of these interactions can be represented as

$$\mathbf{F} = \mathbf{F}_A + \mathbf{F}_R = -\frac{2\Gamma}{3R_1} f_A \Phi(p) + \frac{\varepsilon \chi R_1 \varphi_1^2}{2} f_R, \tag{13.105}$$

where f_A and f_R are dimensionless forces of molecular and electrostatic interactions equal to

$$f_A = \frac{1}{(\mathsf{k}+1)} \left\{ \frac{s}{(s^2-4)(1+\mathsf{k})^2} [8\mathsf{k} - (1+\mathsf{k})^2(s^2-4)] \right.$$
$$\left. + \frac{s(1+\mathsf{k})^2}{(s^2(1+\mathsf{k})^2 - 4(1+\mathsf{k})^2)} [8\mathsf{k} + (1+\mathsf{k})^2 s^2 - 4(1-\mathsf{k})^2] \right\};$$

$$f_R = \frac{\mathsf{k}}{(\mathsf{k}+1)} \frac{e^{-a}}{1-e^{-2a}} [2\alpha - (1+\alpha^2)e^{-a}].$$

Multiplier $\Phi(p)$ characterizes electromagnetic retardation of molecular interaction

$$\Phi(p) = \begin{cases} \dfrac{1}{1+1.77p} & \text{at } p \le 0.57; \\[2ex] -\dfrac{2.45}{60p} + \dfrac{2.17}{180p^2} - \dfrac{0.59}{420p^3} & \text{at } p > 0.57. \end{cases}$$

Here we have introduced the following parameters: Γ is Hamaker constant; ε is permittivity of external liquid; χ is the Debye inverse radius, $\chi = 1/\lambda_D$; λ_D is the thickness of electric double layer (see (11.98)); $\alpha = \phi_2/\phi_1$; ϕ_1 and ϕ_2 are the potentials of the respective drop surfaces; $\mathsf{k} = R_2/R_1$ is the ratio of their radii; $s = 2r/(R_1+R_2)$ is dimensionless distance between the drop's centers (a dimensionless clearance between drops $\Delta = s - 2$); $a = 0.5\chi R_1(1+k)(s-2)$; $p = 2\pi(r - R_1 - R_2)\lambda_L$; λ_L is the London wavelength.

Putting the expressions for h, D_T and F into the equation (13.82) for diffusion flux of drops of type 2 toward the drop of type 1, we get the following relation for the dimensionless flow:

$$\mathrm{J} = \frac{2\lambda_0^2 J}{\pi v_e R_1^3 n_{20}} = (1+\mathsf{k})^3 \varphi_1; \tag{13.106}$$

$$\varphi_1^{-1} = \int_2^\infty \frac{h^2(s)}{s^2(s-\beta)^2} \exp\left[\frac{1+\bar{\mu}}{\mathsf{k}(2+3\bar{\mu})} \int_s^\infty \frac{h(x)}{(x-\beta)^2} \right.$$
$$\left. \times (-S_A \Phi(p) f_A(x) + S_R \tau f_R(x))\, dx \right] ds.$$

Here the following dimensionless parameters are introduced:

$S_A = \dfrac{2\Gamma\lambda_0^2}{3R_1^3\pi\nu_e^2\rho_e}$ – the parameter of molecular interaction;

$S_R = \dfrac{\varepsilon\phi_1^2\lambda_0^2}{3R_1^3\pi\nu_e^2\rho_e}$ – the parameter of electrostatic interaction;

$\tau = R_1\chi = \dfrac{R_1}{\lambda_D}$ – the ratio of drop radius to the thickness of the electric double layer enveloping it ($\tau \gg 1$);

$\bar{\mu} = \dfrac{\mu_i}{\mu_e}$ – the ratio of viscosity factors of internal and external liquids;

$$p = \gamma(1+\mathrm{k})(s-2); \quad \gamma = \pi\frac{R_1}{\lambda_L}; \quad \beta = \frac{2(1-\mathrm{k})^2}{1+\mathrm{k}^2-\mathrm{k}}.$$

To determine diffusion flux, it is convenient to represent expression (13.106) as a system of two ordinary differential equations:

$$\frac{df_1}{dx} = -\frac{(1+\bar{\mu})h(x)}{\mathrm{k}(2+3\bar{\mu})(x-\beta)^2}\left[-S_A\Phi(p)f_A(x) + S_R\tau f_R(x)\right];$$

$$\frac{df_2}{dx} = -\frac{h^2(x)}{x(x-\beta)^2}\exp[f_1(x)], \tag{13.107}$$

$$f_1(\infty) = f_2(\infty) = 0, \quad (2 < x < \infty).$$

Its solution gives $f_2(2)$, and according to (13.106), is equal to the required flux or coalescence frequency

$$\mathrm{J} = f_2(2). \tag{13.108}$$

The system of the equations (13.107) is solved numerically. Below the calculation results are presented.

Coalescence frequency J depends on dimensionless parameters k, $\bar{\mu}$, S_A, S_R, τ, γ, α. The parameter k characterizes relative sizes of interacting drops; $\bar{\mu}$ is the viscosity ratio of drops and ambient liquid; S_A and S_R are the forces of molecular attraction and electrostatic repulsion of drops; τ is the relative thickness of electric double layer, which depends, in particular, on concentration of electrolyte in ambient liquid; γ is the electromagnetic retardation of molecular interaction; α is relative potential of surfaces of interacting drops. Let us estimate the values of these parameters. For hydrosols, the Hamaker constant is $\Gamma \sim 10^{-20}$ J. For viscosity and density of external liquid take $\nu_e \sim 10^{-5}$–10^{-6} m^2/s, $\rho_e \sim 10^3$ kg/m^3. Other parameters are: $\chi \sim 10^7$–10^8 m^{-1}, $\phi_1 \sim 20$ mV, $\lambda_L \sim 10^3$ Å, $\lambda_0 \sim 10^{-4}$ m. Then for drops with radii within the range $(1-100)$ µm, we have $S_A \sim 10^{-8}$–10^{-2}, $S_R \sim 10^{-3}$–10^2 m, $\tau \sim 10$–10^4, $\gamma \geq 3$. The value of parameter α remains uncertain, since at present there are no data to determine dependence of drop surface potentials on their sizes, χ, and properties of liquids. In fact, the case

$\alpha = 1$ has been considered in all publications, therefore main attention will be given to this case.

Depending on values of specified parameters, the regimes of fast and slow coalescence (or, as is commonly said, coagulation) should be distinguished. Consider them successively.

Coagulation of drops with only molecular force of attraction F_A is referred to as fast. In the above-introduced notations, this case corresponds to the zero value of parameter S_R. Denote the appropriate diffusion flux as J_A. The influence of hydrodynamic and molecular forces on coalescence frequency can be determined by comparing flux J_A with J_0 and neglecting the interaction forces. The last flux can be obtained from (13.82) at $F = 0$, $h = h_0$ and $D_T = D_{T0}$:

$$J_0 = \beta^2 (1 + k)^3 \left[\frac{4 - \beta}{2(2 - \beta)} + \frac{2}{\beta} \ln\left(1 - \frac{\beta}{2}\right) \right]^{-1}. \tag{13.109}$$

Analysis of expression (13.109) shows, that the drops of identical size have greatest coalescence frequency, – by about two orders of magnitude greater than collision frequency of drops with ratio of radiuses $k = 0.1$. The account taken of repulsion forces results in slowing down of coagulation rate, therefore such a process is referred to as slow coagulation. By analogy with the theory of Brownian coagulation, the stability factor of disperse system at turbulent coagulation can be introduced; it is the ratio of coagulation frequency in the absence of electrostatic repulsion force ($F_R = 0$) to that in the presence of this force:

$$F = J_A / J. \tag{13.110}$$

In the case of fast coagulation we have $F = 1$, while at slow coagulation $F > 1$. Note, that the parameter C in formula (13.83) is

$$C = \frac{3}{8} J_A. \tag{13.111}$$

It follows that in contrast to elementary models of coagulation, C is not constant, but depends on k, $\bar{\mu}$, S_A, γ, α.

Let us now turn to analysis of the calculation results.

13.7.1 Fast Coagulation

Consider first the coalescence of drops under the action of a molecular attraction force in the absence of electrostatic repulsion ($S_R = 0$). Dependence of coalescence frequency J_A on ratio of drop radii k for various values of viscosity ratio $\bar{\mu}$, and with the electromagnetic retardation neglected ($\gamma = 0$), is shown in Fig. 13.25. The coalescence frequency increases with relative size of drops and with decrease of their relative viscosity. This can be explained by fall of viscous

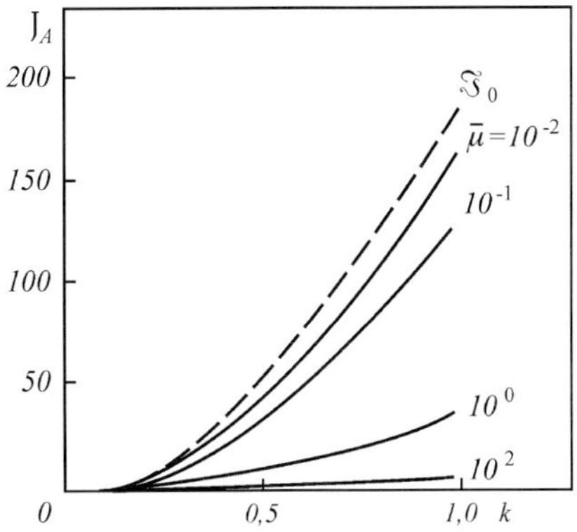

Fig. 13.25 Dependence of coalescence frequency of drops J_A on their relative size k and viscosity ratio of internal and external liquids $\bar{\mu}$: $S_A = 10^{-2}$; $S_R = 0$; $\gamma = 0$; J_0 is the coalescence frequency with hydrodynamic and molecular forces neglected.

resistance in the thin layer of external liquid dividing approaching drops, with decrease of $\bar{\mu}$. In the same figure, the dashed line shows dependence J_0 on k, calculated from the formula (13.109) without taking into account hydrodynamic resistance and molecular forces. Influence of the hydrodynamic resistance and molecular interaction of drops on coalescence frequency is illustrated in Fig. 13.26, where the dependence of fluxes ratio J_A/J_0 on k and $\bar{\mu}$ is shown. Despite the hydrodynamic resistance of drops, small values of $k \leq 0.2$ and $\bar{\mu} \leq 0.1$ at rather big values of the parameter of molecular interaction $S_A \geq 10^{-2}$ are possible

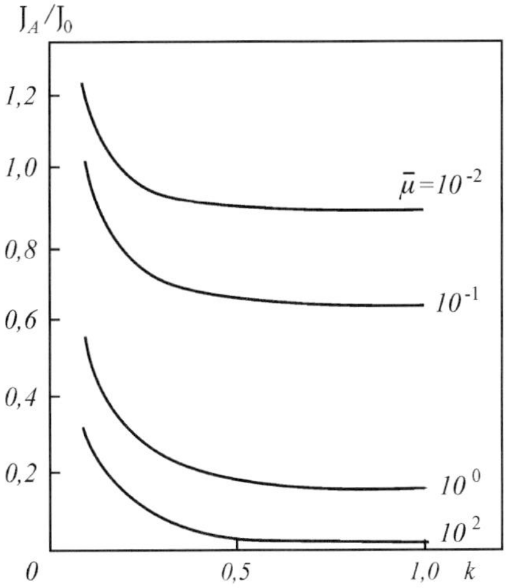

Fig. 13.26 Dependence J_A/J_0 on k and $\bar{\mu}$. Notations are the same as in Fig. 13.25.

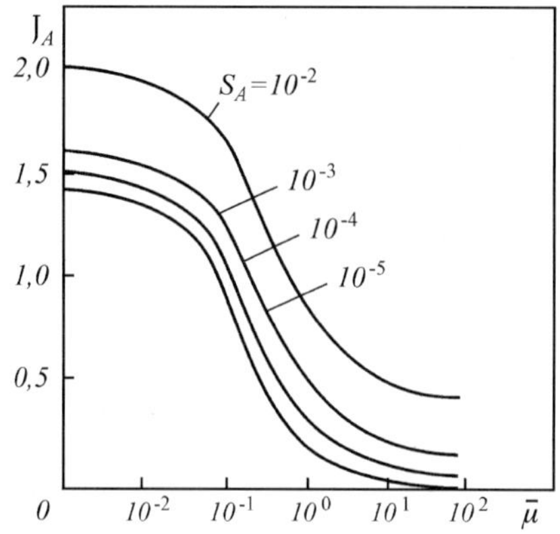

Fig. 13.27 Influence of parameter of molecular interaction S_A on coalescence frequency of drops at $S_R = 0$; $\gamma = 0$; $k = 0.1$.

when $J_A > J_0$. The influence of parameter S_A on coalescence frequency is shown in Fig. 13.27.

It should be noted that sensitivity of coalescence frequency to change of S_A decreases with increase of k. Thus, for drops of equal size ($k = 1$) noticeable difference in values J_A for $S_A = 10^{-2}$ and $S_A = 10^{-6}$ is observed only at $\bar{\mu} \geq 1$. Similar influence on J_A is exerted by electromagnetic retardation of molecular interactions (Fig. 13.28). Decrease of S_A and increase of k weaken influence γ on the coalescence frequency. The parameter C, introduced by (13.111), appears to be 1.05 for values $k = 1$, $\bar{\mu} = 10^2$, $S_A = 10^{-2}$ and $\gamma = 0$.

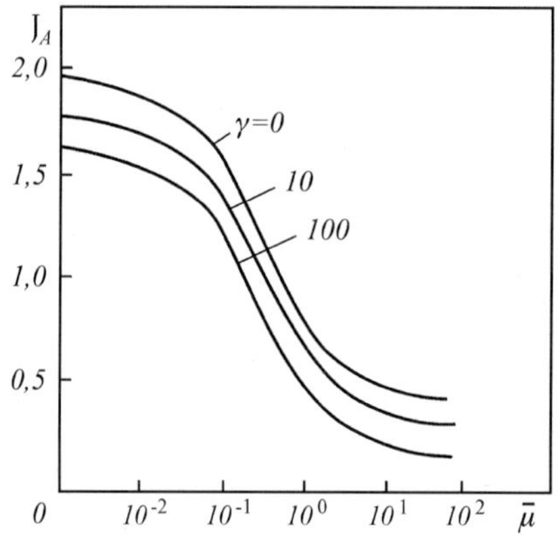

Fig. 13.28 Influence of parameter of electromagnetic retardation of molecular interaction γ on coalescence frequency of drops at $S_A = 10^{-2}$; $S_R = 0$; $k = 0.1$.

13.7.2
Slow Coagulation

Consider now coalescence taking into account the molecular and electrostatic forces. Of the greatest interest is dependence of stability factor F on parameters k, $\bar{\mu}$, S_A, S_R, τ, γ, α, and definition of the criterion for transition from slow to fast coagulation. In Section 11.5 the condition of transition from slow to fast coagulation was considered within the framework of DLVO theory of identical colloid particles without taking into account hydrodynamic interaction

$$V_A + V_R = 0; \quad \frac{dV_A}{dr} + \frac{dV_R}{dr} = 0, \tag{13.112}$$

where V_A and V_R are potentials of molecular and electrostatic interactions of drops (see (11.93) and (11.99)).

The condition (13.112) corresponds to the maximum on the potential energy curve of particle's interaction, where the curve is tangent to the abscissa axis. If we introduce the dimensionless potentials v_A and v_R,

$$V_A = -\frac{\Gamma}{6}\mathrm{v}_A(s); \quad V_R = \frac{\varepsilon r_1 \psi_1^2}{dr}\mathrm{v}_R(s), \tag{13.113}$$

the condition (13.112) can be rewritten as

$$\tau \mathrm{v}_A f_R - 2\mathrm{v}_R f_A = 0; \quad \frac{S_A}{S_R} - 0.5\frac{\mathrm{v}_A}{\mathrm{v}_R} = 0, \tag{13.114}$$

from which it follows, that transition from slow to fast coagulation depends on two parameters: τ and $S_R/S_A = 3R_1\varepsilon\phi_1^2/2\Gamma$. The straight line (Fig. 13.29, dashed

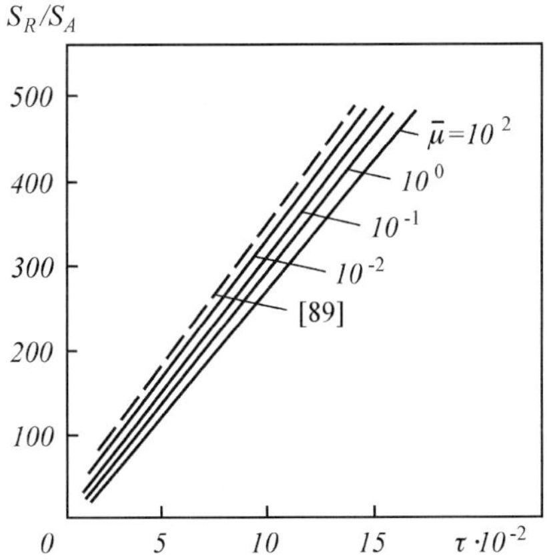

Fig. 13.29 The boundary of regions of fast and slow coagulation.

line) divides the plane $(S_R/S_A, \tau)$ into two areas: one of slow $F > 1$ (above the curve), and the other one of fast $F = 1$ (under the curve) coagulation.

In has been shown in [32] that taking account of hydrodynamic interaction in turbulent coagulation of identical rigid particles, leads to about the same result. The method used in obtaining this result is based on asymptotic estimation of an integral of the type (13.106) by study of behavior of the integrand in the vicinity of $s = 2$. Since for rigid particles at $s \to 2$ the factor of hydrodynamic force behaves as $h(s) \sim 1/(s-2)$, the basic contribution to the integral comes from the integrand's values in the region $s \sim 2$. However, for drops with mobile surface, at $s \to 2$ one has $h(s) \sim 1/(s-2)^{1/2}$, and the integrand in the vicinity of $S = 2$ does not make a major contribution to the value of integral. Therefore the integral should be determined numerically and by doing so one finds the values of parameters S_R/S_A and τ at which F becomes equal to 1 (see Fig. 13.29).

In Figures 13.30–13.33, the dependencies of the stability factor F on parameters t, $\bar{\mu}$, S_A, S_R are shown. Note that, in contrast to the Brownian coagulation, the dependence $F(\tau)$ has non-monotonic character. Its characteristic feature is the presence of maximum, whose position does not depend on either $\boldsymbol{k}$ or $\bar{\mu}$. The maximum value of F is attained in the region of higher values τ with increase of parameter of electrostatic interaction S_R and decrease of parameter of molecular interaction S_A. The appearance of a maximum is explained by mutual influence of hydrodynamic, molecular, and electrostatic forces on the character of drops' approach at small gaps between surfaces. While hydrodynamic force grows monotonously with shrinking of gap, total force of molecular and electrostatic interactions changes non-monotonously with both – s and τ. The maximum value F is reached at $\tau(s-2) = 1.28$. Experimental studies of turbulent coagulation of par-

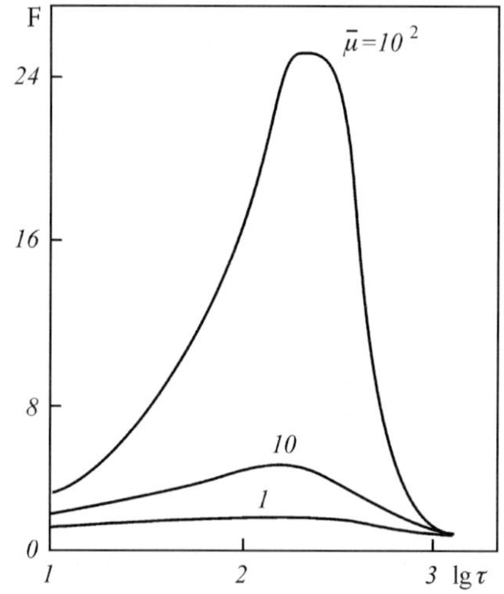

Fig. 13.30 Dependence of the stability factor on τ and $\bar{\mu}$ for $S_A = 10^{-2}$; $k = 1$; $S_R = 2$; $\gamma = 0$; $\alpha = 1$.

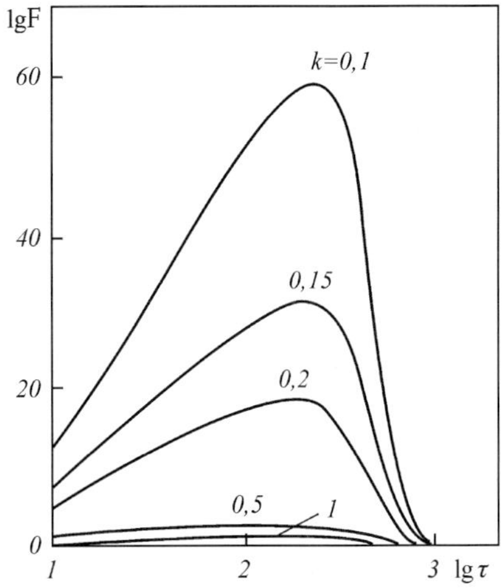

Fig. 13.31 Dependence of stability factor on τ and k for $S_A = 10^{-2}$; $S_R = 2$; $\bar{\mu} = 10^2$; $\gamma = 0$; $\alpha = 1$.

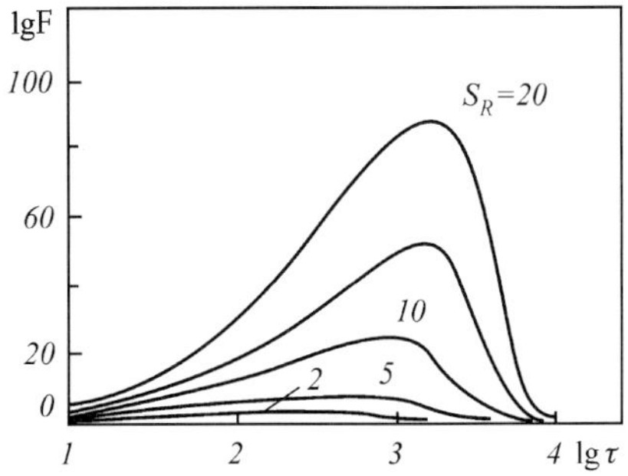

Fig. 13.32 Dependence of stability factor on τ and S_A for $S_R = 2$; $\bar{\mu} = 10^2$; $\gamma = 0$; $\alpha = 1$.

ticles, carried out in [20], do not confirm non-monotonic character of dependence $F(\tau)$. To understand this discrepancy, it is necessary to determine the range of values of τ within which the experiment was carried out. In [20] the dependence of stability factor on volume concentration of anticoagulant (electrolyte added in water flow), rather than on τ, has been determined. The parameter t depends on concentration of electrolyte C (mole/litre) as follow s [51]

$$\tau = 3 \cdot 10^7 zR\sqrt{C}, \tag{13.115}$$

where z is the anticoagulant valence, and R is radius of particles measured in cm.

In the experiment with an anticoagulant a mixture of 1 mole/l HCl and 5.6 mole/l $CaCl_2$, has been used; the average size of particles was 0.6 μm, volume concentration of anticoagulant varied from 1% up to 10%. The data correspond

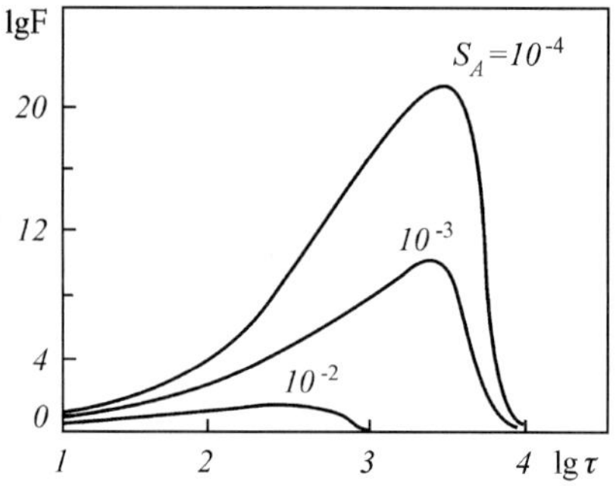

Fig. 13.33 Dependence of stability factor on τ and S_R for $S_A = 10^{-2}$; $k = 1$; $\bar{\mu} = 10^2$; $\gamma = 0$; $\alpha = 1$.

to values of τ from 850 up to 2700. From these data one can estimate parameters S_A and S_R. Experiment was carried out with a pipe of diameter $d = 2.54 \cdot 10^{-2}$ m, the average volume flow rate of liquid was $Q = 1.14 \cdot 10^{-3}$ m^3/s. For water $\nu_e = 10^{-6}$ m^2/s, $\rho_e = 10^3$ kg/m^3, $\varepsilon = 80\varepsilon_0$. These values give $\lambda_0 = 7 \cdot 10^{-5}$ m. For the latex particles used in experiment, $\Gamma = 10^{-21}$ J and $S_A = 5 \cdot 10^{-3}$. To determine parameter S_R, it is necessary to know surface potential of particles. From experimental dependence $F(W_D)$, where W_D is volume concentration of anticoagulant, the unique value $\phi_1 = 13$ mV ($S_R = 1.7$) has been determined, for which the theoretical dependence gives the best approach to experimental data (Fig. 13.34).

It should be noted, that the considered range of τ at $S_A = 5 \cdot 10^{-3}$ and $S_R = 1.7$ falls into the area of decrease of function $F(\tau)$. The maximum value of F on the theoretical curve is attained at $\tau \sim 400$, corresponding to concentration of electrolyte $W_D \sim 0.005$. The experiments were carried out at concentrations $W_D \geq 0.01$,

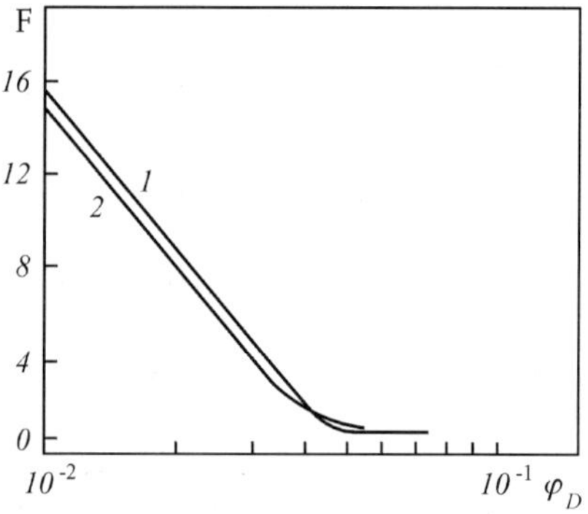

Fig. 13.34 Dependence of stability factor on concentration of anticoagulant: 1 – experiment [33]; 2 – calculations at $S_A = 5 \cdot 10^{-3}$; $S_R = 1.7$; $k = 1$; $\bar{\mu} = 10^3$; $\gamma = 0$; $\alpha = 1$.

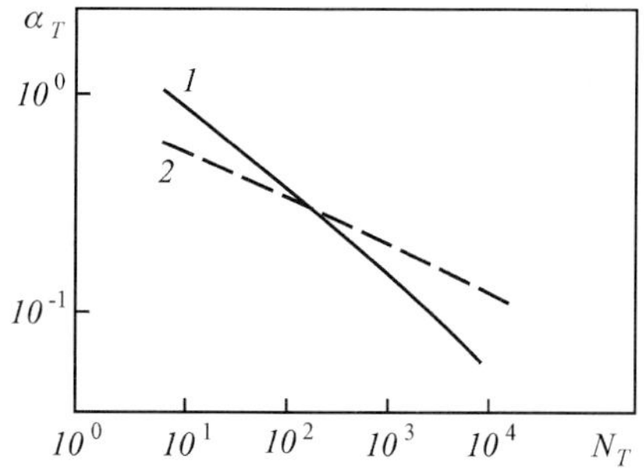

Fig. 13.35 Coagulation rate in agitation: 1, 2 – diffusion and shear model.

therefore the most interesting region of curve $F(\tau)$ cannot be compared with experiment. Besides, small size of particles used in the experiments, makes the possibility of turbulent coagulation of particles unlikely, since at such small sizes of particles the Brownian coagulation can dominate over, or be, at least, of the same order as the turbulent coagulation [19].

The diffusion model of turbulent coagulation is applicable to homogeneous and isotropic turbulent flow. A developed turbulent flow of emulsion in a pipe, in the vicinity of the flow core, can be considered as isotropic. However, the turbulent motion of liquid in a stirrer is neither homogeneous nor isotropic. Therefore the applicability of diffusion model to process of coagulation in agitator is not established with certainty.

Let us compare the coagulation rate in an agitator given by the diffusion model with the rate found in [110] in the model of shear coagulation and confirmed by the experimental results [109]. In Fig. 13.35, on the vertical axis is plotted the ratio α_T of the two frequencies: coagulation frequency of rigid particles of identical size, obtained by trajectory analysis in local shear flow with average shear rate $\dot{\gamma} = (4\varepsilon_0/15\pi\nu_e)^{1/2}$ with the account taken of molecular and hydrodynamic interaction between the particles; and coagulation frequency found in [105] by using a simple model of shear coagulation and neglecting molecular and hydrodynamic forces. On the horizontal axis, the dimensionless parameter $N_T = 6\pi\mu_e R_1^3\dot{\gamma}/\Gamma$ is plotted. For the model of turbulent coagulation, using the above-mentioned dimensionless parameters, one obtains $S_A = 1.16/N_T$, $\alpha_T = 0.15J$. A comparison of values of α_T calculated in the models of shear and turbulent coagulations shows that determination of the coagulation rate in the agitator involves a noticeable error, especially in the regions of small and large values of the parameter of molecular interaction of particles.

Finally, consider the case of interacting drops having viscosities μ_1 and μ_2 that are different from the viscosity of the ambient liquid. It should be treated in the same way as the previous case where the drops have equal internal viscosities, except that we must change the expression for the factor of hydrodynamic resistance h. To this end, consider two drops of types 1 and 2 which move with absolute velocities u_1 and u_2 under the action of turbulent pulsations. In the inertia-less approximation, the equation of motion for the drops has the form

$$F = u_1 h_{11} + u_2 h_{12} = u_1 h_{21} + u_2 h_{22}. \tag{13.116}$$

Introduce the resistance factor of drop 2 into its motion relative to the other drop

$$h = F/(u_1 + u_2). \tag{13.117}$$

Then from (13.116) and (13.117) there follows:

$$h = \frac{h_{11}h_{22} - h_{12}h_{21}}{h_{11} + h_{22} - h_{12} - h_{21}}. \tag{13.118}$$

In the case of non-hindered motion, when the drops are relatively far apart

$$h = h^{(0)} = 6\pi\mu_e \frac{R_2(2 + 3\bar{\mu}_1)(2 + 3\bar{\mu}_2)}{(2 + 3\bar{\mu}_1)(3 + 3\bar{\mu}_2) + \mathsf{k}(2 + 3\bar{\mu}_2)(3 + 3\bar{\mu}_1)}. \tag{13.119}$$

We have introduced new designations $\bar{\mu}_2 = \mu_2/\mu_e$, $\bar{\mu}_1 = \mu_1/\mu_e$.

The factors of hydrodynamic resistance h_{ij} depend on the relative distance s between the drops. Paper [43] solves the problem of slow central motion of two drops in a liquid where the two drops and the liquid all have different viscosities. The factors h_{ij} are found in the infinite series form. The appoximate solution of a similar problem is obtained by method of reflection in [46], and h_{ij} are found as power series in ratios R_1/r and R_2/r, which may be considered as asymptotic expressions for factors h_{ij} at $s \gg 2$. An asymptotic expression for h is obtained in [39] for small values of the gap between the drops at $s \to 2$:

$$h = h_\delta = \frac{\pi^2 \mathsf{k}^{1/2}(\bar{\mu}_1 + \bar{\mu}_2)}{16(1 + \mathsf{k})^2 (s - 2)^{1/2}} + 0[(s - 2)^{1/2}]. \tag{13.120}$$

In the limiting case when one of the drops is a rigid particle ($\bar{\mu}_1 = \infty$),

$$h_\delta = \frac{\mathsf{k}}{2(1 + \mathsf{k})^3 (s - 2)} + \frac{9\pi^2 \mathsf{k}^{1/2} \bar{\mu}_2}{64(1 + \mathsf{k})^2 (s - 2)^{1/2}} + 0[(s - 2)^{1/2}]. \tag{13.121}$$

The asymptotic expressions (13.120) and (13.121) are valid at distances s that obey the corresponding inequalities

$$\delta^{1/2}|\ln \delta| \ll \frac{1}{\max(\bar{\mu}_1 \bar{\mu}_2)} \quad \text{and} \quad \delta^{1/2}|\ln \delta| \ll \frac{1}{\bar{\mu}_2}$$

where $\delta = (1 + \mathsf{k})(s - 2)/2\mathsf{k}$.

To avoid bulky calculations, take

$$h = h^{(0)} + h_\delta \tag{13.122}$$

as the expression for h. It reflects the behavior of h at $s \to 0$ and at $s \to \infty$. Calculations show that the approximate expression (13.122) correctly describes the behavior of h for the values $0 < \mathsf{k} < 1$ and $0 \le (\bar{\mu}_1\bar{\mu}_2) \le 1$.

Introducing the same dimensionless variables as we did earlier, we get the following expression for the diffusion flux of drops of type 2 toward drop 1:

$$\mathsf{J} = \frac{2\lambda_0^2 J}{\pi v_e R_1^3 n_{20}} = (1+\mathsf{k})^3 \varphi_1, \tag{13.123}$$

$$\varphi_1^{-1} = \int_2^\infty \frac{h^2(s)}{s^2(s-\beta)^2} \exp\left(\frac{(2+3\bar{\mu}_1)(1+\bar{\mu}_2) + (2+3\bar{\mu}_2)(1+\bar{\mu}_1)}{\mathsf{k}(1+\mathsf{k})(2+3\bar{\mu}_1)(2+3\bar{\mu}_2)} \times \int_s^\infty \frac{h(s)}{(x-\beta)^2} (-S_A \Phi(p) f_A(x) + S_R \tau f_R(x))\, dx \right) ds.$$

The dependence of the coagulation frequency on the viscosity of the smaller drop $\bar{\mu}_2$ for various viscosities $\bar{\mu}_1$ of the larger drop is depicted in Fig. 13.36.

Since the expression (13.122) for h holds in the region $0 \le (\bar{\mu}_1\bar{\mu}_2) \le 1$ for drops of finite viscosity, and $0 \le \bar{\mu}_2 \le 1$, $\bar{\mu}_1 = \infty$ for interaction of a drop with rigid particle; on the graph, the intermediate area is shown by dashed line. Coalescence frequency becomes smaller as we increase the viscocities of interacting drops. The smaller the size difference between the drops, the more sensitive their coalescence frequency will be to a change in the internal viscosity.

Introduce the parameter Λ equal to the ratio of the coalescence frequency of drops of different viscosities to the coalescence frequency of drops of equal viscos-

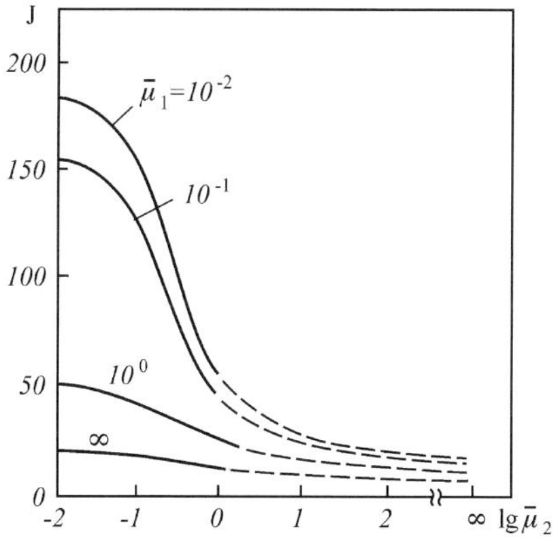

Fig. 13.36 Dependence of the coagulation frequency of drops on the relative viscosities $\bar{\mu}_1$ and $\bar{\mu}_2$ for $k = 1$; $S_A = 10^{-2}$; $S_R = 0$.

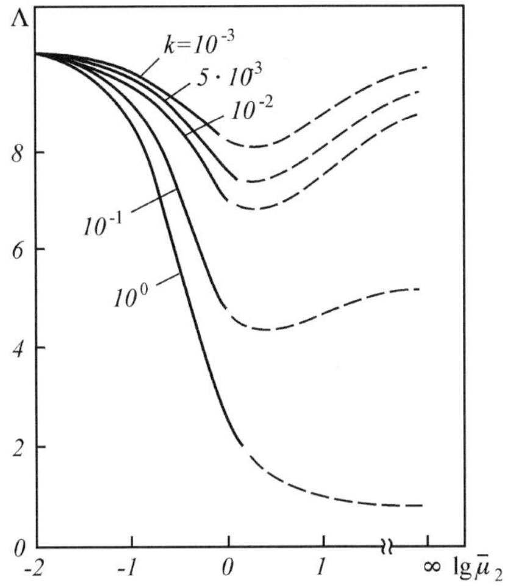

Fig. 13.37 Dependence of Λ on $\bar{\mu}_2$ and k for $\bar{\mu}_1 = 10^{-2}$; $S_A = 10^{-2}$; $S_R = 0$.

ities. This parameter characterizes the effect of the difference in drop viscosities on the coalescence frequency, which is illustrated by Fig. 13.37.

The results shown in Figs. 13.36 and 13.37 correspond to the fast coalescence ($S_R = 0$). For slow coalescence, we must introduce the stability factor F, as we did earlier. The dependence of F on the parameter τ for drops of different internal viscosities is shown in Fig. 13.38. Just as for drops of equal viscosities, this dependence is not monotonous and has a maximum. As the viscosity of drops gets larger, the maximum is shifted to the region of larger values of τ. The dependence of the critical value of τ on S_R/S_A is linear, just as for the drops of equal viscosities. The effect of parameters S_A and S_R on the coalescence frequency and the stability factor has the same character as for drops of equal viscosities.

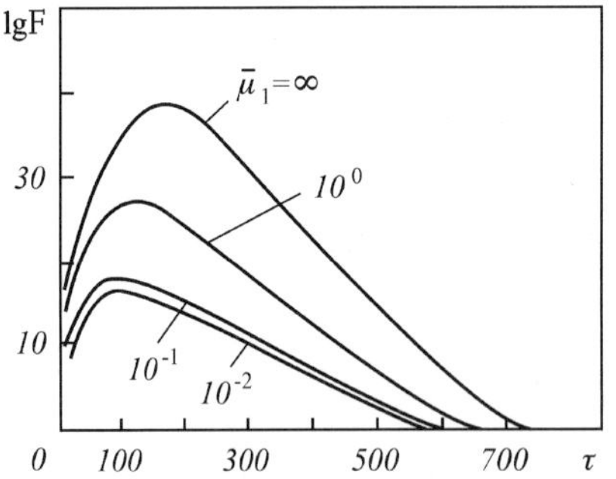

Fig. 13.38 Dependence of the stability factor F on τ and $\bar{\mu}_1$ for $\bar{\mu}_2 = 1$; $k = 0.1$; $S_R = 2$; $S_A = 10^{-2}$.

13.8
Coalescence of Conducting Emulsion Drops in a Turbulent Flow in the Presence of an External Electric Field

Let us find the collision frequency of conducting uncharged spherical drops in a turbulent flow of a dielectric liquid in the presence of a uniform external electric field. Just as before, we assume a developed flow, with drop sizes smaller than the inner scale of turbulence. We assume the drops to be undeformed, which is possible if the external electric field strength E_0 does not exceed the critical value E_{cr}, and the size of drops is sufficiently small. Under these conditions, the factor of mutual diffusion of drops of two types 1 and 2 with regard to hydrodynamic interaction is given by (13.86), while h and $h^{(0)}$ are given by the expressions (13.85) that apply to drops with a completely retarded surface. We must also take into account molecular and electric interaction forces acting on the drops.

Consider the relative motion of two drops of radii R_1 and R_2 under the action of hydrodynamic, electric, and molecular forces. Introduce a spherical system of coordinates (r, θ, Φ) attached to the center of the larger drop 1, the angle θ being measured from the direction of the external electric field strength vector $\boldsymbol{E}_0$. The expression (12.46) corresponds to a radial electric force acting on drop 2. If both drops are uncharged, $q_1 = q_2 = 0$ and

$$F_{2r} = \varepsilon E_0^2 R_1 R_2 (\hat{f}_1 \cos^2\theta + \hat{f}_2 \sin^2\theta). \qquad (13.124)$$

Note that the force F_{2r} is represented in the form that is symmetrical with respect to the drop radii. The factors $\hat{f}_i$ are expressed via the factors f_i appearing in (12.46) as follows

$$\hat{f}_i(\mathrm{k}, \Delta) = f_i[\mathrm{k}\Delta(1+\mathrm{k}^{-1})], \quad i = 1, 2; \qquad (13.125)$$

$$\mathrm{k} = \frac{R_2}{R_1}; \quad \Delta = \frac{r - R_1 - R_2}{R_1 + R_2}.$$

The total force acting on the drop is equal to the sum of molecular and electric forces:

$$F_2(R_1, R_2, r, \theta) = F_{2A}(R_1, R_2, r) + F_{2r}(E_0, R_1, R_2, r, \theta). \qquad (13.126)$$

The force of molecular interaction is taken in the approximate form (13.84)

$$F_{2A} = \frac{\Gamma R_1 R_2}{6(R_1 + R_2)^3} \frac{1}{\Delta^2}. \qquad (13.127)$$

In contrast to drop coalescence in the absence of electric field with a spherically-symmetrical interaction force between the drops that was considered in Section 13.6, in the present case, the electric field causes a dependence of the interaction force on the orientation of the drop pair relative to the vector $\boldsymbol{E}_0$, and consequently, this force possesses an axial, rather than spherical, symmetry. As a result, there appears a partial derivative of the second order with respect to θ in the diffusion equation (11.36). This not only complicates the solution, but also introduces the possibility of rotation of the drop pair relative to the electric field under the conditions of turbulent motion of the ambient liquid. Therefore we try to estimate the diffusion flux by using the following procedure [22].

First, determine the diffusion flux $J_0(0)$ in a unit solid angle, assuming that the interaction force is purely radial. Then integrate the resulting expression over the surface of a sphere whose radius equals the radius of coagulation $R = R_1 + R_2$, taking into account the dependence of the interaction force on the angle θ between the line of centers of the drop pair and the electric field. The flux so determined can be considered as the first approximation of the total diffusion flux J of drops of radius R_2 toward the drop of radius R_1. Carrying out the indicated steps, we get

$$J_0(0) = \frac{v_e}{\lambda_0^2}(R_1 + R_2)\bar{\varphi}_1(\theta)n_0,$$

$$J = 4\pi(R_1 + R_2)^2 \int_0^{\pi/2} j_0(0) \sin\theta \, d\theta. \tag{13.128}$$

where n_0 is the number concentration of drops 2, and $\bar{\varphi}_1(\theta)$ is determined similarly to (13.87).

Taking into account the expressions (13.85) for h, (13.86) for D_T, and (13.126) for the interaction force between the drops, and introducing the same dimensionless variables as in Section 13.6 we arrive at the following equations:

$$J = 4\pi n_{20}\frac{(R_1 + R_2)^3 v_e}{\lambda_0^2}\varphi_1. \tag{13.129}$$

$$\varphi_1 = \int_0^{\pi/2} \bar{\varphi}_1(\theta) \sin\theta \, d\theta.$$

$$(\bar{\varphi}_1(\theta))^{-1} = \int_0^{\infty} \frac{(\Delta + \Delta_1)^2}{\Delta^2(1+\Delta)^2(\Delta + \Delta_2)^2} \exp\left[-S_A \int_\Delta^{\infty} \frac{(\Delta_1 + z)\,dz}{z^3(z + \Delta_2)^3}\right.$$
$$\left. - S_E \int_\Delta^{\infty} \frac{(\Delta_1 + z)(f_1 \cos^2\theta + f_2 \sin^2\theta)}{z(z + \Delta_2)^2} dz\right],$$

$$\Delta_1 = \frac{R_1 R_2}{(R_1 + R_2)^2}; \quad \Delta_2 = \frac{R_1 R_2}{R_1^2 + R_2^2 - R_1 R_2};$$

$$S_A = \frac{\Gamma\lambda_0^2}{36\pi\rho_e v_e^2 (R_1 + R_2)^3}; \quad S_E = \frac{\varepsilon E_0^2 \lambda_0^2}{6\pi\rho_e v_e^2}.$$

It is possible to estimate the diffusion flux J in another way, which is based on the natural assumption that the orientation of the drop pair relative to the electric field changes repeatedly as the drops approach each other. This will lead to a "spreading out" of the concentration profile n_2, and in order to determine the force, one should average it over the angle θ. For this purpose, plug into (13.82) the expressions (13.126) and (13.124) for the force acting on drop 2 and average the result over the sphere surface

$$\langle F_2(R_1, R_2, \Delta)\rangle = \int_0^{\pi} F_2(R_1, R_2, \Delta, \theta)\, \sin\theta\, d\theta$$
$$= F_{2A}(R_1, R_2, r) + \langle F_{2r}(E_0, R_1, R_2, r, \theta)\rangle, \qquad (13.130)$$
$$\langle F_{2r}(E_0, R_1, R_2, r, \theta)\rangle = \frac{1}{3}\varepsilon E_0^2 R_1 R_2(\hat{f}_1 + 2\hat{f}_2).$$

The expression for the flux J looks in this case like (13.129), but the formula for φ_1 will change to

$$\varphi_1^{-1} = \int_0^{\infty} \frac{(\Delta+\Delta_1)^2}{\Delta^2(1+\Delta)^2(\Delta+\Delta_2)^2} \exp\left[-S_A \int_{\Delta}^{\infty} \frac{(\Delta_1+z)\,dz}{z^3(z+\Delta_2)^3}\right.$$
$$\left. - S_E \int_{\Delta}^{\infty} \frac{(\Delta_1+z)(\hat{f}_1+2\hat{f}_2)}{3z(z+\Delta_2)^2} dz\right]. \qquad (13.131)$$

Numerical calculations have shown that the flux given by the formula (13.129) for both of the above-considered methods leads to different expressions for φ_1, with the maximum divergence of 5% in a wide range of parameters S_A and S_E. Since the formula (13.131) requires less calculations, this is the one that was used. In Fig. 13.39, the dependence of φ_1 on the ratio of drop radii $\boldsymbol{k}$ is shown for various values of parameters S_A and S_E. It follows from this dependence that the influence of the electric field on the collision frequency diminishes with the increase of S_A and with the decrease of S_E. Since S_E does not depend on drop sizes, and S_A increases as the drops get smaller, the influence of the electric field on the collision frequency of drops in a turbulent flow becomes smaller with the decrease of the drop radius.

The effect of the electric field on the collision frequency of drops is characterized by the ratio of diffusion fluxes with and without the presence of an electric field

$$\xi = \frac{J(S_A, S_E)}{J(S_A, 0)}.$$

The dependence of ξ on S_A at $S_E = 1$ is shown in Fig. 13.40.

For the characteristic parameter values typical for a turbulent flow of water-oil emulsions $\varepsilon \sim 2\varepsilon_0$, $E_0 \sim 0.9$ kV/cm, $\nu_e \sim 10^{-5}$ m^2/s, $\lambda_0 \sim 10^{-3}$ m, $\rho \sim 900$ kg/m^3, $\Gamma \sim 10^{-20}$ J, we get $S_E \sim 1$, and for identical drops of radius $R \sim 10^{-5}$ m $= 10$ μm, we get $S_A \sim 10^{-7}$. At these parameter values, $\xi \sim 30$. It means that

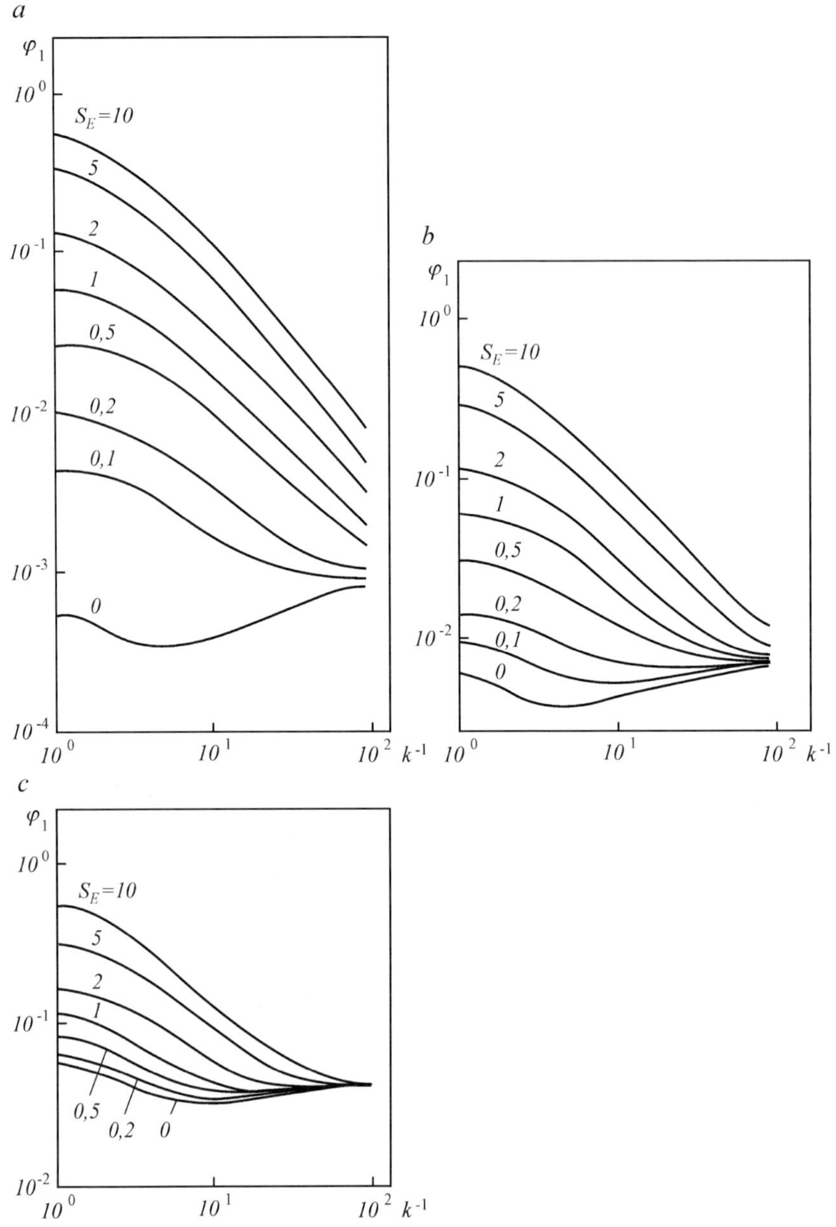

Fig. 13.39 Dependence of φ_1 on k and S_E for S_A: a – 10^{-8}; b – 10^{-6}; c – 10^{-4}.

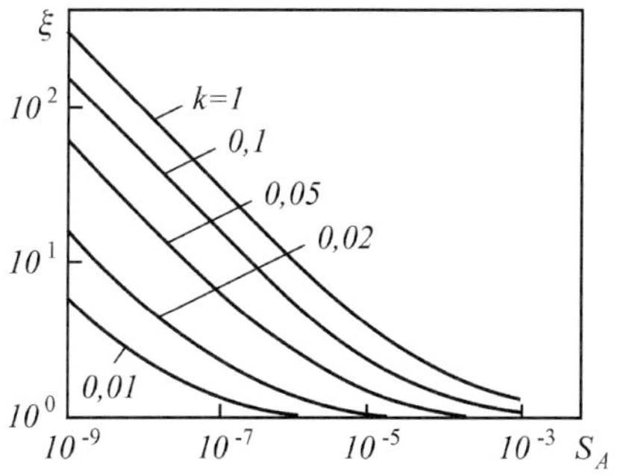

Fig. 13.40 Dependence of ξ on S_A and k for $S_E = 1$.

in the considered case, the presence of the electric field increases the frequency of collisions by a factor of 30. Since $S_A \sim R^{-3}$, an increase of the drop radius causes a substantial increase of the coalescence frequency.

However, despite such effect, even sufficiently strong electric fields are not able to fully counterbalance the influence of hydrodynamic interaction between the approaching drops on their coalescence frequency in a turbulent flow. To estimate the influence of hydrodynamic interaction, consider the ratio of fluxes without and with such interaction, as we did in Section 13.7.

$$\frac{J_0}{J} = \frac{3}{\varphi_1} \tag{13.132}$$

At $S_E \sim 5$, the ratio is $J_0/J \sim 10$, and at $S_E \sim 1$, it equals $J_0/J \sim 50$. So, even in sufficiently strong electric fields, where the paramenter of electrohydrodynamic interaction S_E has a rather large value, the force of hydrodynamic resistance acting on the pair of drops with a fully retarded surface as they approach each other reduces their collision frequency by an order of magnitude as compared to non-interacting drops. Note that in the case of drops with mobile surface, the role of hydrodynamic interaction is not so pronounced, since at small gaps between the drops, the hydrodynamic resistance grows as $1/\Delta^{1/2}$, as opposed to $1/\Delta$ for rigid particles.

By processing the results of numerical calculations, we can get the following approximate expression for the flux (13.129):

$$J = 0.24\pi S_E \nu_e \lambda_0^{-2} (R_1^{1/3} R_2^{2/3} + R_2^{1/3} R_1^{2/3}) n_{20}. \tag{13.133}$$

Finally, we can write the expression for the coagulation kernel $K(V, \omega)$ of the kinetic equation (11.1). Since the diffusion flux of drops of radius R_2 toward the drop of radius R_1 at the unit concentration $n_{20} = 1$ has the meaning of the kernel of coagulation, the replacement of radiuses by volumes in (13.133) gives

$$K(V, \omega) = 0.01\varepsilon E_0^2 \nu_e^{-1} \rho_e^{-1} (V^{1/9}\omega^{2/9} + \omega^{1/9} V^{2/9})^3 \tag{13.134}$$

13.9
Kinetics of Emulsion Drop Coalescence in a Turbulent Flow

First, consider the dynamics of integration of drops with fully retarded surfaces in the absence of electric field. If drops are uncharged, the kernel of the kinetic equation is given by the relation (13.100), namely

$$K(V,\omega) = 16\left(\frac{\Gamma}{3\pi\rho_e\lambda_0^2}\right)^{1/2}(V^{1/3}\omega^{1/6} + \omega^{1/3}V^{1/6}). \tag{13.135}$$

Let us use the method of moments to solve the kinetic equation (11.1). Then, in view of (13.135), we obtain from (11.15) the following equation for zeroth moment, which has the meaning of the number concentration of drops

$$\frac{dm_0}{dt} = -\frac{1}{2}G[m_{1/3}(t)m_{1/6}(t) + m_{1/6}(t)m_{1/3}(t)], \tag{13.136}$$

where $G = 16[\Gamma/(3\pi\rho_e\lambda_0^2)]^{1/2}$.

There are fractional moments appearing in the right-hand side of (13.136). To determine them, recall the method of interpolation of fractional moments through the integer ones (see Section 11.1). Using two-point interpolation, one can express the fractional moments in terms of two moments m_0 and m_1 according to (11.20). Then

$$\frac{d\widehat{m}_0}{dt} = G\frac{m_{1/3}(0)m_{1/6}(0)}{m_0(0)}\widehat{m}_0^{3/2}(t)\widehat{m}_1^{1/2}(t). \tag{13.137}$$

We have introduced the dimensionless moments

$$\widehat{m}_0(t) = \frac{m_0(t)}{m_0(0)}; \quad \widehat{m}_1(t) = \frac{m_1(t)}{m_1(0)},$$

where $m_k(0)$ is the value of k-th moment at the initial moment of time.

Since only the coalescence is being considered, the volume concentration of drops does not change with time and $\widehat{m}_1(t) = 1$. Then the solution of Eq. (13.137) with the initial condition $\widehat{m}_0(t) = 1$ has the form

$$\widehat{m}_0(t) = \frac{m_0(t)}{m_0(0)} = \left[1 + 8G\frac{m_{1/3}(0)m_{1/6}(0)}{m_0(0)}t\right]^{-2} \tag{13.138}$$

Let us estimate the initial values of the fractional moments from two-point interpolation. Then

$$m_{1/3}(0)m_{1/6}(0) \approx [m_0(0)]^{3/2}m_1^{1/2}. \tag{13.139}$$

Now, from (13.138) and (13.139), we derive the number concentration and the average volume of drops as functions of time

$$\bar{m}_0(t) \approx \{1 + 8G[m_0(0)]^{1/2} m_1^{1/2} t\}^{-2} \tag{13.140}$$

$$\frac{V_{av}(t)}{V_{av}(0)} = \frac{1}{\bar{m}_0(t)} \approx \{1 + 8G[V_{av}(0)]^{-1/2} m_1 t\}^{-2}. \tag{13.141}$$

Here we have taken into account the fact that $V_{av}(0) = m_1/m_0(0)$.

It is now possible to estimate the characteristic time of coalescence, that is, the time it takes for the average volume of drops to increase by a factor of e

$$T_A \sim \frac{(\sqrt{e} - 1)\, V_{av}^{1/2}(0)}{128 m_1} \left(\frac{3\pi \rho_e \lambda_0^2}{\Gamma}\right)^{1/2}. \tag{13.142}$$

Let $m_1 = 0.01$; $\lambda_0 = 5 \cdot 10^{-4}$ m; $\Gamma = 10^{-20}$ J; $\rho_e = 850$ kg/m^3; $V_{av} = 4 \cdot 10^{-15}$ m^3 (drops have the radius of 10 μm). Then the characteristic time of drop integration is $T_A \sim 100$ s. Since the proper account of hydrodynamic resistance reduces the diffusion flux for given parameter values by two orders of magnitude (see Section 13.6) as compared to the case of turbulent coalescence with no hydrodynamic resistance, the characteristic time of coalescence increases by the same factor.

Consider now the dynamics of integration of conducting drops in the external electric field of strength E_0, provided that E_0 does not exceed the critical value at which the approaching drops may split. In this case the kinetic equation kernel is determined by the expression (13.134)

$$K(V, \omega) = G_1 (V^{1/9} \omega^{2/9} + \omega^{1/9} V^{2/9})^3, \tag{13.143}$$

where $G_1 = 0.001 \varepsilon E_0^2 \nu_e^{-1} \rho_e^{-1}$.

The equation for zeroth moment m_0 then becomes

$$\frac{d\bar{m}_0}{dt} = -G_1 \frac{m_{1/3}(0) m_{2/3}(0) + 3 m_{4/9}(0) m_{5/9}(0)}{m_0(0)} \bar{m}_0(t) \bar{m}_1(t). \tag{13.144}$$

It has the solution

$$m_0(\tau) = m_0(0) e^{-B\tau}, \tag{13.145}$$

where $\tau = G_1 m_1 t$;

$$B = \frac{m_{1/3}(0) m_{2/3}(0) + 3 m_{4/9}(0) m_{5/9}(0)}{m_0(0)}.$$

It is now possible to estimate the characteristic time of coalescence analogously to the previous case

$$T_E \sim \frac{1}{G_1 m_1 B} = \frac{25 \mu_e}{\varepsilon E_0^2 m_1}. \tag{13.146}$$

Let us take $m_1 = 0.01$; $\mu_e = 5 \cdot 10^{-3}$ kg/m·s; $E_0 = 1$ kV/cm; $\varepsilon \sim 2\varepsilon_0$. Then $T_E \sim 6$ s. If we increase the electric field strength to 2 kV/cm, with all other conditions unchanged, this time will decrease to 1.5 s.

We can estimate the characteristic time necessary for the integration of drops in the electric field to come to completion, in the same manner as we did in Section 13.2

$$t_{coal} \sim \frac{25\mu_e}{\varepsilon E_0^2 m_1} \ln\left[\frac{1.75\Sigma^3}{\varepsilon E_0^2 V_{av}(0)}\right]. \tag{13.147}$$

So, for drops of initial radius 10 μm at $E_0 = 2$ kV/cm, all other parameters being the same as earlier, we have $t_{coal} \sim 6$ s.

Compare the characteristic time of drop integration in a turbulent flow T_E^{turb} with the characteristic time of gravitational sedimentation of the emulsion T_E^{grav} (the electric field is present in both cases):

$$\frac{T_E^{grav}}{T_E^{turb}} \sim 2 \div 3. \tag{13.148}$$

Note that in contrast to the estimation made in Section 13.2, the estimation of characteristic times in the present section is carried out over the average volume, instead of the radius. Keep in mind that the characteristic time of change of the average drop radius is approximately 3 times greater than the characteristic time of change of the average volume.

References

1 Loginov V. I., Dewatering and Desalting of Oils, Chemistry, Moscow, 1979 (in Russian).

2 Voloshuk V. M., Kinetic Theory of Coagulation, Hydrometeoizdat, Moscow, 1984 (in Russian).

3 Hahn G. J., Shapiro S. S., Statistical Models in Engineering, Wiley, 1967.

4 Sinaiski E. G., Separation of two-phase multi-component mixture in oil-gas field equipment, Nedra, Moscow, 1990 (in Russian).

5 Gradshtejn I. S., Ryszik, Tables of Integrals, Sums, Series, and Products, Nauka, Moscow, 1971 (in Russian).

6 Ruckenstein E., Condition for the Size Distribution of Aerosols to be Self-Preserving, J. Colloid Interface Sci., 1975, Vol. 50, No. 3, p. 508–518.

7 Jeffreys G. V., Davis G. A., Coalescence of Droplets and Dispersions (in Recent Advances in Liquid-Liquid Extraction) (Ed. K. Hanson), Pergamon Press, Oxford, 1971, 255–322.

8 Smoluchowski M., Coagulation of Colloids, ONTI, Moscow, 1936 (in Russian).

9 Fuks N. A., Mechanics of Aerosols, Acad. Sci. USSR, Moscow, 1955 (in Russian).

10 Friedlander S. K., Smoke, Dust and Haze: Fundamentals of Aerosol Behavior, Wiley, Interscience, 1977.

11 Hocking L. M., Mutual hydrodynamic Interference in Flow around two Spheres, Quart. J. Roy. Meteorol. Sci., 1959, Vol. 85, p. 363.

12 Krasnogorskaja N. V., Collision Calculations for Particles of comparable Sizes, Papers Acad. Sci. USSR, 1964, Vol. 154, No. 12 (in Russian).

13 Happel J., Brenner H., Low Reynolds Number Hydrodynamics, Prentice-Hall, 1965.

14 Smirnov L. P., Deryaguin B. V., On inertialess electrostatic Deposition of Aerosol Particles on a Sphere at Flow of viscous Fluid around the Sphere, Colloid J., 1967, No. 3 (in Russian).

15 Entov V. M., Kaminskiy V. A., Sinaiski E. G., On Capture of fine Drops by a large one in an electric Field, Fluid Dyn., 1973, No. 5. p. 61–68 (in Russian).

16 Lapiga E. J., Sinaiski E. G., On Capture of small charged Particles by a large uncharged Particle in an external electric Field, Colloid J., 1978, Vol. 40, No. 2 (in Russian).

17 Zhang X., Davis R. H., The Rate of Collision due to Brownian and gravitation Motion of small Drops, J. Fluid Mech., 1991, Vol. 230, p. 479–504.

18 Einstein A., Smoluchowski M., Brownian Motion, ONTI, Moscow, 1936 (in Russian).

19 Levich V. G., Physicochemical Hydrodynamics, Prentice-Hall, Englwood Cliffs, N.J., 1962.

20 Delichatsios M. A., Probstein R. F., Coagulation in Turbulent Flow: Theory and Experiment, J. Colloid Interface Sci., 1975, Vol. 51, p. 394–405.

21 De Boer G. B. J., Hoedamakers G. F. M., Thones D., Coagulation in Turbulent Flow. P. I, Chem. Eng. Res. and Des., 1989, Vol. 67, No. 3, p. 301–307.

22 Entov V. M., Kaminskiy V. A., Lapiga E. J., Calculation of Emulsion Coalescence Rate in a Turbulent Flow, Fluid Dyn., 1976, No. 3, p. 47–55 (in Russian).

23 Sinaiski E. G., Rudkevich A. M., Coagulation of Bubbles in viscous Liquids in a Turbulent Flow, Colloid J., 1981, Vol. 43, No. 2, p. 369–371 (in Russian).

24 Sinaiski E. G., Coagulation of Droplets in a Turbulent Flow of viscous Liquid, Colloid J., 1993, Vol. 55, No. 4, p. 91–103 (in Russian).

25 Sinaiski E. G., Brownian and Turbulent Coagulation of Drops in viscous Liquids and Emulsion Stability, Colloid J., 1994, Vol. 56, No. 1, p. 105–112 (in Russian).

26 Sinaiski E. G., Coagulation of Drops of different Viscosity in a Turbulent Flow of viscous Liquid, Colloid J., 1994, Vol. 56, No. 2, p. 222–225 (in Russian).

27 Tunitski A. N., Kaminskiy V. A., Timashev S. F., Methods of physico-chemical kinetics, Chemistry, Moscow, p. 1972 (in Russian).

28 Deryaguin B. V., Muller V. M., On slow Coagulation of hydrophobic Colloids, Paper Acad. Sci. USSR, 1967, Vol. 176, No. 4, p. 869–872 (in Russian).

29 Spielman O. A., Viscous Interactions in Brownian Coagulation, J. Colloid Interface Sci., 1970, Vol. 33, p. 562–571.

30 Honig E. P., Roebersen G., Wiersema P. H., Effect of hydrodynamic Interaction on the Coagulation of hydrophobic Colloids, J. Colloid Interface Sci., 1971, Vol. 36, p. 91–104.

31 Batchelor G. K., Brownian Diffusion of Particles with hydrodynamic Interaction, J. Fluid Mech., 1976, Vol. 74, p. 1–29.

32 Kaminskiy V. A., On Kinetics of Turbulent Coagulation, Colloid J., 1976, Vol. 38, No. 5, p. 907–912 (in Russian).

33 Hinze J. O., Turbulence, McGraw-Hill, New York, 1959.

34 Monin A. S., Yaglom A. M., Statistical Fluid Mechanics: Mechanics of turbulence (in 2 Vol.), MIT Press, Cambridge, 1971, 1975.

35 Schlichting H., Grenzschicht-Theorie, Karisruhe, G. Braun Verlag, 1964.

36 Sedov L. I., Mechanics of Continuous Media (in 2 Vol.), Nauka, Moscow, 1973.

37 Leontovich M. L., Introduction to Thermodynamics. Statistical Physics, Nauka, Moscow, 1983 (in Russian).

38 Russel W. D., Brownian Motion of small Particles suspended in Liquids, Ann. Rev. Fluid Mech., 1981, Vol. 13, p. 425–455.

39 Zinchenko A. Z., To calculation of hydrodynamic Interaction of Drops at small Reynolds Numbers, Appl. Math Mech., 1978, No. 5, p. 955–959.

40 Davis M. H., The slow Translation and Rotation of two unequal Spheres in a viscous Fluid, Chem. Eng. Sci., 1969, Vol. 24, p. 1769–1776.

41 O'Neil M. E., Majumdar S. R., Asymmetric slow viscous Fluid Motion caused by the Translation or Rotation of two Spheres, P. I, II, ZAMP, 1970, Vol. 21, p. 164.

42 Wacholder E., Weihs D., Slow Motion of a Fluid Sphere in the Vicinity of another Sphere or a Plane Boundary, Chem. Eng. Sci., 1972, Vol. 27, p. 1817–1827.

43 Haber S., Hetsroni G., Solan A., On the low Reynolds Number Motion of two Droplets, J. Multiphase Flow, 1973, Vol. 1, p. 57–71.

44 Rushton E., Davies G. A., The slow unsteady Settling of two Fluid Spheres along their Line of Centres, Appl. Sci. Res., 1973, Vol. 28, p. 37–61.

45 Reed L. D., Morrison F. A., The slow Motion of two touching Fluid Spheres along their Line of Centres, J. Multiphase Flow. 1974, Vol. 1, p. 573–583.

46 Hetsroni G., Haber S., Low Reynolds Number Motion of two Drops submerged in unbounded arbitrary Velocity Field, J. Multiphase Flow, 1978, Vol. 4, p. 1–17.

47 Zinchenko A. Z., Slow asymmetric Motion of two Drops in a viscous Medium, Appl. Math. Mech., 1980, No. 1, p. 30–37 (in Russian).

48 Fuentes Y. O., Kim S., Jeffrey D. J., Mobility Functions for two unequal viscous Drops in Stokes Flow, P. 1: Axisymmertric Motions, P. II: Non-axisymmertric Motions, Phys. Fluids, 1988, Vol. 31, p. 2445–2455; Phys. Fluids. A, 1989, Vol. 1, p. 61–76.

49 Davis R. H., Schonberg J. A., Rallison J. M., The Lubrication Force between two viscous Drops, Phys. Fluids A, 1989, Vol. 1, p. 77–81.

50 Zinchenko A. Z., Calculation of close Interaction of Drops in view of internal Circulation and Effects of Slippage, Appl. Math. Mech., 1981, Vol. 45, p. 759–763.

51 H. R. Kruyt (Ed.), Colloid Science. Vol. 1: Irreversible System, Elsevier, S. Amsterdam, 1952.

52 Hogg R., Healy T. W., Fuerstenau D. W., Mutual Coagulation of Colloidal Dispersions, Trans. Faraday Soc., 1966, Vol. 62, p. 1638–1651.

53 Hamaker H. C., The London-Van der Waals Attraction between spherical Particles, Physica, 1937, Vol. 4, p. 1058–1978.

54 Shenkel J. N., Kitchener J. A., A Test of the Deryaguin-Verwey-Overbeek Theory with a colloidal Suspension, Trans. Faraday Soc., 1960, Vol. 56, p. 161–173.

55 Melcher J. R., Taylor G. I., Electro-hydrodynamics: a Review of the Role of interfacial Shear Stress, Ann. Rev. Fluid. Mech., 1969, Vol. 1, p. 111–146.

56 O'Konski C. T., Thacher H. C., The Distortion of Aerosol Droplets by an electric Field, J. Phys. Chem., 1953, Vol. 57, p. 955–958.

57 Garton C. G., Krasucki Z., Bubbles in insulating Liquids: Stability in an electric Field, Proc. Roy. Soc. Lond. A, 1964, Vol. 280, p. 211–222.

58 Taylor G. I., Disintegration of Water Drops in an electric Field, Proc. Roy. Soc. Lond. A, 1964, Vol. 280, p. 383–397.

59 Rosenklide C. E., A dielectric Fluid Drop in an electric Field, Proc. Roy. Soc. Lond. A, 1969, Vol. 312, p. 473–494.

60 Miskis M. J., Shape of a Drop in an electric Field, Phys. Fluids, 1981, Vol. 24, p. 1967–1972.

61 Adornato P. M., Brown R. A., Shape and Stability of electrostatically levitated Drops, Proc. Roy. Soc. Lond. A, 1983, Vol. 389, p. 101–117.

62 Basaran J. A., Scriven L. E., Axisymmetric Shape and Stability of charged Drops in an electric Field,

Phys. Fluids A., 1989, Vol. 1, p. 799–809.
63 Allan R. S., Mason S. G., Particle Behavior in Shear and electric Fields, I. Deformation and Burst of Fluid Drops, Proc. Roy. Soc. Lond. A, 1962, Vol. 267, p. 45–61.
64 Taylor G. I., Studies in Electrohydrodynamics. I. The Circulation produced in a Drop by an electric Field, Proc. Roy. Soc. Lond. A, 1966, Vol. 291, p. 159–166.
65 Torza S., Cox R. G., Mason S. G., Electrohydrodynamic Deformation and Burst of Liquid Drops, Phil. Trans. R. Soc. Lond., 1971, Vol. 269, p. 295–319.
66 Ajaya O. O., A Note on Taylor's Electrohydrodynamic Theory, Proc. Roy. Soc. Lond. A, 1978, Vol. 364, p. 499–507.
67 Feng J. Q., Scott T. C., A computational Analysis of Electrohydrodynamics of a leaky dielectric Drop in an electric Field, J. Fluid Mech., 1966, Vol. 311, p. 289–326.
68 Shkadov V. J., Shutov A. A., Deformation of Drops and Bubbles in an electric Field, Fluid Dyn., 2002, No. 5, p. 54–66 (in Russian).
69 Thornton J. D., The Application of electrical Energy to chemical and physical Rate Processes, Rev. Pure Appl. Chem., 1968, Vol. 18, p. 197–218.
70 Harker J. H., Ahmadzadeh J., The Effect of electric Field on Mass Transfer from falling Drops, Int. J. Heat Mass Transfer, 1974, Vol. 17, p. 1219–1225.
71 Bailes P. J., Solvent Extraction in an electric Field, Industr. Engng. Chem. Process Des. Dev., 1981, Vol. 20, p. 564–570.
72 Baird M. H. I., Electrostatic Effects on Extraction. In Handbook of Solvent Extraction (Ed. I. C. Lo, M. H. I. Baird, C. Hanson), John Wiley and Sons, 1983, Vol. 20, p. 268–269.
73 Scott T. C., Use of electric Fields in Solvent Extraction, A Review and Prospects, Sep. Purif. Methods, Vol. 18, p. 65–109.
74 Weatherley L. R., Electrically enhanced Extraction. Science and Practice of Liquid-Liquid Extraction (Ed. I. D. Thornton), Oxford University Press, 1968, Vol. 2, p. 407–419.
75 Platsinski K. J., Kerkhof P. J. H. M., Electric Field driven Separations: Phenomena and Applications, Sep. Sci. Technol., 1992, Vol. 27, p. 995–1021.
76 He W., Baird M. H. I., Chang J. S., The Effect of electric Field on Mass Transfer from Drops dispersed in a viscous Liquid, Can. J. Chem. Eng., 1993, Vol. 71, p. 366–376.
77 Sherwood J. D., Breakup of Fluid Droplets in electric and magnetic Fields, J. Fluid. Mech., 1988, Vol. 188, p. 133–146.
78 Landau L. D., Lifshitz E. M., Electrodynamics of continuous Media, GITTL, Moscow, 1957 (in Russian).
79 Ausman E. L., Brook M., Distortion and Disintegration of Water Drops in strong Electric Fields, J. Geophys. Res., 1967, Vol. 72, p. 6131–6141.
80 Brazier-Smith P. R., Stability and Shape of isolated Pair of Water Drops in an electric Field, Phys. Fluids, 1971, Vol. 14, p. 1–6.
81 Stone H. A., Dynamics of Drop Deformation and Breakup in viscous Fluids, Ann. Rev. Fluid. Mech., 1994, Vol. 26, p. 65–102.
82 Nigmatullin P. I., Dynamics of multiphase Mediums (Vol. 1), Nauka, Moscow, 1987 (in Russian).
83 P. Sherman (Ed.), Emulsion Science, Academic Press, London, New York, 1968.
84 Loginov V. I., Dynamics of Breakup of a Fluid Drop in a turbulent Flow, Appl. Mech. Techn. Phys., 1985, No. 4, p. 66–73 (in Russian).
85 Tichonov V. I., Overshoots in random Processes, Nauka, Moscow, 1970 (in Russian).
86 Kolmogorov A. N., On the logarithmic normal Law of Particle Distribution during Breakup, Papers Acad. Sci. USSR, 1941, Vol. (31) No. 2, p. 99–101 (in Russian).

87 Golizin G. S., Fluctuations of Energy Dissipation in locally isotropic Turbulent Flow, Paper Acad. Sci.: USSR, 1962, Vol. 144. No. 3, p. 520–523 (in Russian).
88 Morse P. M., Feshbach H., Methods of theoretical Physics (in 2 Vols), Vol. 2, McGraw-Hill, New York, 1953.
89 Korn G. A., Korn T. M., Mathematical Handbook, McGraw-Hill, New York, 1968.
90 Allan R. S., Mason S. G., Effects of electrical Field on Coalescence in Liquid – Liquid Systems, Trans. Faraday Soc., 1961, Vol. 57, p. 2027.
91 Landau L. D., Lifshitz E. M., Field Theory, Nauka, Moscow, 1973.
92 Smythe W. E., Static and Dynamic Electricity, Mc Graw-Hill, New York, 1939.
93 Lebedev N. N., Skal'skaja I. P., Ufland J. S., Collection of Problems on mathematical Physics, GITTL, Moscow, 1955 (in Russian).
94 Vigovskoy V. P., Redistribution of Charges at Contact of conducting Spheres in an electric Field, Colliod J., 1981, Vol. 43, No. 5, p. 967.
95 Sneddon I., Fourier Transforms, McGraw-Hill, New York, 1951.
96 Russel A., Two charged Spherical Conductors Interaction Forces, Proc. Phys. Soc., 1922, Vol. 35, p. 10.
97 Soo S. L. A., Dynamics of charged Suspensions. Intern. reviews in aerosol physics and chemistry (Vol. 2), Pergamon Press, Oxford – New York, 1971, p. 61–149.
98 Lebedev N. N., The Force acting on a conducting Ball placed into the Field of a Flat Capacitor, J. Techn. Phys., 1962, Vol. 32, No. 2, p. 375 (in Russian).
99 Levchenko D. N., Bergshtein V. V., Hudjakova A. D., Nikolaeva N. M., Oil-Water Emulsions and Methods of Preventing Emulsion Breakdown, Chemistry, Moscow, 1967 (in Russian).
100 Lapiga E. J., Loginov V. I., On the Kernel of the Kinetic Equation of Coalescence, Fluid Dyn., 1980, p. 32–38 (in Russian).
101 Anisimov B. F., Emeljanenko V. G., The Coalescence Criterion for Drops of Water-in-Oil Emulsion in a uniform electric Field, Colloid J., 1977, Vol. 39, No. 3, p. 528 (in Russian).
102 Hochraider D., Hidy G. M., Zebel G., Creeping Motion of charged Particles around a Cylinder in electric Field, J. Colloid Interface Sci., 1969, Vol. 30, No. 1.
103 Van Dyke M., Perturbation Methods in Fluid Mechanics, Academic Press, New York, London, 1964.
104 Camp T. R., Stein P. C., Velocity Gradients and internal Work in turbulent Clouds, J. Boston Soc. Civ. Eng., 1943, Vol. 30, p. 219.
105 Smoluchowski M., Versuch einer Mathematischen Theorie der Koagulations Kinetik kolloider Lösungen, Z. Phys. Chem., 1917, Vol. 92, p. 129–168.
106 Saffman P. G., Turner J. S., On the Collision of Drops in turbulent Clouds, J. Fluid. Mech., 1956, Vol. 1, p. 16–30.
107 Pnueli D., Gutfinger C., Fichman M., A turbulent-brownian Model for Aerosol Coagulation, Aerosol Sci. Technol, 1991, Vol. 14, p. 201–209.
108 Abrahamson J., Collision Rates of small Particles in a vigorously turbulent Fluid, Chem. Eng. Sci., 1975, Vol. 30, p. 1371–1379.
109 Higashitani K., Yamauchi K., Matsuno Y., Hosokawa G., Turbulent Coagulation of Particles dispersed in a viscous Fluid, J. Chem. Eng. Japan, 1983, Vol. 16, p. 299–304.
110 Higashitani K., Ogawa R., Hosokawa G., Matsuno Y., Kinetic Theory of Shear Coagulation for Particles in a viscous Fluid, J. Chem. Eng. Japan, 1983, Vol. 16, p. 299–304.

VI
Gas–Liquid Mixtures

A gas–liquid mixture is a biphasic medium, in which the gas represents the continuous phase and the disperse phase is constituted by liquid drops. The natural gas of gas-condensate fields is in this condition as it enters plants for the dual preparation of gas and condensate (see Part I). Gas and condensate preparation involve the following processes: separation of condensate droplets, water droplets and vapor, and heavy hydrocarbons from the gas, as well as stabilization of separated condensate, i.e. removal of light hydrocarbons and neutral components from the gas. The separation of a condensate from a gas is performed in a gas separator, extraction of water vapor (gas dewatering) and heavy hydrocarbons from a gas is performed in absorbers using special liquid absorbents, and stabilization of condensate is achieved in dividers or deflationers, which are similar in design to oil-and-gas separators. Typical low-temperature separation (LTS) and low-temperature absorption (LTA) technological schemes of plants for the dual preparation of gas and condensate are depicted in Figs. 1.1 and 1.2. The present section deals with the basic processes relating to the preparation of gas and condensate for transportation: separation of gas-condensate mixtures in separators, absorptive dehydration of gas, absorptive extraction of heavy hydrocarbons from gas, and prevention of hydrate formation in gas.

Separation of Multiphase, Multicomponent Systems. E. G. Sinaiski and E. J. Lapiga

ISBN: 978-3-527-40612-8

14
Formation of a Liquid Phase in a Gas Flow

The flow drawn from the inlet pipe of a separator (Fig. 14.1) represents a biphasic gas-liquid mixture. The liquid phase in the inlet pipe is in the form of drops of various sizes suspended in the gas stream and as a thin film on the pipe wall. The volume content of liquid phase in the gas stream is insignificant (about 100 g in 1 m^3 of gas, which corresponds to a volume concentration of liquid of 10^{-2} m^3/m^3 at a pressure of about 100 atm), and there is a high stream velocity in the pipe (up to 10 m s^{-1}). Therefore, the majority of the mass of liquid will be in suspension. The flow in the pipe is turbulent in character, resulting in intensive coagulation of drops and breakage of the large droplets formed. As a result, there is a dynamic balance between these two processes in the flow, which forms the equilibrium distribution of drops over radii, $n_0(R)$. Besides coagulation and drop breakage, the interphase mass-exchange (evaporation or condensation of drops) can affect the establishment of the distribution $n_0(R)$. In a real process of biphase flow, the pressure and temperature can change, for example, upon passage through throttles, heat exchangers, or turbo-expanders. As a result, the thermodynamic balance is disrupted, which leads to nucleation of the liquid phase and to its further growth in conditions of over-saturation. Therefore, if these devices are located prior to a separator in the technological scheme, then it is necessary to take into account the interphase mass exchange. Further devices, passage through which may disrupt the thermodynamic balance between the gas and liquid phases, are referred to as devices of preliminary condensation, or in abbreviated form, DPC, since the end result of these devices is not simply to lower the temperature, but to create the conditions for the deposition of the condensate from the gas.

From the above it follows that two cases are possible: either a DPC is absent or is far enough from the separator (this distance will be estimated later), or a DPC is sufficiently close to the separator that it cannot be ignored. In the first case, the size distribution of drops in the flow in the inlet pipeline is established only due to the processes of coagulation and breakage of drops, whereas in the second case very fine drops are formed downstream of the DPC. The size of these drops changes not only due to coagulation and breakage, but also due to intensive mass exchange under the supersaturation conditions of the mixture.

Separation of Multiphase, Multicomponent Systems. E. G. Sinaiski and E. J. Lapiga

ISBN: 978-3-527-40612-8

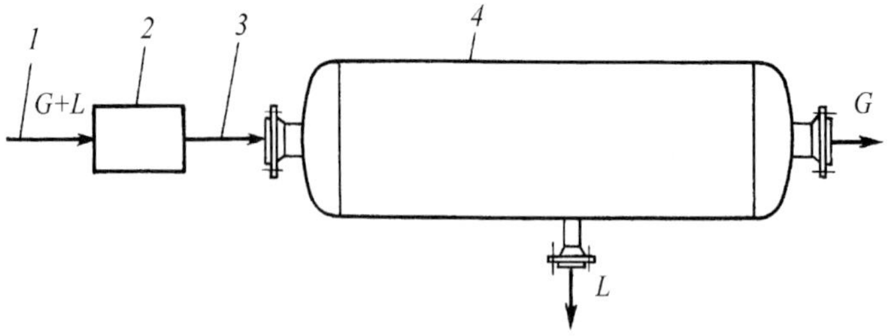

Fig. 14.1 Element of the technological scheme: device of preliminary condensation (DPC) – inlet pipeline – separator: 1, 3: inlet pipeline; 2: DPC; 4 – separator; G: gas; L: liquid.

Thus, drops of two kinds can be formed ahead of a separator, since there are two different mechanisms for their formation. Below, we consider both mechanisms of drop formation, allowing assessment of the parameters needed to determine the separation efficiency of a gas-liquid mixture in a gas separator.

14.1 Formation of a Liquid Phase in the Absence of Condensation

In the considered case, the basic mechanisms of formation of droplets in the turbulent gas flow are processes of coagulation and breakage of drops. These two processes proceed simultaneously. As a result, the size distribution of the drops is established. Assuming homogeneity and isotropy of the turbulent flow, this distribution looks like a logarithmic normal distribution [1]:

$$n(R) = \frac{n_* R_1}{\sigma R} \exp\left(-\frac{\ln^2(R/R_1)}{2\sigma^2}\right), \tag{14.1}$$

where $n_* = 3W \exp(-2.5\sigma^2)/4\pi\sqrt{2\pi}R_{\alpha v}^4$, $R_1 = R_{\alpha v} \exp(-0.5\sigma^2)$, W is the volume concentration of the liquid phase, σ^2 is the dispersion of the distribution, and $R_{\alpha v}$ is the average drop radius.

To define the average radius, it is necessary to consider the processes of coagulation and breakage of the drops. Drops in a turbulent gas flow are only liable to breakage when their radius exceeds some critical value [2]. Drops of radius less than the critical value can only coagulate. Actually, there is no precise border between drops that can coagulate and break since these processes are not strictly defined. Therefore, it is more correct to say that the smaller a drop radius below the critical value, the lower the probability of breakage of the drop in question. If the drop radius exceeds the critical value, the drop in question will with high probability be broken, forming, as a rule, two drops of similar size, along with several small drops referred to as daughter drops or satellites. Therefore, if one estimates the average drop size after a dynamic balance of the coagulation and breakage processes has been established, this size may be taken as being the critical radius.

Breakage of drops in a turbulent gas flow occurs due to the inertial effect caused by a significant difference of density of liquid and gas, and also due to difference of pulsation velocities, i.e. velocities of turbulent pulsations flowing around a drop, at opposite ends of the drop. Breakage of a drop thus occurs due to deformation of its surface.

A drop of density ρ_L, suspended in a turbulent flow of gas with density $\rho_G \ll \rho_L$, is only partially entrained by the gas. Therefore, gas flow around the drop has a relative velocity of the order:

$$u \sim \frac{1}{\sqrt{3}} \varepsilon_0^{1/3} \left(\frac{\rho_L}{\rho_G}\right)^{1/3} \left(\frac{V}{S}\right)^{1/3} \left(\frac{2}{k_f}\right)^{1/3} \tag{14.2}$$

where ε_0 is the specific dissipation of energy, V and S are the volume and cross-section of the drop, respectively, and k_f is the coefficient of aerodynamic resistance of the drop, which is equal to 0.4.

At this velocity, the dynamic thrust acting on the drop is given by:

$$Q \sim \frac{k_f \rho_G}{2} \left(\frac{\rho_L}{\rho_G}\right)^{2/3} \varepsilon_0^{2/3} R^{2/3} \tag{14.3}$$

Besides the dynamic thrust Q caused by the particle inertia in the gas flow, the thrust caused by a change of the pulsation velocity over the drop length also acts on the drop:

$$Q_P \sim k_f \frac{\rho_G (u_1^2 - u_2^2)}{2} \tag{14.4}$$

where u_1 and u_2 are the gas velocities at points separated by a distance of $2R$.

The process of deformation and breakage of a drop is caused by small-scale pulsations λ (see Section 11.3) since large-scale pulsations vary little over distances of the order of the diameter of a drop. The velocity u_λ of such pulsations depends on λ being greater or smaller than the internal scale of turbulence, λ_0 (Kolmogorov scale):

$$\lambda_0 = \frac{d}{\mathrm{Re}^{3/4}} \tag{14.5}$$

where d is the pipe diameter, $\mathrm{Re} = \dfrac{\rho_0 U d}{\mu_0}$ is the Reynolds number of the flow in the pipe, U is the average flow velocity, and μ_0 is the dynamic viscosity of the gas.

For typical values, $d = 0.1$ m, $U = 10$ m s^{-1}, $\rho_G = 40$ kg m^{-3}, $\mu_G = 1.2 \times 10^{-5}$ Pa·s, $\lambda_0 = 1.5 \times 10^{-6}$ m $= 1.5$ μm. The average drop size in the flow, as a rule, is more than this value, and therefore it makes sense to consider only pulsations with $\lambda > \lambda_0$. For such pulsations, see Eq. (11.46):

$$u_\lambda = (\varepsilon_0 \lambda)^{1/3} \tag{14.6}$$

Taking $\lambda = 2R$ in Eq. (14.6) and substituting the obtained expression into Eq. (14.4), one obtains:

$$Q_P \sim 0.5 k_f \rho_G (2R\varepsilon_0)^{2/3} \tag{14.7}$$

From Eqs. (14.3) and (14.7), it follows that:

$$\frac{Q}{Q_P} \sim \left(\frac{\rho_L}{\rho_G}\right)^{2/3} \tag{14.8}$$

This approximation implies that $Q \gg Q_P$ and that the main contribution to the process of drop breakage is that of the dynamic thrust Q. The drop does not break until the dynamic thrust is counterbalanced by the force of surface tension. Therefore, a condition of drop balance is

$$0.5\rho_L^{2/3}\varepsilon_0^{2/3}R^{2/3} = \frac{2\Sigma}{R} \tag{14.9}$$

Eq. (14.9) allows estimation of the critical radius of a drop, i.e. the radius above which a drop has a high probabilityof breakage.

If ε_0 in Eq. (14.9) is expressed in terms of parameters of flow

$$\varepsilon_0 = \nu_0^3/\lambda_0^4 \tag{14.10}$$

and utilizing Eq. (14.5), one obtains an expression for the critical drop radius:

$$R_{cr} \sim \left(\frac{\Sigma}{k_f \rho_G}\right)^{3/5} \left(\frac{\rho_G}{\rho_L}\right)^{2/5} \frac{d^{2/5}}{U^{6/5}} \tag{14.11}$$

If we enter a dimensionless parameter named the Weber number:

$$\mathrm{We} = \frac{\text{dynamic thrust}}{\text{force of surface tension}} = \frac{\rho_G U^2 d}{\Sigma}. \tag{14.12}$$

Equation (14.11) then transforms to

$$\frac{R_{cr}}{d} = k_f^{-3/5}\mathrm{We}^{-3/5}\left(\frac{\rho_G}{\rho_L}\right)^{2/5} \tag{14.13}$$

representing a connection between the three dimensionless parameters R_{cr}/d, We, and ρ_G/ρ_L. For values $d = 0.1$ m, $U = 10$ m s^{-1}, $\rho_G = 40$ kg m^{-3}, $\rho_L = 800$

kg m^{-3} and $\Sigma = 5 \times 10^{-3}$ N m^{-1}, one obtains We $\sim 10^5$, $\rho_G/\rho_L \sim 5 \times 10^{-2}$, and $R_{cr}/d \sim 10^{-3}$. Hence, in the considered case, the critical radius of the drop is of the order of $R_{cr} \sim 100$ μm.

As already noted, Eq. (14.13) applies for drops suspended in a homogeneous and isotropic turbulent flow. The movement of a gas-liquid mixture may be considered as homogeneous and isotropic in the bulk flow, but near a wall this assumption does not hold. Besides, drops can be blown away from a liquid film at the wall, but other drops may be deposited at the wall. These phenomena result in a change of Eq. (14.13). Measurements of the average steady-state size of drops of gas-liquid mixtures in a turbulent flow in a pipe, as described in ref. [3], have shown that the average drop size is described by the following relationship:

$$\frac{R_{cr}}{d} = 0.12\mathrm{We}^{-3/7}\left(\frac{\rho_G}{\rho_L}\right)^{1/7}. \tag{14.14}$$

The difference between Eq. (14.14) and Eq. (14.13) concerns essentially only the numerical factor, since the exponents differ only marginally. Since $k_f^{-3/5} \sim 1.5$ and in Eq. (14.14) the numerical factor is equal 0.12, Eq. (14.14) gives an average steady-state drop radius an order of magnitude smaller than that given by Eq. (14.13).

Thus, Eq. (14.14) will be used for the average steady-state radius of drops formed in the pipeline ahead of a separator.

14.2 Formation of a Liqid Phase in the Process of Condensation

If devices are present ahead of a separator that are capable of varying the gas temperature and pressure, small drops may originate (nucleate) in the flow. Among such devices are throttles, heat exchangers, and turbo-expanders. Since the mechanism of formation of a liquid phase in these devices is essentially identical, the case of a throttle is considered as a representative example.

The mechanism of liquid phase (fog) formation behind a throttle placed in a supply pipe before a separator is based on the process of adiabatic expansion of the gas mixture. In this process, the volume of the mixture increases and the pressure and temperature of the vapor decrease, since the work of expansion is done by the internal energy of the gas. The saturated vapor pressure is reduced by the temperature decrease and this results in increased vapor supersaturation. The degree of supersaturation, s, expresses the ratio of the vapor pressure in the gas, p_v, to the saturated vapor pressure, $p_{v\infty}$, above a flat surface of the same liquid:

$$s = \frac{p_v}{p_{v\infty}}, \quad p_{v\infty} = \exp\left(C - \frac{E}{T}\right) \tag{14.15}$$

Here, T is the absolute temperature, C is a constant, and $E = 0.12M_l l$, where M_l is the molecular mass of the condensable vapor and l is the specific heat of evaporation.

Above a convex surface, which drops have, the saturated vapor pressure is higher than above a flat surface due to the capillary pressure, p_{cap}. The latter increases with reduction of the drop radius, because $p_{cap} = 2\Sigma/R$, where Σ is the coefficient of surface tension of the drop. Therefore, the necessary condition for condensation of the vapor in a gas volume is supersaturation of the vapor, allowing compensation for the increased pressure.

If the formation of drops occurs on condensation nuclei, then the process is called heterogeneous condensation. Such nuclei can be particles of impurities, pollution, droplets, etc. If the formation of drops occurs as a result of vapor condensation on spontaneously forming germs, then the process is called homogeneous condensation.

The process of homogeneous condensation consists of three stages: formation of supersaturated vapor, formation of germs, condensation of vapor on the surfaces of the germs and their further growth. Condensation of a vapor begins only at a certain supersaturation, called the critical value, given by:

$$s_{cr} = \exp\left[1.74 \times 10^7 \frac{M_L}{\rho_L} \left(\frac{\Sigma}{T}\right)^{3/2}\right] \tag{14.16}$$

where ρ_L is the density of the liquid to be condensed.

The process of heterogeneous condensation occurs in the presence of nuclei of condensation. If these nuclei are small drops, than at the establishment of a balance at the surface of drops the Kelvin formula applies:

$$\ln s = \frac{2\Sigma M_L}{AT\rho_L R} \tag{14.17}$$

where A is the gas constant and R is the drop radius.

From Eq. (14.17), knowing the drop radius R, it is possible to find the equilibrium value of the partial pressure of the vapor above the drop as opposed to the vapor pressure $p_{v\infty}$ above a flat surface.

For homogeneous condensation in a gas mixture, at pressures exceeding the saturation pressure, germs of liquid phase originate at a rate determined by the Frenkel formula [4]:

$$I = 1.82 \times 10^{26} \frac{\alpha}{s\rho} (M_L\Sigma)^{1/2} \left(\frac{p}{T}\right)^2 \exp\left[-1.76 \times 10^{16} (\ln s)^{-2} \frac{M_L^2\Sigma}{\rho^2 T^3}\right]^3 \tag{14.18}$$

Here, α is the condensation factor, which gives an indication of the fraction of vapor molecules which, after impact on the liquid surface, remain on it. The literature contains experimental α values for some pure liquids, although the data of different authors for the same liquids show discrepancies. Therefore, reliable data for the condensation factor are apparently lacking. In this connection, it will be assumed that $\alpha = 1$, which is equivalent to assuming that all vapor molecules remain on the drop surface after contacting it.

From the above, it follows that a necessary condition for the formation of a condensed phase is the operation of such processes that increase the supersaturation of mixtures, for example, as a result of temperature lowering or an increase of the gas pressure. In practice, usually both the temperature and pressure change simultaneously, and hence the supersaturation of the mixtures changes. For example, in adiabatic expansion of a gas, the pressure and temperature fall. The first reduces while the second increases the supersaturation. However, on reducing the temperature, the saturation pressure falls much more sharply and this leads to increased supersaturation of mixtures. Supersaturation increases sharply only in the initial stage of the condensation process. During condensation, the temperature of the mixture is increased a little, and in time, values of partial vapor pressure far from the drop and on its surface level off due to reduction of the vapor concentration. Since the flow of vapor to the drop surface decreases, supersaturation also decreases. Finally, the process of formation of germs comes to a halt. Figure 14.2 illustrates the characteristic change of vapor supersaturation with time.

The rate of condensation at the drop surface is determined by the diffusion of the vapor to the surface, and so at a high rate of supersaturation change, ds/dt (for example, in processes causing a rapid increase of vapor supersaturation and associated with a small volume concentration of drops) the diffusion rate may be insufficient for a levelling off of the vapor pressure over the entire volume.

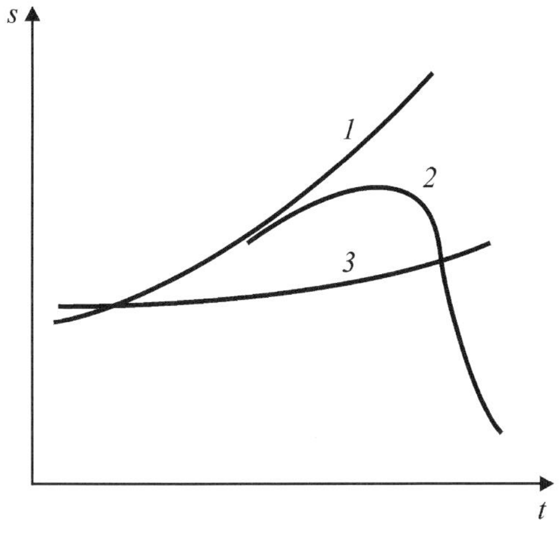

Fig. 14.2 Time dependence of vapor supersaturation: 1: without and 2: with regard to liquid-phase nucleation; 3: critical supersaturation.

Thus, the vapor pressure at the drop surface can differ markedly from the vapor pressure in the bulk of the mixture. As a result, significant supersaturation arises. This leads to intensive germ formation in the initial stage of the process.

Thus, depending on the ratio between the rate of vapor supersaturation and the concentration of droplets (nuclei of condensation), two limiting cases are possible:

1) The rate of vapor diffusion to a drop surface is high and the critical value of supersaturation is not attained so germs are not formed.

2) The rate of vapor diffusion is low; supersaturation exceeds the critical value and intensive germ formation occurs at a rate determined by Eq. (14.18).

A way to achieve cooling and supersaturation of mixtures is adiabatic expansion of a gas due to a throttle effect. In the realm of gas and gas-condensate fields, low-temperature separation (LTS) schemes are used, which include a combination of technological processes directed towards cooling of the production from wells to the requisite temperatures to facilitate separation. The LTS scheme is applied after the gas has been purged of most of the mass of liquid phase in the entrance separator, practically without changes of pressure and temperature. LTS is subsequently applied to condense water vapor and heavy hydrocarbons (HC). LTS entails the following processes.

1. Gas in a heat exchanger or refrigerator is cooled to the lowest possible temperature at which gas hydrates are not formed. Gas condensate and water are thereby isolated from the mixture. Thereafter, water and gas condensate are separated from the gas in separators. At the exit of the separator, a hydrate inhibitor is added to the gas flow.
2. Throttling of the gas, in the course of which the gas pressure and temperature are decreased without doing work and without heat exchange. Throttling is achieved with the aid of a choke restrictor.
3. Adiabatic expansions doing external work, resulting in a decrease in temperature. This process is carried out in a turbo-expander.

Throttling and adiabatic expansion of gas in turbo-expanders occur as a result of gas flow in pipes and channels of variable cross-section. To determine the temperature, pressure, and degree of supersaturation of a mixture in these devices it is necessary to carry out appropriate gas-dynamic calculations. An approach to performing such calculations, based on ref. [5], is outlined below.

Let us assume that in a given section of a pipe there is velocity U_1, pressure p_1, temperature T_1, gas density ρ_{G1}, and cross-sectional area S_1. A function for the change of the cross-sectional area with pipe length $S(x)$ is also specified.

Consider a cross-section of pipe, S_2. The gas-dynamic parameters in this section are determined by one-dimensional conservation equations of mass flow rate, momentum, energy, and equation of state.

The conservation equation of the mass flow rate of gas is

$$\rho_{G1} U_1 S_1 = \rho_{G2} U_2 S_2 \tag{14.19}$$

The conservation equation of energy is:

$$Q + \frac{p_1}{\rho_{G1}} - \frac{p_2}{\rho_{G2}} = L + L_{fr} + g(z_2 - z_1) + E_2 - E_1 + \frac{U_2^2 - U_1^2}{2} \tag{14.20}$$

Here, Q is the heat supplied to 1 kg of gas, L is the work done by 1 kg of gas, L_{fr} is the work done by frictional forces on 1 kg of gas, z is the height of an arrangement of the considered section, and E is the internal energy of the gas.

We may represent Q as a sum of heat fluxes:

$$Q = Q_S + Q_i \tag{14.21}$$

where Q_{S} is the heat flux from the outside due to heat exchange through the pipe wall, and Q_{i} is the heat flux from within due to transformation of frictional forces into heat.

The obvious equality $Q_{\text{i}} = L_{\text{fr}}$ is assumed. The equation of state of the gas is taken in elementary form, assuming the gas to be ideal:

$$p = A\rho_G T. \tag{14.22}$$

The appropriate thermodynamic formulae for an ideal gas are:

$$c_p = c_v + Ah, \quad h = E + AT = c_p T. \tag{14.23}$$

Here, h is the gas enthalpy, and c_{p} and c_{v} are specific thermal capacities at constant pressure and volume, respectively. From Eq. (14.20), it follows that:

$$Q_S - L = h_2 - h_1 + g(z_2 - z_1) + \frac{U_2^2 - U_1^2}{2}. \tag{14.24}$$

If the pipe is horizontal, then the change in the potential energy can be neglected and Eq. (14.24) becomes

$$Q_S - L = h_2 - h_1 + \frac{U_2^2 - U_1^2}{2}. \tag{14.25}$$

In a special case, when there is no heat exchange with the environment ($Q_S = 0$) and the gas does not do mechanical work ($L = 0$), instead of Eq. (14.25) one may write:

$$\frac{U_2^2 - U_1^2}{2} = h_2 - h_1 = c_{p1} T_1 - c_{p2} T_2. \tag{14.26}$$

If the gas velocity in the pipe changes only slightly ($U_1 \approx U_2$), then from Eq. (14.25) it follows that:

$$Q_S - L = c_{p1} T_1 - c_{p2} T_2. \tag{14.27}$$

In particular, at $Q_S = 0$:

$$c_{p1} T_1 = c_{p2} T_2 - L. \tag{14.28}$$

A similar correlation considering the velocity change may be written as:

$$c_{p1} T_1 = c_{p2} T_2 - L + \frac{U_2^2 - U_1^2}{2}. \tag{14.29}$$

From the last two equations, it follows that the temperature of the gas decreases even in the absence of heat exchange through the pipe wall due to the mechanical work done by the gas and the change in kinetic energy.

The momentum conservation equation for the cylindrical fluid tube is:

$$dp + \rho_G U \, dU = -\frac{dF_{fr}}{S} - \frac{dF}{S}, \tag{14.30}$$

where F_{fr} is the friction force applied to a lateral surface of the fluid tube in the direction opposite to the gas pressure, and F is the external force. If these forces are absent or do not change, then Eq. (14.30) takes the simple form:

$$p_1 + \frac{\rho_G U_1^2}{2} = p_2 + \frac{\rho_G U_2^2}{2} \tag{14.31}$$

and is called the Bernoulli equation.

Thus, from the conservation equations of mass, momentum, and energy, it follows that mechanical, thermal, and geometrical factors exert an influence on the velocity of the gas flow, and also on the distributions of pressure, temperature, and density. Next, we consider these influences in more detail within the framework of the assumption of gas ideality.

From the conservation equation of mass flow rate for the gas:

$$G = \rho_G U S = \text{const.} \tag{14.32}$$

it follows that:

$$\frac{dG}{G} = \frac{dS}{S} + \frac{d\rho_G}{\rho_G} + \frac{dU}{U} \tag{14.33}$$

and from the equation of state one obtains:

$$\frac{dp}{\rho_G} = A\left(dT + T\frac{d\rho_G}{\rho_G}\right) \tag{14.34}$$

Substituting $d\rho_G/\rho_G$ in Eq. (14.34) by using Eq. (14.33), one obtains:

$$\frac{dp}{\rho_G} = A\,dT + AT\left(\frac{dG}{G} - \frac{dS}{S} - \frac{dU}{U}\right) \tag{14.35}$$

From the conservation equation of momentum, Eq. (14.30), it follows that:

$$\frac{dp}{\rho_G} = -U\,dU - dL - dL_{fr} \tag{14.36}$$

From the last two equations, we obtain:

$$A\,dT + \frac{a^2}{k}\left(\frac{dG}{G} - \frac{dS}{S}\right) + \left(U^2 - \frac{a^2}{k}\right)\frac{dU}{U} + dL + dL_{fr} = 0 \tag{14.37}$$

where $a^2 = (kAT)^{1/2}$ is the velocity of sound in an ideal gas, and $k = c_p/c_v$ is a constant of the adiabatic curve.

The energy equation, Eq. (14.25), may also be represented by:

$$dQ_S = dh + \frac{dU^2}{2} + dL = \frac{kA}{k-1}dT + U\,dU + dL \tag{14.38}$$

using the following correlation for the enthalpy:

$$dh = d(c_p T) = \frac{kA}{k-1}dT \tag{14.39}$$

Substituting temperature in Eq. (14.37) with the aid of Eq. (14.38) gives:

$$(M^2 - 1)\frac{dU}{U} = \frac{dS}{S} - \frac{dG}{G} - \frac{dL}{a^2} - \frac{k-1}{a^2}dQ_S - \frac{k}{a^2}dL_{fr} \tag{14.40}$$

where $M = U/a$ is the Mach number.

The term on the left-hand side of Eq. (14.40) changes sign at the transition through the velocity of sound. Therefore, the influence of various effects on gas flow under subsonic and supersonic conditions is reversed. At subsonic flow ($M < 1$), flow acceleration ($dU > 0$) may be caused by a narrowing of the channel ($dS < 0$), by supply of an additional mass of gas ($dG > 0$), as a result of work done by the gas ($dL > 0$), or by the supply of heat ($dQ_S > 0$). The same effects in a supersonic flow ($M > 1$) result in retardation. Hence, to regulate the velocity of gas through the critical velocity ($M = 1$) one should, for example, first narrow the channel and then expand it, or first supply heat and then remove it, etc. In the following, we consider each effect of practical interest in turn.

Geometrical effect (nozzle, throttle)

Suppose that $dG = dQ = dL = dL_{fr} = 0$ and $dS \neq 0$. From Eq. (14.40), it follows that:

$$(M^2 - 1)\frac{dU}{U} = \frac{dS}{S} \tag{14.41}$$

At $M < 1$, dU and dS are opposite in sign. Hence, to increase the gas velocity to value of $M = 1$, the cross-sectional area, S, should first be reduced to S_{cr}. Then, to obtain $M > 1$, one should increase this area.

The ratio of cross-sectional area to the appropriate critical value is

$$\frac{S}{S_{cr}} = \frac{[1 + 0.5(k-1)M^2]^{(k+1)/2(k-1)}}{M(0.5k + 0.5)^{(k+2)/2(k-1)}} \tag{14.42}$$

For a mixture of hydrocarbon gases, $k = 1.2$ and

$$\frac{S}{S_{cr}} = \frac{(1 + 0.1M^2)^{5.5}}{1.55M} \tag{14.43}$$

The functional dependence of Eq. (14.43) is shown in Fig. 14.3.

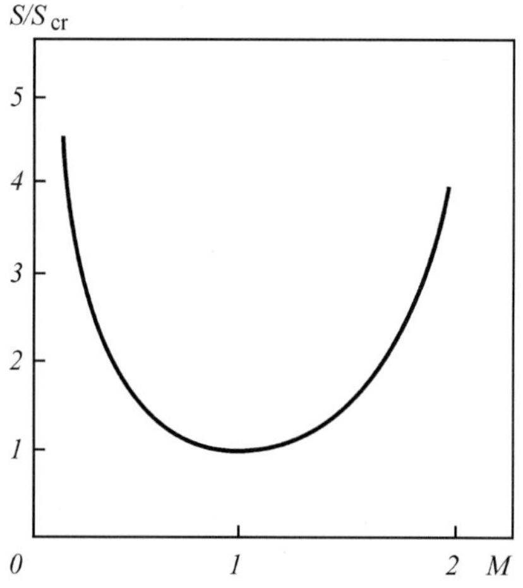

Fig. 14.3 Dependence of S/S_{cr} upon Mach number.

If the velocity, density, pressure, and temperature, as well as the function $S(x)$ are known, then values of these parameters for any cross-section may be obtained from the following system of equations:

$$\begin{aligned}
&\rho_G US = \rho_{G1} U_1 S_1, \quad p\rho_G^{-k} = p_1\rho_{1G}^{-k}, \\
&p = A\rho_G T, \quad c_p T = c_{p1} T_1 - \frac{U^2 - U_1^2}{2}.
\end{aligned} \tag{14.44}$$

The pressure and temperature drop along the nozzle or throttle lead to supersaturation of the mixture. Condensation of water vapor or heavy hydrocarbons in the flow is then possible. Since the residence time of the mixture in the nozzle or throttle is short, fine drops (fog) formed will not have enough time to increase in size to any appreciable extent. Therefore, the main growth of drops should occur in the flow beyond the exit of the nozzle or throttle, where the outflow of the gas jet forms in space as in a divergent nozzle.

Mechanical effect. In the considered case, it is accepted that $dG = dQ_S = dL = dL_{fr} = 0$ and $dS \neq 0$. Thus, from Eq. (14.39), it follows that:

$$(M^2 - 1)\frac{dU}{U} = -\frac{dL}{a^2} \tag{14.45}$$

If $dL > 0$, which corresponds to a case whereby the gas flow does work, for example, on a turbine wheel in a turbo-expander, then in subsonic flow ($M < 1$) the gas is accelerated, but in supersonic flow ($M > 1$) it is slowed down. If energy is supplied to the gas ($dL < 0$), as for example in a compressor, then in subsonic flow the gas is decelerated, whereas in supersonic flow it is accelerated. The basic property of a mechanical nozzle (turbine compressor) is that upon passage of a gas through it, the gas flow velocity is increased, and the pressure, density, and temperature are decreased according to the expressions:

$$\frac{M_2}{M_1} = \left(\frac{T_1}{T_2}\right)^{0.5(k+1)/(k-1)} = \left(\frac{p_1}{p_2}\right)^{0.5(k+1)/k} = \left(\frac{\rho_{G1}}{\rho_{G2}}\right)^{0.5(k+1)} \tag{14.46}$$

Heat effect. It is accepted that $dS = dG = dL = dL_{fr} = 0$ and $dQ \neq 0$. From Eq. (14.39), it follows that:

$$(M^2 - 1)\frac{dU}{U} = -\frac{k-1}{a^2}dQ_S \tag{14.47}$$

Comparison of this equation with Eq. (14.45) shows that since $k > 1$ these equations are identical. Therefore, the influences of heat supply ($dQ_S > 0$) and heat removal ($dQ_S < 0$) in the subsonic flow are to accelerate or decelerate the flow, respectively. In supersonic flow, the reverse is true.

Thus, the aforementioned analysis has shown that upon passage of gas mixture flows through devices of preliminary condensation, a temperature decrease may be expected, favorable for the condensation of water vapor or heavy hydrocarbon vapors.

Consider a gas mixture containing a component that may be condensed as a result of a temperature decrease. By p_{v1} and p_v we designate the partial pressures of this component before and after expansion. Consider passage of the mixture through a throttle. According to Eq. (14.43):

$$p_v = p_{v1}\left(\frac{T}{T_1}\right)^{k/(k-1)} = p_{v1}\left(\frac{\rho_G}{\rho_{G1}}\right) \tag{14.48}$$

Substituting Eq. (14.47) into the expression for supersaturation s, Eq. (14.15), one obtains:

$$s = \exp\left[E\left(\frac{1}{T}-\frac{1}{T_1}\right)\right]\frac{p_v}{p_{v1}} \quad \text{or} \quad \frac{s}{s_1} = \left(\frac{T}{T_1}\right)^{k/(k-1)} \exp\left[\left(\frac{1}{T}-\frac{1}{T_1}\right)\right] \tag{14.49}$$

where p_1 and p are the values of pressure before and during the expansion.

If the variation of p and T over the length of the device is known, then from Eq. (14.48) one can determine the supersaturation s. Then (if it is certain that $s > s_{cr}$) one can derive the rate of formation of germs of a new phase I according to Eq. (14.18). Knowing the residence time t of the mixture in the device, it is easy to determine the numerical and volume concentrations of drops formed in the device as a result of mixture expansion:

$$N = \int_o^t I\,dt, \quad W = \int_o^t V_n I\,dt \tag{14.50}$$

where V_n is the volume of the droplets formed.

The critical radius of germs may be found from the condition of the equality of pressures at the droplet surfaces:

$$p = p_{vp} + \frac{2\Sigma}{R} \tag{14.51}$$

where p_{vp} is the vapor pressure above a droplet surface:

$$p_{vp} = p_\infty \exp\left(\frac{2\Sigma V_L}{RkT}\right) \tag{14.52}$$

Here, V_L is the volume per one molecule of liquid, and k is the Boltzmann constant ($k = 1.3 \times 10^{-29}$ J C^{-1}).

Substitution of Eq. (14.51) into Eq. (14.50) gives an equation defining the radius of the droplet formed:

$$p - \frac{2\Sigma}{R} = p_\infty \exp\left(\frac{2\Sigma V_L}{RkT}\right) \tag{14.53}$$

The further growth of drops is caused by the process of vapor diffusion to the surfaces of the droplet.

15
Coalescence of Drops in a Turbulent Gas Flow

Droplets in motion in a turbulent flow of gas in a pipe are subject to breakage and coagulation (coalescence). These processes occur simultaneously and as a result a dynamic balance is established between them, determined by some distribution of drops with average radius R_{av}.

Assume that drops with radius $R > R_{av}$ will undergo breakage, whereas drops with $R < R_{av}$ will coalesce.

The radius of drops formed in devices of preliminary condensation is much smaller than R_{av}, as determined by the parameters of gas flow in the pipe. Therefore, after passage through the DPC, the drops are formed as a result of two processes. The first process is condensate growth of drops and the second is their coagulation until their radii become equal to R_{av}. From this point on, the average drop size does not change. Hence, having determined the time required to reach this point, it is possible to find the distance at which it is necessary to incorporate the DPC before a separator to enable the drops to acquire their maximum size. First, however, it is necessary to estimate how condensation and coagulation influence the formation of drops. Consider first the rate of drop formation due to coagulation.

In a turbulent flow there are two chief mechanisms of drop coagulation [2], that of turbulent diffusion and that of inertia. The inertial mechanism is based on the assumption that turbulent pulsations do not completely entrain the drop. As a result, relative velocities attained by drops due to turbulent pulsations depend on their masses. The difference in the pulsation velocities of drops of various radii causes their approach and leads to an increase of collision probability. The mechanism of turbulent diffusion is based on an assumption of full entrainment of drops by turbulent pulsations with scales, playing the chief role in the mechanism of approach of drops. Since drops move chaotically under the action of turbulent pulsations, their motion is similar to the phenomenon of diffusion and can be characterized by a coefficient of turbulent diffusion.

The distribution of drops over volumes $n\ (t, P, V)$ at a point in space P and at a moment in time t, in approximation of pair interactions, valid for a low volume concentration of drops, is described by the following kinetic equation:

$$\frac{\partial n}{\partial t} + \nabla \cdot (\boldsymbol{u} n) + \frac{\partial}{\partial V}\left[n \left(\frac{\partial V}{\partial t} \right)_m \right] = I_k + I_b + I_a + I_m + I_n \tag{15.1}$$

Separation of Multiphase, Multicomponent Systems. E. G. Sinaiski and E. J. Lapiga

ISBN: 978-3-527-40612-8

where $\boldsymbol{u}$ is the velocity of a drop with volume V, and dV/dt is the rate of change of drop volume as a result of mass exchange with the surrounding gas (evaporation, condensation). The terms on the right-hand side of Eq. (15.1) represent rates of change of distribution due to the processes of coagulation, breakage, ablation, and deposition at the border of the flow (for example, at a pipe wall) and nucleation. Thus, the rate of formation of drops with volume V as a result of coagulation is given by:

$$I_k = \frac{1}{2}\int_0^V K(\omega, V-\omega)n(\omega, t, P)n(V-\omega, t, P)\, d\omega$$
$$- n(V, t, p)\int_0^\infty K(\omega, V)n(\omega, t, P)\, d\omega \tag{15.2}$$

The rate of formation of drops with volume V due to breakage is given by:

$$I_b = \int_V^\infty f(\omega, V)\boldsymbol{g}(\omega)n(\omega, t, P)\, d\omega - n(V, t, p)\boldsymbol{g}(V) \tag{15.3}$$

The rate of formation of drops with volume V due to nucleation is:

$$I_k = I\delta(V - V_k) \tag{15.4}$$

Reliable data for defining the deposition and ablation of drops at a border of the flow are unfortunately absent.

Here, the following designations are introduced: $K(\omega, V)$ is the collision frequency of drops with volumes V and ω at unit concentration; $f(\omega, V)$ is the number of drops of volume ω formed by the breakage of drops of volume V; $\boldsymbol{g}(V)$ is the probability of breakage of a drop with volume V, I is the rate of formation of germs determined by Eq. (14.18), $\delta(x)$ is the delta function, and V_n is the volume of a germ determined by Eq. (14.52).

The collision frequency and the breakage probability are determined by consideration of the interactions between drops in a field of turbulent pulsations. Their definition represents an independent problem. The problem of drop interactions in an emulsion will be considered in detail in Section V.

The mechanism of drop coagulation depends on the conditions of mixture flow. In laminar flow, the coagulation is caused by the approach of drops due to different velocities of their motion or in the non-uniform field of velocities of an external medium, or on sedimentation in the gravity field. In a turbulent flow, the approach of drops occurs due to chaotic turbulent pulsations. In comparison with the laminar flow, the number of collisions of drops in unit time increases. Any, even insignificant, mixing of the flow increases the collision frequency.

Significant deformation of the drop surface precedes the process of drop breakage. This deformation is caused by the action on the surface of stresses from external and internal flow, for which significant gradients of velocity and dynamic pressure are necessary. In a laminar flow, velocity gradients arise in the flow near the walls, therefore the basic drop breakage is observed in the wall boundary layer where the flow has sheared character. In a turbulent flow, velocity gradients arise in the vicinity of drops when they are flowed around by small-scale eddies. Therefore, the size of drops capable of breakage in the turbulent flow is lower than in the laminar flow. Hereinafter, for simplicity, it will be supposed that the critical size of drops, given by Eq. (14.14), divides drops into those that are capable of breakage (with $R > R_{cr}$) and those that are incapable of breakage (with $R < R_{cr}$). The latter can only coagulate.

15.1 Inertial Mechanism of Coagulation

Consider a drop with radius R_2. The number of times that this drop encounters a drop with radius R_1 in unit time due to the inertial mechanism in turbulent flow may be estimated by means of the following expression [2]:

$$\omega_{12} = \pi(R_1 + R_2)^2 (R_1^2 - R_2^2) \frac{\rho_L \varepsilon_0^{3/4}}{\rho_G \nu_G^{5/4}} \tag{15.5}$$

where ε_0 is the specific energy of dissipation, ν is the coefficient of kinematic viscosity, and n_1 is the number of drops with radius R_1 in unit volume.

Since $\varepsilon_0 = U^3/d$, one may write:

$$\omega_{12} = \pi(R_1 + R_2)^2 (R_1^2 - R_2^2) \frac{\rho_L U^{9/4}}{\rho_G \nu_G^{5/4} d^{3/4}} n_1. \tag{15.6}$$

In the considered case, the coagulation constant, to be entered in Eq. (15.1), is:

$$K(V, \omega) = \frac{81}{256} (V + \omega)^2 |V^2 - \omega^2| \frac{\rho_L U^{9/4}}{\rho_G \nu_G^{5/4} d^{3/4}}. \tag{15.7}$$

Let us estimate the growth rate of drops, assuming them to be identical. The coagulation frequency, Eq. (15.6), must then be averaged over drop sizes:

$$\langle \omega_{12} \rangle = \langle \pi(R_1 + R_2)^2 (R_1^2 - R_2^2) \rangle \frac{\rho_L U^{9/4}}{\rho_G \nu_G^{5/4} d^{3/4}} n_1. \tag{15.8}$$

The value of the average size certainly depends on the form of drop distribution over their sizes. Although there is not a fixed form of distribution, it is possible to obtain an estimate using:

$$\langle (R_1 + R_2)^2 (R_1^2 - R_2^2) \rangle \sim R_{av}^4. \tag{15.9}$$

Now it is easy to write an equation of balance for drop number n:

$$\frac{dn}{dt} = -\frac{1}{2}\pi R_{\alpha v}^{4} \frac{\rho_L U^{9/4}}{\rho_G \nu_G^{5/4} d^{3/4}} n^2. \tag{15.10}$$

The factor of 1/2 appears on the right-hand side of Eq. (15.10) since in calculating the number of collisions the interactions of identical drops are taken into account twice.

Let us introduce the volume content of the drops:

$$W = \frac{4}{3}\pi R_{\alpha v}^{3} n. \tag{15.11}$$

Then, Eq. (15.10) takes the form:

$$\frac{d}{dt}\left(\frac{R_{\alpha v}}{R_{\alpha v}^{0}}\right) = \frac{W}{8}\left(\frac{R_{\alpha v}}{R_{\alpha v}^{0}}\right) R_{\alpha v}^{0} \frac{\rho_L U^{9/4}}{\rho_G \nu_G^{5/4} d^{3/4}}. \tag{15.12}$$

where $R_{\alpha v}^{0}$ is the initial radius of the drops.

The solution of Eq. (15.12) is:

$$\frac{R_{\alpha v}}{R_{\alpha v}^{0}} = \frac{1}{1 - t/t_{in}}, \qquad t_{in}^{-1} = \frac{W}{8} R_{\alpha v}^{0} \frac{\rho_L U^{9/4}}{\rho_G \nu_G^{5/4} d^{3/4}}. \tag{15.13}$$

The time t_{in} can be considered as the characteristic time of drop integration due to the inertial mechanism. It should be expressed not in terms of volume, W, but in terms of mass content of the liquid phase, $q = W\rho_{\text{L}}$:

$$t_{in} = \left(\frac{q}{8} R_{\alpha v}^{0} \frac{\rho_L U^{9/4}}{\rho_G \nu_G^{5/4} d^{3/4}}\right)^{-1}. \tag{15.14}$$

For typical values of $q = 0.2$ kg m^{-3}, $\rho_{\text{G}} = 40$ kg m^{-3}, $d = 0.25$ m, $\nu_{\text{G}} = 3 \times 10^{-7}$ m^2 s^{-1}, $U = 0.4$ m s^{-1}, and $R_{\alpha v} = 5 \times 10^{-5}$ m, one obtains $t_{\text{in}} \sim 3$ s. Thus, the drop radius increases twofold in 1.5 s.

Note that in calculations of such factors, the hydrodynamic resistance of drop motion, as well as forces of molecular interactions, are not taken into account. Hereinafter, it will be shown that correct allowance for these forces significantly increases the characteristic time of drop coagulation.

15.2 Mechanism of Turbulent Diffusion

We now consider the pulsations that can force two drops with radii R_1 and R_2 to approach each other. Consider two limiting cases, $R_2 \ll R_1$ and $R_2 \sim R_1$, without taking into account the resistance of the environment. It is obvious that in the first case the drops will be forced by pulsations of magnitudes $\lambda \sim r - R_1$, and in the second case by $-\lambda \sim r$, where r is the center-to-center distance of the considered drops. Using these limiting correlations, and also the conditions of symmetry $\lambda(r, R_1, R_2) = \lambda(r, R_2, R_1)$, we derive the following estimation for magnitudes of pulsation that would be able to pull together drops of any radius:

$$\lambda \sim r - R_1 - R_2 + \frac{R_1 R_2 (R_1 + R_2)}{R_1^2 + R_2^2 - R_1 R_2} \tag{15.15}$$

The coefficient of turbulent diffusion without consideration of the hindered motion of drops is given by [2]:

$$D_{T0} = \frac{\mu_0}{\rho_G \lambda_0^2} \lambda^2 \tag{15.16}$$

where λ_0 is the internal scale of turbulence. Substitution of Eq. (15.15) into Eq. (15.16) gives:

$$D_{T0} = \frac{\mu_0}{\rho_G \lambda_0^2} \left(r - R_1 - R_2 + \frac{R_1 R_2 (R_1 + R_2)}{R_1^2 + R_2^2 - R_1 R_2} \right)^2 \tag{15.17}$$

The collision frequency of drops of radius R_2 with a drop of radius R_1 is equal to the diffusion flux J_i determined by solution of the stationary diffusion equation. This equation in the spherical-symmetric case is:

$$\frac{1}{r^2} \frac{d}{dr} \left(r^2 D_T \frac{dn_2}{dr} \right) = 0 \tag{15.18}$$

The boundary conditions correspond to instant absorption of drops with radius R_2 by the drop of radius R_1 upon their contact:

$$n_2 = 0 \quad \text{when } r = R_1 + R_2 \tag{15.19}$$

and of constant concentration of drops with radius R_2 far from the drop with radius R_1:

$$n_2 = n_{20} \quad \text{when } r \to \infty \tag{15.20}$$

Solution of Eqs. (15.18)–(15.20) yields an expression for the flux:

$$J_T = 4\pi \left(r^2 D_T \frac{dn_2}{dr} \right)_{r = R_1 + R_2} = 4\pi \left(\int_{R_1 + R_2}^{\infty} \frac{dr}{r^2 D_T(r)} \right)^{-1}. \tag{15.21}$$

Substituting Eq. (15.17) for the coefficient of turbulent diffusion into Eq. (15.21) and taking into account that $\lambda_0 = d/\mathrm{Re}^{3/4}$, one obtains:

$$J_T = 4\pi n_{20}\alpha^2 \left(\frac{U^3}{\nu_G d}\right)^{1/2} \left(\frac{1}{R_1+R_2-\alpha} + \frac{1}{R_1+R_2} + \frac{2}{\alpha}\ln\left|\frac{R_1+R_2-\alpha}{R_1+R_2}\right|\right)^{-1}, \tag{15.22}$$

where $\alpha = R_1 + R_2 - \dfrac{R_1 R_2 (R_1+R_2)}{R_1^2 + R_2^2 - R_1 R_2}$.

In particular, for drops of identical size ($R_1 = R_2 = R$), one may write:

$$J_T = 96\pi n_0 R^3 \left(\frac{U^3}{\nu_G d}\right)^{1/2}. \tag{15.23}$$

Considering Eq. (15.22) in terms of volumes of drops rather than radii and taking into account that the core of the kinetic equation of coagulation $K(V,\omega)$ is equal to the diffusion flux of drops with unit concentration, one obtains:

$$K(V,\omega) = 3\alpha^2 \left(\frac{U^3}{\nu_G d}\right)^{1/2} \times \left(\frac{V^{2/3}+\omega^{2/3}}{(V\omega)^{1/3}(V^{1/3}+\omega^{1/3})} + \frac{2}{\alpha}\ln\left|\frac{(V\omega)^{1/3}}{V^{2/3}+\omega^{2/3}-(V\omega)^{1/3}}\right|\right)^{-1}. \tag{15.24}$$

We now estimate the growth rate of drops due to the mechanism of turbulent diffusion. For simplicity, drops are taken to be of identical size. The collision frequency is then equal to:

$$\omega_{11} = 36 n^2 V \left(\frac{U^3}{\nu_G d}\right)^{1/2}. \tag{15.25}$$

Since $W = Vn$, the balance equation of drop number takes the form:

$$\frac{dn}{dt} = -\omega_{11} = -36 W n \left(\frac{U^3}{\nu_G d}\right)^{1/2}, \quad n(0) = n_0. \tag{15.26}$$

The solution of Eq. (15.26) is

$$n = n_0 \exp\left[-36 W \left(\frac{U^3}{\nu_G d}\right)^{1/2} t\right]. \tag{15.27}$$

Thus, the average drop radius increases with time according to the law:

$$R_{av} = R_{av}^0 \exp\left[12 W \left(\frac{U^3}{\nu_G d}\right)^{1/2} t\right], \tag{15.28}$$

The characteristic time of drop enlargement is thus equal to:

$$t_T \sim \left[12W\left(\frac{U^3}{\nu_G d}\right)^{1/2}\right]^{-1}. \tag{15.29}$$

Compare the characteristic times of drop integration due to the inertial mechanism and the mechanism of turbulent diffusion:

$$\frac{t_{in}}{t_T} \sim 96\frac{\rho_G}{\rho_L}\frac{d}{R^0_{\alpha v}}\mathrm{Re}^{-3/4}. \tag{15.30}$$

For gas-liquid mixtures, characteristic of natural gases, one may take: $\rho_G/\rho_L \sim 0.05$, $d/R^0_{\alpha v} \sim 10^5$, $\mathrm{Re} \sim 10^6$. The time ratio then has the order $t_{in}/t_T \sim 10$. This means that the basic integration mechanism of small drops in a turbulent flow is the mechanism of turbulent diffusion.

Notice that in deriving Eq. (15.29), forces of the hydrodynamic and molecular interaction of drops were not taken into account. Therefore, it should be expected that the characteristic time of drop integration would be less the real time.

The assessment of hydrodynamic and molecular interactions of drops can be made in the same manner as previously described for emulsions in Part V. Upon approach of the drops to each other under the action of turbulent pulsations up to distances smaller than λ_0, they are subject to significant resistance from the environment, and the force of molecular attraction leads to collision and coalescence of the drops. If the basic mechanism of drop coagulation is that of turbulent diffusion, the coefficient of turbulent diffusion depends on the coefficient of hydrodynamic resistance [see Eqs. (11.70), (11.72), and (11.74)] and hence on the relative distance between the approaching drops:

$$D_T(r) = D_{T0}H^2(t), \tag{15.31}$$

$$H(t) = \left(1 + \frac{R_1R_2}{(r-R_1-R_2)(R_1+R_2)}\right)^{-1},$$

where D_{T0} is the coefficient of turbulent diffusion of drops in free motion, determined by Eq. (15.17). Recall the meaning of the coefficient $H(t)$. When drops are far apart ($r \gg R_1 + R_2$), it is possible to assume that they move independently of one another (in accordance with Stokes' law), that the coefficient of resistance is equal to $h_0 = 6\pi\mu_G R_2$, and that $H = 1$. When drops approach, the force of resistance grows in an inversely proportional manner with respect to the distance between their surfaces (if the surface of the drops is completely retarded):

$$h = 6\pi\mu_G\frac{R_1R_2}{R_1+R_2}H(r),$$

and the force of the molecular interaction (see Part V) is approximately equal to:

$$F_A = \frac{\Gamma R_1 R_2}{6(R_1 + R_2)(r - R_1 - R_2)^2}. \tag{15.32}$$

Here, Γ is the Hamaker constant, which is equal to 5×10^{20} J for an aerosol.

For drops commensurable ($R_1 \sim R_2$) and strongly differing in size ($R_1 \gg R_2$), the diffusion flux with due regard for hydrodynamic and molecular forces is given by Eqs. (13.91) and (13.96), and appropriate coagulation frequencies are given by Eqs. (13.97) and (13.98). Both limiting expressions, as well as values of coagulation frequencies in the region of intermediate drop sizes, are approximated by the following expression [6]:

$$K(V,\omega) \sim 5.65\alpha(V\omega)^{1/4} + b(V^{2/3} + \omega^{2/3} - 1.6(V\omega)^{1/3}), \tag{15.33}$$

where $a = 12S_m^{1/2}\frac{\nu_G}{\lambda_0^{1/2}}$; $b = \frac{3^{3/2}}{\Gamma(4/3)}S_m^{1/3}\frac{\nu_G}{\lambda_0^{1/2}}$; $S_m = \frac{\Gamma}{27\rho_G\nu_G^2\lambda_0}$; and $\Gamma(x)$ is a so-called gamma function.

We also highlight another successful approximation used earlier for the assessment of drop coagulation in emulsions (see Part V):

$$K(V,\omega) = G(V^{1/3}\omega^{1/6} + V^{1/6}\omega^{1/3}), \tag{15.34}$$

where $G = 16(\Gamma/3\pi\rho_G\lambda_0^2)^{1/2}$.

15.3
Coalescence of a Polydisperse Ensemble of Drops

Estimations of characteristic times of drop integration, as described in the previous paragraph, related to the elementary case of monodisperse drop distribution without regard for the forces of hydrodynamic and molecular interactions. We now consider the kinetics of integration of a polydisperse ensemble of drops in view of these forces.

The temporal change of the size distributions of drops, $n(V,t)$, under the assumption of a spatially uniform distribution and with regard to drop coagulations alone is described by the kinetic equation of coagulation, which follows from Eq. (15.1):

$$\frac{\partial n}{\partial t} = \frac{1}{2}\int_0^V K(\omega, V-\omega)n(\omega,t)n(V-\omega,t)\,d\omega - n(V,t)\int_0^\infty K(\omega,V)n(\omega,t)\,d\omega. \tag{15.35}$$

Equation (15.35) represents a nonlinear integro-differential equation. For a coagulation kernel of the form of Eq. (15.33), its solution can be obtained by numerical or approximate methods. We confine ourselves to an approximate solution since this allows us to obtain a rather simple analytical solution. In Part V, we presented an approximate method for the solution of a kinetic equation by the method of moments, and we noted the existence of two variants of this method, namely the parametric method and the method of fractional moments. The method is based on transformation of the Eq. (15.35) to an infinite system of equations for distribution moments, m_k:

$$m_k = \int_0^\infty V^k n(V,t)\,dV, \quad (k = 0, 1, 2, \ldots). \tag{15.36}$$

The first two moments have simple physical meaning: m_0 is the numerical concentration of drops, and m_1 is the volume concentration of drops:

$$m_0 = N(t) = \int_0^\infty n(V,t)\,dV, \quad m_1 = W(t) = \int_0^\infty Vn(V,t)\,dV, \tag{15.37}$$

Note that the average drop volume is equal to:

$$V_{av}(t) = \frac{m_1}{m_0}, \tag{15.38}$$

In addition to integer moments (k is an integer), fractional moments appropriate to the non-integer values k are also possible. Of the fractional moments, $m_{1/3}$ and $m_{2/3}$ have physical meaning, characterizing the average radius and the average surface area of drops in unit volume of mixture (area of interface):

$$\begin{aligned} m_{1/3} &= \int_0^\infty V^{1/3} n(V,t)\,dV = \left(\frac{4\pi}{3}\right)^{1/3} R_{av} m_0, \\ m_{2/3} &= \int_0^\infty V^{2/3} n(V,t)\,dV = \left(\frac{1}{36\pi}\right)^{1/3} S_{av}. \end{aligned} \tag{15.39}$$

Sometimes, instead of the drop distribution over volumes $n(V,t)$ it is convenient to use the drop distribution over radii $n(R,t)$ (such a distribution is given by Eq. (14.1)). These two distributions can be connected with the help of the equality $n(V,t)\,dV = n(R,t)\,dR$, which gives an indication of the number of drops with volume or radius, respectively, in the interval $(V, V + dV)$ or $(R, R + dR)$.

Then

$$n(R,t) = (36\pi)^{1/3} V^{2/3} n(V,t) \quad \text{or} \quad n(V,t) = \frac{1}{4\pi R^2} n(R,t) \tag{15.40}$$

If we multiply both parts of Eq. (15.35) by V^i and then integrate with respect to V between 0 and ∞, using the symmetry of the core $K(V,\omega) = K(\omega, V)$, we obtain the following equation for moments m_i:

$$\frac{\partial m_i}{\partial t} = \int_0^\infty \int_0^\infty K(\omega, V)\left(\frac{1}{2}(V+\omega)^i - V^i\right) n(\omega, t) n(V, t)\, d\omega\, dV,$$

$$(I = 0, 1, 2 \ldots). \tag{15.41}$$

To close the system of equations represented by Eq. (15.41) it is necessary to express the right-hand part in terms of moments. To this end, the coagulation kernel should have a special form (for example, the form of a homogeneous polynomial of degrees V and ω), or it is necessary to accept that the distribution conforms to a certain class (for example, a logarithmic normal distribution or a gamma distribution). The first method is called the method of fractional moments, and the second one the parametric method.

Suppose that a drop size distribution conforms to a gamma distribution:

$$n(V, t) = \frac{W}{V_0^2} \frac{(k+1)}{k!} \left(\frac{V}{V_0}\right)^k \exp\left(-\frac{V}{V_0}\right). \tag{15.42}$$

Here, V_0 and k are distribution parameters associated with the average volume $V_{\alpha v}$ and the dispersion σ^2 through the correlations:

$$V_{\alpha v} = \frac{V_0}{k+1}, \quad \sigma^2 = \frac{V_0^2}{k+1}. \tag{15.43}$$

In the parametric method, it is assumed that the distribution with time remains as a gamma distribution and that the parameter V_0 changes with time. If only drop coagulation is taken into account, there is constancy of the volume concentration W, i.e. m_1. When limited to a two-moment approximation, one obtains Eqs. (11.30) for $n(V, t)$, V_0, and the numerical concentration of drops N. Substitution of Eqs. (15.40) and (15.30) into these equations and elimination of V_0 leads to an ordinary differential equation for $N(t)$:

$$\frac{dN}{dt} = -\frac{N^2}{2(k!)^2}\left[a\Omega_{k1}\frac{W^{1/2}}{N^{1/2}(k+1)^{1/2}} + b\Omega_{k2}\frac{W^{2/3}}{N^{2/3}(k+1)^{2/3}}\right], \tag{15.44}$$

where $\Omega_{k1} = 5.65\Gamma^2(k+5/4)$ and $\Omega_{k2} = 2\Gamma(k+5/3)\Gamma(k+1) - 1.6\Gamma^2(k+4/3)$.

Values of Ω_{k1} and Ω_{k} for different values of k are given below:

k	0	1	2	3
Ω_{k1}	4.64	7.25	36.69	387.49
Ω_{k2}	0.53	0.74	3.70	39.38

Since the average volume of drops is of special interest, expressing Eq. (15.44) in terms of $V_{av} = W/N$ rather than N and taking into account $W = \text{const.}$, one obtains the following equation for V_{av}:

$$\frac{dV_{av}}{dt} = \frac{W}{2(k!)^2}\left[a\Omega_{k1}\frac{V_{av}{}^{1/2}}{(k+1)^{1/2}} + b\Omega_{k2}\frac{V_{av}{}^{2/3}}{(k+1)^{2/3}}\right], \quad V_{av}(0) = V_{av}^0, \tag{15.45}$$

The solution of Eq. (15.45) with the initial condition $V_{av}(0) = V_{av}^0$ is:

$$\frac{t}{6} = \frac{u^2 - u_0^2}{2B} - \frac{C(u - u_0)}{B^2} + \frac{C^2}{B^2}\ln\left(\frac{C + Bu}{C + Bu_0}\right), \tag{15.46}$$

where

$$u = V_{av}^{1/6}, \quad u_0 = (V_{av}^0)^{1/6}, \quad B = \frac{bW\Omega_{k2}}{2(k!)^2(k+1)^{2/3}}, \quad C = \frac{aW\Omega_{k1}}{2(k!)^2(k+1)^{1/2}}.$$

Under conditions of turbulent coagulation, the main contribution to the diffusion flux or to the coagulation frequency is made by drops of commensurable sizes (this was shown in Part V for drop coagulation in emulsions). The second term on the right-hand side of Eq. (15.45) is small in comparison with the first one. Neglecting it, the solution may be written in explicit form as:

$$\frac{V_{av}}{V_{av}^0} = \left(1 + \frac{t}{t_1}\right)^2, \quad t_1 = \frac{4(k!)^2(k+1)^{1/2}(V_{av}^0)^{1/2}}{aW\Omega_{k1}}. \tag{15.47}$$

As earlier, consider t_1 to be the characteristic coagulation time of a polydisperse ensemble of drops, caused by the mechanism of turbulent diffusion due to the forces of hydrodynamic and molecular interactions. This time should be estimated. For typical values of the flow, $\rho_G = 40$ kg m^{-3}, $\lambda_0 = 5 \times 10^{-6}$ m, $\mu_G = 1.2 \times 10^{-5}$ Pa·s, $W = 5 \times 10^{-4}$ m^3/m^3 and distribution parameters of $R_{av}^0 = 10^{-6}$ m, $k = 3$, one obtains $1/t_1 = 0.257$ s^{-1}. Thus, a twofold increase in drop radius occurs in a time t of ~7 s. This time is almost two orders of magnitude higher than for a monodisperse distribution without regard to hydrodynamic and molecular forces. Such a big difference in characteristic times is undoubtedly caused not by taking into account the polydispersivity of the distribution, but as a result of considering the interaction forces.

Consider now coagulation caused by the inertial mechanism. In this case, consideration of molecular and hydrodynamic forces results in the following coagulation frequency [7]:

$$K(V,\omega) = 4\pi\gamma\left[\left(\frac{V}{\omega}\right)^{1/3} + \left(\frac{\omega}{V}\right)^{1/3}\right]^2, \tag{15.48}$$

where $\gamma = \Gamma/6\pi\mu_G$.

Note that Eq. (15.48) applies for drops of greatly differing sizes. For the inertial mechanism of coagulation, this case is of greatest interest, because for drops of commensurable size the basic mechanism of coagulation is turbulent diffusion.

Proceeding in the same way as in the case considered above, the following equation is obtained describing the temporal change of the average volume of drops:

$$\frac{dV_{av}}{dt} = 4\pi\gamma A_k W, \quad A_k = \frac{1}{(k!)^2}[\Gamma(k+5/3)\Gamma(k+1/3) + \Gamma^2(k+1)]. \tag{15.49}$$

From this equation, it follows that the average volume of drops increases according to the linear law:

$$\frac{V_{av}}{V_{av}^0} = \left(1 + \frac{t}{t_1}\right)^2, \quad t_2 = \frac{V_{av}^0}{4\pi\gamma A_k W}. \tag{15.50}$$

Clearly, the linear law of drop growth applies only to the initial stage of coagulation.

We now estimate the characteristic time of drop integration, t_2, due to the inertial mechanism. For values of $k = 3$, $W = 4 \times 10^{-4}$ m^3/m^3, $\Gamma = 5 \times 10^{-20}$ J, $\mu_G = 1.2 \times 10^{-5}$ Pa·s, and $R_{av} = 10^{-6}$ m, one obtains $A_3 = 2.28$ and $1/t_2 = 0.34$ c^{-1}. The drop radius increases twofold in 20 s. Thus, in the considered case, the basic mechanism of drop coagulation is that of turbulent diffusion.

Let us consider now the class of logarithmic normal distributions. Let the distribution of drops over volumes be:

$$n(V,t) = \frac{N}{3\sqrt{2\pi}V \ln\sigma} \exp\left(-\frac{\ln^2(V/V_0)}{18\ln^2\sigma}\right). \tag{15.51}$$

Next, we introduce moments of distribution into Eq. (15.51):

$$m_i = \frac{N}{3\sqrt{2\pi}\ln\sigma}\int_0^\infty V^{i-1}\exp\left(-\frac{\ln^2(V/V_0)}{18\ln^2\sigma}\right)dV, \tag{15.52}$$

taking advantage of the approximation Eq. (15.34) for the coagulation kernel and substituting it in Eq. (15.41). Taking further, $i = 1, 2, 3$, one obtains the following system of equations for the first three moments m_0, m_1, and m_2:

$$\frac{dm_0}{dt} = -Gm_{1/3}m_{1/6}, \quad \frac{dm_1}{dt} = 0$$

$$\frac{dm_2}{dt} = G\left(-\frac{1}{2}m_{13/5}m_{1/6} + m_{10/3}m_{7/6}\right.$$

$$\left. + m_{7/3}m_{13/6} - \frac{1}{2}m_{25/6}m_{1/3} + m_{19/6}m_{4/3}\right) \tag{15.53}$$

We then enter fractional moments into the right-hand parts of the obtained system of equations. Therefore, for solving the system of equations it is necessary to express these moments in terms of integers. To this end, one should take advantage of the property [8] of integrals of the form of Eq. (15.52) in allowing evaluation of m_k through m_1 and the parameters V_0 and σ of the distribution according to Eq. (15.51):

$$m_k = m_1 V_0^{k-1} \exp\left(\frac{9}{2}(k^2 - 1) \ln^2 \sigma\right). \tag{15.54}$$

Evaluating fractional moments through integer ones with the help of Eq. (15.54) and substituting in Eq. (15.53), one obtains:

$$\frac{dm_0}{dt} = -\frac{Gm_1^2}{V_0^{3/2}} \exp\left(-\frac{67}{8} \ln^2 \sigma\right), \quad \frac{dm_1}{dt} = 0,$$

$$\frac{dm_2}{dt} = Gm_1^2 V_0^{5/2}\left[-\frac{1}{2} \exp\left(\frac{605}{8} \ln^2 \sigma\right) + \exp\left(\frac{377}{8} \ln^2 \sigma\right)\right.$$

$$\left. + \exp\left(\frac{293}{8} \ln^2 \sigma\right) - \frac{1}{2} \exp\left(\frac{557}{8} \ln^2 \sigma\right) + \exp\left(\frac{297}{8} \ln^2 \sigma\right)\right]. \tag{15.55}$$

Taking into account the correlations:

$$m_0 = \frac{m_1}{V_0} \exp\left(-\frac{9}{2} \ln^2 \sigma\right), \quad m_2 = m_1 V_0 \exp\left(\frac{27}{2} \ln^2 \sigma\right),$$

one obtains a closed system of equations for V_0 and σ. Having determined these unknown values, the average drop volume takes the form:

$$V_{av}(t) = V_0(t) \exp(1.5 \ln^2 \sigma(t)). \tag{15.56}$$

The system of equations, Eq. (15.55), corresponds to a three-moment approximation.

One may also take into account the following moments, but this results in complication of the system of equations. To estimate the integration rate of drops, one may take advantage of the two-moment approximation corresponding to the assumption that the dispersion does not change, but only V_0 does. Then, retaining only the first two equations in system Eq. (15.55), one obtains:

$$N(t) = N(0) \exp\left(-\frac{9}{2} \ln^2 \sigma\right)\left[1 + \frac{GWt}{2(V_{av}^0)^{1/2}} \exp\left(-\frac{41}{8} \ln^2 \sigma\right)\right]^2, \tag{15.57}$$

$$V_{av}(t) = V_{av}^0\left[1 + \frac{GWt}{2(V_{av}^0)^{1/2}} \exp\left(-\frac{41}{8} \ln^2 \sigma\right)\right]^2. \tag{15.58}$$

From Eq. (15.58), it follows that the characteristic coagulation time may be estimated as

$$t_k \sim \frac{2(V_{av}^0)^{1/2}}{GW} \exp\left(\frac{41}{8} \ln^2 \sigma\right). \tag{15.59}$$

Note one characteristic property. After sufficient time:

$$V_{av}(t) \sim \frac{G^2 W^2 t^2}{4}\left[\exp\left(-\frac{41}{8}\ln^2\sigma\right)\right]^2, \tag{15.60}$$

which means that when drops are sufficiently enlarged, they "forget" their initial volume.

If data on the form of drop distributions over volumes are not at hand, one may use the method of approximation of fractional moments. This method is applied to the coagulation of drops with the kernel of Eq. (15.34). If we limit ourselves to the first two moments, m_0 and m_1, then:

$$\frac{dm_0}{dt} = -Gm_{1/3}m_{1/6}, \quad \frac{dm_1}{dt} = 0 \tag{15.61}$$

$$m_0(0) = N_0, \quad m_0 = W = \text{const.}$$

Approximation of fractional moments through integer ones according to Eq. (11.20) in accordance with a two-point scheme gives:

$$\begin{aligned} \frac{m_{1/3}(t)}{m_{1/3}(0)} &= \frac{m_0^{2/3}(t)\; m_1^{1/3}(t)}{m_0^{2/3}(0)\; m^{1/3}(0)} = \frac{m_0^{2/3}(t)}{m_0^{2/3}(0)}, \\ \frac{m_{1/6}(t)}{m_{1/6}(0)} &= \frac{m_0^{5/6}(t)\; m_1^{1/6}(t)}{m_0^{5/6}(0)\; m^{1/6}(0)} = \frac{m_0^{5/6}(t)}{m_0^{5/6}(0)}. \end{aligned} \tag{15.62}$$

Here, it is taken into account that m_1 remains constant during coagulation. Substituting Eq. (15.62) in Eq. (15.61), one obtains:

$$\frac{dm_0}{dt} = -Gm_{1/3}(0)m_{1/6}(0)\left(\frac{m_0}{m_0(0)}\right)^{3/2}, \tag{15.63}$$

The solution is

$$m_0(t) = m_0(0)\left(1+\frac{t}{t_k}\right)^{-2}, \quad t_k = \frac{2}{G}[m_0(0)m_1(0)]^{-1/2}. \tag{15.64}$$

Thus, the average drop volume changes with time according to:

$$V_{av}(t) = V_{av}(0)\left(1+\frac{t}{t_k}\right)^2. \tag{15.65}$$

Over relatively long times, this becomes:

$$V_{av}(t) \sim \frac{64 W^2 \Gamma t^2}{3\pi\rho_G\lambda_0^2}. \tag{15.66}$$

16
Formation of a Liquid Phase in Devices of Preliminary Condensation

During the motion of a two-phase, multi-component mixture in a pipe under conditions of slow change of pressure and temperature, a thermodynamic balance has time to become established. Equilibrium concentrations of components in each phase and volume fractions of phases can be determined by the use of equations for the vapor-liquid balance (see Section 5.7). In devices for the dual preparation of gas and condensate, after the first separation step, in which the bulk of the liquid droplets are separated from the gas, the gas proceeds to the second separation step. In the scheme of low-temperature separation (LTS), devices of preliminary condensation (throttle, heat exchanger, or turbo-expander) are placed before the second step. As a result of a sharp drop in pressure and temperature upon passage through these devices, the thermodynamic balance in the mixture is disrupted. This phenomenon results in extensive formation of a liquid phase through interphase mass exchange at the surface of drops until a new phase balance is established, corresponding to the new values of pressure, temperature, and phase composition. In this chapter, we describe the methods allowing determination of the rate of condensation growth of drops and the quantity of liquid phase that will be formed in devices of preliminary condensation under non-equilibrium conditions.

16.1
Condensation Growth of Drops in a Quiescent Gas–Liquid Mixture

Consider a gas mixture consisting of two components with mole fractions y_{10} and y_{20} at pressure p_1 and temperature T_1. At some point in time, taken to be the initial time, the pressure and temperature change abruptly and become equal to p_2 and T_2, which then remain constant. Thus, the thermodynamic balance that existed in the system at p_1 and T_1 will be disturbed, and at the new values p_2 and T_2 drops of liquid phase with average radius $R_0 = R_n$ will appear in the mixture due to homogeneous condensation. These drops grow in time until a new equilibrium appropriate to the values p_2 and T_2 is established.

Assume that the drops formed consist only of component 1 and that during condensation growth of the drop only component 1 is condensed at the drop sur-

Separation of Multiphase, Multicomponent Systems. E. G. Sinaiski and E. J. Lapiga

ISBN: 978-3-527-40612-8

face, i.e. component 2 is neutral and does not participate in the process of mass exchange between the phases. The surrounding liquid is taken to be quiescent. The basic mechanism of the supply of component 1 to the drop surface is thus diffusion. In addition, it is accepted that thermodynamic equilibrium is established much more rapidly at the interface than in the bulk volume; the characteristic time of diffusion processes in the gas phase is usually much shorter than in the liquid phase since the ratio of diffusion coefficients $D_{iG}/D_{iL} \gg 1$. Therefore, one may consider a quasi-stationary problem, in which the concentration distribution of components in the gas phase has stationary character. This means that mole concentrations of gas y_i depend only on the distance to the drop center, while parameters of the drop (mass and radius) are time-dependent. The problem of drop dynamics in a gas is similar to that considered in Section 6.9.

The flux of the i-th component in the gas phase may be expressed in terms of mole fractions of components y_i in the following manner:

$$J_i = -C_G D_{i2G} \frac{dy_i}{dr} + y_i J, \quad (i = 1, 2), \tag{16.1}$$

where $J = \sum_i J_i$ is the total flux, r is the distance from the center of the drop, and C_G is the mole density of gas (mol m^{-3}).

From the condition of flux conservation:

$$\frac{1}{r^2}\frac{d}{dr}\left(r^2 J_i\right) = 0 \tag{16.2}$$

it follows that:

$$J_i = J_{iw}\frac{R^2}{r^2}, \quad J = J_w \frac{R^2}{r^2}, \tag{16.3}$$

where $R(t)$ is the current radius of the drop, and the index w designates appropriate values at the drop surface.

Substitution of Eq. (16.3) into Eq. (16.1) gives equations describing the temporal change of the concentration of a component in the gas:

$$C_G D_{12G} \frac{r^2}{R^2}\frac{dy_i}{dr} - y_i J_w = -J_{iw}. \tag{16.4}$$

We set concentrations at the drop surface as:

$$y_i = y_{iw} \quad \text{when } r = R \tag{16.5}$$

There is local thermodynamic equilibrium at the interface. Therefore, the values y_{iw} are defined by the appropriate equilibrium relationships. The basic expression is that for an ideal gas mixture:

$$y_{iw} = \frac{p_{is}(T_2)}{p_2}, \tag{16.6}$$

where $p_{\rm is}(T_2)$ is the saturation pressure of the i-th component of the vapor over the liquid surface at temperature T_2. This temperature is taken to be constant.

If the volume concentration of drops is low ($W \ll 1$), then to a first approximation the influence of the neighboring drops (hindrance) may be neglected and we may consider the behaviour of an isolated drop surrounded by an infinite volume of gas. The second boundary condition is then that of constant concentration far from the drop:

$$y_i \to y_{i\infty} \quad \text{when } r \to \infty \tag{16.7}$$

Values $y_{i\infty}$ are considered as given.

The condition expressed by Eq. (16.7) allows us to write the solution of Eq. (16.4) as:

$$y_i = \frac{J_{iw}}{J_w} + \left(y_{i\infty} - \frac{J_{iw}}{J_w}\right) \exp\left(-\frac{R^2 J_w}{C_G D_{12G} r}\right). \tag{16.8}$$

Using the condition expressed by Eq. (16.5), one obtains

$$J_{iw} = J_w \frac{J_{iw} - y_{i\infty} \exp(-RJ_w/C_G D_{12G})}{1 - \exp(-RJ_w/C_G D_{12G})}. \tag{16.9}$$

The condition of neutrality of component 2 gives $J_{2\rm w} = 0$. Then, from Eq. (16.9), it follows that:

$$J_{iw} = \frac{C_G D_{12G}}{R} \ln\left(\frac{1 - y_{1\infty}}{1 - y_{1w}}\right). \tag{16.10}$$

The drop radius, R, and the molar concentration $y_{1\infty}$ far from the drop in Eq. (16.10) are time-dependent. The balance equation of the drop volume is:

$$\frac{dV}{dt} = -4\pi R^2 v_{1L} J_{1w}, \tag{16.11}$$

where v_{1L} is the molar volume of component 1, which determines the drop radius. Since $J_{1\rm w} = J_{\rm w}$, substituting Eq. (16.10) into Eq. (16.11), one obtains:

$$\frac{dV}{dt} = -4\pi \left(\frac{3V}{4\pi}\right)^{1/3} v_{1L} C_G D_{12G} \ln\left(\frac{1 - y_{1\infty}}{1 - y_{1w}}\right). \tag{16.12}$$

We can now derive an equation for $y_{1\infty}(t)$. Let the drops be characterized by a distribution over volumes $n(V,t)$. If we consider only the change of distribution due to condensation growth of drops, then from the kinetic equation, Eq. (15.1), it follows that:

$$\frac{\partial n}{\partial t} + \frac{\partial}{\partial V}\left(n\left(\frac{dV}{dt}\right)_m\right) = 0. \tag{16.13}$$

Numerical and volume concentrations of drops are determined by the formulae:

$$N = \int_0^\infty n(V,t)\,dV, \quad W = \int_0^\infty Vn(V,t)\,dV, \tag{16.14}$$

in the considered case of $N = \text{const}$.

Considering now the mass balance of the i-th component, one obtains equations that describe the temporal change of $y_{1\infty}$:

$$\frac{dy_{1\infty}}{dt} = \frac{4\pi}{C_G}\int_0^\infty R^2 J_w n(V,t)\,dV, \tag{16.15}$$

Substituting the expression for J_w, Eq. (16.10), one obtains:

$$\frac{dy_{1\infty}}{dt} = 4\pi\left(\frac{3}{4\pi}\right)^{1/3} D_{12G}\ln\left(\frac{1-y_{1\infty}}{1-y_{1w}}\right)\int_0^\infty V^{1/3}n(V,t)\,dV, \tag{16.16}$$

Thus, Eqs. (16.12), (16.13), and (16.16) form a closed system for the definition of $V(t)$, $n(V,t)$, and $y_{1\infty}$. They should be solved under the initial conditions:

$$n(V,t) = n_0(V), \quad V(0) = V_0, \quad y_{1\infty}(0) = y_{10}. \tag{16.17}$$

We solve the obtained system by the method of moments. Let us introduce moments of the i-th order:

$$m_k = \int_0^\infty V^k n(V,t)\,dV, \quad (k = 0, 1, 2, \ldots). \tag{16.18}$$

Recall that $m_0 = N$ and $m_1 = W$.

We multiply Eq. (16.13) by V^i and then integrate the result in terms of V from 0 to ∞. Limiting ourselves to the two first moments, after simple transformations we obtain:

$$\frac{dm_0}{dt} = 0, \quad \frac{dm_1}{dt} = \int_0^\infty \frac{dV}{dt}n(V,t)\,dV. \tag{16.19}$$

Using Eq. (16.12), we find that:

$$\frac{dm_1}{dt} = -4\pi\left(\frac{3}{4\pi}\right)^{1/3} \upsilon_{1L} D_{12G} C_G \ln\left(\frac{1-y_{1\infty}}{1-y_{1w}}\right) m_{1/3}. \tag{16.20}$$

Introduction of the moments allows us to rewrite Eq. (16.16) as:

$$\frac{dy_{1\infty}}{dt} = 4\pi\left(\frac{3}{4\pi}\right)^{1/3} D_{12G} \ln\left(\frac{1-y_{1\infty}}{1-y_{1w}}\right) m_{1/3}. \tag{16.21}$$

The equation for the radius of a drop follows from Eq. (16.12):

$$\frac{dR}{dt} = -\frac{C_G D_{12G} \upsilon_{1L}}{R} \ln\left(\frac{1-y_{1\infty}}{1-y_{1w}}\right). \tag{16.22}$$

Equations (16.20)–(16.22) can be solved in the case of an infinitely dilute mixture when either the first component or the second one is present in the gas mixture in only a small amount. Of the utmost interest is the case of a low content of the first, i.e. condensing, component. For processes occurring in a throttle, heat exchanger, or turbo-expander, water vapor or heavy hydrocarbon vapors are condensed. In natural gas, these components are, as a rule, only present in small amounts and therefore the approximation of an infinitely dilute mixture is justified.

The assumption made allows us to assume that $y_{iw} \ll 1$ and $y_{i\infty} \ll 1$. Then

$$\ln\left(\frac{1-y_{1\infty}}{1-y_{1w}}\right) \approx -(y_{1\infty} - y_{1w}) = -\Delta y_1. \tag{16.23}$$

and the system of equation becomes:

$$\frac{dm_1}{dt} = 4\pi\left(\frac{3}{4\pi}\right)^{1/3} \upsilon_{1L} D_{12G} C_G m_{1/3} \Delta y_1. \tag{16.24}$$

$$\frac{d(\Delta y_1)}{dt} = -4\pi\left(\frac{3}{4\pi}\right)^{1/3} D_{12G} C_G m_{1/3} \Delta y_1. \tag{16.25}$$

$$\frac{dR}{dt} = \frac{C_G D_{12G} \upsilon_{1L}}{R} \Delta y_1. \tag{16.26}$$

The fractional moment $m_{1/3}$ appears in the right-hand terms of Eqs. (16.24) and (16.25). This should be expressed in terms of integer moments. For this purpose, interpolation formulae, Eq. (11.20), are used for two-point interpolation:

$$\hat{m}_{1/3}(t) = \hat{m}_1(t). \tag{16.27}$$

Here, non-dimensional moments $\hat{m}_i = m_i(t)/m_i(0)$ are introduced.

Since the zero moment is the numerical concentration of drops, $\hat{m}_0(t) = 1$ and:

$$\hat{m}_{1/3}(t) = \hat{m}_1^{1/3}(t) \tag{16.28}$$

The last correlation allows transformation of Eq. (16.24) into

$$\frac{d\hat{m}_1}{d\tau} = \frac{\Delta y}{\Delta y_1(0)}\hat{m}_1^{1/3} \tag{16.29}$$

where $\tau = 3\upsilon_{1L}C_G D_{12G}\Delta y_1(0)t/R^2(0)$.

Having expressed Δy_1 in terms of $\hat{m}_1$ through Eq. (16.25) and having substituted this into Eq. (16.29), one obtains the equation:

$$\frac{d\hat{m}_1}{d\tau} = (a - b\hat{m}_1)\hat{m}_1^{1/3}, \quad \hat{m}_1(0) = 1 \tag{16.30}$$

$$b = \frac{m_1(0)}{\upsilon_{1L}C_G\Delta y_1(0)}, \qquad a = 1 + b.$$

The solution to this equation is found in implicit form:

$$\tau = -\frac{\lambda^3 - 1}{\lambda}\left(0.5\ln\frac{(\hat{m}_1^{1/3} - \lambda)^2(1 + \lambda + \lambda^2)}{(\hat{m}_1^{2/3} + \lambda\hat{m}_1^{1/3} + \lambda^2)(1 - \lambda)^2} + \sqrt{3}\left(\arctan\frac{2\hat{m}_1^{1/3} + \lambda}{\lambda\sqrt{3}} - \arctan\frac{2 + \lambda}{\lambda\sqrt{3}}\right)\right) \tag{16.31}$$

where $\lambda = (a/b)^{1/3}$.

Equation (16.31) allows determination of the change of the volume content of drops with time, $m_1(t)$. As $\tau \to \infty$, the system tends to reach an equilibrium state. At this, $\Delta y_1 \to 0$ and from the relationship:

$$m_1(t) + \upsilon_{1L}C_G\Delta y_1 = m_1(0) + \upsilon_{1L}C_G\Delta y(0), \tag{16.32}$$

which stems from Eqs. (16.24) and (16.25), it follows that $\tilde{m}_1(\tau) \to \lambda^3$. In particular, having determined $\tilde{m}_1$ and having used the condition of constancy of numerical concentration of drops, it is possible to find the average volume of drops:

$$V_{av}(t) = V_{av}(0)\tilde{m}_1(\tau), \tag{16.33}$$

where $V_{av}(0)$ is the initial average volume of the drops.

Since the system tends towards equilibrium as $\tau \to \infty$, to estimate the characteristic time of equilibrium establishment, τ_{eq}, it make sense to define it as the time after which the volume contents of drops will differ from the appropriate

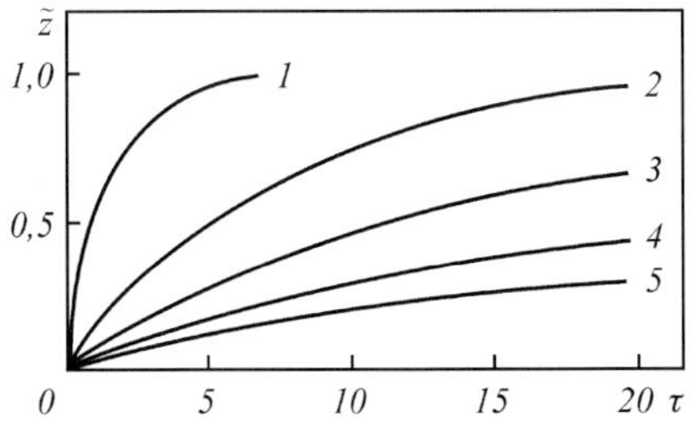

Fig. 16.1 Dependence of $\tilde{z}$ upon τ for various values of λ: 1: 1.5; 2: 3; 3: 5; 4: 7; 5: 10.

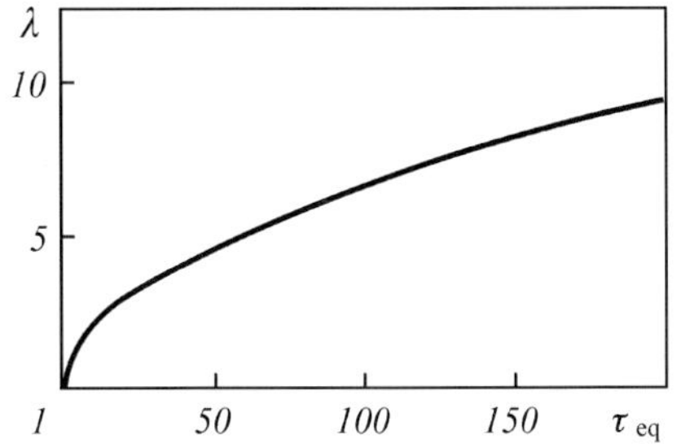

Fig. 16.2 Relationship between λ and the dimensionless time of establishment of balance τ_{eq}.

equilibrium value $m_{1eq} = \lambda^3$ by 10%. In Fig. 16.1, the dimensionless parameter $z = (V_{\alpha v}/V_{\alpha v}(0))^{1/3}$, which describes the average drop radius, is plotted against time τ for various values of the parameter λ. For convenience, the value $\tilde{z} = (z-1)/(\lambda-1)$ is plotted as the ordinate. The relationship between λ and the characteristic time of equilibrium establishment in the system τ_{eq} is plotted in Fig. 16.2.

The solution obtained as described above is valid at small values of volume content of the liquid phase. The average distance between drops is proportional to $m_1^{1/3}$, and therefore with the growth of m_1 the average distance between drops decreases, which results in the need to take into account the mutual influence of drops, in other words, to take into account hindrance. To this end, the boundary condition in the bulk gas needs to be formulated, proceeding from the cellular model (see Section 10.7):

$$y_1 = y_{1\infty}, \quad \text{when } R = R_c \tag{16.34}$$

where $R_c = R(t)/m_1^{1/3}(t)$ is the radius of a cell equal to the average distance between drops.

Returning to Eq. (16.4) and using this boundary condition, one obtains:

$$\begin{aligned} y_i &= \frac{J_{iw}}{J_w} + \frac{(y_{i\infty} - y_{iw}) \exp(-R^2 J_w / C_G D_{12G} r)}{\exp(-R^2 J_w / C_G D_{12G} R_c) - \exp(-R J_w / C_G D_{12G})}, \\ \frac{J_{iw}}{J_w} &= \frac{y_{iw} \exp(-R^2 J_w / C_G D_{12G} R_c) - y_{i\infty} \exp(-R J_w / C_G D_{12G})}{\exp(-R^2 J_w / C_G D_{12G} R_c) - \exp(-R J_w / C_G D_{12G})}. \end{aligned} \tag{16.35}$$

Taking $J_{2w} = 0$ in Eq. (16.35), one finds:

$$J_w = \frac{C_G D_{12G}}{R(1 - R/R_c)} \ln\left(\frac{1 - y_{1\infty}}{1 - y_{1w}}\right). \tag{16.36}$$

In an approximation for an infinitely dilute mixture:

$$J_w \approx \frac{C_G D_{12G}}{R(1 - R/R_z)} \Delta y_1. \tag{16.37}$$

The equations for Δy_1 and m_1 become:

$$\frac{dm_1}{dt} = 4\pi \left(\frac{3}{4\pi}\right)^{1/3} \upsilon_{1L} D_{12G} C_G m_{1/3} \Delta y_1 (1 - m_1^{1/3})^{-1}, \tag{16.38}$$

$$\frac{d(\Delta y_1)}{dt} = -4\pi \left(\frac{3}{4\pi}\right)^{1/3} D_{12G} m_{1/3} \Delta y_1 (1 - m_1^{1/3})^{-1}. \tag{16.39}$$

Expressing the fractional moment $m_{1/3}$ through integer moments m_0 and m_1, one obtains:

$$\frac{d\hat{m}_1}{dt} = \frac{(a - b\hat{m}_1)\hat{m}_1^{1/3}}{(1 - \hat{m}_1 m_1(0))^{1/3}}, \quad \hat{m}_1(0) = 1 \tag{16.40}$$

The solution is:

$$\begin{aligned} \tau = &-\frac{\lambda^3 - 1}{\lambda} \left(0.5 \ln \frac{(\hat{m}_1^{1/3} - \lambda)^2 (1 + \lambda + \lambda^2)}{(\hat{m}_1^{2/3} + \lambda \hat{m}_1^{1/3} + \lambda^2)(1 - \lambda)^2} \right. \\ &\left. + \sqrt{3}\left(\arctan \frac{2\hat{m}_1^{1/3} + \lambda}{\lambda\sqrt{3}} - \arctan \frac{2 + \lambda}{\lambda\sqrt{3}}\right)\right) \\ &+ (\lambda^3 - 1) m_1(0) \ln\left(1 - \frac{b\hat{m}_1}{1 + b}\right), \end{aligned} \tag{16.41}$$

where $\lambda = (1 + \upsilon_{1L} C_G \Delta y_1(0)/m_1(0))^{1/3}$, $b = (\lambda^3 - 1)^{-1}$.

Compare Eqs. (16.31) and (16.41), which relate to the non-hindered (free) and hindered cases, respectively. It can be seen that, for the same values of parameters, τ determined from Eq. (16.41) is less than τ determined from Eq. (16.31). Hence, the taking into account of hindrance results in faster establishment of equilibrium in the system.

Consider now the case when, as a result of a pressure and temperature change, there is homogeneous condensation throughout the volume. We designate with I the rate of formation of germs of the new phase (number of drops formed in unit volume in unit time). The volume of the incipient drops V_n is taken to be constant. Then, in the right-hand part of Eq. (16.13), a term describing the incipiency rate of the liquid phase appears:

$$\frac{\partial n}{\partial t}+\frac{\partial}{\partial V}\left(n\frac{dV}{dt}\right)=I\delta(V-V_n). \tag{16.42}$$

In the considered case, the system of equations for the first two moments becomes:

$$\frac{dm_0}{dt}=I,\quad \frac{dm_1}{dt}=\int_0^{\infty} n\frac{dV}{dt}\,dV+IV_n. \tag{16.43}$$

From the first equation, it follows that:

$$m_0=m_0(0)+\int_0^1 I\,dt. \tag{16.44}$$

The second equation transforms into

$$\frac{dm_1}{dt}=4\pi\left(\frac{3}{4\pi}\right)^{1/3} v_{1L}D_{12G}C_G m_{1/3}\Delta y_1+IV_n. \tag{16.45}$$

Expressing the fractional moment $m_{1/3}$ through integer moments m_0 and m_1, and having taken advantage of Eq. (16.44), one obtains:

$$\hat{m}_{1/3}=\hat{m}_1^{1/3}\left(1+\frac{1}{m_0(0)}\int_0^1 I\,dt\right)^{2/3} \tag{16.46}$$

In contrast to the case considered before, here it is necessary to take into account the dependence of y_{1w} on time, since in accordance with the correlation of Eq. (16.6) the pressures p_{1s} and p are changed upon expansion of the gas. Then, C_G also changes, and Eq. (16.25) converts to:

$$\frac{d(\Delta y_1 C_G)}{dt}+4\pi\left(\frac{3}{4\pi}\right)^{1/3} D_{12G}C_G m_{1/3}\Delta y_1+\frac{d(y_{1w}C_G)}{dt}=IV_nC_L. \tag{16.47}$$

Entering the dimensionless variables

$$\Delta\tilde{y}_1 = \frac{\Delta y_1}{\Delta y_1(0)}, \quad \tilde{y}_{1w} = \frac{y_{1w}}{\Delta y_1(0)}, \quad \tilde{C}_G = \frac{C_G}{C_{G0}},$$

$$\tau = \frac{3C_{G0}D_{12G}t}{R^2(0)C_L}, \quad \tilde{I} = \frac{IV_n C_L^3 R^2(0)}{3C_{G0}^2 D_{12G}(\Delta y_1(0))^2},$$

Eq. (16.47) takes the form:

$$\frac{d(\Delta\tilde{y}_1\tilde{C}_G)}{d\tau} + \upsilon_{1L}D_{12G}C_{G0}\Delta y_1(0) + C_G\tilde{m}_{1/3}\Delta\tilde{y}_1$$
$$+ \upsilon_{1L}D_{12G}C_{G0}\Delta y_1(0)\frac{d(\tilde{y}_{1w}\tilde{C}_G)}{d\tau} = \tilde{I}. \tag{16.48}$$

The equation for the first moment transforms into:

$$\frac{d\hat{m}_1}{d\tau} - \Delta\tilde{y}_1\hat{m}_1^{1/3}\hat{m}_0^{2/3} = \tilde{I}(\tau) \tag{16.49}$$

From the last two equations, it follows that:

$$\frac{d}{d\tau}(\hat{m}_1 m_1(0) + \upsilon_{1L}C_{G0}\tilde{C}_G\Delta y_1(0)\Delta\tilde{y}_1) + \upsilon_{1L}C_{G0}\Delta y_1(0)\frac{d}{d\tau}(\tilde{C}_G\tilde{y}_1) = \tilde{I} \tag{16.50}$$

or

$$\hat{m}_1 m_1(0) + \upsilon_{1L}C_{G0}\tilde{C}_G\Delta y_1(0)\Delta\tilde{y}_1 + \upsilon_{1L}C_{G0}\tilde{C}_G y_{1w}(0)\tilde{y}_{1w}$$
$$= m_1(0) + \upsilon_{1L}C_{G0}\Delta y_1(0) + \upsilon_{1L}C_{G0}y_{1w}(0) + \int_0^\tau \tilde{I}\,d\tau. \tag{16.51}$$

By expressing Δy_1 by way of Eq. (16.51) and substituting the result into Eq. (16.49), one obtains an equation for the volume content of the liquid phase:

$$\frac{d\hat{m}_1}{d\tau} - \left(1 + \frac{1}{m_1(0)}\int_0^\tau \tilde{I}\,d\tau\right)^{2/3}(a - b\hat{m}_1)\hat{m}_1^{1/3} = \tilde{I}, \tag{16.52}$$

where $b = \dfrac{m_1(0)}{\upsilon_{1L}C_{G0}\Delta y_1(0)}$, $a = b + \dfrac{y_{1\infty}(0) - y_{1w}(\tau)}{\Delta y_1(0)} + \dfrac{1}{\upsilon_{1L}C_{G0}\Delta y_1(0)}\int_0^\tau \tilde{I}\,d\tau$.

Thus, Eq. (16.52) expresses the temporal change of the volume content of the liquid phase during homogeneous condensation of drops of identical volume V_n with intensity I for a given change in pressure and temperature.

If the origination of a liquid phase does not occur, then the change in pressure and temperature will result in a change of the volumetric concentration of already available liquid phase. Thus, it is necessary to set $\tilde{I} = 0$ in Eq. (16.52), and the equation reduces to:

$$\frac{d\hat{m}_1}{d\tau} = (a - b\hat{m}_1)\hat{m}_1^{1/3}, \quad \hat{m}_1(0) = 1, \tag{16.53}$$

where $b = \dfrac{m_1(0)}{v_{1L}C_{G0}\Delta y_1(0)}$, $a = b + \dfrac{y_{1\infty}(0) - y_{1w}(\tau)}{\Delta y_1(0)}$.

Equation (16.31) is also needed, although its use is inconvenient. Calculations show that the following expression gives a good approximation in the interval $1 \leq \lambda \leq 7$:

$$\tilde{z} = \frac{z-1}{\lambda - 1} = 1 - \exp(-\alpha\tau), \quad \alpha = \frac{0.331}{\lambda - 1} - 0.026 \tag{16.54}$$

16.2
Condensation Growth of Drops in a Turbulent Flow of a Gas–Liquid Mixture

The expressions derived in the previous section may be applied to a mixture at rest or under weak agitation. Actually, a gas flow with condensate drops suspended in it, as in elements of field equipment, especially in pipelines, is characterized by extensive turbulence, which results in strong mixing and equalization of the concentrations of the components in the gas phase. Characteristic values of parameters of gas flow in a pipeline are as follows: $U \sim 10$–$50\ \text{m s}^{-1}$, $d \sim 0.2$–1 m, $\rho_G \sim 50\ \text{kg m}^{-3}$, $\mu_G \sim 10^{-5}$ Pa·s. The associated Reynolds number, Re, is $\sim 10^7$ to 2.5×10^8, and the inner scale of turbulence is $\lambda_0 \sim (0.1$–$5.6) \times 10^{-6}$ m $= 0.1$–5.6 µm.

Gas flow in equipment elements of greater size, for example in separators, is characterized by velocities and diameters of $U \sim 0.1$–$1\ \text{m s}^{-1}$, $d \sim 1$–2 m. Thus, Re $\sim 5 \times 10^5$ to 10^7 and $\lambda_0 \sim 5.6$–53 µm.

The inner scale of turbulence, λ_0, defines the character of hydrodynamic and mass-exchange processes in areas in which size is greater or smaller than λ_0. Since processes in the vicinity of drops are of greatest interest, the size of these regions is commensurable with drop sizes. Let R_{av} be the average radius of an ensemble of drops under consideration. The character of the processes then depends on the ratio R_{av}/λ_0.

Consider the mass-exchange of a drop with radius R suspended in a turbulent gas flow. At the initial moment in time, the composition of the drop is specified in the form of a mass concentration ρ_{i0L} (kg m^{-3}) and a gas concentration ρ_{i0G}, where $i = 1, 2 \ldots, s$, and s is the number of components. Two simplifying assumptions are made:

1. The thermodynamic balance is established much faster at the drop surface than in the gas volume. There is then local thermodynamic balance at the interface, allowing determination of equilibrium values of the component concentrations in both phases using equations of vapor-liquid balance (see Section 5.7).

2. The basic mechanism of component delivery to the interface in the gas phase is transfer by turbulent pulsations characterized by a scale of λ. The amount of component flux, J_i, to a drop (or from it) depends on the ratio R_{av}/λ_0 and is equal to [2]:

$$J_i = \begin{cases} 8\left(\dfrac{2}{3\sqrt{3}}\right)^{1/3} D_{imG}^{2/3}\left(\dfrac{\rho_L}{\rho_G}\right)^{1/3} \dfrac{R^2 U^{3/4}}{v_G^{5/12} d^{1/4}} \Delta\rho_{iG}, & \text{at } R \ll \lambda_0, \\ 4\pi\sqrt[3]{2}\left(\dfrac{\rho_L}{\rho_G}\right)^{1/3} \dfrac{UR^{7/3}}{d^{1/3}} \Delta\rho_{iG}, & \text{at } R \gg \lambda_0, \end{cases} \tag{16.55}$$

where D_{imG} is the coefficient of binary diffusion for the i-th component of the gas, ρ_L and ρ_G are the densities of the gas and liquid, respectively, U is the average flow velocity, v_G is the kinematic viscosity of the gas, and $\Delta\rho_{iG} = \rho_{iG} - \rho_{iGw}$ is the difference in mass concentrations of the i-th component of the gas in the bulk flow and at the interface.

Two expressions for the diffusion coefficient of a component in a turbulent flow are indicative of multiple mechanisms for its delivery to the drop surface. In particular, at $R \gg \lambda_0$ turbulent diffusion dominates (as opposed to molecular diffusion).

3. The basic change in concentration, ρ_{iG}, occurs in a thin layer at the drop surface. Outside of this layer (in the bulk flow), the concentration ρ_{iG} is homogeneous and changes only with time.

4. The distribution of components in the liquid phase (within a drop) is homogeneous and changes only with time.

After a time t_{eq} has elapsed (characteristic time of balance establishment in the mixture), the process of mass-exchange practically ceases, that is, ρ_{iG} tends towards ρ_{iGw}.

Consider a distribution of drops over volumes $n(V,t)$. The mass balance of the i-th component in unit volume of the mixture in the gas phase can then be written as:

$$\frac{d\rho_{iG}}{dt} = -\int_0^\infty J_i n(V,t)\, dV, \tag{16.56}$$

To this equation should be added the equation of drop mass balance:

$$\frac{d}{dt}(\rho_L V) = \sum_{i=1}^{s} J_i, \tag{16.57}$$

and the equation describing the change of $n(V,t)$ owing to the change of drop volume due to condensation mass-exchange alone:

$$\frac{\partial n}{\partial t} + \frac{\partial}{\partial V}\left(n\frac{dV}{dt}\right) = 0, \tag{16.58}$$

Since $\rho_L = \sum_i \rho_{iL}$, then instead of Eq. (16.57), one can write:

$$\frac{d}{dt}\left(\sum_{i=1}^{s} \rho_{iL} V\right) = \sum_{i=1}^{s} J_i, \tag{16.59}$$

Here, ρ_{iL} denotes the concentration of the i-th component averaged over the entire drop volume. Then, from the condition of mass balance for the i-th component in the drop:

$$\frac{d}{dt}(\rho_{iL} V) = J_i \tag{16.60}$$

the change of ρ_{iL} with time may be determined.

Thus, having determined ρ_{iGw} and having specified the initial concentrations of components in the gaseous and liquid phases, ρ_{iG} and ρ_{iL}, respectively, as well as the initial values of the volume, V_0, and the distribution over volumes, $n_0(V)$, from Eqs. (16.56), (16.58)–(16.60) in relation to Eq. (16.55), one can ascertain the condition of the mixture at any point in time, as well as the characteristic time of balance establishment. However, before proceeding to solve the problem at hand, it is first necessary to ascertain which of the two expressions for J_i in Eq. (16.55) should be used. For this purpose, it is necessary to estimate the average radius of the drops.

A distribution of drops is formed in the intensely turbulent flow in a pipeline, where the average drop radius, R_{av}, is given by:

$$R_{av} = 0.09d\left[\frac{1}{U}\left(\frac{2\Sigma}{\rho_L d}\right)^{1/2}\right]^{6/7}\left(\frac{\rho_G}{\rho_L}\right)^{1/7} \tag{16.61}$$

We can now estimate possible values of R_{av}. If we consider the flow of a gas-liquid mixture in a pipeline, then $d \sim 0.2$–1 m, $\rho_L \sim 750$ kg m^{-3}, $\rho_G \sim 50$ kg m^{-3}, $\Sigma \sim 5 \times 10^{-3}$ N m^{-1}, $U \sim 10$–50 m s^{-1}, and $R_{av} \sim (10\text{–}100) \times 10^{-6}$ m $= 10$–100 μm. Comparison with values of λ_0 found earlier shows that in a pipeline $\lambda_0 < R_{av}$. For processes in a separator, we may have $d \sim 1$–2 m and $U \sim 0.1$–1 m s^{-1} and estimation of the average radius of drops shows that in this case both $\lambda_0 < R_{av}$, and $\lambda_0 > R_{av}$ are possible.

Consider first the case $\lambda_0 < R_{av}$. From Eq. (16.55), it follows that:

$$J_i = 5.04\pi\left(\frac{\rho_L}{\rho_G}\right)^{1/3}\frac{R^{7/3} U \Delta\rho_{iG}}{d^{1/3}} \tag{16.62}$$

In the elementary case of a monodisperse mixture, Eqs. (16.56) and (16.58)–(16.60) are reduced to:

$$\frac{d\rho_{iG}}{dt} = -\frac{3W_0 J_i}{4\pi R_0^3}, \tag{16.63}$$

$$\frac{d}{dt}\left(\sum_{i=1}^{s} \rho_{iL} V\right) = \sum_{i=1}^{s} J_i, \tag{16.64}$$

$$\frac{d(\rho_{iL} V)}{dt} = J_i, \tag{16.65}$$

Substituting Eq. (16.62) into the first two equations and considering that to a first approximation $\rho_L = \text{const.}$, one obtains:

$$\frac{d(\Delta\rho_{iG})}{dt} = -E\Delta\rho_i \left(\frac{V}{V_0}\right)^{7/9}, \tag{16.66}$$

$$\frac{d(V/V_0)}{dt} = \frac{E}{\rho_L W_0}\left(\frac{V}{V_0}\right)^{7/9} \sum_{i=1}^{s} \Delta\rho_{iG}, \tag{16.67}$$

where $E = 5.04\pi(3/4\pi)^{7/9}(\rho_L/\rho_G)^{1/3} W_0 U/V_0^{2/9} d^{1/3}$.

Summing the first equation over i, adding to the second equation, and integrating the resulting expression, one obtains:

$$\sum_{i=1}^{s} \frac{\Delta\rho_{iG}}{\rho_L} = \sum_{i=1}^{s} \frac{\Delta\rho_{iG}(0)}{\rho_L} + W_0 - W_0 \frac{V}{V_0}. \tag{16.68}$$

Substitution of Eq. (16.68) into Eq. (16.67) yields:

$$\frac{Et}{W_0} = \int_0^{V/V_0} \frac{dz}{z^{7/9}\left(\sum_i \Delta\rho_{iG}(0)/\rho_L + W_0 - W_0 z\right)}. \tag{16.69}$$

The integral in Eq. (16.69) does not undertake in quadratures, therefore one should proceed in the following way. From Eqs. (16.63)–(16.65), it follows that the characteristic time of the change in drop volume V is greater than the characteristic time of the change in the concentration of components ρ_{iG} in the gas phase. This means that when we consider the temporal change of ρ_{iG}, the value of R can be taken to be constant, equal to R_0. Then, to a first approximation, it is possible to write:

$$\frac{V - V_0}{V_0} = \frac{1}{\rho_L W_0} \sum_{i=1}^{s} (\rho_{iG0} - \rho_{iGw})(1 - e^{-At}), \tag{16.70}$$

where $A = 3.78(\rho_L/\rho_G)^{1/3} W_0 U/R_0^{2/3} d^{1/3}$.

Otherwise, the solution of Eqs. (16.66) and (16.67) would be obtained in implicit form:

$$\tau = -\left(\frac{V}{V_0}\right)^{-7/9} \ln\left[1 + \frac{W_0 \rho_L}{\sum_i \Delta\rho_{iG}(0)}\left(1 - \frac{V}{V_0}\right)\right],$$

where $\tau = Et$.

The relationships obtained allow estimation of the maximum value of the relative increase in drop volume:

$$\left(\frac{V - V_0}{V_0}\right)_{max} = \frac{\rho_G}{\rho_L W_0 M_G} \sum_{i=1}^{s} M_i(y_{i0} - y_{iw}) \equiv B, \tag{16.70a}$$

where $y_i = \rho_{iG} M_G / \rho_G M_i$ and $x_i = \rho_{iL} M_L / \rho_L M_i$ are mole fractions and M_i, M_G, and M_L are the molecular masses of components in the gaseous and liquid phases. For characteristic parameter values of $W_0 \sim 10^{-4}$–10^{-5} m^3/m^3, $\rho_G/\rho_L \sim 10^{-1}$, $M_i/M_G \sim 0.25$, and $y_{i0}/y_{iw} \sim 10^{-2}$–$10^{-3}$, one obtains $B \sim 1.25$–26. Hence, under certain conditions, the change in drop volume owing to condensation growth can be significant and may exert a noticeable influence on the separation rate of liquid phase from the gas in the separator.

Since drops of average radius are formed in the pipeline, then for R_0 in Eq. (16.70) it is necessary to use R_{av}, which is defined according to Eq. (16.61). The value of the parameter A depends on where the process of drop mass-exchange is considered, i.e. in a pipeline or in a separator. In the first case, d corresponds to the diameter of the pipeline, and:

$$A = A_p = \frac{14.3 U^{4/7} \rho_L^{5/7} W_0}{d^{5/7} \Sigma^{2/7} \rho_G^{3/7}} \tag{16.71}$$

If the process occurs in a separator, then:

$$A = A_s = \frac{14.3 U_s^{4/7} \rho_L^{5/7} W_0 U_s}{d^{8/21} \Sigma^{2/7} \rho_G^{3/7} D^{1/3}} \tag{16.72}$$

Here, $U_s = U(d/D)^2$ is the average velocity of the flow in the separator and D is the diameter of the separator.

We consider first the process in the pipeline. Since the exponent of the exponential function in Eq. (16.70) does not depend on the properties of the components, the characteristic time of balance establishment is identical for all components and is equal to:

$$t_{eq} \sim A_p^{-1} \sim 0.07 \frac{d^{5/7} \Sigma^{2/7} \rho_G^{3/7}}{U^{11/7} \rho_L^{5/7} W_0} \tag{16.73}$$

From Eq. (16.73), it follows that an increase in the flow velocity and a reduction in the pipe diameter will result in a decrease in the time of balance establishment.

The distance at which the balance is established is given by:

$$L_{eq} = d\frac{0.07}{W_0}\left(\frac{\Sigma^2 \rho_G^3}{U^4 \rho_L^5 d^2}\right)^{1/7} \tag{16.74}$$

For characteristic values of a pipeline, $\Sigma \sim 5 \times 10^{-3}$ N m^{-1}, $\rho_G \sim 50$ kg m^{-3}, $U \sim 30$ m s^{-1}, $\rho_L \sim 750$ kg m^{-3}, $d \sim 0.1$ m, $W_0 \sim 10^{-4}$, $\rho_{G0}/\rho_{L0} \sim 0.1$, one obtains $L_{eq} \sim 4.5$ m and $t_{eq} \sim 0.15$ s.

If the process occurs in a separator, then Eqs. (16.73) and (16.74) are slightly modified:

$$t_{eq} \sim 0.07\frac{D^{1/3}\Sigma^{2/7} d^{8/21} \rho_G^{3/7}}{U^{4/7} \rho_L^{5/7} W_0 U_s} \tag{16.75}$$

$$L_{eq} \sim d\frac{0.07 D^{1/3}\Sigma^{2/7} d^{8/21} \rho_G^{3/7}}{W_0 U^{4/7} \rho_L^{5/7}} \tag{16.76}$$

The solution obtained corresponds to a monodisperse distribution of drops without regard for coagulation. We consider a possible solution taking into account a polydisperse distribution and the coagulation of drops.

To take into account a continuous distribution of drops over volumes, it is necessary to take advantage of Eqs. (16.56) and (16.62), from which it follows that:

$$\frac{d\rho_{iG}}{dt} = -\Lambda \Delta\rho_{iG} m_{7/9}, \quad \Lambda = 4\pi\sqrt[3]{2}\left(\frac{3}{4\pi}\right)^{7/9}\left(\frac{\rho_L}{\rho_G}\right)^{1/3}\frac{U}{d^{1/3}} \tag{16.77}$$

where $m_{7/9}$ is the fractional moment of the order 7/9.

Equations for the first two moments, m_0 and m_1, can be derived in the ordinary way from the kinetic equation, Eq. (16.48):

$$\frac{dm_0}{dt} = 0, \quad \frac{dm_1}{dt} = \int_0^\infty n(V,t)\frac{dV}{dt}dV. \tag{16.78}$$

Taking into account that

$$\frac{dV}{dt} = \Lambda \sum_{i=1}^{s} \frac{\Delta\rho_{iG}}{\rho_L} V^{7/9},$$

from Eq. (16.78) we obtain:

$$\frac{dm_1}{dt} = \Lambda \sum_{i=1}^{s} \frac{\Delta\rho_{iG}}{\rho_L} m_{7/9}. \tag{16.79}$$

Next, we express the fractional moment in terms of integer values using a two-point scheme of interpolation, $m_{7/9} = m_0^{2/9} m_1^{7/9}$. Substituting the result in Eqs. (16.77) and (16.78) and taking into account that in the considered case the numerical concentration of drops m_0 remains constant, we obtain:

$$\Lambda t m_0^{2/9} = \int_{m_1(0)}^{m_1} \frac{dm_1}{m_1^{7/9}\left(\sum_i \Delta\rho_{iG}(0)/\rho_L + m_1(0) - m_1\right)}. \tag{16.80}$$

This expression gives in implicit form the dependence of m_1 on t, which is similar to the correlation of Eq. (16.69).

It is possible to simultaneously take into account condensation growth and the coagulation of drops if we enter into the right-hand part of Eq. (16.58) the collision term:

$$\frac{\partial n}{\partial t} + \frac{\partial}{\partial V}\left(n \frac{dV}{dt}\right) = I_k. \tag{16.81}$$

Thus, Eq. (16.77) does not change and the equations for the first two moments become

$$\frac{dm_0}{dt} = -G m_{1/3} m_{1/6}, \quad \frac{dm_1}{dt} = \sum_{i=1}^{s} \frac{\Delta\rho_i}{\rho_L} m_{7/9}. \tag{16.82}$$

Expressing fractional moments in terms of integer values and entering the following dimensionless variables:

$$\tau = \frac{\Lambda W_0^{7/9} t}{V_0^{2/9}}, \quad \widehat{m}_0 = \frac{m_0}{m_0(0)}, \quad \widehat{m}_1 = \frac{m_1}{m_1(0)}, \quad \gamma = \frac{G}{\Lambda V_0^{5/18}},$$

$$\tilde{\rho}_{iG} = \frac{\rho_{iG}}{\rho_L}, \quad \Delta\tilde{\rho}_{iG} = \frac{\Delta\rho_{iG}}{\rho_L},$$

one obtains:

$$\frac{d\widehat{m}_0}{d\tau} = -\gamma \widehat{m}_0^{3/2} \widehat{m}_1^{1/2} W_0^{5/9}, \quad \frac{d\widehat{m}_1}{d\tau} = \widehat{m}_0^{2/9} \widehat{m}_1^{7/9} \sum_{i=1}^{s} \Delta\tilde{\rho}_{iG},$$

$$\frac{d(\Delta\tilde{\rho}_{iG})}{d\tau} = -W_0 \Delta\tilde{\rho}_{iG} \widehat{m}_0^{2/9} \widehat{m}_1^{7/9}, \tag{16.83}$$

$$\widehat{m}_0(0) = \widehat{m}_0(0) = 1, \quad \Delta\tilde{\rho}_{iG}(0) = \Delta\tilde{\rho}_{iG0}.$$

Having solved the system of equations, Eqs. (16.83), one obtains the temporal change of the concentration of a gas component and also the change of the disperse structure of the mixture. This structure includes the numerical concentration of drops $m_0 = N$, the volume content of drops $m_1 = W$, and the average volume of drops $m_1/m_0 = V_{av}$.

We now proceed to the second case, when $R_{av} < \lambda_0$. The diffusion flux is thus equal to

$$J_i = A_i \frac{V_0}{W_0}\left(\frac{V}{V_0}\right)^{2/3} \Delta\rho_{iG}, \quad A_i = 2.35 \frac{D_{imG}^{2/3} U^{3/4} W_0 \rho_L^{1/3}}{v_G^{5/12} d^{1/4} V_0^{1/3} \rho_G^{1/3}} \tag{16.84}$$

Substitution of Eq. (16.84) into Eqs. (16.63)–(16.65) gives:

$$\frac{d(\Delta\rho_{iG})}{dt} = -A_i \left(\frac{V}{V_0}\right)^{2/3} \Delta\rho_{iG}, \quad \frac{d(V/V_0)}{dt} = -\sum_{i=1}^{s} A_i \frac{\Delta\rho_{iG}}{\rho_L W_0}\left(\frac{V}{V_0}\right)^{2/3}. \tag{16.85}$$

Coefficients A_i differ from one another only by the coefficients of diffusion D_{imG}. As they have the same order of magnitude for different components in the gas phase, for simplicity they may be taken to be identical ($A_i = A =$ const.). Then, Eqs. (16.85) yield an exact solution. If we enter a dimensionless time $\tau = At$ and proceed from mass concentration to mole fraction, then this solution is:

$$\tau = \frac{1}{a^2}\left(0.5 \ln \frac{(1-a)^3 (z^3 - a^3)}{(1-a^3)(z-a)^3} - \sqrt{3}\left(\arctan \frac{2+a}{a\sqrt{3}} - \arctan \frac{2z+a}{a\sqrt{3}}\right)\right) \tag{16.86}$$

where $z = R/R_0$, $a^3 = 1 + \rho_G \sum M_i \Delta y_{i0} / \rho_L M_G W_0$, M_i and M_G are the molecular masses of components and the gas, y_i is the mole fraction of the i-th component, and $\Delta y_{i0} = y_{i0} - y_i$ is the difference in mole fractions at p_1, T_1 and p_2, T_2, respectively.

Equation (16.86) in implicit form gives the dependence of the dimensionless drop radius R/R_0 on the dimensionless time τ. Since $z \to a$ as $\tau \to \infty$ and $z \to 1$ as $a \to 1$, then instead of z it is convenient to enter $z_1 = (z-1)/(a-1)$. From Eq. (16.86), it follows that a change of drop size depends on two parameters, the initial volume content of the liquid phase W and parameter $\beta = \rho_G \sum_i M_i \Delta y_{i0} / \rho_L M_G$, describing the deviation of the mixture composition from equilibrium. In Fig. 16.3, the dependence of z_1 on τ for various values of the parameter a is shown. It can be seen that with increasing volume content of drops W_0, the growth time and the final volume of drops decrease. For example, for $W_0 \sim 7 \times 10^{-4}$ m^3/m^3, the drop radius increases 1.4-fold in a time of $t = 1.5$ s, while for $W_0 \sim 7 \times 10^{-5}$ m^3/m^3 it increases in 2.7 times in a time of $t = 6$ s. Since the mixture entering the separator contains about 100 g m^{-3} of liquid phase (condensate) at a pressure of 100 MPa, which corresponds to $W_0 \sim 7 \times 10^{-4}$ m^3/m^3, and the characteristic dwell time of the mixture in the

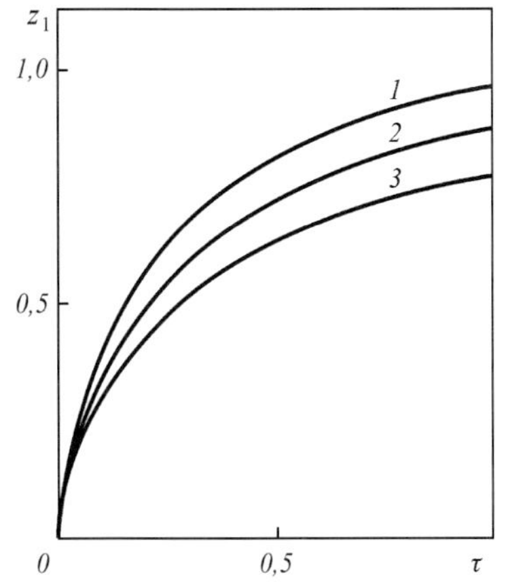

Fig. 16.3 Dependence of z_1 upon τ for different values of a: 1: 2.71; 2: 2.15; 3: 1.44.

separator is about 20 s, it is possible to conclude that the mixture in the separator is under conditions of phase balance.

Equation (16.86) is bulky and inconvenient for calculations. With sufficient accuracy, it may be approximated by the expression:

$$z = 1 + (a-1)(1 - e^{-B\tau}), \quad B = 4.38(a-1) - (a-1)^2 \tag{16.87}$$

We consider now in tandem the taking into account of the polydispersiveness of drop distribution and the probability of drop coagulation.

One takes into account the continuous distribution of drops over volumes in the same manner as in the case of $R_{av} > \lambda_0$; the only difference is the choice of expression for the diffusional flux J_i. The concentration distribution in the gas phase is thus described by the equations:

$$\frac{d\rho_{iG}}{dt} = -\Lambda_i \Delta\rho_i m_{2/3}, \quad \Lambda_i = 2.25 D_{imG} \left(\frac{\rho_L}{\rho_G}\right)^{1/3} \frac{U_s^{3/4}}{v_G^{5/12} D^{1/4}} \tag{16.88}$$

Equation (16.88) is written for the case of a process occurring in a separator, and therefore in Λ_i depends on the flow velocity U_s and the diameter D of the separator.

The equation for moment m_1 (the moment m_0 remains constant) will become:

$$\frac{dm_1}{dt} = \sum_i \Lambda_i \frac{\Delta\rho_i}{\rho_L} m_{2/3}. \tag{16.89}$$

Transition from a fractional moment to an integer one allows us to rewrite Eqs. (16.88) and (16.89) as

$$\frac{d\rho_{iG}}{dt} = -\Lambda_i \Delta\rho_i m_0^{1/3} m_1^{2/3}, \quad \frac{dm_1}{dt} = \sum_i \Lambda_i \frac{\Delta\rho_i}{\rho_L} m_0^{1/3} m_1^{2/3}. \tag{16.90}$$

Let us accept for simplicity that diffusion coefficients D_{imG} are identical ($\Lambda_i = \Lambda$). The system of equations then has an exact solution:

$$\Lambda \left(\frac{W_0}{V_0}\right)^{1/3} t = \frac{0.5}{W_0} \left(0.5 \ln \frac{(1-a)(z^3 - a^3)}{(1-a^3)(z-a)} - \sqrt{3}\left(\arctan \frac{2+a}{a\sqrt{3}} - \arctan \frac{2z+a}{a\sqrt{3}}\right)\right) \tag{16.91}$$

$$\Lambda = 2.25 D_{mG} \left(\frac{\rho_L}{\rho_G}\right)^{1/3} \frac{U_s^{3/4}}{v_G^{5/12} D^{1/4}}$$

where $z = R/R_0$ and $a^3 = 1 + \rho_G \sum_i M_i \Delta y_{i0} / \rho_L M_G W_0$.

Simultaneously taking into account polydispersiveness and the coagulation of the drops, the following system of equations is obtained:

$$\begin{aligned} \frac{d\tilde{\rho}_{iG}}{d\tau} &= -\Delta\tilde{\rho}_{iG} \widehat{m}_0^{1/3} \widehat{m}_1^{2/3} W_0, \\ \frac{d\widehat{m}_1}{d\tau} &= \sum_{i=1}^{s} \Delta\tilde{\rho}_{iG} \widehat{m}_0^{1/3} \widehat{m}_1^{2/3}, \\ \frac{d\widehat{m}_0}{d\tau} &= -\gamma \widehat{m}_0^{3/2} \widehat{m}_1^{1/2}. \end{aligned} \tag{16.92}$$

Here, the following dimensionless variables are introduced:

$$\tau = \Lambda t / V_0^{1/3}, \quad \Delta\tilde{\rho}_{iG} = \Delta\rho_{iG}/\rho_L, \quad \gamma = G/\Lambda V_0^{1/6},$$
$$\widehat{m}_0 = m_0/m_0(0), \quad \widehat{m}_1 = m_1/m_1(0).$$

16.3 Enlargement of Drops During the Passage of a Gas–Liquid Mixture Through Devices of Preliminary Condensation

In the scheme of low-temperature separation, a device of preliminary condensation (DPC) is placed before the separator. This may be a throttle, a heat-exchanger, or a turbo-expander, which serves to decrease the temperature of the mixture. Upon passage of a gas-condensate mixture through these devices, the phase balance established by the flow of the mixture in the pipeline is disrupted. As a result, there will be probable formation of a liquid phase due to the process of nucleation, and transition of components from one phase to another owing to the processes of mass-exchange, evaporation, and condensation. Violation of the thermodynamic balance of phases is caused by a change in pressure p and temperature T. The main interest concerns changes of these parameters whereby the size of condensate drops is increased, since this facilitates their detachment from the gas in a separator.

In the previous section, the process of drop integration was considered as a result of a change in thermobaric conditions and simple formulae were derived allowing the estimation of increments of drop volumes (see, for example, Eq. (16.87)). It is obvious that, in general, the size of drops can both increase and decrease. Growth of drops occurs when the condition $a > 1$ is met, or when:

$$\sum_i M_i \Delta y_{i0} > 0, \tag{16.93}$$

where Δy_{i0} is the difference in the equilibrium mole fractions of the i-th component at p_1, T_1 and p_2, T_2.

Since $\sum_i M_i y_{i0} = M_{G0}$ and $\sum_i M_i \Delta y_{iw} = M_{Gw}$, then the inequality, Eq. (16.93), may be written in the form:

$$M_{G0} > M_{Gw}. \tag{16.94}$$

Thus, for better integration of drops it is necessary to vary the pressure and temperature so that the inequality, Eq. (16.94), is satisfied. In the following paragraphs, processes occurring in various devices are considered. The discussion is limited to some assessment of the role of various DPCs and their efficiency as devices of preliminary integration of condensate drops ahead of a separator. Therefore, without considering the behavior of gas-liquid mixtures in DPCs, suppose that during the flow to the DPC pipeline the gas, containing condensate drops, at p_1, T_1 is in thermodynamic equilibrium with the liquid. In a certain section of the pipeline the pressure and temperature abruptly change and become equal to p_2 and T_2 $(p_2 \leq p_1, T_2 < T_1)$. Further along the mixture flow, thermodynamic equilibrium is re-established, but corresponding to values p_2 and T_2. The equilibrium conditions of the mixture can be determined with the help of vapor-liquid balance equations based on the Peng–Robinson equation of state. As an example, we consider $T_1 = 20$ °C, three values of pressure $p_1 = 10, 7$, and 4 MPa, and three different gas-liquid mixtures distinguished by their condensate mass contents of $q = 100, 500$, and 2000 g m^{-3}. Table 16.1 shows the general

Table 16.1

q	CO_2	N_2	C_1	C_2	C_3	C_4	F_1	F_2	F_3	F_4	F_5
100	0.0015	0.0458	0.7909	0.0866	0.0353	0.0145	0.0115	0.0066	0.0037	0.0021	0.0015
500	0.0018	0.0421	0.7266	0.0826	0.0393	0.0198	0.0306	0.0260	0.0164	0.0092	0.0066
2000	0.0011	0.0325	0.5544	0.0716	0.0506	0.0341	0.0806	0.0767	0.0500	0.0281	0.0203

Table 16.2

p_1	CO_2	N_2	C_1	C_2	C_3	C_4	F_1	F_2	F_3	F_4	F_5
$q = 100$ g m^{-3}											
10	0.0015	0.0478	0.8156	0.0847	0.0317	0.0110	0.0058	0.0015	0.0003	0.0	0.0
7	0.0015	0.0478	0.8174	0.0854	0.0316	0.0105	0.0047	0.0009	0.0001	0.0	0.0
4	6.0015	0.0474	0.8148	0.0867	0.0329	0.0112	0.0047	0.0007	0.0001	0.0	0.0
$q = 500$ g m^{-3}											
10	0.0017	0.0511	0.8288	0.0756	0.0265	0.0084	0.0059	0.0016	0.0003	0.0	0.0
47	0.0018	0.0501	0.8290	0.0783	0.0271	0.0080	0.0046	0.0010	0.0001	0.0	0.0
	0.0019	0.0487	0.8217	0.0830	0.0300	0.0090	0.0045	0.0008	0.0008	0.0	0.0
$q = 2000$ g m^{-3}											
10	0.0009	0.0627	0.8435	0.0568	0.0213	0.0069	0.0059	0.0016	0.0003	0.0	0.0
7	0.0010	0.0579	0.8442	0.0620	0.0224	0.0066	0.0046	0.0010	0.0001	0.0	0.0
4	0.0012	0.0529	0.8328	0.0725	0.0275	0.0077	0.0045	0.0008	0.0001	0.0	0.0

mixture compositions at the indicated values of q. It is accepted that the mixture consists of two neutral components CO_2 and N_2, four first hydrocarbonic components C_1, C_2, C_3, C_4, as well as residual C_{5+}, which is separated into five fractions F_1, F_2, F_3, F_4, F_5.

Table 16.2 shows the compositions of the gas phase at various pressure p_1 before a DPC for different values of q.

Compositions of the gas phase beyond the DPC are listed in Table 16.3.

Using the data listed in the tables, it is easy to find the greatest possible radius of the enlarged drops after passage through the DPC. We shall consider a case involving the use of a heat exchanger as the DPC. As an input to the heat exchanger, one supplies a gas-liquid mixture at temperature $T = 20$ °C, and pressure $p_1 = 4, 7$, and 10 MPa. The pressure in the heat-exchanger remains practically constant ($p_2 = p_1$). As a result of cooling of the mixture in the heat exchanger, the phase equilibrium is disrupted, a portion of the components passes from the liquid phase into the gas phase, and another portion goes from the gas phase into the liquid phase. The drop size thus increases according to the law given by Eq. (16.87):

$$\frac{R}{R_0} = 1 + (a-1)(1 - e^{-B\tau}), \quad B = 4.38(a-1) - (a-1)^2 \qquad (16.95)$$

where $a^3 = 1 + \rho_G \sum_i M_i \Delta y_{i0} / \rho_L M_G W_0$.

From Eq. (16.95), it can be seen that upon the attainment of equilibrium, beyond the heat exchanger (as $\tau \to \infty$), the drop radius becomes equal to aR.

Table 16.3

T_1	CO_2	N_2	C_1	C_2	C_3	C_4	F_1	F_2	F_3	F_4	F_5
$q = 100$ g m^{-3}											
$P_2 = 10$ MPa											
10	0.0015	0.0483	0.8199	0.0836	0.0303	0.0101	0.0049	0.0011	0.0002	0.0	0.0
0	0.0014	0.0489	0.8248	0.0820	0.0287	0.0090	0.0040	0.0009	0.0001	0.0	0.0
−10	0.0014	0.0497	0.8300	0.0800	0.0269	0.0080	0.0034	0.0009	0.0001	0.0	0.0
−20	0.0013	0.0506	0.8349	0.0773	0.0250	0.0072	0.0029	0.0006	0.0001	0.0	0.0
$p_2 = 7$ MPa											
10	0.0015	0.0482	0.8224	0.0843	0.0300	0.0093	0.0036	0.0006	0.0001	0.0	0.0
0	0.0014	0.0488	0.8284	0.0826	0.0279	0.0078	0.0026	0.0004	0.00	0.0	0.0
−10	0.0014	0.0496	0.8355	0.0800	0.0251	0.0063	0.0018	0.0002	0.00	0.0	0.0
−20	0.0013	0.0507	0.8438	0.0762	0.0217	0.0048	0.0012	0.0001	0.00	0.0	0.0
$p_2 = 4$ MPa											
10	0.0015	0.0477	0.8194	0.0861	0.0316	0.0098	0.0033	0.0004	0.0	0.0	0.0
0	0.0015	0.0482	0.8250	0.0850	0.0296	0.0082	0.0022	0.0002	0.0	0.0	0.0
−10	0.0015	0.0487	0.8317	0.0833	0.0269	0.0063	0.0013	0.0001	0.0	0.0	0.0
−20	0.0014	0.0495	0.8400	0.0806	0.0232	0.0045	0.0008	0.0001	0.0	0.0	0.0
$q = 500$ g m^{-3}											
$p_2 = 10$ MPa											
10	0.0016	0.0522	0.8357	0.0728	0.0242	0.0073	0.0047	0.0021	0.0002	0.0	0.0
0	0.0016	0.0534	0.8425	0.0695	0.0218	0.0062	0.0038	0.0009	0.0001	0.0	0.0
−10	0.0015	0.0560	0.8491	0.0656	0.0195	0.0053	0.0031	0.0007	0.0001	0.0	0.0
−20	0.0014	0.0565	0.8584	0.0610	0.0163	0.0039	0.0020	0.0004	0.0001	0.0	0.0
$p_2 = 7$ MPa											
10	0.0017	0.0510	0.8367	0.0755	0.0244	0.0066	0.0034	0.0007	0.0	0.0	0.0
0	0.0016	0.0520	0.8448	0.0720	0.0214	0.0053	0.0024	0.0004	0.0	0.0	0.0
−10	0.0015	0.0534	0.8532	0.0674	0.0182	0.0040	0.0017	0.0003	0.0	0.0	0.0
−20	0.0014	0.0551	0.8618	0.0619	0.0151	0.0031	0.0012	0.0002	0.0	0.0	0.0
$p_2 = 4$ MPa											
10	0.0018	0.0493	0.8294	0.0810	0.0276	0.0073	0.0031	0.0005	0.0	0.0	0.0
0	0.0017	0.0501	0.8377	0.0782	0.0243	0.0058	0.0020	0.0002	0.0	0.0	0.0
−10	0.0017	0.0510	0.8466	0.0744	0.0205	0.0041	0.0013	0.0002	0.0	0.0	0.0
−20	0.0016	0.0520	0.8464	0.0694	0.0166	0.0029	0.0008	0.0001	0.0	0.0	0.0
$q = 2000$ g m^{-3}											
$p_2 = 10$ MPa											
0	0.0008	0.0695	0.8561	0.0479	0.0160	0.0048	0.0038	0.0009	0.0001	0.0	0.0
−10	0.0007	0.0744	0.8600	0.0432	0.0138	0.0040	0.0030	0.0007	0.0001	0.0	0.0
−20	0.0006	0.0811	0.8613	0.0386	0.0118	0.0033	0.0025	0.0006	0.0001	0.0	0.0

Table 16.3 *(continued)*

T_1	CO_2	N_2	C_1	C_2	C_3	C_4	F_1	F_2	F_3	F_4	F_5
$p_2 = 7$ MPa											
0	0.0009	0.0627	0.8614	0.0519	0.0159	0.0041	0.0025	0.0005	0.0	0.0	0.0
−10	0.0008	0.0660	0.8687	0.0462	0.0130	0.0031	0.0018	0.0003	0.0	0.0	0.0
−20	0.0007	0.0704	0.8742	0.0403	0.0105	0.0024	0.0013	0.0002	0.0	0.0	0.0
$p_2 = 4$ MPa											
0	0.0011	0.0560	0.8545	0.0625	0.0192	0.0044	0.0021	0.0003	0.0	0.0	0.0
−10	0.0010	0.0579	0.8649	0.0562	0.0152	0.0032	0.0014	0.0002	0.0	0.0	0.0
−20	0.0008	0.0602	0.8747	0.0492	0.0118	0.0022	0.0009	0.0001	0.0	0.0	0.0

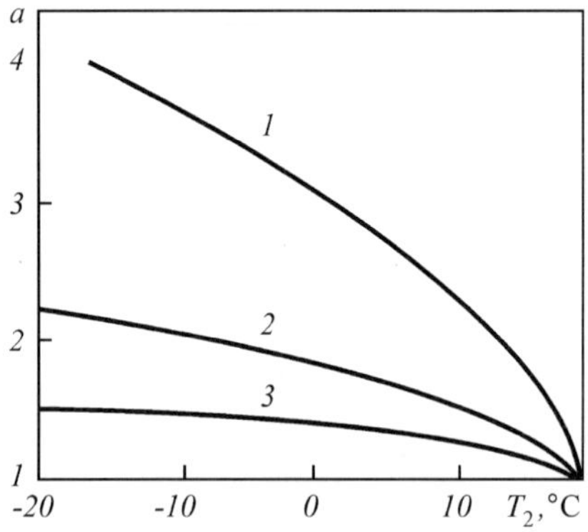

Fig. 16.4 Dependence of a upon T_2 for different values of q (kg m^{-3}) at $p_1 = p_2 = 10$ MPa; $T_1 = 20$ °C: 1: $q = 0.1$; 2: $q = 0.5$; 3: $q = 3$.

Hence, the parameter a determines the factor by which the initial drop radius is increased. This parameter depends on p_1, T_1, p_2, T_2, the initial composition of the mixture, and the condensate factor q. These dependences are shown graphically in Figs. 16.4 and 16.5.

At fixed values of pressure and temperature at the input of the heat exchanger and fixed gas condensate factor, the value of a increases with reduction of the temperature output. The value of q strongly influences the growth of drops. A reduction of q results in an increased degree of enlargement of the drops, albeit in a longer time. This may be explained as follows. The volume content of drops is determined by $W = q/\rho_L$. So, for the considered values of $q = 100, 500$, and 2000 g m^{-3} and $\rho_L = 700$ kg m^{-3}, we have $W = 1.4 \times 10^{-4}, 7 \times 10^{-4}$, and 2.8×10^{-3}. The parameter W determines the average distance between drops in the mixture.

For the indicated values of the parameters, these distances are equal to 20 R, 11.2 R, and 7 R, respectively. Hence, reduction of W results in an increase in the

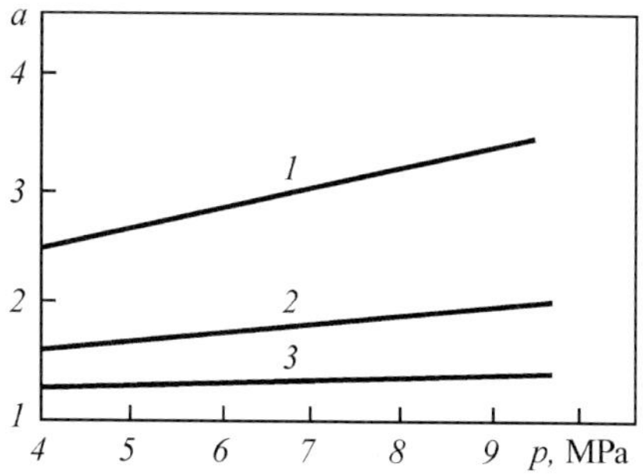

Fig. 16.5 Dependence of a upon p for different values of q (kg m^{-3}) at $T_1 = 20$ °C; $T_2 = -10$ °C: 1: $q = 0.1$; 2: $q = 0.5$; 3: $q = 3$.

average distance between drops and, consequently, to a decrease in their numerical concentration. Thus, the time needed for balance establishment in the system increases, which creates more favorable conditions for drop growth. An increase in pressure also promotes drop growth. The characteristic time of drop integration for $p_1 = p_2 = 10$ MPa, $T_1 = 20$ °C, $T_2 = 0$ °C, $q = 100$ g m^{-3} is equal to $t = 1.4$ s. A change of T_2 affects t only slightly, while an increase in q reduces t markedly.

If a throttle is used as the DPC, the following conditions are satisfied: $p_2 < p_1$, $T_2 < T_1$. So, for values of $p_1 = 10$ MPa, $T_1 = 20$ °C, $T_2 = 7$ MPa, $T_2 = 10$ °C, $q = 100$ g m^{-3}, the drop radius is increased threefold after passage through the throttle. For the same parameter values in a heat exchanger, the drop radius is increased fourfold. Hence, a heat exchanger has an advantage over a throttle in that it allows the growth of larger drops. Besides, an essential feature of a throttle as a device for drop condensation is that there is a sharp contraction of pipe cross-section, so that the resulting increase in flow velocity leads to breakage of the drops suspended in the flow and hence to a reduction in their average size.

In the following section, we show that a multitude of fine droplets, as formed by nucleation in a gas upon passage through a throttle, undergo extensive coagulation. As a result, the growth of drops after the throttle is chiefly due to coagulation rather than condensation.

16.4
Formation of a Liquid Phase in a Throttle

Throttling of a gas is carried out by means of a nozzle placed in a delivery pipeline before a separator. The gas flow velocity in a nozzle consisting of a conic throttling orifice with a narrowed cross-section is increased, and therefore the pressure and temperature of the gas decrease.

Let the pressure, temperature, and gas velocity before the throttle be p_1, T_1, and u_1, and let those at the output be p_2, T_2, and u_2 $(p_2 < p_1, T_2 < T_1, u_2 > u_1)$. For simplicity of the calculations, it is accepted that the gas is composed of a mixture of two gases, which, by convention, are referred to as gas and vapor. Their fractional pressures are designated as p_G and p_v. Reduction of the pressure and temperature in a throttle can induce condensation of vapor. We consider conditions under which this may occur. The key parameter describing the probability and the rate of liquid-phase nucleation is the supersaturation, s, which is equal to the ratio of the vapor pressure p_v to the saturation pressure $p_s(T)$ at temperature T (see Eq. (14.15)):

$$s = \frac{p_v}{p_s(T)}, \tag{16.96}$$

Here, $p_v = p_2 y_v$ is the partial pressure of the vapor, and p_s is the saturation pressure [9]:

$$p_s = p_c \exp\left[h\left(1 - \frac{T_c}{T}\right)\right], \quad h = \frac{\ln(p_c)}{T_c/T_b - 1}, \tag{16.97}$$

T_c is the critical temperature and T_b is the boiling point.

A necessary condition of vapor condensation is that the inequality $s > 1$ has to be met. However, this condition alone is not enough. A further requirement is that supersaturation exceeds the critical value:

$$s > s_{cr} = \exp\left[1.74 \times 10^7 \frac{M_L}{\rho_L}\left(\frac{\Sigma}{T}\right)^{3/2}\right] \tag{16.98}$$

Here, M_L and ρ_L are the molecular mass and the density of the condensable component, and Σ is the surface tension coefficient of the liquid phase. Since Σ depends on the pressure and temperature (see Section 17.1), s_{cr} also depends on these parameters. Thus, if the values p_2 and T_2 after the throttle are such that $1 < s < s_{cr}$, then there is no formation of a new phase in the throttle. Consider next the case $s > s_{cr}$. The value of the partial vapor pressure can be found if the mole fraction of the vapor, y_v, is known. To estimate this, we take advantage of the correlation $p_{v2} = p_2 y_v$ and Eq. (16.97) for p_s. If we take into account that $p_{v1} = p_s(T_1)$, then the condition for the absence of homogeneous condensation becomes:

$$\exp\left[hT_c\left(\frac{1}{T_2} - \frac{1}{T_1}\right)\right] < \frac{p_2}{p_1} < \exp\left[1.74 \times 10^7 \frac{M_L}{\rho_L}\left(\frac{\Sigma}{T}\right)^{3/2} - hT_c\left(\frac{1}{T_2} - \frac{1}{T_1}\right)\right] \tag{16.99}$$

Thus, if a gas mixture enters a throttle at p_1, T_1, and then p_2 and T_2 satisfy the inequality Eq. (16.99), a liquid phase does not appear in the flow. Under such conditions, condensation can only occur in the presence of nuclei of condensation, for example, already available condensate droplets in the flow ahead of the throttle. If the parameters of the throttle are such that p_1, T_1, p_2, and T_2 satisfy the inequality $s > s_{cr}$, then in the flow in the throttle, and possibly beyond it, a new phase is formed, the extent of which depends on the degree of mixture supersaturation.

We now consider characteristic values of the parameters. As an example, consider a methane/propane mixture. Let the pressure and temperature in the throttle fall from $p_1 = 12$ MPa, $T_1 = 293$ K to $p_2 = 7.5$ MPa, $T_2 = 269$ K. The saturation pressure for the propane changes from 0.796 MPa to 0.419 MPa. The equilibrium mole fractions of the propane corresponding to these values are $y_{v1} = 0.066$ and $y_{v2} = 0.056$. Considering that before the throttle $s = 1$, one obtains $s = 1.18$ beyond the throttle.

Upon formation of the liquid phase during homogeneous condensation there is heat evolution Q_{in} characterized by a heat of condensation l (kJ kg^{-1}). Therefore, the temperature in the throttle does not decrease as sharply as in the absence of condensation. To correct for calorification, it is necessary to take into account Q_{in} in the energy balance equation. If it is accepted that germs are formed practically instantly, then Q_{in} will consist of two additive terms. The first characterizes the heat evolution upon the formation of germs, and the second the heat evolution upon further condensation growth of drops:

$$Q_{in} = Q_{in1} + Q_{in2} \tag{16.100}$$

The flow of gas in the throttle is considered to be one-dimensional and stationary. The x-axis is directed along the throttle axis and $x = 0$ corresponds to the entrance section. The system of equations describing the flow in the throttle includes the conservation equation of gas mass flow rate:

$$\frac{d(\rho_G u S)}{dx} = 0, \tag{16.101}$$

the energy equation in the absence of physical work done by the gas and heat exchange with the environment:

$$\rho_G u \frac{d(c_p T)}{dx} = u \frac{dp}{dx} + Q_{in}, \tag{16.102}$$

the equation of motion:

$$\rho_G u \frac{du}{dx} = -\frac{dp}{dx}, \tag{16.103}$$

and the equation of state for the gas:

$$p = z \frac{A \rho_G T}{M_G}. \tag{16.104}$$

Here, ρ_G is the gas density; u, p, and T are the velocity, pressure, and temperature of the gas, respectively, S is the cross-sectional area of the throttle, c_p is the specific thermal capacity of the gas, Q_{in} is the heat evolved during condensation, z is the coefficient of gas compressibility, A is the gas constant, and M is the molecular mass of the gas.

The energy of condensation consists of the energy liberated by liquid-phase nucleation and by condensation growth of the formed droplets:

$$Q_{in} = Il\rho_L V_n + M_L l \int 4\pi R^2 J_{1w} n(V, x)\, dV, \tag{16.105}$$

where V_n is the germ volume, l is the heat of condensation, I is the rate of liquid-phase nucleation, ρ_L and M_L are the density and molecular mass of the liquid phase, J_{1v} is the flux of the condensed vapor on a drop of volume V, and $n(V, x)$ is the distribution of drops over volumes in a section of length x.

It is necessary to supplement the listed equations with further equations obtained in Section 16.3, which describe the dynamics of a drop and the change $n(V, x)$ due to condensation growth and coagulation. Suppose that in the course of a nucleation only propane is condensed, while methane remains unaffected.

Let us consider the model drop growth in the presence of a neutral component (see Section 16.1). The system of equations describing the behavior of drops is then:

$$\frac{\partial n}{\partial t} + \frac{\partial}{\partial V}\left(n\frac{dV}{dt}\right) = I\delta(V - V_n) + I_k, \tag{16.106}$$

$$\frac{dV}{dt} = -4\pi R^2 v_{1L} J_{1w} = 4\pi\left(\frac{3V}{4\pi}\right)^{1/3} \rho_G D_{12G}\Delta y v_{1L}, \tag{16.107}$$

$$\frac{d(y_{1\infty}\rho_G)}{dt} = -4\pi\left(\frac{3V}{4\pi}\right)^{1/3} \rho_G D_{12G}\Delta y \int_0^\infty V^{1/3} n\, dV - I\rho_L V_n \tag{16.108}$$

$$V_n = \frac{4\pi}{3}\left(\frac{2\Sigma}{p}\right)^3, \quad y_{1\infty}(0) = \frac{p_s(T_1)}{p_1}, \quad y_{1w} = \frac{p_s(T)}{p}, \quad \Delta y = y_{1\infty} - y_{1w}, \tag{16.109}$$

Applying the method of moments to Eqs. (16.106)–(16.109), one obtains:

$$\frac{dm_0}{dt} = I - Gm_{1/3}m_{1/6}, \quad \frac{dm_1}{dt} = 4\pi\left(\frac{3}{4\pi}\right)^{1/3} \rho_G D_{12G}\Delta y v_{1L} m_{1/3} + IV_n, \tag{16.110}$$

$$\frac{d(\Delta y\rho_G)}{dt} = -4\pi\left(\frac{3}{4\pi}\right)^{1/3} \rho_G D_{12G}\Delta y m^{1/3} - \frac{d(y_{1w}\rho_G)}{dt} - I\rho_L V_n \tag{16.111}$$

Next, we express the fractional moments in terms of integer values:

$$m_{1/3} = m_0^{2/3} m_1^{1/3}, \quad m_{1/6} = m_0^{5/6} m_1^{1/6}. \tag{16.112}$$

Equations (16.110) and (16.111) thus become:

$$\frac{dm_0}{dt} = I - Gm_0^{3/2}m_1^{1/2}, \tag{16.113}$$

$$\frac{dm_1}{dt} = 4\pi\left(\frac{3}{4\pi}\right)^{1/3}\rho_G D_{12G}\Delta y v_{1L} m_0^{2/3} m_1^{13} + IV_n, \tag{16.114}$$

$$\frac{d(\Delta y \rho_G)}{dt} = -4\pi\left(\frac{3}{4\pi}\right)^{1/3}\rho_G D_{12G}\Delta y m_0^{2/3} m_1^{1/3} - \frac{d(y_{1w}\rho_G)}{dt} - I\rho_L V_n \tag{16.115}$$

$$m_0(0) = m_1(0) = \Delta y(0) = 0. \tag{16.116}$$

To these equations one should add Eqs. (16.101)–(16.105), and subsequent transformation leads to:

$$\rho_G u S = \rho_{G1} u_1 S_1, \tag{16.117}$$

$$\rho_G u \frac{d(c_p T)}{dx} = -\rho_G u^2 \frac{du}{dx} + \frac{32\pi\Sigma^2 l \rho_L I}{3p^3} + 4\pi\left(\frac{3}{4\pi}\right)^{1/3}\rho_G D_{12G} M_L \Delta y m_0^{2/3} m_1^{1/3}, \tag{16.118}$$

$$\rho_G u \frac{du}{dx} = -\frac{dp}{dx}, \tag{16.119}$$

$$p = z\frac{A\rho_G T}{M_G}. \tag{16.120}$$

Besides, there are the following expressions:

$$I_n = 1.82 \times 10^{26}\pi\left(\frac{p}{T}\right)^2 (M_L\Sigma)^{1/2}(s\rho_L)^{-1} y_{1\infty}^2 \times \exp\left(-1.76\times 10^{16}\left(\frac{M_L}{\rho_L}\right)^2\left(\frac{\Sigma}{T}\right)^3 \ln^{-2} s\right) \tag{16.121}$$

$$s = \frac{y_{1\infty} p}{p_s(T)} = 1 + \frac{\Delta y}{y_{1w}}, \quad p_s = p_c \exp\left(h\left(1 - \frac{T_c}{T}\right)\right), \quad h = \frac{\ln(p_c)}{T_c/T_b - 1}$$

$$\Sigma = 10^{-3}(21.04 - 3.74p + 0.37p^2 - 0.015p^3) \tag{16.122}$$

$$z = (0.4 \lg T_r + 0.73)^{p_r} + 0.1p_r \tag{16.123}$$

$$l = \frac{AT_{r1}}{M_1}(7.08(1 - T_{r1})^{0.354} + 10.95\omega(1 - T_{r1})^{0.456}) \tag{16.124}$$

It is necessary to take into account that Σ, z, and l depend on the temperature and pressure. The dependence of Σ on pressure can be obtained by approximation of the experimental data in the interval $p = 1$–12 MPa (see Section 17.1):

$$\Sigma = 10^{-3}(21.04 - 3.74p + 0.37p^2 - 0.015p^3) \tag{16.125}$$

The dependence of Σ on temperature for hydrocarbon systems in the considered pressure interval is rather weak and hence is not taken into account.

The dependence of z on p and T is derived in ref. [10]:

$$z = (0.4 \lg T_r + 0.73)^{p_r} + 0.1 p_r \tag{16.126}$$

where T_r and p_r are the reduced pressure and the reduced temperature (see Section 5.7).

The heat of condensation is determined by the formula [11]:

$$l = \frac{AT_{r1}}{M_1}\left(7.08(1 - T_{r1})^{0.354} + 10.95\omega(1 - T_{r1})^{0.456}\right) \tag{16.127}$$

The subscript index 1 denotes a condensed component, while ω is an eccentricity factor indicating the deviation of the molecular form from spherical. For methane, ω is 0.08; for ethane 0.098; for propane 0.152, etc. For propane at $T_{r1} = 0.79$, $T_{c1} = 390$ K, $M_1 = 44$ kg kmol^{-1}, $l = 360$ kJ kg^{-1}.

Before proceeding to offer a solution to a general problem, we will try to estimate the volume of liquid phase formed on the basis of a simplified statement. Assume that coagulation and condensation growth of drops in the throttle are absent and that the flow is adiabatic. The distribution of flow parameters is then described by the following equations:

$$\begin{gathered} \rho_G u S = \rho_{G1} u_1 S_1, \\ p\rho_{G1}^k = p_1 \rho_G^k, \quad T_1 p_1^{(1-k)/k} = T p^{(1-k)/k} \\ \rho_G u \frac{d(c_p T)}{dx} = -\rho_G u^2 \frac{du}{dx} + V_n \rho_L l I, \end{gathered} \tag{16.128}$$

where k is the adiabatic exponent.

We can then transform the last equation to the form:

$$\frac{\rho_{G1} u_1 S_1}{S} \frac{d}{dx}(c_p T + 0.5u^2) = V_n \rho_L l I$$

which, after integration, reduces to

$$(c_p T - c_{p1} T_1) + 0.5(u^2 - u_1^2) = \frac{\rho_L l}{\rho_{G1} u_1} \int_0^x \frac{S(x)}{S_1} V_n I \, dx \tag{16.129}$$

Substitution of Eqs. (16.109) and (16.121) for V_n and I, respectively, into Eq. (16.129) results in

$$(c_pT - c_{p1}T_1) + 0.5(u^2 - u_1^2) = \varphi(x) \tag{16.130}$$

$$\varphi(x) = 1.29 \times 10^{27} \frac{\pi l M_L^{1/2} \Sigma_1^{7/2} y_{1\infty}^2}{\rho_L u_1 T_1^2 p_1} \int_0^x \frac{S}{S_1} \left(\frac{\Sigma}{\Sigma_1}\right)^{7/2} \frac{p}{p_1} \left(\frac{T}{T_1}\right)^2 s^{-1}$$
$$\times \exp\left(-1.76 \times 10^{16} \left(\frac{M_L}{\rho_L}\right)^2 \left(\frac{\Sigma_1}{T_1}\right)^3 \left(\frac{\Sigma}{\Sigma_1}\right)^3 \left(\frac{T_1}{T}\right)^3 \ln^{-2} s\right) dx$$

Reducing Eqs. (16.128) with respect to Eq. (16.130) yields a form convenient for further use:

$$\frac{T}{T_1} = \left(\frac{S}{S_1}\right)^2 \left(1 + 2\frac{\varphi(x/L)}{u_1^2} + 2c_{p1}T_1 \frac{1 - c_pT/c_{p1}T_1}{u_1^2}\right)^{(1-k)/k}$$
$$\frac{u}{u_1} = \frac{S_1}{S}\left(\frac{T}{T_1}\right)^{1/(1-k)}, \quad \frac{p}{p_1} = \left(\frac{T_1}{T}\right)^{1/(1-k)}, \quad \frac{\rho_G}{\rho_{G1}} = \frac{u_1 S_1}{uS}, \tag{16.131}$$

where L is the throttle length.

If the change of the throttle cross-sectional area is described by a function $S(x)$, and if the values of the input parameters are p_1, u_1, T_1, and ρ_{G1}, from Eq. (16.131) one can find the distribution of these parameters over the length. Of the utmost interest is the volumetric content of the liquid phase W, since the rates of coagulation and condensation growth of drops depend on this value. In the considered approach, this parameter is equal to:

$$W = \frac{1}{QL} \int_0^L IV_n \, dx, \tag{16.132}$$

where Q is the gas flow rate and L is the throttle length.

With reference to Eq. (16.121), the last relationship may be rewritten as:

$$W(x) = 1.29 \times 10^{27} \frac{\pi M_L^{1/2} \Sigma_1^{7/2}}{\rho_L u T_1^2 p_1} \int_0^x \left(\frac{\Sigma}{\Sigma_1}\right)^{7/2} \frac{p}{p_1} \left(\frac{T}{T_1}\right)^2 s^{-1} y_{1\infty}^2$$
$$\times \exp\left(-1.76 \times 10^{16} \left(\frac{M_L}{\rho_L}\right)^2 \left(\frac{\Sigma}{T}\right)^3 \ln^{-2} s\right) d(x/L) \tag{16.133}$$

Let us accept that $p_1 = 12$ MPa, $\rho_{G1} = 100$ kgm^{-3}, $\Sigma_1 = 3.54 \times 10^{-3}$ N m^{-1}, $M_L = 44$ kg kmol^{-1}, $\rho_L = 700$ kg m^{-3}, $T_1 = 293$ K. Changes in temperature T, pressure p, saturation pressure p_s, coefficient of surface tension Σ, supersaturation s, volume of germs V_n, and rate of liquid phase formation I are summarized in Table 16.4. The value of the parameter $2c_{p1}T_1/u_1^2$ is 500.

From the presented results, it follows that formation of liquid phase occurs in a small section of the throttle, in which the contraction changes from 0.3 to 0.24. The velocity in the latter section is equal to 5.3 u_1.

Let the radius of the throttle cross-section change lengthwise according to a linear law from $r_1 = 0.05$ m at the input to $r_2 = 0.0245$ m at the output. If

Table 16.4

S/S_1	1	0.45	0.3	0.27	0.25	0.24
T, K	293	290	285	281.3	277.2	271
p, MPa	12	11.3	10.2	9.6	8.7	7.5
p_s, MPa	0.8	0.76	0.67	0.6	0.53	0.44
S	1	1.012	1.041	1.095	1.12	1.17
Σ, N m^{-1}	3.5×10^{-3}	4.4×10^{-3}	5.4×10^{-3}	6×10^{-3}	6.6×10^{-3}	7.5×10^{-3}
V_n, m^3	8.2×10^{-28}	1.9×10^{-27}	4.9×10^{-27}	7.7×10^{-27}	1.4×10^{-26}	3×10^{-26}
IV_n, m^3 s^{-1}	0	0	0	0.9×10^{-3}	2.9×10^{-2}	2×10^{-1}

the length of the throttle is $L = 0.1$ m, then the radius of the section where $S/S_1 = 0.3$ is 0.027 m. Accordingly, the length over which the liquid phase is formed is 0.008 m. It is then easy to estimate the volume content of liquid at the output of the throttle:

$$W \sim \frac{1}{QL}\int_{0.092}^{0.1} IV_n\,dx \sim 10^{-5} \tag{16.134}$$

Such a small volume content of liquid is caused by neglect of the drop growth due to vapor condensation and coagulation. Below, it will be shown that, in spite of the short times of the processes, taking these factors into account increases W by two orders of magnitude.

We now proceed to a solution of the general problem and consider the system of equations, Eqs. (16.113)–(16.124). The following dimensionless variables are introduced:

$$\tilde{p} = \frac{p}{p_1},\quad \tilde{T} = \frac{T}{T_1},\quad \tilde{\rho} = \frac{\rho_G}{\rho_{G1}},\quad \tilde{u} = \frac{u}{u_1},\quad \tilde{S} = \frac{S}{S_1},\quad \xi = \frac{X}{L},\quad \tilde{\Sigma} = \frac{\Sigma}{\Sigma_1},$$

$$\tilde{m}_0 = m_0 V_n(0),\quad \gamma = \frac{u_1^2}{c_{p1}T_1},\quad \delta = \frac{LI(0)}{\rho_{G1}u_1 c_{p1}T},\quad \tilde{V}_n = \frac{V_n}{V_{n1}},$$

$$\varepsilon = \frac{4\pi(3/4\pi)^{1/3} lLD_{12G}M_L}{c_{p1}u_1T_1V_{n1}^{2/3}},\quad I_1 = 1.92\times 10^{27}\frac{\pi M_L^{1/2}\Sigma_1^{7/2}}{\rho_L T_1^2 p_1},\quad \tilde{I} = \frac{I}{I_1},$$

$$\lambda = \frac{\rho_{G1}u_1^2}{p_1},\quad \beta = 1.76\times 10^{16}\left(\frac{M_L}{\rho_L}\right)^2\left(\frac{\Sigma_1}{T_1}\right)^3,\quad \tilde{z} = \frac{z}{z_1},\quad a = \frac{LIV_n}{u_1},$$

$$f = a\frac{\rho_L}{\rho_{G1}},\quad b = \frac{G_1 L}{V^{1/2}u_1},\quad \tilde{c}_p = \frac{c_p}{c_{p1}},\quad C = \frac{4\pi(3/4\pi)^{1/3}LD_{12G}v_{1L}\rho_{G1}}{u_1V_n^{2/3}},$$

$$\tilde{p}_c = \frac{p_{c1}}{p_1},\quad \tilde{T}_c = \frac{T_{c1}}{T_1},\quad d = \frac{4\pi(3/4\pi)^{1/3}D_{12G}}{u_1V_n^{2/3}},$$

$$G_1 = \left(\frac{16\Gamma}{3\pi\rho_{G1}\lambda_0^2}\right)^{1/2},\quad \lambda_0 = \frac{2r_1}{\mathrm{Re}^{3/4}},\quad \mathrm{Re} = \frac{2r_1u_1\rho_{G1}}{\mu_G}.$$

In view of the new variables, the system of equations will become:

$$\begin{aligned}
&\tilde{\rho}\tilde{u}\tilde{S} = 1,\\
&\tilde{\rho}\tilde{u}\frac{d(\tilde{c}_p\tilde{T})}{d\xi} = -\tilde{\gamma}\rho^2\tilde{u}^2\frac{d\tilde{u}}{d\xi} + \delta\tilde{I} + \varepsilon\tilde{\rho}\Delta y\tilde{m}_0^{2/3}m_1^{1/3},\\
&\tilde{I} = \frac{\tilde{\Sigma}_1^{7/2}y_{1\infty}^2}{\tilde{T}_1^2\tilde{p}s}\exp\left(-\frac{\beta\tilde{\Sigma}^3}{\tilde{T}^3\ln^2 s}\right),\\
&\lambda\tilde{\rho}\tilde{u}\frac{d\tilde{u}}{d\xi} = -\frac{d\tilde{p}}{d\xi},\\
&\tilde{p} = \tilde{z}\tilde{\rho}\tilde{T},\\
&\tilde{u}\frac{d\tilde{m}_0}{d\xi} = a\frac{\tilde{I}}{\tilde{V}_n} - b\frac{\tilde{m}_0^{3/2}m_1^{1/3}}{\tilde{\rho}^{1/2}\tilde{S}^{7/8}},\\
&\tilde{u}\frac{d\tilde{m}_1}{d\xi} = a\tilde{I}_n + C\tilde{\rho}\Delta y\tilde{m}_0^{2/3}m_1^{1/3},\\
&\tilde{\rho}\tilde{u}\frac{d(\Delta y)}{d\xi} = -d\Delta y\tilde{m}_0^{2/3}m_1^{1/3}\tilde{\rho} - \tilde{u}\frac{d(\tilde{\rho}y_{1w})}{d\xi} - \tilde{u}\Delta y\frac{d\tilde{\rho}}{d\xi} - f\tilde{I}\tilde{V}_n,\\
&s = 1 + \frac{\Delta y}{y_{1w}},\quad y_{1w} = \frac{\tilde{p}_c}{\tilde{p}}\exp\left(h\left(1 - \frac{\tilde{T}_c}{\tilde{T}}\right)\right),\\
&\tilde{z} = (0.4\lg T_r + 0.73)^{p_r} + 0.1p_r\\
&\tilde{\Sigma} = \frac{(21.04 - 3.74(10^{-6}p_1\tilde{p}) + 0.37(10^{-6}p_1\tilde{p})^2 - 0.015(10^{-6}p_1\tilde{p})^3)}{(21.04 - 3.74(10^{-6}p_1) + 0.37(10^{-6}p_1)^2 - 0.015(10^{-6}p_1))}
\end{aligned}\tag{16.135}$$

Note that in the last equation p_1 has dimension $[p_1] = \text{Pa}$, in contrast to Eq. (16.122), in which pressure has dimension MPa. The factor of 10^{-6} appears for this reason.

The system of equations, Eq. (16.135), describes the variation of the gas and thermodynamic parameters of the flow in a conical confuser with a given change in cross-sectional area with length, $S(x)$, and conditions at the input of:

$$\tilde{p}(0) = \tilde{T}(0) = \tilde{\rho}(0) = \tilde{u}(0) = 1,\quad \tilde{m}_0(0) = m_1(0) = \Delta y(0) = 0, \tag{16.136}$$

Solution of these equations gives values $\tilde{p}$, $\tilde{T}$, $\tilde{\rho}$, $\tilde{u}$, $\tilde{m}_0$, m_1, and s in any section in which the flow velocity has not achieved the velocity of sound, i.e. when the condition $M < 1$ is met, where M is the Mach number.

Equations (16.135) with conditions according to Eq. (16.136) are then solved numerically. As an example, we consider a binary mixture consisting of methane (90%) and propane (10%). The mean molecular mass of this mixture is $M_G = 18.84$ kg kmol^{-1}. Its thermophysical properties can be determined with the aid of methods for multicomponent mixtures [9–11]. In particular, for mixtures, the critical values of temperature and pressure (referred to as pseudocritical values) and the eccentricity factor may be determined from the following expressions:

$$T_{pc} = \sum_i T_{ic}\, y_i, \quad p_{pc} = \sum_i p_{ic}\, y_i, \quad \omega_{pc} = \sum_i \omega_i\, y_i,$$

where T_i, p_i, ω_i are the respective values of the pure components, and y_i is the mole fraction of components in the mixture.

For the calculations, the following values of the parameters are chosen: $T_1 = 293$ °C, $\rho_L = 590$ kg m^{-3}, $c_p = 3.74$ kJ kg^{-1} K^{-1}, $M_L = 44$ kg kmol^{-1}, $D_{12G} = 10^{-7}$ m^2 s^{-1}, $\Gamma = 5 \times 10^{-20}$ J, $L = 0.1$ m, and $r = 0.1$ m. Pressures at the entrance are taken to be: $p_1 = 12, 10, 8,$ and 6 MPa, with velocities u_1 of 5, 10, 30, and 50 m s^{-1}. The radius of the confuser decreases according to a linear law, and so the cross-sectional area was set as:

$$\tilde{S} = (1 - 0.5\xi)^2$$

To compare the influence of various factors on drop growth, the following problems are solved successively:

1. The gas-dynamic problem. Changes only of the gas-dynamic parameters $\tilde{p}$, $\tilde{u}$, $\tilde{\rho}$, and $\tilde{T}$ were considered. Coagulation and phase transitions were ignored. Equations (16.135) at $b = C = \delta = \varepsilon = 0$ were integrated in the subsonic region. In Table 16.5 are listed values of $\tilde{p}$, $\tilde{u}$, $\tilde{\rho}$, $\tilde{T}$, and s in the immediate vicinity of the critical section at $p_1 = 12$ MPa. For comparison with the case of taking into account coagulation and phase transitions, numerical m_0 and volume concentrations of drops m_1 determined from the found values of the gas-dynamic parameters are also shown in this table.
2. Taking into account drop coagulation without regard for phase transitions. In this case, Eqs. (16.135) are integrated at $\varepsilon = C = 0$. In comparison with the first case, only the

Table 16.5

u_1, m s^{-1}	50	30	10	5
ξ_{cr}	1	1.222	1.55	1.682
$\tilde{p}$	0.629	0.629	0.627	0.618
$\tilde{u}$	6.11	10.08	30.25	61.31
$\tilde{\rho}$	0.655	0.655	0.653	0.645
$\tilde{T}$	0.96	0.96	0.96	0.959
$\tilde{m}_0 \times 10^4$	1.64	1.6	1.53	1.5
$m_1 \times 10^5$	2.89	2.85	2.79	2.59
S	1.38	1.38	1.39	1.4

numerical concentration of drops m_0 changes. It becomes less, and so the average drop volume $V_{av} = m_1/m_0$ is increased approximately 10^2-fold compared to the first case (at $p_1 = 12$ MPa and $u_1 = 50$ m s^{-1}).

3. Taking into account phase transitions without regard for drop coagulation. In this case, it is necessary to set $b = 0$ in Eqs. (16.135). Comparison with the first case shows that the volume content of drops increases ($m_1 = 8.08 \times 10^{-3}$ *cf.* 2.59×10^{-5} at $p_1 = 12$ MPa and $u_1 = 5$ m s^{-1}), and that the numerical concentration of drops decreases much more. The average volume of drops ($\tilde{V} = V_{av}/V_{av}(0) = 8 \times 10^9$ *cf.* 200) and the supersaturation of the mixture increase accordingly. Comparison of cases 2 and 3 shows that the basic growth of drops in the throttle is caused by phase transitions (condensation).
4. The general case when drop coagulation and phase transitions are taken into account. Of greatest interest is the dependence of the volume content of the liquid phase m_1 at the output of the throttle upon velocity u_1 and pressure p_1 at the input (Fig. 16.6). At a fixed pressure p_1, the value of m_1 increases with increasing velocity u_1 and this increase is more pronounced the higher the pressure p_1. The increase of pressure at a constant value of u_1 also results in an increase of m_1. Taking into account coagulation and phase transitions considerably increases the average drop volume at the output of the throttle (Fig. 16.7) (10^9–10^{10}-fold). It should be noted that the main change in the gas-dynamic parameters occurs in a very small area near the critical section.

The depicted results are obtained on the assumption that before the throttle the condensed component (vapor) is under saturation conditions, i.e. $\Delta\gamma_1(0) = 0$

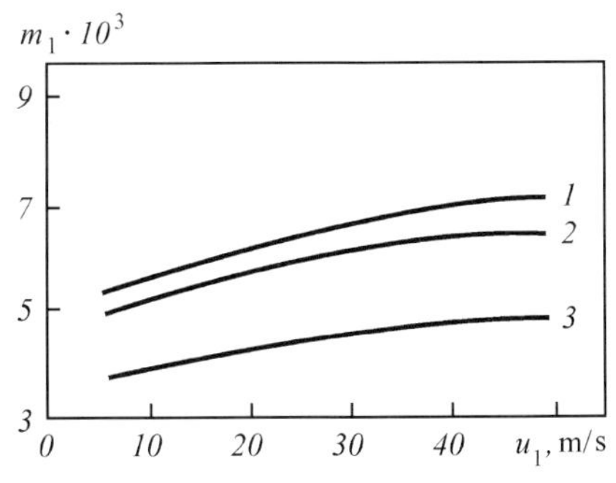

Fig. 16.6 Dependence of m_1 upon u_1 at various values of p_1 (MPa): 1: 12; 2: 10; 3: 6.

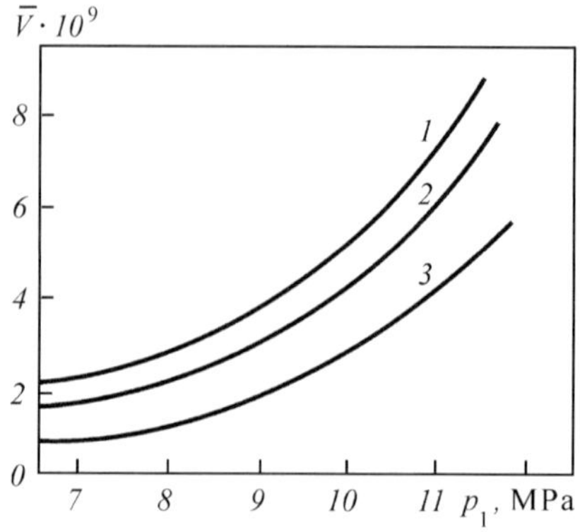

Fig. 16.7 Dependence of $\bar{V}$ upon p_1 for various u_1 (m s^{-1}): 1: 50; 2: 30.5; 3: 10.

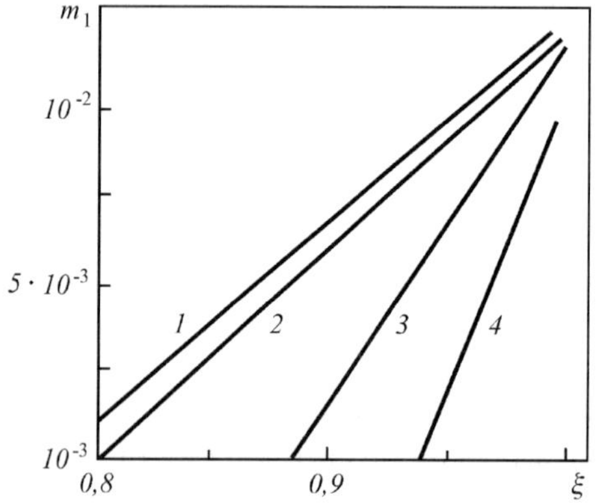

Fig. 16.8 Change of the volumetric contents of the liquid phase m_1 with length of a throttle for the different values of $\Delta y_1(0)$: 1: 0; 2: -10^{-4}; 3: -5×10^{-3}; 4: -10^{-3}.

and $s(0) = 1$. If this component in the gas before the throttle has not yet achieved saturation conditions ($\Delta y_1(0) < 0$), then under certain conditions condensation cannot take place in the throttle (Fig. 16.8). The further away from saturation conditions the vapor is, the greater the lateral displacement of the region in which vapor condensation occurs from the entrance section and, above a certain value $\Delta y_1(0)$, condensation in the throttle ceases.

Thus, calculations show that condensation in a throttle occurs only at certain values of the parameters. If condensation occurs, then the principal contribution to the enlargement of drops stems from condensation growth rather than coagulation. This is so because the characteristic time of condensation is small in comparison with the characteristic time of coagulation. Hence, it follows that behind the throttle the thermodynamic balance will be established very quickly and further agglomeration of drops up to the greatest possible size will occur due to coagulation.

16.5
Fomation of a Liquid Phase in a Heat-Exchanger

In contrast to a throttle, in a heat-exchanger the process of condensation occurs at essentially constant pressure. The gas is cooled during its flow through the tubes of an evaporator. The walls of the tubes are cooled externally by means of a liquid or gaseous refrigerant down to a temperature of -18 °C or below. The primary goal here is to define the amount of the liquid phase condensed in the refrigerating device.

As was shown in the previous section, the formation of a liquid phase in the condensation of heavy hydrocarbons begins under conditions whereby the inequality $s > s_{cr}$ is met. Solving this inequality with respect to temperature, it is possible to determine the temperature, T_{cr}, at which vapor condensation begins. Since the temperature at the entrance is higher than T_{cr}, in the flow through the heat-exchanger the temperature of the gas decreases and in some section becomes equal to T_{cr}. Beyond this section, very small droplets appear in the gas, which then increase in size due to phase transitions at their surfaces and coagulation with other droplets. Extensive nucleation and vapor condensation results in calorification of the heat of condensation, which can slightly increase the temperature of the gas. Since the volume content of the liquid phase is small, and the intensity of gas cooling over the length of the heat-exchanger is sufficiently great, to a first approximation the heat of condensation can be neglected. Therefore, the temperature distribution over the pipe length can be determined independently, and then from the known temperature distribution one may determine the quantity of liquid phase formed.

Suppose that the gas at the entrance has temperature T_0, the temperature of the wall is T_w, and the velocity of the gas is U. Consider first the case when gas cooling occurs due to heat conductivity. The temperature change is then described by the equation:

$$U\frac{\partial T}{\partial x} = \frac{\lambda_G}{\rho_G c_p}\frac{1}{r}\frac{\partial}{\partial r}\left(r\frac{\partial T}{\partial r}\right), \tag{16.137}$$

provided that

$$T(0,r) = T_0, \quad T(x,r_c) = T_w, \quad \left(\frac{\partial T}{\partial r}\right)_{r=0} = 0, \tag{16.138}$$

where λ_G is the thermal conductivity coefficient of the gas, c_p is the isobaric heat capacity of the gas, r is the current radius, r_c is the radius of the tube, the x-axis is directed along the tube center line, and $x = 0$ corresponds to the input section.

The solution of Eq. (16.137) with the conditions of Eq. (16.138) is:

$$\Theta = \frac{T - T_0}{T_w - T_0} = 1 - 2\sum_{k=1}^{\infty} \frac{J_0(p_k r/r_c)}{p_k J_1(p_k)} e^{-p_k^2 \tau}, \tag{16.139}$$

where $\tau = \lambda_G x / r_c^2 U \rho_G c_p$; J_0 and J_1 are Bessel functions; p_k are roots of the equation $J_0(p_k) = 0$, where $p_1 = 2.4$, $p_2 = 5.52$, $p_3 = 8.65$, $p_4 = 11.79$, etc. The corresponding values of $J_1(p_k)$ are equal to $J_1(p_1) = 0.519$, $J_1(p_2) = -0.34$, $J_1(p_3) = 0.27$, $J_1(p_4) = -0.23$, etc.

Next, we introduce temperature averages over a tube cross-section:

$$\bar{T} = 2r_c^{-2} \int_0^{r_c} Tr\, dr, \tag{16.140}$$

Substituting Eq. (16.139) into Eq. (16.140), one obtains:

$$\frac{\bar{T}}{T_0} = \frac{T_w}{T_0} + 4\left(1 - \frac{T_w}{T_0}\right) \sum_{k=1}^{\infty} p_k^{-2} e^{-p_k^2 \tau},$$

or

$$\Theta = \frac{T - T_0}{T_w - T_0} = 4\sum_{k=1}^{\infty} p_k^{-2} e^{-p_k^2 \tau} \tag{16.141}$$

The series on the right-hand side of Eq. (16.141) converges quickly at $\tau > 0$, and therefore it is possible to retain only the first term:

$$\frac{\bar{T}}{T_0} \approx 0.694\left(1 - \frac{T_w}{T_0}\right) e^{-5.76\tau} \tag{16.142}$$

Since the tubes through which the cooled gas moves are sufficiently narrow ($2r_c \ll L$), for temperature one may use Eq. (16.141) or Eq. (16.142).

The dependence $\Theta_1(\tau)$ is shown in Fig. 16.9. Using Eq. (16.142), it is possible to determine some design parameters of the heat-exchanger. The most rational design of the heat-exchanger (see Part I) is a shell-and-tube heat-exchanger, consisting of a bundle of N parallel tubes with radius r_c and length L, located in a casing of radius R_k. If the radius of the supply pipe to the heat-exchanger is R_t, then from the conservation condition of the gas flow rate, under the assumption that the gas density is constant, it follows that:

$$R_t^2 U_t = N r_c^2 U = \frac{Q_G}{\pi}, \tag{16.143}$$

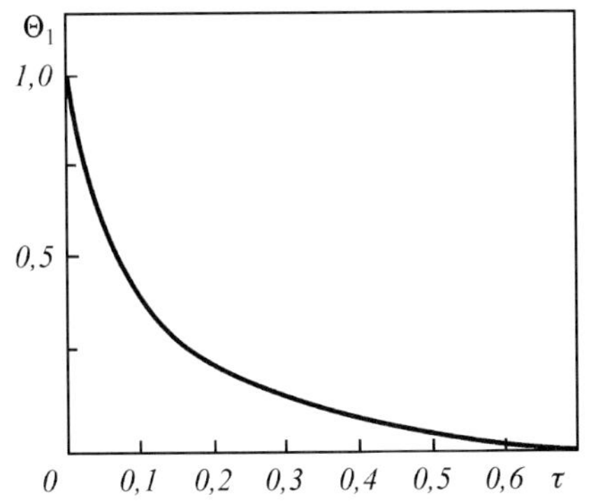

Fig. 16.9 Dependence of temperature Θ_1 upon τ.

where U_t and U are the gas velocity in the supply pipe and in the tube of the heat-exchanger, respectively, and Q_G is the volume flow rate of gas at pressure p and temperature T. The gas flow rate Q_n is commonly given at normal conditions. The connection between Q_G and Q_n is evident:

$$Q_G = \frac{zTp_n Q_n}{T_n p}, \tag{16.144}$$

where T_n and p_n are the normal temperature and pressure ($T_n = 293$ K, $p_n = 0.1$ MPa), and z is the gas-compressibility factor.

Consider the parameter t contained in Eq. (16.141). Using Eq. (16.143), it can be written as:

$$\tau = \frac{\pi \lambda_G N x}{Q_G \rho_G c_p}. \tag{16.145}$$

Now, Eqs. (16.144) and (16.145) allow the determination of the number of tubes in the heat-exchanger:

$$N = \frac{\rho_G c_p \tau z T p_n Q_n}{\pi \lambda_G L T_n p}. \tag{16.146}$$

The key parameters of a heat-exchanger are the number of tubes in a bundle N, their radius r_c and length L. If L and r_c are specified, then one can determine N for a given pressure p_0 and temperature T_0 at the entrance, temperature of the wall T_w, gas temperature at the exit T_1, gas flow rate Q_n, and radius of the supply pipe. Preset values allow the determination of the value of τ from Fig. 16.9 and then the value of N from Eq. (16.146).

Consider an example. Suppose that it is necessary to cool a gas from $T_0 = 293$ K down to $T_1 = 263$ K. The wall temperature of the cooled tubes is maintained at

$T_w = 253$ K, the pressure at the entrance is $p_0 = 10$ MPa, and the gas flow rate is $Q_n = 10^4\ m^3\ day^{-1} = 0.121\ m^3\ s^{-1}$. Other parameters are $\rho_G = 100\ kg\ m^{-3}$, $c_p = 3.26\ kJ\ kg^{-1}\ K^{-1}$, $\lambda_G = 1.26 \times 10^{-4}\ W\ m^{-1}\ K^{-1}$, and $L = 6$ m. From Fig. 16.9 one obtains a value of $\tau = 0.175$ and from Eq. (16.146) it follows that $N = 210$.

The described method of calculation is suitable for flows with relatively low velocities, where there is heat transfer due to conduction. At high gas velocities, the flow in the cooling tubes is turbulent, and so heat transfer occurs basically as a result of extensive cross-mixing of the flow, which averages out the temperature across the cross-section of the tube.

Let us consider a gas flow in a cylindrical tube of radius r_c. The temperature at the entrance is T_0, and the wall temperature is maintained constant such that $T_w < T_0$. We consider a flow velocity U and a temperature T averaged over the tube cross-section, and accept that the heat loss through the pipe wall and into the environment is determined by the formula:

$$Q_w = k_T(T - T_w)St, \tag{16.147}$$

where k_T is the heat-transfer coefficient of the tube wall, S is the area of the tube surface, and t is time. The value of k_T is calculated by:

$$\frac{1}{k_T} = \frac{1}{\alpha_1} + \frac{1}{\alpha_2} + \frac{h}{\lambda_w}, \tag{16.148}$$

where α_1 and α_2 are coefficients of heat emission from the cooled gas to the wall and from the wall to the environment, respectively, h is the wall thickness, and λ_w is the coefficient of heat conductivity of the wall. Generally speaking, k_T depends on the temperature T, but taking this dependence into account is difficult. Besides, this dependence will not strongly influence the average temperature of the gas. Therefore, k_T can be taken to be constant.

We set the x-axis along the center line of the tube and express the heat balance in the cylindrical volume of radius r_c and length dx. Owing to convective heat transfer, in unit time a quantity of heat equal to $\pi r_c^2 \rho_G c_p UT$ enters this volume, but a quantity of heat equal to $\pi r_c^2 \rho_G c_p U(T + dT)$ leaves. Hence, the influx of heat into the considered volume in unit time is equal to $\pi r_c^2 \rho_G c_p U\, dT$. On the other hand, a quantity of heat, which, according to Eq. (16.147) is equal to $2\pi r_c k_T(T - T_w)\, dx$, will be lost from the considered volume through the pipe walls in unit time. As a result of nucleation and further condensation growth of drops, an energy E is generated, given by:

$$E = \left(V_n \rho_L l I + M_L l \int_0^\infty 4\pi R^2 J_{1w} n(V, x)\, dV \right) 2\pi r_c\, dx \tag{16.149}$$

Equating heat fluxes, we obtain:

$$r_c \rho_G c_p U\, dT = -2k_T(T - T_w)\, dx + E.$$

Considering that $V_n = 32\pi\Sigma^3/3p^3$, and invoking the algebraic expression for the flux of the condensed component J_{1w} (see Section 16.1), one obtains:

$$\rho_G c_p U \frac{dT}{dx} = -\frac{2k_T(T - T_w)}{r_c} + \frac{32\pi\Sigma^3 l\rho_L I}{3p^3} + 4\pi\left(\frac{3}{4\pi}\right)^{1/3} M_L l\rho_L D_{12G}\Delta y m_0^{2/3} m_1^{1/3}, \tag{16.150}$$

If we neglect the heat of condensation, then the solution of Eq. (16.150) under the condition $T(0) = T$ is

$$\Theta_1 = \frac{T - T_w}{T_0 - T_w} = \exp(-\beta\xi), \tag{16.151}$$

where $\beta\xi = 2\pi k_T r_c Nx/c_p\rho_G Q_G$.

Because the surface area of the heat-exchanger is $S = 2\pi r_c LN$, Eq. (16.151) can be written as:

$$\Theta_1 = \exp\left(-\frac{k_T Sx}{\rho_G c_p Q_G L}\right), \tag{16.152}$$

Having set the temperature at the exit of the heat exchanger as T_1, from Eq. (16.152) we can determine the area of heat exchange:

$$S = \frac{\rho_G c_p Q_G}{k_T} \ln\left(\frac{T_0 - T_w}{T_1 - T_w}\right) \tag{16.153}$$

Consider now the flow of a binary gas/vapor mixture in a heat exchanger with regard to condensation and coagulation. The relevant system of equations is similar to Eqs. (16.135). The main difference lies in the constancy of the tube cross-section, i.e. $\tilde{S} = 1$. Thus, we have:

$$\tilde{\rho}\tilde{u} = 1,$$

$$\tilde{\rho}\tilde{u}\frac{d(\tilde{c}_p\tilde{T})}{d\xi} = -\gamma\tilde{\rho}^2\tilde{u}^2\frac{d\tilde{u}}{d\xi} + \delta\tilde{I} + \varepsilon\tilde{\rho}\Delta y_0\tilde{m}_0^{2/3}m_1^{1/3} - \beta_1(\tilde{T} - \tilde{T}_w),$$

$$\tilde{I} = \frac{\tilde{\Sigma}_1^{7/2}y_{1\infty}^2}{\tilde{T}_1^2\tilde{p}s}\exp\left(-\frac{\beta\tilde{\Sigma}^3}{\tilde{T}^3\ln^2 s}\right),$$

$$\lambda\tilde{\rho}\tilde{u}\frac{d\tilde{u}}{d\xi} = -\frac{d\tilde{p}}{d\xi},$$

$$\tilde{p} = \tilde{z}\tilde{\rho}\tilde{T},$$

$$\tilde{u}\frac{d\tilde{m}_0}{d\xi} = a\frac{\tilde{I}}{\tilde{V}_n} - b\frac{\tilde{m}_0^{3/2}m_1^{1/3}}{\tilde{\rho}^{1/2}}, \tag{16.154}$$

$$\tilde{u}\frac{d\tilde{m}_1}{d\xi} = a\tilde{I}_n + C\tilde{p}\Delta y\tilde{m}_0^{2/3}m_1^{1/3},$$

$$\tilde{p}\tilde{u}\frac{d(\Delta y)}{d\xi} = -d\Delta y\tilde{m}_0^{2/3}m_1^{1/3}\tilde{p} - \tilde{u}\frac{d(\tilde{p}y_{1w})}{d\xi} - \tilde{u}\Delta y\frac{d\tilde{p}}{d\xi} - f\tilde{I}\tilde{V}_n,$$

$$s = 1 + \frac{\Delta y}{y_{1w}}, \quad y_{1w} = \frac{\tilde{p}_c}{\tilde{p}}\exp\left(h\left(1 - \frac{\tilde{T}_c}{T}\right)\right),$$

$$\tilde{z} = (0.4\lg T_r + 0.73)^{p_r} + 0.1p_r,$$

$$\tilde{\Sigma} = \frac{(21.04 - 3.74(10^{-6}p_1\tilde{p}) + 0.37(10^{-6}p_1\tilde{p})^2 - 0.015(10^{-6}p_1\tilde{p})^3)}{(21.04 - 3.74(10^{-6}p_1) + 0.37(10^{-6}p_1)^2 - 0.015(10^{-6}p_1))}$$

Besides the dimensionless parameters entered in Eq. (16.135), here are added:

$$\tilde{T}_w = \frac{T_w}{T_1}, \quad \beta_1 = \frac{2\pi k_T r_c NL}{c_p \rho_{G1} U_1 Q_G}.$$

The system of equations, Eqs. (16.154), is solved under the initial conditions:

$$\tilde{T}(0) = \tilde{p}(0) = \tilde{u}(0) = 1, \quad \tilde{m}_0(0) = m_1(0) = \Delta y(0) = 0. \tag{16.155}$$

Let us consider an example of a numerical solution of Eqs. (16.154) under the conditions of Eq. (16.155) for a shell-and-tube heat-exchanger consisting of 281 tubes with diameter $2r_c = 0.025$ m and length $L = 12$ m. The heat-transfer coefficient is $k_T = 0.42$ kW m^{-2} K^{-1}. At the entrance of the heat-exchanger, the incoming binary gas mixture contains 90% methane and 10% propane. The gas flow rate is $Q_G = 0.7$ m^3 s^{-1} at a pressure of $p_1 = 5$ MPa and temperature $T_1 = 293$ K. Additionally, $T_w = 253$ K, $c_p = 2.77$ kJ kg^{-1} K^{-1}, $\rho_{G1} = 40$ kg m^{-3}, and $U_1 = 9$ m s^{-1}. Figures 16.10 and 16.11 show the changes of volume content of the liquid phase, m_1, and the deviation of the mole concentration of the condensing com-

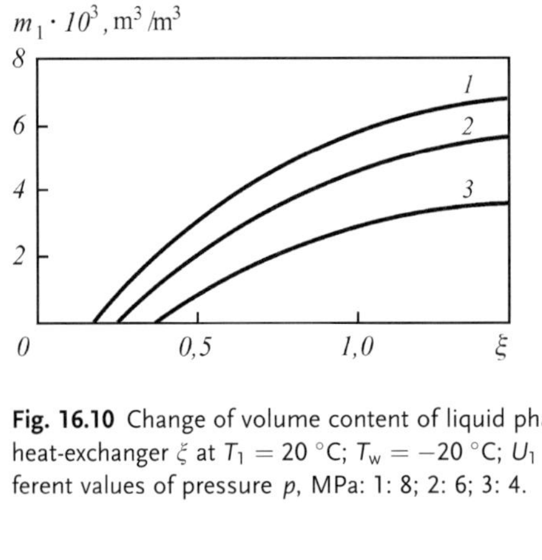

Fig. 16.10 Change of volume content of liquid phase m_1 with length of heat-exchanger ξ at $T_1 = 20$ °C; $T_w = -20$ °C; $U_1 = 9$ m s^{-1}; for different values of pressure p, MPa: 1: 8; 2: 6; 3: 4.

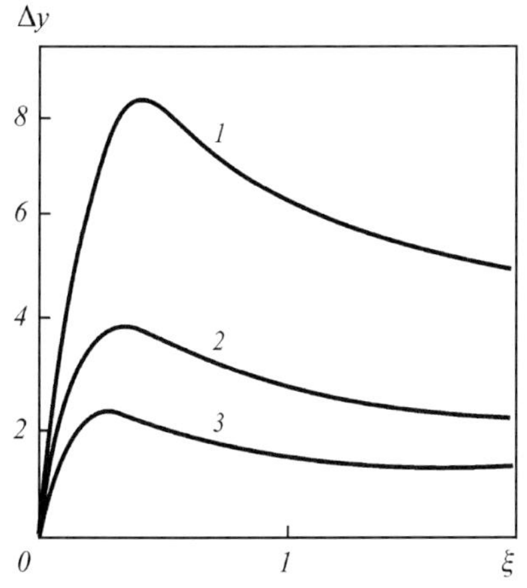

Fig. 16.11 Change of deviation of mole concentration of condensing component in gas phase Δy with length of heat-exchanger ξ at $T_1 = 20\ °C$; $T_w = -20\ °C$; $U_1 = 9\ m\ s^{-1}$ for different values p, MPa: 1: 4; 2: 6; 3: 8.

ponent (propane), Δy, from the equilibrium values with the length of heat-exchanger for various values of pressure, p. Nucleation of drops begins at some distance from the entrance. The higher the gas pressure, the shorter this distance. A pressure increase from 4 to 8 MPa increases the volume of the liquid phase almost twofold. A reduction of the temperature in the heat-exchanger leads first to prompt supersaturation of the mixtures (increased Δy), and subsequently there is extensive vapor condensation, which reduces Δy.

In conclusion, it should be noted that the considered method is limited to the condition of constant wall temperature of the heat-exchanger tubes. Actually, the temperature T_w can change with the length of the tubes. For its definition, it is necessary to solve the external thermal problem, i.e. to determine the temperature distribution in the intertubular space. This avoids the main difficulties. The solution procedure can become a lot more complicated, especially in cases where the flow of the cooling agent is in the opposite direction to the direction of the cooled gas flow.

17
Surface Tension

17.1
Physics of Surface Tension

The phenomenon of surface tension reveals itself in many processes that one encounters not only in engineering, but also in everyday life. It is enough to mention the following examples: the evolution of soap bubbles as they are formed, then go up, and finally burst; the rise of liquid in a capillary tube above the level of liquid in the tank into which the tube was plunged; disintegration of liquid into drops when it flows out of a jet through a thin orifice nozzle, or atomizer; the printing process in jet printers; a thin liquid layer sticking to the surface of a body when the body is removed from liquid; the behavior of a drop on a flat solid surface – it can remain spherical or spread over the surface depending on the forces of interaction between liquid, solid surface, and air. Of particular importance if the role of surface tension in the formation of drops during disintegration of liquid jets, and in the splitting of large drops in a flow of gas-liquid mixture.

Liquid consists of molecules in the state of motion that are kept at relatively small distances from each other by Van der Waals's forces of molecular attraction. Since liquid can be treated as a continuous medium (this is the basic hypothesis of continuum), a small element of this medium can be characterized by the presence of some average molecular field. In this field, the net attractive force acting on an arbitrary molecule in a small element of continuous medium is zero on the average (forces in all directions balance each other).

This assumption is no longer true at the interface, for example on the boundary between liquid and gas, because molecular motion in a liquid is more constrained than in a gas, where interaction between molecules is relatively weak. The interaction of molecules in a liquid makes it harder for them to escape from it through the interface boundary (evaporation). Therefore, molecules located at the surface of the liquid are acted upon by forces directed along the surface or inward (Fig. 17.1).

As a result, the interface experiences a tensile force and behaves as membrane seeking to reduce its size. It is necessary to point out that representation of the interface as a smooth surface of the meniscus type where the concentration of molecules suffers a discontinuity is an idealization. Actually, this surface has

Separation of Multiphase, Multicomponent Systems. E. G. Sinaiski and E. J. Lapiga

ISBN: 978-3-527-40612-8

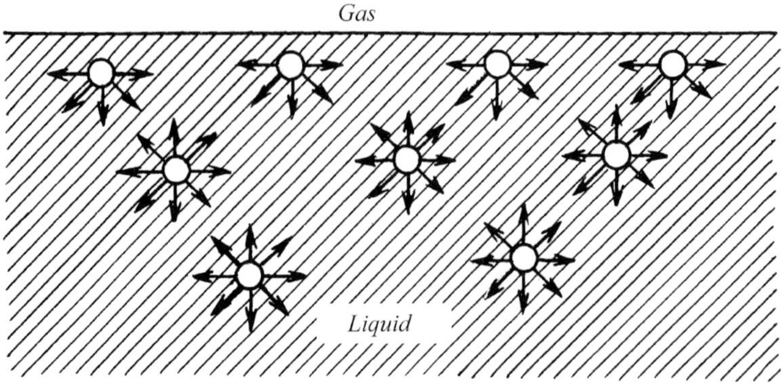

Fig. 17.1 Gas-liquid interface.

thickness of about 100 nanometers, and the concentration of molecules in it changes continuously from the value characteristic of the liquid phase to the value characteristic of the gas phase.

The behavior of the interface can be described using energy approach. Since molecules of liquid at the interface experience smaller attraction from adjacent molecules than the molecules in the bulk of the liquid, the energy of attraction acting on one molecule of liquid at the interface is less than the same energy in the bulk of the liquid. Because of this, the energy of a surface molecule constitutes only a part of the energy typical of internal molecules. As far as the free energy of the system tends to the minimum, the surface area of the liquid tends to decrease.

Designate by Σ the force acting on a unit length and aspiring to reduce the surface area. Then, introducing the free energy F [12],

$$F = E - TS, \tag{17.1}$$

one gets the following expression for Σ:

$$\Sigma = \frac{\partial F}{\partial s}, \tag{17.2}$$

where s is the surface area.

It follows from (17.2) that $\Sigma > 0$ when F decreases together with s. The quantity Σ is called the coefficient of surface tension. The dimensionality of Σ is N/m. Table 17.1 provides values of S for some pure liquids in equilibrium with their vapor at normal temperature.

The coefficient of surface tension decreases the with an increase of temperature, so

$$\frac{\partial \Sigma}{\partial T} < 0. \tag{17.3}$$

Table 17.1

Material	Σ (mN/m)
Water	72.88
Nitromethane	32.66
Gasoline	28.88
Methanol	22.50
Mercury	486.5

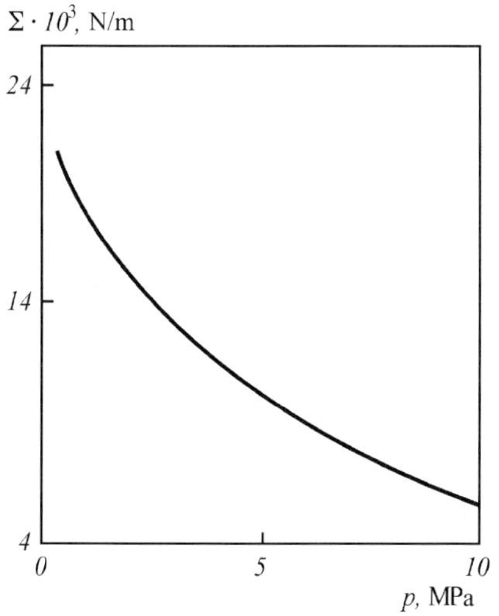

Fig. 17.2 Dependence of surface tension coefficient on pressure for natural gas condensate.

The diminution of Σ with an increase of T occurs practically linearly and is approximated by the number $-0.1\ \mathrm{N}/{}^0\mathrm{K}$. The surface tension of natural hydrocarbon liquids depends both on temperature and pressure. On Fig. 17.2, a typical dependence of Σ upon pressure is shown for natural gas condensate.

It is known that adsorption of a surface-active substance (surfactant) on the interface results in a formation at this surface of an oriented monolayer that lowers the surface tension. Typical water solutions of surfactants contain organic molecules with long hydrocarbon tails and polar heads [13]. Hydrocarbons are practically insoluble in water, and water is a highly polar liquid. Figure 17.3 shows how molecules of ideal surfactant are adsorbed on the water surface. The polar heads of molecules penetrate into water, while hydrocarbon tails remain in the gaseous medium. Formation of the monolayer requires a relatively small number of molecules.

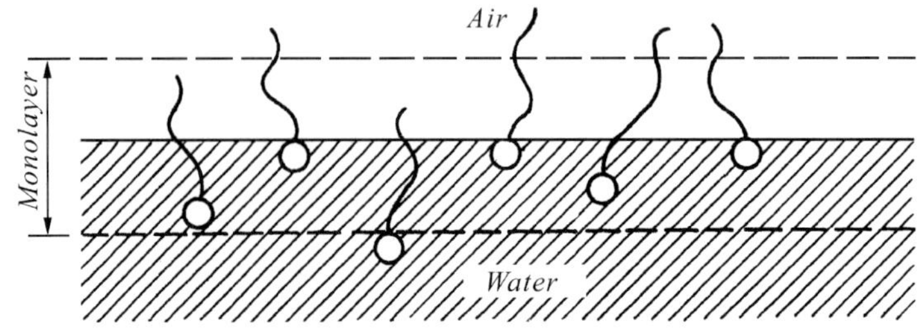

Fig. 17.3 Adsorption of surfactant on an interface.

Let's introduce the concept of surface concentration of surfactant (measured in g mole/m^2) [2]. If the environment is a binary solution with the volume molar concentration C of surfactant dissolved in it, then the quantities Γ and C are related by the Gibbs equation (aka the Gibbs isotherm):

$$\Gamma = -\frac{C}{AT}\frac{\partial \Sigma}{\partial C}, \tag{17.4}$$

where A is the gas constant.

We conclude from (17.4 that an increase of surfactant concentration C in a solution causes reduction of the surface tension, provided that the surfactant is adsorbed at the surface, i.e. $\Gamma > 0$. If the concentration of surfactant in the solution is small, then molecules of the surfactant behave independently. The increase in C leads to their aggregation and formation of micelles. Micelles are formations whose form can be approximately taken as spherical. They contain from 50 to 100 molecules.

The polar heads of molecules are surrounded by double layers. They are in the water while hydrocarbonic "tails" are amassed in the area where there is no water. Micelles are formed when surfactant concentration C exceeds critical value C_{cr}, known as the concentration of micelle formation. The increase of C over C_{cr} does not change S, it results only in the additional formation of micelles [14].

Let us now proceed to the kinetics of the adsorption process. The characteristic time necessary for the establishment of equilibrium value of Σ at the liquid surface is estimated by $t \sim l^2/D$, where D is the diffusion coefficient of surfactant molecules, l is the molecular size. This is the typical time during which surface molecules are swapped with their neighboring molecules in the liquid. For characteristic values $D \sim 10^{-9}$ m^2/s and $l \sim 0.3$ nm (the diameter of water molecule), it equals $t \sim 10^{-10}$ s. Hence, the equilibrium value of Σ is established practically instantaneously.

Another characteristic time is the average time spent by a molecule at the surface when the liquid is at equilibrium with its vapor, i.e. under the condition that the number of evaporating molecules is equal to the number of condensing molecules. For water vapor, the diffusion coefficient is $D \sim 10^{-6}$ m^2/s, and the thickness of vapor layer is $l \sim 0.1$ μm. It is then possible to estimate the frequency of collision of vapor molecules with the surface, $D/l^2 \sim 10^8\ \text{s}^{-1}$. If the fraction of molecules capable of being condensed is 0.1, then the characteristic lifetime of a molecule at the interface is $\sim 10^{-7}$ s. This is more than the characteristic time for the establishment of Σ, but it is rather small nevertheless.

The reasoning carried out for a liquid – gas interface can be repeated for the interface, separating two non-mixing liquids. The value of surface tension for the liquid – liquid interface lies between Σ_1 and Σ_2, that is, between the surface tension values of either liquid.

Up till now we considered only flat interfaces. The surface tension aspires to bend the surface. As a result, there is a negative pressure jump as we go through the interface from the side where the centre of curvature is located. The expression for the pressure jump is called the Laplace-Young equation:

$$\Delta p = \Sigma\left(\frac{1}{R_1} + \frac{1}{R_2}\right), \tag{17.5}$$

where R_1 and R_2 are the main curvature radii of the surface.

In the specific case of spherical drops or bubbles, $R_1 = R_2 = R$, and

$$p_{cap} = \Delta p = p_i - p_e = \frac{2\Sigma}{R}, \tag{17.6}$$

where p_i, p_e are the internal and external pressures, and R is the drop's radius.

The form of Eq. (17.6) reflects the fact that at the onset of thermodynamic balance, the surface energy should be minimal, and for a given volume of the liquid, the minimal surface area corresponds to the area of a sphere. It follows from (17.6) that when $R \to \infty$ (the flat surface), $\Delta p \to 0$.

Go on now to the behavior of liquid on a solid surface. Unlike molecules of liquid or vapor, molecules of the solid surface are motionless. If we replace the area occupied by gas on Fig. 17.1 with a solid surface (for example, glass), then liquid molecules at the interface experience greater attraction from molecules of the solid body than from the molecules in the bulk of the liquid. Thus, liquid molecules adhere to the solid surface. The solid surface, as opposed to the liquid one, does not happen to be absolutely homogeneous (clean). The presence of surface irregularities and various impurities influences the surface tension of liquid. Therefore, attraction forces from the molecules of a real solid body can be smaller than might be expected for the ideal smooth surface.

A liquid drop put on a solid surface interacts not only with the surface of the solid body, but also with the surrounding gas (Fig. 17.4). Two cases are possible: the drop spreads out freely over the surface of solid body, or it retains the form of a drop, whose surface forms a finite angle θ with the solid surface at points of contact, as shown in Fig. 17.4. The force Σ, applied to the free surface at the points of contact, and acting on the surface, is directed tangentially to the interface. We can then identify several forces acting on the contour that is formed by the points of contact between the solid body and the drop: force Σ_{SG} between mol-

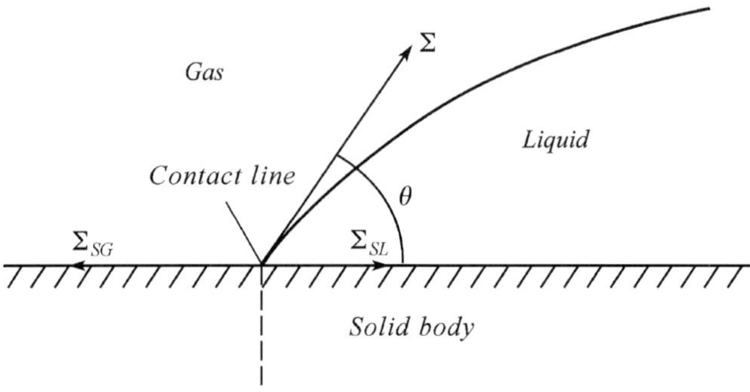

Fig. 17.4 Liquid drop on a solid surface.

ecules of solid body and the gas; force Σ_{LG} between molecules of solid body and the liquid, and the force $\Sigma\cos\theta$, which represents a projection of liquid-gas surface tension force on the solid body plane. From the equilibrium condition for the contact surface, there follows Young's equation:

$$\Sigma\cos\theta = \Sigma_{SG} - \Sigma_{SL}, \tag{17.7}$$

or

$$k = \frac{\Sigma_{SG} - \Sigma_{SL}}{\Sigma} = \cos\theta. \tag{17.8}$$

The value k is called the coefficient of moistening. It is obvious that $-1 \le k \le 1$. If the contact angle $\theta = 0$, then $k = 1$ and the liquid drop completely spreads over the solid surface. In this case the surface is called completely wettable. For $0 < k < 1$, the contact angle is $0 < \theta < \pi/2$. In this case the surface of solid body is called wettable, or more precisely, partially wettable. Another limiting case corresponds to $k = -1$, i.e. $\theta = \pi$. Then the solid surface is called completely non-wettable. It follows from (17.8) that this case corresponds to large values of Σ_{SL} that are close to Σ. At $-1 < k < 0$, the angle is $\pi/2 < \theta < \pi$. In this case the surface is called non-wettable. A good example is a drop of mercury on a glass, for which the contact angle is $\theta \approx 140°$.

Now rewrite (17.8) in the form

$$k = 1 + \frac{S}{\Sigma}, \quad S = \Sigma_{SG} - \Sigma_{SL} - \Sigma. \tag{17.9}$$

The parameter S is called the coefficient of spreading. If $S > 0$, then $k > 1$. In this case the line of contact cannot be in an equilibrium state.

The equilibrium condition for the line of contact (17.7) is applicable if the line of contact is immobile. In problems that involve motion of liquid over a solid surface it is necessary to consider a movable line of contact [15]. In such problems, the concept of a dynamic line of contact or angle of contact is introduced.

The condition for the line of contact is a boundary condition of the corresponding hydrodynamic problem and as such, it should take into account the force of viscous resistance in addition to the forces of surface tension.

17.2 Capillary Motion

The phenomenon of capillarity is defined as the property of a liquid to form a finite or zero angle at the contact with a solid body. In practice, the capillarity phenomenon is manifested as the ability of a liquid to rise or drop in very thin capillary tubes or in a porous medium characterized by its microcapillary internal structure. In this context, the capillary motion is understood to be a motion induced by the action of surface forces. A most basic demonstration of the phenomenon of capillarity is the rise of liquid in a vertical capillary tube immersed with one end in a tank with liquid, with the other end opened into the atmosphere. In this experiment, the liquid can rise to a height greater than the height of the liquid in the tank. Other examples are the spreading of drops on blotting paper and the moistening flow of liquid over a solid surface. In these cases the surface tension is the driving force of the flow.

Surface tension cannot be constant over the entire interface surface because the surface tension coefficient depends on the concentration of surfactant adsorbed at the surface, and on the temperature and the electric charge of the surface. Any change of Σ results in a corresponding change in the balance of forces acting on the interface and therefore it may cause a liquid flow. Indeed, the boundary conditions at the interface between two immiscible liquids are the equalities of normal and tangential stresses. The continuity condition for normal stresses includes pressures, which at the curved surface are connected by the Laplace-Young equation (17.5). If the surface tension has a gradient tangential to the interface, then the continuity of tangential stresses demands that viscous stresses should change along the surface, and thus, the velocity field in the liquid phase and the form of the interface should change accordingly.

The subject of the present section is the capillary motion in a vertical narrow tube [2]. The moistening flow and the flow due to a surface gradient of surface tension will be considered, respectively, in Sections 17.3 and 17.5.

Consider the rise of a liquid in a vertical capillary cylindrical tube with radius a, one end of which is immersed into a tank with liquid, and the other end is opened into the atmosphere (Fig. 17.5).

In a narrow tube with a circular cross section, the shape of the liquid meniscus approximates a spherical segment with constant radius of curvature $R = a/\cos\theta$, where θ is a static contact angle. The deviation of the meniscus shape from a sphere can be caused by the influence of gravity. The ratio of hydrostatic gravitational force to the surface tension force is characterized by a dimensionless pa-

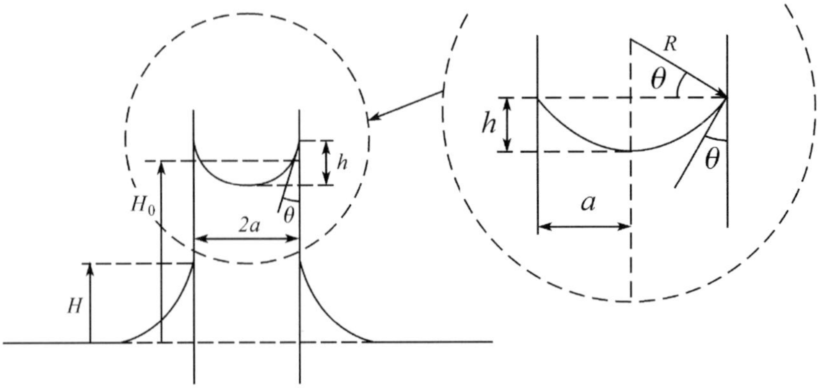

Fig. 17.5 The rise of liquid in a vertical capillary.

rameter called the Bond number:

$$\mathrm{Bo} = \frac{\text{force of gravity}}{\text{surface tension force}} = \frac{\rho g L^2}{\Sigma}, \tag{17.10}$$

where L is the characteristic linear size.

At $\mathrm{Bo} \gg 1$, the surface tension force can be neglected in comparison with gravity. At $\mathrm{Bo} \ll 1$, on the other hand, gravity is immaterial.

Looking at the region occupied by the meniscus, one should take $L = h$. Then the requirement that gravity should not cause a significant deformation of the spherical shape of the meniscus leads to $\mathrm{Bo} \ll 1$, i.e. $h^2 \rho g \ll \Sigma$, from which we get

$$h \ll \left(\frac{\Sigma}{\rho g}\right)^2 = \Delta_k \tag{17.11}$$

The quantity Δ_k has the dimensionality of length and is called the capillary length.

Consider now the equilibrium condition for a liquid column with height H_o. From the Laplace-Young condition (17.5) it follows that the pressure drop at the meniscus equals $\Delta p = p_a - p_L = 2\Sigma \cos\theta/a$, where p_L is the average pressure in the liquid at the interface, and p_a is the atmospheric pressure. The hydrostatic pressure at the tube entrance (in the tank) is equal to $p_L + \rho g H_0 \approx p_a$. It follows from these two expressions that

$$\rho g H_0 = \frac{2\Sigma \cos\theta}{a}. \tag{17.12}$$

This gives us the equilibrium height of the liquid column:

$$H_0 = 2\left(\frac{\Sigma}{\rho g}\right)\frac{\cos\theta}{a} = 2\frac{\Delta_k^2}{a}\cos\theta. \tag{17.13}$$

Depending on the value of the contact angle, the height of column varies in the interval

$$-2\frac{\Delta_k^2}{a} < H_0 < 2\frac{\Delta_k^2}{a}.$$

In particular, for a column of mercury, $\theta \approx 140^\circ$, and the column does not rise, but drops below the level of mercury in the tank.

It follows from (17.13) that the smaller the radius of the capillary, the higher (or lower) the liquid can rise (or descend). Thus, for water, $\Sigma = 73$ mN/m, and in glass capillary with $a = 0.1$ mm we have $H = 0.15$ m. On the other hand, measuring the height of a raising (or descending) liquid column in the capillary, it is possible to determine the value of the surface tension coefficient with high accuracy.

Let us now determine how fast the liquid will rise. We assume that the flow has a Poiseuille velocity profile (see (6.27)) with $u_{\max} = 2U$ and $h = \dfrac{a}{\sqrt{2}}$. Then

$$u = \frac{dH}{dt} = \frac{a^2}{8\mu}\frac{\Delta p}{H}, \tag{17.14}$$

where $H \le H_0$ is the distance between the surface of the liquid in the tank and the free surface in the capillary at some moment t. It is well known that a Poiseuille velocity profile is established in a developed flow after $t \gg a^2/\nu$, so our assumption is only an approximation to the actual velocity profile. The quantity Δp appearing in (17.14) can be considered as the difference between the hydrostatic pressures at the final height H_0 and at the current height H. Then, taking into account (17.12), we have

$$\Delta p = \rho g H_0 - \rho g H = \frac{2\Sigma\cos\theta}{a} - \rho g H. \tag{17.15}$$

and, substituting (17.15) in (17.14), we obtain

$$\frac{\mu}{\Sigma}\frac{dH}{dt} = \frac{1}{8}\left(2\frac{a}{H}\cos\theta - \frac{\rho g a^2}{\Sigma}\right). \tag{17.16}$$

Here the parameter $\rho g a^2/\Sigma = \mathrm{Bo}$ is the Bond number for the given capillary radius. The other dimensionless parameter follows from the form of the left-hand side of (17.16). Designate the characteristic velocity by U. Then it is possible to introduce the capillary number,

$$\mathrm{Ca} = \frac{\text{viscous force}}{\text{force of the surface tension}} = \frac{\mu U}{\Sigma}. \tag{17.17}$$

The solution of Eq. (17.16) under the condition of $H = 0$ at $t = 0$ has the form

$$t = t_c\left(\ln\left(\frac{1}{1 - H/H_0}\right) - \frac{H}{H_0}\right), \tag{17.18}$$

where we have introduced the parameter

$$t_c = \frac{8\mu H_0}{\rho g a^2} = 8\frac{\mu}{\Sigma}\frac{H_0}{\mathrm{Bo}}. \tag{17.19}$$

Since $H \to H_0$ at $t \to \infty$ and the first term in the right-hand side of (17.18) is much greater than the second, it is possible to write approximately

$$\frac{H}{H_0} \approx 1 - e^{-t/t_c}. \tag{17.20}$$

Then time t_c can be considered as the characteristic time of ascent of the liquid column. For water in the example considered above ($H_0 = 0.15$ m, $a = 0.1$ mm), we get $t_c = 12$ s. Notice that $a^2/\nu \sim 10^{-2}$ s and the assumption $t_c \gg a^2/\nu$ for which the solution was obtained is satisfied. In this example, Bond's number is equal to $\mathrm{Bo} = 1.3 \cdot 10^{-3}$, which indicates (see (17.10)) that gravity is immaterial as compared to surface tension.

17.3 Moistening Flows

A moistening flow is defined as the flow of a thin layer of a liquid over a solid surface. A review of publications in this field is contained in [16]. For most practical applications, the primary goal is to force a liquid to spread with a thin layer over as large a surface as feasible. Though the stresses due to surface forces are small in comparison with the pressure difference that is needed to overcome viscous forces, it is still necessary to account for them, given a fixed flow rate of liquid.

The hydrodynamic problem of a moistening flow is the problem of finding a stationary solution that relates the film's thickness, the velocity of the film's flow, the liquid's properties, and the parameters describing the film's geometry. The two-dimensional character of the flow, the presence of an unknown free boundary, and the nonlinearity of the problem are responsible for the greater part of mathematical difficulties that complicate the solution. There is also an important related problem of examining the stability of this flow.

One classical problem is to find the thickness of the moistening film on a plate that is extracted from a tank filled with viscous liquid [2]. We shall limit ourselves to consider the flow only for small capillary numbers $\mathrm{Ca} \ll 1$. We also assume that the thickness of the film δ_f, that adheres to the plate after its extraction

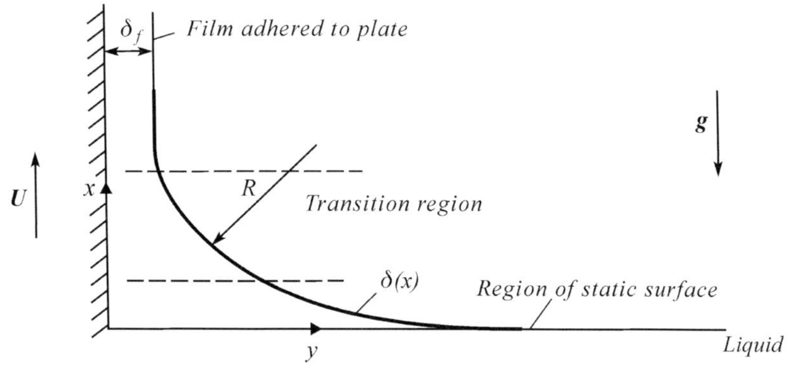

Fig. 17.6 Extraction of a plate from a viscous liquid.

from the tank is small in comparison with the characteristic linear size of the flow (Fig. 17.6).

The viscosity force is important for the flow in the film region but negligible in the region of static liquid surface in the tank, where the shape of the meniscus is determined by the balance of surface tension and hydrostatic pressure. The transition from the static surface to the surface of the established film occurs through an intermediate surface curved in the shape of a meniscus. Since the established film has a small thickness, the film surface should be completely moistening and the contact angle should be equal to zero. If R is the curvature radius of the meniscus in the transition area, then the characteristic linear scale of the flow in this area is equal to R. Hence, the condition that the film thickness must be infinitesimal reduces to the inequality $\delta \ll R$. The flow in the film can be regarded as almost parallel to the surface of the plate.

Consider the flow in the film. Since we are mostly interested in the thickness of the established film, non-stationary terms in the equations can be neglected.

The stationary state condition means that $t \gg t_{eq} = R^2/\nu$, where t_{eq} is the time it takes to establish equilibrium. We further assume that the Reynolds number $\mathrm{Re} = \rho UR/\mu \ll 1$, which allows us to drop inertial terms in the Navier–Stokes equations. In addition, we assume gravitational effects to be small. This means that the Bond number is $\mathrm{Bo} = \rho g R^2/\Sigma \ll 1$.

The assumptions we have made allow us to reduce the equations of motion to the balance condition of the pressure gradient and the gradient of shear stresses. If the surface tension coefficient remains constant at the interface, then the capillary pressure inside the liquid is equal to

$$-\Delta p = \frac{\Sigma}{R} = \frac{\Sigma\delta''}{(1+\delta'^2)^{3/2}} \approx \Sigma\delta''. \tag{17.21}$$

In the derivation of (17.21), we used the fact that $R_1 = \infty$, $R_2 = R = (1+\delta'^2)^{3/2}/\delta''$ is the curvature radius of the surface $\delta = \delta(x)$. In addition, in the film's flow region, $(\delta')^2 \ll 1$.

In view of the assumptions made thus far, the Navier–Stokes equations are reduced to the equality of the viscous and surface forces

$$\Sigma \frac{d^3\delta}{dx^3} + \mu \frac{\partial^2 u}{\partial y^2} = 0, \tag{17.22}$$

where u is the velocity in the film along the x axis.

Taking the boundary conditions, which stipulate that adhesion occurs at the plate,

$$u = U \quad \text{at } y = 0, \tag{17.23}$$

and applying them on the free surface of the film,

$$\mu \frac{\partial u}{dy} = 0, \quad \text{for } y = \delta(x) \tag{17.24}$$

we obtain, after integrating (17.22) over y,

$$u = U - \frac{\Sigma}{\mu} \frac{d^3\delta}{dx^3} \left(\frac{y^2}{2} - \delta y \right). \tag{17.25}$$

Eq. (17.25) allows us to find the flow rate of the liquid in the film

$$Q = \int_0^{\delta(x)} u\, dy = U\delta + \frac{\Sigma}{\mu} \frac{d^3\delta}{dx^3} \frac{\delta^3}{3}. \tag{17.26}$$

Far away from the surface of the reservoir, the conditions $\delta \to \delta_f = \text{const}$ and

$$Q \to U\delta_f. \tag{17.27}$$

must hold for the film. Equating (17.26) and (17.27), we find that

$$\delta^3 \delta'' + 3\text{Ca}\delta = 3\text{Ca}\delta_f. \tag{17.28}$$

Using dimensionless variables $\eta = \delta/\delta_f$, $\xi = x/(3\text{Ca})^{1/3}\delta_f$, we can rewrite Eq. (17.28) in the form

$$\eta^3 \eta''' = 1 - \eta. \tag{17.29}$$

To solve Eq. (17.29), one needs three conditions, and one extra condition is needed in order to determine δ_f. Details of the solution are adduced in [2]. The thickness δ_f of the film at the surface is equal to

$$\frac{\delta_f}{\Delta_k} = 0.946(\text{Ca})^{2/3}. \tag{17.30}$$

Since $\delta'' = 1/R$ and $\Delta_k = \sqrt{2}/\delta''$, Eq. (17.30) reduces to

$$\frac{\delta_f}{R} = 0.643(3\mathrm{Ca})^{2/3}. \qquad (17.31)$$

The solution holds under conditions $\delta_f/R \ll 1$ and $\mathrm{Ca} \ll 1$. A more detailed asymptotic analysis of the solution for the above-considered problem can be found in [17]. The formal procedure of matching asymptotic expansions demonstrates that the solution in the form (17.30) is applicable for $\mathrm{Ca} \to 0$. In the case of finite values of $\mathrm{Ca} < 1$, the expression (17.30) is modified [17]:

$$\frac{\delta_f}{\Delta_k} = 0.946(\mathrm{Ca})^{2/3}(1 - 0.113\mathrm{Ca}^{1/3} + \cdots). \qquad (17.32)$$

Another problem, similar in character to the one considered above, is the problem of cavitation in a thin layer between two rotating cylinders [18]. Rotation of two cylinders leads to an outflow of viscous liquid from the tank, which is accompanied by the formation of a thin moistening layer at the surface of the cylinders. The clearance between the surfaces of cylinders is small; therefore the flow in the formed film is driven by the forces of viscosity and surface tension. In the clearance between cylinders, the liquid flows parallel to the surface. The character of the flow is same as in a film at the surface of a plate being removed from the tank.

Another similar problem is the problem of extrusion by air of a viscous liquid from a capillary (Fig. 17.7) [19]. The flow is analogous to the flow in the problem of a lubricating layer between rotating cylinders. To convince ourselves in this similarity, it is necessary to attach the coordinate system to a bubble moving with constant velocity U. Then, in its proper system, the bubble is motionless, and the walls of the capillary move with the velocity U to the left. Near the front of the bubble, the surface of the liquid forms a meniscus-like interface, which then transfigures into a film flow at the surface of the capillary. For small values of Re and Ca, the equation of motion reduces to a balance of viscous and capillary

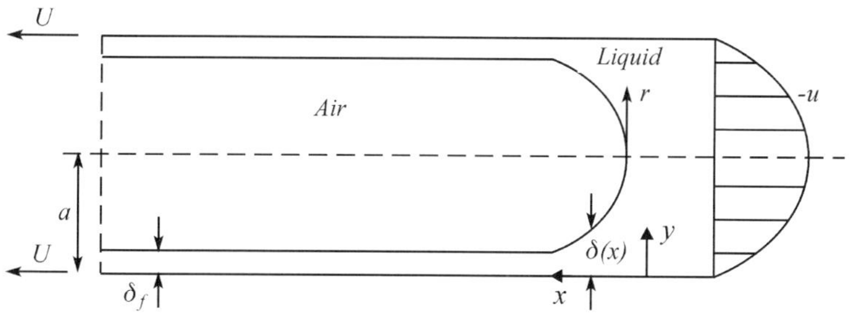

Fig. 17.7 Extrusion of viscous liquid from capillary by air.

forces. In the transition region near the front of the bubble, the thickness of the film layer changes as

$$\delta = \frac{1}{2}\frac{x}{a} + 1.79(3\text{Ca})^{2/3}a. \tag{17.33}$$

17.4
Waves at the Surface of a Liquid and Desintegration of Jets

In the previous paragraph we discussed equilibrium shapes of the interface in certain moistening flow. There is one aspect still missing in that discussion, namely, the behavior of perturbations on the interface. If particles on the surface of the liquid experience a small displacement due to some random fluctuation, so that the surface of the liquid becomes deformed, deviating from its equilibrium shape, there appear forces that try to return the surface to its equilibrium form. They can be caused by two factors. First, the deformation of the surface results in a corresponding increase of its free energy. As a consequence, there appear capillary forces that try to reduce the total surface, restoring the shape to equilibrium. Second, if the liquid is placed in a gravitational field and the surface is flat, the force of gravity will tend to smooth any perturbations, returning the surface to the original flat shape. Under the action of these two forces, fluid particles displaced from the equilibrium position will try to return to it. However, due to inertia, they will pass the equilibrium position, again evoking the returning forces, and so on. Hence, any random fluctuation will create waves on the surface of liquid.

If the main factor causing the occurrence of waves is surface tension, then such waves are called capillary waves. In the case of predominance of gravitational forces we can talk about gravitational waves. An illustration of the first case is the outflow of a liquid jet from a nozzle into the air. At some distance from the nozzle, the surface of the liquid jet gets covered by waves, and then the jet disintegrates into drops (the process of jet "breakage" takes place). The second case is exemplified by ordinary waves on the surface of the water.

Consider flat sinusoidal waves with small amplitude. Let λ be the length and a – the amplitude of the wave. Waves arise at the interface between air and water and propagate in the positive direction of the x axis with the velocity c (Fig. 17.8). Assume water to be a non-viscous, incompressible liquid. In the absence of viscous dissipation, the wave amplitude can be taken as constant. Denote by $\zeta(x, t)$ the vertical perturbation on the interface whose equilibrium position is horizontal. Then ζ can be written as

$$z = \zeta(x, t) = a \sin(\omega t - kx), \tag{17.34}$$

where the wave number k and the frequency ω are related to the period τ and the wavelength λ by the formulas

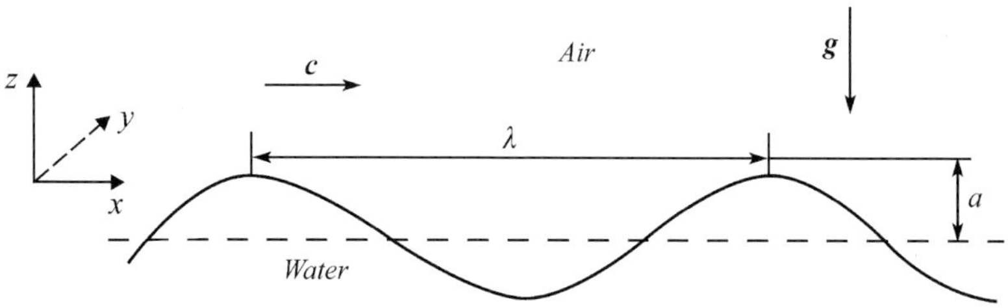

Fig. 17.8 Waves on a water surface.

$$\omega = \frac{2\pi}{\tau}, \quad k = \frac{\omega}{c} = \frac{2\pi}{\lambda}, \quad \lambda = c\tau. \tag{17.35}$$

The zero viscosity of the liquid means that the amplitude can neither decrease nor increase, but the wave can be dispersive, i.e. ω can depend on k. It is shown in [2] that perturbations with small amplitude as compared to the wavelength ($a \ll \lambda$) are described by the linearized equation

$$\rho \frac{\partial \bar{\mathbf{u}}}{\partial t} = -\nabla p_e, \tag{17.36}$$

where $\bar{\mathbf{u}}$ is the propagation velocity of perturbations, $p_e = p - \rho g z$ is the total pressure.

Denote the atmospheric pressure by p_a. Then the hydrostatic pressure of liquid in the equilibrium state is

$$p_0 = p_a - \rho g z. \tag{17.37}$$

In a perturbed state,

$$p = p_0 + (p_e)_g, \tag{17.38}$$

where $(p_e)_g$ is the pressure perturbation.

At the interface, the condition $p = p_a$ should be satisfied. Then from (17.37) and (17.38), there follows:

$$(p_e)_g = \rho_L g \zeta. \tag{17.39}$$

The curvature of the interface results in the appearance of capillary pressure

$$(p_e)_g = \frac{\Sigma}{R} \approx -\Sigma \frac{\partial^2 \zeta}{\partial x^2}. \tag{17.40}$$

Here we have taken into account the Laplace-Young equation (17.5), subject to the conditions $R_1 = \infty$, $R_2 = -\Sigma\zeta_{xx}$ that apply to small perturbations (see (17.21)).

It follows from (17.34) that $\zeta_{xx} = -k^2\zeta$ and

$$(p_e)_g = \Sigma k^2 \zeta. \tag{17.41}$$

The excessive pressure in the liquid near the interface is the sum of capillary and hydrostatic pressures:

$$p_e = \rho\left(g + \frac{k^2\Sigma}{\rho_L}\right)\zeta = \rho\zeta g_{\text{eff}}. \tag{17.42}$$

It follows from (17.42) that the effect of surface tension is equivalent to an increase of hydrostatic pressure that would occur if the gravity constant changed from g up to g_{eff}.

If the force of surface tension is much stronger than gravity, then the waves are capillary. This condition can be written as $k^2\Sigma/\rho \gg g$, or, in view of the equality $k = 2\pi/\lambda$, as

$$\lambda \ll 2\pi\left(\frac{\Sigma}{\rho g}\right)^{1/2} = 2\pi\Delta_k. \tag{17.43}$$

In another limiting case, $k^2\Sigma/\rho \ll g$, or

$$\lambda \gg 2\pi\left(\frac{\Sigma}{\rho g}\right)^{1/2} = 2\pi\Delta_k, \tag{17.44}$$

the perturbations give rise to gravity waves.

Thus the capillary length Δ_k is a characteristic linear scale that divides the wavelength spectrum into the capillary ($\lambda \ll \Delta_k$) and the gravitational ($\lambda \gg \Delta_k$) regions. For instance, the capillary length of water ($\Sigma = 73$ mN/m, $\rho = 1000$ kg/m^3) is $1.7 \cdot 10^{-2}$ m. Thus we have we have $\lambda \ll 10^{-2}$ м (short wavelength, high frequency) for capillary waves, and $\lambda \gg 10^{-2}$ m (long wavelengths, low frequencies) for gravitational waves.

It follows from Eq. (17.36) and from the continuity equation $\nabla \cdot \boldsymbol{u} = 0$ that the perturbation of the velocity field is has zero divergence and curl ($\nabla \times \boldsymbol{u} = 0$), so it must be a potential field $\boldsymbol{u} = \nabla\phi$, with the potential ϕ satisfying the Laplace equation $\Delta\phi = 0$. Therefore Eq. (16.36) can be rewritten as

$$\rho\nabla\left(\frac{\partial\phi}{\partial t}\right) = -\nabla p_e$$

or

$$\rho\frac{\partial\phi}{\partial t} = -p_e + F(t),$$

where $F(t)$ is an arbitrary function of time. It can be set to zero, because if it were non-zero, it would be possible to introduce a new potential

$$\phi' = \phi + \int F\, dt,$$

which allows to eliminate $F(t)$ from the equation. Thus,

$$p_e = -\rho \frac{\partial \phi}{\partial t}. \tag{17.45}$$

And now, using the condition (17.42) for the liquid, one obtains a condition that should hold at the interface

$$\left(\frac{\partial \phi}{\partial t}\right)_{z=0} = -g_{\text{eff}}\zeta. \tag{17.46}$$

Note that small value of the amplitude ($a \ll \lambda$) allows us to formulate the condition (17.46) not for $z = \zeta$, but for $z = 0$, i.e. the boundary condition at the perturbed surface is "shifted" to the unperturbed surface, which corresponds to the linearization of the boundary condition.

The second boundary condition at the interface follows from the kinematic condition of equality between the vertical velocity of fluid particles and the vertical velocity of surface motion

$$\frac{D\zeta}{Dt} = w \quad \text{for } z = \zeta. \tag{17.47}$$

Since $D\zeta/Dt = \partial\zeta/\partial t + \mathbf{u} \cdot \nabla\zeta$, then, linearizating this operator, neglecting the higher-order term $\mathbf{u} \cdot \nabla\zeta$ and applying this condition to the unperturbed surface $z = 0$, we obtain

$$\frac{\partial \zeta}{\partial t} = \left(\frac{\partial \phi}{\partial z}\right)_{z=0}. \tag{17.48}$$

Other boundary conditions depend on the geometry of the problem. Thus, if we consider the waves on the surface of a liquid in very deep reservoir, then, assuming that far away from the surface, perturbations are absent, we have

$$\phi \to \text{const}, \quad \text{at } z \to -\infty. \tag{17.49}$$

In this one can seek for the solution of the Laplace equation in the form

$$\phi = Ae^{kz} \cos(\omega t - kx). \tag{17.50}$$

It is easy to verify that (17.50) satisfies the conditions (17.46) and (17.48) when

$$\omega^2 = kg_{\text{eff}}, \quad A = \omega a \tag{17.51}$$

and under the condition that perturbations at the liquid surface look like (17.34). Perturbations decay exponentially as $z \to -\infty$, and the characteristic depth of attenuation is $1/k = \lambda/2\pi$.

The propagation velocity of surface waves is equal

$$c = \frac{\omega}{k} = \left(\frac{g_{\text{eff}}}{k}\right)^{1/2} = \left(\frac{\lambda g_{\text{eff}}}{2\pi}\right)^{1/2} = \left(\frac{2\pi\Sigma}{\rho\lambda} + \frac{g\lambda}{2\pi}\right)^{1/2}. \tag{17.52}$$

The wavelength corresponding to the velocity minimum is

$$c_{\text{min}} = \left(\frac{g\lambda_{\text{min}}}{\pi}\right)^{1/2}, \quad \lambda_{\text{min}} = 2\pi\left(\frac{\Sigma}{\rho g}\right)^{1/2} = 2\pi\Delta_k. \tag{17.53}$$

For water, it equals $l_{\text{min}} = 1.7 \times 10^{-2}$ m and $c_{\text{min}} = 0.23$ m/s.

For capillary waves, we get from (17.52)

$$c_{cap} = \left(\frac{2\pi\Sigma}{\rho\lambda}\right)^{1/2}. \tag{17.54}$$

This correlation between the velocity of propagation of capillary waves and the coefficient of surface tension is frequently is used in experimental measurements of Σ.

If we neglect liquid viscosity, we come to the conclusion that amplitudes of plane surface waves do not change. If we account for viscosity and for the change of the surface tension coefficient, we have to conclude that surface waves get attenuated. The problem of suppression of waves by surfactants will be discussed in the next section.

Now let us consider the waves formed on the free surface of a cylindrical liquid jet. These waves cause the jet to ultimately disintegrate into drops. Therefore we are interested in the conditions at which small capillary perturbations can increase the amplitude, resulting in breakup of the jet. From the physical viewpoint, disintegration of a slowly moving cylindrical jet occurs for the following reason. If a jet accidentally becomes thicker in some random cross section, the surface tension tries to reduce the free surface and thus to minimize the free energy. Simple geometrical considerations show [20] that the surface of a cylinder with a given volume will become smaller if this volume is broken into spherical drops whose radius will be in 1.5 times greater than the radius of the cylinder. Indeed, let us consider the cylinder of radius R and length L. Its volume and surface area are equal to $V_c = \pi R^2 L$, $S_c = 2\pi RL$, so that $V_c = S_c R/2$. Let this cylinder break up into n spheres of radius r. Then the volume of the spheres and their surface are equal to $V_s = 4\pi R^3 n/3$, $S_S = 4\pi R^2 n = 3V_s/r$. It follows from the condition $V_s = V_c$ that $S_s = 3V_c/r = 3S_c R/2r = 1.5S_c R/r$, or $r/R = 1.5S_c/S_s$. From the condition $S_c/S_s > 1$, it follows that $r > 1.5R$. In addition, it is possible to show that if a jet disintegrates into $n \geq 2$ identical spherical drops, then the average distance between the drops will be greater than $1.5rn/(n-1)$ with $1 < n/(n-1) \leq 2$.

The reason for the disintegration of the jet is the instability of perturbations of the cylindrical shape of the jet's surface. This instability induces a significant decrease of the jet's cross section and ultimately the breakup of the jet and release of the free energy. Since the total surface area of the drops thus formed is less than the surface area of the cylindrical jet, the surface energy decreases. All this is the consequence of the axial symmetry of the jet. In the case of a flat jet, no such phenomena will occur.

There are many various kinds of jet breakup. Consider only those that result in a disintegration of the jet into spherical drops due to the action of capillary forces. We assume the velocity of the jet's outflow into the air to be small. If this velocity is large enough, then stability of the jet's surface will be influenced by the dynamics of the environment, in particular, the forces of viscous friction and the change of pressure. Therefore, breakup of high-velocity jets has a different character. Indeed, a jet flowing out into a gas with a small velocity breaks up into rather large drops, while at a high velocity of outflow it breaks up into a set of fine drops of various sizes.

The theory of hydrodynamic stability for small perturbations of low-velocity jets is relatively simple. If perturbations are not small in comparison with the jet's radius, it is necessary to take into account nonlinear effects that considerably complicate the problem. A detailed discussion of this question can be found in the review [21].

The problem of linear stability of an incompressible nonviscous liquid in the form of an infinitely long cylinder of circular cross section surrounded by air was first considered by Rayleigh [22]. This work and the subsequent works [23, 24] on hydrodynamic stability include four stages. The first stage is the determination of parameters of the basic undisturbed flow: fields of velocities, pressures and temperatures. The next stage is the assumption about the insignificance of perturbations of these parameters and linearization of equations and boundary conditions. The result is a homogeneous linear system of equations in partial derivatives. The coefficients of this system can depend on spatial coordinates, but they do not depend on time. In the third stage, we look for an elementary solution for a chosen initial perturbation. Usually we seek for a solution in the form of a complex Fourier-representation of periodic functions. For example, one can look for an elementary solution in the form of a normal mode

$$\phi(x, z, t) = \Phi(z) \exp(i\omega t - kx), \tag{17.55}$$

where Φ and ω are complex numbers and k is a real number.

The real part ϕ represents a one-dimensional wave that propagates in the direction of the x axis. This wave can grow, attenuate or remain constant with time. The latter case is realized, for example, in the expression (17.50). Let us set $\omega = \omega_r + i\omega_i$. Then at $\omega_i < 0$ the time-dependent index of the exponential in (17.55) is positive and the value of ϕ increases with time all the way to infinity. It means that perturbations grow with time and are unstable. For $\omega_i > 0$, the index of the exponential function is negative and perturbations decrease with time. It means that perturbations are steady. At $\omega_i = 0$ the quantity ϕ either does not depend on t (if $\omega_r = 0$), or is a periodic function of time t. In this case we talk about "neutral stability".

It is usually convenient to represent perturbations as a superposition of normal modes, each of which can be considered independently because the equations are linear. In this case the primary problem is to find the spectrum of normal modes, which will then give you the perturbation.

The last stage is to find the eigenfunctions $\Phi(x)$ and the corresponding complex wave velocities c. During this procedure the linear system of partial differential derivatives is reduced to a linear system of ordinary differential equations.

The method used to research stability of small perturbations of a cylindrical jet of non-viscous incompressible fluid is similar to the previously discussed method of handling the small perturbation problem. The main difference from the case of a plane surface is the axial symmetry of the problem, which consequently features the characteristic linear size a (radius of the jet). In a coordinate system moving with the jet's velocity, the jet itself is motionless. Let us neglect gravity and take into account only the force of surface tension. Then the pressure along the jet is equal to $p = p_a + \Sigma/a$ (since $R_1 = \infty$, $R_2 = a$). Now proceed to linearize the equations and the boundary conditions. Assume the perturbations of the flow to be small and consider the equation of motion. After linearization, i.e. after rejecting the second order terms, one obtains Eq. (17.36) for velocity perturbations and full pressure. Since the flow is a potential one, any perturbation of the velocity potential satisfies the Laplace equation $\Delta\phi = 0$, which in a cylindrical coordinate system (r, θ, z) is written as

$$\frac{1}{r}\frac{\partial}{\partial r}\left(r\frac{\partial \phi}{\partial r}\right) + \frac{1}{r^2}\frac{\partial^2 \phi}{\partial \theta^2} + \frac{\partial^2 \phi}{\partial z^2} = 0. \tag{17.56}$$

Let the radial displacement of the jet's surface for a small perturbation be

$$r = \zeta(z, \theta, t). \tag{17.57}$$

The linearized boundary condition at the jet's surface is similar to (17.40) for a flat surface. For an axisymmetric problem, this condition becomes

$$(p_e)_g = -\Sigma\left(\frac{\partial^2 \zeta}{\partial z^2} + \frac{1}{a^2}\frac{\partial^2 \zeta}{\partial \theta^2} + \frac{\zeta}{a^2}\right)_{r=a}. \tag{17.58}$$

Similarly to (17.48), one can write the kinematic condition at the jet's surface:

$$\frac{\partial \zeta}{\partial t} = \left(\frac{\partial \phi}{\partial r}\right)_{r=a}. \tag{17.59}$$

Note that conditions (17.58) and (17.59) are applied to the non-perturbed jet surface $r = a$, just as in the plane-surface problem.

Thus, Eq. (17.56) and the conditions (17.58) and (17.59) represent a linearized equation with boundary conditions.

Proceed now to determine the perturbations. Perturbations will be sought in the form that follows from the general representation of the normal mode (17.55)

$$\phi(r, \theta, z, t) = \Phi(r)e^{\beta t} \cos(kz + n\theta), \tag{17.60}$$

where k is the axial wave number, n is a positive integer number, β is the coefficient of amplification or attenuation.

Substituting the expression (17.60) into Eq. (17.56) and separating variables, one obtains a Bessel equation for $\Phi(r)$,

$$\frac{d^2\Phi}{dr^2} + \frac{1}{r}\frac{d\Phi}{dr} - \left(\frac{n^2}{r^2} + k^2\right)\Phi = 0, \tag{17.61}$$

whose general solution is

$$\Phi = AI_n(kr) + BK_n(kr), \tag{17.62}$$

where I_n and K_n are the modified Bessel functions of the first and the second kind, respectively. Since K_n has a singularity at $r \to \infty$, the requirement that Φ must be finite means that $B = 0$. Therefore,

$$\Phi = AI_n(kr). \tag{17.63}$$

Now, using the kinematic condition (17.59) and the general formula for Φ, we obtain from (17.60) and (17.63) an expression for the radial displacement of the perturbed surface:

$$\zeta = \frac{AkI_n'(kr)}{\beta} e^{\beta t} \cos(kz + n\theta). \tag{17.64}$$

Substitution of (17.64) into (17.58) gives us the full pressure p_e. Furthermore, by using the Bernoulli equation (17.45) $p_e + \rho\partial\phi/\partial t = 0$, one can eliminate the constant A. The result is a dispersive equation that relates parameters of the perturbation ϕ (17.60):

$$\beta^2 = \left(\frac{\Sigma}{\rho_L a^3}\right)\frac{\alpha I_n'(\alpha)}{I_n(\alpha)}(1 - \alpha^2 - n^2), \quad \alpha = ak. \tag{17.65}$$

Eq. (17.65) determines the stability conditions for the jet. It should be noted that $\beta^2 \sim \Sigma/\rho a^3$, whereas in the plane case involving capillary waves at the surface of a deep reservoir we had $\omega^2 \sim \Sigma/\rho\lambda^3$ according to (17.52). These perturbations also could be derived from (17.73). Indeed, consider the case of $\alpha = ak \gg 1$ or $a \gg \lambda$, i.e. perturbations whose wavelength is small in comparison with the jet's radius. It then follows from the properties of the Bessel function that $I_n'(\alpha)/I_n(\alpha) \approx 1$. Putting $\beta = i\omega$ into (17.65), one finds that $\omega^2 = k^3\Sigma/p \sim \Sigma/\rho\lambda^3$, which corresponds to neutrally stable capillary waves on the surface of a deep water reservoir.

The most important solutions are the ones with $\beta^2 > 0$, which corresponds to unstable perturbations. Since $I_n'(\alpha)/I_n(\alpha) > 0$ for all $\alpha \neq 0$, it follows from (17.65) that $\beta^2 < 0$ for all α if $n \neq 0$ and for all $|\alpha| \geq 1$ if $n = 0$. For these values of α and n, the factor β is imaginary. It means that waves at the jet's surface do not intensify and are not subject to damping. We conclude from the form of surface perturbations (17.64) that for $n \neq 0$, perturbations do not possess axial symmetry (i.e. they depend on θ), while the case $n = 0$ corresponds to axially symmetric perturbations. It follows from the above that for asymmetric perturbations, the jet is always stable. On the other hand, it can be seen from (17.65) that $\beta^2 > 0$ for $-1 < \alpha < 1$ at $n = 0$. Since $\alpha = ak = 2\pi a/\lambda$, this condition means that $\lambda \geq 2\pi a$, i.e. the jet is unstable with respect to axially symmetric perturbations if the wavelength $\lambda = 2\pi/k$ of these perturbations exceeds the perimeter of the circular section $2\pi a$ for an unperturbed jet.

From (17.65), we can find the maximum value of the amplification factor β_{max} as a function of α for $n = 0$. For example,

$$\beta_{max} = 0.343\left(\frac{\Sigma}{\rho_L a^3}\right)^{1/2} \quad \text{at } \alpha = 0.697. \tag{17.66}$$

This value corresponds to the fastest-growing perturbation with wavelength

$$\lambda_{max} = \frac{2\pi}{k_{max}} = \frac{2\pi a}{0.697} = 9.02a. \tag{17.67}$$

According to Rayleigh's hypothesis, it is just this mode that is responsible for the jet's disintegration.

The results obtained so far correspond to the case of a stationary jet, i.e. the jet whose motion is not taken into account. Nevertheless, they can be used for the estimation of the distance L_B at which the jet, moving with a constant velocity U, will be broken. Let

$$L_B = Ut_B. \tag{17.68}$$

Here t_B is the time during which the fastest-growing mode will increase the amplitude of the perturbation from a small initial value up to the value of about

$2\pi a$, equal to perimeter of the unperturbed cross section of the jet. An increase of the initial amplitude by a factor of e results in $\beta_{\max} t_B = 1$, or

$$t_B = \frac{1}{\beta_{\max}} = 2.92\left(\frac{\Sigma}{\rho a^3}\right)^{-1/2}. \qquad (17.69)$$

An increase of the initial amplitude by a factor of 100 ($\beta_{\max} t_B = 4.61$) leads to

$$t_B = \frac{1}{\beta_{\max}} = 13.4\left(\frac{\Sigma}{\rho a^3}\right)^{-1/2}. \qquad (17.70)$$

For instance, for a water jet with the diameter of 5 mm, we get $(\rho a^3/\Sigma)^{1/2} = 4.14 \cdot 10^{-2}$ s. If the jet's velocity is equal to $U = 0.1$ m/s, it can become unstable and be broken very quickly ($L_B \sim 10^{-2}$ m).

The wavelength (17.67) of these perturbations that was found earlier can be considered as the distance between drops into which the jet will split. A drop of diameter d will be formed from the liquid contained in the cylinder of length $\lambda_{\max}$. Therefore

$$\frac{\pi d^3}{6} = \pi a^2 \lambda_{\max}. \qquad (17.71)$$

Substituting $\lambda_{\max}$ from (17.67), we can find the diameter of newly formed drops:

$$d = 3.78a. \qquad (17.72)$$

17.5
Flow Caused by a Surface Tension Gradient – The Marangoni Effect

As was already mentioned in Section 17.1, a change of surface tension Σ along the interface causes additional tangential stresses (Marangoni stresses) at this surface, and consequently gives rise to forces acting on the two media adjoining to either side of the surface. These forces, together with the forces of viscous friction, change the velocity and pressure fields in these media. In particular, if the interface divides two quiescent liquids or a liquid and a gas, then a change of Σ may produce a motion of the liquid. This phenomenon is referred to as the Marangoni effect.

Σ can also change along the normal to the interface, resulting in a special kind of instability called the Marangoni instability. This kind of instability is similar to convective instability that arises in a medium when the density gradient is directed against the gravity force (the top layers of liquid are heavier than the bottom layers).

A model that pictures the interface as a thin layer with high viscosity (the Boussinesq model [2]) has gained widespread acceptance. For such a layer, one has to write down a special phenomenological correlation that connects the surface stress tensor with the rate of surface strain tensor [25]. Paper [26] uses this model to examine the problem of the influence of surfactants on the dynamics of a free interface.

Let us consider only the change of Σ in the direction tangential to the interface. The change of surface tension can be caused by a change of temperature, a change of surfactant or impurities concentration, and also by the presence of electric charge or electrostatic potential on the surface. Depending on the mechanism that causes the change of Σ, the corresponding flow is called thermo-capillary, diffusive-capillary or electro-capillary flow.

One can adduce many examples of flows induced by the gradient of surface tension. It is sufficient to recall a phenomenon well known from the ancient times – the damping of ocean waves that takes place when oil is poured into the ocean, spreading with a thin layer over the water surface. Another familiar phenomenon is the "dancing" of a ball of solid camphor on the water surface. All these phenomena are caused by a change of Σ.

Thus, in order to solve the hydrodynamic problem of liquid motion in view of the change of Σ at the interface, we should first find out the distribution of substance concentration, temperature and electric charge over the surface. These distributions, in turn, are influenced by the distribution of hydrodynamic parameters. Therefore the solution of this problem requires utilization of conservation laws – the equations of mass, momentum, energy, and electric charge conservation with the appropriate boundary conditions that represent the balance of forces at the interface: the equality of tangential forces and the "jump" in normal forces which equals the capillary pressure. In the case of Boussinesq model, it is necessary to know the surface viscosity of the layer. From now on, we are going to neglect the surface viscosity.

The change of surface tension coefficient along the interface gives rise to a tangential force directed from the smaller values of Σ to the larger values:

$$\boldsymbol{f} = \nabla_s \Sigma. \tag{17.73}$$

Denote through $\boldsymbol{f}^a$ and $\boldsymbol{f}^b$ the forces with which liquids a and b, divided by the surface, are acting on the interface. Then

$$\boldsymbol{f}^a = \boldsymbol{n} \cdot \boldsymbol{T}^a + \boldsymbol{n}p^a, \quad \boldsymbol{f}^b = -\boldsymbol{n} \cdot \boldsymbol{T}^b - \boldsymbol{n}p^b. \tag{17.74}$$

Here $\boldsymbol{T}$ is the viscous stress tensor, $\boldsymbol{n}$ is the normal vector to the interface, directed for the sake of definiteness towards liquid b.

Taking into account (17.73) and (17.74), we can write the force equilibrium conditions, expressed in projections on the tangential line (s) and the normal (n) to the surface, in the form

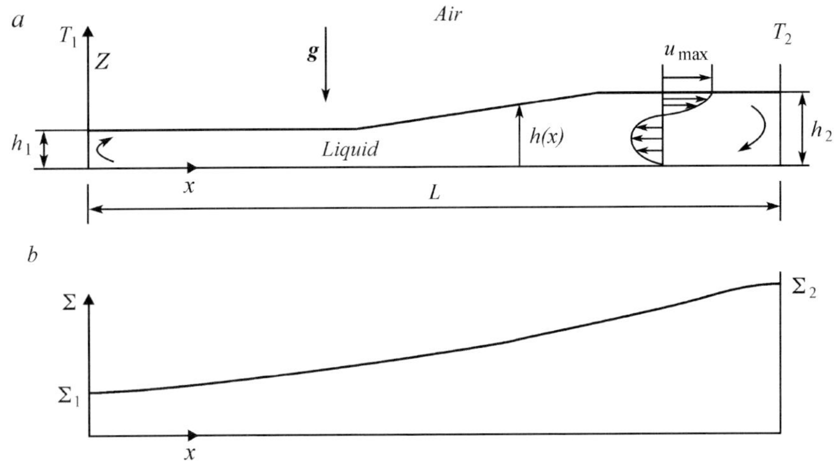

Fig. 17.9 Thermo-capillary motion of liquid in a shallow rectangular tray (*a*) and the change of surface tension coefficient along the tray (*b*).

$$\boldsymbol{f}_s^a + \boldsymbol{f}_s^b + \nabla_s \Sigma = 0, \tag{17.75}$$

$$f_n^a + f_n^b = \Sigma \left(\frac{1}{R_1} + \frac{1}{R_2} \right). \tag{17.76}$$

As an example, consider the thermo-capillary motion of a thin layer of liquid in a shallow rectangular tray (Fig. 17.9) [2, 27, 28]. The change of surface tension is caused by the difference of temperatures ($T_1 > T_2$) at the lateral walls of the tray. Hence there is a temperature gradient along the free surface, which, in turn, creates a surface tension gradient. Recall that for a liquid, $\partial\Sigma/\partial T < 0$. Let's assume that the size of the tray in the direction normal to the XOZ plane and the length L of the tray along the x-axis are much greater than the height h of the liquid layer. Also, the region of non-uniform flow is limited to a small area near the walls of the tray, and the flow can be considered as two-dimensional. The condition $h \ll L$ allows us to assume in the first approximation that the flow is almost parallel to the x-axis and that the height of the tray is roughly the same along the entire length.

Thus, there is a small vertical component of velocity (small as compared to its longitudinal counterpart) present in the liquid. Suppose the Reynolds number of the arising flow is small. To estimate Re, let's take h_1 as the characteristic linear size and u_{max} as the characteristic velocity, where u_{max} is the maximum longitudinal velocity on the free surface that is induced by the surface tension gradient. Since the shape of the surface changes slowly along the x-axis, we can approximate $|\nabla_s\Sigma| \approx d\Sigma/dx$. From the condition (17.75), it follows that $\mu(\partial u/\partial y)_s = d\Sigma/dx$. Inasmuch as $(\partial u/\partial y)_s \sim u_{max}/h_1$, we have

$$u_{\max} \sim \frac{h_1}{\mu}\frac{d\Sigma}{dx} \sim \frac{h_1}{\mu}\frac{(\Sigma_2 - \Sigma_1)}{L}. \tag{17.77}$$

Let's now find Reynolds number:

$$\mathrm{Re} = \frac{u_{\max} h_1}{\nu} = \left(\frac{h_1}{L}\right)^2 \frac{(\Sigma_2 - \Sigma_1)}{\rho\nu^2} \quad L \ll 1. \tag{17.78}$$

If the condition (17.78) is satisfied, then, taking into consideration the obvious inequalities $\partial^2 u/\partial x^2 \ll \partial^2 u/\partial z^2$, $v \ll u$ and the fact that the curvature of the liquid surface is very small, the equations of motion in the noninertial approximation, are written as

$$\frac{\partial p}{dx} = \mu \frac{\partial^2 u}{\partial z^2}, \quad \frac{\partial p}{dx} = -\rho g. \tag{17.79}$$

Let's integrate the continuity equation

$$\frac{\partial u}{dx} + \frac{\partial v}{\partial z} = 0 \tag{17.80}$$

over z from 0 to h, and note that $v = 0$ at $z = 0$ and $v \approx 0$ at $z = h$. As a result, we obtain a conservation equation for flow rate in the liquid layer

$$\int_0^{h(x)} u(z)\, dz = 0. \tag{17.81}$$

When deriving (17.81) we used the fact that $h(x)$ is a slowly varying function of x, i.e. $|h'(x)| \sim h/L \ll 1$. It should be noted that the inequality $v \ll u$ is true everywhere, except for regions near the edges of the layer, where the vertical component of the velocity v can be of the same order or even greater than u.

Let's complement Eqs. (17.79)–(17.81) with the boundary conditions at the bottom of the tray,

$$u = 0 \quad \text{at } z = 0, \tag{17.82}$$

and at the interface (see (17.75)),

$$\mu \frac{\partial u}{\partial z} = \frac{d\Sigma}{dx} \quad \text{at } z = h(x). \tag{17.83}$$

Viscous stresses due to the air are neglected. Furthermore, as the curvature of the surface is small, normal gradients at the surface are absent and the condition (17.76) reduces to the equality of pressures:

$$p = p_a \quad \text{at } z = h(x), \tag{17.84}$$

where p_a is the atmospheric pressure.

Integrating (17.79) over z and using conditions (17.82) and (17.83), one gets:

$$\mu u = \left(\frac{d\Sigma}{dx} - h\frac{\partial p}{\partial x}\right) z + \frac{1}{2}\frac{\partial p}{\partial x} z^2. \tag{17.85}$$

Now use the expression for hydrostatic pressure

$$p = p_a + \rho g(h - z). \tag{17.86}$$

which leads to

$$\frac{\partial p}{\partial x} = \rho g \frac{dh}{dx}. \tag{17.87}$$

Using (17.85) together with (17.87) and (17.81), we find that

$$\Sigma - \Sigma_1 = \frac{\rho g}{3}(h^2 - h_1^2), \tag{17.88}$$

where $\Sigma = \Sigma_1$ and $h = h_1$ at $x = 0$.

The condition (17.88) relates the changes in $h(x)$ and $\Sigma(x)$. Now we can use (17.85) to find the longitudinal velocity

$$u = \frac{z}{2\mu}\left(\frac{3z}{2h} - 1\right)\frac{d\Sigma}{dx} \tag{17.89}$$

and its maximum value, which is attained on the free surface

$$u_{\max} = \frac{h}{4\mu}\frac{d\Sigma}{dx}. \tag{17.90}$$

The velocity is equal to zero at the height $z = 2h/3$, and has a maximum at $z = h/3$. Let Reynolds number be equal to $\mathrm{Re} = h_1 u_{\max}/v \sim 0.1$. Then

$$h_1^2 \ll \frac{0.4\rho v^2}{d\Sigma/dx}. \tag{17.91}$$

The dependence $\Sigma(T)$ is almost linear for a liquid. Thus, for water, $d\Sigma/dT \approx -0.15$ mN/mq·K. Therefore, if $dT/dx = -100$ K/m, then $d\Sigma/dx = 1.5$ mN/m^2.

If $v = 10^{-6}$ m^2/s, $\rho = 10^3$ kg/m^3, $\Sigma_1 = 73$ mN/m, and h_1 is defined with a due regard to (17.91), then Bond's number is $\mathrm{Bo} = \rho g h_1/\Sigma_1 = 3.6 \cdot 10^{-3}$. The small value of Bo testifies to the primary influence of surface tension forces in comparison with the gravity force.

So, the estimation just performed shows that in the above-mentioned problem of a one-dimensional thermo-capillary flow of a thin layer, surface tension plays the mane role in formation of the flow.

If the change in surface tension on the interface is caused by the adsorption of a surfactant from the solution, then it is also necessary to determine the surfactant concentration near the interface.

Denote through Γ the molar concentration (mole/m^2) of surfactant at the interface. Then the equation describing the change of Γ, looks like a convective diffusion equation that takes into account the delivery of matter from the liquids which are divided by the interface. Making the assumption that each liquid is a binary solution, the diffusion equation can be derived in the same manner as in Section 4.4. Suppose that chemical reactions are absent, the diffusion is governed by Fick's law, and the diffusion coefficients are constant. Then the equation of surfactant diffusion at the interface has the form [2]

$$\frac{\partial \Gamma}{\partial t} + \nabla_s \cdot (\Gamma \boldsymbol{u}_s) = D_s \nabla_s \Gamma + [D \nabla C] \cdot \boldsymbol{n}, \tag{17.92}$$

where $\boldsymbol{u}_s$ is the average molar rate of transport of the surfactant along the interface, D_s and D are, respectively, the binary diffusion coefficients at the surface and in the solution, C is the molar concentration of surfactant in the solution, $\boldsymbol{n}$ is the external normal to the surface, and the square brackets stand for the jump of the quantity enclosed by brackets in the process of transition through the interface.

Eq. (17.92) shows that the change of surfactant concentration at the interface may occur due to a local change of Γ, a convection transfer of surfactant along the interface, surface diffusion, and delivery of matter from the solution. Implied in the last term is the approximation of adsorption-desorption kinetics, i.e. the assumption that the process of adsorption-desorption occurs much faster than diffusion. To solve Eq. (17.92), it is necessary to know the distributions of velocities and concentrations in solutions. All of this considerably complicates the problem. Solutions of certain problems of a similar kind can be found in [2, 20].

Let us dwell on some important practical problems, the possible methods of solution and the final results.

One of ways to enhance the recovery oil from a of porous reservoir is by pumping foam and surfactant into the reservoir. The complexity of the problem lies in the need to account for all factors that influence the form of gas bobbles and for the tendency of the bubbles to block microscopic capillaries of the porous medium.

Therefore the primary goal is to force the bubbles to move inside the microcapillaries, pushing out the oil and dragging the oil film that covers the capillary walls down the capillary. This problem is similar to the problem of motion of a long (but finite) bubble inside a capillary filled with viscous liquid that was already mentioned in Section 17.3 (see Fig. 17.7). The difference is that it is neces-

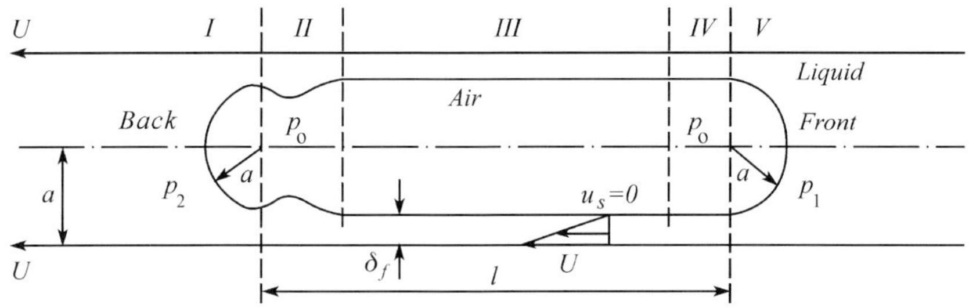

Fig. 17.10 Extrusion of viscous liquid from a capillary by air in the presence of a surfactant: *I*, *V* – back and front hemispherical caps; *II*, *IV* – back and front transitive regions; *III* – region of cylindrical moistening film.

sary to take into account the adsorption of surfactant on the bubble's surface, which causes a change of the surface tension coefficient. Consequently, the hydrodynamic problem must be supplemented with equations of surfactant transfer (17.92) from the liquid to the interface and also with the conditions (17.75) and (17.76). The presence of a surfactant leads to an additional pressure drop pushing the bubble and enhancing the velocity of its motion through the capillary by 2 to 4 orders of magnitude as compared to the bubble's velocity in the absence of a surfactant [20].

Fig. 17.7 depicted the motion of a semi-infinite bubble inside a capillary in the absence of a surfactant. Consider now the case when a surfactant is present in the liquid. Let's switch to a coordinate system where the bubble is at rest and the capillary wall moves from the right to the left with the velocity U (Fig. 17.10). We are mostly interested in two parameters that can be calculated and determined experimentally: the thickness δ_f of the liquid film between the bubble and the capillary wall, and the additional pressure drop Δp inside the liquid, caused by the presence of the bubble.

The detailed analysis of this problem can be found in papers [29–31].

The motion of a long bubble in a capillary tube filled with viscous liquid in the absence of a surfactant is analyzed in [19]. In this paper, it is shown that if the Reynolds number is small and gravity is neglected, the flow depends exclusively on one dimensionless parameter, namely, the capillary number $\text{Ca} = \mu U/\Sigma$, where μ is the liquid's viscosity, U is the velocity of the bubble, and Σ is the surface tension coefficient at the gas – liquid interface. When Ca takes asymptotically small values (i.e. $\text{Ca} \to 0$), the flow can be divided into five regions as shown in Fig. 17.10. A hemispherical cap is formed at each edge of the bubble. The pressure and shape of the cap are controlled by capillary forces only. The hemispherical caps are connected with the cylindrical regions by transitional regions. It has been shown that in the cylindrical regions, the thickness of the moistening film and the additional pressure drop are determined by the expressions

$$\delta_f/a = 1.34\text{Ca}^{2/3}, \quad \Delta p = (9.4\text{Ca}^{2/3}\Sigma)/a. \tag{17.93}$$

Numerous experiments have been performed [32, 33] to test the expressions (17.93). The biggest challenge in conducting these experiments is to ensure that the interface is clean from surfactants and impurities. Consequently, in each experiment there were traces of pollution and surfactants. In all experiments with clean interface the thickness of the moistening film was greater than in the formula (17.93), and the difference was increasing with the increase of Ca. Digressions from the formula were also detected in measurements of Δp, which was higher than suggested by (17.93). In particular, it turned out that Δp depends on the size (length) of the bubble [34]. Similar results were obtained from experiments that involved the addition of surfactants into the liquid [35–37]. At first, pressure grew as the bubble length increased, but then it tended asymptotically to a constant that was much greater than what would be expected for a clean surface.

Experiments have shown that adsorption of a surfactant on the interface creates additional stresses that block the motion of long bubbles inside capillaries. Therefore the presence of these stresses necessitates the creation of a larger pressure drop to push bubbles through the capillary. It also leads to an increase in the thickness of the moistening film at the capillary wall.

Theoretical works [29–31, 38–42] rely on the method of small perturbation and numerical methods to investigate the influence of surfactants on the motion of long bubbles and drops inside capillaries. For infinitely long bubbles, the presence of a surfactant leads to an increase of pressure in the front part of the bubble and of the moistening film's thickness by a factor of $4^{2/3}$ as compared to the case of a pure liquid. The influence of the finite value of length of relatively long bubbles was investigated in the paper [42], where it was shown that the additional pressure drop depends on the bubble length.

Let us estimate the additional pressure drop caused by the presence of surfactant in the liquid. Although in reality the front and rear parts of the bubble's surface have slightly different shapes (the curvature radius of the back "cap" is greater than that of the front "cap"), for simplicity's sake, we consider them as identical hemispheres with radii equal to the radius of the capillary. The length of the liquid film confined between the wall and the bubble is l. The presence of a surfactant in the liquid flowing around the motionless bubble, leads to a transfer of the surfactant to the bubble surface via convective diffusion. This gives rise to a non-uniform distribution of the surfactant at the bubble surface. The surfactant is pushed to the back of the bubble and accumulates there. An increased surfactant concentration caused a reduction of Σ. Therefore, Σ decreases from the front to the rear of the bubble. As a result, the pressure in the rear becomes higher than in the front the bubble, and the difference $p_2 - p_1$ should increase the velocity of the bubble's motion.

Let's make use of some results obtained while solving the problem of the bubble's motion in the absence of surfactants. In this case, the tangential stress at the bubble's surface vanishes because the viscosity of gas is negligible as compared to liquid. For small values of the capillary number $\mathrm{Ca} \ll 1$, the change in the thickness of the layer near the bubble's front is given by the formula (17.31). Then the curvature of the intermediate region of the front meniscus is estimated as

$$1/R_1 + 1/R_2 = \delta'' + 1/a = 2(1 + 1.79(3\mathrm{Ca})^{2/3})/a.$$

Here R_l is the curvature radius of the bubble's meniscus in its normal cross section, and R_2 is the curvature radius of the axial cross section. Obviously, the contribution to the axial pressure drop is due to the second term. Therefore, according to Eq. (17.5), the pressure drop at the bubble's front is $\Delta p_1 = 3.58 \cdot (\Sigma/a)(3\mathrm{Ca})^{2/3}$. The change of curvature in the rear part of the bubble, as was already mentioned, differs from that in the front part. The pressure drop in the rear of the bubble is equal to $\Delta p_2 = 0.94 \cdot (\Sigma/a)(3\mathrm{Ca})^{2/3}$. As a result, the pressure drop between the rear and front parts of the bubble turns out to be

$$p_2 - p_1 = (9.4\mathrm{Ca}^{2/3}\Sigma)/a. \tag{17.94}$$

Now, let a surfactant be present in the liquid. To simplify the analysis, we assume that the surfactant is spreading over the bubble's surface with the velocity $\boldsymbol{u}_s$ and only as a result of surface convection. Then it follows from (17.92) that $\nabla_s \cdot (\Gamma \boldsymbol{u}_s) = 0$. From this equation and the continuity equation $\nabla_s \cdot (\Gamma \boldsymbol{u}_s) = 0$, we obtain

$$(\nabla_s \Gamma) \cdot \boldsymbol{u}_s = 0. \tag{17.95}$$

Since $\nabla_s \Gamma$ is parallel to $\boldsymbol{u}_s$, it follows from this equation that $\boldsymbol{u}_s = 0$. It means that the there is no surfactant transfer over the bubble surface, i.e. the surface is fully retarded and the bubble moves like a solid body, which considerably increases the tangential stress at the bubble surface. The balance of forces in the direction of the x axis on the bubble's surface can be written approximately as

$$\pi a^2 (p_2 - p_1) = 2\pi a l \tau_s = (2\pi a l \mu U)/\delta_f. \tag{17.96}$$

where τ_s is the tangential stress.

Let's estimate the thickness δ_f. The coefficient of surface tension Σ is assumed to be a slowly varying function of x, and the velocity profile in the film is assumed to be linear and satisfying conditions

$$u = 0 \quad \text{at } y = \delta(x), \qquad u = -U \quad \text{at } y = 0. \tag{17.97}$$

From the condition (17.26) stating that flow rate in the film should remain constant, one obtains

$$Q = \frac{1}{2} U \delta_f. \tag{17.98}$$

To determine δ_f, one should use Eq. (17.22) with conditions (17.97). The solution is carried out by the same method as in Section 17.3, and the outcome will be

$$\frac{\delta_f}{a} = 0.643(6\mathrm{Ca})^{2/3}. \qquad (17.99)$$

A substitution of (17.99) into (17.96) yields the pressure difference between the two ends of the bubble:

$$p_2 - p_1 = 9.42\mathrm{Ca}^{1/3}\frac{\Sigma}{a}\frac{l}{a}. \qquad (17.100)$$

Note that $p_2 - p_1$ actually stands for the difference of capillary pressures between the front and the rear of the bubble, since in according to (17.6),

$$p_2 - p_1 = (p_2 - p_0) - (p_1 - p_0) = -\frac{2\Sigma_2}{a} + \frac{2\Sigma_1}{a} = \frac{2}{a}(\Sigma_1 - \Sigma_2). \qquad (17.101)$$

Equating (17.100) with (17.101), we can find the maximum difference of surface tension coefficients at the surface of the bubble:

$$\frac{\Sigma_1 - \Sigma_2}{\Sigma} = 0.471\mathrm{Ca}^{1/3}\frac{l}{a}. \qquad (17.102)$$

Note that relation (17.101) was obtained in the approximation $\mathrm{Ca} \ll 1$ and it also involves the assumption that the change Σ along the x-axis is small, which allowed us to neglect the $d\Sigma/dx$ term in the equation of motion.

Now, compare the formulas (17.94) and (17.101). The pressure difference between the two ends of the bubble in the presence of a surfactant is proportional to $\mathrm{Ca}^{1/3}l/a$, while in the absence of a surfactant it is proportional to $(\mathrm{Ca})^{2/3}$. For instance, if $\mathrm{Ca} \sim 10^{-6}$, then, accounting for the surfactant's presence, we get a pressure difference that is two orders of magnitude greater than it would be without the surfactant. But the comparison of the values of δ_f in both cases (see (17.39) and (17.99)) shows that they differ insignificantly.

So, when a surfactant is present, the velocity of the bubble's motion relative to the capillary wall will be much greater than in the absence of the surfactant.

Another case that is important in many applications involves the motion of a drop in a liquid containing a surfactant, which can be adsorbed at the drop's surface [2]. The motion of the drop results in that, due to the constant stretching of the surface, the surface density of adsorbed surfactant molecules in the front part of the surface will be smaller than in the case of the drop's equilibrium with the solution. In the rear part of the drop, the surface density will exceed the density at equilibrium. Because, unlike the surface of a solid particle, the surface of a drop is mobile, the motion of the liquid will cause surfactant molecules to drift to the rear part of drop and accumulate there. Accumulation of surfactant results in a decreased surface tension in the rear part of the drop. On the other hand, the increase of surfactant concentration in the rear part leads to the appearance of a surface diffusion flux in the opposite direction – from the rear to the front. This flux tries to oppose the surface transfer of the surfactant and in this way it impedes further accumulation of surfactant in the rear part of the drop.

In a specific case, when the rate of surfactant exchange between the external liquid and drop's surface is restrained by the adsorption-desorption process, the concentration of surfactant on the drop's surface is close to the equilibrium value. The drop sinks in the liquid with Stokes' velocity under the action of gravity without getting deformed. The expression for the drop's velocity was derived in [2] and is written as

$$U = 3U_0 \frac{\mu_e + \mu_i + \gamma}{2\mu_e + 3\mu_i + 3\gamma}. \tag{17.103}$$

where $U_0 = \dfrac{2(\rho_i - \rho_L)ga^2}{9\mu_e}$ is the Stokes velocity of a solid particle, $\gamma = \dfrac{2\Gamma_0}{3\alpha a}\dfrac{\partial \Sigma}{\partial \Gamma}$, Γ_0 is the equilibrium surface concentration of the surfactant, α is the proportionality coefficient between the rate of the adsorption-desorption process and the deviation of the surface concentration of surfactant Γ' from its equilibrium value ($j^* = -\alpha\Gamma'$); μ_i, ρ_i and μ_e, ρ_e are, respectively, viscosity and density of the drop and viscosity and density of the external liquid. Let's consider the coefficient γ. Rewrite it in the form

$$\gamma = \frac{2\Gamma_0}{3\alpha a}\frac{\partial \Sigma}{\partial \Gamma} = \frac{2\Gamma_0}{3\alpha a}\frac{\partial \Sigma}{\partial C_0}\frac{\partial C_0}{\partial \Gamma}, \tag{17.104}$$

where C_0 is the surfactant concentration in a region far off from the drop.

In accordance with Gibbs' formula (17.4),

$$\frac{\partial \Sigma}{\partial C_0} = \frac{\Gamma_0 AT}{C_0},$$

therefore for $\Gamma = \Gamma_0 + \Gamma'$ and $\Gamma' \ll \Gamma_0$,

$$\gamma = \frac{2\Gamma_0^2 AT}{3\alpha a C_0}\frac{\partial C_0}{\partial \Gamma_0}. \tag{17.105}$$

The derivative $\partial C_0/\partial \Gamma_0$ in the right-hand part is dependent on the mechanism of the adsorption process. Usually it is assumed that C_0 and Γ_0 are related by the Langmuir adsorption isotherm [13],

$$\Gamma_0 = \frac{kC_0}{1 + kC_0/\Gamma_\infty}, \tag{17.106}$$

where Γ_∞ is the limiting surface concentration that corresponds to the case of the surfactant completely suffusing the drop's surface. Here $kC_0/\Gamma_\infty \gg 1$, meaning that $\Gamma_0 \to \Gamma_\infty$.

At $\gamma \gg \mu_e + \mu_i$, the formula (17.103) turns into the Stokes formula, which corresponds to a completely retarded surface of the drop, and at $\gamma \ll \mu_e + \mu_i$ it transforms into the Hadamard-Rybczynski formula. In the general case, accounting for the surfactant leads to a decrease in the velocity of the drop's descent.

When the surfactant's adsorption at the drop surface is characterized by the rate of its delivery from the volume to the surface, i.e. by the process of convective diffusion, it becomes necessary to solve the appropriate diffusion problem with the conditions $C = C_0$ far away from the drop and $C = C_1$ at the drop's surface. Here C_1 is the concentration of surfactant in the solution in equilibrium with the surface concentration of surfactant Γ. In this case the velocity of drop's descent is determined by the formula (17.103), but with a different value of γ [2]. For instance, if the value of Γ doesn't differ too much from the equilibrium value Γ_0 that corresponds to the concentration C_0, then

$$\gamma = \frac{2AT\Gamma_0\delta}{3DRC_0},$$

where δ is the average thickness of the boundary diffusion layer.

One kind of dependence of Σ from Γ is represented by the so-called equation of interface state [43]

$$\Sigma = \Sigma_\infty + (\Sigma_r - \Sigma_\infty)\Sigma_1(\Gamma/\Gamma_\infty), \quad (0 \le \Gamma \le \Gamma_\infty), \tag{17.107}$$

where Σ_r is the surface tension of pure liquid, and $\Sigma(\Gamma/\Gamma_\infty)$ is a function that depends on the properties of the surfactant. In accordance with [44], this function is equal to

$$\Sigma_1(x) = (\alpha + 1)(1 + \theta(\alpha)x)^{-3} - \alpha, \quad \theta(\alpha) = (1 + 1/\alpha)^{1/3} - 1. \tag{17.108}$$

The parameter α depends on the surfactant's properties. For instance, for $\alpha \ll 1$, we have $\Sigma_1 \sim \Sigma_\infty$, and for $\alpha \gg 1$, we have $\Sigma_1 \sim 1 - \Gamma/\Gamma_\infty$.

Let us examine yet another example of a surfactant influencing surface tension: the suppression of waves on the water surface by a layer of oil spread over the water. In Section 17.4, we have considered capillary waves on a clean water surface without regard for the viscosity of the liquid. When a layer of oil is present, it is obviously necessary to take viscosity into account. In addition, the presence of the surfactant changes the surface tension Σ of the interface, resulting in the appearance of a surface gradient.

First, we consider the influence of the liquid's viscosity on the damping of plane capillary waves on deep water. Suppose the liquid has low viscosity so that viscous effects only manifest themselves inside a thin boundary layer near the interface. Hence, outside the boundary layer, the liquid flow is potential, and the potential is described by the Laplace equation, while the liquid flow near the surface is described by the boundary layer equations with the accompanying condition that the tangent viscous stress at the free interface must be zero. The solution of this problem can be found in [2]. The main difference from the case of a non-viscous liquid is the appearance of a coefficient of the form $\exp(-\beta_1 t)$ in the expression for perturbations of the vertical displacement of the interface (here $\beta_1 = (2k)k\nu$ and k is the wave number). Hence the presence of viscosity leads to the penetration of perturbations into the liquid to the depth $\sim 1/k$ below the interface, and to exponential attenuation of these perturbations with time.

Suppose now that there is a surfactant on the surface. Just as before, we limit ourselves to the case when the transfer of the surfactant along the surface occurs exclusively via convection. Then Eq. (17.92) is true, and for small perturbations of the surface's shape, it is possible to take $u_s = 0$ at $z = 0$. This condition means that the interface is retarded by the surfactant and the velocity in the viscous layer changes more significantly than in the case of a clean surface. Thus the problem becomes similar to the one describing the behavior of perturbations in a viscous boundary layer of an infinite liquid constrained by a solid elastic surface that executes small oscillations [45]. Oscillations of the interface give rise to harmonic oscillations of the longitudinal pressure gradient $\partial p/\partial x$ with frequency ω and to the corresponding velocity perturbation in the liquid. The resultant viscous boundary layer has the thickness

$$\delta \sim 2\left(\frac{2v}{\omega}\right)^{1/2}. \tag{17.109}$$

The damping of perturbations at the liquid surface with time has exponential character, with the damping coefficient

$$\beta_2 = \frac{1}{2}\left(\frac{\omega}{2\nu}\right)^{1/2} k\nu. \tag{17.110}$$

Let's now form the ratio of damping coefficients obtained for perturbations with and without the surfactant.

$$B = \frac{\beta_2}{\beta_1} \sim \left(\frac{\omega}{\nu k^2}\right)^{1/2} = \left(\frac{c\lambda}{2\pi\nu}\right)^{1/2}, \tag{17.111}$$

where λ is the wavelength and c is the wave velocity.

Let's estimate the obtained results. Take $\lambda_{\min}$ and $c_{\min}$ from (17.53) to be the wavelength and the wave velocity. For water ($\nu = 10^{-6}\ \mathrm{m^2/s}$), the outcome is $\lambda_{\min} = 1.7 \cdot 10^{-2}$ m, $c_{\min} = 0.23$ m/s and $B = 25$. Therefore, $B \gg 1$ and $\beta_1 \ll \beta_2$.

Thus, the presence of a surfactant on the interface facilitates a faster damping of waves on the liquid surface. The damping caused by the surfactant is more intense than the damping due to viscosity.

17.6
Pulverization of a Liquid and Breakup of Drops in a Gas Flow

Pulverization of liquid occurs in the process of outflow of a liquid jet under high pressure into a gas medium. This process is of interest to us because of its numerous technical applications. Examples include the injection of combustible liquid (fuel) that occurs in heating systems, gas turbines, diesel and rocket engines; painting a surface by using a pulverizer; water sprinklers used in agriculture; and many other processes in various fields having to do with disintegration of liquids, such as medicine and meteorology. Many devices have been designed with the purpose of pulverizing a liquid to smallest-size droplets.

The importance of the problem of liquid pulverization has led to the emergence of a separate branch of science and engineering out of numerous research projects in this field [46].

It was shown in Section 17.4 that in the case of slow outflow of liquid into a motionless gas, small perturbations of the jet surface produce instability resulting in the jet disintegrating into drops of roughly the same size. In industrial plants, the outflow of liquid from nozzles, atomizers and other devices occurs with high velocity. Because of the perturbations of the liquid-gas interface, which are caused by the interaction of the jet with the walls of devices and with the gas environment, the jet quickly becomes unstable even at small distances from the point where the liquid enters the volume of gas. The jet disintegrates into a set of drops of different sizes. The spectrum of drops thus formed depends on the parameters of the atomizer, the jet hydrodynamics, the aerodynamics of the surrounding gas, the exchange of heat and mass between the liquid and gaseous phases, and on the properties of both the liquid and the gas. The flow regime (which can be laminar or turbulent) has a great influence on the formation of the droplet spectrum. At the present time, there is no perfect theory that would allow us to determine such parameters as the size distribution of drops formed in the process of jet disintegration, the hydro- and aerodynamic characteristics of the flow, etc. Therefore the prevalent method of jet pulverization research is by experiments which aim to find the dispersiveness of the formed gas-liquid mixture and the flow structure, and to check some results of simplified theoretical models.

There are numerous ways to produce jets. The most important requirement is the high velocity of the liquid relative to the gas, which guarantees a finely-dispersed atomization of jets. The liquid can be input into the gas in using a number of techniques, in particular, by injection through the nozzle directly into the bulk of the gas either parallel to or against the gas stream. To achieve a high velocity of liquid outflow, it is necessary to create a big pressure drop inside the spraying device. During the input of liquid into the gas flow through a small orifice at high pressure, the energy of compression converts into kinetic energy. As result, the liquid flows out from the nozzle with a high velocity. Let us give some typical values of the outflow velocity. For hydrocarbon gas mixture, the pressure drop of 0.14 MPa in the atomizer results in the outflow velocity of about 19 m/s, if we disregard the losses due to friction. If we increase the pressure drop to 5.5 Mpa, this will increase the outflow velocity to 117 m/s.

Devices that spray liquids are called atomizers. They are classified as direct-jet atomizers, which create a cylindrical jet, and centrifugal atomizers, which create a swirling jet with a gas core inside. A thin liquid cylindrical layer (sheet) is formed at the exit of a centrifugal atomizer and then expands in the radial direction as the jet moves farther away from the outflow orifice and quickly breaks up into small drops. The expansion angle may range from 30° to 180°. There is a wide variety of atomizers of the above-mentioned types, whose construction depends on the type of the working liquid, on the expected operating conditions, and on the required dispersiveness of the resulting drops. Various kinds of atomizers are presented in [46], which also discusses their advantages and disadvantages and the methods of operation.

The spraying of a liquid relies on two basic processes: the loss of stability of the jet lowing out from the atomizer, with the formation of relatively large drops, and a further breakup of these drops into smaller ones. The loss of stability of a cylindrical jet was discussed (for the elementary case) earlier in Section 17.4. We now consider the second process, namely, the breakup of a large drop.

We start with the most elementary process – the formation of a drop, which flows out slowly from a small aperture with diameter d_0 (tap, pipette etc.) under the influence of the gravity force. The diameter D of the drop, whose moment of "birth" is simply the instant when the drop detaches itself from the aperture, can be found if we equate the weight $\pi\rho_L g D^3/6$ of the drop to the surface tension force $\pi d_0 \Sigma$ acting on the perimeter of the outlet aperture:

$$D = \left(\frac{6d_0\Sigma}{\rho_L g}\right)^{1/3}. \tag{17.112}$$

Using the data adduced in Section 17.1, we find for the water flowing out from the aperture with $d_0 = 1$ mm, the emerging drops have diameter $D = 3.6$ mm. A reduction of the aperture size to $d_0 = 10$ μm will result in the new drop diameter of $D = 784$ μm.

A more thorough analysis shows that the diameter of a drop that falls down after getting detached from a horizontal moistened surface is equal to

$$D = 3.3\left(\frac{\Sigma}{\rho_L g}\right)^{1/3}. \tag{17.113}$$

For water drops, the formula (17.113) gives $D = 9$ mm.

Thus, if the outflow occurs at a slow rate (the liquid is dripping), the balance of two forces – gravity and surface tension – lies at the core of the drop formation process. The size of the emerging drops is rather big. It is obvious that an analogous approach would be completely unsuitable for the majority cases of jet and drop breakup, since they are characterized by high outflow velocities and small size of the emerging drops ($\sim$ 1–100 μm). The gravity force does not have any appreciable influence on the process of drop formation during disintegration of high-velocity liquid jets.

Now, consider the droplet formation resulting from the breakup of relatively large drops in a gas flow. The drop is subjected to an external aerodynamic force (pressure p_G). It should be counterbalanced by the internal pressure of the drop p_i and the capillary pressure p_σ:

$$p_i = p_\sigma + p_G = \text{const.} \tag{17.114}$$

Recall that for a spherical drop,

$$p_\sigma = \frac{4\Sigma}{D}. \tag{17.115}$$

The drop remains stable as long as the change of p_0 is compensated by the change of p_σ so that p_i remains constant. If $p_G \gg p_\sigma$, which is true for large drops, then $p_i \approx p_G$. In this case the change of p_G cannot be compensated and the drop gets deformed and splinters into smaller drops until the increase of p_σ compensates p_G. Such reasoning leads to the concept of critical drop size. The more the drop size exceeds the critical value, the greater is the probability of drop breakup and the shorter the time during which it's likely to happen. It is easy to see that the critical size is given by the maximum value of the diameter. If this value is exceeded, the drop will break apart.

The shape that can be assumed by a drop in a gas flow depends on the character of the flow in the vicinity of the drop, as well as the properties of both the liquid and the gas, such as density, coefficient of viscosity and coefficient of surface tension. The three basic kinds of drop deformations (Fig. 17.11) are described in [46].

1. The drop looks like an oblate spheroid (Fig. 17.11, a). At further deformation the drop becomes a torus and then stretches and splits into fine droplets.
2. The drop looks like a long cylindrical body with rounded ends (cigar) (Fig. 17.11, b). Then the drop breaks into smaller droplets.
3. Local deformations of the drop result in irregular dents and ledges. The protruding parts eventually separate and form smaller droplets (Fig. 17.11, c).

The first shape of the drop is formed under the influence of pressure and viscous stresses acting on the drop surface as the gas is engaged in both translational and rotational flow around the drop. In a shear flow, the drop will take the second shape. The third shape is characteristic of irregular gas flows, for example,

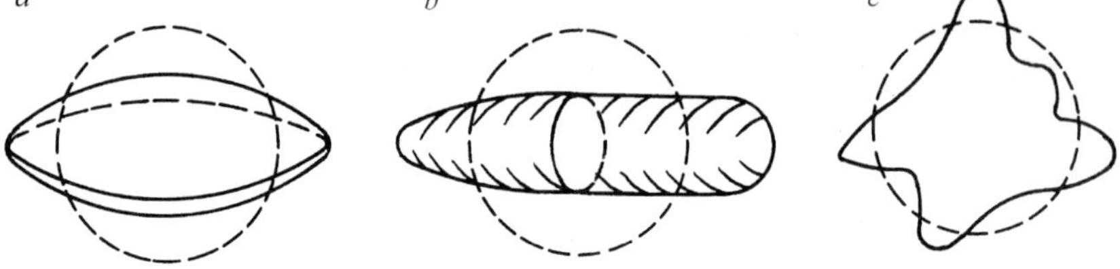

Fig. 17.11 Basic shapes of drops that get deformed in a gas flow.

a turbulent flow. Most often, a large drop splits into two drops of roughly the same size and several drops of smaller sizes. The latter drops are called satellites.

To determine the shape of the drop surface it is necessary to write the balance of forces (dynamical pressure, viscous and surface forces) acting on the drop. If the drop has a low viscosity, it is sufficient to write the balance of aerodynamic resistance force having the order $\rho_G U_G^2$, and surface tension force having the order Σ/D. The ratio of these forces forms a dimensionless parameter

$$\mathrm{We} = \frac{\text{aerodynamic resistance}}{\text{surface tension force}} = \frac{\rho_G U_G^2 D}{\Sigma}, \tag{17.116}$$

called the Weber number. Denote by $S = \pi D^2/4$ the largest cross section of the drop that is perpendicular to the gas velocity U_G. Then the equality of aerodynamic and surface tension forces becomes

$$\frac{\pi D^2}{4} C_D 0.5 \rho_G U_G^2 = \pi \Sigma D, \tag{17.117}$$

where C_D is the resistance coefficient dependent on Re.

The equality (17.117) can be considered as the initial condition at which the drop starts to deform and disintegrate. It can be rewritten as

$$\mathrm{We}_{cr} = \frac{8}{C_D}, \tag{17.118}$$

where We_{cr} is the critical value of the Weber number; when We_{cr} is exceeded, the drop breaks apart.

Eq. (17.117) allows us to estimate the maximum size of a stable drop:

$$D_{\max} = \frac{8\Sigma}{C_D \rho_G U_G^2} \tag{17.119}$$

and the critical velocity of gas for a drop of diameter D (again, the drop will split when the critical velocity is exceeded)

$$U_{cr} = \left(\frac{8\Sigma}{C_D \rho_G D} \right)^{1/2}. \tag{17.120}$$

When we account for viscous stress in the formula for the balance of forces at the drop surface, this results in the appearance of an additional dimensionless parameter called the Ohnezorge number [47]:

$$\mathrm{Oh} = \frac{\text{inner viscosity force}}{\text{surface tension force}} = \frac{\mu_L}{(\rho_L \Sigma D)^{1/2}} = \left(\frac{\rho_G}{\rho_L} \right)^{1/2} \frac{\mathrm{We}^{1/2}}{\mathrm{Re}_L}, \tag{17.121}$$

where μ_L is the viscosity coefficient of the drop.

Then, in view of the correction for internal viscosity, the critical Weber number that corresponds to the drop's state prior to breakup is equal to

$$\mathrm{We}_{cr} = \mathrm{We}_{cr}^{(0)} = (1 + f(\mathrm{Oh})), \tag{17.122}$$

where $\mathrm{We}_{cr}^{(0)}$ is the critical Weber number obtained without regard for viscosity, as, for example, Eq. (17.118).

There exists the following empirical correlation for the formula (17.122) [46]

$$\mathrm{We}_{cr} = \mathrm{We}_{cr}^{(0)} = (1 + 14\mathrm{Oh}). \tag{17.123}$$

Of special interest is the process of drop breakup in a turbulent gas flow. An expression for the dynamic thrust acting on the drop surface in an isotropic turbulent flow of gas is obtained in [2]:

$$Q \sim \frac{C_D \rho_G}{2} \left(\frac{\rho_L}{\rho_G}\right)^{2/3} \frac{{\rho_L}^{2/3}}{\rho_G^{2/3}} R^{2/3}, \tag{17.124}$$

where ρ_L and ρ_G are the liquid and gas densities, R is the drop radius, ε is the specific dissipation of energy in a turbulent gas flow ($\varepsilon \sim \rho_G U_G^3/L$), L is the characteristic linear size of the flow region, for example the diameter of the pipe.

Equating the dynamic thrust Q to the surface tension force for the drop, $2\Sigma/R$, one obtains the formula for the maximum stable size of the drop:

$$D_{\max} \sim \left(\frac{\Sigma}{C_D \rho_L}\right)^{3/5} \left(\frac{\rho_L}{\rho_G}\right)^{2/5} \frac{L^{2/5}}{U_G^{6/5}}. \tag{17.125}$$

or

$$D_{\max} = C\left(\frac{\Sigma}{\rho_G}\right)^{3/5} \varepsilon^{-2/5}. \tag{17.126}$$

According to [48], the constant $C = 0.725$ and $\mathrm{We}_{\mathrm{cr}}$ is

$$\mathrm{We}_{cr} = \frac{2\rho_G}{\Sigma} \varepsilon^{2/3} D_{\max}^{5/3} = 1.18. \tag{17.127}$$

There can exist three possible desintegration regimes for a jet emerging from an atomizer, depending on the relative significance of gravitational, inertial, viscous and surface forces [46]:

1. The influence of the surrounding gas is negligible, and the velocity of jet outflow is small ($\mathrm{Re}_L \ll 1$). This process leads to formation of relatively large drops. This is the case of Rayleigh instability described earlier in Section 17.4.

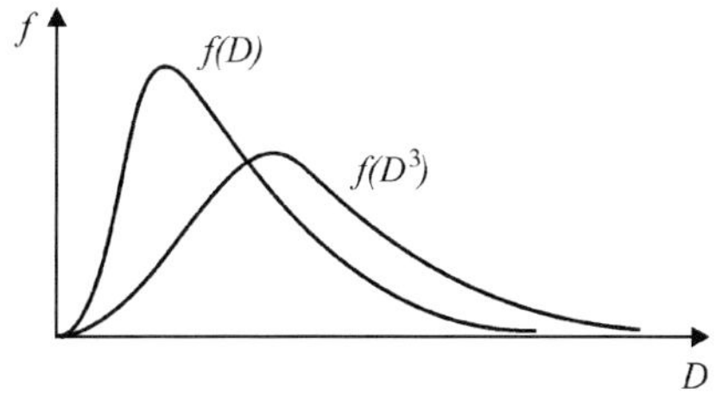

Fig. 17.12 Distribution of drops over sizes in a liquid jet flowing out from an atomizer.

2. Desintegration occurs at finite numbers Re_L, and the influence of the surrounding gas cannot be neglected. A spectrum of droplets is formed.
3. Disintegration 3 occurs at $\mathrm{Re}_L \gg 1$. For all practical purposes, disintegration starts immediately from the moment the liquid enters the gas. A spectrum of droplets is formed in this case; the average diameter of drops is much smaller than the diameter of the jet.

The spectrum of droplets formed when the jet disintegrates is characterized by distribution of drops over diameters $f(D)$ or over volumes $f(V)$, where $V = \pi D^3/6$.

The typical shape of these distributions is shown in Fig. 17.12. The distributions of drops over sizes that have been obtained in experiments can be approximated by some known theoretical distributions [49], which include:

– the normal distribution

$$f(D) = \frac{1}{\sqrt{2\pi}\sigma_n} \exp\left(-\frac{1}{2\sigma_n^2}(D - \bar{D})^2\right), \tag{17.128}$$

where $\bar{D}$ is the expectation value (average diameter), σ_n^2 is the variance;

– the logarithmically normal (lognormal) distribution

$$f(D) = \frac{1}{\sqrt{2\pi} D\sigma_g} \exp\left(-\frac{1}{2\sigma_g^2}(\ln D - \ln \bar{D})^2\right), \tag{17.129}$$

– the Nakayama-Tonosava distribution

$$f(D) = aD^p \exp(-bD^q), \tag{17.130}$$

where a, p, q are empirical constants.

Currently, the most frequently used distribution is the Rosin-Ramler distribution

$$f(D) = qbD^{q-1} \exp(-bD^q), \tag{17.131}$$

where q and b are the empirical constants of the distribution.

The greater the value of q in the distribution (17.130), the closer the distribution gets to a monodispersive one. At $q \to \infty$ we get a monodispersive distribution of drops. For most jets, $1.5 < q < 4$, although for some centrifugal atomizers the value of q can be as high as 7.

Some additional data about the breakup of drops can be found in [50–53].

18
Efficiency of Gas-Liquid Separation in Separators

At the present time, there is a great variety of separator designs (see Section 2.1). Nevertheless, they can be grouped into two basic classes according to the physical principles of gas-liquid mixture separation: gravitational and inertial separators.

Gravitational separators are big horizontal or vertical containers, in which separation of phases occurs due to the force of gravity. Since the size of drops entering the separator from the supply pipeline is small, the efficient removal of these drops from the gas flow due only to gravity will take a rather long time. Consequently, gravitational separators should have large sizes. In inertial separators, the separation of phases occurs due to inertial forces as the mixture flows around various obstacles (jalousies, grids, strings etc.) or spins in centrifugal devices.

In modern separator constructions, both principles are usually applied. Therefore any separator, as a rule, consists of two parts: the sedimentation section and the end section. In the sedimentation section, drop sedimentation occurs due to gravity, and in some constructions with tangential gas input sedimentation occurs together with deposition due to the centrifugal force. The end section is equipped with various drop-catching devices, for example, with centrifugal branch pipes, mesh and string nozzles orifices etc. In these devices, separation of drops from the gas occurs due to inertia forces. Fine cleaning of gas from very small droplets is achieved by using filter cartridges, which are cylinders with fibrous stuffing through which the mixture is filtered.

The quality of separation of gas-liquid mixture in separators depends on the velocity (flow rate) of gas, thermobaric conditions, physical and chemical properties of the phases, geometrical parameters and, most importantly, dispersiveness of the liquid phase (distribution of drops over sizes and the parameters of this distribution) that goes into the separator together with the gas flow from the delivery pipeline. The key parameter of this distribution is the average size of drops that are formed in the gas flow in the delivery pipeline. Therefore, it depends on the parameters of the pipeline and also on the parameters of the device of preliminary condensation (DPC), which, as a rule, is placed at some distance from the separator (see technological schemes in Chapter 1). Therefore, the efficiency of gas-liquid mixture separation is influenced not only by the parameters of the separator, but also by the particular details of the technological circuit before the separator.

Separation of Multiphase, Multicomponent Systems. E. G. Sinaiski and E. J. Lapiga

ISBN: 978-3-527-40612-8

The main parameter that serves as a measure of separation quality is the coefficient of efficiency (CE), which is equal to the ratio of the volume of the liquid phase settled in the separator to the quantity of the liquid phase contained in the mixture at the separator entrance:

$$\eta = \frac{W_s}{W_0} = \frac{W_0 - W_1}{W_0} = 1 - K. \tag{18.1}$$

Here W_0, W_1, W_s are, respectively, the volume concentration of liquid at the entrance to the separator, the volume concentration at the exit, and the fraction of volume that settles in the separator; K is the coefficient of ablation.

Denote by $n_0(R)$ the distribution of drops over radii at the entrance to the separator, and by $n_1(R)$ – the distribution at the exit. Then

$$W_0 = \int_0^\infty \frac{1}{3}\pi R^3 n_0(R)\,dR, \quad W_1 = \int_0^\infty \frac{4}{3}\pi R^3 n_1(R)\,dR.$$

Let's define the "transfer function" of separator as the ratio of the two size distributions of drops – at the exit and at the entrance:

$$\Phi(R) = \frac{n_1(R)}{n_0(R)}. \tag{18.2}$$

Then the expression (5.1) can be represented as

$$\eta = 1 - \int_0^{R_m} R^3\Phi(R)n_0(R)\,dR \Big/ \int_0^\infty R^3 n_0(R)\,dR. \tag{18.3}$$

where R_{m} is the minimum radius of drops.

If the separator consists of several sections connected in series, with each section characterized by its own transfer function $\Phi_i(R)$, than the transfer function of the separator is

$$\Phi(R) = \prod_{i=1}^{N} \Phi_i(R). \tag{18.4}$$

From formulas (18.3) and (18.4) we can find the CE of a separator consisting of several sections connected in series:

$$\eta = 1 - \prod_{i=1}^{N} K_i, \tag{18.5}$$

where K_i is the ablation coefficient of i-th section equal to

$$K_i = \int_0^{R_m} R^3 \Phi_i(R) n_0(R)\, dR \Big/ \int_0^{R_m} R^3 n_0(R)\, dR, \tag{18.6}$$

If the separator sections are connected in parallel, multiplication in formulas (18.4) and (18.5) gets replaced by summation.

Thus, in order to determine the CE of the separator, it is necessary to know the size distribution $n_0(R)$ of drops entering the separator, and also the transfer function $\Phi_i(R)$ of each section.

Inasmuch as the CE essentially depends on the average size of the drops, any influence on the gas-liquid mixture that results in the augmentation of drops increases the CE of the separator. There are two basic mechanisms of drop augmentation: coagulation and condensation.

The first mechanism is based on the interaction of drops, specifically, on their collision and subsequent merging. The second mechanism is based on the exchange of mass between the liquid and gaseous phases as a result of violation of phase balance during the passage of mixture through the DPC. The rates of drop augmentation due to condensation and coagulation depend on such factors as the initial sizes of the drops, their volume concentrations, thermobaric conditions, properties of phases, and flow parameters. If a DPC is placed before the separator, then in addition to the drops of condensate already present in the flow, very small droplets are formed, whose size grows rapidly as they move in the supply pipeline from the DPC to the separator due to the simultaneous processes of condensation and coagulation. Since each of these processes is characterized by a certain time, the knowledge of how the radius of drops changes with time enables us to determine the optimal distance from the separator, at which we should place the DPC in order to maximize the CE of the separator.

The technical registration certificate of the separator always specifies its efficiency, which is usually quite high, but it does not indicate under what conditions it can be achieved. Therefore the primary goal is to find the CE for a given separator design as a function of pressure, temperature, gas flow rate, and construction parameters.

The subject of the present chapter is the methods of calculating CE's of elementary gravitational separators. It also analyzes the influence of parameters of the supply pipeline and the DPC on the CE. Because the input of the mixture is usually carried out in such a way that the longitudinal velocity profile at the entrance is non-uniform in its cross section, and besides, the drops can also change their size, the derivation of the expression for CE is carried out taking into account the non-uniformity of the velocity profile and a possible change of the radius of drops. The calculation methods used in this chapter for gravitational separators will be applied again in chapter 19 to determine the CE's of sedimentation sections of separators that also contain additional droplet-catcher sections of various designs.

18.1
The Influence of Non-Uniformity of the Velocity Profile on the Efficiency Coefficient of Gravitational Separators

The gas-liquid mixture arriving at the separator entrance contains a small volume concentration of the liquid phase ($W_0 \ll 1$). It means that the liquid phase has practically no influence on the velocity distribution inside the flow. It is also possible to neglect the mutual interaction of drops, i.e. the constrained character of their motion. Suppose that the velocity profile $u_0(y)$ is given at the entrance to the separator. Direct the x-axis along the axis of the separator, and the y-axis perpendicular to the separator axis. For simplicity's sake, let us take a separator with a rectangular cross section. The influence of the walls' curvature in the case of the circular cross section will be studied later. The equations of motion for a drop of radius R in the approximation that neglects inertia have the form:

$$\frac{dx}{dt} = u_x, \quad \frac{dy}{dt} = u_y - u_s, \tag{18.7}$$

where u_x and u_y are the components of the gas velocity, and u_s is the drop descent velocity, that is, Stokes velocity

$$u_s = \frac{2\Delta\rho g R^2}{9\mu_G}, \quad \Delta\rho = \rho_L - \rho_G. \tag{18.8}$$

It is assumed, that drops radius does not change during motion, i.e. condensation and coagulation of drops are not taken into account.

Since the gas velocity is small in comparison with the speed of sound, the gas is incompressible and its density ρ_G is constant. In this case the continuity equation yields

$$u_x = \frac{\partial\psi}{\partial y}, \quad u_y = -\frac{\partial\psi}{\partial x}. \tag{18.9}$$

where ψ is a function of the flow.

It follows from Eqs. (18.7) and (18.9) that

$$\frac{dy}{dx} = \frac{-\partial\psi/\partial x + u_s}{\partial\psi/\partial y}. \tag{18.10}$$

and from Eq. (18.10), we find:

$$d\psi = -u_s\, dx$$

or

$$\psi(x, y) = C - u_s x. \tag{18.11}$$

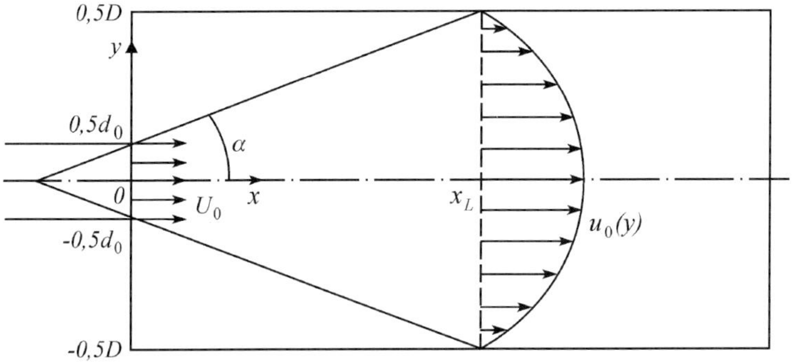

Fig. 18.1 Inflow of mixture into the separator.

If the function $\psi(x, y)$ has been found by solving the appropriate hydrodynamic problem, then the expression (18.11) represents the equation of the drop trajectory. The constant C is determined by the initial position of the drop at the separator entrance. As far as the diameter of the branch pipe for gas input is much smaller than the cross-sectional diameter of the separator, the gas flows into the separator as an expanding jet with the opening angle $2\alpha \approx 28^\circ$ (Fig. 18.1). Then the distance at which the jet reaches the wall of the separator equals

$$X_L \approx 2d_0\left(\frac{D}{d_0} - 1\right), \tag{18.12}$$

where d_0 and D are the diameters of the gas input branch pipe and the separator. In particular, at $D \gg d_0$, which is nearly always true, $X_L \approx 2D$.

Thus, in the region $x > X_L$ the profile $u_0(y)$ of longitudinal velocity is established.

As a rule, a special device (baffle-plate, damper etc.) is placed at the separator entrance. Its purpose is to increase the opening angle of the jet, reducing the stagnant zone between the separator wall and the jet boundary. Thereby a stable profile of longitudinal velocity is established at a small distance from the separator entrance. Consider the sedimentation of drops in this region. The average velocity of the flow is equal to

$$U = \frac{1}{D}\int_0^D u_0(y)\,dy.$$

Denote through $y_0(R)$ the height of the layer that will be depleted of all drops with radius R along the entire length L of the separator. Then the transfer function will be equal to

$$\Phi(R) = \int_0^{y_0(R)} u_0(y)\,dy = 1 - \frac{\psi(0, y_0)}{u_s D}. \tag{18.13}$$

We took into account expressions (18.2) and (18.11) in deriving (18.13). We have also used the inequality $X_L \ll L$ which allowed us to shift the condition formulated for $x = X_L$ to another point $x = 0$. To determine $\psi(0, y_0)$, one should use the expression (18.11). With this purpose in mind, let us consider a trajectory that begins at the point $(0, y_0)$ and ends at the point $(L, 0)$. Then

$$\psi(0, y_0) = \psi(L, 0) + Lu_s. \tag{18.14}$$

Physically, the difference between two values of the flow function ψ_1 and ψ_2 is equal to the gas flow rate between the stream-lines $\psi = \psi_1$ and $\psi = \psi_2$. Since the total linear gas flow rate is equal to

$$Q = \int_0^D u_0(y)\, dy,$$

and the flow function is determined up to a constant, the condition $\psi(x, 0) = 0$ should be satisfied at the bottom wall of the separator. Then we have $\psi(x, D) = Q$. Because of this, we also get $\psi(L, 0) = 0$ and then, from the formula (18.14), there follows $\psi(0, y_0) = Lu_s$.

Now the transfer function of the separator can be found from expression (18.2):

$$\Phi(R) = \begin{cases} 1 - \left(\dfrac{R}{R_m}\right)^2 & \text{at } R \le R_m, \\ 0 & \text{at } R > R_m, \end{cases} \tag{18.15}$$

where $R_m = (9DQ_G\mu_G/2\Delta\rho g V)^{1/2}$; Q_G is the volume gas flow rate and V the volume of the sedimentation section.

If the drop radius changes in the course of the motion as $R = R(x, R_0)$, where R_0 is the initial radius of drop at the separator entrance, then the formula (8.11) gets altered:

$$\psi = \psi(0, y_0) - \int_0^L u_s(x, R_0)\, dx. \tag{18.16}$$

The transfer function in this case is equal to

$$\Phi(R_0) = 1 - \frac{1}{DU}\int_0^L u_s(x, R_0)\, dx. \tag{18.17}$$

The efficiency coefficient is determined by the ratio (18.3), but the integration should be carried out over the initial drops radius R_0.

Summarizing, we should make the following observation: the expression for the efficiency coefficient clearly implies that for a horizontal gravitational separator, this coefficient does not depend on the velocity profile – assuming that the gas flow rate is constant.

18.2 The Efficiency Coefficient of a Horizontal Gravitational Separation

We model the separator as a volume of rectangular cross section with length L and height D_h. The distribution of drops over radii $n_0(R)$ is given at the entrance. The physical meaning of $n_0(R)$ is that $n_0(R)\,dR$ is equal to the number of drops whose radius belongs to the interval $(R, R+dR)$, and the number of drops and their volume in a unit volume of the mixture are, respectively,

$$N_0 = \int_0^\infty n_0(R)\,dR, \quad W = \int_0^\infty V n_0(R)\,dR, \quad V = \frac{4}{3}\pi R^3. \tag{18.18}$$

Consider sedimentation of drops in the gravitational field, supposing that drops move independently from each other. Direct the x-axis along the separator axis; the y-axis will be vertical, i.e. directed against the force of gravity. The equations of motion for the drop then become

$$\begin{aligned} \frac{4\pi R^3 \rho_L}{3}\frac{dv_x}{dt} &= 0.5 f \rho_G \pi R^2 |\mathbf{u}_r| (U_h - v_x), \\ \frac{4\pi R^3 \rho_L}{3}\frac{dv_y}{dt} &= \frac{4\pi R^3 g \Delta\rho}{3} - 0.5 f \rho_G \pi R^2 |\mathbf{u}_r| v_y. \end{aligned} \tag{18.19}$$

Here $\mathbf{u}_r = (U_h - v_x, v_y)$ is the velocity of the drop relative to the gas; $v_x = dx/dt$ and $v_y = dy/dt$ are the components of the drop's velocity; U_h is the gas velocity; f is the resistance coefficient equal to

$$f = \begin{cases} 24\mathrm{Re}^{-1} & \text{at } \mathrm{Re} < 2, \\ 18.5\mathrm{Re}^{-0.6} & \text{at } 2 < \mathrm{Re} < 500, \\ 0.44 & \text{at } \mathrm{Re} > 500, \end{cases} \tag{18.20}$$

$\mathrm{Re} = 2Rr_G u_r/\mu_G$ is the Reynolds number of the drop.

It is necessary to set the position and the velocity of the drop at the initial moment

$$x(0) = 0, \quad y(0) = y_0, \quad (dx/dt)_{i=0} = U, \quad (dy/dt)_{i=0} = 0. \tag{18.21}$$

Solving equations (18.19) with the conditions (18.21), we can obtain all possible trajectories for a drop with radius R. By trying all possible values of y_0, it is possible to find the minimum radius R_m of drops that can be settled in the separator. The meaning of R_m is that all drops with radius $R > R_m$ will be removed from the gas flow and settled in the separator. In order to determine R_m, it is necessary to select from all trajectories the one that starts at the point $(0, D_h)$ and ends at the point $(L, 0)$.

In the general case, the equations (18.19) must be solved numerically. However, the numerical procedure is inconvenient to us because later on, we will want to use the results to determine the CE of the separator. Let us try to simplify the equations. First, we estimate the terms that appear in these equations.

Rewrite the first equation (18.19) as

$$\varepsilon \frac{d(v_x/U_h)}{d(t/t_c)} = \frac{|\mathbf{u}_r|}{U_h}\left(1 - \frac{v_x}{U_h}\right), \tag{18.22}$$

where $t_c = L/U_h$ is the characteristic residence time (i.e. time spent by the drop inside the separator), and $\varepsilon = 8R\rho_L/3f\rho_G t_c U_h$ is a dimensionless parameter.

For gas-liquid mixtures in a separator, we can make the following estimates: $R/L \leq 10^{-5}$; $\rho_L/\rho_G \sim 10^{-2}$; $f \sim 1$ and $\varepsilon \leq 10^{-2}$. At these values, it follows from (18.22) that $v_x \approx U_h$. Note that as the drop size increases, the ratio R/L and the parameter ε increase as well and the inequality $\varepsilon \ll 1$ can be violated. Thus, for $R/L \sim 10^{-3}$ we get $\varepsilon \sim 1$ (for example, $L \sim 1$ m, $R \sim 10^{-3}$ m). However, drops this big cannot be formed in the turbulent flow in a pipeline. And even if they exist, their number is very small. Besides, large drops are easily separated from the gas in the settling section. So, the assumption $\varepsilon \ll 1$ means that the inertia of the longitudinal motion of drops can be neglected. At the same time, $|\mathbf{u}_r| \approx v_y$ and the second equation (18.2) can be rewritten as

$$\frac{1}{g}\frac{dv_y}{dt} = \frac{\Delta\rho}{\rho_L} - \frac{3f}{8Rg\rho_L}v_y^2. \tag{18.23}$$

The estimation of the inertial term in the left-hand side of Eq. (18.23) can be carried out in a similar way. Since $dv_y/dt \sim v_y^2/D_h$, the ratio of the inertial force to the resistance force is equal to

$$\varepsilon_1 = \frac{1}{g}\frac{dv_y}{dt} : \left(\frac{3f}{8Rg\rho_L}v_y^2\right) \sim \varepsilon\frac{L}{D_h}.$$

When $\varepsilon < 10^{-2}$ and $L/D_h \sim 3$, we have $\varepsilon_1 \leq 3 \cdot 10^{-2}$.

Thus, when the inequalities $\varepsilon \ll 1$ and $\varepsilon_1 \ll 1$ are true, the inertial terms in equations (18.19) can be neglected. Then it follows from (18.19) that

$$v_x = U_h, \quad v_y^2 = \frac{8Rg\Delta\rho}{3\rho_G f}. \tag{18.24}$$

Since, according to (18.20), f depends on v_y, the relation (18.24) is a nonlinear equation with respect to v_y. Let us introduce a dimensionless parameter

$$\mathrm{Ar} = \frac{8R^3\rho_G^2 g\Delta\rho}{3\rho_G\mu_G^2}, \tag{18.25}$$

called the Archimedes number. According to [54], the expression for the resistance coefficient of the particle can be written as

$$f = \frac{4Ar}{3\mathrm{Re}^2}, \tag{18.26}$$

where, according to the formula (18.20), the Reynolds and Archimedes numbers are connected by the following approximate dependence:

$$\mathrm{Re} = \frac{\mathrm{Ar}}{18 + 0.575\mathrm{Ar}^{1/2}}. \tag{18.27}$$

Now, using expressions (18.25) and (18.26), we can determine the sedimentation velocity of a drop with radius R from (18.24):

$$v_y = \frac{\mu_G \mathrm{Ar}}{2R\rho_G(18 + 0.575\mathrm{Ar}^{1/2})}. \tag{18.28}$$

Taking the inertialess approximation and neglecting the Archimedes force, we see that the equations of motion for the drop reduce to

$$\frac{dx}{dt} = U_h, \quad \frac{dy}{dt} = -v_y, \quad x(0) = 0, \quad y(0) = y_0. \tag{18.29}$$

Assuming U_h to be constant, one obtains

$$y = y_0 - \frac{v_y x}{U_h}. \tag{18.30}$$

Assuming in (18.30) $y_0 = D_h$, $x = L$, $y = 0$ and using the relation (18.28), we get the following equation for the minimum drop radius

$$\frac{\mathrm{Ar}}{18 + 0.575\mathrm{Ar}^{1/2}} = \frac{2RU_h D_h \rho_G}{L\mu_G}. \tag{18.31}$$

Denote the minimum radius of drops by R_{ms}, provided that they descend with Stokes velocity (see the formula (18.16)):

$$R_{ms} = \left(\frac{9U_h D_h \mu_G}{2\Delta\rho g L}\right)^{1/2}. \tag{18.32}$$

Then, if we introduce a dimensionless minimum radius of the drop $r_m = R_\mathrm{m}/R_\mathrm{ms}$, Eq. (18.31) will change to

$$r_m^2 \approx 1 + 0.28\beta^{1/4} r_m^{3/2}, \quad \beta = U_h^3 D_h^3 \rho_G^2 / L\mu_G.$$

For the characteristic values of parameters entering β, we have β ranging between 1 and 10. This condition allows us to find an approximate solution of the equation

$$r_m^2 \approx 1 + \frac{0.28\beta^{1/4}}{2 + 0.42\beta^{1/4}}. \qquad (18.33)$$

Now we can proceed to determine the CE of the separator. We start with the case of a monodisperse mixture with drops of the same size R_{av} at the entrance. The volume concentration of liquid at the entrance W_0 (for any given values of pressure and temperature, W_0 can be determined from the equations of vapor-liquid balance) is given. To find the amount of the liquid phase settled in the separator, it is sufficient to consider a drop whose final position at the exit will be $(0, \mathrm{L})$ and to determine its initial position at the entrance cross section. From (18.30), we find $y_0 = v_y L/U_h$. Now it is easy to find the volume concentration of the liquid phase at the exit $W_1 = (1 - y_0)W_0$ and the CE of the separator:

$$\eta = y_0/D_h. \qquad (18.34)$$

Substituting here the expressions for y_0 and v_y from (18.28), one gets:

$$\eta = \frac{L\mathrm{Ar}}{2R_{av}\mathrm{Re}(18 + 0.575\mathrm{Ar}^{1/2})}. \qquad (18.35)$$

Denote by $t = L/U_{\mathrm{h}}$ the residence time of the mixture in the separator, and by $t_{\mathrm{d}} = 2R_{\mathrm{av}} \cdot \mathrm{Re}(18 + 0.575\mathrm{Ar}^{1/2})/U_{\mathrm{h}}Ar$ – the time it takes for the drop of radius R to leave the layer whose height is D. Then (18.35) takes the simplest form:

$$\eta = t/t_d, \quad t \le t_d. \qquad (18.36)$$

Thus, for a monodisperse mixture, the CE depends linearly on the residence time of the mixture in the separator. Consider now the case when the gas-liquid mixture arriving at the entrance has a continuous size distribution of drops $n_0(R, y)$ and the volume concentration

$$W_0 = \int_0^\infty \frac{4}{3}\pi R^3 n_0(R, y)\, dR.$$

Since there are no drops with radius $R > R_{\mathrm{m}}$ at the exit, the volume concentration of the liquid phase at the exit is

$$W_1 = \int_0^{R_m}\int_{y_0}^{D_h} \frac{4}{3}\pi R^3 n_0(R, y)\, dy\, dR.$$

Now it is possible to determine the CE of the separator:

$$\eta_h = 1 - \frac{4\pi}{3W_0} = \int_0^{R_m}\int_{y_0}^{D_h} R^3 n_0(R, y)\, dy\, dR, \quad y_0 = v_y\frac{L}{U_h}. \qquad (18.37)$$

Take the distribution (14.1), which is formed in the separator supply pipeline, to be the initial distribution of drops. Then, considering the distribution at the entrance as uniform over the cross section, we obtain from the expression (18.37)

$$\eta_h = \frac{\exp(-3\sigma^2)}{\sqrt{2\pi}\sigma W_0} \int_0^{R_m} R^2 \left(1 - \frac{y_0}{D}\right) \exp\left(-\frac{\ln^2 R/R_1}{2\sigma^2}\right) dR. \tag{18.38}$$

Introduce the following dimensionless variables:

$$z = \frac{R}{R_{av}}, \quad z_1 = \frac{R_1}{R_{av}} = \exp(-0.5\sigma^2), \quad \tau = \frac{t}{t_{ds}} = \frac{Lu_{s0}}{U_h D_h},$$

$$\mathrm{Ar} = \frac{8R^3 \rho_G g \Delta\rho}{\mu_G^2}, \quad \mathrm{Ar}_{av} = \frac{8R_{av}^3 \rho_G g \Delta\rho}{\mu_G^2}, \quad z_m = \frac{R_m}{R_{av}}.$$

Here t is the residence time of mixture in the separator; t_{ds} is the time it takes for the drop with radius R_{av} to leave the layer of height D_h if this drop moves with the Stokes velocity $u_{s0} = 2g\Delta\rho R_{av}^3/9\mu_G$.

Written in new variables, the expression (18.38) for the CE will take the form

$$\eta_h = 1 - \frac{\exp(-3\sigma^2)}{\sqrt{2\pi}\sigma} \int_0^{z_m} z^2 \left(1 - \frac{\tau z^2}{1 + 0.032\mathrm{Ar}_{av}^{1/2} z^{3/2}}\right) \exp\left(-\frac{\ln^2 z/z_1}{2\sigma^2}\right) dz \tag{18.39}$$

Looking at (18.33), we see that the number z_m appearing in (18.39) equals

$$z_m = 1 + \left(\frac{0.28\beta^{1/4}}{2 + 0.42\beta^{1/4}}\right)\tau^{-1/2}, \quad \beta = 0.00017\tau^{-3}\mathrm{Ar}_{av}^2. \tag{18.40}$$

The transfer function of our separator is equal to

$$\Phi_h(R) = \begin{cases} 1 - \dfrac{z^2\tau}{1 + 0.032Ar^{1/2}z^{3/2}} & \text{at } R \le R_m, \\ 0 & \text{at } R > R_m. \end{cases} \tag{18.41}$$

So, the expression (18.39) gives the dependence of the CE of a horizontal gravitational separator on dimensionless parameters ρ, σ, τ, z_1 and Ar_{av} which describe the dispersiveness of the flow, the geometrical proportions of the separator, and also the hydrodynamic, physical and chemical parameters of the mixture. It should be noted that the variance σ^2 of the radial distribution of drops that forms in the separator supply pipeline changes within small limits ($0.4 \le \sigma \le 0.5$) for a large velocity range as shown in [3]; therefore, in the further calculations this variance will be taken to be 0.45.

In Fig. 18.2, the dependence of η on the parameters τ and Ar_{av} is shown. As an example, consider the CE calculation for a separator at the following parameter

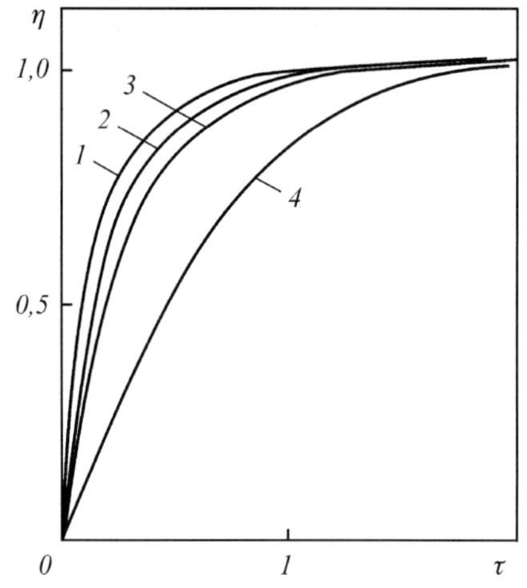

Fig. 18.2 Dependence of the efficiency η of a horizontal gravitational separator on τ for different values Ar_{ar} at $\sigma = 0.5$: 1 – 1; 2 – 10; 3 – 10^2; 4 – 10^3.

values: $D_h = 1.6$ m, $L = 3$ m, $p = 5$ MPa, $T = 250$ °K, $\rho_L = 750$ kg/m^3, $\rho_G = 40$ kg/m^3, $\mu_G = 1.2 \cdot 10^{-5}$ Pa·s. Fig. 18.3 shows the dependence of the CE on the gas flow rate Q_G at normal conditions for various values of the diameter d of the separator supply pipeline. From this dependence, we see that the CE decreases with the increase of the gas flow rate and with the decrease of the pipeline diameter.

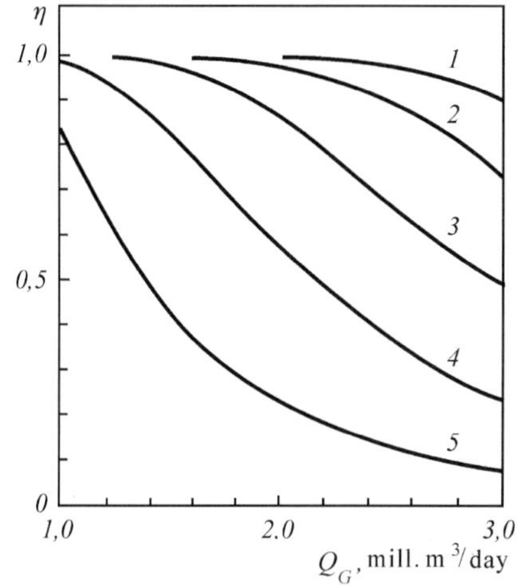

Fig. 18.3 Dependence of η of a horizontal gravitational separator on the gas flow rate Q_G for different values d, m: 1 – 0.35; 2 – 0.3; 3 – 0.25; 4 – 0.2; 5 – 0.15.

The latter correlation has a simple explanation: as d gets smaller, the average drop radius ($R_{av} = R_{cr}$, where R_{cr} is given by the formula (14.14)) decreases, so the sedimentation velocity in the separator decreases as well. On the horizontal axis, we plotted the gas flow rate under the normal conditions, which is a standard convention. Therefore, all other things being equal, an increase in gas pressure or a decrease in temperature will result in a reduced gas flow rate under the operating conditions. The velocity of the gas will be reduced and as a consequence, the CE of the separator will increase. However, such a dependence of the CE on pressure is not characteristic of all separators. Later on, it will be shown that for some types of separators the dependence CE(p) is not a monotonous function.

18.3 The Efficiency Coefficient of Vertical Gravitational Separators

Consider the separation of liquid from gas in a vertical gravitational separator where the flow is directed against the force of gravity. Obviously, the drops that reach the exit of the separator will be the ones that satisfy the inequality $u_s < U_v$, where u_s is the sedimentation velocity of a drop with the radius R, and U_v is the flow velocity, which is assumed to be constant. The minimum size of drops is determined from the condition $u_s = U_v$. Let us invoke the expression (18.28) for velocity u_s. Then in order to determine the minimum drop radius, one should solve the following equation:

$$\frac{\mathrm{Ar}}{2R(18 + 0.575\mathrm{Ar}^{1/2})} = \frac{U_v \rho_G}{\mu_G}. \tag{18.42}$$

A comparison of equations (18.31) and (18.42) for the minimum drop radius in horizontal and vertical separators shows that if the equality $U_v/U_h = D_h/D_v$ holds (here the bottom indexes v and h correspond to the parameters of horizontal and vertical separators), then, all other things being equal, the minimum radii of drops in both separators coincide. But this certainly does not mean that their CE's should be the same, because the ratio of liquid phase volume concentrations at the exit of the two separators is

$$W_1^{(h)}/W_1^{(v)} = \int_0^{R_m^{(h)}} \int_{y_0^{(h)}}^{\infty} R^3 n_0(R, y)\, dy\, dR \bigg/ \int_0^{R_m^{(v)}} \int_{y_0^{(v)}}^{\infty} R^3 n_0(R, y)\, dy\, dR < 1.$$

If we take (14.1) as the initial distribution and consider it to be uniform at the entrance, the CE of a vertical separator can be represented as

$$\eta_h = 1 - \frac{\exp(-3\sigma^2)}{\sqrt{2\pi}\sigma} \int_0^{z_m} z^2 \exp\left(-\frac{\ln^2 z/z_1}{2\sigma^2}\right) dz. \tag{18.43}$$

$z_m = r_m(U_v/u_{s0})^{1/2}$, $r_m = R_m/R_{ms}$, $R_{ms} = (9\mu_0 U_v/2g\Delta\rho)^{1/2}$, $\beta = U_v^3 \rho_G^2/\mu_G g\Delta\rho$, and r_{m} is determined by the expression (18.33).

The CE of a vertical gravitational separator depends on parameters σ, β, U/u_{s0} and Ar_{av}. The character of dependence of η on each of these parameters is the same as for a horizontal separator. However, other things being equal, the same value of CE can be attained in vertical separators at much smaller gas flow rates, then in horizontal ones. Another noteworthy fact is that, according to (18.43), η does not depend on the height L of the separator. This contradicts the experimental results, according to which for a given velocity of gas and with all other thing being equal, the amount of condensate separated the from gas increases with the height L. The higher the gas velocity U, the more noticeable this phenomenon becomes. Experiments also suggest that, starting from a certain height, the separator CE practically does not change. It means that the process responsible for this effect operates for only a limited time. As time goes on, it gradually disappears.

One of the mechanisms that influence the amount of condensate separated from gas is coagulation of drops. In the process of coagulation, the average volume of a drop increases. Fine drops, which sedimentation velocity $u_s < U$, move upwards with the flow and can coagulate in the process of motion. The rate of coagulation depends on the volume concentration of drops. Since the main bulk of the liquid is enclosed in large drops, which are separated early on, near the separator entrance, the volume concentration of drops in the ascending flow is small and therefore drop coagulation occurs rather slowly. Let t_k be the characteristic time of coagulation. Then, if the residence time of the mixture in the separator $t = L/U < t_k$, the change of height L will influence CE; this won't happen if $t > t_k$. An increase in the gas velocity U, on the one hand, increases the minimum radius of drops and thus worsens separation efficiency, and on the other hard, intensifies coagulation of drops, resulting in a dependence of the amount of settled condensate on L. All these theoretical conclusions are in good agreement with the dependence of CE on L and U, obtained from experiments.

The growth of drops in the flow can also occur due to condensation, but this can only happen in the absence of phase equilibrium. It was shown earlier that phase equilibrium develops much faster than dynamic equilibrium. Therefore, condensation-driven growth of drops in the separator is only possible if the DPC is located in the immediate vicinity of the separator, for instance, directly at the entrance.

Later on we shall be estimating the effect of drop coagulation and condensation on the value of CE.

If we disregard these processes, the key parameters influencing the CE of a vertical gravitational separator will be the gas flow rate at normal conditions Q, pressure p, temperature T, and also the diameters of the separator D and the supply pipeline d. The character of these dependences is the same as for horizontal separators. But for the same parameters values, the CE of a vertical separator is smaller than that of a horizontal one. Fig. 18.4 shows the dependence of CE on Q and d.

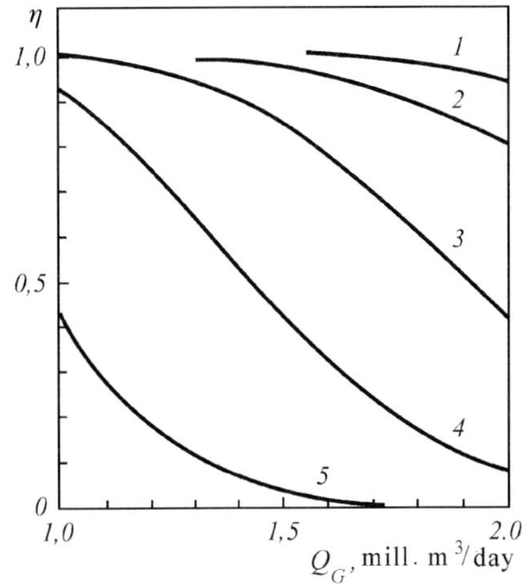

Fig. 18.4 Dependence of the efficiency coefficient for a vertical gravitational separator η on the gas flow rate Q_G for different values of d, m: 1 – 0.35; 2 – 0.3; 3 – 0.25; 4 – 0.2; 5 – 0.15.

If we assume that the drops are setting with the Stokes velocity, then the transfer function of the vertical gravitational separator may be written in the simplest form

$$\Phi_v(R) = \begin{cases} 1 & \text{at } R < R_m, \\ 0 & \text{at } R \geq R_m. \end{cases} \tag{18.44}$$

18.4
The Effect of Phase Transition on the Efficiency Coefficient of a Separator

As mentioned in the previous section, the condensation growth of drops can affect the CE value. Drops arriving at the separator entrance from the supply pipeline can grow by condensation provided that the thermodynamic balance of phases is violated at the entrance. This can happen, for instance, if we place one of the devices for preliminary condensation directly at the entrance.

Let the pressure and temperature p_2, T_2 in the separator differ from the corresponding values p_1, T_1 in the supply pipeline. As the mixture moves in the pipeline, the phase balance corresponding the values p_1 and T_1 is established. This balance is violated in the separator, resulting in the condensation growth of drops. In Section 16.1, we studied two possible mechanisms of condensation growth of drops: the diffusion mechanism that works under the weak mixing condition, and the turbulent one. Consider both these mechanisms successively.

The diffusion mechanism of growth in a binary mixture leads to the following expression that describes the change of the drop radius (see (16.54)):

$$R/R_0 = 1 + (\lambda - 1)(1 - e^{-\alpha\tau}). \tag{18.45}$$

where R_0 is the initial radius of the drop, for example, the average radius of the drops formed in the supply pipeline; $\lambda = (1 + \rho_G \Delta y(0)/\rho_L W_0)^{1/2}$, ρ_G and ρ_L are the densities of the gas and the liquid; $\Delta y(0)$ is difference of the equilibrium molar fractions of the condensed component respectively at p_1, T_1 and p_2, T_2; W_0 is the initial volume concentration of the liquid phase in the gas; $\alpha = 0.331/(\lambda - 1) - 0.026$, $\tau = 3\rho_G D_{12} \Delta y(0) t/\rho_L R_0^2$; D_{12} is the coefficient of binary diffusion of the condensed component in the gas.

Fist, consider the CE of a horizontal separator. For simplicity's sake, we assume that drops are settling with the Stokes velocity $u_s(R)$, which is equal to

$$u_s = u_{s0}(1 + (\lambda - 1)(1 - e^{-\alpha\tau}))^2, \quad u_{s0} = \frac{2g\Delta\rho R_0^2}{9\mu_G}. \tag{18.46}$$

Suppose that the longitudinal gas velocity U in the separator is constant. Then, plugging the variable $x = Ut$ into (18.46) and using the expression (18.17), we get the following expression for the transfer function of the separator

$$\Phi_h(R) = \begin{cases} 1 - \dfrac{u_{s0} L I(R_0)}{UD} & \text{at } R < R_m, \\ 0 & \text{at } R \geq R_m, \end{cases} \tag{18.47}$$

$$I(R_0) = \lambda^2 - \frac{2\lambda}{\alpha\tau_L}(\lambda - 1)(1 - e^{-\alpha\tau_L}) + \frac{(\lambda - 1)^2}{2\alpha\tau_L}(1 - e^{-2\alpha\tau_L}),$$

$$\tau_L = \frac{3\rho_G D_{12} \Delta y(0) L}{\rho_L U R_2^2}.$$

In the absence of condensation growth of drops ($\alpha = 0$), from (18.47) there follows the already-familiar formula (18.15).

Let's estimate the magnitudes of the terms entering the right-hand side of the expression for $I(R_0)$. For the characteristic values $L = 3$ m; $D = 1$ m; $U = 0.2$–0.5 m/s; $\mu_G = 10^{-5}$ Pa·s; $\rho_L = 750\ \text{kg/m}^3$; $\rho_G = 50\ \text{kg/m}^3$; $R_0 = 10^{-5}$ m; $D_{12} = 10^{-7}$ m²/s; $\Delta y(0) = 10^{-2}$; $W_0 = 10^{-4}$ one gets $\lambda = 2$; $\alpha = 0.3$; $\tau_L = 30$ and $\alpha\tau_L = 9$. The obtained value of the parameter $\alpha\tau_L$ allows us to neglect the second and third terms in the expression for $I(R_0)$. As a result, the formula (18.47) becomes:

$$\Phi_h(R) = \begin{cases} 1 - \lambda^2 \dfrac{u_{s0} L}{U D_h} & \text{at } R < R_m, \\ 0 & \text{at } R \geq R_m. \end{cases} \tag{18.48}$$

To determine the CE of the separator, we first have to find the minimum drop radius R_m. It can be found from the expression (18.16) into which we should plug

$y_0 = D_h$, $\psi = 0$ and $\psi(0, D_h) = UD_h$. The result is:

$$\int_0^L u_s\,dx = UD_h. \tag{18.49}$$

Substituting the formula (18.46) for u_s, integrating the resulting expression and taking into account the estimations made above, we find the minimum radius of drops:

$$R_m = R_{m0}/\lambda, \tag{18.50}$$

where $R_{m0} = (9\mu_G D_h U_h/2g\Delta\rho_L)^{1/2}$ is the minimum radius drops would have if we disregarded the condensation growth.

Thus, accounting for phase transformations results in a decrease ($\lambda > 1$, condensation) or an increase ($0 < \lambda < 1$, evaporation) of the minimum radius of drops by a factor of λ.

To determine the CE of a separator, one needs to substitute into Eq. (18.37) the appropriate expressions for R_m and $\Phi(R_0)$. Introducing dimensionless variables $z = R_0/R_{av}$, $z_1 = R_1/R_{av}$, $\tau = Lu_{s0}/U_h D_h$, $u_{s0} = 2g\Delta\rho R_{av}^2/9\mu_G$, we obtain

$$\eta_h = 1 - \frac{\exp(-3\sigma^2)}{\sqrt{2\pi}\sigma}\int_0^{z_m} z^2(1-\lambda^2\tau z^2)\exp\left(-\frac{\ln^2 z/z_1}{2\sigma^2}\right)dz, \tag{18.51}$$

where $z_m = \tau^{1/2}/\lambda$.

Fig. 18.5 illustrates the dependence of η on the dimensionless residence time τ of the mixture in a separator for various values of λ that describes the degree of

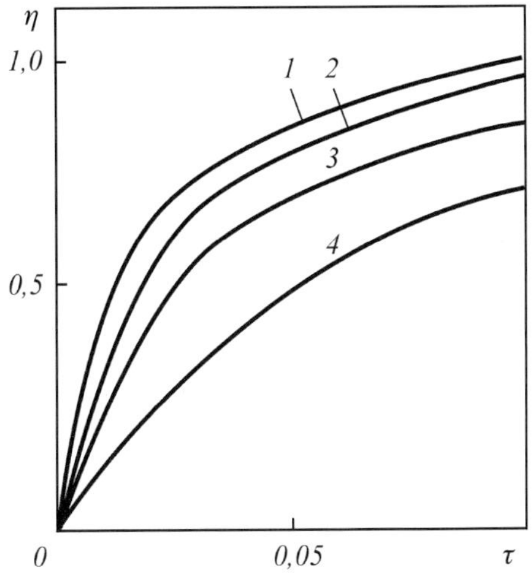

Fig. 18.5 Dependence of η of a horizontal gravitational separator on τ for λ: 1 – 2; 2 – 1.7; 3 – 1.4; 4 – 1.

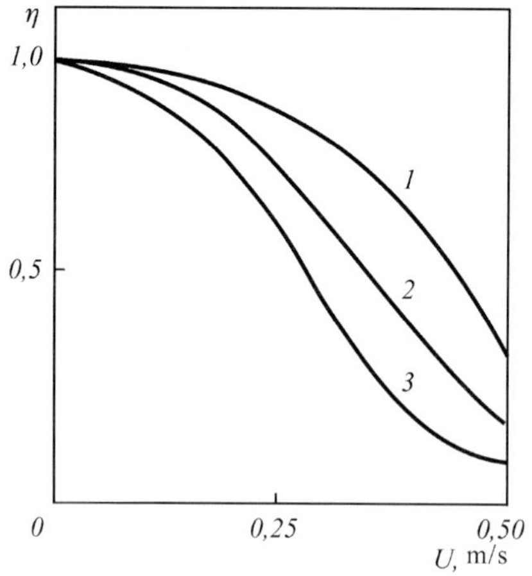

Fig. 18.6 Dependence of η of a horizontal gravitational separator on U for different values of λ: 1 – 2; 2 – 1.4; 3 – 1.

saturation of the mixture. The case $\lambda = 1$ corresponds to the absence of phase transitions. It can be seen from the figure that an increase of λ results in the efficiency growth, because the size of drops increases. The dependence of CE on the flow velocity U in a separator is presented in Fig. 18.6 for the following values of parameters: $p = 5$ MPa; $D = 1$ m; $\mu_G = 10^{-5}$ Pa·s; $\rho_G = 40$ kg/m^3; $\rho_L = 750$ kg/m^3; $L = 3$ m.

Determination of the CE for a vertical gravitational separator can be carried out in a similar way. The expression for η looks as follows:

$$\eta_v = 1 - \frac{\exp(-3\sigma^2)}{\sqrt{2\pi}\sigma} \int_0^{z_m} z^2 \exp\left(-\frac{\ln^2 z/z_1}{2\sigma^2}\right) dz, \tag{18.52}$$

where $z_m = (U/u_{s0})^{1/2}/\lambda$.

The dependence of η on U/u_{s0} for different values of λ is shown in Fig. 18.7.

Consider now the augmentation of drops due to phase transitions in a turbulent flow. In Section 16.1, the expression (16.70) for the change of the average drop volume V was derived for the case $R_{av} > \lambda_0$:

$$\frac{V}{V_0} = 1 + \frac{1}{\rho_L W_0} \sum_i (\rho_{iG0} - \rho_{iGw})(1 - e^{-A_c t}), \tag{18.53}$$

where $A_c = 14.3 \cdot U_t^{4/7} \rho_L^{5/7} W_0 U_h / d^{8/21} D_h^{1/3} \Sigma^{2/7} \rho_G^{3/7}$; U_t and U_h are the flow velocities in the supply pipeline and in the separator; W_0 is the volume concentration of the liquid phase at the separator entrance; d and D_h are the pipeline and separator diameters; Σ is the surface tension coefficient of drops.

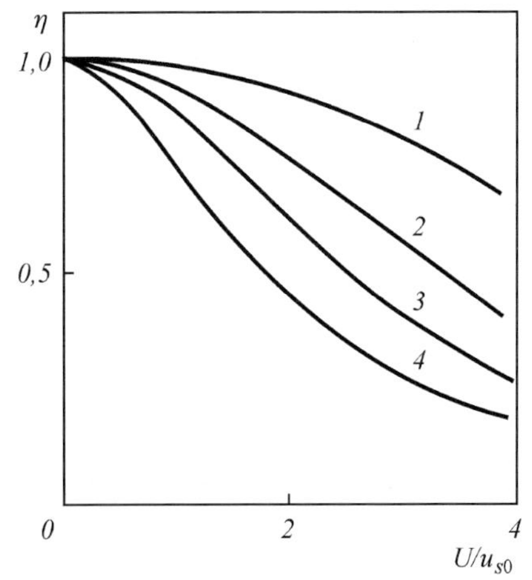

Fig. 18.7 The dependence of η of a horizontal gravitational separator on U/u_{s0} for different values of λ: 1 – 2; 2 – 1.4; 3 – 1.2; 4 – 1.

We shall restrict ourselves to considering the sedimentation of drops in a horizontal gravitational separator. If the size of drops does not change, and they are settling with Stokes velocity, then the transfer function is determined by the formula (18.15). Let the size of drops grow according to the law (18.53). Then the trajectories of averaged motion of drops are determined by the equation

$$\frac{dy}{dx} = -\frac{D}{L}\left(\frac{R}{R_{m0}}\right)^2 = -\frac{D_h}{L}\left(\frac{R_0}{R_{m0}}\right)^2 (1 + B(1 - e^{-A_c x/U_h}))^{2/3}, \tag{18.54}$$

with the initial condition $y(0) = y_0$. Here we denoted $B = \sum_i (\rho_{iG0} - \rho_{iGw})/\rho_L W_0$, R_{m0} is the minimum radius of the drops that don't grow.

The solution of Eq. (18.54) has the form

$$y = y_0 - \frac{D_h}{L}\left(\frac{R_0}{R_{m0}}\right)^2 C, \tag{18.55}$$

$$C = \frac{3U}{A_c}\left(-0.5(z^2 - 1) + 0.5\alpha^2\left(0.5 \ln \frac{(1-\alpha)^3(z^3 - \alpha^3)}{(1 - \alpha^3)(z - \alpha)^3} + \sqrt{3}\left(\arctan\left(\frac{2+\alpha}{\alpha\sqrt{3}}\right) - \arctan\left(\frac{2z+\alpha}{\alpha\sqrt{3}}\right)\right)\right)\right),$$

$$\alpha = (1 + B)^{1/3}, \quad z = (1 + B(1 - e^{-A_c x/U_h}))^{1/3}.$$

To determine the minimum radius of drops, one needs to plug $y_0 = D_h$, $y = 0$ and $x = L$ into (18.55). The result is

Table 18.1 The values of parameter λ for various W_0 and U.

W_0, m^3/m^3	U, m/s					
	0.1	0.2	0.3	0.4	0.5	0.6
10^{-5}	2.27	2.27	2.5	2.68	2.82	2.93
$5 \cdot 10^{-5}$	1.68	1.81	1.88	1.93	1.95	1.97
10^{-4}	1.47	1.57	1.55	1.56	1.56	1.56
$5 \cdot 10^{-4}$	1.15	1.15	1.15	1.15	1.15	1.15
10^{-3}	1.04	1.04	1.04	1.04	1.04	1.04

$$R_m = \frac{R_{m0}}{\lambda}, \quad \lambda = \left(\frac{A_c L}{3 U_h C_L}\right)^{-1/2}. \tag{18.56}$$

Here C_L is the value of the parameter C at $x = L$.

Thus, the accounting for phase transitions boils down to the introduction of the correction factor λ^{-1} in the expression for the minimum drop radius. The CE of a separator is determined by the formula (18.51), in which R_m should be represented by the expression (18.56). What we're mostly interested in is the dependence of λ on the gas velocity U_h in the separator and the volume concentration W_0 of the liquid phase at the entrance. The values of λ are given in Table 18.1 for $d = 0.1$ m; $D_h = 1$ m; $\Sigma = 5 \cdot 10^{-3}$ N/m; $\rho_L = 750$ kg/m^3; $L = 3$ m; $\Delta y(0) = 10^{-2}$.

The dependence of the CE on U_h for different values of W_0 is shown in Fig. 18.8. For a fixed flow velocity U_h, the CE decreases with the growth of W_0 because as W_0 gets larger, the average distance between drops decreases, the phase equilibrium is established faster, and the growth of drops slows down.

If $R_{av} < \lambda_0$, then in the above expressions, it is necessary to replace A_c by A_i, which is determined by the equality (16.84). The dependence of η on U_h and W_0 in the current case practically coincides with the case $R_{av} > \lambda_0$.

Thus, as shown by the calculations above, the condensation growth of drops can exert a noticeable influence on the CE of a separator if there exists a sufficiently strong oversaturation of the mixture (Δy not less than 10^{-2}) and if the volume concentration W_0 of the liquid phase is small (no more than $5 \cdot 10^{-4}$ m^3/m^3). This is possible if the condensing device is located almost at the entrance to the separator. However, in this case the removal of small drops formed in the device from the flow is performed with an extremely low efficiency. Therefore, when calculating the CE of a separator, the condensation growth of drops can be neglected.

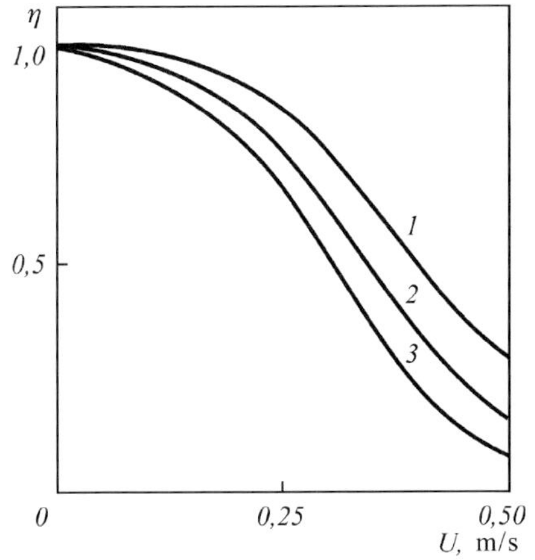

Fig. 18.8 The dependence of η of a horizontal gravitational separator on U for different values of W_0, m^3/m^3: 1 – 10^{-5}; 2 – $5 \cdot 10^{-5}$; 3 – absence of phase transitions.

18.5
The Influence of Drop Coalescence on the Efficiency Coefficient of a Separator

Coagulation of drops in a turbulent gas flow was considered in Section 15.2. The change of the average drop volume with time is described by

$$\frac{V}{V_0} = (1 + \beta t)^2, \quad \beta = 8 W_0 \left(\frac{\Gamma}{3\pi \rho_G V_0 \lambda_0^2} \right)^{1/2}, \tag{18.57}$$

where V_0 is the initial drop volume; Γ is the Hamaker constant; λ_0 is the internal scale of turbulence. Consider coagulation of drops in a horizontal gravitational separator. Let the drops at the separator entrance have the same volume equal to the average volume V_{av} of drops formed in the supply pipeline.

Consider the elementary volume containing N drops of volume V_{av} at the entrance and observe this volume, assuming that drops don't leave it. As the volume is moving in the separator, the drops inside the volume coagulate, the number of drops decreases, the total volume of drops is conserved, and the average drop volume increases according to the expression (18.57).

Note that the residence time of the drops inside the separator depends on the initial position of our volume in the entrance cross section. To determine the CE of the separator, we use correlations (18.3) and (18.17). Suppose the drops settle with the Stokes velocity. We then obtain:

$$\eta_k = \frac{\alpha}{D_h U_h} \int_0^L V^{2/3}\, dx, \quad \alpha = \left(\frac{3}{4\pi} \right)^{2/3} \frac{2g\Delta\rho}{9\mu_G}. \tag{18.58}$$

Substituting the expression (18.57) here, we have

$$\eta_k = \frac{3\alpha V_0^{2/3}}{7D_h\beta}\left(\left(1+\frac{\beta L}{U_h}\right)^{7/3} - 1\right). \tag{18.59}$$

Neglecting coagulation, we find the CE in this approximation:

$$\eta_k = \frac{\alpha V_0^{2/3} L}{U_h D_h}. \tag{18.60}$$

Let's compose the ratio of the efficiency coefficients presented above:

$$\varepsilon = \frac{\eta_k}{\eta_h} = \frac{3U_h}{7L\beta}\left(\left(1+\frac{\beta L}{U_h}\right)^{7/3} - 1\right). \tag{18.61}$$

It follows from there that the influence of drop coagulation on the CE of the separator is determined by the parameter β_L/U_h. We are interested in dependence of the CE on the volume concentration of the liquid phase W and on the gas velocity U_h.

Using expressions for β, λ_0 and $V_0 = V_{av}$, one finds that $\beta L/U_h \sim U_h W$. For the characteristic parameter values $D_h = 1$ m; $d = 0.1$ m; $\rho_G = 50$ kg/m^3; $\rho_L = 750$ kg/m^3; $\Gamma = 5 \cdot 10^{-20}$; $\Sigma = 10^{-2}$ N/m; $\mu_G = 10^{-5}$ Pa·s; $L = 3$ m; $W \leq 10^{-3}$ m^3/m^3; $U \leq 1$ m/s, we get $\beta L/U \sim 0.3$. The small value of this parameter allows us to take

$$\varepsilon \approx 1 + \frac{2\beta L}{3U_h}. \tag{18.62}$$

Thus, taking drop coagulation into account, we must increase the CE by the quantity equal to $2\beta L\eta/3U_h$. The main factors influencing the CE are the flow velocity U_h and the volume concentration of liquid W. In the example presented above, coagulation exerts a noticeable influence on the CE only at $W \geq 10^{-3}$ m^3/m^3 and $U \geq 1$ m/c. However, at such flow velocities, the CE of a gravitational separator is practically zero and the correction for coagulation can not produce any noticeable increase. It is necessary keep in mind that, since the volume concentration is defined as the volume of the liquid phase per 1 m^3 of mixture at operating conditions, the value of W depends on pressure and temperature, namely,

$$W = W_n \frac{293p}{zT}. \tag{18.63}$$

where z is the gas compressibility, W_n is the volume concentration of liquid under normal conditions.

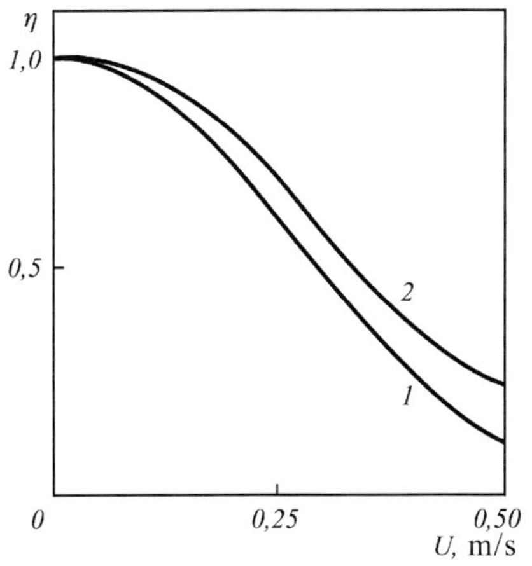

Fig. 18.9 The dependence of η of a horizontal gravitational separator on gas velocity U: 1 – without coagulation; 2 – with coagulation at $W = 5 \cdot 10^{-3}$ m^3/m^3.

This means that at fixed values of W_n (it is exactly this parameter that is used in practice; it is called the condensate factor), the volume concentration of the liquid phase W increases with the growth of pressure and with the decrease of temperature. At the same time the coagulation rate of drops grows as well. Fig. 18.9 shows the dependence of CE of a horizontal gravitational separator on the flow velocity U_h in the separator with and without regard for the coagulation of drops. In gravitational separators, the velocity does not exceed 0.3 m/s, so we can conclude from the estimations above that coagulation has no significant effect on the separator CE, and can be ignored in the course of calculations.

18.6
The Effect of Curvature of the Separator Wall on the Efficiency Coefficient

Up till now, in order to simplify CE calculations, the separator was modeled by a container with a rectangular cross section. It is evident that such approximation will not have any effect on the CE of a vertical device. However, a similar statement is not obvious for a horizontal separator.

Consider sedimentation of drops in a horizontal with round cross section. Let the gas flow in a pipe of diameter D with a velocity that is uniform over the cross section and equal to the mass-averaged velocity U_h. "Chop" the pipe's cross section into m vertical strips of the same width A. The height z of each strip depends on its distance δ from the pipe's axis:

$$z = 2(0.5D^2 - \delta^2)^{1/2}. \tag{18.64}$$

Table 18.2 Efficiency of a horizontal separator.

Velocity U, m/s	0.086	0.13	0.23	0.3	0.4	0.5
Rectangular cross section	0.99	0.97	0.64	0.4	0.2	0.12
Round cross section	0.99	0.97	0.69	0.46	0.25	0.14

Now break the interior region of the pipe into vertical layers by planes parallel to the pipe's axis. Each of the layers can be considered a horizontal separator of rectangular cross section, whose CE η_1 is easy to determine. To determine the CE of all separators, it is necessary to integrate η over δ. The results of CE calculation for a horizontal gravitational separator with rectangular and circular cross sections are presented in Table 18.2 for the following values of parameters: $p = 5$ MPa; $\mu_G = 1.2 \cdot 10^{-5}$ Pa·s; $\rho_G = 40$ kg/m^3; $\rho_L = 750$ kg/m^3; $d = 0.1$ m; D = 1 m; $\Sigma = 10^{-2}$ N/m.

It follows from the given results that wall curvature causes an increase of the separator's CE, but this increase is insignificant.

Thus, the curvature of walls may noticeably influence the CE of the separator only at rather high flow velocities, where the efficiency of separation is small. Since this velocity interval does not present any practical interest, one can conclude that for practical calculations the separator can be modeled by a container with rectangular cross section.

18.7
The Influence of a Distance Between the Preliminary Condensation Device and the Separator on the Efficiency Coefficient

In the previous paragraphs, we considered the processes occurring in a device of preliminary condensation (DPC), in a supply pipeline, and in a separator. It was shown that separation efficiency essentially depends on the behavior of the gas-liquid mixture in all three elements of the technological scheme. We are now interested in the question about the influence of relative positioning of these elements. In particular, one important parameter is the distance between the DPC and the separator, that is, the length of the connecting pipeline.

At the passage of the gas-liquid mixture through the DPC (throttle, heat-exchanger, or turbo-expander), fine droplets (fog) are formed in the flow as a result of temperature and pressure decrease. The size of these droplets in the fog is much smaller than the minimum radius of drops typical for a separator. If the DPC is placed at the separator entrance, the droplets formed there will not be captured by the separator. For the separation to be efficient, they have to be enlarged. Coagulation of drops begins already in the DPC. However, the short residence time in the DPC does not allow them to be enlarged to the required size, and the further coagulation of drops takes place in the pipeline connecting the DPC with the separator. As was shown earlier, the basic mechanism of drop aggregation in the pipeline is coagulation in a turbulent gas flow. To achieve the desired augmentation, the drops should be given a sufficiently long time depending on parameters of the flow, diameter of the pipeline, thermobaric conditions and the phase properties. On the other hand, drops cannot increase in size indefinitely, since there is a critical drop size for the given parameters of the flow and the pipeline. Beyond this size, drops begin to disintegrate. Knowing the mechanism of drop coagulation and the greatest possible drop size would help determine such a distance from the DPC to the separator that drops would have just enough time to be augmented up to the largest possible size.

In the process of coagulation, the size of drops increases due according to the law (18.57). For the characteristic values of parameters $\rho_G = 50\ \text{kg/m}^3$; $W = 5 \cdot 10^{-3}\ \text{m}^3/\text{m}^3$; $\Gamma = 5 \cdot 10^{-20}$ J; $\lambda_0 = 10^{-6}$ m, we have $\beta t \gg 1$. This inequality allows us to write the law governing the change of the average drop volume as

$$V \approx \frac{64W^2\Gamma}{3\lambda_0^2\pi\rho_G}t^2. \tag{18.65}$$

It should be noted that at $t \gg 1/\beta$, the size of drops does not depend on the initial size.

We shall demand that the final size of drops must be equal to the average size R_{cr} (see (14.14)) formed in the turbulent flow inside the pipe. Using the expression $V_{av} = 4\pi R_{cr}^3/3$ and the correlation (14.5) for λ_0 and taking $t = L/U$, one gets the required distance from the DPC to the separator:

$$L_0 = 0.021\frac{\mu_G^{3/4}(2\Sigma)^{9/14}d^{31/28}}{\Gamma^{1/2}\rho_L^{6/7}U^{29/28}\rho_G^{1/28}W}. \tag{18.66}$$

Distance L decreases with the increase of flow velocity and with the reduction of pipeline diameter d. It is explained by the fact that an increase of U and a decrease of d results, on the one hand, in a decreased average drop size in the flow, and on the other hand, in an increased rate of drop coagulation.

Let's make an estimate for the possible values of L. Put a throttle before the separator. Behind the throttle, the pressure is $p = 10$ MPa; the temperature is $T = 273\ ^\circ$K; gas velocity in the pipeline is $U = 10$ m/s; the pipeline diameter is $d = 0.4$ m; the densities of liquid and gas are, respectively, $\rho_L = 750\ \text{kg/m}^3$, $\rho_G = 100\ \text{kg/m}^3$; the surface tension coefficient of the liquid is $\Sigma = 5 \cdot 10^{-3}$ N/m; the viscosity coefficient is $\mu_G = 10^{-5}$ Pa·s; the volume concentration of the liquid phase in the mixture is $W = 5.5 \cdot 10^{-3}\ \text{m}^3/\text{m}^3$; the Hamaker constant is $\Gamma = 5 \cdot 10–20$ J. At these values of parameters, the expression (18.66) yields $L = 90$ m. For the same parameter values, a reduction of the pipeline diameter to $d = 0.2$ m results in the decrease of L to 11 m. But this will produce an unwanted side effect – a decreased minimum size of drops settling in the separator, and hence, a decreased separator efficiency.

Thus, the distance from the DPC to the separator should be such that, on the one hand, it is not too large, and on the other hand, the CE of the separator does not show any noticeable decrease.

The data above is in good agreement with the results of experiments [55], where the authors determined the optimal distance between the throttle and the separator at which the CE reaches its maximal value.

19
The Efficiency of Separation of Gas–Liqid Mixtures in Separators with Drop Catcher Orifices

Gravitational separators have low separation efficiencies and small gas productivities. To raise these parameters, it is necessary to increase separator sizes or to equip separators with additional droplet catcher sections, capable of catching the finely dispersed components of gas-liquid mixtures, that have no time to sediment in the settling section of the separator. The designs of droplet catcher sections have been reviewed in Section 2.1. The following types of droplet catcher sections are most commonly used: the jalousie, centrifugal, mesh, and string sections. In these sections, the liquid is separated from the gas, basically, due to inertia, surface, and hydrodynamic forces. It should be noted that the majority of separators equipped with droplet catcher sections are vertical apparatuses.

The separator can be represented as consisting of two sections connected in tandem: the settling section and the end droplet catcher section.

The separation process in such a separator can be described as follows. The gas-liquid mixture enters the separator with a certain distribution of drop radii $n_0(R)$, at the entrance (Fig. 19.1). In the settling section, characterized by a minimum drop radius R_m, all drops with radii $R > R_m$ are separated. In a settling section of horizontal gravitational type, besides these drops, a part of the drops with sizes in the interval $0 < R < R_m$ (shown by the horizontal hatched area) is also separated from the gas. The gas-liquid mixture that remains after the removal of the large drops, proceeds to the end section, which is equipped with droplet catcher orifices with their own characteristic minimum radius, $R_{m1} < R_m$. As a result, drops having radii greater than this value are additionally removed from the distribution (vertical hatching) and at the exit of the separator only a small part of the distribution with $0 < R < R_{m1}$ remains. As will be shown later, the minimum drop radius does not necessarily exist for all types of orifices; therefore, the introduction of R_{m1} is conditional.

If the transfer functions of the sections are known, the coefficients of effectiveness and ablation of the separator can be determined through Eqs. (18.3) and (18.6).

In the previous chapter, expressions were derived for the CE of gravitational separators, which can be considered as the CE of the settling section, η_1. In the present chapter, separation processes in droplet capture orifices are considered and appropriate expressions for their CE, η_2, are derived.

Separation of Multiphase, Multicomponent Systems. E. G. Sinaiski and E. J. Lapiga

ISBN: 978-3-527-40612-8

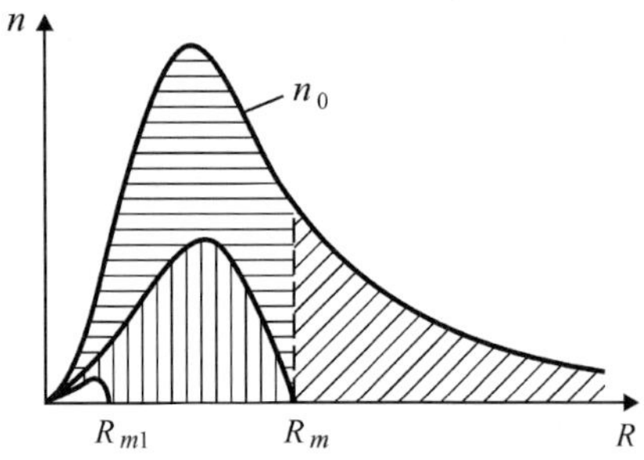

Fig. 19.1 Distribution of drop radii.

19.1 The Efficiency Coefficient of Separators with Jalousie Orifices

Consider a vertical separator consisting of two sections: a gravitational settling section and a droplet catching mist extractor equipped with jalousie orifices aligned with the direction of gravity perpendicular to the plane of Fig. 19.2. The jalousie section consists of packed corrugated plates, separated by a distance of h_0. As a rule, the value h_0 is assumed to be constant. The central angle of the corrugations is $2\varphi_i$, and $i = 0$ corresponds to the angle of the entrance section. This arrangement results in zigzag channels between adjacent plates for the passage of gas. At the entrance of the jalousie packing, the incoming gas has a certain distribution of drop radii. The gas velocity at the entrance is U. We assume, that deposition of drops at the walls of jalousie channels occurs basically due to inertia of the drops; the velocity in the jalousie cross-section is slow, uniform,

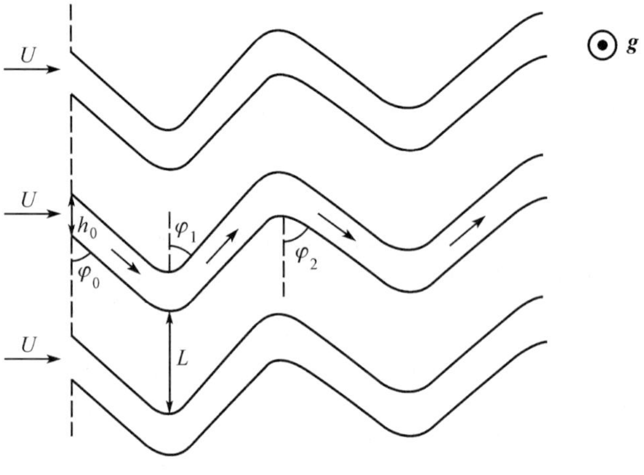

Fig. 19.2 Schematic representation of a jalousie packing.

and parallel to the walls; drops are small, so the resistance is given by the Stokes force.

As analysis of equations of motion for a drop of radius R shows, the jalousie transfer function depends on the Stokes number $S = 2R^2\rho_L U/9L\mu_0$ that describes drop inertia and on the geometrical parameters $a_0 = h_0/L$, $\varphi_0, \varphi_1, \ldots, \varphi_n$, where n is the number of corrugations. Having determined drop trajectories, it is possible to find the transfer function of the jalousie section

$$\Phi(R) = \prod_{i=0}^{n}(1 - A_i). \tag{19.1}$$

Parameters A_i describing the deposition rate of drops of radius R at jalousie walls are equal to

$$A_0 = \frac{S \cot \varphi_0}{a_0}(1 - e^{-1/s}), \quad A_1 = \frac{S \cot \varphi_1}{a_0 - \cot \varphi_0 S(1 - e^{-1/s})}(1 - e^{-1/s}),$$

$$A_i = \frac{S(\cot \varphi_i - \cot \varphi_{i-z})}{(a_0 - (\cot \varphi_{i-1} - \cot \varphi_{i-2})S(1 - e^{-1/s}))}(1 - e^{-1/s}), \quad (i \geq 2).$$

The minimum Stokes number and the appropriate minimum radii of drops for each corrugation are determined from the condition $A_i = 1$. For the minimum radius of drops to decrease as the number of corrugations increases, it is necessary that the angle φ_i decreases with the growth of i. Otherwise, drops will be deposited only at the first two corrugations. At $\varphi_0 > \varphi_1 > \cdots > \varphi_n$, the smallest value of minimum drop radius will achieved in the last corrugation. If the inequality $a_0 < 0.4\ (\cot \varphi_0 + \cot \varphi_{n-1})$ is true, we get the following approximated expression for the minimum Stokes number

$$S_m \approx \frac{a}{\cot \varphi_n + \cot \varphi_{n-1}}. \tag{19.2}$$

For values $h_0 = 0.01$ m, $L = 0.025$ m, $\cot \varphi_i = (6 + i)/12$, $\rho_L = 650$ kg/m^3, $U = 1$ m/s, $\mu_G = 10^{-5}$ Pa·s, and $n = 7$, we have $S_m = 0.19$. Using the expression for the Stokes number, one finds the minimum radius of captured drops: $R_m = 1.8 \cdot 10^{-5}$ m $= 18$ µm. To capture drops of radius 10 µ at the same values of parameters, we must take $n = 35$. A reduction of R_m may be achieved by increasing the velocity U or by changing the geometrical parameters. In particular, a narrower cross-section and a smaller tilt of the wall promotes a more efficient capture of drops. As an example of CE calculation for a separator with jalousie section, consider a vertical separator with the following parameter values: diameter of the separator $D = 1.6$ m; diameter of the supply pipeline $d = 0.3$ m; pressure $p = 13$ MPa; temperature $T = 293$ °K; liquid and gas densities $\rho p_L = 615$ kg/m^3, $\rho_G = 143$ kg/m^3; gas viscosity $\mu_G = 1.73 \cdot 10$ Pa·s. The resultant dependence of the separator CE on the gas flow rate Q under normal conditions is shown on

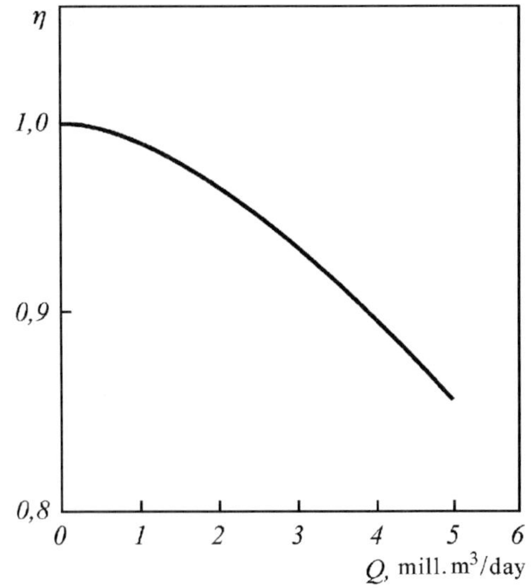

Fig. 19.3 Dependence of the efficiency coefficient of a separator with a jalousie section on the gas flow rate.

Fig. 19.3. A comparison with a similar dependence (see Fig 18.4) shows that a jalousie droplet catcher section increases the separator CE. It should be noted that when the flow velocity exceeds the critical value, this can lead to an ablation of the liquid film formed at jalousie walls, which would lower the separation efficiency.

The critical velocity value is usually established experimentally.

19.2
The Efficiency Coefficient of a Separator with Multicyclone Orifices

An additional section equipped with centrifugal branch pipes (multicyclone), is arranged, as a rule, in the top part of a vertical separator, on a horizontal platform. Each branch pipe is a cylinder, at whose entrance the flow is swirled by a special device. Depending on how this swirling is achieved, branch pipes are classified as tangential and axial. In the first type of branch pipes, the flow is injected tangentially through special slots in the branch pipe wall. In branch pipes of the second type, the flow is swirled at the entrance by a fairing with special blades, installed at a certain angle to the flow.

Consider the flow of a gas-liquid mixture in a branch pipe modeled by a cylinder of radius R_c and length L_c. Assume that gas velocity has two components: the axial component v_z and the tangential component v_φ. The radial component of the gas velocity is small as compared to v_z and v_φ and, in the first approximation, its influence on the process of drop deposition at the branch pipe wall can be neglected. Assuming that the gas flow running into a plate with centrifugal branch pipes is uniform over the separator cross-section, it is possible to find the average longitudinal velocity of gas in a branch pipe

$$v_z = U\frac{S_1}{S_2}. \tag{19.3}$$

Here U is the average velocity of gas in the settling section of the separator, S_1 and S_2 are the working cross-section areas of the settling section and the end section, respectively.

Consider the motion of a drop of radius R in a centrifugal branch pipe. Since large drops have already been separated from the gas in the settling section of the separator, only relatively small drops move in the branch pipe. Therefore, resistance forces acting on the drops obey Stokes's law, and gravity can be neglected. These assumptions enable us to write the equations of drop motion as

$$\begin{aligned}
&\frac{4}{3}\pi R^3\rho_L\frac{du_z}{dt} = -\frac{4}{3}\pi R^3 g\Delta\rho - 6\pi\mu_G R(u_z - v_z),\\
&\frac{4}{3}\pi R^3\rho_L\frac{du_r}{dt} = -6\pi\mu_G R u_r + \frac{4}{3}\pi R^3\rho_L\frac{u_\varphi^2}{r},\\
&\frac{4}{3}\pi R^3\rho_L\left(\frac{du_\varphi}{dt} + \frac{u_r u_\varphi}{r}\right) = -6\pi\mu_G R(u_\varphi - v_\varphi),
\end{aligned} \tag{19.4}$$

where u_z, u_r and u_φ are the components of drop velocity.

Let us estimate the order of magnitude of the inertial terms in the left-hand side of equations (19.4). We have

$$\begin{aligned}
\varepsilon_1 &= \frac{4}{3}\pi R^3\rho_L\frac{du_z}{dt} : 6\pi\mu_G R v_z \sim \frac{2R^2\rho_L v_z}{9L_c\mu_G},\\
\varepsilon_2 &= \frac{4}{3}\pi R^3\rho_L\frac{du_r}{dt} : 6\pi\mu_G R u_r \sim \frac{2R^2\rho_L v_z}{9L_c\mu_G},\\
\varepsilon_3 &= \frac{4}{3}\pi R^3\rho_L\left(\frac{du_\varphi}{dt} + \frac{u_r u_\varphi}{r}\right) : 6\pi\mu_G R u_\varphi \sim \frac{2R^2\rho_L}{9\mu_G}\left(\frac{v_z}{L_c} + \frac{u_r}{r}\right),
\end{aligned}$$

For the characteristic parameter values $\rho_L \sim 750\ \mathrm{kg/m^3}$; $v_z = 5$ m/s; $\mu_G = 10^{-5}$ Pa·s; $L_c \geq 0.3$ m; $R \leq 2\cdot 10^{-5}$ m, we have $\varepsilon_1 = \varepsilon_2 \leq 0.1$. It means that the inertial terms in the left-hand side of the first two equations (19.4) are negligibly small. At the same time, $u_r/r \sim 2R^2 r_L\omega^2/9\mu_G$, where ω is the angular velocity of the swirling flow at the entrance to the branch pipe, and

$$\varepsilon_3 \sim \frac{2R^2\rho_L}{9\mu_G}\left(\frac{v_z}{L_c} + \frac{2R^2\rho_L\omega^2}{9\mu_G}\right).$$

It follows from here that $\varepsilon_3 < 0.1$ at $\omega < 50\ \mathrm{c}^{-1}$.

Thus, at the angular velocity of the swirling flow $\omega \leq 50$ c^{-1}, the equations of motion of drops of radius $R \leq 2 \cdot 10^{-5}$ m (it is precisely such drops that reach the entrance of centrifugal branch pipes), look like

$$6\pi\mu_G R\left(v_z - \frac{dz}{dt}\right) = \frac{4\pi R^3 \Delta\rho g}{3},$$

$$6\pi\mu_G R\frac{dr}{dt} = \frac{4\pi R^3 \rho_L g u_\varphi^2}{3r}, \qquad u_\varphi = v_\varphi, \tag{19.5}$$

$$z(0) = 0, \quad r(0) = r_0 \quad (0 < r < R_c).$$

As it was already explained, centrifugal branch pipes may be of two types: axial and tangential. They differ radically in their method of mixture input. Besides, the profile of the tangential component of gas velocity $v_\varphi(r)$ depends on the design of the branch pipe [56, 57]. We can combine the analysis of branch pipes of both kinds assuming that each of them can be characterized by some average value of the spin angle φ. For axial branch pipes, we can assume it to be equal to the flow spin angle at the entrance φ_0, whereas for tangential branch pipes, the value φ_i may be determined in the following way. Denote by L_{ap}, h_{ap} and n_{ap} the height, the width, and the number of tangential slots in the bottom part of the branch pipe wall.

The average tangential velocity of the gas entering through the slots is equal to

$$\bar{v}_\varphi = \frac{Q}{N n_{ap} h_{ap} L_{ap}}, \tag{19.6}$$

where N is the number of branch pipes; Q is the gas flow rate.

The average flow spin angle φ_{av} can be estimated from the relation

$$\tan \varphi_{av} = \frac{v_\varphi}{v_z} \sim \frac{\pi(R_c^2 - R_t^2)}{n_{ap} h_{ap} L_{ap}}. \tag{19.7}$$

In the numerator of the expression (19.7), it is the area of the ring cross-section of the swirling flow in the branch pipe, and $R_t = n_{ap}h_{ap}/2$ is the internal radius of this cross-section.

It is shown in papers [56, 57] that in axial branch pipes, the tangential component of velocity that depends on the design of the swirler is determined by

$$v_\varphi = Cr^k, \tag{19.8}$$

At $k = -1$, the flow is spinning, obeying the law of circulation constancy (potential rotation); at $k = 0$, the spin angle is constant, i.e. does not depend on radius; and at $k = 1$, the rotation obeys the laws of solid body rotation (a quasi-solid rotation). To ensure that rotation obeys (19.8), the blades of the swirler at the exit should have a certain dependence of the spin angle on the radius. The question about the appropriate law for tangential branch pipes remains open and demands an experimental check.

The choice of the appropriate relation from the possibilities (19.8) should be restricted by a conservation condition for the average value of the tangent of the flow spin angle

$$\tan \varphi_{av} = \frac{2}{R_c^2} \int_0^{R_c} r v_\varphi \, dr \tag{19.9}$$

under the assumption that the axial component of velocity remains constant. Then, from the condition (19.9), we can find the constant C that appears in (19.8):

$$C = \frac{(k+2)}{2R_c^2} v_z \tan \varphi_{av}, \quad (k > -2).$$

Consider now the equations of motion of a drop of radius R in the branch pipe:

$$\frac{dz}{dt} = v_z - u_s, \quad \frac{dr}{dt} = \frac{2\rho_L R^2 v_\varphi^2}{9\mu_G r},$$
$$z(0) = 0, \quad r(0) = r_0, \quad u_s = \frac{2g\Delta\rho R^2}{9\mu_G}. \tag{19.10}$$

The critical trajectory of the drop corresponds to the value of r_0 for which the drop exiting the branch pipe lands at the wall at $r = R_c$. The solution of equations (19.10) gives us the critical value of r_0:

$$\left(\frac{r_0}{R_c}\right)^2 = \begin{cases} \left(1 - \dfrac{(1-k)(k+2)^2 \rho_L R^2 v_z \tan^2 \varphi L_c}{9\mu_G R_c^2 (1 - u_{s0}/v_z}\right)^{1/(1-k)} & \text{at } k \neq 1, \\ \exp\left(-\dfrac{\rho_L R^2 v_z L_c \tan^2 \varphi}{2\mu_G R_c^2 (1 - u_{s0}/v_z}\right) & \text{at } k = 1. \end{cases} \tag{19.11}$$

The minimum radius of drops R_{m1} is determined from Eq. (19.11) under the condition $r_0 = 0$. Obviously, R_{m1} exists only at $k < 1$. The coefficient of efficiency (CE) of the branch pipe is equal to

$$\eta_z = 1 - \frac{4\pi}{3W_{01}} \int_0^{R_{m1}} R^3 n_{01}(R) \frac{r_0^2}{R_c^2} dR, \tag{19.12}$$

where W_{01} and $n_{01}(R)$ are the volume concentration and the drop distribution at the entrance to the branch pipe.

Note that in the case when a minimum radius does not exist, the upper limit of integration in (19.12) should be replaced by R_m, where R_m is the minimum radius of drops in the settling section. The volume concentration W_{01} of drops at the entrance to the branch pipe is equal to that at the exit of the settling section, hence

$$W_{01} = \frac{1}{3}\int_0^{R_m} R^3 n_0(R)\, dR. \tag{19.13}$$

Please also note that the longitudinal velocity of the flow in branch pipes is high ($v_z \geq 1$ m/s), therefore for all drops entering the branch pipe, the inequality $u_s \ll v_z$ should be satisfied.

We now consider the various laws governing the swirling of the flow (19.8).

1. The spin angle is not changing with radius ($k = 0$). We have for this case:

$$v_\varphi = v_z \tan\varphi, \quad R_{m1} = \left(\frac{9\mu_G R_c^2}{4\rho_L v_z L_c \tan^2\varphi}\right)^{1/2}; \quad \left(\frac{r_0}{R_c}\right)^2 = \left(1 - \left(\frac{R}{R_{m1}}\right)^2\right).$$

The CE of the branch pipe follows from Eq. (19.12):

$$\eta_z = 1 - \frac{\exp(-3\sigma^2)}{\sqrt{2\pi}\sigma}\int_0^{z_{m1}} z^2\left(1 - \frac{4}{9\gamma_\varphi z^2}\right)\exp\left(-\frac{\ln^2 z/z_1}{2\sigma^2}\right) dz,$$

$$z_{m1} = \left(\frac{9}{4\gamma_\varphi}\right)^{1/2}, \quad \gamma_\varphi = \frac{R_{av} v_z^2 L_c \tan^2\varphi}{\mu_G R_c^2}. \tag{19.14}$$

2. Quasi-solid rotation ($k = 1$). In this case we have:

$$v_\varphi = \frac{3r}{2R_c} v_z \tan\varphi, \quad \left(\frac{r_0}{R_c}\right)^2 = \exp\left(-\frac{R^2 v_z L_c \rho_L \tan^2\varphi}{\mu_G R_c^2}\right).$$

The CE of the branch pipe is equal to

$$\eta_z = 1 - \frac{\exp(-3\sigma^2)}{\sqrt{2\pi}\sigma}\int_0^{z_m} z^2 \exp\left(-\gamma_\varphi z^2 - \frac{\ln^2 z/z_1}{2\sigma^2}\right) dz. \tag{19.15}$$

3. Potential rotation ($k = -1$). In this case,

$$v_\varphi = \frac{R_c}{2r} v_z \tan\varphi, \quad R_{m1} = \left(\frac{9\mu_G R_c^2}{4\rho_L v_z L_c \tan^2\varphi}\right)^{1/2},$$

$$\left(\frac{r_0}{R_c}\right)^2 = \left(1 - \left(\frac{R}{R_{m1}}\right)^2\right)^{1/2}.$$

and the CE of the branch pipe is

$$\eta_z = 1 - \frac{\exp(-3\sigma^2)}{\sqrt{2\pi}\sigma}\int_0^{z_{m1}} z^2\left(1 - \frac{2}{9}\gamma_\varphi z^2\right)^{1/2}\exp\left(-\frac{\ln z^2/z_1}{2\sigma^2}\right) dz,$$

$$z_{m1} = \left(\frac{9}{2\gamma_\varphi}\right)^{1/2}. \tag{19.16}$$

The dependence of the CE of an axial branch pipe on the parameter γ_φ for various laws governing the swirling of the flow (i.e. various values of k) is shown in Fig. 19.4, a. In the region $\gamma_\varphi \gg 1$, the branch pipe with a swirler that ensures the flow rotation with a constant spin angle ($k = 0$), possesses the highest effi-

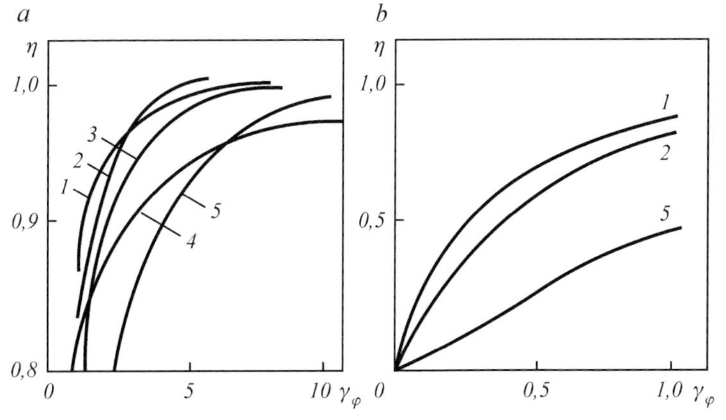

Fig. 19.4 Dependence of the CE of a centrifugal branch pipe (gas cyclone (*a*) and hydrocyclone (*b*) on the parameter γ_φ for various values of *k*: 1 – 1; 2 – 0; 3 – 0.5; 4 – 2; 5 – –1.

ciency. As we increase the parameter γ_φ, the difference in the CE of branch pipes with various types of swirling flow becomes smaller. Note that for the typical parameter values in gas branch pipes (cyclones), the inequality $\gamma_\varphi \gg 1$ is nearly always true. Therefore the CE of gas cyclones is not really sensitive to a change of the type of swirling for the flow.

A different situation arises for hydrocyclones, in which the density of the disperse phase is only moderately greater than the density of the continuous phase. Therefore, for hydrocyclones, $\gamma_\varphi < 1$, and their CE (Fig. 19.4, b) is small and essentially depends on the profile of the tangential flow velocity. From the three laws of swirling considered above, the greatest efficiency is provided by the law of quasi-solid rotation.

Consider now a tangential branch pipe for which the spin angle is estimated from Eq. (19.7). With the same reasoning as before, it is easy to find an expression for the CE:

$$\eta_v = 1 - \frac{\exp(-3\sigma^2)}{\sqrt{2\pi}\sigma} \int_0^{z_{m1}} z^2 \left(1 - \varepsilon^2 - \frac{4}{9}\gamma_{\varphi t} z^2\right) \exp\left(-\frac{\ln^2(z/z_1)}{2\sigma^2}\right) dz, \tag{19.17}$$

where $\varepsilon = R_t/R_c$, $\gamma_{\varphi t} = \gamma_\varphi/(1 - R_t^2/R_c^2)$.

If the tangential branch pipe is designed in such a way that $n_{ap}h_{ap} = 2R_c$, then $\varepsilon = 0$, and the calculation of CE for the tangential branch pipe can be performed in the same way as for the axial one. The only difference is in the definition of the flow spin angle.

Let us now examine the dependence of the CE of a vertical separator equipped with centrifugal branch pipes (multicyclones) on various parameters. The dependence on thermobaric conditions (pressure and temperatures), geometrical parameters (diameter of supply pipeline, length of branch pipes), and also on the gas flow rate presents the greatest interest to us.

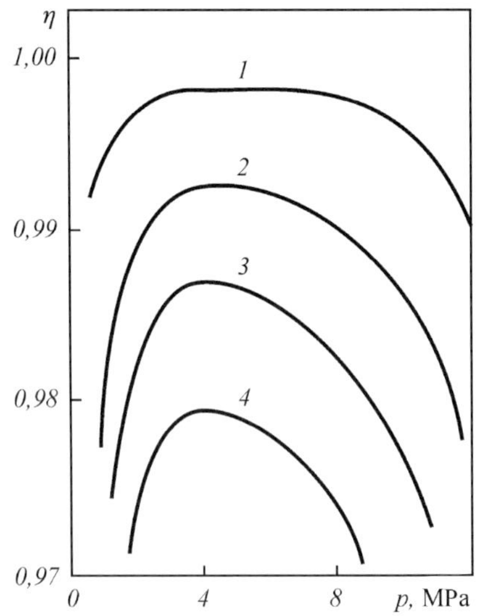

Fig. 19.5 Dependence of η on pressure in a separator with tangential centrifugal branch pipes for different values of the gas flow rate Q, mill. m^3/day ($D = 1.6$ m; $d = 0.35$ m; $T = 0\ ^\circ C$; $N = 92$; $L_c = 0.3$ m; $2R_c = 0.1$ m; $L_{ap} = 0.075$ m; $h_{ap} = 0.005$ m; $n_{ap} = 20$): 1 – 5; 2 – 9; 3 – 12; 4 – 15.

As pressure growth, the separator's CE increases at first, and then decreases. The maximum value of CE is achieved at $p = 3$–4 MPa (Fig. 19.5). An increase in the gas flow rate (taken at normal conditions) results in a reduction of CE (Fig. 19.6). Pressure affects the gas flow rate at operating conditions, the density of the gas, and the coefficient of surface tension of drops. The size of drops entering the separator depends on the above-mentioned parameters. An increase of pressure results in a reduction of the gas flow rate and of the coefficient of surface tension. A reduced gas flow rate, in its turn, leads to an increase of the average radius of drops formed in the supply pipeline. However, the decrease of the gas flow rate reduces the gas velocity and thereby the centrifugal force. Thus, the interaction of these factors leads to a decline of the separator CE together with the gas flow rate. The coefficient of surface tension first noticeably decreases with the growth of pressure up to 9 MPa, then changes slightly at $p \geq 9$ MPa. So the increase of pressure initially reduces the average size of drops, but then exerts a smaller effect on it. This results in a non-monotonous change of the CE. It should be noted that, if we disregard for the settling section, the CE of centrifugal branch pipes has a maximum in the region of low flow rates. Since the CE of a settling section is high in this region, then CE of the separator as a whole decreases monotonously with an increase of the gas flow rate.

A change of temperature has only a small effect on the CE. A change of the supply pipeline diameter produces the same effect as in gravitational separators. An increase in the length of branch pipes will also result in an increased CE of the separator (Fig. 19.7).

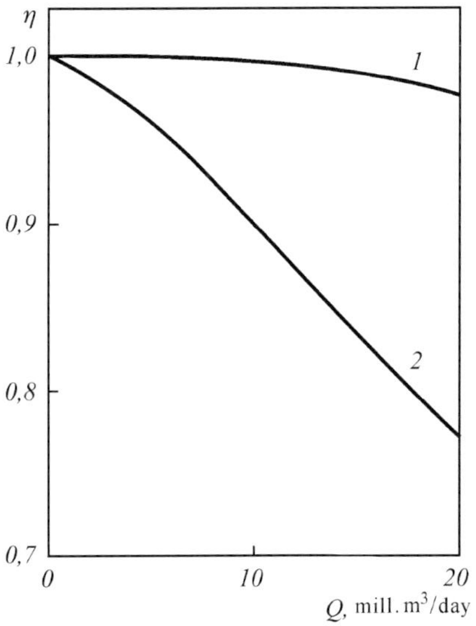

Fig. 19.6 Dependence of η on the gas flow rate Q, mill. m^3/day in a separator with tangential centrifugal branch pipes for different values d, m (p = 10 MPa, other parameters are the same as in Fig. 19.5): 1 – 0.4; 2 – 0.3.

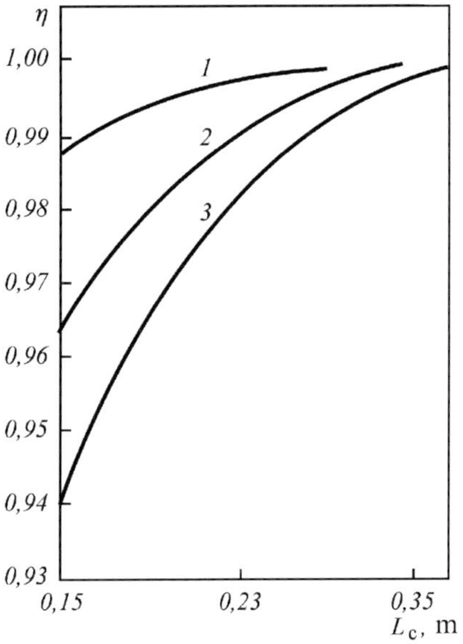

Fig. 19.7 Dependence of η of a separator with tangential centrifugal branch pipes on the branch pipe length L_c for different values of the gas flow rate Q, mill. m^3/day ($D = 0.45$ m; $d = 0.15$ m; $T = 10$ °C; $N = 7$; $2R_c = 0.1$ m; $p = 3$ MPa): 1 – 5; 2 – 9; 3 – 12; 4 – 15.

19.3
The Efficiency Coefficient of a Separator with String Orifices

The body of a string droplet catcher consists of a set of separating packages (sections), each of which is composed of rows of frames with stretched threads made from wire, nylon, etc. The threads are oriented parallel to gravity and perpendicular to gas flow. Frames and packages are placed in series.

As before, we understand the coefficient of efficiency (CE) to be the ratio of the amount (volume) of the liquid phase caught by a droplet catcher, to the amount of the liquid phase at the entrance

$$\eta = \frac{Q_s}{Q_0} = 1 - \frac{Q_1}{Q_0} = 1 - K, \qquad (19.18)$$

where K is the coefficient of ablation.

The value of η depends on gas velocity, volume concentration and dispersiveness of the liquid phase, geometrical size of the droplet catcher section, pressure, temperature etc.

Denote by K_i the ablation coefficient of the i-th frame. Since frames in the package are connected in series, according to (18.5), the CE of a package consisting of N frames is

$$\eta_s = 1 - \prod_{i=1}^{N} K_i. \qquad (19.19)$$

Similarly, the CE of a block containing M packages is equal to

$$\eta_s = 1 - \prod_{j=1}^{M}\left(\prod_{i=1}^{N} K_{ij}\right). \qquad (19.20)$$

Here K_{ij} is the CE of i-th frame in j-th package.

If all frames and packages are identical, then $K_{ij} = K$ and the CE of the block is

$$\eta_s = 1 - K^{MN}. \qquad (19.21)$$

So, to determine the CE of a string droplet catcher section, it is sufficient to know the CE of a single frame K.

A frame is a row of parallel cylindrical strings of diameter $d_1 = 0.3$–0.5 mm spaced at $l = 2$–3 mm apart. Since $d_1/2l \ll 1$, the neighboring strings do not exert a marked effect on the gas flow pattern around a single string, so it will suffice to consider drop deposition on a single cylinder, after which we can go on to consider the entire frame.

So, let us consider the capture of drops of a given size by a single cylinder (Fig. 19.8). The mechanism of drop deposition depends on flow conditions, and also on

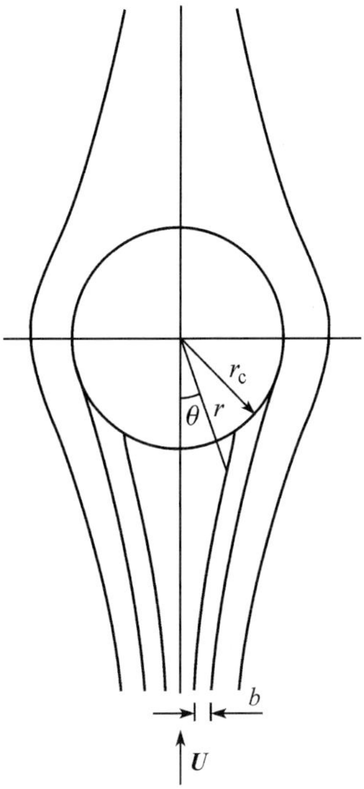

Fig. 19.8 Deposition of drops on a cylinder.

the ratio of the drop radius R to the cylinder radius $r_c = d_1/2$. First, we estimate Reynolds number for the gas flow around the string. For the characteristic values $U = 2.5\text{–}10$ m/s; $r_c = (1.5\text{–}2.5) \cdot 10^{-4}$ m; $\mu_G = 10^{-5}$ Pa·s; $\rho_G = 50$ kg/m^3 we obtain Re = 375–2500.

The stream of gas flowing around the cylinder can have a laminar character, but behind the cylinder, there can be a turbulent trace caused by the ablation of the boundary layer. High values of Re mean that the capture of drops by the cylinder has inertial character and can be considered within the framework of the inertial mechanism described in Section 10.5. As the gas flows around the cylinder, the drop trajectories, due to the inertia of drops, deviate from the streamlines, and drops are deposited on the cylinder surface. The bigger the drop size, the larger these deviations are. For relatively large drops, their trajectories are close to straight lines. For the capture of large drops, the efficiency coefficient of a single cylinder is determined by the formula

$$\eta_z = (r_z + R)/(0.5l + r_c). \tag{19.22}$$

On the other hand, very small drops have small inertia, so they follow the streamlines. In this case, the capture efficiency of such drops, which also takes their adhesion into account, is equal to

$$\eta_z = 2r_c/l. \tag{19.23}$$

The actual value of capture efficiency lies between these two numbers, and can be determined by considering the motion of drops as the flow goes around the cylinder.

Consider the deposition of drops of the same radius from the gas when the flow is transversal relative to the cylinder, and the gas flows around the cylinder with the velocity U on a large distance from the cylinder. Assume that drops are sufficiently small, so the resistance will be given by the Stokes force. Then drop trajectories are described by the equation

$$\frac{4\pi}{3} R^3 \rho_L \frac{d^2 \boldsymbol{r}}{dt^2} = 6\pi\mu_G R\left(\boldsymbol{U} - \frac{d\boldsymbol{r}}{dt}\right), \tag{19.24}$$

where $\boldsymbol{r}$ is the radius-vector of the drop center.

Taking projections on the radial and tangential directions, we can reduce this equation to

$$\begin{aligned} \frac{d^2 r}{dt^2} &= r\left(\frac{d\theta}{dt}\right)^2 + \frac{9\mu_G}{2R^2\rho_L}\left(U_r - \frac{dr}{dt}\right), \\ r\frac{d^2\theta}{dt^2} &= -2\frac{dr}{dt}\frac{d\theta}{dt} + \frac{9\mu_G}{2R^2\rho_L}\left(U_\theta - r\frac{d\theta}{dt}\right). \end{aligned} \tag{19.25}$$

Introduce the following dimensionless variables:

$$x = \frac{r}{r_z}, \quad \tau = \frac{U_\infty t}{r_z}, \quad S = \frac{4U\rho_L R^2}{9\mu_G d_1}, \quad v_r = \frac{U_r}{U_\infty}, \quad v_\theta = \frac{U_\theta}{U_\infty}.$$

In new variables, equations (19.25) will turn into

$$\begin{aligned} \frac{d^2 x}{d\tau^2} &= x\left(\frac{d\theta}{d\tau}\right)^2 + S^{-1}\left(v_r - \frac{dx}{d\tau}\right), \\ x\frac{d^2\theta}{d\tau^2} &= -2\frac{dx}{d\tau}\frac{d\theta}{d\tau} + S^{-1}\left(v_\theta - x\frac{d\theta}{d\tau}\right), \\ z(0) &= x_0, \quad \theta(0) = \theta_0 \end{aligned} \tag{19.26}$$

In the right-hand side of the system of equations (19.26), we see a dimensionless parameter S – the Stokes number. It is well known [2, 58] that drops of a given size will not reach the cylinder surface for all values of S. There exists such a critical value S_{cr} that deposition at the cylinder is possible only for $S > S_{cr}$.

For a potential flow around the cylinder, the approximate solution of equations (19.26) gives $S_{cr} = 0.0625$. However, at $Re \gg 1$, a viscous boundary layer of thickness $\delta \sim r_c/\sqrt{Re}$ is formed near the cylinder surface, which causes a stronger curving of streamlines near the surface than we would expect for a potential flow. As a result, trajectories are pushed away from the surface. This means that the number of droplets reaching the surface in a unit time decreases. We can account for the velocity field in the presence of a boundary layer by using Blasius' solution [59].

Drop trajectories are obtained by integration of equations (19.26) for different coordinate values of the initial position of drops. All trajectories can be grouped in two categories: trajectories ending at the cylinder surface, and trajectories that go around the cylinder. The trajectory dividing the first category from the second is called the limiting trajectory. It is spaced at the distance b from a straight line that runs through the cylinder axis parallel to the gas flow velocity far away from the cylinder. It is obvious that the flux of drops captured by the cylinder is proportional to b. Let's call the ratio $G = b/r_c$ the dimensionless radius of the cross-section of capture of drops of radius R by the cylinder. Numerical integration of equations (19.26) allows us to obtain the dependence of G on Stokes number S (Fig. 19.9).

If the velocity field corresponds to a potential flow around the cylinder, the obtained dependence (curve 2) is in good agreement with the approximate solution [2] (curve 1) in the region $S > 1$. At smaller values of the Stokes number, the difference becomes noticeable, and the critical value equals $S_{cr} = 0.1$. The influence of the boundary layer (curve 3) is insignificant at $S \geq 10$, while at $S < 1$ it becomes essential. In particular, $S_{cr} = 0.25$. Hence, the minimum radius of drops that are captured by the cylinder will be greater than the one expected for a potential flow, once we account for the boundary layer.

The dependence of G on S can be approximated by the following expression:

$$G = \left(\frac{S}{S + A}\right)^2, \tag{19.27}$$

The constant A has the following values: 0.35 for a potential flow, approximate solution; 0.44 for a potential flow, numerical solution; 0.66 for a flow with a boundary layer, numerical solution.

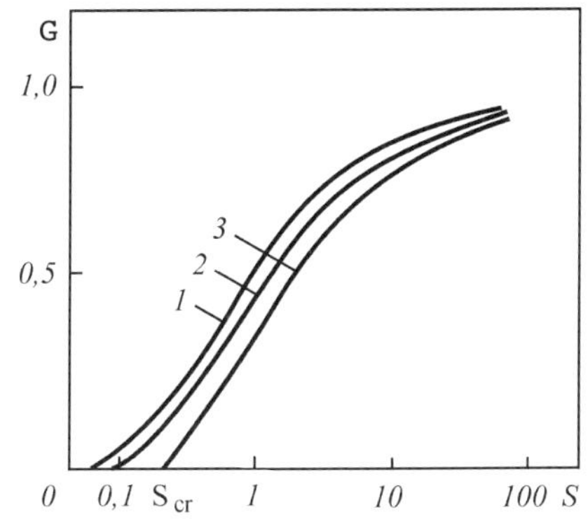

Fig. 19.9 Dependence of G on Stokes number S: 1 – potential flow, approximate solution; 2 – potential flow, numerical solution; 3 – flow with the formation of a boundary layer, numerical solution.

Now let us study the capture of droplets by a frame. The coefficient of ablation of the frame is

$$K = 1 - \frac{2G}{l/r_c + 2}. \tag{19.28}$$

If all drops have the same radius, the CE of the package is equal to:

$$\eta_p = 1 - \left(1 - \frac{2G}{l/r_c + 2}\right)^{MN}. \tag{19.29}$$

For the values of parameters typical for string orifices, the conditions $2b/(l + 2r_c) \ll 1$ and MN $\gg 1$ are satisfied. At the same time,

$$\eta_p \approx 1 - \exp\left(\frac{2MNG}{l/r_c + 2}\right). \tag{19.30}$$

Let's take an example. Let $\rho_L = 750\ \text{kg/m}^3$, $\mu_G = 10^{-5}$ Pa·s, $R = 1.5 \cdot 10^{-6}$ m, $r_c = 5 \cdot 10^{-4}$ m, $l = 3 \cdot 10^{-3}$ m, $N = 10$ and M $= 4$. Then S $= 0.5$, G $= 0.25$, $\eta_p = 0.92$.

If we have a gas-liquid mixture with the volume concentration of liquid $W_0 = 5 \cdot 10^{-4}\ \text{m}^3/\text{m}^3$ at the entrance of the string droplet catcher, then at the exit, we have $W_1 = 4 \cdot 10^{-5}\ \text{m}^3/\text{m}^3$. The dependence of η on gas velocity U is shown in Fig. 19.10. For the chosen values of parameters, the critical velocity, i.e. the velocity at which S $=$ S$_{cr}$, is equal to $U_{cr}^{(1)} = 1.56$ m/s. It means that when $U < U_{cr}^{(1)}$, the efficiency of the droplet catcher section is practically equal to zero. Keep in mind that if drop have different radii, then the critical Stokes number will be determined by the average radius of drops.

The existence of the critical Stokes number means that strings will fail to capture drops whose radius is less than the critical radius,

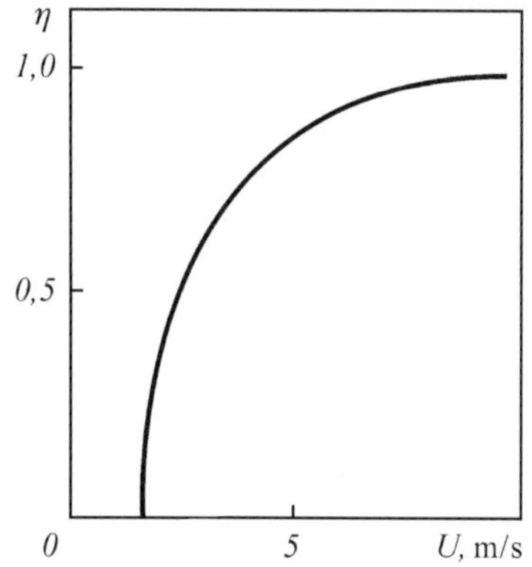

Fig. 19.10 Dependence of η of a string orifice on gas velocity.

$$R_{cr} = 0.75\left(\frac{\mu_G d_1}{2U\rho_L}\right)^{1/2}. \tag{19.31}$$

From (19.31), it follows that an increase in the flow velocity reduces the minimum radius of droplets that can be captured, and thereby increases the capture efficiency (CE) of the string droplet catcher. On the other hand, an increase in the flow velocity can destabilize the liquid film formed on the string surface, stripping drops away from the surface, and thus, reducing the CE.

Thus, a string orifice is characterized by two critical velocities. The first critical velocity limits the size of drops being captured, and is found from the condition $S_{cr} = 0.25$. The second critical velocity corresponds to the beginning of a secondary ablation of drops from the surface of the liquid film that is formed on strings. We can estimate the value of the second critical velocity.

To this purpose, consider a transversal flow of gas (with drops suspended in it) around a vertical cylindrical string. Drops are deposited at the string, forming a thin layer of liquid that flows down the string. The layer thickness increases in the direction of gravity. At some distance from the top end of the string, the thickness of the film achieves the critical value (for a given gas velocity), at which stability of the film will be disturbed. At this point destruction of the film and ablation of drops from its surface becomes possible. Direct the x-axis from the top end of the string downwards, parallel to gravity. Denote the thickness of the film by h. The velocity with which liquid flows down the string can be estimated by equating the gravity force and the force of viscous friction. We obtain as a result:

$$u_x = \frac{\rho_L g h^2}{3\mu_L}. \tag{19.32}$$

To determine the critical height h of the string, we should write the liquid flow rate conservation condition: the flow rate through the film cross-section is equal to the amount of liquid deposited from the gas flow on the string segment from our cross-section all the way to top,

$$2r_c h u_x = 2bUWx. \tag{19.33}$$

Substitution of the previous expression into (19.33) gives

$$h = \left(\frac{3b\mu_L UWx}{r_c\rho_L g}\right)^{1/3}. \tag{19.34}$$

To determine the critical thickness h_{cr} at which the film breakdown is possible, consider the forces acting on the film in the case of transversal flow. There is a dynamic pressure from the gas, which causes film accumulation in the rear part of the string. Forces of viscous friction at the wall and forces of surface tension at the free surface try to impede the breakdown of the film. Therefore, for the breakdown to happen, the following inequality should be satisfied:

$$0.5\rho_G U^2 h \geq \Sigma + \frac{\pi\mu_L r_c U}{2h}. \tag{19.35}$$

From (19.35), we can find the critical film thickness:

$$h_{cr} = \frac{\Sigma}{\rho_G U^2}\left(1+\left(1+\frac{2\pi\rho_G r_c \mu_L U^3}{\Sigma^2}\right)^{1/2}\right). \tag{19.36}$$

Substituting into (19.36) the expression for h from (19.34), we get the distance x_{cr} from the top end of the string to the point where the film can be destabilized by the oncoming flow:

$$x_{cr} = \frac{\rho_L g r_c}{3\mu_L UWb}\left(\frac{\Sigma}{\rho_G U^2}\left(1+\frac{2\pi\rho_G r_c \mu_L U^3}{\Sigma^2}\right)^{1/2}\right)^3. \tag{19.37}$$

Hence, in order to prevent secondary ablation at the given values of parameters that appear in the right-hand side of the equation (19.37), the height of the string should not exceed h_{cr}.

With an increase in the gas velocity, the critical height of the string becomes smaller. Therefore one raw of strings is insufficient for efficient removal of drops from the gas flow. If we increase the number of string rows, then, as velocity goes above the critical value, the next rows will capture those drops that were not captured by the previous row. Since each row of strings catches from the flow a part of the remaining liquid (causing a consecutive reduction of the volume concentration W of liquid in the gas, and an increase of x_{cr} for each of the subsequent rows), the critical velocity of gas with grow as the number of strings increases.

Consider N rows of strings placed in series. Assume that the distance between the rows is large in comparison with the diameter of strings. In this case, any perturbation of the velocity field that might occur as the gas flows over any one row, will not influence the next row. Let W_0 be the volume concentration of liquid before the string droplet catcher, and $x_{cr}^{(1)}$ be critical height of the first row of strings. Adopt the following model for the subsequent calculations. In the string region where $0 < x < x_{cr}^{(1)}$, some drops are deposited on the string, while at $x_{cr}^{(1)} < x < H$ (where H is height of the string), all drops are swept away by the gas flow. Under this assumption, the second row of strings is facing the flow that has the following volume concentration of drops:

$$W = \begin{cases} W_1 & \text{at } 0 < x < x_{cr}^{(1)}, \\ W_0 & \text{at } x_{cr}^{(1)} < x < H, \end{cases} \tag{19.38}$$

where

$$W_1 = W_0\left(1 - \frac{bx_{cr}^{(1)}}{H(l+2r_c)}\right). \tag{19.39}$$

Then the conservation condition for the liquid phase flow rate becomes:

$$2r_c h u_x = 2bU(x_{cr}^{(1)} W_1 + (x - x_{cr}^{(1)}) W_0). \tag{19.40}$$

Using expressions (19.32) and (19.39), we can find the critical height of the second row of strings from (19.40):

$$x_{cr}^{(2)} = x_{cr}^{(1)} + x_{cr}^{(1)} \left(1 - \left(1 - \frac{bx_{cr}^{(1)}}{H(l + 2r_c)}\right)\right). \tag{19.41}$$

Repeating this reasoning for the subsequent rows, we can find the critical height for any row of strings:

$$x_{cr}^{(k+1)} = x_{cr}^{(1)} + x_{cr}^{(k)} \left\{1 - \prod_{i=1}^{k} \left(1 - \frac{bx_{cr}^{(i)}}{H(l + 2r_c)}\right)\right\}, \quad (k \geq 1). \tag{19.42}$$

We see from the last formula that the distance from the top end of the string to the point of the film breakdown gets larger as the number of string rows increases. For the given flow velocity and string height values, the number of rows should be chosen in such a way that no film breakdown occurs at the last string row, in other words, we should impose the condition

$$\frac{x_{cr}^{(1)}}{H} + \frac{x_{cr}^{(N+1)}}{H} \left\{1 - \prod_{i=1}^{N-1} \left(1 - \frac{bx_{cr}^{(i)}}{H(l + 2r_c)}\right)\right\} \geq 1. \tag{19.43}$$

The number of strings rows chosen according to this inequality, will ensure the operation of string droplet catcher without ablation of droplets.

Determination of the critical gas flow rate and the CE of a separator equipped with a string droplet catcher orifice requires a preliminary determination of gas-liquid flow parameters in the entire system, including the supply pipeline, the settling section of the separator, and the droplet catcher orifice – in the same way as it was done earlier for a separator with a centrifugal orifice. The outcome is the following pair expressions for the CE's of a vertical and a horizontal separator

$$\eta_v = 1 - \frac{\exp(-3\sigma^2)}{\sqrt{2\pi}\sigma} \int_0^{z_{mv}} z^2 \exp\left(-\frac{N}{f + 2}\left(\frac{S_{av} z^2}{S_{av} z^2 + 0.66}\right)^2 - \frac{\ln^2(z/z_1)}{2\sigma^2}\right) dz, \tag{19.44}$$

$$\eta_h = 1 - \frac{\exp(-3\sigma^2)}{\sqrt{2\pi}\sigma} \int_0^{z_{mh}} z^2 \exp\left(1 - \frac{\tau z^2}{1 + 0.032 \mathrm{Ar}_{av}^{1/2} z^{3/2}}\right)$$
$$\times \exp\left(-\frac{N}{f + 2}\left(\frac{S_{av} z^2}{S_{av} z^2 + 0.66}\right)^2 - \frac{\ln^2(z/z_1)}{2\sigma^2}\right) dz, \tag{19.45}$$

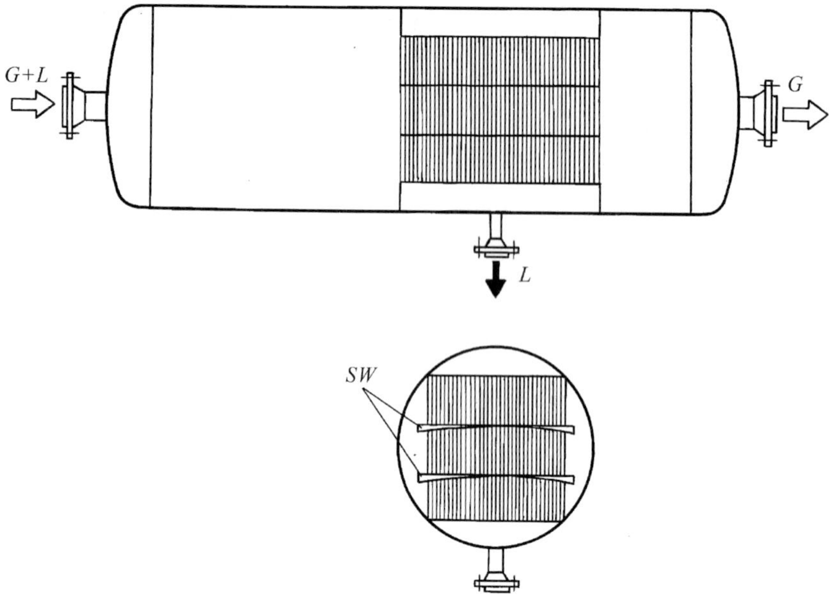

Fig. 19.11 Arrangement of string orifices in a horizontal separator: *G* – gas; *L* – liquid; *SW* – sectionalized partitions.

where x_{mv} and z_{mh} are the non-dimensional minimum radii of drops in the settling sections of vertical and horizontal separators, respectively; $S_{av} = 2U_{st}\rho_L R_{av}^2/9\mu_G r_c$; U_{st} is the velocity in the string droplet catcher; Ar_{av} is Archimedes' number, calculated from the average radius R_{av} of drops formed in the supply pipeline; $z_1 = \exp(-\sigma^2)$; τ is a non-dimensional residence time of the mixture in the settling section of the separator; $f = l/r_c$.

The string section can be placed both in horizontal and vertical separators. In a horizontal separator, it is convenient to place the section vertically in the cross-section of the body (Fig. 19.11). To maintain conditions that prevent the film breakdown, the string droplet catcher should be subdivided into sections, according to the calculated critical height of strings. The placement of a droplet catcher section into a vertical separator involves reorganization of the gas flow. This can result in a reduction of the working cross-section of the string section and an increase of the flow velocity in the string section above the critical value. One promising way to increase productivity and efficiency of a vertical separator is to install the string droplet catcher at the separator exit, as shown in Fig. 19.12. Thanks to its small dimensions, the string block has a number of useful applications such as re-equipment of separators that have low efficiency and/or productivity, treatment of oil gas, and treatment of the gas going toward the flare.

CE calculations for separators equipped with string orifices have shown their high efficiency for gas flow rates in the sub-critical region. Therefore, we should

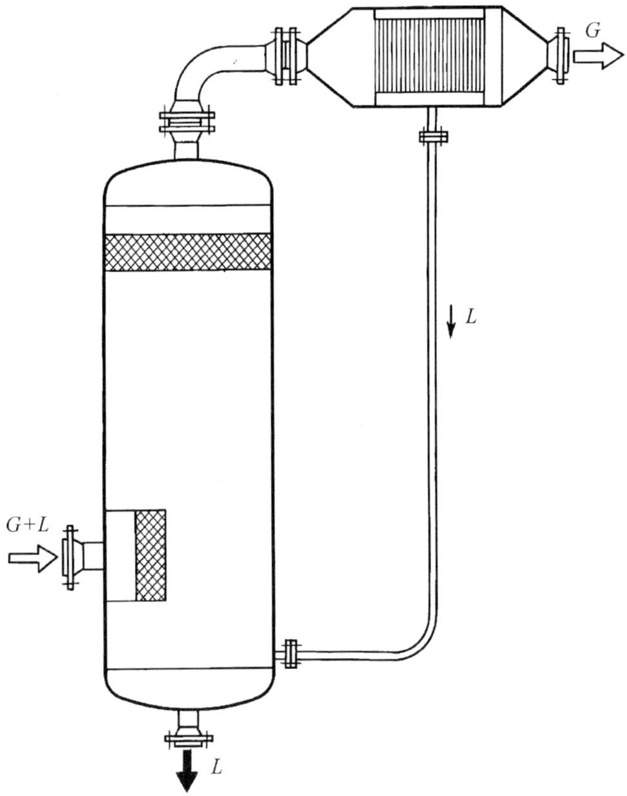

Fig. 19.12 Placement of string orifice at the exit of a vertical separator.

try to determine the critical gas flow rate and its dependence on various parameters.

Consider a horizontal separator with a string orifice located in its cross section (Fig. 19.11). Let the cross-sectional area of the orifice be equal to 2/3 of the working cross-sectional area of the separator. Choose the following parameter values: $D = 1$ m; $d = 0.15$ m; $p = 5$ MPa; $T = 280$ °K; $\rho_L = 750$ kg/m^3; $\rho_G = 50$ kg/m^3; $\Sigma = 10^{-2}$ N/m; $\mu_L = 10^{-3}$ Pa·s; $d_1 = 5 \cdot 10^{-4}$ m; $l = 2.5 \cdot 10^{-3}$ m. The orifice consists of several sections. All strings have the same height H. Fig. 19.13, a shows the dependence of the critical gas flow rate Q_{cr} (where Q_{cr} is reduced to the normal conditions), on the number N of string rows for various values of H, when the condensate factor equals $W_0 = 100$ cm^3/m^3. The critical flow rate increases with the decrease of H and W_0, and with the increase of the number N of rows. The influence of pressure on the critical gas flow rate is illustrated in Fig. 19.13.

There is a pressure (about 9 MPa) at which the permissible gas flow rate is maximized. The viscosity and the volume concentration of liquid to be separated

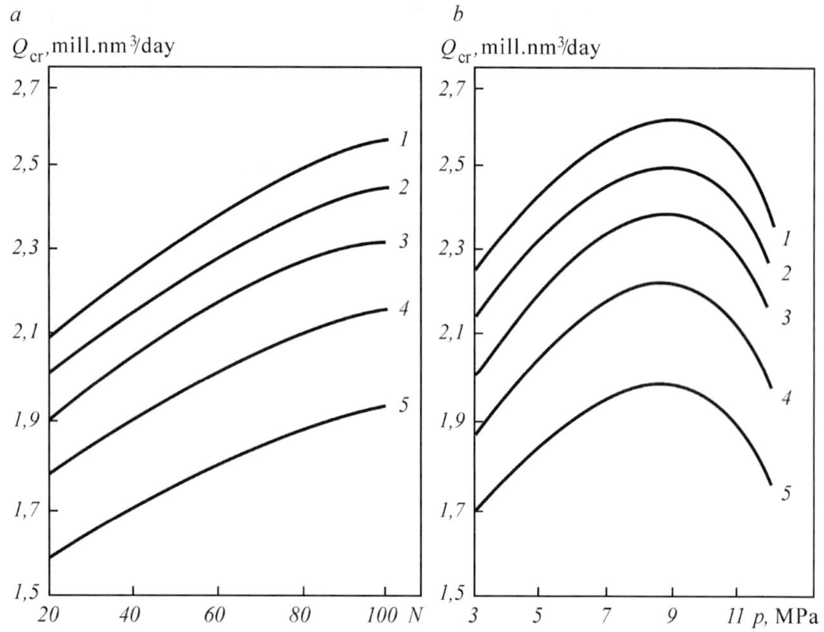

Fig. 19.13 Dependence of the critical gas flow rate Q_{cr} in a horizontal separator with string orifices on the number N of string rows (*a*) and the pressure (*b*) for different values of the string height H, m: 1 – 0.141;

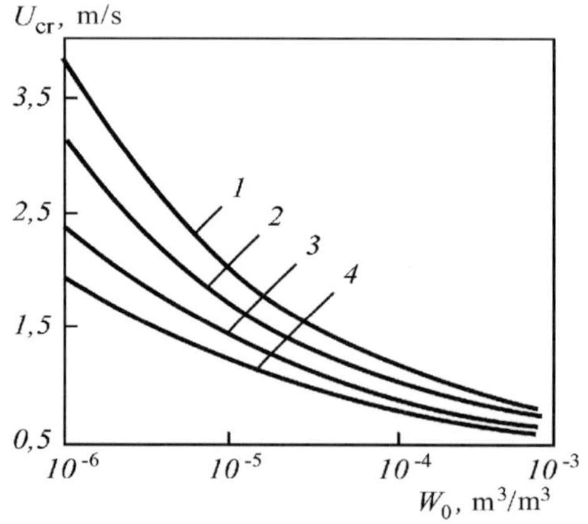

Fig. 19.14 Dependence of the critical gas velocity U_{cr} in a horizontal separator with a string orifice on the condensate factor W_0:

Number of the curve in the figure	1	2	3	4
N	100	40	100	40
H	0.141	0.141	0.707	0.707

both have a noticeable influence on the value of the critical gas flow rate. Smaller μ_L means that the liquid will flow down the strings at a faster rate, so the film will be thinner, and therefore the critical flow rate will be higher. Fig. 19.14 shows the dependence of the critical gas velocity in a separator on the condensate factor W_0. As W_0 gets lower, the critical velocity increases. This means that string orifices can be used with the greatest efficiency in the separation of gas with a low concentration of liquid. Note that usually the condensate factor is given assuming the normal conditions, while in the calculation formulas presented above, W is taken at the operating conditions. This fact should be kept in mind when comparing theoretical results with experimental ones.

19.4 The Efficiency Coefficient of a Separator with Mesh Orifices

Mesh droplet catcher orifices are made from hose knitted grid and placed in vertical and horizontal separators in the device cross-section.

Deposition of drops on grid wires is similar to deposition of drops on strings, which was considered in the previous section. As opposed to string orifices, the internal structure of mesh orifices is characterized by a chaotic distribution of cells over the orifice thickness. A mesh orifice can be considered as a porous highly permeable medium with the following intrinsic parameters: specific free volume ε_V (m^3/m^3); thickness of the mesh bundle H; average size of grid cells $l \times l$; diameter of wires d_W.

A mesh bundle will be modeled by N parallel flat layers of grid, with a constant distance h between them. Specific volume ε_V may be expressed in terms of the geometrical parameters of the grid:

$$\varepsilon_v = 1 - \frac{\pi d_w^2}{2lh}. \tag{19.46}$$

Now we can determine specific surface area a of the grid, the distance h between layers, and the number N of layers

$$a = \frac{4(1-\varepsilon_v)}{d_w}, \quad h = \frac{\pi d_w^2}{2l(1-\varepsilon_v)}, \quad N = \frac{2Hl(1-\varepsilon_v)}{\pi d_w^2}. \tag{19.47}$$

For the characteristic values $d_w = 3 \cdot 10^{-4}$ m, $\varepsilon_V = 0.975$ m^3/m^3, $l = 3 \cdot 10^{-3}$ m, $H = 0.1$ m we have $a = 330$ m^2/m^3, $h = 1.8 \cdot 10^{-3}$ m, and $N = 55$.

Each layer of the grid will be modeled by two frames with stretched strings located perpendicularly to each other. We can then use the results from the previous section. If all drops in the oncoming flow running into the mesh bundle have the same size, then the CE of the mesh droplet catcher is

$$\eta = 1 - \exp\left(-\frac{4Nb}{l + 2r_c}\right). \tag{19.48}$$

where r_c is the radius of wires; $b = r_c G$, the quantity G is given by the formula (19.27). For a continuous distribution of drop sizes, the CE of a separator with a mesh droplet catcher is determined by the expressions (19.44) and (19.45), in which we must substitute *2N* instead of *N*.

Mesh orifices, as well as string ones, are characterized by a critical velocity or critical flow rate; when it's exceeded, the secondary ablation of drops takes place. The mechanism of ablation depends on the orientation of the orifice with respect to the direction of gravity. First, consider a vertical orifice. At drops are deposed on the grid wires, the liquid flows downwards. However, the possibility of transverse flow – along the horizontal wires of the grid results in that the flow of liquid effectively doubles. Hence, the thickness of the film that is formed on the vertical wires increases as compared to the string orifice case, and accordingly, the critical velocity of a mesh orifice is less than that of a string orifice. To determine the critical velocity, it is possible to use results of the previous section, but every instance of *W* in the appropriate formulas should be replaced by *2W*.

Next, consider the critical velocity for a horizontal mesh orifice. The following expression is usually taken to represent the critical velocity:

$$u_{cr} = K\left(\frac{g\Sigma(\rho_L - \rho_G)}{\rho_G^2}\right)^{1/4}, \tag{19.49}$$

where *K* is an empirical constant.

However, there are many factors that the formula (19.49) doesn't take into account, in particular, the structure of the mesh bundle. Therefore, in order to determine *K*, we have to perform numerous experiments with various mesh orifices.

Let's take the following model. The mesh bundle is considered as a highly permeable porous medium. Gas flows through microchannels whose effective diameter is

$$d_{eff} = \frac{4\varepsilon_v}{a} = \frac{\varepsilon_v d_w}{1 - \varepsilon_v}. \tag{19.50}$$

Gas velocity is equal to the gas flow rate divided by the total area of microchannel cross-sections $S_{ch} = \varepsilon_v S/\varepsilon_k$, where S is the working area of the orifice, $a_k = 1/h$ is a coefficient that helps account for the curvature of the channels.

Drops that are ablated from the surface of wires move through microchannels together with the gas. In the process of motion, they experience an additional splitting. The average radius of drops inside microchannels can be estimated from the formula (14.14), in which d_{eff} should be taken for the diameter. If the velocity of ascending gas flow in the orifice exceeds the velocity of drop sedimentation, then drops will be carried away from the orifice. Equating both velocities, one can find the critical gas velocity, which marks the point where the secondary ablation of drops begins

Table 19.1

p, atm	ρ_G, kg/m³	ρ_L, kg/m³	μ_G, Pa·s	Σ, N/m	K_{exp}	K_{th}
21	16.6	760	$1.2 \cdot 10^{-5}$	$2 \cdot 10^{-2}$	0.56	0.52
25	19.7	996	$1.3 \cdot 10^{-5}$	$6.7 \cdot 10^{-2}$	0.46	0.49
35.2	27.3	695	$1.5 \cdot 10^{-5}$	$1.4 \cdot 10^{-2}$	0.66	0.57
36	27	720	$1.6 \cdot 10^{-5}$	$1.5 \cdot 10^{-2}$	0.63	0.55
50	49	704	$1.8 \cdot 10^{-5}$	$1.2 \cdot 10^{-2}$	0.72	0.74

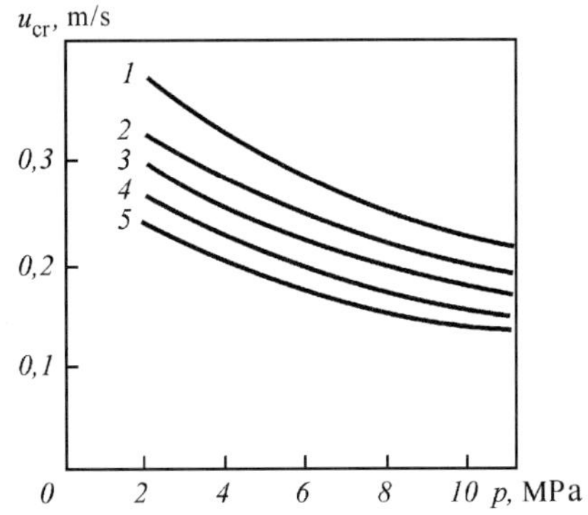

Fig. 19.15 Dependence of critical velocity u_{cr} of a vertical separator with a horizontal mesh orifice on pressure p for different values of specific grid surface area a, m²/m³ ($\varepsilon = 0.075$; $D = 1$ m; $d = 0.15$ m; $d_w = 0.00035$ m; $T = 7$ °C; $H = 0.15$ m): 1 – 400; 2 – 600; 3 – 800; 4 – 1000; 5 – 1200.

$$u_{cr} = \frac{0.0625\pi\varepsilon_v d_w^2 \Sigma^{6/19} \rho_G^{2/19}}{l^2(1-\varepsilon_v)\rho_L^{8/19}} \left(\frac{\Delta\rho g}{\mu_G}\right)^{7/19} \left(\frac{\varepsilon_v d_w}{1-\varepsilon_v}\right)^{8/19}. \tag{19.51}$$

The expression (18.51) can be brought to the form (19.49) and then the appropriate factor K can be found from it. Theoretical and experimental values of this factor are given in Table 19.1.

The formula (19.51) shows that the key parameters influencing u_{cr} are the pressure p (Fig. 19.15), on which the factors Σ, μ_G and ρ_G depend, and also the parameters ε_V, d_w, and l, which characterize the structure of the mesh orifice. The critical velocity of the mesh orifice decreases with an increase of pressure and of the wire diameter (or the grid surface area). Although horizontal mesh orifices have much lower critical velocity than vertical ones, they also have one advantage: they don't need to be sectionalized, as vertical orifices do. Unlike velocity, the critical flow rate Q_{cr}, is not a monotonous function of pressure (Fig. 19.16). The

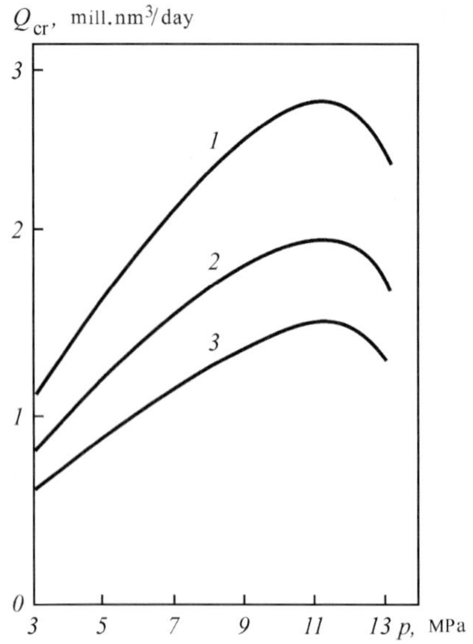

Fig. 19.16 Dependence of critical gas flow rate Q_{cr} of a vertical separator with a horizontal mesh orifice on pressure p for different values of ε (other parameters are the same as in Fig. 19.15): 1 – 0.98; 2 – 0.975; 3 – 0.97.

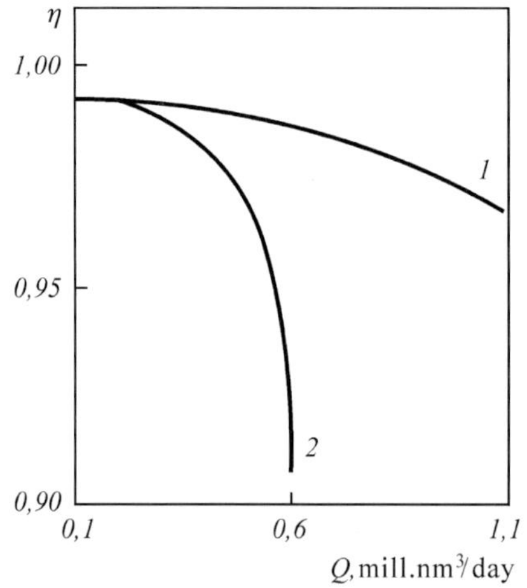

Fig. 19.17 Dependence of η of a vertical separator on tgas flow rate Q (p = 11 Mpa; $T = 5$ °C; other parameters are the same as in Fig. 19.15): 1 – with horizontal mesh orifice; 2 – without mesh orifice.

greatest value of Q_{cr} is achieved at $p \approx 11$ MPa. Therefore, mesh separators should be used at high pressure. Because the pressure falls in the course of exploitation, the productivity of mesh separators decreases as well.

By equipping separators with mesh orifices, we can substantially improve their CE. Figure 19.17 shows the dependence of the CE of a mesh separator on the gas flow rate. For comparison, the figure also shows the same dependence for a gravitational separator without a mesh orifice.

20
Absorption Extraction of Heavy Hydrocarbons and Water Vapor from Natural Gas

One of the methods of extracting water vapor and hydrocarbons (which form a gas condensate) from natural gas is by absorption. Liquid absorbent is injected into the gas flow that enters the field equipment devices designed for physical processing of hydrocarbon raw material. In Section 2.2, we studied the devices and technologies that are used in absorption extraction of heavy hydrocarbons from gas, and in gas dewatering. It was mentioned that two radically different methods exist: input of the absorbent directly into the gas flow under the concurrent flow condition (gas and absorbent move in the same direction); and input of the absorbent against the gas flow (countercurrent flow condition). The first method is realized when the absorbent is injection directly into the gas-transporting pipeline (this method is called the intratube absorption), or into the contact section of a direct-flow spray absorber. The second method is realized in vertical plate absorbers, where gas enters the bottom part of the device, and absorbent is delivered into the top part and then flows down successively from one contact plate to another. In the latter case, at each plate both concurrent and countercurrent flows are possible – depending on the design of contact plates.

In the process of extraction of heavy hydrocarbons, a liquid saturated with heavy hydrocarbons can be taken as the absorbent, for example, petroleum, solar or transformer oil, weathered condensate etc. For the absorption dehydration of gas, glycols are used: diethylene glycol (DEG) or triethylene glycol (TEG).

Consider successively the calculation of absorption processes for intratube absorption, in concurrent spray absorbers, and, finally, in countercurrent plate absorbers.

20.1
Concurrent Absorption of Heavy Hydrocarbons

An absorbent is introduced in the form of fine drops into the flow of a multicomponent gas mixture moving in a pipe of constant cross section. Let us suppose that all drops have the same radius R, and the composition of the liquid phase is given either by mass concentrations of components ρ_{i0} (kg/m^3), or by their molar fractions x_{i0}. Pressure p and temperature T of gas in the pipe are given, and they

Separation of Multiphase, Multicomponent Systems. E. G. Sinaiski and E. J. Lapiga

ISBN: 978-3-527-40612-8

are different from the values p_a and T_a at which the absorbent is prepared. It means that the initial molar fractions x_{i0} correspond to the equilibrium values at p_a and T_a, to the initial composition of the mixture, and to the volume concentration W_0 of the absorbent. For the values p and T, i.e. at the moment when the absorbent enters the gas flow, the balance between phases is violated, triggering a mass exchange between the liquid and gas phases, which continues until a new equilibrium is established. Consider the process of mass exchange in a simplified, quasi-stationary formulation (see Section 16.2) under the condition of local thermodynamic equilibrium at the drop surface.

Let ρ_{iL} be the current mass concentration of i-th component in the drop, ρ_{iGw} – mass concentration of i-th component in the gas at the drop boundary, ρ_{iG} – mass concentration of i-th component in bulk of the gas flow. Since the volume concentration of absorbent in the gas is small, it is safe to assume that in course of mass exchange, drops of absorbent do not exert any noticeable influence on each other.

Equations describing the change of component concentrations in liquid and gaseous phases, and the change of the drop volume V, are similar to Eqs. (16.63) to (16.65):

$$\frac{d\rho_{iG}}{dt} = -\frac{W_0 J_i}{V_0}, \quad \frac{d(V\rho_{iL})}{dt} = J_i, \quad \frac{dV}{dt} = \sum_i \frac{J_i}{\rho_L}, \tag{20.1}$$

where W_0 is the initial volume concentration of the liquid phase; V_0 and V are, respectively, the initial and the current volume of the drop; J_i is the mass flux of i-th component, which, according to (16.55), equals

$$J_i = 4\pi\sqrt[3]{2}\left(\frac{\rho_L}{\rho_G}\right)^{1/3} \frac{UR^{7/3}\Delta\rho_i}{d^{1/3}}. \tag{20.2}$$

Here ρ_L and ρ_G are the phase densities; U is the gas velocity in the tube; $\Delta\rho_i = \rho_{iG} - \rho_{iGw}$; d is the tube diameter.

For the initial conditions, we take $\rho_{iG} = \rho_{iG0}$, $\rho_{iL} = \rho_{iL0}$, $V = V_0$, where ρ_{iG0} are mass concentrations of components in the gas flow before the input of the absorbent, and ρ_{iL0} are mass concentrations of components in a freshly prepared absorbent.

The absorbent can be introduced into the flow in several ways. The most widespread method is the input of absorbent through atomizers that inject a finely-dispersed spray in or against the direction of the flow. Absorbent can also be injected through special apertures in the pipe wall, so that the liquid jet enters the pipe perpendicular to the flow and practically at once becomes disintegrated by the rapidly moving gas, producing a drop spectrum. Since the flow is turbulent, in the process of breakup and coagulation the size distribution of drops stabilizes. The resulting distribution is approximated by distribution (14.1) with the average radius (14.14). In the process of mass exchange with the gas, the size of drops increases until their radius exceeds the stable radius R_{cr}, which is the characteristic radius for the given flow parameters in the pipe, and then they with split with a high probability. This allows us to assume that in the process of mass exchange in the pipe, the size of drops does not change on the average, i.e. $V = V_0 = V_{av}$. Then the system of equations will simplify and turn into

$$\frac{dy_i}{dt} = -E(y_i - y_{iw})W_0, \quad \frac{dx_i}{dt} = \frac{E\rho_G M_L(y_i - y_{iw})}{\rho_L M_G}, \tag{20.3}$$

where $E = 3.78U\rho_L^{1/3}/d^{1/3}\rho_G^{1/3}R_{av}^{2/3}$, and x_i, y_i are molar fractions of the components in liquid and gaseous phases.

For given values of p, T, x_{i0} and y_{i0}, the quantities y_{iw} remain constant. Also, the molecular weight M_0 and the density ρ_G of the gas both change insignificantly, therefore the solution of equations (20.3) looks like

$$y_i = y_{iw} + (y_{i0} - y_{iw})e^{-\tau}, \tag{20.4}$$

where $\tau = EW_0t$.

It follows from the formula (20.4) that the characteristic time for the establishment of equilibrium in the gaseous phase is estimated by

$$t_{eq}^{(G)} \sim \frac{1}{EW_0} = 0.264\frac{R_{av}^{2/3}d^{1/3}}{W_0U}\left(\frac{\rho_L}{\rho_G}\right)^{1/3}, \tag{20.5}$$

where R_{av} is the average radius of drops formed in the turbulent flow in the pipe.

From the second equation (20.3), it is possible to estimate the characteristic time for the establishment of equilibrium in the liquid phase:

$$t_{eq}^{(L)} \sim \frac{W_0\rho_L M_G}{\rho_G M_L}t_{eq}^{(G)}. \tag{20.6}$$

Since $W_0\rho_L M_G/\rho_G M_L \ll 1$, we see that $t_{eq}^{(L)} \ll t_{eq}^{(G)}$. Hence, the equilibrium in a liquid phase is established much faster than in the gaseous phase, and the time it takes to establish phase equilibrium in the mixture is estimated by $t_{eq}^{(G)}$. A substitution of the expression (14.11) for R_{av} into (20.5) yields the characteristic time for the establishment of phase equilibrium:

$$t_{eq} \sim 0.06\frac{d^{5/7}\Sigma^{2/7}\rho_G^{3/7}\rho_L^{2/7}}{qU^{11/7}}, \tag{20.7}$$

where q is the specific mass flow rate of the absorbent (kg per m^3 of gas at operating conditions).

An important parameter is the distance at which a phase equilibrium is established:

$$L_{eq} \sim 0.06\frac{d^{5/7}\Sigma^{2/7}\rho_G^{3/7}\rho_L^{2/7}}{qU^{4/7}}, \tag{20.8}$$

From the formula (20.8), it follows that the length at which absorption occurs becomes smaller with the increase of absorbent flow rate q and velocity U, and also with the reduction of pipe diameter d. An increase of the flow rate results in an increased volume concentration W_0 of the absorbent, and hence in a larger phase contact surface area. Similarly, a velocity increase and reduction of the pipe diameter will reduce the size of drops in the gas stream, thus increasing the contact surface area.

For the characteristic parameter values $d = 0.4$ m, $p = 5$ MPa, $T = 293$ °K, $\Sigma = 10^{-3}$ N/m, $\rho_L = 750$ kg/m^3, $\rho_G = 40$ kg/m^3, $q = 0.1$ kg/m^3 and $U = 15$ m/s corresponding to the gas flow rate $Q_G = 10$ mill. m^3/day under normal conditions or $Q_G = 2$ m^3/s at the operating conditions, we have $L_{eq} \approx 0.32$ m. A reduction of the gas flow rate up to 1 mill. m^3/day, with other parameters remaining the same, results in $L_{eq} \approx 3.73$ m. In practice, these estimations allow us to decide at what distances fresh absorbent should be injected into the gas flow during multistage absorption. The exhausted absorbent should be separated for from the gas after each stage of contact.

Especially important to us is the amount of heavy hydrocarbons C_{3+} and C_{5+} extracted from the gas, in terms of percentages of the initial concentration:

$$m_{C_{k+}} = \frac{\left(\sum_{i=k}^{N} \rho_{iG0} - \sum_{i=k}^{N} \rho_{iG}\right)}{\sum_{i=k}^{N} \rho_{iG0}} 100 \tag{20.9}$$

or molar fractions:

$$m_{C_{k+}} = \left\{ 1 - \frac{M_{G0}\rho_G \sum_{i=k}^{N} y_i M_i}{M_G \rho_{G0} \sum_{i=k}^{N} y_{i0} M_i} \right\} 100. \tag{20.10}$$

Substituting into (20.10) the correlations (20.4), we obtain:

$$m_{C_{k+}} = (m_{C_{k+}})_{eq}(1 - e^{\tau}), \tag{20.11}$$

where $(m_{C_{k+}})_{eq}$ are equilibrium values of $m_{C_{k+}}$ at $\tau \to \infty$, which are equal to

$$(m_{C_{k+}})_{eq} = \left\{ 1 - \frac{M_{G0}\rho_G \sum_{i=k}^{N} y_{iw} M_i}{M_G \rho_{G0} \sum_{i=k}^{N} y_{i0} M_i} \right\} 100. \tag{20.12}$$

The values y_{iw} entering the expression (20.12) should be determined from the equations of liquid-vapor equilibrium, using the unified equation of state (see Section 5.7).

Thus, calculations for the absorption process in a gas flow should be carried out as follows.

1. Set the initial compositions of absorbent x_{i0} and gas y_{i0}, and also the absorbent flow rate q at operating conditions, or the volume concentration of the absorbent $W_0 = q/\rho_L$.
2. Calculate the composition of the resultant gas-liquid mixture:

$$\eta_i = (1-\alpha_0)\,y_{i0} + \alpha x_{i0}, \qquad \alpha_0 = \frac{0.02404q}{\sum_i M_i x_{i0} - 0.02404q}; \qquad (20.13)$$

3. For the given values of p and T, calculate the liquid-vapor equilibrium for the mixture with the composition (20.13), and also the density and the molecular weights of liquid and gaseous phases, using the Peng-Robinson equation of state (see Section 5.7);
4. Determine the dimensionless time τ and the characteristic time T_{eq} and length L_{eq}.
5. Find the percentage of heavy hydrocarbons C_{3+} and C_{5+} extracted from the gas.

As an example, let's carry out the calculation for intratube absorption in a flow of gas-liquid mixture, whose composition is given in Table 20.1.

The components that appear in C_{5+} are grouped into fractions that are designated through F_i in the table.

The results of calculation of the equilibrium values $(m_{C_{k+}})_{eq}$ are reported below. The calculation was carried out for nine values of pressure and five values of temperature that varied, respectively, in the intervals from 4 to 12 MPa and from 253 to 293 °K, and for five values of absorbent flow rate that varied from 0.1 to 2 kg/m^3. The results are presented in Figs. 20.1 and 20.2.

Pressure p and temperature T exert a strong influence on the efficiency of extraction of heavy hydrocarbons from gas. With the growth of pressure up to 6–7 Mpa, the extraction of hydrocarbons of C_{3+} group grows 1–2% at temperatures below 263 °K and 10–20% at temperatures above 263 °K. For pressures above 7 Mpa, extraction falls sharply: by 30–70% at $T = 293$ °K and 15–95% at $T = 253$ °K. The greater the absorbent flow rate q, the smaller is the reduction of $(m_{C_{3+}})_{eq}$. The value $(m_{C_{5+}})_{eq}$ increases by 3–5% with the growth of pressure up to 6 MPa, and then it decreases by 82–85% at 293 °K, and by 50–90% at 253 °K.

As temperature grows, the values $(m_{C_{3+}})_{eq}$ and $(m_{C_{5+}})_{eq}$ also diminish, with the exception of the isobar 12 MPa at $q = 0.1$ kg/m^3. At the other values of q, we observe a reduction of the degree of hydrocarbons extraction with the growth of temperature for all pressure values.

Table 20.1 The initial composition of the gas-liquid mixture.

Component	Concentrations of components (molar fraction)		Molecular mass of components M_i (kg/kmole)
	In a gas, y_{i0}	In a liquid, x_{i0}	
CO_2	0.0015	0.0050	44.01
N_2	0.0469	0.00010	28.01
C_1	0.8091	0.00500	16.04
C_2	0.0878	0.03678	30.01
C_3	0.0341	0.08619	44.1
C_4	0.0130	0.07970	58.12
F_1	0.0063	0.24093	80
F_2	0.0012	0.23984	100
F_3	0.0001	0.15798	128
F_4	0	0.08895	172
F_5	0	0.06403	240

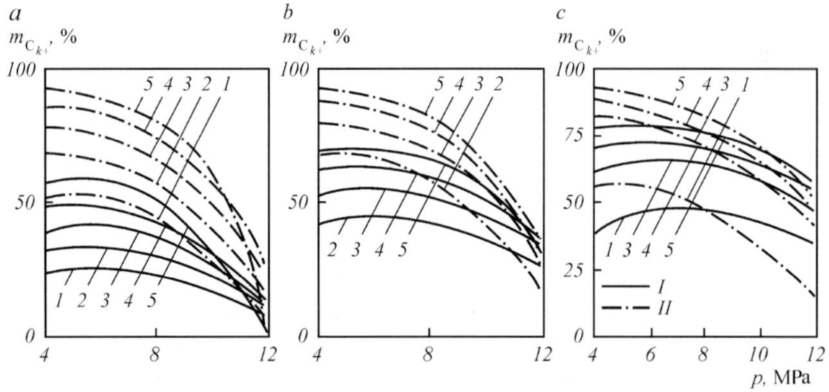

Fig. 20.1 Dependence of the efficiency of extraction of heavy hydrocarbons $m_{C_{k+}}$ from gas on pressure, for various values of T, °C: a, b, c – q accordingly 0.1, 0.5, and 2 kg/m^3; 1 – 20; 2 – 10; 3 – 0; 4 – −10; 5 – −20; I – C_{3+}; II – C_{5+}.

With an increase of q up to 1–1.2 kg/м^3, the value $(m_{C_{3+}})_{eq}$ grows at first, and then remains stable; on the other hand, the value $(m_{C_{5+}})_{eq}$ practically does not change.

The peculiar features of heavy hydrocarbon extraction elucidated above (the complex dependence of the degree of extraction on various conditions) can be explained by phase transitions occurring during the contact of the absorbent

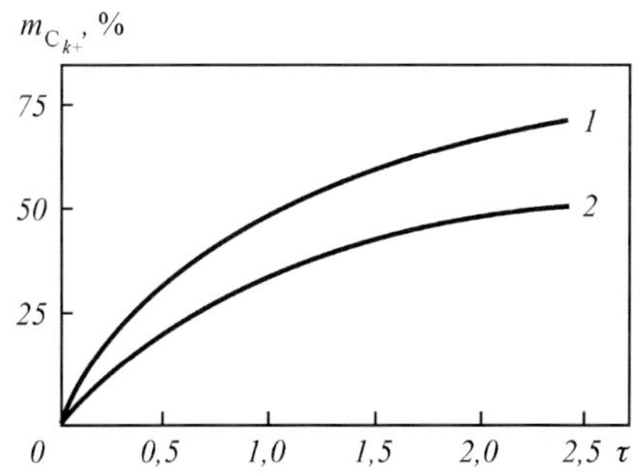

Fig. 20.2 Dependence of the efficiency of extraction of heavy hydrocarbons $m_{C_{k+}}$ from gas on time ($p = 6$ MPa; $T = 0$ °C; $q = 0.5$ kg/m^3): 1 – C_{5+}; 2 – C_{3+}.

with the gas flow. As can be seen form expression (20.12), the difference between the values y_{iw} and y_{i0} exerts a predominant effect on $(m_{C_{k+}})_{eq}$. The ratio $M_{G0}\rho_G/M_G\rho_{G0} \sim 1$. For example, as the pressure changed from 4 to 12 MPa at $q = 0.1$ kg/m^3, the molar fractions y_{iw} increased by 1 or 2 orders of magnitude. A similar picture is observed for the other values of q.

Thus, the greatest degree of heavy hydrocarbons extraction from the gas flow is achieved at pressures 4–6 MPa, and the lower the temperature, the greater is the extraction efficiency. An increase in volume concentration W of drops (or in the absorbent flow rate q) causes an essential increase in $m_{C_{k+}}$ only in the range of q from 0.1 kg/м^3 to 1.2 kg/м^3. In the indicated ranges of p, q and for temperatures below 263 °K, the degree of heavy hydrocarbons extraction can be as high as 50–60% for $(m_{C_{3+}})_{eq}$, and 85–90% for $(m_{C_{5+}})_{eq}$.

If the degree of extraction fails to attain a required value for the given p, T, q, x_{i0} and y_{i0}, then the process of absorption must be carried out in several stages (multistage absorption). At each stage, we have to deliver a fresh absorbent to the gas flow, while removing the exhausted absorbent from the gas. The arrangement of such a multistage process in the pipeline presents severe difficulties. This is why multistage absorption is performed in special multistage absorbers operating in concurrent conditions.

20.2
Multistage Concurrent Absorption of Heavy Hydrocarbons

Consider now the multistage concurrent absorption calculations, as exemplified by the spray absorber and by absorption inside the pipeline.

One possible design of a direct-flow spray absorber is shown schematically in Fig. 20.3. Each stage of the absorber consists of a contact chamber and a mesh separator placed in tandem. To simplify calculations, assume that the separator operates with a 100% efficiency, i.e. dry gas comes out from the contact stage.

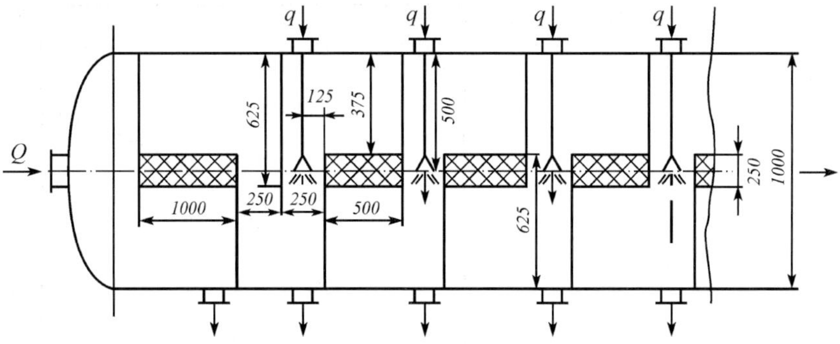

Fig. 20.3 Direct-flow multistage spray absorber.

Table 20.2 Initial compositions of the gas and the absorbent (molar fraction, %).

Component	Gas, y_{i0}	Absorbent, x_{i0}
CO_2	0.15	0.00
C_1	93.42	0.00
C_2	4.99	0.00
C_3	0.83	0.00
C_4	0.39	0.00
C_5	0.22	0.00
F_1	0.00	22.00
F_2	0.00	10.00
F_3	0.00	10.00
F_4	0.00	20.00
F_5	0.00	38.00

Gas composition at the entrance of the absorber, and composition of the fresh absorbent (straw oil) is given in Table 20.2. Fractions into which hydrocarbons C_{6+} are grouped are designated by F_i.

For the given values of p, T, and Q_G, it is possible to determine the length of the contact zone, by demanding, for example, that molar concentrations at the exit differ by 1% from their corresponding equilibrium values. Using expression (20.4), we then obtain:

$$L_k = 0.29 \frac{d^{5/7} \Sigma^{2/7} \rho_G^{3/7} \rho_L^{2/7}}{q U^{4/7}}, \tag{20.14}$$

If the fresh absorbent delivered to each stage has the same composition and flow rate, then the length of each stage is also the same and is given by the formula (20.14). Since all the used absorbent is removed from the gas in the separator, the composition of the gas entering n-th stage is equal to that at the exit of $(n-1)$-th stage:

$$y_{i0,n} = y_{iw,n-1} + 0.01(y_{i0,n-1} - y_{iw,n-1}), \tag{20.15}$$

The calculation for each stage is carried out by the method described in the previous section, until the degree of extraction $m_{C_{k+}}$ will achieve the required value. In this manner, the necessary quantity of contact stages can be determined.

The calculation of the absorber represented in Fig. 20.3 was carried out for the range of pressures from 2 to 12 MPa, for the range of temperatures from -10 to 25 °C, and for the range of absorbent flow rates q from 0.01 to 0.04 m^3 per thousand m^3 of gas. The gas flow rate is equal to $Q_0 = 1.2$ mill. m^3/day. The results of calculation are shown in Fig. 20.4. The dependence of $m_{C_{k+}}$ on pressure has a non-monotonous character. The greatest degree of extraction is achieved in the pressure range $p = 4$–6 MPa.

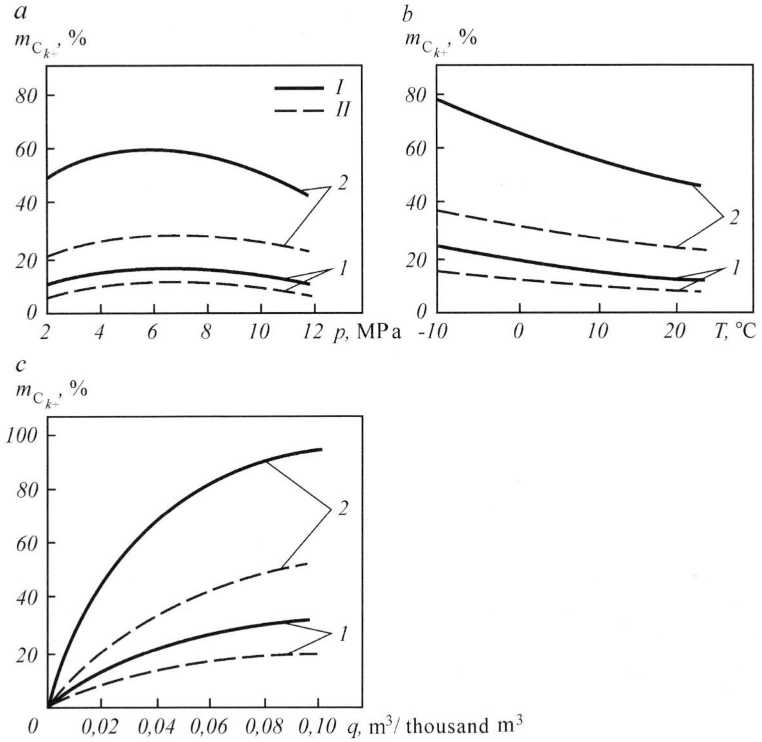

Fig. 20.4 Dependence of the coefficient of extraction (aka extraction factor) of heavy hydrocarbons $m_{C_{k+}}$ from gas on pressure (*a*), temperature (*b*), and absorbent flow rate (*c*): 1 – one stage; 2 – eight stages; *I* – C_{5+}; *II* – C_{3+}.

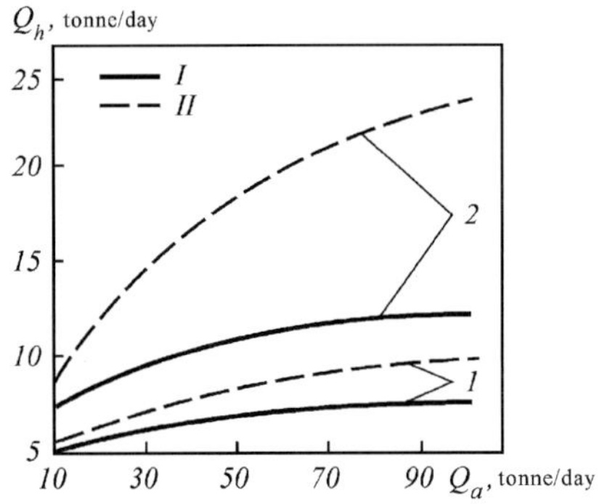

Fig. 20.5 Dependence of the amount Q_h of heavy hydrocarbons extracted from the gas on the absorbent flow rate Q_a: 1 – one stage; 2 – eight stages; I – C_{5+}; II – C_{3+}.

The amount of hydrocarbons extracted from gas (ton/day) can be determined by the formula:

$$Q_h = 44.64 Q_G \sum_{k+} (y_{i0} - y_i) M_i. \tag{20.16}$$

Fig. 20.5 shows the dependence of Q_h on the total amount of delivered absorbent Q_a (ton/day) and on the number of contact stages. It can be seen from these dependences that solar oil is an efficient absorbent for certain thermobaric conditions. However, on some gas-condensate fields in the North, its use is restricted by the difficulty of transporting large amounts of solar oil. A more recent proposal is to use weathered condensate as an absorbent.

As an example, consider the calculation of concurrent absorption as applied to a gas-condensate field with low reservoir pressure and positive temperatures of gas at the entrance to the DCPG. Such thermobaric conditions make it possible to carry out intrapipe absorption even with such a low-efficient absorbent as weathered condensate. To increase the condensate recovery in the process of absorption, we can deliver the absorbent into gas flow in the pipeline before the separators at the gas gauging station. Before we start to calculate the absorption process, it is necessary to determine the composition of weathered condensate that gets collected after the first stage of separation. Denote by p_B and T_B the pressure and temperature in the divider where the condensate weathering takes place. The composition of the condensate collected after the first step of separation, and also the composition of the reservoir gas are given in Table 20.3.

At the gauging station, gas from the delivery pipeline is distributed over 16 separators. The process can be organized in two ways. The first way calls for a single-stage input of weathered condensate into gas before each separator. The second way relies on two-stage absorption, which is arranged by connecting a pair of separators in series, weathered condensate being input before each separator.

Table 20.3 Initial composition of the gas and the absorbent (molar fraction %).

Component	Condensate	Gas	Component	Condensate	Gas
N_2	0.36	4.68	C_6	15.22	0.65
C_1	18.3	79.6	C_7	14.96	0.43
C_2	11.39	9.38	C_8	9.11	0.22
C_3	13.83	3.56	C_9	0.88	0.02
C_4	3.44	0.43	C_{10}	1.37	0.03
C_5	11.16	1.02			

Calculation of the absorption process consists of the following stages:

1. Determine compositions of the gas and the condensate after the first stage of separation;
2. Determine composition of the weathered condensate at the given values p_B and T_B;
3. Determine compositions of the gas and the condensate after their mixture before the gauging station.
4. Determine the amount of heavy hydrocarbons belonging to C_{3+} and C_{5+} groups that are additionally extracted from the gas.

The calculation results for a single-stage absorption are shown in Fig. 20.6. It is worth to note that at certain values of pressure and temperature, some components are desorbed from the liquid to the gaseous phase (this corresponds to negative values of m_{ex}). Thus, for condensate weathered at $T_B = 20$ °C and $p_B = 0.1$ MPa, desorption is observed at $T \geq 15$ °C and $p \leq 1.2$ MPa. A reduction of T_B and an increase of p_B results in desorption of heavy hydrocarbons at higher values of p and lower values of T.

A two-stage scheme of absorption allows us to extract additional amounts of hydrocarbons, as compared to a single-stage scheme, provided the absence of desorption. For instance, at $T_B = 20$ °C; $p_B = 0.1$ MPa; $T = 10$ °C; and $p = 3$ Mpa, a single-stage input of 100 liter/thousand m3 of weathered condensate allows for the extraction of 15.7 ton/day of C_{3+} and 5.71 ton/day of C_{5+}. A two-stage input scheme allows for the extraction of 16.7 ton/day of C_{3+} and 6.1 ton/day of C_{5+} using the same amount of absorbent (50 liter/thousand m^3 at each stage).

The calculations performed above have shown that weathered condensate can be used as absorbent in the extraction of heavy hydrocarbons. The efficiency of extraction depends strongly on condensate weathering conditions, thermobaric conditions, quantity of the entered absorbent, and also on the number of contact steps.

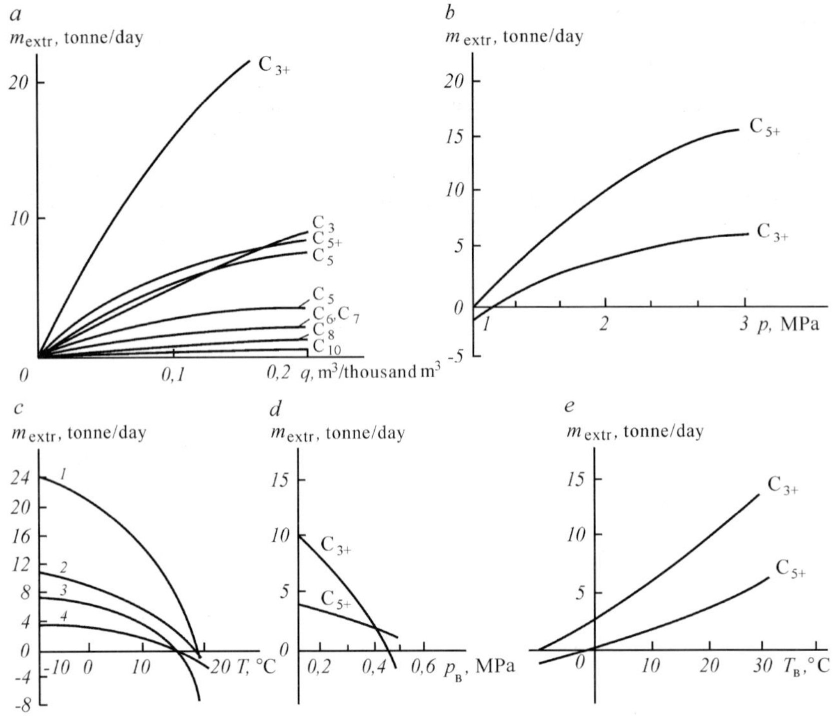

Fig. 20.6 Dependence of m_{ex} on $q(a)$, $p(b)$, $T(c)$, $p_B(d)$, $T_B(e)$ in a technological scheme for single-stage absorption: $a - T_B = 20$ °C; $p_B = 0.1$ MPa; $T = 10$ °C; $p = 3$ MPa; $b - q = 0.1$ m^3/thousand m^3; $T = 10$ °C; $T_B = 20$ °C; $p_B = 0.1$ MPa; $c - T_B = 20$ °C; $p_B = 0.1$ MPa; $p = 3$ MPa; 1 – C_{3+}; $q = 0.1$ m^3/thousand m^3; 2 – C_{3+}; $q = 0.03$ m^3/thousand m^3; 3 – C_{5+}; $q = 0.1$ m^3/thousand m^3; 4 – C_{5+}; $q = 0.03$ m^3/thousand m^3; $d - q = 0.1$ m^3/thousand m^3; $T = 10$ °C; $p = 2$ MPa; $T_B = 20$ °C; $e - q = 0.1$ m^3/thousand m^3; $T = 10$ °C; $p = 2$ MPa; $p_B = 0.1$ MPa.

20.3 Counter-Current Absorption of Heavy Hydrocarbons

Consider now the absorption extraction of heavy hydrocarbons under the conditions of counter-current flow in a column absorber presented schematically in Fig. 20.7. Gas of a given composition $y_0 = (y_{01}, y_{02}, \ldots, y_{0n})$, where y_{0i} is the molar fraction of i-th component, with the flow rate Q_{G0}, enters the bottom part of the column. At the same time, an absorbent with composition $\mathrm{x}_0 = (x_{01}, x_{02}, \ldots, x_{0n})$ and flow rate q_0 enters the top part of the column. The number of contact stages is equal to N. Each stage is equipped with a perforated plate operating in the ablation regime. This means that the liquid is not collected on the plate, but exists in a dispersed state in the inter-plate space. Each contact stage contains a separation device, for example, a mesh droplet catcher, in which the exhausted absorbent is separated from the gas and directed toward the next plate.

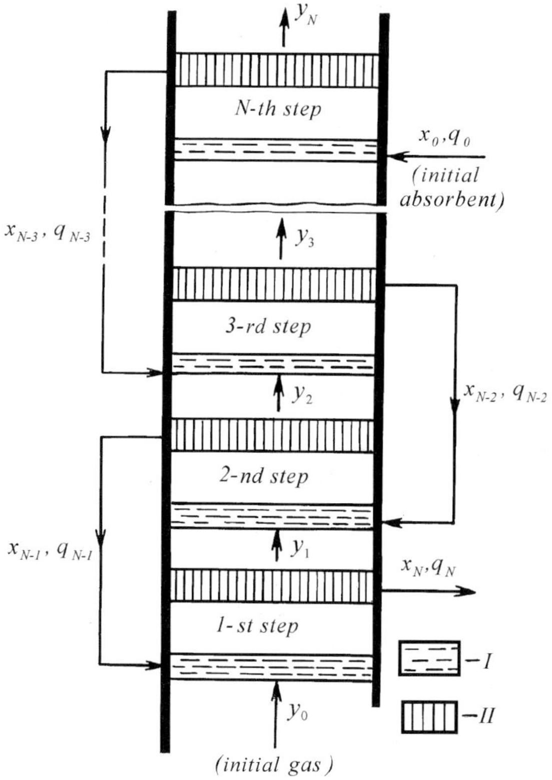

Fig. 20.7 Schematic image of a column plate absorber: *I* – contact plate; *II* – separation device.

We assume, for simplicity, that the droplet catcher completely separates the liquid from the gas.

The average velocity of gas motion in a contact stage is equal to

$$U = \frac{4Q_{Gp}}{\pi D^2} = \frac{0.4TzDQ_{Gn}}{\pi D^2 293p}. \tag{20.17}$$

Here Q_{Gp} and Q_{Gn} are, respectively, gas flow rates (million m^3/day) under operatibg and normal conditions; z is the compressibility factor of the gas; p and T are pressure and temperature; D is the diameter of working section of the column. So, at $Q_{Gn} = 10$ mill. m^3/day; $D = 2.4$ m; $p = 8$ MPa; $T = 253$ °K and $z = 0.7$ we have $U = 0.2$ m/s. Reynolds number is $\mathrm{Re} = \rho_G UD/\mu_G = 5 \cdot 10^7$. The appropriate internal scale of turbulence is $\lambda_0 = 5$ μ. Knowing the average velocity of ascending gas flow, it is possible to determine the maximum size of drops that are carried away with the flow, by using Stokes formula:

$$R_m = \left(\frac{9\mu_G U}{2(\rho_L - \rho_G)g} \right)^{1/2}. \tag{20.18}$$

For values $\mu_G = 2 \cdot 10^{-5}$ Pa·s, $\rho_L = 650\ \text{kg/m}^3$, $\rho_G = 100\ \text{kg/m}^3$ and $U = 0.2$ m/s we have $R_m = 56\ \mu$. Thus, the radius of drops moving together with the flow in the inter-plate space does not exceed R_m. The size of these drops can change during motion because of the mass exchange with the gas and the drop coagulation. Each of these processes is characterized by its own time. Thus, the characteristic time of mass exchange is equal to:

$$t_{eq}^{(1)} \sim 0.26 \frac{R^{2/3} D^{1/3}}{WU} \left(\frac{\rho_L}{\rho_G} \right)^{1/3}, \tag{20.19}$$

where W is the volume concentration of absorber in the gas, which equals

$$W = \frac{29.2 \cdot 10^{-4} qp}{zT}. \tag{20.20}$$

Here q is the absorbent flow rate (liters per 1000 m^3 of gas at normal conditions); p is the pressure (MPa). So, for values $q = 220$ liter/thousand m^3; $p = 8$ MPa; $z = 0.7$, $T = 253$ °K, $R = 4 \cdot 10^{-5}$ m, $D = 2.4$ m, $U = 0.2$ m/s, $\rho_G = 100\ \text{kg/m}^3$, and $\rho_L = 650\ \text{kg/m}^3$ we get $W = 2.5 \cdot 10^{-2}\ \text{m}^3/\text{m}^3$ and $t_{eq}^{(1)} = 0.2$ s.

Thus, small sizes of drops and their relatively high volume concentration result in a quick establishment of phase equilibrium in the inter-plate space.

The characteristic time of drop augmentation due to their coagulation is estimated by the following expression:

$$t_{eq}^{(2)} \sim \left(\frac{\pi^2 \rho_G \lambda_0^2 R_{av}^2}{16 \Gamma W^2} \right)^{1/2}, \tag{20.21}$$

Here R_{av} is the average drop radius; $\Gamma = 5 \cdot 10^{-20}$ J is Hamaker's constant.

For the above-considered example, at $R_{av} \sim 50\ \mu$ we have $t_{eq}^{(2)} \sim 2.5$ s. Thus, $t_{eq}^{(2)} \gg t_{eq}^{(1)}$.

The distance between contact plates should be chosen from the condition that both phase and dynamic equilibrium must be established during the motion of the mixture in the inter-plate space. For this purpose, the residence time of the mixture should be equal to $t_{eq}^{(2)}$, and the distance between contact plates – to

$$L_k \geq U t_{eq}^{(2)} = \frac{0.4 T z D Q_{Gn}}{\pi D^2 293 p} \left(\frac{\pi^2 \rho_G \lambda_0^2 R_{av}^2}{16 \Gamma W^2} \right)^{1/2}, \tag{20.22}$$

The fulfillment of condition (20.22) guarantees the establishment of phase equilibrium in the inter-plate space, while also allowing us to reduce the ablation of the used absorbent.

Proceed now to calculate the mass exchange process in the absorber. Denote the gas composition at the entrance of j-th contact stage by y_{j-1}; the gas flow rate at j-th stage by Q_j; the composition of the liquid at the entrance to j-*th* stage by x_{N-j}; the liquid flow rate at j-th stage by Q_{N-j}. At each contact stage, these phases mix together, the composition of the mixture is determined from the condition of vapor-liquid equilibrium at the giver values of p and T, molecular masses of components, and flow rates of phases.

The conditions of thermodynamic equilibrium of phases for a multi-component system can written as a system of equations

$$p^{(G)} = p^{(L)}; \quad T^{(G)} = T^{(L)}; \quad f_i^{(G)} = f_i^{(L)}. \tag{20.23}$$

Here $i = (1, 2, \ldots, n)$ is the index of the component; G and L designate gas and liquid phases, and f_i is the fugacity of i-th component.

To determine fugacities of components in each phase, it is necessary to use the combined Peng-Robinson equation of state (see Section 5.7).

The system of equations (20.23) complemented by the mass balance equations, has been solved numerically. As a result, equilibrium values of molar concentrations of components, and the properties of phases have been obtained. As an example, consider the calculation of the column absorber equipped with ten perforated contact plates, working in the ablation regime. The following parameter values were chosen: $p = 7.65$ MPa; $T = -24$ °C; $Q_G = 10$ mill. m^3/day; $q = 217$ liter/thousand m^3; $D = 2.4$ m. Initial compositions of the gas and the absorbent are shown in Table 20.4.

For the chosen values of parameters, the estimation of the minimal distance between contact plates gives $L_k = 0.5$ m. Note that in the actual absorber, this distance was equal to 0.6 m.

The amounts of heavy hydrocarbons extracted from the gas at each stage and over the entire apparatus are given in Table 20.5.

Table 20.4 Initial compositions of the gas and the absorbent (molar fraction, %).

Components	Gas	Absorbent	Components	Gas	Absorbent
CO_2	0.71	0.76	C_5	0.32	5.96
N_2	0.32	0.03	C_6	0.20	7.69
C_1	90.37	30.68	C_7	0.11	8.53
C_2	5.64	8.35	C_8	0.06	8.22
C_3	1.69	6.09	C_9	0.02	4.91
C_4	0.52	4.39	C_{10+}	0.01	14.39

Note: properties of the fraction C_{10+}: boiling temperature 510.1 °K; molecular mass 205 kg/kmole.

Table 20.5 Efficiency of heavy hydrocarbon extraction from the gas.

Stage number	Concentration in the gas, g/m³		Component output, ton/day		Liquid phase litre/thousand m³		Extraction of liquid litre/thousand m³
	C_{3+}	C_{5+}	C_{3+}	C_{5+}	Entrance	Exit	
1	42.31	6.45	286.00	198.79	286.67	346.40	77.78
2	39.46	5.46	29.84	10.13	277.62	286.67	9.11
3	38.24	5.35	12.67	1.19	273.64	277.62	4.05
4	37.49	5.37	7.88	−0.15	271.16	273.64	2.57
5	36.93	5.40	5.96	−0.33	269.23	271.16	2.01
6	36.42	5.44	5.29	−0.32	267.47	269.23	1.83
7	35.86	5.47	5.91	−0.27	265.42	267.47	2.14
8	35.06	5.50	8.47	−0.25	262.15	265.42	3.31
9	33.67	5.54	14.75	−0.25	255.37	262.15	6.83
10	30.98	5.60	31.96	0.33	216.95	255.37	38.42
Over the entire column			408.73	208.86			148.06

20.4 Gas Dehydration in Concurrent Flow

The process of dehydration of gas by absorption does not differ radically from absorption extraction of heavy hydrocarbons from the gas. The only difference is that another liquid – glycol (diethyleneglycol-DEG or triethyleneglycol – TEG), which has the tendency to absorb water vapors from gas, is used as the absorbent. The solubility of hydrocarbons in glycols is small in comparison with the solubility of water. Therefore in the first approximation, we can assume that, as the liquid glycol comes into contact with natural gas, only water vapor participates in the mass-exchange process. The equilibrium in the system "glycol–natural gas" at the given pressure p and temperature T is established after a certain time t_{eq}. The equilibrium concentration of moisture in the gas can be found by using approximate formulas or diagrams of the type presented in Fig. 20.8 [60].

On gas fields with small debit, direct-flow spray absorbers that consist of several stages connected in series (see Fig. 2.19, a) are frequently used for the dehydration of gas from moisture, as the gas is being prepared for transportation. Each stage consists of a contact chamber and a separator. The absorbent (DEG) with the flow rate q is injected through the atomizer into the contact chamber.

Because the size of the drops formed in the process of spraying depends on their velocity relative to the gas stream, injection is usually performed against the gas flow, which promotes formation of smaller drops due to secondary breakup. At first, drops move against the flow for a while; then the flow drags them along.

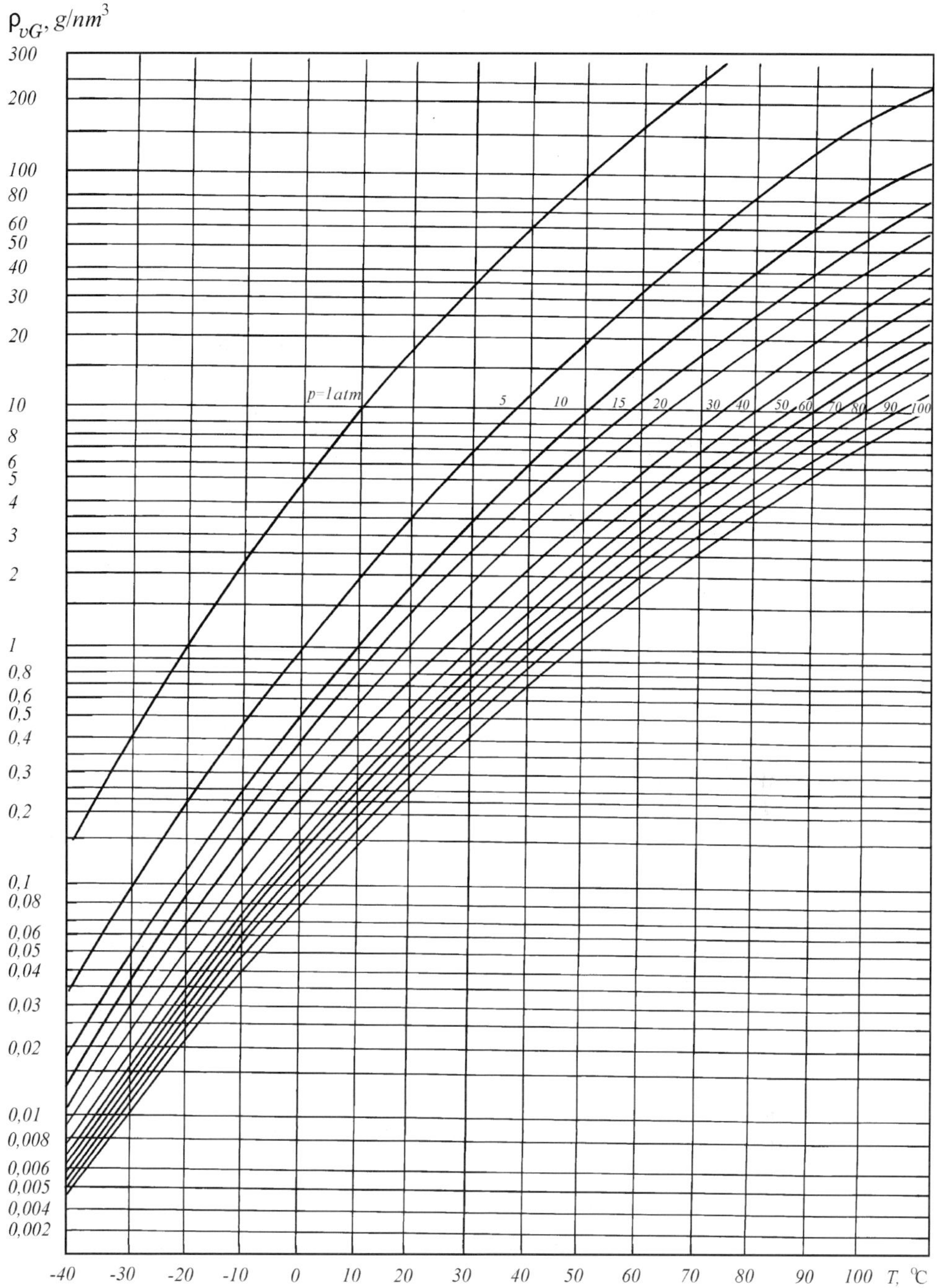

Fig. 20.8 Equilibrium concentration of water vapors in natural gas.

During their contact with the gas, drops absorb the water vapor contained in the gas. After that, the gas-liquid flow enters the separator, where the liquid phase gets separated from the gas. In order to determine the parameters of any single stage of the process, it is necessary to know the dynamics of the absorption process, and the efficiency of drop capture in the separator. The present section will consider the dynamics of mass exchange between DEG drops and a humid gas. For simplicity, it is supposed that the separator completely catches all drops, and that all stages of the absorber are identical.

The most common method of calculating the process of gas dewatering is based on the assumption that a phase equilibrium exists between the absorbent (highly concentrated solution of DEG) and the gas containing water vapors [60]. This method allows us to determine the equilibrium concentrations of water in the vapor and liquid phases, and, knowing the initial concentration, to determine the maximum quantity of moisture that can be extracted from the gas at the given values of p and T. If an identical amount of the absorbent is delivered at each stage, then the obtained data enable us to find the number of theoretical contact stages for the required degree of gas dewatering at the exit from the absorber. Since the equilibrium occurs in time t_{eq}, by specifying the deviation from the equilibrium at each contact stage, it is possible to find the number of working stages.

Consider the absorption of moisture from a gas flow by absorbent drops. Let gas containing water vapor with mass concentration ρ_{G0} $(\mathrm{kg/m^3})$ arrive at the entrance to the contact chamber. Denote through ρ_{G1} the concentration of vapor in the gas at the exit from the contact chamber. It should be noted that mass concentration values are taken at operating conditions. Define the dehydration factor as the ratio between the amount of water vapor extracted from the gas in the contact chamber, and the amount of water vapor at the entrance to the chamber:

$$\eta = 1 - \frac{\rho_{G1}}{\rho_{G0}}. \qquad (20.24)$$

Sometimes the degree of the gas dehydration is characterized by the parameter

$$\varphi = \frac{y_0 - y_1}{y_0 - y_{eq}}, \qquad (20.25)$$

where y_0 and y_1 are the molar fractions of moisture in the gas at the entrance and at the exit from the contact chamber, and y_{eq} is the equilibrium (for given values of p, T) molar fraction of moisture in the gas above the DEG solution at the entrance.

The parameters introduced above are connected by the formula

$$\varphi = \frac{\eta}{1 - 803.5 K x_0/\rho_{G0}}, \qquad (20.26)$$

where K is the equilibrium constant and x_0 is the molar fraction of water in the absorbent.

The primary goal consists in the determination of the dependence of η on the parameters describing the gas and the absorbent, and on geometrical dimensions of the contact chamber. Direct the x-axis along the contact chamber axis, so that $x = 0$ corresponds to the entrance cross-section. Assume that the absorbent is atomized in the gas flow and forms drops of identical radius equal to the stable radius of drops in a turbulent flow:

$$R_{av} = 0.09 D_k \left(\frac{2\Sigma}{U_r^2 \rho_L D_k} \right)^{3/7} \left(\frac{\rho_G}{\rho_L} \right)^{1/7}. \tag{20.27}$$

Here D_k is the diameter of the contact chamber, U_r is the velocity of the liquid flowing out from the atomizer relative to the gas; ρ_G and ρ_L are the densities of the gas and the absorbent. Usually the absorbent flow rate is 8–14 liters on 1000 m^3 of gas, which corresponds to the volume concentration of liquid $W = 8 \cdot 10^{-4}$–$1.4 \cdot 10^{-3}\ m^3/m^3$. Thus the average distance between drops is equal to $l \sim R/\sqrt[3]{W} \approx 10R$ so the restrained character of the flow can be neglected.

The mass exchange for a drop suspended in a turbulent flow, occurs due to the delivery of substance to the drop surface by turbulent pulsations, and also via molecular diffusion. As it was shown in Section 16.2, the expression for the mass flux of the substance at the drop surface depends on the ratio between the drop radius and the internal scale of turbulence $\lambda_0 = D_k/\mathrm{Re}^{3/4}$, where D_k is the diameter of the working cross-section of the absorber, and Re is Reynolds number. For the characteristic parameter values $\rho_G = 50\ \mathrm{kg/m^3}$; $U = 3\ \mathrm{m/s}$; $D_k = 0.4\ \mathrm{m}$; $\mu_0 = 10^{-5}$ Pa·s, we have $\mathrm{Re} = 6 \cdot 10^6$ and $\lambda_0 = 5 \cdot 10^{-6}$ m. Since the drop size is $R \sim 2 \cdot 10^{-5}$ m, the inequality $\lambda_0 < R$ holds, and the delivery of substance to the drop surface is mostly performed by turbulent pulsations. Thus the mass flux on the drop surface is equal to (see (16.55))

$$J = 5.04\pi \left(\frac{\rho_L}{\rho_G} \right)^{1/3} R^{7/3} U \Delta\rho D_k^{-1/3}, \tag{20.28}$$

where $\Delta\rho = \rho_{vG} - \rho_{wG}$, ρ_{vG} and ρ_{wG} are, respectively, mass concentrations of moisture in the bulk of the gas flow, and near the external drop surface.

We assume, as usual, that there is a local thermodynamic equilibrium at the drop surface. Then the value ρ_{wG} is equal to the equilibrium concentration of moisture above the surface of the solution. At thermodynamic equilibrium of the system "absorbent – natural gas – water vapor", the mass concentration ρ_{wG} of water vapor is determined by the pressure p, temperature T, and the mass fraction α_w of glycol in the absorbent solution. Usually [61], the dependence $\rho_{wG}(p, T, \alpha_w)$ is presented as a pair of dependencies – $\rho_{wG}(p, T_t)$ where T_t is the dew-point temperature, and $\alpha_w(T_t, T)$. The first dependence looks as follows:

$$\rho_{wG} = 4.67 \cdot 10^{-6} (\exp(0.0735 T_t - 0.00027 T_t^2)) p^{-1} + 0.0418 \exp(0.054 T_t - 0.0002 T_t^2). \tag{20.29}$$

The second dependence is given in [61] in a graphic form. But it can be approximated by the expression:

$$A(1-\alpha_w)^B = T_t + 273, \tag{20.30}$$

$$A = 277.0793 + 1.1972T, \quad B = 0.033258 + 0.000297T$$

in the range 5 °C $< T <$ 35 °C; $0.9 < \alpha_w < 0.995$. The temperature T that appears in (20.29) and (20.30) is measured in °C, p – in MPa, and ρ_{wG} – in g/m^3.

Denote through N the number of drops in a unit volume of the mixture. Then the equation describing the change of moisture concentration in the gas assumes the form:

$$\frac{d\rho_{vG}}{dt} = -NJ. \tag{20.31}$$

Express N through volume concentration W of the absorbent and the drop volume:

$$N = \frac{3W}{4\pi R^3}. \tag{20.32}$$

Usually the absorbent flow rate q is specified in liters per 1000 м^3 under normal conditions. Then W is determined by the correlation (20.20). From (20.31) and (20.32), there follows

$$\frac{d\rho_{vG}}{dt} = -\frac{3JW}{4\pi R^3}, \quad \rho_{vG}(0) = \rho_{0G}. \tag{20.33}$$

The change of mass concentration of moisture inside the drop is described by the diffusion equation:

$$\frac{\partial \rho_{vL}}{\partial t} = D_L \frac{1}{r^2}\frac{\partial}{\partial r}\left(r^2 \frac{\partial \rho_{vL}}{\partial r}\right), \quad (0 < r < R) \tag{20.34}$$

with the following initial and boundary conditions:

$$\rho_{vL}(r,0) = \rho_{0L}, \quad \rho_{vL}(R,t) = \rho_{wL}, \quad (\partial \rho_{vL}/\partial r)_{r=0} = 0, \tag{20.35}$$

where D_L is the diffusion coefficient of water in the absorbent ($D_L = 1.9 \cdot 10^{-10}$ m^2/s); ρ_{0L} the initial concentration of water in the drop.

Eqs. (20.33) and (20.34) should be complemented by the equations of mass balance – at the drop surface:

$$J = 4\pi R^2 D_L \left(r^2 \frac{\partial \rho_{vL}}{\partial r}\right)_{r=R} \tag{20.36}$$

and for the entire drop:

$$J = \frac{d}{dt}\left(\frac{4}{3}\pi R^3 D_L \rho_{vL}\right). \tag{20.37}$$

The mass of moisture in the gas is small in comparison with the mass of the arriving absorbent; therefore the increase of the drop volume will be insignificant, so in the first approximation, the volume change will be neglected and the radius R of the drop will be considered constant and equal to the average radius R_{av}, formed during atomization in a turbulent gas flow. In view of this, the system of equations describing the change of ρ_{vG} and ρ_{vL} with time becomes:

$$\begin{aligned}
&\frac{\partial \rho_{vL}}{\partial t} = D_L \frac{1}{r^2}\frac{\partial}{\partial r}\left(r^2 \frac{\partial \rho_{vL}}{\partial r}\right),\\
&\frac{d\rho_{vG}}{dt} = -3.78\left(\frac{\rho_L}{\rho_G}\right)^{1/3} R^{-2/3} U(\rho_{vG} - \rho_{wG}) D_k^{-1/3},\\
&D_L\left(\frac{\partial \rho_{vL}}{\partial r}\right)_{r=R} = 1.26\left(\frac{\rho_L}{\rho_G}\right)^{1/3} R^{7/3} U(\rho_{vG} - \rho_{wG}) D_k^{-1/3},\\
&\rho_{wG} = f(\alpha_w, p, T),\\
&\rho_{vL}(r,0) = \rho_{0L}, \quad \rho_{vL}(R,t) = \rho_{wL}, \quad (\partial \rho_{vL}/\partial r)_{r=o} = 0, \quad \rho_{vG}(0) = \rho_{0G}.
\end{aligned} \tag{20.38}$$

The system of equations (20.38) allows for an analytical solution, subject to the condition of linear dependence:

$$\rho_{wG} = a(p,T) - b(p,T)\alpha_w. \tag{20.39}$$

Such linear dependence gives quite a good approximation of the real dependence for some ranges of p and T.

Instead of the mass concentration ρ_{vL}, we can consider the mass fraction α of glycol in the water solution of the absorbent, which is connected with ρ_{vL} by the correlation $\rho_{vL} = \rho_w(1-\alpha)$, where ρ_w is the density of water. Then we can rewrite (20.38) and (20.39) in dimensionless variables

$$z_0 = \frac{\rho_{vG} - \rho_{0G}}{\rho_{0G}}, \quad z_L = \frac{\alpha - \alpha_0}{\alpha_0}, \quad \tau = \frac{D_L t}{R^2}, \quad \xi = \frac{r}{R}, \quad \gamma = \frac{\rho_w \alpha_0}{\rho_{0G}},$$

$$\Lambda = 1.26\left(\frac{\rho_L}{\rho_G}\right)^{1/3} R^{4/3} U D_k^{-1/3} D_L^{-1}, \quad \beta = \frac{a - b\alpha_0}{\rho_{0G}}, \quad \delta = \frac{b\alpha_0}{\rho_{0G}},$$

where ρ_L is the density of glycol; α_0 is the initial mass fraction of glycol in the absorbent; U is the average flow velocity in the absorber.

In the new variables, Eqs. (20.38) take the form:

$$\frac{\partial z_L}{\partial \tau} = \frac{1}{\xi^2}\frac{\partial}{\partial \xi}\left(\xi^2 \frac{\partial z_L}{\partial \xi}\right), \quad (0 < \xi < 1),$$

$$\frac{dz_G}{d\tau} = -3W\Lambda(z_G - z_{wG}),$$

$$z_{wG} + 1 = \beta - \delta z_{wL}, \quad \Lambda(z_G - z_{wG}) = -\gamma\left(\frac{\partial z_L}{\partial \xi}\right)_{\xi=1}, \tag{20.40}$$

$$\left(\frac{\partial z_L}{\partial \xi}\right)_{\xi=0}, \quad z_L(\xi, 0) = 0, \quad z_L(1,\tau) = z_{wL}, \quad z_G(0) = 0.$$

Applying the Laplace transform to Eq. (20.40), we obtain:

$$z_G = -\varepsilon(1-\beta)\left\{\frac{1}{3+\varepsilon} - 2\sum_{k=1}^{\infty}\frac{(p_k - \tan p_k)\exp(-p_k^2\tau)}{p_k(2\varepsilon + \nu p_k^2) + \varepsilon \tan p_k(p_k^2 - \lambda p_k^4 - 2)}\right\}, \tag{20.41}$$

where p_k are the roots of the transcendental equation,

$$\tan p_k = \frac{\varepsilon p_k(1 - \lambda p_k^2)}{p_k^2 + \varepsilon(1 - \lambda p_k^2)}, \tag{20.42}$$

Here we have introduced non-dimensional parameters

$$\varepsilon = 3W\rho_w/b, \quad \lambda = 0.26D_L(D_k\rho_G/\rho_L R^4)^{1/3}/WU, \quad \nu = b\alpha_0/(a - b\alpha_0).$$

At $\tau \to \infty$, the solution (20.41) tends to the equilibrium solution, which is equal to

$$z_G^{(eq)} = -\frac{\varepsilon(1-\beta)}{3+\varepsilon}. \tag{20.43}$$

Physically, the quantity $-z_G$ is equal to the fraction of moisture extracted from the gas, therefore the efficiency of gas dewatering can be characterized by the parameter

$$\eta = z_G = -\varepsilon(1-\beta)\left\{\frac{1}{3+\varepsilon} - 2\sum_{k=1}^{\infty}\frac{(p_k - \mathrm{tg} p_k)\exp(-p_k^2\tau)}{p_k(2\varepsilon + \nu p_k^2) + \varepsilon \mathrm{tg} p_k(p_k^2 - \lambda p_k^4 - 2)}\right\}, \tag{20.44}$$

A more simple solution may be obtained if we take the mass concentration $\bar{\rho}_{vL}$ averaged over the absorbent volume, and the corresponding mass fraction of glycol $\bar{\alpha}$, and substitute them into the equations (20.38). This results in the following system of equations:

$$\frac{d\rho_{vG}}{dt} = -\frac{WJ}{V}, \quad \frac{d\bar{\rho}_{vL}}{dt} = \frac{J}{V},$$

$$\bar{\rho}_{vL} = \rho_w(1-\bar{\alpha}), \quad \rho_{vG} = a - b\bar{\alpha}, \quad V = \frac{4\pi}{3}R^3, \tag{20.45}$$

$$\rho_{vG}(0) = \rho_{0G}, \quad \bar{\alpha}(0) = \alpha_0,$$

whose solution is:

$$\bar{\alpha} = \alpha_0 - \frac{A}{B}(1 - e^{-B\tau_1}), \quad \rho_{vG} = \rho_{0G} W(\alpha_0 - \bar{\alpha}), \tag{20.46}$$

$$A = 1 - \frac{a - b\alpha_0}{\rho_{0G}}, \quad B = \frac{\rho_w W + b}{\rho_{0G}}, \quad \tau_1 = 3.78 L_k \left(\frac{\rho_{0G}}{\rho_w}\right)\left(\frac{\rho_L}{\rho_{vG} D_k R^2}\right)^{1/3},$$

Here L_k is the length of the contact zone.

A further enhancement of the obtained solution (20.46) can be made, if we take into account the real dependence of ρ_{wG} on p, T, and $\bar{\alpha}$. The resultant equation is:

$$\frac{d\bar{\alpha}}{d\tau_1} = -\left(1 - \frac{\rho_w W}{\rho_{0G}}(\alpha_0 - \bar{\alpha}) - \frac{\rho_{wG}}{\rho_{0G}}\right), \quad \bar{\alpha}(0) = \alpha_0, \tag{20.47}$$

$$\rho_{vG} = \rho_{0G} + \rho_w W(\bar{\alpha} - \alpha_0),$$

which must be complemented by the correlations (20.29) and (20.30).

The given equations allow us to calculate the residual moisture concentration in the gas at the exit from a contact stage. This residual concentration should be taken as the initial condition for the calculation of the next contact stage. The dew-point T_d of the dewatered gas can be found either from the graphic dependence of ρ_{vG} on p and T (Fig. 20.8), or by solving Eq. (20.29) with respect to T_t.

As an example, consider the calculation of a direct-flow multistage spray absorber whose construction and test results are described in [62]. The absorber consists of four successive contact stages. The absorbent is a highly concentrated water solution of DEG, delivered to each stage with different flow rates q, and after each stage the exhausted absorbent is separated from the gas in a separator. The experiments were carried out at the pressure $p = 2.1$ MPa and the temperature of 25 °C. The gas flow rate is $Q_G = 1$ mill. м^3/day; the moisture concentration in the gas at the entrance is $\rho_{0G} = 1.3$ g/m^3 under normal conditions. Theoretical and experimental results are presented in Table 20.6.

You can see from these results that the values obtained using equations (20.47) show the best agreement with the experimental data. Fig. 20.9 shows the dependence of the mass concentration ρ_{vG} of moisture in the gas at the exit from each

Table 20.6 Direct-flow spray absorber: the concentration of moisture in the gas after each contact stage.

Parameters	Contact stages			
	I	II	III	IV
q, kg/thousand m^3	14	16.5	13	18
ρ_{vG}, g/m^3:				
Equation (20.41)	0.33	0.1	0.053	0.041
Equation (20.46)	0.32	0.095	0.052	0.041
Equation (20.47)	0.35	0.114	0.067	0.041
Experiment [62]	0.34	0.125	0.067	0.038

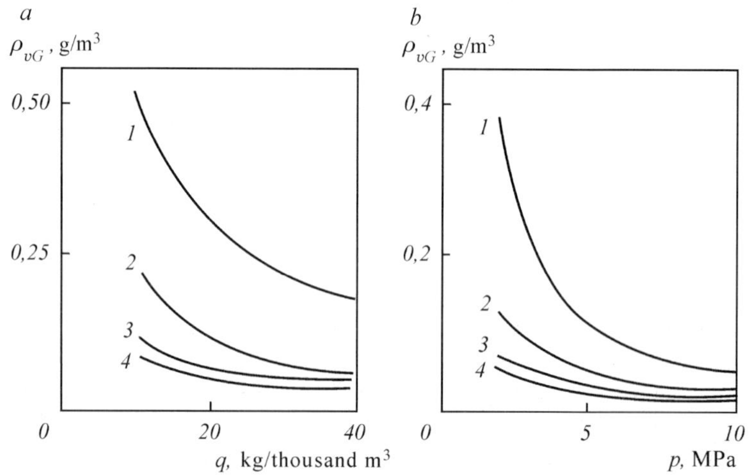

Fig. 20.9 Dependence of moisture mass concentration ρ_{vG} in the gas at the exit from a contact stage on the quantity of injected absorbent (*a*) and the pressure (*b*): 1 – 4 – stage numbers.

contact stage on the absorbent flow rate (assumed to be the same in each stage), and on the pressure.

The length of the contact chamber is connected to the dimensionless time by the correlation $L_k = R^2 U\tau/D_L$ or $L_k = (\rho_B/\rho_{0G})(\rho_{vG} D_k R^2/\rho_L)^{1/3}\tau_1/3.78$. In order to find L_k, it is necessary to know τ or τ_1. They can be found in one of the following two methods, depending on the requirements imposed on the absorption process. In the first case, we should demand that after the time τ, the amount of moisture in the gas at the contact chamber exit must comprise S% of the appropriate equilibrium value. Then the value of τ will be determined from the condition

$$1 - \eta = 0.01S\left(1 - \varepsilon\frac{1-\beta}{3+\varepsilon}\right). \qquad (20.48)$$

In the second case, the restriction is imposed on the mass concentration $\bar{\alpha}$ of glycol in the exhausted absorbent. Then the second equation (20.47) gives us:

$$\eta = \frac{\rho_w W(\alpha_0 - \bar{\alpha})}{\rho_{0G}}. \qquad (20.49)$$

Thus, Eqs. (20.47) and (20.48) allow us to find the dimensionless time τ or τ_1, and thereby, the required length of the contact stage.

The moisture concentration in the gas and the efficiency η of gas dehydration at each contact stage are connected by the formula

$$\rho_{vG} = \rho_{0G}(1 - \eta). \qquad (20.50)$$

The efficiency of dehydration in a multistage direct-flow absorber with N consecutive contact stages connected in series is

$$\eta_N = 1 - \prod_{i=1}^{N}(1 - \eta_i). \qquad (20.51)$$

If all stages are identical and their efficiency is equal to η, then

$$\eta_N = 1 - (1 - \eta)^N. \qquad (20.52)$$

If the efficiency of each stage has been determined, then, having set the required degree of gas dehydration or the dew-point temperature, we can calculate the number N of contact stages from (20.51) or (20.52).

In the above-considered example, the obtained length of the contact chamber, $L_k = 0.54$ m, corresponds to the requirement that at each stage, the deviation from the equilibrium at the exit amounts to 10%. If the required dew-point of the gas is 8 °C, one contact stage will be sufficient, while if we want to have the dew-point at 15 °C, three stages will be required, and the length of each stage should be 0.54 m.

20.5 Gas Dehydration in Counter-Current Absorbers with High-Speed Separation-Contact Elements

Recently, column absorbers for gas dehydration have begun to adopt high-speed direct-flow centrifugal separation-contact elements with tangential input of gas and recirculation of the absorbent (see Fig. 2.17). These elements are installed on horizontal plates in vertical counter-flow devices. The absorbent delivered to the top (highly concentrated water solution of DEG) flows downward from plate to plate. The layer of absorbent at each plate is supported at some height that, generally speaking, can vary for different plates. The absorbent enters the separation-contact element through a special tube. There, it flows out of the tube into a spinning gas flow, in which the liquid disintegrates, and the formed fine drops are picked up by the flow and thrown against the wall of the element. As a result, there are two processes occurring simultaneously inside the element: mass exchange between the drops and the gas, and separation of drops from the gas.

Since the absorbent flows continuously from plate to plate and is also repeatedly diluted by the absorbent coming out of the elements, the state of the absorbent layered on the plates can be characterized by residence times: inside the element $t_e = L_e/u_e$, and at the plate $t_i = V/Q_a$, where V is the absorbent volume at the plate, L_e is the height of the element, Q_a is the absorbent flow rate, and u_e is the average vertical drift velocity of the gas inside the element (i.e. velocity along the symmetry axis of the element). A comparison of these two times for the characteristic parameter values shows that $t_e \ll t_t$. This means as it dwells at the plate, the absorbent has enough time to make numerous contacts with the gas inside the elements installed on this plate.

The key parameter describing the absorbent state is the recirculation number, defined as the ratio of the absorbent flow rate q_a through the contact elements to the general absorbent flow rate Q_a in the absorber.

$$n = \frac{q_\alpha}{Q_\alpha}. \tag{20.53}$$

In practical terms, the parameter n determines how many times a volume V of the absorbent will pass through elements located at one plate during its residence time t_t at the plate. The primary goal is to find out how the concentrations of glycol in the absorbent solution and of moisture in the gas at the exit from the element depend on the number n of recirculations.

Assume that a uniform mixing of the absorbent occurs on the plate, so that glycol concentration in the absorbent solution changes only with time. Suppose further that the ablation of liquid from centrifugal elements by the gas flow is neglibibly small. The last assumption is valid only if the elements operate in the subcritical regime.

Let α_α and $\bar{\alpha}_\alpha$ be the mass fractions of glycol in the absorbent solution, respectively, within the element, and inside the layer on the plate; and $\bar{\alpha}_{\alpha 0}$ and $\alpha_{\alpha 1}$ be the corresponding values in the absorbent solution coming to our plate from the previous plate, and at the exit from the "previous" element. The equation describing the variation of α_α with time is similar to (20.47) and has the form

$$\frac{d\alpha_\alpha}{d\tau_1} = -\left(1 - \frac{\rho_w W}{\rho_{0G}}(\alpha_\alpha - \bar{\alpha}_\alpha) - \frac{\rho_{wG}}{\rho_{0G}}\right), \quad \alpha_\alpha(0) = \bar{\alpha}_\alpha, \tag{20.54}$$

where $\tau = 3.78u_e t\left(\frac{\rho_{0G}}{\rho_w}\right)\left(\frac{\rho_\alpha}{\rho_G d_e R_{\alpha v}^2}\right)^{1/3}$, is the non-dimensional time; ρ_{0G} is the mass concentration of water vapor in the gas at the entrance to the element; ρ_B, ρ_α, and ρ_G are, respectively, the densities of water, absorbent and gas; R is the average radius and W – the volume concentration of absorbent drops in the gas flow inside the elements; d_e is the diameter of elements; ρ_{vG} is the equilibrium concentration of water vapor (at the current values of p, T) above the absorbent solution with glycol concentration α_α.

For the average radius R of drops formed during the disintegration of the jet that flows into the separation contact element, we take the average radius $R_{\alpha v}$ of drops formed in the turbulent flow (see Eq. (20.27), into which we must plug the corresponding parameters of the element).

Eq. (20.54) describes the change of the mass fraction of glycol in the absorbent solution inside the contact element in the dimensionless time interval $0 < \tau < \tau(t_e)$.

The balance-of-mass relation for the absorbent at the plate, accounting for the recirculations and for the absorbent flowing from plate to plate, gives us the following equation for $\bar{\alpha}_\alpha$

$$\frac{d\bar{\alpha}_\alpha}{d\tau} = -\beta((n+1)\bar{\alpha}_\alpha - \bar{\alpha}_{\alpha 0} - n\alpha_{\alpha 1}) \quad (0 < \tau < \tau_1), \quad \bar{\alpha}_\alpha(0) = \bar{\alpha}_{\alpha 0}, \tag{20.55}$$

Here $\beta = Q_\alpha L_e/\tau_e V u_c = 1/\tau_t(t_t)$ is the dimensionless residence time of the absorbent at the plate.

Eq. (20.55) describes the change of the mass fraction $\bar{\alpha}_\alpha$ of glycol in the absorbent solution on the plate in the dimensionless time interval $0 < \tau < \tau_t$.

Eqs. (20.54) and (20.55) must be supplemented with the relation connecting the equilibrium concentration ρ_{wG} of water vapor in the gas with p, T, and α_α, which, according to [61], looks like:

$$\rho_{wG} = 10^6 p_w\left(\frac{749}{p} + B\right), \quad p_w = p_w^* x\gamma, \tag{20.56}$$

$$B = \exp(0.06858(0.01T)^4 - 0.3798(0.01T)^3 + 1.06606(0.01T)^2 - 2.00075(0.01T) + 4.2216),$$

$$p_w^* = \exp(-0.60212(0.01T)^4 + 1.475(0.01T)^3 - 2.97304(0.01T)^2 + 7.19863(0.01T) + 6.41465),$$

$$\gamma = \exp\left(-\frac{2.303}{273+T}\left(\frac{mx}{1-x} + n_1\right)^{-2}\right), \quad x = \frac{1-\alpha_\alpha}{1-\alpha_\alpha + 18.02\alpha_\alpha/M_\alpha},$$

Here x and γ are the molar fraction of water in the absorbent solution and the activity factor; m and n_1 are constants (for DEG, $m = 0.0245$ and $n = 0.137$); M_α is the molecular mass of the absorbent.

The expression (20.56) is cumbersome and inconvenient for further use. For a solution with the mass fraction of DEG in the range $0.9 < \alpha_\alpha < 1$, we can approximate (20.56) with a high accuracy by the following expression:

$$\rho_{wG} = a(1 - \alpha_\alpha),$$

$$a = 3.95 \cdot 10^{-6} p_w^* \left(\frac{749}{p} + B\right) \exp\left(-\frac{115}{273 + T}\right). \tag{20.57}$$

Thus, the system of equations (20.54), (20.55), and (20.57) describes the time dynamics of DEG concentration in the absorbent layer at the plate, with a proper account for the absorbent recirculations in the contact elements.

The change of water vapor concentration in the gas that is being dewatered is described by the second equation (20.47)

$$\rho_{vG} = \rho_{0G} - \rho_w W(\alpha_{\alpha 1} - \bar{\alpha}_\alpha), \tag{20.58}$$

where ρ_{vG} is the moisture concentration of gas at the entrance to the contact element.

To solve the obtained system of equations, one first needs to determine the number n of recirculations, the volume concentration W of the absorbent in the gas, and the volume V of the absorbent at the plate.

The number of recirculations depends on the element design, the number N_e of elements at the plate, flow rate Q_G of the gas, flow rate Q_a of the absorbent, and also on the excess height H of the absorbent layer at the plate above the point where the absorbent is injected into the element. Processing of the element test data has produced the following empirical dependence

$$n = 3.456 \cdot 10^{-3} N_e \sqrt{2gH}(Q_G Q_a)^{-1}. \tag{20.59}$$

The flow rates appearing in this expression have the following dimensionality: $[Q_G]$ = million m^3/day; $[Q_a]$ = m^3/thousand m^3.

The volume of the absorbent on the plate is

$$V = 0.25\pi H(D^2 - N_e d_e^2), \tag{20.60}$$

where D is the plate diameter, d_e is the diameter of contact elements.

The volume concentration of absorbent in the gas flow is given by the expression

$$W = \frac{q_a}{Q_G}, \tag{20.61}$$

Note that flow rates entering this expression should be taken at the operating conditions (m^3/s).

As the characteristic time of change of glycol concentration in the absorbent inside the element is small as compared to that of glycol concentration at the plate ($\tau_e \ll \tau_1$), this gives us grounds to assume that during the time spent by the absorbent inside the contact element, the mass fraction of glycol in the absorbent at the plate $\bar{\alpha}_\alpha$ does not change noticeably. This simplifies the solution. First, from the equation (20.54), one finds $\alpha_\alpha(\tau)$ at constant $\bar{\alpha}_\alpha$, and then $\bar{\alpha}_\alpha(\tau)$ follows from (20.55). As a result, we obtain

$$\begin{gathered}\alpha_\alpha = A\bar{\alpha}_\alpha + G,\\ \bar{\alpha}_\alpha = \left(\bar{\alpha}_{\alpha 0} - \frac{\bar{\alpha}_{\alpha 0} + nG}{n+1-nA}\right)\exp\left(-\frac{(n+1-nA)\tau}{\tau_t}\right) + \frac{\bar{\alpha}_{\alpha 0} + nG}{n+1-nA},\\ A = \frac{\bar{a}}{\delta+\bar{a}}e^{-(\delta+\bar{a})\tau_1} + \frac{\delta}{\delta+\bar{a}}, \quad G = \frac{\bar{a}-1}{\delta+\bar{a}}\left(1-e^{-(\delta+\bar{a})\tau_1}\right),\\ \delta = \frac{\rho_w W}{\rho_{0G}}, \quad \bar{a} = \frac{a}{\rho_{0G}}.\end{gathered} \tag{20.62}$$

If the absorbent flow rate is given in liter/thousand m^3 under normal conditions, then $W = 10^{-6} n q_a$ (m^3/m^3) and $\delta = 10^{-6} n q_a/\rho_{0G}$.

It follows from (20.62) that at $\tau \to \infty$, the value $\bar{\alpha}_\alpha$ tends to the corresponding equilibrium value $\bar{\alpha}_\alpha^{(eq)}$, which equals

$$\bar{\alpha}_\alpha^{(eq)} = \frac{\bar{\alpha}_{\alpha 0} + nG}{n+1-nA}. \tag{20.63}$$

At the moment $\tau = \tau_t$, which is when the absorbent must leave the plate, the mass fraction of glycol in the exhausted absorbent (provided that the circulation number is large ($n \gg 1$)) will be equal to

$$\begin{aligned}\bar{\alpha}_\alpha \approx &\left(\bar{\alpha}_{\alpha 0} - \frac{\bar{\alpha}_{\alpha 0} + (\bar{a}-1)\rho_{0G}/10^{-3}q_a}{1+\bar{a}\rho_{0G}/10^{-3}q_a}\right)\exp\left(-\frac{1+\bar{a}\rho_{0G}}{10^{-3}q_a}\right)\\ &+ \frac{\bar{\alpha}_{\alpha 0} + (\bar{a}-1)\rho_{0G}/10^{-3}q_a}{1+\bar{a}\rho_{0G}/10^{-3}q_a}.\end{aligned} \tag{20.64}$$

In Fig. 20.10, the calculated mass fraction $\bar{\alpha}_\alpha$ of glycol in the absorbent layer on the plate after n-multiple recycling of the absorbent in separation-contact elements is plotted against the recirculation number n. Calculations were carried out for the following values of the key parameters: $p = 7.5$ MPa; $T = 13.5$ °C; $\rho_{0G} = 0.24$ g/m^3; $\bar{\alpha}_{\alpha 0} = 0.9916$; $Q_G = 12.8$ million. m^3/day; $Q_a = 5.5$ kg/thousand m^3.

It follows from the obtained results that the main change of $\bar{\alpha}_\alpha$ occurs in the interval of values of n from 1 to 7. An increase of the recirculation number above 7 changes $\bar{\alpha}_\alpha$ a little, and $\bar{\alpha}_\alpha$ is then close to the value predicted by the formula (20.64). It means that the absorbent is close to the saturated state.

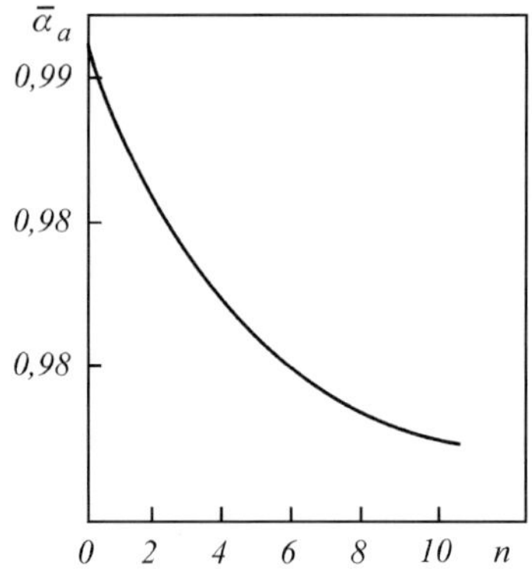

Fig. 20.10 Dependence of glycol mass fraction $\bar{\alpha}_a$ in the absorbent solution on the plate on the number n of recirculations.

Consider now the dehydration process in a counter-flow vertical absorber represented schematically in Fig. 20.7. The gas with mass concentration of moisture ρ_{0G} (kg/м^3 under normal conditions) enters the absorber from the bottom, and the DEG solution with mass fraction $\bar{\alpha}_\alpha$ of DEG enters from the top. At each stage, the corresponding concentrations are denoted through $\rho_{N-i,G}$ and $\bar{\alpha}_i$ $(i = \overline{1, N})$, where N is the number of contact stages. The numbering of stages goes from the top downward. Introduce the following designations: H_i is the height of the absorbent layer at i-th plate above the point where absorbent is injected into the element; Q_G is the gas flow rate under normal conditions; Q_a is the absorbent flow rate in the device; $q_a^{(i)}$ is the flow rate of the absorbent through an element at i-th plate; $n_i = q_a^{(i)}/Q_a$ is the number of recirculations at the plate; $\tau^{(i)} = 3{,}78(\rho_{N-i,G}/\rho_w)u_e t(\rho_\alpha/\rho_w R^2 d_e)^{1/3}$.

Then for each plate we can write a system of equations

$$\frac{d\bar{\alpha}_i}{d\tau^{(i)}} = -\beta((n_i+1)\bar{\alpha}_i - \bar{\alpha}_{i-1} - n_i\alpha_{i-1}), \quad \bar{\alpha}_i(0) = \bar{\alpha}_{i-1},$$

$$\frac{d\alpha_i}{d\tau^{(i)}} = -\left(1 - \frac{\rho_w W_i}{\rho_{N-i+1,G}}(\alpha_i - \bar{\alpha}_i) - \frac{\rho_{i,wG}}{\rho_{N-i+1,G}}\right), \quad \alpha_i(0) = \bar{\alpha}_i, \tag{20.65}$$

$$\rho_{i,wG} = a(1-\alpha_i)\rho_{N-i+1,G}.$$

In order to solve equations (20.65), it is necessary to know the number of recirculations n_i at each plate. Suppose it is determined by the formula (20.59). Our task is complicated by the fact that the values $\bar{\alpha}_0$ and ρ_{0G} are specified at different ends of the absorber. Therefore, the numerical solution is obtained using the method of successive approximations from the top downward. The conditions $\tau_e^{(i)} \ll \tau_i^{(i)}$ let us simplify the solution considerably. Since each approximation assumes that the value $\rho_{N-i+1,G}$ entering (20.65) is known from the previous approximation, the first two equations can be integrated independently from each other. The result is:

$$\alpha_{i-1} = A_i \bar{\alpha}_i + G_i,$$

$$\bar{\alpha}_i = \left(\bar{\alpha}_{i-1} - \frac{\bar{\alpha}_{i-1} + n_i G_i}{n_i + 1 - n_i A_i}\right) \exp\left(-\frac{(n_i + 1 - n_i A_i)\tau^{(i)}}{\tau_t^{(i)}}\right) + \frac{\bar{\alpha}_{i-1} + n_i G_i}{n_i + 1 - n_i A_i}, \quad (20.66)$$

$$A_i = e^{-\bar{a}_i \tau_e^{(i)}}, \quad G_i = (1 - A_i)\left(1 - \frac{\rho_{N-i+1,G}}{a\rho_{N-i,G}}\right).$$

Let us give an example of the calculation of a countercurrent absorber equipped with separation-contact elements, at the following values of the key parameters: $p = 5.5$ MPa; $T = 15$ °C; $Q_G = 10$ mill. m^3/day; $Q_\alpha = 7 \cdot 10^{-3}$ m^3/(thousand.m^3) of gas under normal conditions; $\bar{\alpha}_0 = 0.9915$. Fig. 20.11 shows the dependence of relative moisture concentration ρ_{vG}/ρ_{0G} in a dewatered gas at the absorber exit on the number of contact stages for different values of the recirculation number *n*. We have assumed the number of recirculations to be the same at each stage.

In Table 20.7, the calculated values of the dew-point temperature of dewatered gas and the test results are given for a three-stage absorber.

One important consequence is the reduction of the number of contact stages with the increase of the recirculation number *n*. Thus, the moisture concentration $\rho_{vG}/\rho_{0G} = 0.1$ at $n = 2$ is achieved for a dewatered gas in the six-stage absorber,

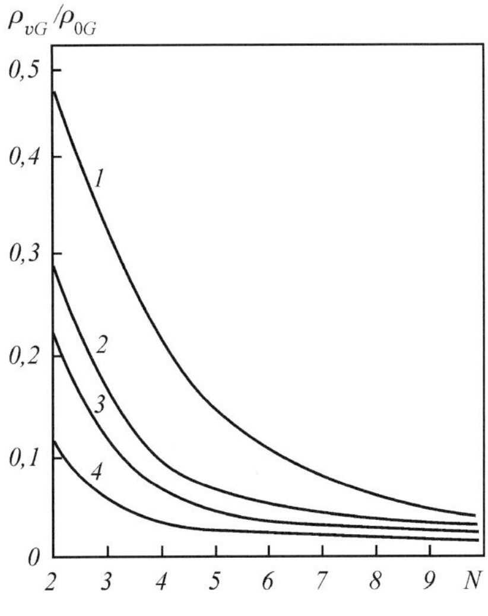

Fig. 20.11 Dependence of the relative moisture concentration ρ_{vG}/ρ_{0G} in a dewatered gas on the number *N* of contact stages for different values of the number of recirculations *n*: *1* – 2.08; *2* – 3.81; *3* – 5.38; *4* – 11.42.

Table 20.7

Gas flow rate, mill. m^3/day	Absorbent flow rate, kg/mill. m^3	Pressure, MPa	Temperature, °C	DEG concentration in the absorbent	Dew point, temperature, °C
12.8	4.92	7.5	13.5	0.9617/0.9619	−19/−20
11.5	4.83	7.0	13.2	0.9651/0.9569	−19/−19
8.5	7.16	7.3	15	0.9596/0.9656	−20/−17

Note: The numerator gives the theoretical value, and the denominator – the operating value.

while if we increase the recirculation number to 11, the same concentration will be achieved in a two-stage absorber. Note that in a real device, the recirculation number can be different at different absorber stages. Since the absorbent at the bottom stages contains an increased (in comparison with the initial absorbent) concentration of water, it makes sense to increase the recirculation number for the bottom stages. This can be achieved by increasing absorbent layer thickness at the plate.

Thus, recirculation of the absorbent at each stage of an absorber allows us to make a more efficient use of its absorbing ability at each contact stage, thereby reducing the number of stages. However, in this context, it is necessary be aware that an increased recycling number n results in a proportional increase of volume concentration W of the liquid phase in the gas (see formulas (20.53) and (20.61)) and, consequently, in an increased ablation of liquid from the element.

21
Prevention of Gas-Hydrate Formation in Natural Gas

In view of putting gas and gas-condensate fields in operation in the complicated conditions of Far North, with the presence in gas of hydrogen sulphide and carbon dioxide, the urgent problem arises of improvement in process and technology of application of hydrate inhibitors in systems for gas production and its preparation for transport.

Hydrates of natural gases are formed in wells, in elements of the field equipment for gas preparation, and also in gas pipelines at certain values of pressure and temperature determined by the phase diagrams of heterogeneous equilibrium. In practice, in order to determine the conditions for the beginning of hydrate formation, one uses graphic methods, calculation method with use of equilibrium constants, semi-graphical method with the Barrer–Stuart equation, and experimental method.

There are simple and mixed hydrates. Simple hydrates represent hydrates of separate gases. These hydrates are formed from molecules of water and hydrator, which can be one of hydrocarbon gases. Hydrators of mixed hydrates and hydrates of natural gases are the gas mixtures, rather than single gases. The composition of mixed hydrates and their component structure changes depending on temperature and partial pressure of components in gas phase.

Methods used for prevention of hydrate formation are dictated by physical and chemical nature of a process. Since equilibrium parameters of hydrates formation depend on partial pressure of water vapor in hydrating medium, any action lowering such pressure reduces the temperature of hydrate formation. In practice, the two ways are used: dehydration of gas from moisture and the input into gas flow of various water-absorbing substances called inhibitors.

An effective and reliable method to prevent hydrate formation is gas dehydration during its preparation before transport in gas pipeline. It is necessary to dewater gas up to the dew-point some degrees below minimally possible temperature in gas pipeline. As hydrate inhibitor is high-concentrated water solution of glycol (DEG). The process of gas dehydration was in detail considered in the previous chapter.

The second method uses the delivery of the inhibitor of hydrate formation directly into gas flow in the pipeline or in elements of field equipment above the so called protected point, i.e. the place in which the expected temperature can be

Separation of Multiphase, Multicomponent Systems. E. G. Sinaiski and E. J. Lapiga

ISBN: 978-3-527-40612-8

lower than the temperature of hydrate formation. Under the action of the inhibitor the structural connections between the water molecules break down, and the energy of their interaction changes. As a result, the pressure of the water vapor in gas phase reduces, and hence the temperature of hydrate formation decreases. The most widespread inhibitor of hydrate formation in considered method is high-concentrated water solution of methanol. Advantages of its employment in comparison with other inhibitors are relatively low cost, high anti-hydrate activity (the greatest reduction of the temperature of hydrate format ion with all other things equal), low viscosity, low temperature of freezing, low solubility in unstable condensate, etc.

By now an extensive experience has been accumulated in using of methanol in gas industry. The big volume of field and laboratory experiments together with theoretical works on thermodynamics of hydrocarbon systems have allowed to elucidate the influence of an inhibitor on the natural gas-water system [61, 63–65].

Existing methods of technological calculations of the inhibition process [65] are based on the assumption that there exists a thermodynamic balance between liquid (inhibitor) and gas (natural gas) phases. Application of this method allows to determine equilibrium values of concentration of water vapor and inhibitor in a gas at given values of pressure, temperature, inhibitor's mass concentration in the solution, composition of gas, and specific flow rate of inhibitor required for given temperature decrease of hydrate formation:

$$\Delta T = T_0 - T_2, \tag{21.1}$$

where T_0 is the temperature of hydrate formation in gas without input of inhibitor, and T_2 is the temperature at the point to be protected.

Let the inhibitor enters gas flow at point 1, where the temperature is T_1 ($T_2 < T_0 < T_1$). If there is a phase balance at point 2, then ΔT is determined by the relation [65]

$$\Delta T = -\tilde{A}\ln(\gamma_{wr}x_{wr}), \tag{21.2}$$

$$\ln\gamma_i = \ln\gamma_i^{\infty}\left[1 + \left(\frac{\ln\gamma_i^{\infty}}{\ln\gamma_j^{\infty}}\right)\left(\frac{x_{ir}}{x_{jr}}\right)\right]^{-2}, \quad (i, j = w, I\ (i \neq j)),$$

$$\gamma_w^{\infty} = 2.2 - \frac{530}{T}, \quad \gamma_I^{\infty} = 3.1 - \frac{715}{T}, \quad (243\ {}^{\circ}\mathrm{K} < T < 320\ {}^{\circ}\mathrm{K}),$$

where x_{wr}, x_{Ir} are molar fractions of water and inhibitor in the solution; γ_w, γ_I are activity factors of water and inhibitor; γ_w^{∞}, γ_I^{∞} are appropriate limiting values of activity factors; $\tilde{A}$ is the empirical constant (for methanol $\tilde{A} = 85$).

After the input of an inhibitor at point 1 the balance in the gas-water vapor system breaks. During the flow of gas with drops of the inhibitor solution from point 1 to point 2, gas temperature T_0, drops' temperature T_L, and also concentrations of water and inhibitor in gas y_w, y_I and in a drop x_w, x_I change. Since the values x_w and x_I figuring in (21.2) correspond to equilibrium values, the primary goal consists in determining the distance between points 1 and 2, or the time t_{eq} of balance establishment in system of gas-inhibitor drops. The value t_{eq} depends on p, T_L, T_0, initial concentrations x_{w0}, x_{I0}, relative flow rate of inhibitor q, measured in kg/10^3 м^3 of gas under normal conditions, gas composition, and parameters describing the flow of gas-liquid mixture. To solve the problem of definition t_{eq} it is necessary to consider the dynamics of mass-exchange between the inhibitor drops and the ambient gas.

One of the widespread methods of bringing the inhibitor into gas flow in the pipeline is spraying through atomizers providing such regime of liquid jet outflow, at which it breaks up into fine drops with various diameters. As a result, the following drop – size distribution is formed at the atomizer's exit [51]:

$$n_d(D) = \begin{cases} \dfrac{N_d\beta^2 \exp(\beta)}{(1+\beta)D_{\max}} \left(\dfrac{D_{\max}}{D}\right)^3 \exp\left(-\beta\dfrac{D_{\max}}{D}\right), & D \leq D_{\max}, \\ 0, & D > D_{\max}, \end{cases} \tag{21.3}$$

where D is the diameter of a drop, N_d is the number of drops in a unit volume, β is the parameter determined by the atomizer's design (for a centrifugal atomizer $\beta = 0.19$; for direct-jet atomizer $\beta = 0.35$), $D_{\max}$ is the maximal drop diameter characteristic of one of the basic assumptions in deriving the distribution (21.3).

Appropriate expressions $D_{\max}$ for atomizers of mentioned types can be found in [51, 66].

It follows from (21.3) that the key parameters of the distribution – the average diameter and the variance – are:

$$\begin{aligned} D_{av} &= \frac{1}{N_d}\int_0^{D_{\max}} Dn(D)\,dD = \frac{1}{1+\beta}D_{\max}, \\ \sigma^2 &= \frac{1}{N_d}\int_0^{D_{\max}} (D - D_{av})^2 n(D)\,dD \\ &= \frac{\beta^2 \exp(\beta)}{1+\beta} D_{\max}^2 \left[-Ei(-\beta) - \frac{\exp(-\beta)}{1+\beta}\right], \end{aligned} \tag{21.4}$$

where $Ei(-\beta)$ is the integral exponential function.

Another important parameter is the volume content of drops W. If the specific flow rate of liquid from the atomizer q_L is given, then $W = q_L/\rho_L$, where ρ_L is the density of liquid, and it follows from (21.3) that

$$N_d = \frac{q_L}{3\rho_L V_{\max}^2[\exp(-\beta) + \beta Ei(-\beta)]}, \quad V_{\max} = \pi D_{\max}^3/6. \tag{21.5}$$

The motion of formed ensemble of drops with gas flow is accompanied by continuous change of drops distribution over sizes; this results from the concurrent processes of mass-exchange between the drops and the gas, coagulation and breakup of drops under action of intensive turbulent pulsations of various scales, and also deposition of drops at pipe walls and droplet detachment from the surface of thin liquid film formed at the internal surface of the pipe.

As the experimental researches of the flow in pipes of gas-liquid mixtures with small content of liquid phase have shown [3], at some distance from the place of appearance of the liquid in gas flow (behind the throttle or the atomizer) the distribution of drops over sizes is stabilized and is well described by lognormal distribution characterized by the average radius:

$$R_{av} = 0.09d\left[\frac{1}{U}\left(\frac{2\Sigma}{\rho_L d}\right)^{1/7}\right]^{6/7}\left(\frac{\rho_G}{\rho_L}\right)^{1/7}, \qquad (21.6)$$

where d is the diameter of the tube, U is the flow-rate average velocity of gas, Σ is the coefficient of surface tension, ρ_G is the density of gas, ρ_L is the density of liquid.

For prevention of hydrate formation, an inhibitor is injected into natural gas – the high-concentration water solution of methanol in amount no more than 1 kg on 1000 m^3 of gas at normal conditions. It corresponds to the volume content of liquid $W \leq 5 \cdot 10^{-5}\ m^3/m^3$ at characteristic pressure 5 MPa. Such small value W allows to neglect the influence of drops on dynamic parameters of gas and to consider dynamics of drop ensemble under action of the flow undisturbed by disperse phase.

An important problem is the research of change of the initial drop spectrum described by distribution (21.3) under the influence of mass-exchange (evaporation of methanol, condensation of water vapor), coagulation, breakup and deposition of drops at the pipe wall. As of now, it is not possible to take into account the flux of drops from the wall (ablation), because at the present time there are no reliable relevant data in literature.

The dispersiveness of the flow is characterized by distribution of drops over volume n, which is (in approximation of two-particle interactions in spatially homogeneous case) described by the kinetic equation

$$\frac{\partial n}{\partial t} + \frac{\partial}{\partial V}\left(n\frac{dV}{dt}\right) = I_k + I_b - I_d. \qquad (21.7)$$

Here dV/dt is the rate of a drop volume change due to mass-exchange with the ambient gas. Terms in the right part characterize the rate of distribution change due to coagulation, breakup, and deposition at the wall, and have the following form (see (11.119)):

$$I_k = \frac{1}{2}\int_0^V K(\omega, V-\omega)n(\omega,t)n(V-\omega,t)\,d\omega - n(V,t)\int_0^\infty K(V,\omega)n(\omega,t)\,d\omega,$$

$$I_b = \int_0^\infty P(V,\omega)f(\omega)n(\omega,t)\,d\omega - n(V,t)f(V), \qquad (21.8)$$

$$I_d = \frac{n}{\bar{t}_l},$$

where $\bar{t}_l$ is the average lifetime of a drop in the flow.

Consider in succession the mass-exchange between drops and hydrocarbon gas, coagulation and breakup of drops, and their deposition at walls of the pipe, which will allow to determine dV/dt, I_k, I_b and I_d.

Consider first the behavior of an isolated drop in quiescent gas, and then the process of mass-exchange in an ensemble of drops in the quiescent gas and in the turbulent gas flow in the pipe. Then the behavior of spectrum of drops injected into a gas flow through atomizers will be analyzed.

21.1
The Dynamics of Mass Exchange between Hydrate-Inhibitor Drops and Hydrocarbon Gas

Let at some point 1 of the gas pipeline an inhibitor in the form of small droplets is injected in the gas flow. The way of injection will not be discussed, we will only assume that the total volume of drops is small in comparison with the volume of gas, so drops do not exert noticeable influence on the gas flow. Drops of the inhibitor consist of high-concentration water-methanol solution.

After the input of inhibitor at point 1 the balance in the system gas–water vapor breaks. During the flow of gas with inhibitor drops from point 1 to a point of interest, the concentration of water and methanol in gas y_{w}, y_m and in drop x_{w}, x_m, temperatures of gas T_G and drops T_L undergo change. The primary goal consists in determining the change of listed parameters as functions of time, and finding the characteristic time t_{eq} of balance establishment in system gas-inhibitor drops. The value t_{eq} depends on pressure p, temperatures of liquid and gas T_L, T_G, initial concentration of components in liquid x_{w0}, x_{m0}, relative flow rate of the inhibitor q measured in $\mathrm{kg}/1000\ \mathrm{m}^3$ of gas under normal conditions, composition of gas, and parameters describing specific properties of the gas-liquid mix flow. To solve the problem of determining t_{eq}, it is necessary to consider dynamics of mass-exchange of inhibitor drops with ambient hydrocarbon gas.

We begin with considering behavior of an isolated drop in quiescent gas, and then will focus on mass-exchange of drop ensemble in quiescent gas and in turbulent flow of gas in the pipe.

Consider a spherical liquid drop in a boundless gas medium. The liquid phase represents a solution consisting of inhibitor–water solution of methanol. Denote molar fractions of water and methanol in a drop through x_w and x_m. The gas phase represents multi-component mixture including natural gas with molar concentration of components y_i, as well as water and methanol vapor y_w and y_m, respectively. At the initial moment of time $t = 0$ drop and gas temperatures T_{L0} and T_{G0}, molar concentration of liquid x_{w0}, x_{m0} and gas y_{i0}, y_{w0}, y_{m0} phases, and drop radius R_0 are specified.

Make the following assumptions: the gas is motionless; the drop does not move relative to gas; there exists local thermodynamic balance at the liquid-gas interface; the pressures in gas and liquid phases are equal and constant; natural gas is considered neutral. This means that the gas does not dissolve in the liquid phase, while the water and methanol transfer through the interface remains possible. The characteristic time of the heat-mass-transfers in the gas phase is small in comparison with the characteristic time in the liquid phase. This assumption allows to formulate the problem in quasi-stationary approximation: the component concentration and temperature distributions in gas are stationary and depend only on distance r from the center of the drop, while component concentration and temperature in the liquid phase change in time and are homogeneous within the drop volume; the natural gas is considered as one component (pseudo-gas) with properties determined by known rules of averaging for multi-component mixtures [9]. The molar concentration of pseudo-gas is designated y_{nG}; the mass transfer of components in gas is caused by mechanism of molecular diffusion characterized by binary diffusion factor D_{im}; cross effects are neglected.

The problem of drop evaporation into the ideal gas medium in a similar formulation was considered in Sec. 6.9. The equations describing the mass and heat transfer in the gas phase, for a stationary process with spherical symmetry are:

$$\frac{d}{dr}(r^2 J_i) = 0, \quad \frac{d}{dr}(r^2 E) = 0, \quad (i = w, M, nG) \tag{21.9}$$

where J_i is the specific molar flux of i-th component, and E is the specific heat flux.

Mass and heat fluxes in the multi-component gas are:

$$J_i = -\frac{\rho_G D_{im}}{M_G}\frac{dy_i}{dr} + y_i \sum_j J_j, \quad (i, j = w, M, nG),$$
$$E = -\lambda_G \frac{dT_G}{dr} + \sum_j J_j H_{Gj}, \tag{21.10}$$

where ρ_G, M_G, λ_G, H_{Gj} are the density, the molecular mass, gas thermal conductivity and molar enthalpy of i-th gas component.

The boundary conditions for Eqs. (21.9) and (21.19) are:

$$J_i = J_{iB}, \quad E = E_B, \quad y_i = y_{iB}, \quad T_G = T_B. \tag{21.11}$$

The component concentration, temperature, and conditions of limitation of fluxes far from the drop are given as:

$$y_i \to y_{i\infty}, \quad T_G \to T_\infty, \quad |J_i| \text{ and } |E| < C \quad \text{at } r \to \infty. \tag{21.12}$$

According to the fourth assumption, the mass flux of natural gas at the drop surface is equal to zero:

$$J_{nGB} = 0, \quad \text{at } r = R, \tag{21.13}$$

Equations describing change with time of component concentrations, temperature and volume of the drop, can be derived from the conditions of mass, energy and volume equilibrium:

$$\begin{aligned}
&\frac{d}{dt}(Nx_i) = -4\pi R^2 J_{iB}, \quad \sum_j x_j = 1, \quad i, j = w, M, \\
&\frac{dN}{dt} = -4\pi R^2 \sum_j J_{jB}, \\
&\frac{dV}{dt} = -4\pi R^2 \sum_j \mathrm{v}_{Lj} J_{jB}, \\
&\frac{d}{dt}\left(N\sum_j x_j(H_{Lj} - pV_{Lj})\right) = -4\pi R^2 \sum_j H_{GjB} J_{iB} - p\frac{dV}{dt}.
\end{aligned} \tag{21.14}$$

Here $N = V/\sum_j \mathrm{v}_{Lj}x_j = V/\mathrm{v}_{Lm}$ is the number of moles in the drop; v_{Lj}, v_{Lm} are molar volumes of j-th component and the mixture of liquid phase; R, V are the radius and volume of the drop; H_{Lj} is molar enthalpy of j-th component of the liquid phase.

Systems of equations (21.9), (21.10) and (21.14) should be supplemented by expressions for the enthalpy:

$$H_{Gj} = H^0_{Gj} + c_{pGj}(T_G - T^0), \quad H_{Lj} = H^0_{Lj} + c_{pLj}(T_L - T^0) \tag{21.15}$$

and restraints should be imposed on the temperatures of gas and liquid and molar enthalpies at the interface surface:

$$T_L = T_B, \quad H_{Gjw} - H_{Ljw} = l^0_j + (c_{pGj} - c_{pLj})(T_L - T^0), \tag{21.16}$$

where c_{pGj}, c_{pLj} are specific molar heat capacities of components of gas and liquid phases; H^0_{Gj}, H^0_{Lj}, l^0_j are molar values of the enthalpy and the heat of vaporization of appropriate components at reference temperature T°.

The initial conditions for Eq. (21.14) are:

$$x_i(0) = x_{i0}, \quad T_L(0) = T_{L0}, \quad R(0) = R_0. \tag{21.17}$$

For completion of the system (21.9)–(21.17) it is necessary to add relations connecting concentrations in both phases at the interface. The assumption of thermodynamic balance at the interface allows to use results of works [64, 65] according to which equilibrium values of mass concentration of water vapor C_{WB} and methanol vapor C_{MB} in gas are equal to

$$C_{WB} = \frac{2.93 \cdot 10^{-4} \gamma_w x_w}{T_L} \exp\left(18.304 - \frac{3.816 \cdot 10^3}{T_L - 46.13} + \frac{18p}{AT_L} + \frac{2\beta_m p}{zAT_L + \alpha_m \beta_m p}\right),$$

$$C_{MB} = \frac{3.9 p_s \gamma_M x_M}{zT_L} \exp\left\{-\frac{2p}{AT_L}\left(\delta - \frac{\upsilon_{LMN}}{2}\right)\right\}, \qquad (21.18)$$

$$\alpha_m = \sum_j \alpha_j y_{jB}, \quad \beta_m = \sum_j \beta_j y_{jB}.$$

Here z is gas compressibility factor; A gas constant; δ empirical coefficient; υ_{LMN} molar volume of methanol under normal conditions; p_s saturation pressure for methanol vapor determined by Antuan's formula:

$$p_s = 1.33 \cdot 10^{-4} \exp\left(18.95 - \frac{3626.55}{T_L - 34.29}\right).$$

The values of factors α_j and β_j are taken in accord with experimental data for the moisture content of pure gases: for CH_4 $\alpha = 0.725$, $\ln \beta = 6.87 - 0.0093 T_L$; for C_2H_6 $a = 0.6$, $\ln \beta = 7.19 - 0.008 T_L$; for CO_2 and C_{3+} $\alpha = 0.568 - 0.0008 T_L$, $\ln \beta = 8.85 - 0.0117 T_L$; for H_2S $\alpha = 0.56 - 0.0009 T_L$, $\ln \beta = 8.44 - 0.0091 T_L$.

The empirical factor δ is determined by the formula

$$\delta = \begin{cases} -\exp(7.83 - 0.01 T_L) & \text{at } 280\ {}^\circ\text{K} < T_L \le 340\ {}^\circ\text{K}, \\ -\exp(8.103 - 0.011 T_L) & \text{at } 243\ {}^\circ\text{K} \le T_L \le 280\ {}^\circ\text{K}. \end{cases}$$

It should be noted that pressure p figuring in (21.18) is measured in MPa.

Using the connection between molar and mass concentrations

$$y_{iB} = \frac{C_{iB}}{M_i \sum_j (C_{jB}/M_j)}. \qquad (21.19)$$

and substituting it in (21.18), one finds necessary dependences y_{iB} from x_j and T_L.

Thus, the system of equations (21.9), (21.10), (21.14), (21.15), (21.19) with boundary and initial conditions (21.11)–(21.13), (21.16), (21.17) and relations (21.18) describe the dynamics of heat-mass exchange of the water-methanol drop in motionless gas within the framework of the above-made assumptions.

Equations (21.9), (21.10) with conditions (21.11)–(21.13) could be solved independently from the others. Introducing dimensionless variables

$$\bar{R} = \frac{R}{R_0}, \quad \tau = \frac{3\upsilon_{Lw0}\rho_G D_{wm} t}{M_G R_0^2}, \quad \bar{T}_L = \frac{T_L}{T_{L0}}, \quad \bar{T}_G = \frac{T_G}{T_{L0}}, \quad \bar{T}^0 = \frac{T^0}{T_{L0}},$$

$$\bar{d}_{im} = \frac{D_{im}}{D_{wm}}, \quad \bar{\upsilon}_{Lm} = \frac{\upsilon_{Lm}}{\upsilon_{Lw0}}, \quad \bar{\upsilon}_{Li} = \frac{\upsilon_{Li}}{\upsilon_{Lw0}},$$

$$I_i = \frac{J_i R_0 M_G}{\rho_G D_{wm}} \quad \bar{c}_{pLm} = \frac{c_{pLm}}{c_{pLw}}, \quad \bar{l}_i^0 = \frac{l_i^0}{l_{iw}^0},$$

$$\bar{c}_{pGi} = \frac{c_{pGi}}{c_{pGw}}, \quad \bar{c}_{pLi} = \frac{c_{pLi}}{c_{pLw}}, \quad \varepsilon = \frac{c_{pGw}\rho_G D_{wm}}{\lambda_G M_G}, \quad \mu = \frac{c_{pGw}}{c_{pLw}}, \quad \nu = \frac{l_w^0}{T_{L0} c_{pGw}},$$

we obtain system of equations describing the change of the dimensionless concentrations of components, temperatures, and the drop radius:

$$\frac{dx_i}{d\tau} = \frac{\bar{\upsilon}_{Lm}}{\bar{R}}\left(x_i \sum_j \bar{\upsilon}_{Lj} I_{jB}\right), \quad (i = w, M),$$

$$\frac{d\bar{T}_L}{d\tau} = \frac{\mu\bar{\upsilon}_{Lm}}{\varepsilon\bar{R}\bar{c}_{pLm}}\left\{\frac{a(\bar{T}_G - \bar{T}_L)}{\bar{R}(\exp(a) - 1)} - \varepsilon\nu\sum_j I_{iB}\left[\bar{l}_j^0 + \frac{1}{\nu\mu}(\mu\bar{c}_{pGj} - \bar{c}_{pLj})(\bar{T}_L - \bar{T}^0)\right]\right\},$$

$$\frac{d\bar{R}}{d\tau} = -\frac{1}{3}\sum_j I_j - I_{iB}, \quad (i = w, M). \tag{21.20}$$

$$I_{jB} = I_B \frac{y_{jB} - y_{j\infty}\exp(-\bar{R}I_B/\bar{d}_{jB})}{1 - \exp(-\bar{R}I_B/\bar{d}_{jB})}, \quad (j = w, M, nG),$$

$$I_B = \frac{\bar{d}_{nGB}}{\bar{R}} \ln\left(\frac{1 - y_{w\infty} - y_{m\infty}}{1 - y_{wB} - y_{mB}}\right), \quad a = \varepsilon\bar{R}\sum_j I_{jB}\bar{c}_{pGj},$$

$$\bar{\upsilon}_{Lm} = \sum_j x_j \bar{\upsilon}_{Lj}, \quad \bar{c}_{pLm} = \sum_j x_j \bar{c}_{pLj}.$$

To these equations, we must add the initial conditions

$$x_i(0) = x_{i0}, \quad \bar{T}_L(0) = \bar{R}(0) = 1, \quad (i = w, M), \tag{21.21}$$

and the conditions (21.18), (21.19), transformed to dimensionless form.

The system of equations (21.19)–(21.21) is solved numerically. As a result, the changes in time of water and methanol concentrations in a drop, temperature and radius of drops are determined. Thermo-physical properties of gas and liquid phases involved in equations can be determined by methods given in [9]. Calculations were carried out for various pressures, initial temperatures of the drop and gas, and initial concentrations of methanol in the inhibitor solution. The composition of gas used in calculations depends on p, T and should be determined in advance from the equations of vapor-liquid equilibrium. Thus, for $p = 8$ MPa and $T = 313$ °K, the following composition is obtained (molar fractions): $N_2 = 0.81$; $CO_2 = 0.22$; $CH_4 = 96.97$; $C_2H_6 = 1.74$; $C_3H_8 = 0.16$; $i - C_4 = 0.07$; $n - C_4 = 0.03$.

Shown in Fig. 21.1 is the dependence of molar concentration of water in the drop on dimensionless time τ for various values of pressure and initial mass content of methanol 70% and 95%. Eventually the methanol completely evaporates

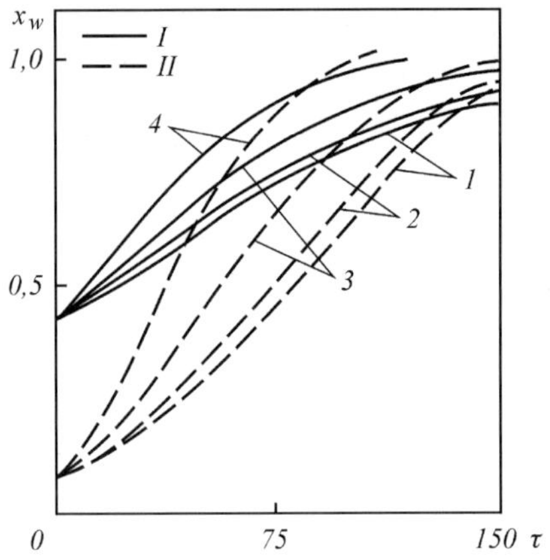

Fig. 21.1 Molar concentration of water in the drop x_w as a function of τ for different pressures and initial methanol mass concentrtions ($T_{L0} = 313$ *K*; $T_{G0} = 288$ *K*): *1* – 12 MPa; *2* – 8 MPa; *3* – 4 MPa; *4* – 2 MPa; *I* – $x_{M0} = 70\%$; *II* – $x_{M0} = 95\%$.

from the drop. Denoted as τ_{eq}, is the characteristic time of equilibrium establishment, determined by given deviation x_w from appropriate equilibrium value.

For the case of 1% deviation, the dependence τ_{eq} on pressure is depicted in Fig. 21.2. It follows from calculations, that within the interval 50–100% of initial concentrations of methanol, the time τ_{eq} determined by this method practically does not depend on x_{m0}.

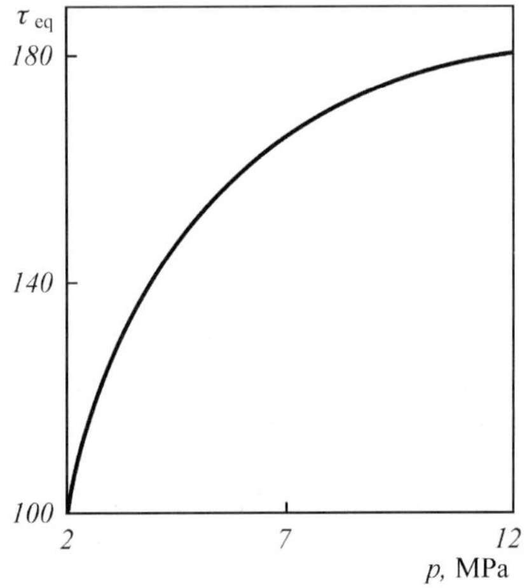

Fig. 21.2 The characteristic time of equilibrium establishment in the drop τ_{eq} as a function of pressure p ($T_{L0} = 313$ *K*; $T_{G0} = 288$ *K*; $x_{M0} = 95\%$).

The transition to dimensional quantities gives us for the value $\tau_{eq} = 100$ the corresponding time $t = 25$ s for a drop with the initial radius $R_0 = 10^{-4}$ m = 100 μm.

Since $t \sim R_0^2 \tau$, the decrease of R_0 to $2 \cdot 10^{-5}$ m lowers t to 1s. The data shown in Figs. 21.1 and 21.2 correspond to the temperature of gas $T_0 = 313$ °K and the initial temperature of the drop $T_{L0} = 288$ °K. As time goes on, the temperature of the drop grows all the way to the gas temperature, and the characteristic heating time for the drop τ_T is much shorter than τ_{eq}. Of general interest is the time change of molar concentrations of water and methanol in the gas phase on the drop surface. The value of y_{wB} grows monotonously and tends to the corresponding saturation value. The molar concentration of methanol y_{wB} first grows sharply and then achieves its maximum value during the time τ_T. Once the drop temperature becomes equal to that of the gas, the flux of evaporating methanol decreases, which results in the reduction of y_{wB} in the surface-adjacent layer. Calculations carried out for various initial gas and drop temperatures have shown the dynamics of mass exchange between the drop and the gas is mostly influenced by the gas temperature, rather than the initial temperature difference between the gas and the drop. This can be explained by the fact that the characteristic time to establish temperature equilibrium is one order of magnitude smaller than the characteristic time to establish concentration equilibrium. Since the evaporation of methanol occurs with higher intensity than condensation of water vapor at the drop, the size of the drop decreases with time. An increase of pressure reduces the equilibrium drop size.

Consider the mass-exchange process for an ensemble of drops with identical initial radii R_0. Denote by W_0 and n the initial volume and number concentrations of drops, and by q – the mass flow rate of the inhibitor, given in kg per thousand m^3 of gas under normal conditions. The newly-introduced quantities are related by

$$W_0 = \frac{4}{3}\pi R_0^3 n = \frac{2.93pq}{zT_G\rho_L}. \tag{21.22}$$

Assume that there is no breakup or coagulation of drops. It means that n remains constant.

The average distance between drops is estimated by the formula $L_{av} \sim R/\sqrt[3]{W}$, therefore any increase of the pressure or the inhibitor flow rate reduces L_{av}. When this happens, neighboring drops begin to exert noticeable influence on the exchange of mass between the drop and the gas. In is case we talk about "hindered" mass-exchange of drops.

Consider the mass-exchange of a fixed drop with the ambient gas. Leave all previously made assumptions unchanged. In this case, Eqs. (21.9), (21.19), and (21.14) still hold. Conditions (21.11), (21.13) and (21.16) at the drop surface also hold, while conditions (21.12) that apply far away from the drop should be reformulated for the distance $r = R_z = R/2\sqrt[3]{W}$. Besides, in the case under consideration, the values W, $y_{i\infty}$, and T_G change with time.

Equations describing $y_{i\infty}$ and T_G can be obtained from the equations of mass and energy balance for a unit volume of the gas-liquid mixture, which, when written in the dimensionless variables introduced above, take the form:

$$\frac{dy_{i\infty}}{d\tau} = -\varphi W_0 \bar{R}^2 \frac{I_{iB} - y_{i\infty} \sum_j I_{jB}}{1 - W},$$

$$\frac{d\bar{T}_G}{d\tau} = -\frac{\varphi \bar{R}^2 W_0}{\varepsilon(1 - W)\bar{c}_{pGm}} \left\{ \frac{a(\bar{T}_G - \bar{T}_L)}{\bar{R}(\exp(a(1 - \bar{R}/\bar{R}_z)) - 1)} - \varepsilon v \sum_j I_{jB} \left[I_j^0 + \frac{1}{v\mu}(\mu \bar{c}_{pGj} - \bar{c}_{pLj})(\bar{T}_L - \bar{T}^0) \right] \right\}, \tag{21.23}$$

$$a = \varepsilon \bar{R} \sum_j I_{jB} \bar{c}_{pGj},$$

$$W = \bar{R}^3, \quad y_{i\infty}(0) = y_{i0}, \quad \bar{T}_G(0) = \bar{T}_{G0},$$

where $\overline{W} = W/W_0$, $\varphi = M_G/\upsilon_{Lw0}\rho_G$, $\bar{R}_1 = R_2/R_0$, $\bar{T}_{G0} = T_{G0}/T_{L0}$.

Of basic interest in the case under consideration is the influence of the inhibitor flow rate q on the dynamics of mass-exchange between drops and the gas.

Fig. 21.3 shows the change of x_w with time for various values of q. The increase of the inhibitor flow rate amplifies the effect of "hindered" mass-exchange process and, as consequence, decreases the intensity of evaporation of methanol. The time change of the moisture content in the gas has a non-monotonous character (Fig. 21.4).

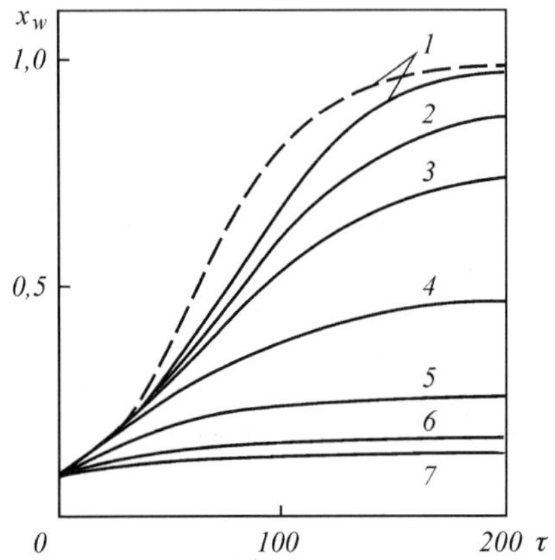

Fig. 21.3 Dependence of molar concentration of water in the drop on τ for different inhibitor flow rates q (kg/thousand m^3 at normal conditions): 1 – 1; 2 – 3; 3 – 5; 4 – 10; 5 – 20; 6 – 50; 7 – 100; $p = 8$ MPa; $T_{L0} = 313$ K; $T_{G0} = 288$ K; $x_{M0} = 95\%$.

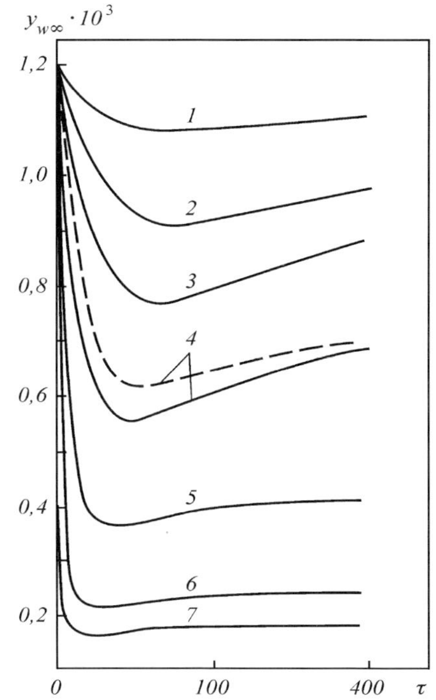

Fig. 21.4 Moisture concentration in the gas y_{w0} as a function of time τ for different inhibitor flow rates q (kg/thousand. m^3) at normal conditions: $1 - q = 1$; $2 - 3$; $3 - 5$; $4 - 10$; $5 - 20$; $6 - 50$; $7 - 100$; $p = 8$ MPa; $T_{L0} = 313$ K; $T_{G0} = 288$ K; $x_{M0} = 95\%$.

As inhibitor flow rate increases, the minimum value of y_{w0} becomes more pronounced and is shifted toward the region of smaller values of τ. The characteristic relaxation time of the system, which is determined by moisture concentration in the gas phase, decreases with an increase of the inhibitor flow rate (Fig. 21.5). The change of q also affects the established size of the drops. This size increases with the growth of q, because each drop releases less methanol into the to gas. But since the number of drops of a given radius increases as q grows, the total amount of methanol transferred into the gaseous phase increases. The end result in an increased molar concentration of methanol and a reduced concentration of water vapor in the gas. The temperature of hydrate formation correspondingly decreases. The heat exchange of drops with the ambient gas doesn't much affect the characteristic relaxation time of the system and the appropriate equilibrium values of water and methanol concentrations in the liquid and gaseous phases. To support this conjecture, we can carry out calculations without using the energy equation (the second equation in (21.20)). The results are depicted in Fig. 21.3 and 21.4 by the dotted line.

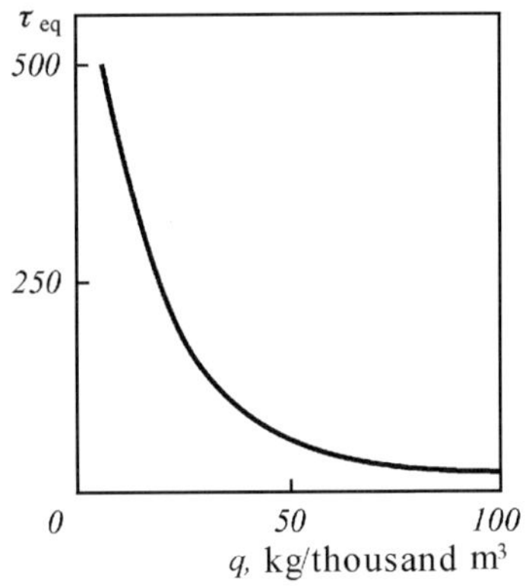

Fig. 21.5 Dependence of the characteristic relaxation time of the system τ_{eq} on q ($p = 8$ MPa; $T_{L0} = 313$ K; $T_{G0} = 288$ K; $x_{M0} = 95\%$).

The flow of biphasic mixture in wells, pipes, and elements of the field equipment is characterized by a developed turbulence, which results in intensive agitation of the mixture and a faster heat and mass exchange of drops with the gas flow. In addition, turbulence causes breakup and coagulation of drops. For the turbulent flow of a biphasic mixture with a small content of the liquid phase, there emerges a distribution of drops with the average radius

$$R_{av} = 0.12d\left(\frac{\rho_G}{\rho_L}\right)^{4/7} We^{-3/7}. \tag{21.24}$$

where d is the diameter of the pipe, $We = \rho_G U^2/\Sigma$ is Weber's number, U is the average gas velocity, Σ is the surface tension coefficient of the liquid phase.

The estimation of R_{av} for characteristic parameter values shows that $R_{av} \gg \lambda_0$, where $\lambda_0 = d/\mathrm{Re}^{3/4}$ is the internal scale of turbulence. In a turbulent flow, both heat and mass exchange of drops with the gas are intensified, as compared to a quiescent medium. The delivery of substance and heat to or from the drop surface occurs via the mechanisms of turbulent diffusion and heat conductivity. The estimation of characteristic times of both processes, with the use of expressions for transport factors in a turbulent flow, has shown that in our case of small liquid phase volume concentrations, the heat equilibrium is established faster then the concentration equilibrium. In this context, it is possible to neglect the difference of gas and liquid temperatures, and to consider the temperatures of the drops and the gas to be equal. Let us keep all previously made assumptions, and in addition to these, assume that initially all drops have the same radius (21.24). Then the mass-exchange process for the considered drop is described by the same equations as before, in which the molar fluxes of components at the drop surface will be given by the appropriate expressions for diffusion fluxes as applied to particles suspended in a turbulent flow (see Section 16.2). In dimensionless variables (the bottom index "0" denotes a paramenter value at the initial conditions),

$$\tau_1 = \frac{\sqrt[3]{2}\rho_L^{1/3}\rho_G^{2/3}U\upsilon_{Lw}t}{M_G d^{1/3}R_{av}^{2/3}}, \quad \bar{R} = \frac{R}{R_{av}}, \quad \varepsilon = \frac{M_G}{\rho_G \upsilon_{Lw}},$$

$$\upsilon_{Lj} = \frac{\upsilon_{Lj}}{\upsilon_{Lw}}, \quad \bar{\upsilon}_{Lm} = \frac{\upsilon_{Lm}}{\upsilon_{Lw}}, \quad \Delta y_j = y_{j\infty} - y_{jB}, \quad \zeta_j = \frac{M_j}{M_G}$$

the equations describing the change of component concentrations for the drop and the gas become:

$$\frac{dx_w}{d\tau_1} = -\frac{3\bar{\upsilon}_{Lm}}{\bar{R}^{2/3}}\{x_w(\zeta_w\Delta y_w + \zeta_M\Delta y_M) - \zeta_w\Delta y_w\},$$

$$x_M = 1 - x_w,$$

$$\frac{dy_{i\infty}}{d\tau_1} = -\frac{3\varepsilon W_0\bar{R}^{7/3}}{1-W}\{\zeta_i\Delta y_i - y_{i\infty}(\zeta_w\Delta y_w + \zeta_M\Delta y_M)\}, \quad (i = w, \text{M}), \tag{21.25}$$

$$\frac{d\bar{R}}{d\tau_1} = \bar{R}^{1/3}(\bar{\upsilon}_{Lw}\Delta y_w + \bar{\upsilon}_{LM}\Delta y_M),$$

$$x_w(0) = x_{w0}, \quad x_M(0) = x_{M0}, \quad \bar{R}(0) = 1, \quad y_{w\infty}(0) = x_{w0}, \quad y_{w\infty}(0) = x_{w0}.$$

Consider as an example the mass exchange for water-methanol solution drops that were input into a turbulent gas flow in a pipe at the following parameter values: pipe diameter $d = 0.4$ m; pressure $P = 8$ MPa; gas temperature $T_G = 313$ °K; inhibitor flow rate $q = 1$ kg/thousand nm^3; initial mass concentration of methanol in the solution $x_{M0} = 95\%$. Fig. 21.6 shows the dependence of the characteristic relaxation time of the system, t_{eq}, on the flow rate Q of the gas under normal conditions. The decrease of t_{eq} with the growth Q is explained by the reduction of the average drop size. All other factors being equal, the time t_{eq} grows with the increase of the pipe diameter and the pressure. Knowing τ_{eq}, it is

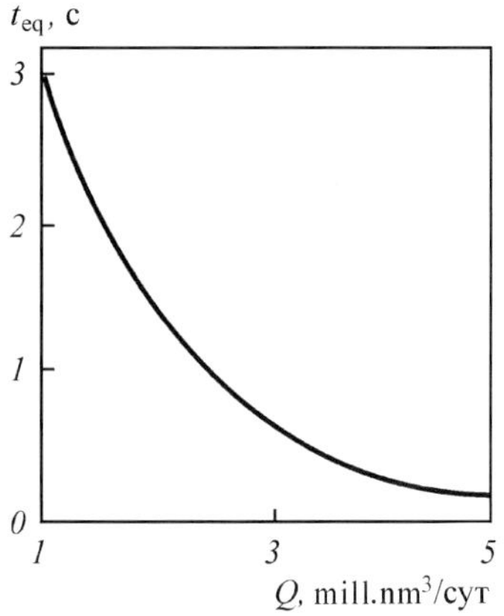

Fig. 21.6 Dependence of the characteristic relaxation time of the system, τ_{eq} on the gas flow rate Q.

easy to find the distance L from the solution entrance point to the point where equilibrium is established. Thus, for the example considered above, for the flow rate of gas $Q = 5$ mill. nm^3/day, we get $L = 2$ m. The length L increases with the fall of gas temperature. Reducing the temperature to 288 °K, we get an increase of L to 12 m.

21.2
Evolution of the Spectrum of Hydrate-Inhibitor Drops Injected into a Turbulent Flow

The most efficient method of inputting hydrate inhibitor into the flow of natural gas in the pipeline is by spraying with the help of atomizers. This results in the formation of a spectrum of drops with distribution (21.3) inside the flow. The average size of the resultant drops is smaller than the stable size that is characteristic of the turbulent flow. During the motion, drops change their size as a result of mass exchange with the gas, and also coagulation and breakup. Besides, drops can be deposited at the pipe wall and then swept away from the surface of the liquid film that forms on the wall.

Let us consider the change of the initial drop spectrum, which is described by distribution (21.3), under the influence of mass exchange (evaporation of methanol or water), coagulations, breakup, and deposition of drops on the pipe wall. As was mentioned earlier, it is not feasible to take into account the flux of drops detached from the wall, because of the absence of reliable data.

The evolution of the volume distribution of drops, $n(V, t)$, is described by the kinetic equation (21.7), which can only be solved if we know the expressions for the rates of change of $n(V, t)$ in the right-hand side of the equation.

Appearing in the expression for I_k is the coagulation kernel $K(\omega, V)$, which is also known as the coagulation constant. It gives the collision frequency of drops in terms of volumes ω and V and can be determined by studying the relative motion of two drops under the action of various interaction forces: gravitational, hydrodynamic, or molecular. The analysis of various mechanisms of drop coagulation that was carried out in Section 13.6 shows that for the currently considered process, turbulent diffusion makes the overwhelming contribution to the drop coagulation rate in a turbulent flow. Thus, the coagulation kernel is

$$K(V, \omega) = G(V^{1/3}\omega^{1/6} + V^{1/6}\omega^{1/3}), \tag{21.26}$$

which takes into account the hydrodynamic and molecular interaction forces. Here $G = 16(\Gamma/3\pi\rho_G\lambda_0^2)^{1/2}$; Γ is Hamaker's constant ($\Gamma = 5 \cdot 10^{-20}$ J for aerosols); $\lambda_0 = d/\mathrm{Re}^{3/4}$ is the inner scale of turbulence; Re is the Reynolds number of the gas flow.

The rate of change of n that is due to the breakup of drops is expressed in terms of the breakup frequency $f(V)$ of drops whose volume lies in the interval $(V, V + dV)$ and the probability $P(V, \omega)$ of drop with volume $(V, V + dV)$ being formed after the breakup of a drop of volume $(\omega, \omega + d\omega)$. A model of drop breakup was considered in Section 11.7, using the assumption that the breakup of a single drop is completely determined by the fluctuations of energy dissipation in its vicinity. It was remarked that, irrespective of the initial spectrum, after a certain time the distribution of drops becomes lognormal as a result of drop breakup. To determine the breakup frequency $f(V)$, one has to estimate the minimum radius $R_{\min}$ of drops that are subject to breakup in a turbulent flow. This size can be determined theoretically, by comparing the forces acting on the drop that can result in a significant deformation of its surface. In [67] the expression for $R_{\min}$ is obtained from the comparison of viscous friction and capillary forces, and in [2] – from the comparison of dynamic pressure and capillary forces. The motion of drops in the gas does not evoke a great viscous friction force, therefore the second model is more preferable and it makes sense to represent $R_{\min}$ by the expression

$$R_{\min} = 2.64 \frac{\Sigma^{3/5}}{\rho_G^{1/5} \bar{\varepsilon}^{2/5}}, \tag{21.27}$$

where $\bar{\varepsilon} = \rho_G U^3/d$ is the average specific dissipation of energy in the turbulent flow.

The breakup frequency for a drop with volume V was obtained in Section 11.7:

$$f(V) = \frac{1}{2.03\sqrt{2\pi}T_0}\gamma(x), \quad \gamma(x) = \frac{1}{\sqrt{2\pi}\phi(x)}\exp(-x^2/2), \tag{21.28}$$

$$x = -a\ln(bV/V_{\min}), \quad V_{\min} = 4\pi R_{\min}^3/3, \quad \phi(x) = \frac{1}{\sqrt{2\pi}}\int_{-\infty}^{x}\exp\left(-\frac{y^2}{2}\right)dy.$$

Here $T_0 = \sqrt{\mu_0/\bar{\varepsilon}}$ is the correlation time constant of random process $\varepsilon(t)$; $\bar{\varepsilon}(t)$ is the average value of energy dissipation in the vicinity of the drop; μ_G is the gas viscosity; parameters a and b are determined by the dependence of $R_{\min}$ on $\bar{\varepsilon}$. In particular, for the dependence (21.27), we have $a = 1.44$ and $b = 1.22$.

It follows from (21.28) that $f(V)$ grows monotonously with V. For drop volumes that are small in comparison with the minimum volume, $V \ll V_{\min}$ ($x \gg 1$), the frequency of breakup tends to zero as $(2\pi)^{-1/2}\exp(-x^2/2)$, while for $V \gg V_{\min}$ ($x \ll 1$), the frequency of breakup is a linear function $f \sim -x^2$.

As demonstrated by experimental studies of drop breakup dynamics [68], a drop usually splits into two drops of about the same size and a number of satellites with considerably smaller sizes. It is therefore permissible to assume that during breakup, the drop will split into two drops of the same volume. Thus, the probability of breakup can be written as

$$P(V,\omega) = 2\delta\left(V - \frac{\omega}{2}\right), \tag{21.29}$$

where $\delta(x)$ is the delta-function.

Substituting (21.29) into the expression for I_b and using the properties of the delta-function, we get:

$$I_b = 4f(2V)n(2V,t) - f(V)n(V,t). \tag{21.30}$$

The lifetime $\bar{t}_l$ of a drop of volume V in a gas flow enters the expression for the rate of change of the drop distribution, I_d, that takes place because of the deposition of liquid on the pipe wall. The lifetime can be interpreted as the time it takes for the drop of volume V to reach the pipe wall. This time is averaged over all possible initial positions of the drop in the cross section of the flow. To determine $\bar{t}_l$, we should first estimate the velocity of transversal motion of the drop in the gas flow. As shown in [69], the transversal drift of drops inside the core of the turbulent flow occurs due to the gravitational force, and the drift velocity is equal to

$$U_s = t_r g, \tag{21.31}$$

where t_r is the drop relaxation time, which for highly- and coarsely dispersed particles equals $t_r = \rho_L D^2/18\mu_G$, g is the acceleration of gravity.

In a thin wall layer, the velocity of particles' migration to the wall is determined by the semi-empirical relationship

$$u_m = \begin{cases} 7.25 \cdot 10^4 \left(\dfrac{\tau_+}{1+\omega_+\tau_+} \right)^2 u_d & \text{at } \dfrac{\tau_+}{1+\omega_+\tau_+} < 16.6, \\ 0.2u_d & \text{at } \dfrac{\tau_+}{1+\omega_+\tau_+} \geq 16.6, \end{cases} \tag{21.32}$$

where $t_+ = D^2\rho_L u_d^2/18\nu_G^2\rho_G$ is the dimensionless drop relaxation time; $\omega_+ = 20\nu_G\rho_G/d\rho_L u_d$ is the dimensionless frequency of large-scale pulsations; $u_d = U(k_f/8)^{1/2}$ is the dynamic velocity of the gas; k_f is the coefficient of friction, ν_G is the kinematic viscosity of the gas.

The estimation of velocities of gravitational sedimentation and migration of drops shows that $u_s \ll u_m$. It means that the lifetime of a drop is determined by time of its residence in the core of the flow. After averaging t_l over all initial positions in the cross section of the flow core, we obtain:

$$\bar{t}_l = \frac{36\, d\mu_G}{\pi g \rho_L D^2}, \tag{21.33}$$

Thus, the kinetic equation (21.7) together with the appropriate equations (21.8), (21.26), (21.28)–(21.30), which should be complemented by equations (21.25) that define the change of drop volume due to the mass exchange, describe the dynamics of the initial volume distribution of drops $n_d(V)$ behind the atomizer. This distribution is connected to the diameter distribution $n_d(D)$ by the equality

$$n_d(V) = \frac{2}{\pi D^2} n_d(D). \tag{21.34}$$

In the general case, the obtained system of equations can be solved numerically. The estimation of characteristic times of processes involved allows us to simplify both the system of equations and the method of solution.

The characteristic times of mass-exchange and coagulation of drops of the average volume V_{av} are estimated by the expressions:

$$t_m \sim \frac{M_G d^{1/3} V_{av0}^{2/9}}{\rho_L^{1/3} \rho_G^{2/3} U v_{Lw}}, \qquad t_k \sim \frac{V_{av1}^{1/2}}{G W_0}.$$

Here V_{av0} and V_{av1} are the average volumes of the drop before and after the completion of the mass-exchange process.

For characteristic values $M_G = 20$ kg/kmole, $d = 0.5$ m, $V_{av0} = 10^{-15}$ m^3, $V_{av1} = 3 \cdot 10^{-17}$ m^3, $\rho_L = 10^3$ kg/m^3, $\rho_G = 50$ kg/m^3, $U = 15$ m/s, $v_{Lw} = 2 \cdot 10^{-5}$ m^3/mole, $W_0 = 5 \cdot 10^{-5}$ m^3/m^3, we get $t_m/t_k \sim 10^{-4}$. This means that when considering the mass-exchange process, we can neglect drop coagulation. Besides, if the average radius of drops forming at the exit from the atomizer is smaller then the minimum radius of drops that can experience breakup, we can also neglect the breakup of drops. For the chosen values of parameters we see that $\bar{t}_l \geq t_k$, therefore the sedimentation of drops on the pipe wall may be ignored as well.

Thus, for drops in a gas flow, the mass-exchange process happens much faster then the other processes. The evolution of the initial distribution (21.3) is described by the equation

$$\frac{\partial n}{\partial t} + \frac{\partial}{\partial V}\left(n \frac{dV}{dt}\right) = 0, \quad n(V,0) = n_d(V). \tag{21.35}$$

which, together with the system of equations describing the mass-exchange of drops, allows to investigate the dynamics of mass exchange between a polydispersive ensemble of drops and a hydrocarbon gas.

Note that in the process of mass exchange, the number concentration of drops $N = \int_0^\infty n(V,t)\, dV$, is conserved, while the volume concentration and the average size of drops decrease owing to methanol evaporation being more intensive than water vapor condensation.

Replace the distribution (21.3) by the logarithmic normal volume distribution of drops:

$$n_d(V) = \frac{N_d}{a\sqrt{\pi} V} \exp\left\{\left(-\frac{\ln^2(V/V_*)}{\alpha^2}\right)\right\}, \tag{21.36}$$

having chosen the parameters of this distribution in such a way that both distributions have the same average diameters and variances. Then the parameters a and V_* will be equal to

$$\alpha = \{18 \ln[-(1+\beta)\exp(\beta) Ei(-\beta)]\}^{1/2},$$

$$V_* = \frac{\pi D_*^3}{6} = \frac{\pi}{6} D_{\max}^3 \left\{ \frac{\beta}{(1+\beta)[-(1+\beta)\cdot\exp(\beta)\cdot Ei(-\beta)]^{1/2}} \right\}^3, \tag{21.37}$$

Consider the case when the volume distribution of drops changes only due to the mass exchange of drops with the ambient gas.

Suppose at the initial moment the distribution has the form (21.3). It is characterized by three parameters – N, β, and $D_{\max}$. By its physical meaning, N is the number of drops in a unit volume of the gas-liquid mixture; β is a parameter describing the design of the atomizer, which is equal to 0.35 for a direct-flow atomizer and 0.19 for a centrifugal atomizer. The expression for $D_{\max}$ was derived for a direct-flow atomizer [51]:

$$\left(\frac{\Sigma}{\rho_L u^2 D_{\max}}\right)\left(1+\frac{10^6 \mu_L^2}{\rho_L D_{\max}}\right)^{1/12}\left(1-0.5\frac{\rho_G}{\rho_L}\right) = 4.8\cdot 10^{-5} \tag{21.38}$$

and for a centrifugal atomizer [66]:

$$\frac{D_{\max}}{h} = 8.5\left(\frac{\Sigma}{\Delta p_d h}\right)^{1/3}\left(1+\frac{10^6 \mu_L}{\rho_L \Sigma h}\right)^{1/10}\left(1-0.5\frac{\rho_G}{\rho_L}\right) \tag{21.39}$$

by processing the data from pneumatic spraying experiments.

Here Σ is the surface tension coefficient of the liquid; ρ_L and ρ_G are the liquid and gas densities; u is the velocity of liquid outflow from the atomizer; μ_L is the dynamic viscosity of the liquid; Δp_d is the pressure drop across the atomizer; h is the thickness of the liquid sheet at the cut of the atomizer nozzle.

The parameter N can be expressed in terms of specific mass flow rate of liquid q_L, maximum volume of drops $V_{\max}$, and β:

$$N = C(\beta)\frac{q_L}{\rho_L V_{\max}}, \quad C(\beta) = \frac{1+\beta}{\beta[1+\beta\cdot\exp(\beta)\cdot Ei(-\beta)]}, \tag{21.40}$$

where $Ei(-\beta)$ is an integral exponential function.

In addition to the distribution of drops over diameters (21.3) let us introduce their distribution over volumes:

$$n_d(V) = n_d(D)\frac{dD}{dV} = n_d(D)\frac{2}{\pi D^2}. \tag{21.41}$$

Substituting (21.3) into (21.41), we obtain:

$$n_d(V) = \begin{cases} N_d\left(\dfrac{V_{\max}}{V}\right)^{5/3}\exp\left[-\beta\left(\dfrac{V_{\max}}{V}\right)^{1/3}\right] & \text{at } V \le V_{\max}, \\ 0 & \text{at } V \le V_{\max}, \end{cases} \tag{21.42}$$

$$N_d = \frac{q_L}{3\rho_L V_{\max}^2[\exp(-\beta)+\beta Ei(-\beta)]}.$$

Consider a unit volume of hydrocarbon gas. Suppose at the initial moment it contains a homogeneously sprayed water-methanol solution of the total volume q_L/ρ_L consisting of drops whose distribution is given by (21.42). The gas is characterized by the molar fractions of components y_i ($i = w, M$ for water vapor and methanol and $i = \overline{1, S}$ for the other components). The composition of drops is given by the molar fractions of water and methanol x_j ($j = w, M$). The temperature and the pressure are assumed to be identical in both phases.

The dynamics of mass-exchange between a mono-dispersive ensemble of hydrate inhibitor drops and a hydrocarbon gas has been considered in Section 21.1 within the framework of the above-made assumptions in a quasi-stationary approximation. In dimensionless variables:

$$\bar{V} = \frac{V}{V_0}, \quad \tau = \frac{3\upsilon_{Lw0}\rho_G D_w t}{M_G (3V_0/4\pi)^{2/3}}, \quad \bar{D}_i = \frac{D_i}{D_w},$$

$$\bar{\upsilon}_{Lm} = \frac{\upsilon_{Lm}}{\upsilon_{LM0}}, \quad I_i = \frac{J_i M_G (3V_0/4\pi)^{1/3}}{\rho_G D_{wm}}, \quad \overline{W} = \frac{W}{W_0}$$

the system of equations describing the time change of the drop volume and of both water and methanol concentrations inside the drop, assumes the form:

$$\frac{dx_w}{d\tau} = \frac{\bar{\upsilon}_{Lm}}{\bar{V}^{1/3}}\left[x_w \sum_j I_{jB} - I_{wB}\right], \quad \frac{d\bar{V}}{d\tau} = -\frac{1}{2}\bar{V}^{2/3}\sum_j \upsilon_{Lj} I_{jB},$$

$$I_{jB} = I_B \frac{y_{jB}\exp(-\bar{V}^{1/3} I_B 2\sqrt[3]{W}/\bar{D}_j) - y_j \exp(-\bar{V}^{1/3} I_B/\bar{D}_j)}{\exp(-\bar{V}^{1/3} I_B 2\sqrt[3]{W}/\bar{D}_j) - \exp(-\bar{V}^{1/3} I_B/\bar{D}_j)}, \tag{21.43}$$

$$I_B = \frac{\bar{D}_{nGB}}{\bar{V}^{1/3}} \ln\left(\frac{1 - y_w}{1 - y_{wB}}\right), \quad y_{jB} = f(x_{jB}) \quad (i, j = w, M),$$

Here V_0 is the characteristic volume of a drop chosen to be equal to $V_{\max}$; v is the molar volume; D_j is the binary coefficient of diffusion; J_j is the molar flux; M is the molecular mass; the lower indexes L and G denote parameters belonging, respectively, to the liquid and the gaseous phase; 0 and m denote the values taken at the initial moment of time and averaged over composition; B denotes the values taken at the interface.

From the condition of mass equilibrium for a unit volume of gas-liquid mixture with volume distribution of drops $n(V, t)$, there follow equations for the molar concentration of components in the gaseous phase:

$$\frac{dy_i}{d\tau} = \frac{\varphi W_0}{1 - W}\int_0^\infty \bar{V}^{2/3}\left(I_{iB} - y_i \sum_j I_{jB}\right)\bar{n}\, d\bar{V}, \tag{21.44}$$

where $\varphi = M_G q_L/\rho_G \upsilon_{Lm0}\rho_L$, $W_0 = q_L/\rho_L$, $\bar{n} = n3\rho_L V_{\max}^2/q_L$ and $W = W_0 \int_0^\infty \bar{V}\bar{n}\, d\bar{V}$.

By implication, W_0 and W are the initial and current volume concentrations of the liquid phase in a gas-liquid mixture.

To complete the system of equations (21.43) and (21.44), we must add the equation describing the change of $n(V, t)$. Since we only consider the mass exchange of the drops with the ambient gas, this equation looks like

$$\frac{\partial \bar{n}}{\partial \tau}+\frac{\partial}{\partial \bar{V}}\left(\bar{n}\frac{d\bar{V}}{d\tau}\right)=0, \quad \bar{n}=n\frac{V_0^2}{W_0}. \tag{21.45}$$

Thus, Eqs. (21.43)–(21.45) with the initial conditions

$$x_w(0)=x_{w0}, \quad \bar{V}(0)=\bar{V}_0, \quad y_i(0)=x_{i0}, \quad \bar{n}(\bar{V},0)=n_d(\bar{V}) \tag{21.46}$$

describe the time change of water and methanol concentrations in the liquid and gaseous phases and the time change of the volume distribution of drops.

For the case of small volume concentration of drops ($W \ll 1$) Eqs. (21.43) and (21.44) can be solved independently from (21.45); then, from (21.45), we can find $n(\bar{V},t)$. At the same time, Eq. (21.45) can be solved by using expressions for $d\bar{V}/d\tau$ and for the molar fluxes J_{iB}, J_B at the surface of drops. Denoting:

$$\frac{\partial \bar{V}}{\partial \tau}=\bar{V}^{1/3}\mathrm{f}(\tau), \quad \mathrm{f}(\tau)=-\sum_j \bar{\upsilon}_{Lj} I_{jB}\bar{V}^{1/3}, \tag{21.47}$$

and introducing new variables,

$$\xi=\int_0^\tau \mathrm{f}(\tau)\,d\tau, \quad \eta=1.5\bar{V}^{2/3}, \tag{21.48}$$

instead of τ and $\bar{V}$, Eq. (21.45) can be transformed into

$$\frac{\partial}{\partial \xi}\left[\left(\frac{2\eta}{3}\right)^{1/2}\bar{n}\right]+\frac{\partial}{\partial \eta}\left[\left(\frac{2\eta}{3}\right)^{1/2}\bar{n}\right]=0, \tag{21.49}$$

which has a general solution

$$\bar{n}(\eta,\xi)=\left(\frac{2\eta}{3}\right)^{-1/2}F(\eta-\xi). \tag{21.50}$$

Using the initial condition (21.46),

$$\bar{n}(\bar{V},0)=\bar{n}_d(\bar{V})=\left(\frac{2\eta}{3}\right)^{-1/2}F(\eta),$$

one obtains:

$$n(\bar{V},\tau)=\begin{cases} n_d\left(\frac{2\eta}{3}\right)^{-1/2}\left(\frac{2(\eta-\xi)}{3}\right)^{-2}\exp\left(-\beta\left(\frac{2}{3}(\eta-\xi)\right)^{-1/2}\right), \\ \qquad \eta-\xi\le 1.5, \\ 0, \quad \eta-\xi>1.5. \end{cases} \tag{21.51}$$

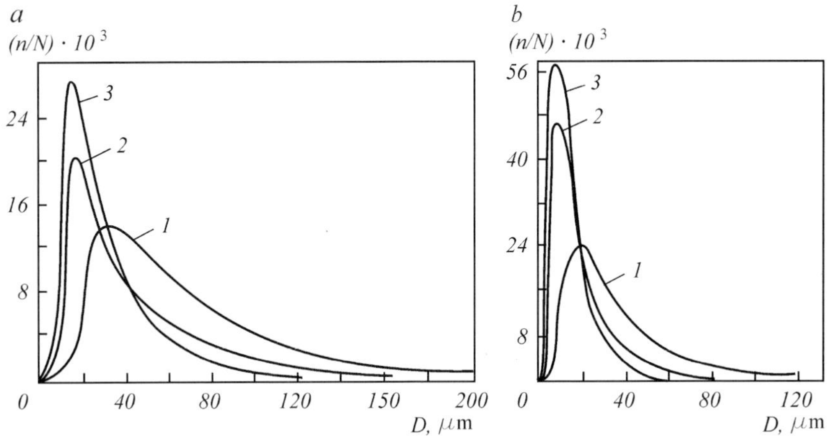

Fig. 21.7 The time change of distribution density of drops over diameters for direct-flow (*a*) and centrifugal (*b*) atomizers: *1* – 1 c; *2* – 15 c; *3* – 100 c.

Fig. 21.7 shows the calculated time changes of the diameter distribution density of drops for a direct-flow and a centrifugal atomizer. The gas and hydrate formation inhibitor parameters are the same as in the example calculation adduced in Section 21.1.

Consider now the mass-exchange of drops in a turbulent flow. Take as the initial distribution the lognormal distribution (21.36) which was used to approximate the initial distribution (21.3).

The dynamics of mass-exchange for a mono-dispersive ensemble of drops suspended in a turbulent flow of hydrocarbon gas was considered in Section 21.1, with certain assumptions being made. A similar approach for the poly-dispersive case with a continuous volume distribution of drops $n(V, t)$ yields the following system of the equations describing the change of molar concentrations of water x_{w} and methanol $x_m = 1 - x_w$ in the liquid phase, components y_i in the gaseous phase, and the drop volume V:

$$\frac{dx_w}{d\tau} = -\frac{3\bar{v}_{Lm}}{\bar{V}^{2/9}}\{x_w(\zeta_w \Delta y_w + \zeta_M \Delta y_M) - \zeta_w \Delta y_w\},$$

$$\frac{dy_i}{d\tau} = -\frac{3\delta}{1-W}\left\{\zeta_i \int_0^{\infty} \bar{V}^{7/9} \Delta y_i \bar{n}\, d\bar{V} - y_i \sum_j \zeta_j \int_0^{\infty} \bar{V}^{7/9} \Delta y_j \bar{n}\, d\bar{V}\right\}, \tag{21.52}$$

$$\frac{d\bar{V}}{d\tau} = 3\bar{V}^{7/9} \sum_j \bar{V}_{Lj} \Delta y_j \zeta_j,$$

$$y_{jB} = f(x_i), \quad (i, j = w, M).$$

Here we introduced the following dimensionless parameters:

$$\tau = \frac{\sqrt[3]{2}\rho_L^{1/3}\rho_G^{2/3}u\upsilon_{Lw}}{M_G D^{1/3}(3V_0/4\pi)^{2/9}}, \quad \bar{V} = \frac{V}{V_0}, \quad \bar{\upsilon}_{Lj} = \frac{\upsilon_{Lj}}{\upsilon_{Lw}}, \quad \bar{n} = nV_0^2/W_0,$$

$$\bar{\upsilon}_{Lm} = \frac{\upsilon_{Lm}}{\upsilon_{Lw}}, \quad \zeta_j = \frac{M_i}{M_G}, \quad \delta = \frac{M_G}{\rho_G \upsilon_{Lw}}, \quad \Delta y_j = y_j - y_{jB},$$

where $\bar{\upsilon}_{Lm} = \sum \upsilon_{Lj}x_j$ is the average molar volume of the liquid phase; M_i is the molecular mass of *i-th* component; V_0 is the initial volume of the drop; the index B corresponds to the value at the interface.

The last equation (21.52), establishing the connection between concentrations of components at the interface, follows from the appropriate equilibrium relations and depends on pressure and temperature.

In addition to Eq. (21.52), it is necessary to impose the initial conditions

$$x_w(0) = x_{w0}, \quad y_i(0) = y_{i0}, \quad \bar{V}(0) = 1 \quad (i = w, M).$$

The dynamics of the initial distribution of drops is described by the equation

$$\frac{\partial \bar{n}}{\partial \tau} + \frac{\partial}{\partial \bar{V}}\left(\bar{n}\frac{d\bar{V}}{d\tau}\right) = 0, \quad \bar{n}(\bar{V}, 0) = \bar{n}_d(\bar{V}). \tag{21.53}$$

As an example, consider the mass-exchange of water-methanol drops with natural gas at temperature $T = 293$ °K and pressure $p = 6$ MПа. The flow rate of the inhibitor is equal to $q_L = 1$ kg on 1000 м^3 of gas under normal conditions. The mass exchange occurs in a pipe of diameter $d = 0.4$ m, the gas flow rate is equal to 10 mill м^3/day. The initial distribution of drops over diameters (21.3) is characterized by the parameters $\beta = 0.19$, and $D_{\max} = 1.25 \cdot 10^{-4}$ m. The parameters of the corresponding logarithmic normal distribution (21.36) are $\alpha = 3.38$; $D_* = 1.3 \cdot 10^{-5}$ m. The system of equations (21.52) and (21.53) has been solved numerically. Fig. 21.8 shows the evolution of the diameter distribution of drops. The average diameter of drops decreases with time – from the initial value $2 \cdot 10^{-5}$ m to $7.9 \cdot 10^{-6}$ m, and the root-mean-square deviation decreases from $1.61 \cdot 10^{-5}$ m to $8.6 \cdot 10^{-6}$ m. The characteristic relaxation time is estimated by the value of 0.2 s. In the same figure, the dashed line shows the logorithmic normal distribution with parameters $\alpha = 3.75$ and $D^* = 5.35 \cdot 10^{-6}$ m which approximates the established distribution.

After the completion of mass-exchange of drops with the gas, any further evolution of the distribution occurs as a result of drop coagulation. Consequently, the average radius of drops increases, and when it reaches the value of the order of the minimum radius of splitting drops, breakup begins to exert a noticeable influence on the change of the volume distribution of drops. Another process taking place at the same time is the deposition of drops on the pipe wall, which, together with breakup, reduces the number of large drops. Thus $n(V, t)$ is described by the equation

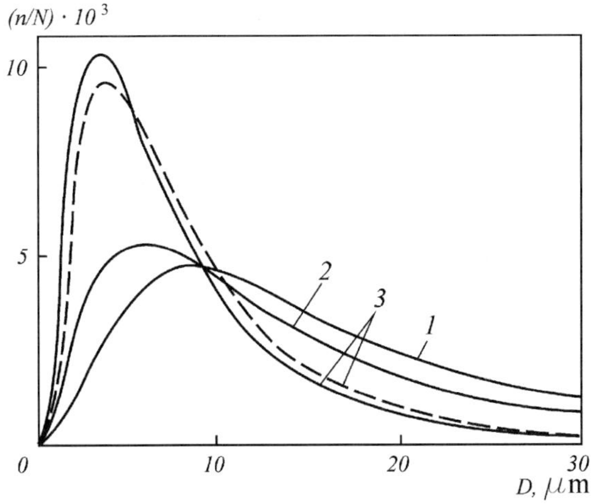

Fig. 21.8 Evolution of hydrate inhibitor drop distribution over diameters during their mass-exchange with hydrocarbon gas: *1* – 0 c; *2* – 0.02 c; *3* – 0.2 c.

$$\frac{\partial n}{\partial \tau} = I_k + I_b - I_d. \tag{21.54}$$

Eq. (21.54) will be solved by the method of moments. Introduce the moments of the distribution:

$$m_k = \int_0^\infty V^k n(V,t)\, dV, \quad (k = 0, 1, 2, \ldots). \tag{21.55}$$

Number and volume concentrations of drops N and W, and the variance dispersion are expressed through first three moments:

$$N = m_0, \quad W = m_1, \quad \sigma^2 = \frac{m_2 m_0 - m_1^2}{m_0}. \tag{21.56}$$

Multiplying both parts of Eq. (21.54) by V_k, integrating the resulting expression over V that ranges from 0 to ∞, and taking into account the symmetry of the coagulation kernel $K(V,\omega)$, we obtain:

$$\frac{dm_k}{dt} = \int_0^\infty \int_0^\infty K(V,\omega)\left[\frac{1}{2}(V+\omega)^k - V^k\right] n(V,t) n(\omega,t)\, dV\, d\omega$$
$$+ (2^{1-k} - 1)\int_0^\infty V^k f(V) n(V,t)\, dV - \left(\frac{6}{\pi}\right)^{2/3} \frac{\pi g \rho_L}{36\, d\mu_G} m_{k+2/3}. \tag{21.57}$$

The uniform power dependence of $K(V,\omega)$ on V and ω (see (21.26)) allows us to express the first term the right-hand side of (21.57) through the moments of the distribution. To perform a similar procedure with the second term, let us make use of the asymptotic behavior of $f(x)$ at $x \ll -1$ and $x \gg 1$, which gives us the right to present the breakup frequency as

$$3.03\sqrt{2\pi}T_0 f(x) = \begin{cases} \dfrac{1}{\sqrt{2\pi}}\exp\left(-\dfrac{x^2}{2}\right) - x & \text{at } x \le 0, \\ \dfrac{1}{\sqrt{2\pi}}\exp\left(-\dfrac{x^2}{2}\right) & \text{at } x > 0. \end{cases} \tag{21.58}$$

The approximation (21.58) provides an accurate estimate of $f(x)$ in the region $x \in (-\infty, -2) \cup (1, \infty)$. On the remaining interval $(-2, 1)$, the maximum error goes as high as 50% at $x = 0$. This, however, does not exert an appreciable influence on the final result. So let us seek the solution of (21.57) among the logarithmic normal distributions of the type (21.36), considering parameters N, α and V_* time-dependent. In such a case the moment of any order can be expressed in terms of m_1, V_* and α as

$$m_k = V_*^{k-1}\exp\left\{\frac{\alpha^2(k^2-1)}{4}\right\}m_1 \tag{21.59}$$

and the integral in the right part of (21.57) is equal to

$$\int_0^\infty V^k f(V) n(V,t)\, dV = \frac{V_{\min}^k m_0}{4.06\pi T_0} I_k. \tag{21.60}$$

where

$$I_k = \frac{1}{\alpha a b^k}\left\{\frac{1}{A}\exp(C_k) + \frac{1}{\sqrt{2}D^2}\exp(F_k)[\exp(-E_k^2) - \sqrt{\pi}E_k(1 - \varnothing(E_k))]\right\},$$

$$A = \left(\frac{1}{2} + \frac{1}{a^2\alpha^2}\right)^{1/2}, \quad C_k = \frac{(2S - k\alpha)^2}{2\alpha^2(2 + a^2\alpha^2)} - \frac{S^2}{\alpha^2}, \quad D = \frac{1}{a\alpha}, \quad E_k = \frac{2S - k\alpha^2}{2\alpha},$$

$$F_k = -k\left(S - \frac{\alpha^2 k}{4}\right)^{1/2}, \quad C_k = \ln\left(\frac{V_{\min}}{bV_*}\right), \quad \varnothing(z) = \frac{2}{\sqrt{\pi}}\int_0^z \exp(-z^2)\, dz,$$

Putting $k = 0; 1; 2$ into (21.57), and using the relations above, we can obtain equations for first three moments:

$$\frac{d\bar{m}_0}{d\tau_1} = -\frac{\bar{m}_0^{95/72}\bar{m}_1^{31/36}}{\bar{m}_2^{13/72}} - \varphi\frac{\bar{m}_0^{2/9}\bar{m}_1^{8/9}}{\bar{m}_2^{1/9}} + \psi I_0 \bar{m}_0,$$

$$\frac{d\bar{m}_1}{d\tau_1} = -\varphi\exp\left(-\frac{5\alpha_0^2}{9}\right)\frac{\bar{m}_1^{37/9}}{\bar{m}_0^{17/9}\bar{m}_2^{11/9}}, \tag{21.61}$$

$$\frac{d\bar{m}_2}{d\tau_1} = 2\exp\left(-\frac{\alpha_0^2}{4}\right)\frac{\bar{m}_1^{67/36}\bar{m}_2^{23/72}}{\bar{m}_0^{13/72}} - \varphi\exp\left(\frac{344\alpha_0^2}{144}\right)\frac{\bar{m}_1^{94/9}}{\bar{m}_0^{50/9}\bar{m}_2^{35/9}}$$
$$-\frac{\psi I_2 \bar{V}_{min}^2}{2}\exp\left(-\frac{\alpha_0^2}{2}\right)\bar{m}_0,$$
$$\bar{m}_0(0) = \bar{m}_1(0) = \bar{m}_2(0) = 1.$$

Here, we introduced the following dimensionless variables:

$$\tau_1 = \frac{GW_0 t}{V_{av}^{1/2}(0)\exp(13\alpha_0^2/144)}, \qquad \bar{m}_k = \frac{m_k}{m_k(0)},$$
$$\varphi = \frac{0.13 g\rho_L V_{av}^{7/6}(0)\exp(5\alpha_0^2/144)}{d\mu_G G W_0},$$
$$\bar{V}_{min} = \frac{V_{min}}{V_{av}(0)}, \quad \bar{V}_* = \frac{V_*}{V_{av}(0)}, \quad \psi = \frac{0.078 V_{av}^{1/2}(0)\exp(13\alpha_0^2/144)}{T_0 G W_0},$$

The numbers W_0 and α_0 stand for the corresponding initial values.

Since I_k depends on a and V_*, we should complement (21.61) with the correlations that follow from (21.56) and allow us to express these parameters through the moments of the distribution

$$\frac{\alpha}{\alpha_0} = \left\{1 + \frac{2}{\alpha_0^2}\ln\left(\frac{\bar{m}_2\bar{m}_0}{\bar{m}_1^2}\right)\right\}^{1/2}, \qquad \bar{V}_* = \frac{\bar{m}_1^2}{\bar{m}_2^{1/2}\bar{m}_0^{3/2}}\bar{V}_*(0), \tag{21.62}$$

Thus, by solving Eqs. (21.61) and (21.62), we can obtain dimensionless moments $\bar{m}_0$, $\bar{m}_1$ and $\bar{m}_2$, which enables us to trace evolution of the volume distribution of drops and the change of the main characteristics of the distribution, such as the average volume of drops V_{av}, the volume concentration of drops W, and the distribution variance σ^2:

$$\frac{V_{av}}{V_{av}(0)} = \frac{\bar{m}_1}{\bar{m}_0}, \quad \frac{W}{W_0} = \bar{m}_1, \quad \frac{\sigma_0}{\sigma^2(0)} = \left\{\frac{\exp(\alpha_0^2/2)\bar{m}_2\bar{m}_0 - \bar{m}_1^2}{\exp(\alpha_0^2/2) - 1}\right\}.$$

There are some special cases, where equations (21.61) can be simplified:

1. Coagulation

Suppose the size distribution of drops changes only due to coagulation. Then, putting into (21.61) $\varphi = \psi = 0$, we get:

$$\frac{d\bar{m}_0}{d\tau_1} = -\frac{\bar{m}_0^{95/72}}{\bar{m}_2^{13/72}}, \quad \bar{m}_1 = 1, \quad \frac{d\bar{m}_2}{d\tau_1} = 2e^{-\alpha_0^2/4}\frac{\bar{m}_2^{23/72}}{\bar{m}_0^{13/72}}, \tag{21.63}$$
$$\bar{m}_0(0) = \bar{m}_2(0) = 1.$$

This leads to

$$\bar{m}_2 = [(1-h)\sqrt{\bar{m}_0} + h]^2/\bar{m}_0, \quad h = 2\exp(-\alpha_0^2/4). \tag{21.64}$$

The parameter h appearing in (21.64) is much smaller than 1, therefore the second moment $\bar{m}_2 \approx 1$ and

$$\begin{aligned}
&\bar{m}_0 \approx (1+0.32\tau_1)^{-3.13}, \quad \frac{V_{av}}{V_{av}(0)} \approx (1+0.32\tau_1)^{3.13}, \\
&\frac{\alpha}{\alpha_0} \approx 1 - \frac{6.16}{\alpha_0^2} n(1+0.32\tau_1), \quad \frac{V_*}{V_*(0)} \approx (1+0.32\tau_1)^{4.69}.
\end{aligned} \tag{21.65}$$

Taking the same parameter values as in the example considered earlier and assuming $\alpha_0 = 3.75$, we obtain $t = 12\tau_1$. We then conclude from (21.65) that the average size of drops will double during the time $t \sim 35$ s.

2. Coagulation and breakup

Suppose the drop distribution evolves in time only due to the coagulation and breakup. Then the volume concentration of drops W or the first moment of the distribution m_1 remains constant, and equations (21.63) become

$$\begin{aligned}
&\frac{d\bar{m}_0}{d\tau_1} = -\frac{\bar{m}_0^{95/72}}{\bar{m}_2^{13/72}} + \psi I_0 \bar{m}_0, \quad \bar{m}_1 = 1, \\
&\frac{d\bar{m}_2}{d\tau_1} = 2e^{-\alpha_0^2/4}\frac{\bar{m}_2^{23/72}}{\bar{m}_0^{13/72}} - \frac{\psi I_2}{2}\bar{V}_{\min}^2 e^{-\alpha_0^2/2}\bar{m}_0.
\end{aligned} \tag{21.66}$$

From these equations, it follows that the moments and parameters of the established distribution are:

$$\begin{aligned}
&\bar{m}_0 = (\psi^{9/2} I_0^{9/2} L^{13/8})^{-1}, \quad \bar{m}_2 = L^2 \bar{m}_0^3, \quad L = I_2 \bar{V}_{\min}^2 / 4I_0 \exp(\alpha_0^2/4), \\
&\frac{V_*}{V_*(0)} = 4.6 \cdot 10^{-3} \exp\left(\frac{31\alpha_0^2}{32}\right) \psi^{27/2} \bar{V}_{\min}^{31/4} I_0^{77/8} L_2^{31/8}, \\
&\frac{\alpha}{\alpha_0} = \left\{1 - \frac{36}{\alpha_0^2}\ln(\psi I_0 L^{1/4})\right\}^{1/2},
\end{aligned} \tag{21.67}$$

The established radius of drops is:

$$\begin{aligned}
R_{av}^{eq} &= \bar{V}_{av}^{1/3} \exp\left[-\frac{(\alpha^2 - \alpha_0^2)}{18}\right] \\
&= 0.236\psi^{7/2} \bar{V}_{\min}^{25/12} I_0^{59/24} L_2^{25/24} \exp\left(-\frac{25\alpha_0^2}{96}\right).
\end{aligned} \tag{21.68}$$

Take the following numerical example. At $d = 0.5$ m, $\mu_G = 10^{-5}$ Pa·s, $\rho_G = 50$ kg/m^3, $\rho_L = 10^3$ kg/m^3, $u = 10$ m/s, $\Sigma = 10^{-2}$ N/m, $W_0 = 5 \cdot 10^{-5}$ m^3/m^3, we get $\alpha = 3$, $\bar{m}_0 = 1.6 \cdot 10^{-3}$, and $R_c^{eq} = 3.4 \cdot 10^{-5}$ m. This value is in good agreement with the results of the experiment [3], which gives for the chosen values of parameters a stable radius value of $R_c^{eq} = 3.6 \cdot 10^{-5}$ m.

3. Deposition on the wall

Taking into account only the sedimentation of drops on the pipe wall enables us to derive an exact solution of the kinetic equation. In our case this equation reduces to

$$\frac{\partial n}{\partial t} = \left(\frac{6}{\pi}\right)^{2/3} \frac{\pi g \rho_L}{36\, d\mu_G} V^{2/3} n, \quad n(V, 0) = n_d(V). \tag{21.69}$$

from which it follows that

$$n = \exp(-\varphi \bar{V}^{2/3} \tau_1) n_d(V). \tag{21.70}$$

The moments of the resultant distribution are equal to:

$$\begin{aligned} m_0 &= N = N(0) \frac{1}{\alpha\sqrt{\pi}} \int_{-\infty}^{\infty} \exp\left\{ -\frac{z^2}{\alpha^2} - h_1 \exp\left(\frac{2z}{3}\right) \right\} dz, \\ m_1 &= N(0) V_*^0 \frac{1}{\alpha\sqrt{\pi}} \int_{-\infty}^{\infty} \exp\left\{ -\frac{z^2}{\alpha^2} - h_1 \exp\left(\frac{2z}{3}\right) + z \right\} dz, \\ m_2 &= N(0) (V_*^0)^2 \frac{1}{\alpha\sqrt{\pi}} \int_{-\infty}^{\infty} \exp\left\{ \frac{z^2}{\alpha^2} - h_1 \exp\left(\frac{2z}{3}\right) + 2z \right\} dz, \end{aligned} \tag{21.70}$$

where $h_1 = \varphi\tau \left(\dfrac{V_*(0)}{V_c(0)}\right)^{2/3}$.

References

1 Kolmogorov A. N., On logarithmic normal law of particle distribution during breakup, Papers Acad. Sci. USSR, 1941, Vol. 31, No. 2, p. 99–101 (in Russian).

2 Levich V. G., Physicochemical hydrodynamics, Prentice-Hall, Englwood Cliffs, N.J., 1962.

3 Guseinov Ch. S., Asaturjan A. S., Determination of drop modal size in two-phase flow, J. Appl. Cemistry, 1977, Vol. 50, No. 4, p. 848–853.

4 Amelin A. G., Theoretical bases of fog formation during vapor condensation Chemistry, Moscow, 1966 (in Russian).

5 Abramovich G. N., Applied Gas Dynamics, Nauka, Moscow, 1976 (in Russian).

6 Loginov V. I., Dewatering and desalting of oils, Chemistry, Moscow, 1979 (in Russian).

7 Deryaguin B. V., Smirnov L. P., On inertialless deposition of particles

from a fluid flow on a sphere due to the action of Van der Waals attraction forces, Studies in surface forces, Nauka, Moscow, 1962 (in Russian).
8 Gradstein I. O., Ryszik I. M., Tables of integrals, sums, series and products, Nauka, Moscow, 1971.
9 Ried R. C., Prausnitz J. M., Sherwood T. K., The properties of gases and liquids, McGraw-Hill, New York, 1977.
10 Gurevich G. R., Brusilovskiy A. I., Reference textbook on calculations of phase state and properties of gas-condensate mixtures, Nedra, Moscow, 1984 (in Russian).
11 Katz D. L., Cornell D., Kobayashi R., et al., Handbook on natural gas engineering, McGraw-Hill, New York, 1965.
12 Prigogine L., Defay R., Chemical thermodynamics, Longmans Green and Co., London, New York, Toronto, 1954.
13 Fridrichsberg D. A., A course in colloid chemistry, Chemistry, Leningrad, 1974, (in Russian).
14 Hiemenz P. O., Principles of colloid and surface chemistry, Marcel Dekker, New York, 1986, 815 p.
15 Dussan V. E. B., On the spreading of liquids on solid surfaces: static and dynamic contact lines, Ann. Rev. Fluid Mech., 1979, Vol. 11, p. 371–400.
16 Ruschak K. J., Coating flows, Ann. Rev. Fluid Mech., 1985, Vol. 10, p. 65–89.
17 Wilson S. D. R., The drag-out problem in film coating, J. Eng. Math., 1982, Vol. 16, p. 209–221.
18 Taylor G. I., Cavitation of a viscous fluid in narrow passages, J. Fluid Mech., 1963, Vol. 16, p. 595–619.
19 Bretherton K. J., The motion of long bubbles in tubes, J. Fluid Mech., 1961, Vol. 10, p. 166–188.
20 Probstein R. F., Physicochemlcal hydrodynamics, Butterworths, 1989.
21 Bogy D. B., Drop Formation in circular Liquid Jet, Ann. Rev. Fluid Mech., 1979, Vol. 11, p. 207–228.
22 Rayleigh D., The theory of sound (in 2 vols.), Dover, New York, 1945.
23 Lin C. C., The theory of hydrodynamic stability, Cambridge Univ. Press, Cambridge, 1955.
24 Drazin P. G., Reid W. H., Hydrodynamic stability, Cambridge Univ. Press, Cambridge, 1981.
25 Slattery J. C., Interfacial transport phenomena, Springer-Verlag, New-York, 1990.
26 Boulton-Stone J. M., The effect of surfactant on bursting gas bubbles, J. Fluid Mech., 1995, Vol. 302, p. 231–257.
27 Yih C.-S., Fluid motion induced by surface tension variation, Phys. Fluids, 1983, Vol. 11, p. 477–480.
28 Pimputkar S. M., Ostrach S., Transient thermocapillary flow in thin liquid layers, Phys. Fluids., 1980, Vol. 23, p. 1281–1285.
29 Ratulowski J., Chang H. C., Marangoni effects of trace impurities on the motion of long gas bubbles in capillaries, J. Fluid Mech., 1990, Vol. 210, p. 303–328.
30 Stebe K. S., Lin S. Y., Maldarelly C., Remobilization surfactant-retarded particle interfaces. I. Stress-free conditions at the interfaces of micellar solutions of surfactants with fast sorption kinetics, Phys. Fluids A, 1991, Vol. 3 (1), p. 3–20.
31 Stebe K. S., Bartes-Biesel D., Marangoni effects of adsorption-desorption controlled surfactants on the leading end of an infinitely long bubble in capillary, J. Fluid Mech., 1995, Vol. 286, p. 25–48.
32 Chen J. D., Measuring the film thickness surrounding a bubble inside a capillary, J. Colloid Interface Sci., 1985, Vol. 109, p. 341–349.
33 Schwartz L. W., Princen H. M., Kiss A. D., On the motion of bubbles in capillary tubes, J. Fluld Mech., 1986, Vol. 172, p. 259–275.
34 Marchessault R. F., Mason S. G., Flow of entrapped bubbles through a capillary, Ind. Engng. Chem., 1960, Vol. 52 (1), p. 79–81.
35 Bartes-Biesel D., Moulai-Mostefa V., Meister E., Effect of surfactant on the flow of large gas bubbles in capillary tubes, Proc. Physico-chem. Hydrod.

Nato Conf. (ed. M. Verlarde), La Rabida, Spain.

36 Hirasaki G., Lawson J. B., Mechanism of foam flow in porous media: apparent Viscosity in smooth Capillaries, Soc. Petr. Engng., 1986, Vol. 25, p. 176–190.

37 Gintey G. M., Radke C. J., The influence of soluble surfactants on the flow of long bubbles through a cylindrical capillary, ACS Symp. Series., 1989, Vol. 396, p. 480–501.

38 Herbolzheimer E., The effect of surfactant on the motion of a bubble in a capillary, AIChE Annual Meetings, 1987, 15–20 November, New York, Paper 68.

39 Chang H. C., Ratulowski J., Bubble transport in a capillary, AICHE Annual Meetings, 1987, 15–20 November, New York, Paper 681.

40 Park C. W., Effects of insoluble surfactants on dip coating, J. Colloid Interface Sci., 1991, Vol. 146, p. 382–394.

41 Borhan A., Mao C. F., Effects of surfactants on the motion of drops through circular tubes, Phys. Fluids A, 1992, Vol. 4 (12), p. 2628–2640.

42 Park C. W., Influence of a soluble surfactants on the motion of a finite bubble in a capillary tube, Phys. Fluids A, 1992, Vol. 4 (11), p. 2335–2346.

43 Jensen O. E., Grotberg J. B., Insoluble surfactant spreading on a thin viscous film: shock evolution and film rapture, J. Fluid Mech., 1992, Vol. 240, p. 259–288.

44 Gaver D. P., Grotberg J. B., The dynamics of a localized surfactant on a thin film, J. Fluid Mech., 1990, Vol. 213, p. 127–148.

45 Landau L. D., Lifshiz E. M., Theoretical physics. V. 6. Hydrodynamics, Nauka, Moscow, 1988.

46 Lefebvre A. H., Atomization and sprays. Hemisphere, 1989.

47 Ohnesorge W., Formation of drops by nozzles and the breakup of liquid jets, Z. Angew Math. Mech., 1936, Vol. 16, p. 335–358.

48 Hinze J. O., Fundamentals of the hydrodynamic mechanics of splitting in dispersion processes, AICHE J., 1955, Vol. 1, No. 3, p. 289–295.

49 Paloposki T., Drop size distribution in liquid sprays, Acta Polltechnica Scandinavica, Mech. Eng. Ser., 1994, No. 114, p. 299–209.

50 Kolmogorov A. N., On breakup of droplets in a turbulent flow, Papers Acad. Sci., 1949, Vol. 66, No. 5, p. 825–829 (in Russian).

51 Troesch H., Atomization of liquid, Chemie–Ingeniur–Technik, 1954, B. 26, N 6, S. 311–320.

52 Loginov V. I., Dynamics of fluid breakage in a turbulent flow, Appl. Mech. Techn. Phys., 1985, No. 4, p. 66–73 (in Russian).

53 Sinaiski E. G., Michaleva G. V., Evolution of hydrate inhibitor droplet spectrum in a turbulent flow of natural gas, J. Appl. Chem., 1993, Vol. 66, No. 3, p. 544–555 (in Russian).

54 Kasatkin A. G., Basic processes and apparatuses in chemical technology, Chemistry, Moscow, 1971 (in Russian).

55 Trivus N. A., Tichonenko N. V., Fedorin E. P., Operating efficiency improvement for low-temperature separators, Gazovoe delo. 1972, No. 1 p. 26–29 (in Russian).

56 Shukin V. K., Halatov A. A., Heat exchange, mass exchange, and hydrodynamics of swirled flows in axially-symmetric channels, Mechanical Engineering, Moscow, 1982 (in Russian).

57 Gupta A. K., Liley D. B., Syred N., Swirl flows, Abacus Press, 1984.

58 Fuks N. A., Mechanics of aerosols, Acad. Sci. USSR, Moscow, 1955 (in Russian).

59 Schlichting H., Grenzschicht–Theorie, G. Braun Verlag, Karlsruhe, 1964.

60 Zhdanova N. V., Haliph A. L., Dehydration of natural gas. Nedra, Moscow, 1975 (in Russian).

61 Guchman L. I., Preparation of gas for long-distance transportation at northern fields, Nedra, Leningrad, 1980 (in Russian).

62 Sun A. M., Gas dehydration by DEG in high-speed, direct-flow spray

absorbers, Gas Industry, 1976, No. 10, p. 56–58 (in Russian).
63 Bik S. S., Makagon Yu. F., Fomina V. I., Gas hydrates, Chemistry, Moscow, 1980 (in Russian).
64 Buchgalter E. B., Methanol and its usage in gas industry, Nedra, Moscow, 1986 (in Russian).
65 Istomin V. A., Jakushev V. S., Gas hydrates in natural conditions, Nedra, Moscow, 1992 (in Russian).
66 Golovkov L. G., Distribution of droplets over size during atomization of liquid by centrifugal atomizers, Eng.-Phys. J., 1964, Vol. 7, No. 11, p. 55–61.
67 Sherman P. (Ed.), Emulsion Science, Academic Press, London, New York, 1968.
68 Jiang Y. J., Umemura A., Law O. K., An experimental investigation on the collision behavior of hydrocarbon droplets, J. Fluid Mech., 1992, Vol. 234, p. 171–190.
69 Mednikov E. P., Turbulent transport and aerosol deposition, Nauka, Moscow, 1981 (in Russian).

VII
Liquid–Gas Mixtures

We shall apply the term "liquid-gas mixtures" to the mixtures in which the liquid forms the continuous phase, while gas forms the disperse phase. An example is the oil-gas mixture that forms as oil moves inside boreholes, elements of field equipment and oil pipelines.

Crude oil contains plenty of low-molecular compounds dissolved in it at high reservoir pressure. Among these compounds are saturated hydrocarbons of paraffin family (methane, ethane, propane etc.), and natural gases (carbon-dioxide, nitrogen, hydrogen sulphide etc). As oil travels through bores and pipelines, thermobaric conditions change (pressure and temperature fall), resulting in violation of phase equilibrium and emission of lightweight components from oil. As a result, liquid-gas or gas-liquid mixture is formed when oil is still flowing through boreholes. Various types of the flow are possible, depending on the ratio between gas and liquid volume concentrations: bubble flow, plug flow, slug flow.

Oil gases at various fields differ by their fractional composition, but, as a rule, the predominant component is methane (from 30 up to 95%). At normal conditions, quantitative ratios between phases of hydrocarbon systems are represented by the quantity called the gas factor and having dimensionality m^3/m^3 or m^3/ton. The gas factor of hydrocarbon systems can vary over a wide range – from low values, measured in ones and tens, to high values, measured in hundreds and thousands m^3 of gas per 1 ton or 1 m^3 of oil. It is obvious that for low gas factors, the mixture is a liquid-gas one, and for high gas factors, we have a gas-liquid mixture. Therefore, in the former case we talk about oil-gas systems, and in the latter case – about gas-oil or gas-condensate systems.

Each oil field is characterized by its own gas factor. Fields with predominantly light oil are characterized by high gas factors. The gas factor can also be high in the case when in the process of field exploitation, gas breaks from the gas cap of the stratum or from gas – bearing horizons into productive wells. There are also fields with very low gas factor.

Oil coming into transporting pipelines has a low gas factor, because most of the oil gas gets separated from oil in separators at the field preparation plant. Separation of oil from gas is an important technological process that is realized in several stages. The first stage accomplishes preliminary separation of the free gas that forms when oil flows through wells and systems of oil gathering. The further

Separation of Multiphase, Multicomponent Systems. E. G. Sinaiski and E. J. Lapiga

ISBN: 978-3-527-40612-8

separation is performed in separators arranged in series. In each consecutive separator the pressure is lowered as compared to its predecessor. Such step-by-step separation provides extraction of certain groups of light hydrocarbons from oil and prevents an instantaneous release of large quantities of gas. Instantaneous separation of gas in large amounts could cause foaming of oil or disrupt the normal operation of equipment.

The phase state of oil-gas systems in equilibrium conditions is determined by the equations of liquid-vapor equilibrium on the basis of the united equation of state (see Section 5.7). Most often, the Peng-Robinson equation of state is used.

There are two ways to perform degassing of oil during its preparation. The first way is differential separation, which involves gradual multistage degassing. At each stage, the pressure decreases and the extracted gas is removed from the system. As a result, the liquid phase is enriched by components with high boiling temperature, as a part of light hydrocarbons gets removed together with the gas. The second way is contact separation, where the pressure is reduced at once, and the extracted gas is not removed from the system, so the total composition of the mixture does not change during this process. The total quantity of extracted gas is smaller for differential separation than for contact separation, because step-by-step reduction of pressure is accompanied by the removal of light hydrocarbons, leading to a higher concentration of heavy hydrocarbons in the remaining oil, which, in turn, reduces the quantity of extracted gas at the further decrease of pressure. In spite of this drawback, differential separation possesses one advantage in comparison with contact separation, namely, that oil is more fully cleansed of lightweight gases. In real practice, of course, separation is never purely differential or contact. The real process can only be close to one or the other method.

The existing methods of calculation of separation processes are based on the assumption that a phase equilibrium exists in the system. Of great practical interest is the dynamics of the process, in particular the determination of characteristic relaxation times for the liquid-gas system. Accounting for the non-equilibrium effects is important in research of separation processes, because the gaseous dispersive phase formed in the oil is not immediately separated from the liquid – it takes some time for gas bubbles to reach the surface. During their motion in the liquid, gas bubbles change their size due to the processes of mass exchange and coagulation. Bubbles that fail to reach the surface during the mixture's residence time in the separator are ablated from the device, thereby diminishing the efficiency of separation.

Thus, the problems arising in liquid-gas systems analysis are analogous to those examined in Chapter VI for gas-liquid systems.

22
Dynamics of Gas Bubbles in a Multi-Component Liquid

If a bubble of radius R_0 was initially placed into a supersaturated (at the given pressure and temperature) mixture, then, due to the concentration difference $\Delta\rho = \rho_{L\infty} - \rho_{Lw}$ between the gas dissolved in the liquid far away from the bubble and the appropriate equilibrium value at the bubble surface, there arises a directional diffusion flux of the dissolved substance toward the surface. At the interface, the transition of substance from the liquid to the gaseous state takes place. The result is the increase in the bubble volume. The growth of the bubble, in its turn, results in the increase of its lift velocity, as well as the increase of convective diffusion flux. The statement of the problem and the basic dynamic equations for a bubble in a solution were described in Section 6.8.

Consider a multicomponent solution. The distribution of concentrations of the dissolved substances is described by the equations of convective diffusion

$$\frac{\partial \rho_{iL}}{\partial t} + u_r \frac{\partial \rho_{iL}}{\partial r} + u_\theta \frac{1}{r}\frac{\partial \rho_{iL}}{\partial \theta} = D_{iL}\frac{1}{r^2}\frac{\partial}{\partial r} r^2 \frac{\partial \rho_{iL}}{\partial r}, \tag{22.1}$$

where u_r and u_θ are the radial and tangential velocity components of the flow that goes around the bubble, r and θ are spherical coordinates, ρ_{iL} are the mass concentrations of dissolved components, D_{iL} is the coefficient of binary diffusion.

The mass balance equation for the gas in the bubble is

$$\frac{d}{dt}\left(\frac{4}{3}\pi R^3 \rho_{iG}\right) = I_{iD} = 4\pi R^2 D_{iL}\left(\frac{\partial \rho_{iL}}{\partial r}\right)_{r=R}, \tag{22.2}$$

where ρ_{iG} is the mass concentration of i-th gas component in the bubble.

The dynamic balance equation for the bubble, known as Rayleigh's equation, looks like

$$\rho_L\left(R\ddot{R} + \frac{3}{2}(\dot{R})^2 + 4\nu_L \frac{1}{R}\dot{R}\right) = p_G - p_\infty - \frac{2\Sigma}{R}. \tag{22.3}$$

Here ρ_L is the liquid phase density; R – the radius of the bubble; ν_L – kinematic viscosity of the liquid; Σ – the coefficient of surface tension, p_G – gas pressure inside the bubble; p_∞ – gas pressure far away from the bubble.

Separation of Multiphase, Multicomponent Systems. E. G. Sinaiski and E. J. Lapiga

ISBN: 978-3-527-40612-8

Suppose that the gas inside the bubble represents an ideal gas mixture. Then

$$p_G = \sum_i p_i, \quad p_i = \frac{AT\rho_{iG}}{M_i}, \tag{22.4}$$

where p is the partial pressure of i-th component in the gas; A is the gas constant; M_i is the molecular mass of i-th component; T is temperature of both the gas and the solution, taken to be constant.

Let us restrict ourselves to the consideration of an infinitely diluted solution. Then, on the basis of Henry's law, we have at the interface boundary the following connection between the concentration ρ_{iwL} of i-th component in the solution and the partial pressure of the same component in the bubble:

$$p_{iG} = B_i \frac{\rho_{iwL}/M_i}{\rho_s/M_s + \sum_i \rho_{jwL}/M_j}, \tag{22.5}$$

Here B_i is Henry's constant; ρ_s, M_s are the density and molecular mass of the solvent.

In addition to Eqs. (22.1) to (22.5), we should take the following initial and boundary conditions:

$$\rho_{iL}(r,0) = \rho_{i\infty L}, \quad \rho_{iL}(\infty,t) = \rho_{i\infty L}, \quad \rho_{iL}(R,t) = \rho_{iwL},$$
$$R(0) = R_0, \quad \dot{R}(0) = 0. \tag{22.6}$$

Eqs. (22.1)–(22.5) with conditions (22.6) describe the dynamics of the bubble in a solution under the condition that the velocity field of the flow that goes around the bubble is given, and that the bubble itself is not subject to deformation.

In view of the nonlinearity of the problem, its solution cannot be presented as a superposition of two solutions: the solution for the growth of a non-lifting bubble, and for the lift of a non-growing bubble. Nevertheless, there are two possible limiting cases when such separation of the problem is appropriate. This is the case of small-sized bubbles, where the lift can be neglected, and the case of large bubbles, whose further growth can be ignored. Under some additional conditions that will be elucidated below, in the first case, the diffusion flux at the bubble surface can be derived from the solution of the problem of diffusion growth of a non-lifting bubble, and in the second case – from the solution of the problem of convective diffusion of a non-growing, lifting bubble.

22.1
Motion of a Non-Growing Bubble in a Binary Solution

The convective diffusion on a spherical particle placed in an infinitely diluted binary solution is has been studied extensively for the case $\mathrm{Re} \leq 1$ [1]. Generally speaking, the motion of a gas bubble differs from that of a solid particle. The basic difference is that the bubble surface is free or, using conventional terminology, unretarded. Therefore the hydrodynamic resistance experienced by a lifting bubble is less than the resistance acting on a solid particle. However, for the bubble moving in a real liquid, its surface is stabilized by impurities which are present in the liquid, including surface active substances. As a result, the mobility of the bubble surface reduces, sometimes all the way to zero, so the bubbles move in the liquid as solid particles. This fact is adduced in the review [2] devoted to the determination of velocity of lifting bubbles.

Restrict ourselves to the problem of a lifting bubble at $\mathrm{Re} \ll 1$. Then the flow moves around the bubble in a viscous regime and the distribution of velocities can be found by solving the problem using Stokesean approximation. For an infinitely diluted solution, the Peclet's diffusion number is Đå$_D \gg 1$ and as the bubble moves, a boundary layer is formed on its surface. It is in this layer that the main change of concentration of the diffusing component occurs. A non-growing bubble is lifting with a constant velocity, so the concentration distribution of a substance dissolved in the liquid is described by the stationary equation of convective diffusion. The solution of the corresponding diffusion problem for a solid particle and for a bubble with non-retarded surface at $\mathrm{Re} \ll 1$ gives the following expressions for the diffusion flux on the particle:

– for a solid particle:

$$I = 7.98(\rho_{1\infty L} - \rho_{1wL}) D_{1L}^{2/3} U^{1/2} R^{4/3}, \tag{22.7}$$

– for a bubble:

$$I = 8\left(\frac{\pi D_{1L} U}{6R}\right)^{1/2} R^2 (\rho_{1\infty L} - \rho_{1wL}). \tag{22.8}$$

Here D_{1L} is the diffusion coefficient of the solute indicated by the index 1; $U = \alpha R^2$ is the lifting (or sedimentation) velocity of the particle (for a solid particle, $\alpha = 2\Delta\rho g/9\mu_L$, while for a bubble with free surface, $\alpha = 2\Delta\rho g/3\mu_L$, where $\Delta\rho = \rho_L - \rho_G$).

Suppose the bubble contains only the gas of component 1. Then the expression for diffusion flux can be used to estimate the increase of the bubble size under the condition that the bubble grows sufficiently slowly. Let the bubble move as a solid particle. Substituting the expression (22.7) for the flux into Eq. (22.2), we will get an equation that tells us how fast the radius of the bubble with completely retarded surface changes with time:

$$\frac{d}{dt}\left(\frac{4}{3}\pi R^3 \rho_{1G}\right) = 7.98(\rho_{1\infty L} - \rho_{1wL}) D_{1L}^{2/3} U^{1/2} R^{4/3}. \tag{22.9}$$

In addition to this equation, we shall use Eq. (22.3), in which we neglect all inertial and viscous terms (this is permissible for relative large bubbles):

$$p_G = p_\infty + \frac{2\Sigma}{R}, \tag{22.10}$$

The equation of state for gas in the bubble is

$$p_{1G} = \frac{AT\rho_{1G}}{M_1}, \tag{22.11}$$

and it follows from Henry's law that

$$p_{1G} = B_1 \frac{\rho_{1wL}/M_1}{\rho_s/M_s + \rho_{1wL}/M_1}. \tag{22.12}$$

Consider the lifting of a bubble with the initial radius R_0 in a solution layer of thickness H. The x-axis follows the direction of gravity, its origin fixed on the layer surface. Introduce the following dimensionless variables:

$$\bar{R} = \frac{R}{R_0}, \quad y = \frac{6D_{1L}^{2/3}\rho_s AT}{\pi R_0^3 \alpha^{2/3} B_1 M_s} x, \quad \varphi = \frac{2\Sigma}{R_0 p_\infty}, \quad C_\infty = \frac{\rho_{1\infty L} M_s B}{\rho_s M_1 p_\infty},$$

$$C_{eq} = \frac{1}{1 - p_\infty/B_1}, \quad a = \frac{\varphi}{C_\infty - C_{eq}}.$$

In these new variables, the solution of Eqs. (22.9)–(22.12) is

$$\bar{R}^3 - 1 + 0.5(2\varphi + 3a)(\bar{R}^2 - 1) + a(2\varphi + 3a)(\bar{R} - 1) + a^2(2\varphi + 3a)\ln\left(\frac{\bar{R} - a}{1 - a}\right) = (C_\infty - C_{eq})(y_0 - y), \tag{22.13}$$

where y_0 is the dimensionless depth at which the bubble was located at the initial moment.

In the particular case of $\varphi \ll 1$, that is, when capillary pressure can be neglected, the solution (22.13) simplifies:

$$\bar{R} \approx (1 + (C_\infty - C_{eq})(y_0 - y))^{1/3}, \tag{22.14}$$

The dimensionless time τ during which the radius of the bubble reaches the value R is defined by the formula

$$\tau = 3\frac{\bar{R} - 1}{C_\infty - C_{eq}} + (2\varphi + 3a)\frac{\ln((\bar{R} - a)/(1 - a))}{C_\infty - C_{eq}}, \tag{22.15}$$

where $\tau = 6\alpha^{1/3} D_{1L} ATt/\pi R_0 B_1 M_1$.

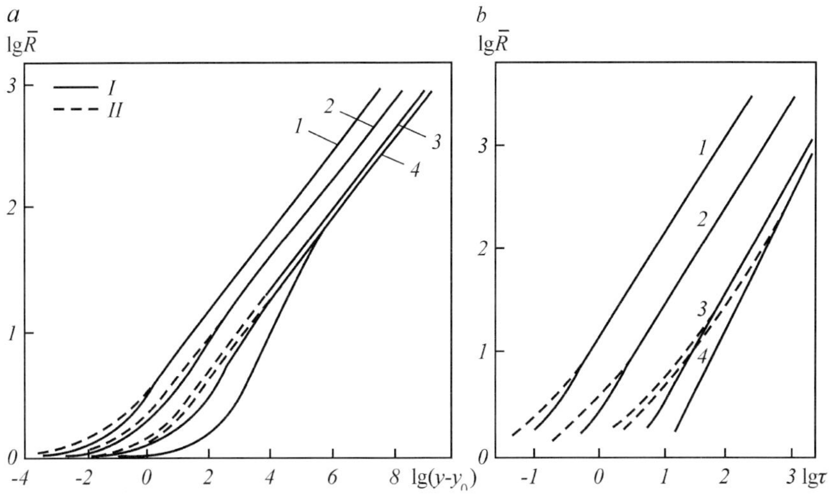

Fig. 22.1 lg $\bar{R}$ vs. lg$(y_0 - y)$ (*a*) and lg $\bar{R}$ vs. lg τ (*b*) for different values of ΔC: *1* – 50; *2* – 10; *3* – 1.5; *4* – 1.05; *I* – exact solution; *II* – without capillary pressure.

At $\varphi \ll 1$, (22.15) reduces to

$$\tau \approx 3\frac{\bar{R} - 1}{C_\infty - C_{eq}}. \tag{22.16}$$

Fig. 22.1 shows the dependence of the dimensionless radius of the bubble $\bar{R}$ on the depth $y_0 - y$ at which the bubble was located at the moment of time τ, and on τ, for various values of mixture oversaturation $\Delta C = C_\infty - C_{eq}$ at $\varphi = 1$. Continuous lines indicate the dependencies found using formulas (22.13) and (22.15), and dotted lines – the dependencies found using formulas (22.14) and (22.16).

From the adduced dependencies, it can be seen that at small values of $\Delta C - \varphi$, the approximate dependencies (22.14) to (22.16) can lead to exaggerated values for the bubble radius. With the increase of $\Delta C - \varphi$, digressions from the true value of $\bar{R}$ shift to the region of smaller values of $y_0 - y$. Thus, at $\Delta C \gg \varphi$ the capillary pressure has a noticeable effect only at the initial stage of bubble growth. At $y_o \geq 10^4$, the radius $\bar{R}_H$ of the bubble that achieves the layer surface $(y = 0)$ is determined to a sufficient accuracy by the formula

$$\bar{R}_H \approx \left(1 + (C_\infty - C_{eq})\,y_H\right)^{1/3}. \tag{22.17}$$

When the inequality $\Delta C y_H \gg 1$ is satisfied, which is nearly always the case, we have

$$\bar{R}_H \approx ((C_\infty - C_{eq})\gamma_H)^{1/3}. \tag{22.18}$$

Here γ_H is the dimensionless height of the liquid layer.

Since $\bar{R}_H = R_H/R_0$ and $\gamma_H \sim R_0^{-3}$, it follows from (22.18) that the size of a bubble that has passed through a layer of height $H \gg R_0$ for all practical purposes does not depend on the initial radius.

Summarizing, we should note that the obtained results are only true in the limiting case when a change in the bubble size has a minor influence on its lift velocity. Later on, it will be shown that this case is possible when the bubble radius surpasses some critical value.

22.2
Diffusion Growth of a Motionless Bubble in a Binary Solution

Consider now the diffusion growth of an isolated motionless bubble in a binary solution. Denote by ρ_{1L} the mass concentration of the dissolved component in the liquid. Suppose the bubble consists only of component 1, the process is spherically symmetrical, and the distribution ρ_{1L} is described by the equation of convective diffusion (22.1), in which should be should take $u_\theta = 0$ and $u_r = (R/r)^2\dot{R}$ under the condition $\rho_{1G} \ll \rho_L$:

$$\frac{\partial \rho_{1L}}{\partial t} + \left(\frac{R}{r}\right)^2 \dot{R}\frac{\partial \rho_{1L}}{\partial r} = D_{1L}\frac{1}{r^2}\frac{\partial}{\partial r}r^2\frac{\partial \rho_{iL}}{\partial r}. \tag{22.19}$$

The balance of mass (22.2) for the bubble reduces to the equation

$$\frac{d}{dt}\left(\frac{4}{3}\pi R^3 \rho_{1G}\right) = I_{1D} = 4\pi R^2 D_{1L}\left(\frac{\partial \rho_{1L}}{\partial r}\right)_{r=R}. \tag{22.20}$$

Neglect all viscous and inertial terms in Eq. (22.3). Then this equation will become

$$p_G = p_\infty + \frac{2\Sigma}{R}. \tag{22.21}$$

In addition to these equations, we should consider the equation of state for the gas in the bubble,

$$p_G = p_{1G} = \frac{AT\rho_{1G}}{M_1}, \tag{22.22}$$

Henry's law,

$$p_{1G} = B_1\frac{\rho_{1wL}/M_1}{\rho_s/M_s + \rho_{1wL}/M_1}, \tag{22.23}$$

and the initial and boundary conditions,

$$\rho_{1L}(r,0)=\rho_{1\infty L},\quad \rho_{1L}(\infty,t)=\rho_{1\infty L},\quad \rho_{1L}(R,t)=\rho_{1wL},$$
$$R(0)=R_0,\quad \dot{R}(0)=0. \tag{22.24}$$

The existence of a thin diffusion boundary layer near the bubble surface allows us to find an approximate solution of the formulated problem. Let us use the method of integral relations, which boils down to selecting a diffusion layer of thickness $\delta \ll R$ in the liquid around the bubble, with the assumption that the change of concentration of the dissolved component from $\rho_{1\infty L}$ up to ρ_{1wL} occurs in this layer. Then following conditions should be satisfied:

$$\rho_{1L}(R+\delta,t)=\rho_{1L\infty},\quad \left(\frac{\partial \rho_{1L}}{\partial r}\right)_{r=R+\delta}=0,\quad \rho_{1L}(R,t)=\rho_{1wL}. \tag{22.25}$$

Conditions (22.25) allow us to seek for the distribution of concentration ρ_{1L} in the form

$$\frac{\rho_{1L}-\rho_{1\infty L}}{\rho_{1wL}-\rho_{1\infty L}}=\begin{cases}(1-x)^2, & \text{at } 0\le x\le 1,\\ 0, & \text{at } x>1,\end{cases} \tag{22.26}$$

where $x=(r-R)/\delta$.

Integrate Eq. (22.19) over r from R up to $R+\delta$. Taking into account expressions (22.26), (22.25), and the condition $\delta/R \ll 1$, we get the following equation, accurate to the terms of the order δ/R:

$$\frac{\delta}{R}=\frac{\rho_{1G}R^3-\rho_{1G0}R_0^3}{R^3(\rho_{1\infty L}-\rho_{1wL})}. \tag{22.27}$$

We can now determine the diffusion flux at the bubble surface:

$$I_{1D}=4\pi R^2 D_{1L}\left(\frac{\partial \rho_{1L}}{\partial r}\right)_{r=R}=8\pi R^4 D_{1L}\frac{(\rho_{1\infty L}-\rho_{1wL})^2}{\rho_{1G}R^3-\rho_{1G0}R_0^3}. \tag{22.28}$$

Substituting the flux into the balance-of-mass equation (22.20) for the bubble, we obtain:

$$\frac{d(\rho_{1G}R^3)}{dt}=6R^4 D_{1L}\frac{(\rho_{1\infty L}-\rho_{1wL})^2}{\rho_{1G}R^3-\rho_{1G0}R_0^3}. \tag{22.29}$$

Now, make a transition to dimensionless parameters:

$$\bar{R}=\frac{R}{R_0},\quad \tau=\left(\frac{\rho_{1\infty L}AT}{M_1 p_\infty R_0}\right)^2 D_{1L}t,\quad a=\frac{\rho_s}{\rho_{1\infty L}},\quad q=\frac{2\Sigma}{R_0 p_\infty},\quad K=\frac{B_1}{p_\infty}.$$

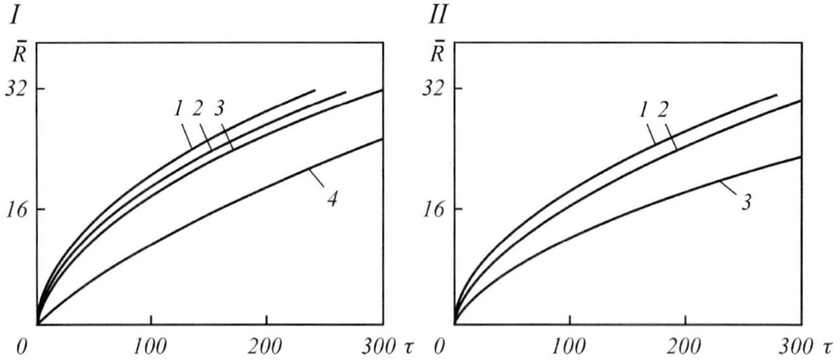

Fig. 22.2 Bubble radius $\bar{R}$ vs. time τ: *I*: $K = 100$; $a = 5$; q: *1* – 0; *2* – 0.5; *3* – 1; *4* – 5; *II*: $q = 1$; $a = 0.5$; – 0; *2* – 0.5; K: *1* – 100; *2* – 10; *3* – 3.

in Eqs. (22.21)–(22.23), (22.27), and (22.29). The equations will transform into

$$\frac{d\bar{R}}{d\tau} = \frac{2\bar{R}_1^3}{(\bar{R}+2q/3)(\bar{R}^3+q\bar{R}^2-1-q)}\left(1-\frac{a(\bar{R}+q)}{(K-1)\bar{R}-q}\right), \quad \bar{R}(0)=1,$$

$$\frac{\delta}{R} = \frac{p_\infty M_1 Z}{AT\rho_{1\infty L}}, \quad Z = \left(1+\frac{q}{\bar{R}}-\frac{1-q}{\bar{R}^3}\right)\left(1-\frac{a(\bar{R}+q)}{(K-1)\bar{R}-q}\right)^{-1}, \tag{22.30}$$

$$\frac{\rho_{1G}}{\rho_{1G0}} = \frac{1+q/\bar{R}}{1+q}.$$

The dependencies $\bar{R}(\tau)$ and $Z(\bar{R})$, obtained by numerical integration of equations (22.30), are shown in Figs. 22.2 and 22.3 for the various values of parameters q and K. At the initial stage, the surface tension, characterized by the parameter q, exerts a noticeable influence on the bubble growth. At first, bubble growth is impeded by surface tension, and the external boundary of the boundary layer is moving away from the bubble surface. Then, as $\bar{R}$ grows, the role of surface tension decreases, the thickness of the boundary layer is stabilized and remains constant in time. At the same time, $Z \to (1+q)^{-1}$ and, as we see from the last equation (22.30), the relative density of gas in the bubble decreases and also remains constant.

Now, compare diffusion fluxes for the two limiting cases considered above. In the case of convective diffusion on a lifting, non-growing bubble with free surface, the diffusion flux equals

$$I = 5.8(\rho_{1\infty L}-\rho_{1wL})\left(\frac{D_{1L}gR^5}{3\nu_L}\right)^{1/2}. \tag{22.31}$$

The diffusion flux on a growing, non-lifting bubble is determined by the expression (22.28). Composing the ratio of these fluxes, we get

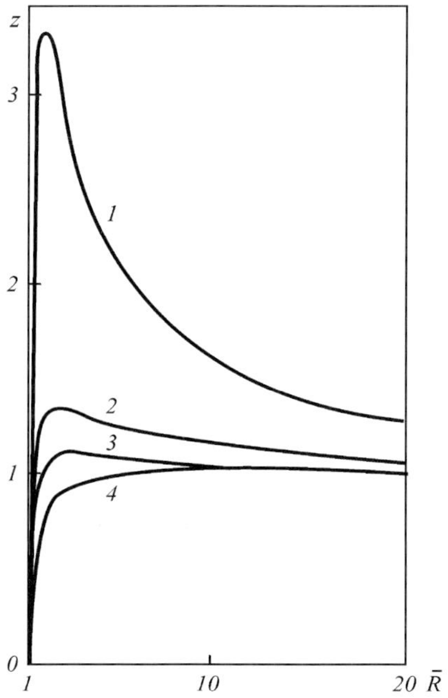

Fig. 22.3 Relative thickness of the diffusion layer z vs. bubble radius $\bar{R}$ for various values of q ($a = 5$; $K = 100$): *1* – 5; *2* – 1; *3* – 0.5; *4* – 0.

$$\frac{I}{I_{1D}} = \frac{6.8}{Z}\left(\frac{\bar{R}_{cr}^{(1)}}{\bar{R}}\right)^{3/2}, \quad \bar{R}_{cr}^{(1)} = \left(\frac{\rho_{1\infty L}^2 A^2 T^2 D_{1L}\nu_L}{M_1^2 p_\infty^2 g R_0^3}\right)^{1/3}. \tag{22.32}$$

Since $Z \sim 1$, it follows from (22.32) that for $\bar{R} \ll 3.16\bar{R}_{cr}^{(1)}$, the diffusion flux is determined by the expression (22.28) for a non-lifting bubble, and for $\bar{R} \gg 3.16\bar{R}_{cr}^{(1)}$ – by the expression (22.31) for a lifting, non-growing bubble.

A similar procedure carried out for a bubble with fully retarded surface produces the following ratio of fluxes:

$$\frac{I}{I_{1D}} = \frac{3.12}{Z}\left(\frac{\bar{R}_{cr}^{(1)}}{\bar{R}}\right)^{3/2}, \quad \bar{R}_{cr}^{(2)} = \left(\frac{\rho_{1\infty L}^3 A^3 T^3 D_{1L}\nu_L}{M_1^3 p_\infty^3 g R_0^3}\right)^{1/3}. \tag{22.33}$$

From this expression, it follows that for $\bar{R}$ satisfying the inequality $\bar{R} \ll 5.12\bar{R}_{cr}^{(2)}$, the diffusion flux is determined by the expression (22.28) for a non-lifting bubble, and for $\bar{R} \gg 5.12\bar{R}_{cr}^{(2)}$ – by the expression

$$I = 7.98(\rho_{1\infty L} - \rho_{1wL})\left(\frac{2D_{1L}^2 g R^6}{9\nu_L}\right)^{1/3}. \tag{22.34}$$

for a lifting, non-growing bubble.

Finally, we want to remind you that our estimations were based on the assumption that diffusion fluxes I_{1D} are equal for the bubbles with free and retarded surface. In reality, surfactants adsorbed at the bubble surface can interfere with the transition of component *1* from the liquid phase into the gaseous phase. This issue will be addressed in Section 22.5.

22.3
The Initial Stage of Bubble Growth in a Multi-Component Solution

Consider the initial stage of growth of a spherical gas bubble with initial radius R_0 in a multicomponent supersaturated solution due to the diffusion of components with low concentration that dissolved in the liquid. Assume that the process is isothermal, and the factors D_{iL}, B, and Z are constants. The bubble is small enough to ignore its lifting inside the liquid. The dynamics of bubble growth is described by Eqs. (22.1)–(22.5), in which we must set $u_U = 0$.

Let us use the approximate expression for the concentration of a dissolved component in the liquid in the vicinity of the bubble surface [3]:

$$\rho_{iwL} = \rho_{i\infty L} - \left(\frac{D_{iL}}{\pi}\right)^{1/2} \int_0^t \frac{R^2(x)(\partial \rho_{iL}/\partial r)_{r=R}}{\left(\int_x^t R^4(y)\,dy\right)^{1/2}} dx. \tag{22.35}$$

Use Eq. (22.2) to express the partial derivative as

$$\left(\frac{\partial \rho_{iL}}{\partial r}\right)_{r=R} = \frac{1}{3R^2 D_{iL}} \frac{d(\rho_{iG} R^3)}{dt}$$

and substitute it into the relation (22.35):

$$\rho_{iwL} = \rho_{i\infty L} - \left(\frac{D_{iL}}{\pi}\right)^{1/2} \frac{M_i B_i}{3ATD_{iL}} \int_0^t \frac{R^2(x)(\partial R^2 \rho_{iwL}/\partial r)_{r=R}}{\left(\int_x^t R^4(y)\,dy\right)^{1/2}} dx. \tag{22.36}$$

Making a change of variables by the formulas

$$u = \int_0^t R^4(y)\,dy, \quad v = \int_0^x R^4(y)\,dy,$$

we can bring Eq. (22.36) to the form

$$\rho_{iwL} = \rho_{i\infty L} - \left(\frac{D_{iL}}{\pi}\right)^{1/2} \frac{M_i B_i}{3ATD_{iL}} \int_0^u \frac{(dR^3 \rho_{iwL}/dv)}{(u-v)^{1/2}} dv. \tag{22.37}$$

Neglecting inertial and viscous forces in (22.3) and taking into account Eqs. (22.4) and (22.5) we get

$$p_\infty + \frac{2\Sigma}{R} = \sum_i p_{iG} = \frac{\sum_i B_i \rho_{iwL}/M_i}{\rho_s/M_s + \sum_j \rho_{jwL}/M_j}. \tag{22.38}$$

The last equation must also be true at the initial moment, so

$$p_\infty + \frac{2\Sigma}{R_0} = \frac{\sum_i B_i \rho_{iwL0}/M_i}{\rho_s/M_s + \sum_i \rho_{jwL0}/M_j}. \tag{22.39}$$

Subtracting Eq. (22.35) from Eq. (22.34), we get

$$\frac{2\Sigma}{R_0}\left(1 - \frac{R_0}{R}\right) = \frac{\sum_i B_i \rho_{iwL0}/M_i}{\rho_s/M_s + \sum_j \rho_{jwL0}/M_j} \times \left\{1 - \frac{\sum_i B_i \rho_{iwL}/M_i \left(\rho_s/M_s + \sum_j \rho_{iwL0}/M_i\right)}{\sum_i (B_i \rho_{iwL0}/M_i)\left(\rho_s/M_s + \sum_j \rho_{jwL}/M_j\right)}\right\}. \tag{22.40}$$

Now, introduce the following dimensionless variables:

$$\bar{R} = \frac{R}{R_0}, \quad x_{iw} = \frac{\rho_{iwL}}{\rho_{i\infty L}}, \quad x_{iw0} = \frac{\rho_{iwL0}}{\rho_{i\infty L}},$$

$$\alpha_i = \frac{B_i \rho_{i\infty L} M_s R_0}{2 M_i \rho_s \Sigma}, \quad \beta_i = \left(\frac{D_{1L}}{D_{iL}}\right)^{1/2} \frac{M_i B_i}{M_1 B_1}, \quad \gamma_i = \frac{\rho_{i\infty L} M_L}{\rho_i M_i}.$$

In these variables, Eqs. (22.37) and (22.40) will become

$$x_{iw} = 1 - \beta_i \int_0^u \frac{(d\bar{R}^3 x_{iw}/dv)}{(u-v)^{1/2}} dv. \tag{22.41}$$

$$1 - \frac{1}{\bar{R}} = \frac{\sum_i \alpha_i x_{iw0}}{1 + \sum_j \gamma_j x_{jw0}} \cdot \left\{1 - \frac{\sum_i \alpha_i x_{iw}\left(1 + \sum_j \gamma_j x_{jw0}\right)}{\sum_i (\alpha_i x_{iw0})\left(1 + \sum_j \gamma_j x_{jw}\right)}\right\}. \tag{22.42}$$

At the initial stage of bubble growth, we seek for the solution in the form

$$\bar{R} = 1 + f, \quad x_{iw} = x_{iw0}(1 + \varphi_i), \tag{22.43}$$

where f and $\varphi_i \ll 1$.

Substituting (22.43) into (22.41) and (22.42) and denoting $G_i = \bar{R}^3 x_{iw}$, we get the following:

$$x_{iw0}(1 + \varphi_i) = 1 - \beta_i \int_0^u \frac{G_i'}{(u - v)^{1/2}} dv,$$

$$f = E\left\{1 - \frac{1}{E} \frac{\sum_i (\alpha_i x_{iw0})(1 + \varphi_i)}{1 + \sum_j \gamma_j x_{jw0}(1 + \varphi_j)}\right\}, \quad E = \frac{\sum_i \alpha_i x_{iw0}}{1 + \sum_j \gamma_j x_{jw0}}.$$

Expanding G_i as a power series of f and φ_i and leaving only the first-order terms, we can write:

$$G_i = x_{iw0}(1 + 3f + \varphi_i).$$

Then the system of equations describing the initial stage of bubble growth takes the form

$$\begin{aligned} &x_{iw0}(1 + \varphi_i) = 1 - \beta_i \int_0^u \frac{G_i'}{(u - v)^{1/2}} dv \\ &f = E \sum_i \psi_i \varphi_i, \quad \psi_i = \frac{\gamma_i x_{iw0}}{1 + \sum_j \gamma_j x_{jw0}} - \frac{\alpha_i}{\sum_j \alpha_j x_{jw0}}. \end{aligned} \tag{22.44}$$

Let us apply the Laplace transfom to Eqs. (22.44), taking into account the fact that the integral in the right-hand side of the first equation is a convolution of two functions. The result is the following system of equations for transforms $\tilde{G}_i(S)$, $\tilde{f}_i(S)$, and $\tilde{\varphi}_i(S)$:

$$x_{iw0}\left(\frac{1}{S} + \tilde{\varphi}_i\right) = \frac{1}{S} - \beta_i \left(\frac{\pi}{S}\right)^{1/2} (S\tilde{G}_i - x_{iw0}),$$

$$\tilde{f} = E \sum_i \psi_i \tilde{\varphi}_i, \quad \tilde{G}_i = x_{iw0}\left(\frac{1}{S} + 3\tilde{f} + \tilde{\varphi}_i\right),$$

whose solution takes the form

$$\tilde{\varphi}_i = \frac{1/S - x_{iw0}/S - 3B_i\sqrt{\pi S} x_{iw0} \tilde{f}}{x_{iw0}(1 + \beta_i \sqrt{\pi S})},$$

$$\tilde{f} = E \sum_i \psi_i \frac{1 - x_{iw0}}{S x_{iw0}(1 + \beta_i \sqrt{\pi S})} \left\{1 + 3E \sum_i \psi_i \beta_i \frac{\sqrt{\pi S}}{1 + \beta_i \sqrt{\pi S}}\right\}^{-1}.$$

We are looking for the solution at small deviations from the initial state $(t \to 0)$. Let us therefore expand the obtained expressions for transforms into power series in the vicinity of the infinity point:

$$\tilde{f} = E \sum_i \psi_i \frac{1 - x_{iw0}}{\beta_i \sqrt{\pi} x_{iw0}} \left\{ 1 + 3E \sum_i \psi_i \right\}^{-1} \frac{1}{S\sqrt{S}} + \cdots$$

$$\tilde{\varphi}_i = \frac{1}{\beta_i \sqrt{\pi}} \left\{ 1 - x_{iw0} - 3E x_{iw0} \beta_i \frac{\sum_j \psi_j (1 - x_{jw0}) / \beta_j x_{jw0}}{1 + 3E \sum_j \psi_j} \right\} \frac{1}{S\sqrt{S}} + \cdots$$

Going back to the original quantities, we obtain:

$$\mathrm{f} = E \sum_i \psi_i \frac{1 - x_{iw0}}{\beta_i \sqrt{\pi} x_{iw0}} \left\{ 1 + 3E \sum_i \psi_i \right\}^{-1} \sqrt{t} + \cdots$$

$$\varphi_i = \frac{2}{\beta_i \sqrt{\pi}} \left\{ 1 - x_{iw0} - 3E x_{iw0} \beta_i \frac{\sum_j \psi_j (1 - x_{jw0}) / \beta_j x_{jw0}}{1 + 3E \sum_j \psi_j} \right\} \sqrt{t} + \cdots$$

Now, from (22.43), we find the evolution of the bubble radius at the initial stage of growth:

$$\bar{R} = 1 + 2E \sum_i \psi_i \frac{1 - x_{iw0}}{\beta_i \pi x_{iw0}} \left\{ 1 + 3E \sum_i \psi_i \right\}^{-1} \sqrt{t} + \cdots \tag{22.45}$$

22.4 Bubble Dynamics in a Multi-Componenet Solution

Consider the dynamics of a bubble in a multicomponent solution under the condition that its lift can be neglected. At $u_\theta = 0$, the system of equations (22.1)–(22.6) can be solved by the same approximate method that used in Section 22.2 for a binary solution. For each component dissolved in liquid, select its own thin diffusion layer with the thickness $\delta_i \ll R$, around the bubble. In each layer, we seek for the distribution of concentration in the form

$$\frac{\rho_{iL} - \rho_{i\infty L}}{\rho_{iwL} - \rho_{i\infty L}} = (1 - x_i)^2, \quad (R \le r \le R + \delta_i), \tag{22.46}$$

and outside the layer, we have

$$\rho_{iL} = \rho_{i\infty L}, \quad (r > R + \delta_i). \tag{22.47}$$

Plug $u_\theta = 0$ into (22.1) and integrate this equation over r from R up to $R + \delta_i$. We then obtain Eq. (22.27) for the layer thickness δ_i, which is accurate to the order of δ_i/R:

$$\delta_i = \frac{\rho_{iG}R^3 - \rho_{iG0}R_0^3}{R^2(\rho_{i\infty L} - \rho_{iwL})}. \tag{22.48}$$

Introduce the following dimensionless variables:

$$\tau = \frac{6\rho_s ATD_{1L}}{B_1 M_L R_0^2}t, \quad \bar{R} = \frac{R}{R_0}, \quad \bar{\rho}_{iG} = \rho_{iG}\frac{AT}{B_i M_i}, \quad x_{iw} = \rho_{iw}\frac{M_s}{M_i\rho_s},$$

$$x_{i\infty} = \rho_{i\infty}\frac{M_s}{M_i\rho_s}, \quad d_i = \frac{D_{iL}}{D_{1L}}, \quad \alpha_i = \frac{B_i M_s}{AT\rho_s}, \quad \gamma = \frac{36\rho_s D_{1L}^2}{p_\infty R_0^2},$$

$$\lambda = \frac{2\mu_L}{3\rho_s D_{1L}}, \quad \mu_i = \frac{M_i}{M_s}, \quad \beta_i = \frac{B_i}{B_1}, \quad q = \frac{2\Sigma}{p_\infty R_0}, \quad K = \frac{B_1}{p_\infty}.$$

In new variables, the system of equations describing (in the currently considered approximation) the diffusion growth of a bubble in a multicomponent solution will become:

$$\frac{d\bar{\rho}_{1G}}{d\tau} = \frac{d_i(x_{i\infty} - x_{iw})^2}{\alpha_i\beta_i(\bar{\rho}_{iG}\bar{R}^3 - \bar{\rho}_{iG0})} - \frac{3\bar{\rho}_{iG}v}{\bar{R}}, \quad v = \frac{d\bar{R}}{d\tau},$$

$$\frac{dv}{d\tau} = \frac{\alpha_1^2\left(K\sum_i \beta_i\bar{\rho}_{iG} - 1 - q/\bar{R} - \gamma\lambda v/\alpha_i\bar{R}\right)}{\gamma\bar{R}\left(1 - \sum_i \mu_i x_{iw}\right)} - \frac{3v^2}{\bar{R}},$$

$$x_{iw} = \frac{\bar{\rho}_{iG}}{1 - \sum_j \bar{\rho}_{jG}}, \tag{22.49}$$

$$\bar{\rho}_{iG}(0) = \bar{\rho}_{iG0}, \quad \bar{R}(0) = 1, \quad v(0) = 0.$$

In order to solve this system of equations, it is necessary to set the initial density values $\bar{\rho}_{iG0}$ of gas components in the bubble. These values cannot be known in advance. The calculations above show that for any initial composition of gas in the bubble, the componential composition of gas is established very quickly and practically does not change afterwards. The initial composition of the solution is conventionally specified by mass percentages Π_i of components. Mole fractions are expressed in terms of mass percentages as follows:

$$x_{i\infty} = \frac{\Pi_i M_s}{M_i\left(100 - \sum_j \Pi_j\right)}. \tag{22.50}$$

Denote by y_{i0} the initial mole fractions of components in the bubble ($\sum_i y_{i0} = 1$). It is obvious that

$$y_{i0} = \frac{\rho_{iG0}/M_i}{\sum\limits_j \rho_{jG0}/M_j}. \tag{22.51}$$

The condition of dynamic equilibrium of the bubble at the initial moment implies

$$\sum_j \frac{AT\rho_{jG0}}{M_j} = p_\infty + \frac{2\Sigma}{R_0}. \tag{22.52}$$

Substituting the expression (22.51) into (22.52) and switching to dimensionless variables introduced earlier, we get the initial conditions for $\bar{\rho}_{iG0}$:

$$\bar{\rho}_{iG0} = \frac{y_{i0}(1+q)}{k\beta_i}. \tag{22.53}$$

As an example, consider the growth of a gas bubble in a solution where two components with mass concentrations $\Pi_1 = 7\%$ and $\Pi_2 = 3.5\%$ are dissolved. The other parameters are: $\alpha_1 = 1.6$; $\alpha_2 = 0.16$; $d_1 = d_2 = 1$; $\mu_1 = 0.05$; $\mu_2 = 0.1$; $x_{1\infty} = 2.5$; $x_{2\infty} = 0.6$; $q = 1$; $K = 100$. The initial composition of the bubble is $y_{10} = 0.99$; $y_{20} = 0.01$. Fig. 22.4 shows the time evolution of the bubble radius and the mass concentrations in the bubble.

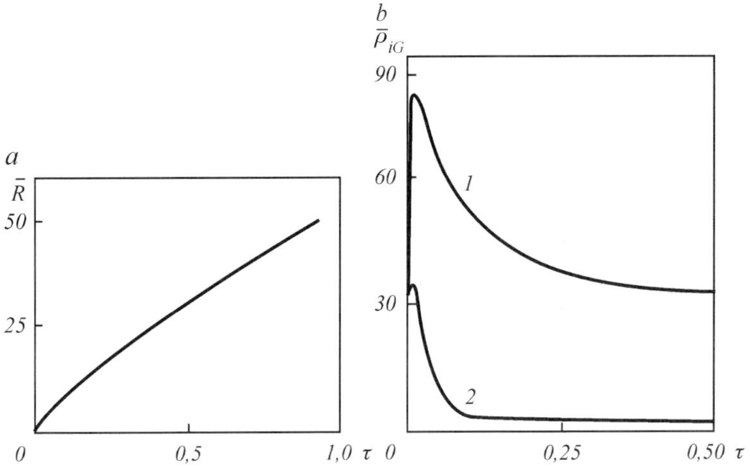

Fig. 22.4 Bubble radius $\bar{R}$ (*a*) and component densities $\bar{\rho}_{iG}$ in the bubble (*b*) as functions of time τ: 1 – $i = 1$; 2 – $i = 2$.

The calculations performed allow us to make the following conclusions:

1. The initial stage of bubble growth is characterized by a rapid increase of the pressure and density of gas. Afterwards, they diminish and then remain practically constant;
2. Inertial and viscous items in the equation of dynamic equilibrium for the bubble are essential only in the initial stage of growth;
3. The surface tension has a noticeable effect on bubble growth. As a result, in the early stage of growth, the diffusion boundary layer gets thicker; then, as time goes on, its thickness decreases and eventually stabilizes;
4. The established fractional composition of gas in the bubble does not depend on the initial composition.

22.5 The Effect of Surfactants on Bubble Growth

In the majority of publications devoted to research of diffusion growth of gas bubbles in a supersaturated solution, the coefficient of surface tension Σ is considered to be constant and the quantity of gas entering the bubble is determined by the diffusion flux of gas dissolved in the liquid at the bubble surface. In reality, the solution, for example, a natural hydrocarbon mixture, will always contain surfactants which, being adsorbed at the interface, reduce Σ on the one hand, and on the other hand, interfere with the transition of dissolved substance from the solution to the bubble. The effect of surfactants on the value of Σ is known [4]. As to the influence of surfactants on the transition of gas from the dissolved state into the gaseous one, this effect is poorly studied.

To examine the influence of surfactants on bubble growth, consider the following problem. A spherical gas bubble is placed in a binary solution, supersaturated at the given pressure and temperature. As a result, there emerges a diffusion flux toward the bubble surface. The process is accompanied by adsorption of a surfactant, which is present in the liquid, on the bubble surface. Assume that the process is diffusion-controlled, i.e. adsorption-desorption of surfactant molecules at the interface occurs sufficiently quickly. In such a case, the surface concentration of surfactant Γ may be taken equal to the equilibrium value [4]:

$$\frac{\Gamma}{\Gamma_\infty} = \frac{k_1 \rho_{2wL}}{1 + k_1 \rho_{2wL}}, \tag{22.54}$$

where Γ_∞ is the surfactant concentration at full saturation (limiting adsorption); k_1 is the constant of equilibrium of the adsorption process; ρ_{2vL} is the surfactant concentration in the solution at the bubble surface.

During the adsorption of surfactant molecules on the gas-liquid interface, a monomolecular layer is formed. At low surface concentrations of the surfactant (far from full saturation), it looks as though the molecules are "floating" on the surface. At higher concentrations, the number of molecules in the surface layer increases. In a fully saturated state, the interface surface is covered by a tightly packed layer of surfactant molecules. This layer will impede the transition of gas molecules dissolved in the liquid through the interface. The fraction of the surface area s that is occupied by the surfactant is proportional to $1 - \Gamma/\Gamma_\infty$. Therefore, it makes sense to assume that the diffusion flux of component 1 dissolved in the liquid is some power function of the surface area occupied by the surfactant:

$$J_1 = 4\pi R^2 D_{1L}\left(1 - \frac{\Gamma}{\Gamma_\infty}\right)^n \left(\frac{\partial \rho_{1L}}{\partial r}\right)_{r=R}. \tag{22.55}$$

One noteworthy attempt to take into account the effect of the surfactant film at the bubble surface on the rate of diffusion growth was made in [5]. It is based on the assumption that the intensities of substance transfer through a surface free from surfactants and through a surface covered by a surfactant are characterized by different diffusion factors. The proposed model introduced a number of parameters, whose determination requires extensive experimental work that involves different surfactants.

The considered process is described by the following system of equations:

$$\begin{aligned}
&\frac{\partial \rho_{iL}}{\partial t} + \left(\frac{R}{r}\right)^2 \dot{R}\frac{\partial \rho_{iL}}{\partial r} = D_{iL}\frac{1}{r^2}\frac{\partial}{\partial r}\left(r^2\frac{\partial \rho_{iL}}{\partial r}\right), \quad (i = 1, 2),\\
&\frac{d}{dt}\left(\frac{4\pi}{3}\rho_G R^3\right) = 4\pi R^3 D_{iL}\left(1 - \frac{\Gamma}{\Gamma_\infty}\right)^n \left(\frac{\partial \rho_{iL}}{\partial r}\right)_{r=R},\\
&\rho_L\left(R\ddot{R} + \frac{3}{2}R\dot{R}^2 + \frac{4\nu_L}{R}\dot{R}\right) = p_G - p_\infty - \frac{2\Sigma}{R},\\
&\frac{d}{dt}(\Gamma R^2) = R^2 D_{2L}\left(\frac{\partial \rho_{2L}}{\partial r}\right)_{r=R},\\
&p_G = \frac{AT\rho_G}{M_1}, \quad p_G = \frac{B_1\rho_{1w}M_2M_L}{M_1M_2\rho_L + M_2M_L\rho_{1w} + M_1M_L\rho_{2w}},\\
&\Sigma = \Sigma_0 - \frac{AT}{F_0}\ln(1 + k_2\rho_{2w}).
\end{aligned} \tag{22.56}$$

This system must be supplemented by the initial and boundary conditions:

$$\begin{aligned}
&\rho_{iL}(0, r) = \rho_{iL\infty}, \quad \rho_{iL}(t, \infty) = \rho_{iL\infty}, \quad \rho_{iL}(t, R) = \rho_{iw}, \quad (i = 1, 2),\\
&\Gamma(0) = 0, \quad R(0) = R_0, \quad R(0) = 0.
\end{aligned} \tag{22.57}$$

To solve the obtained equations, we can use the approximate method of integral relations that was considered earlier. Select two thin diffusion layers around the bubble. These layers contain dissolved gas and the surfactant and have thicknesses $\delta_1 \ll R$ and $\delta_2 \ll R$. The mass concentration distributions inside the layers are equal to

$$\frac{\rho_{1L} - \rho_{1\infty L}}{\rho_{1wL} - \rho_{1\infty L}} = (1 - \xi_1)^2, \quad \xi_1 = \frac{r - R}{\delta_1}, \quad (R \le r \le R + \delta_1),$$
$$\frac{\rho_{2L} - \rho_{2\infty L}}{\rho_{2wL} - \rho_{2\infty L}} = (1 - \xi_2)^2, \quad \xi_2 = \frac{r - R}{\delta_2}, \quad (R \le r \le R + \delta_2), \tag{22.58}$$

while concentrations outside the layers are equal to the corresponding unperturbed values far away from the bubble.

Integration of the first equation (22.56) over r from R to $R + \delta_1$ at $i = 1$ and from R to $R + \delta_2$ at $i = 2$ yields the following expressions, which are accurate the order of δ_i/R:

$$\delta_2 = \frac{3\Gamma}{\rho_{2\infty L} - \rho_{2wL}},$$
$$\frac{d(R^2\Gamma)}{dt} = \frac{D_{2L}R^2(\rho_{2\infty L} - \rho_{2wL})}{3\Gamma},$$
$$\frac{d}{dt}((\rho_{1\infty L} - \rho_{1wL})\delta_1 R^2) = \frac{6D_{1L}R^2(\rho_{1\infty L} - \rho_{1wL})R^2}{\delta_1},$$
$$\frac{d(R^3\rho_G)}{dt} = \frac{6D_{1L}R^2(\rho_{1\infty L} - \rho_{1wL})(1 - \Gamma/\Gamma_\infty)^n}{\delta_1}. \tag{22.59}$$

Introduce the dimensionless variables:

$$\tau = \frac{6\rho_L ATD_{1L}t}{B_1 M_L R_0^2}, \quad \bar{R} = \frac{R}{R_0}, \quad G = \frac{\Gamma}{\Gamma_\infty}, \quad \Delta = \frac{\delta_1}{R}, \quad \bar{\Sigma} = \frac{\Sigma}{\Sigma_0},$$

$$x_{i\infty} = \frac{\rho_{i\infty} M_L}{M_i \rho_L}, \quad x_{iw} = \frac{\rho_{iw} M_L}{M_i \rho_L}, \quad \bar{P} = \frac{p_G}{B_1}, \quad \varphi = \frac{AT}{F_0 \Sigma_0},$$

$$\beta = \frac{\alpha D_{2L} R_0 \rho_L m_2}{\Gamma_\infty}, \quad \gamma = \frac{36\rho_L D_{1L}^2}{p_\infty R_0^2}, \quad \alpha = \frac{B_1 M_L}{\rho_L AT}, \quad q = \frac{2\Sigma}{p_\infty R_0},$$

$$K = \frac{B_1}{p_\infty}, \quad K_i = \frac{k_i M_2 \rho_L}{M_L}, \quad m_i = \frac{M_i}{M_L}, \quad \lambda = \frac{2\mu_L}{3\rho_L D_{1L}}.$$

The equations are then transformed into

$$\frac{d}{d\tau}(\Delta \bar{R}^3 (x_{1\infty} - x_{1w})) = \frac{\alpha \bar{R}(\bar{x}_{1\infty} - x_{1w})}{\Delta},$$

$$\frac{d(P\bar{R}^3)}{d\tau} = \frac{\bar{R}(1 - G)^n (x_{1\infty} - x_{1w})}{\Delta},$$

$$\frac{d(G\bar{R}^3)}{d\tau} = \frac{\beta \bar{R}^2 (x_{1\infty} - x_{1w})^2}{\Delta},$$

$$\bar{\Sigma} = 1 - \varphi \ln(1 + K_2 x_{2w}), \tag{22.60}$$

$$\frac{\gamma}{\alpha^2}(1 + x_{1w} m_1 + x_{2w} m_2)\left(\bar{R}\ddot{\bar{R}} + \frac{3}{2}\dot{\bar{R}}^2\right) = K\bar{P} - 1 - \frac{q\bar{\Sigma}}{\bar{R}} + \frac{\lambda\gamma}{\alpha\bar{R}}\dot{\bar{R}},$$

$$\bar{p} = \frac{x_{1w}}{1 + x_{1w} + x_{2w}}, \quad G = \frac{K_1 x_{2w}}{1 + K_1 x_{2w}},$$

$$\bar{R}(0) = 1, \quad G(0) = \Delta(0) = 0, \quad \bar{P}(0) = \bar{P}_0.$$

The bubble will increase in size if the inequality $KP_0 - 1 - q > 0$ is satisfied, otherwise it will be dissolved in the liquid.

Some results of the numerical solution of Eqs. (22.60) are shown in Figs. 22.5 to 22.7. Dependences on parameters β and n that characterize the influence of the surfactant, and also on the parameter K that determines the influence of pressure, are of greatest interest. At small values of τ, the effect of the surfactant causes a reduction of surface tension, and the bubble grows faster than it would grow in the absence of the surfactant. At $n = 0$, the surfactant layer on the surface does not impede the transition of dissolved substance into the bubble, and the influence of the surfactant is manifest only in the reduction of surface tension. With an increase of n, the rate of bubble growth slows down considerably. An increase of the parameter K (the fall of pressure p_∞) results in enhancement of the bubble growth rate. Fig. 22.6 shows the change of surface concentration of surfactant at $K = 4.45$; $n = 0$; $x_{1\infty} = 1$, and $q = 1$ for various values of β. An in-

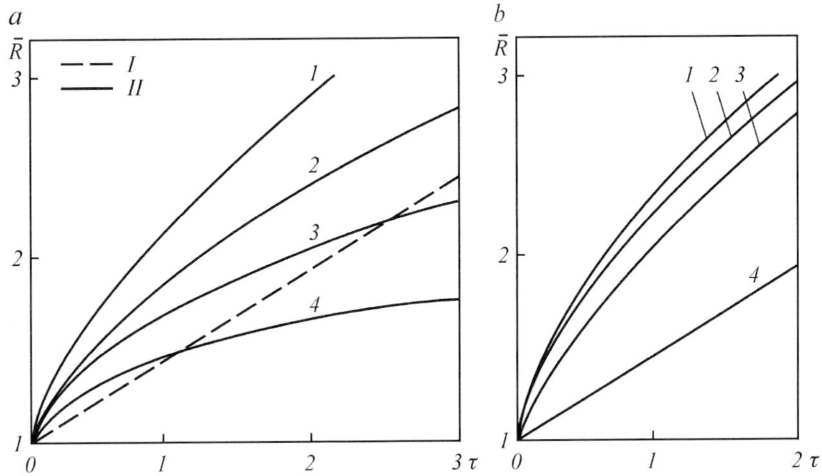

Fig. 22.5 Bubble radius as a function of time ($K = 4.45$; $x_{1w0} = 0.8$; $x_{1\infty} = 1$; $q = 1$): a – n: 1 – 0; 2 – 0.5; 3 – 1; 4 – 2; I – $\beta = 0$ (without surfactant); II – $\beta = 10$; b – $n = 0$; β: 1 – 1000; 2 – 100; 3 – 10; 4 – 0.

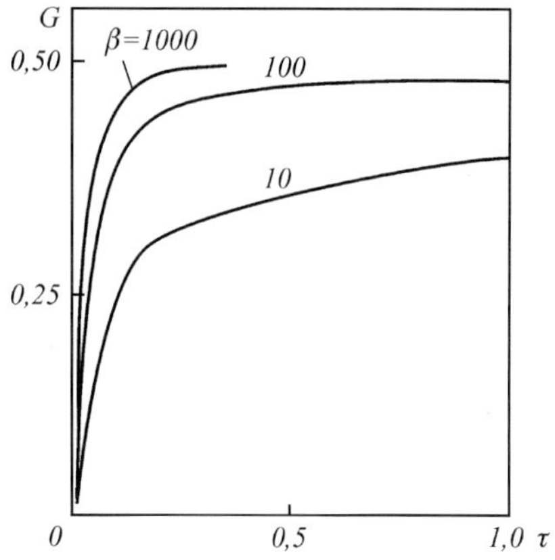

Fig. 22.6 Surface concentrsation G of surfactant vs. time τ for different values of β.

crease of β results in an increase of the bubble size (Fig. 22.5, b) on the one hand, and on the other hand, since the characteristic time of change of the surface concentration of surfactant is $t \sim \beta^{-1}$, it results in a shorter time required to establish the equilibrium value of G. An increase of K for the same values of other parameters will cause G to decrease. The last circumstance is explained by the fact that with the increase of K, the rate of bubble growth increases, while the flux of surfactant at the surface changes insignificantly. Fig. 22.7 shows how the gas density in the bubble changes with time. It can be seen that the main change of ρ_G occurs at the initial stage of growth, when inertial, viscous, and capillary forces are essential.

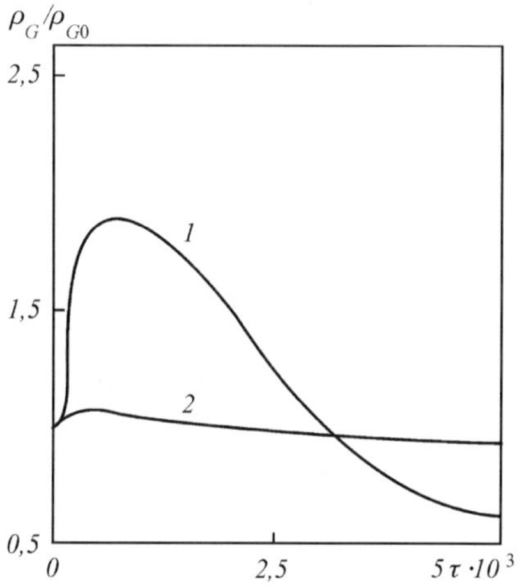

Fig. 22.7 Dependence of gas density in the bubble on time for different values ($K = 4.45$; $n = 0$; $x_{1\infty} = 1$; $q = 1$) of K: 1 – $K = 100$; 2 – $K = 4.45$.

23
Separation of Liquid–Gas Mixtures

Consider a quiescent homogeneous liquid-gas mixture with density ρ_L and viscosity μ_L, occupying a volume V with height of the layer H. The mass concentrations of components dissolved in the liquid are ρ_{iL} $(i = \overline{1,s})$). At the initial moment, the pressure above the surface of the mixture is p_0 and the mixture is in equilibrium. If the pressure above the surface drops abruptly to $p_\infty < p_0$, small bubbles (germs) are formed inside the liquid, which then start to grow as some components dissolved in the liquid diffuse into them. If the layer has a sufficient height, an ancillary growth mechanism emerges – the reduction of hydrostatic pressure as the bubbles come closer to the surface. After they reach the surface, the bubbles burst, releasing the gas contained in them. If the pressure above the mixture surface is kept constant (that is possible if the released gas is removed from the system), then such a process is called differential separation. If the released gas is allowed to accumulate above the surface, the pressure will grow with time. Then the process is called contact separation. In both cases, after some time the solution becomes depleted of dissolved components, and the intensity of cold boiling decreases. After a time t_{eq} boiling practically stops and the system comes into equilibrium.

Let ρ_s be the mass concentration of solvent, and ρ_{iG} – mass concentrations of components in the gaseous phase. Assume the process to be isothermal. Then the continuity equation for components in the solution is

$$\frac{\partial \rho_{iL}}{\partial t} + \nabla \cdot (\rho_{iL} \mathbf{w}_{iL}) = \sum_{\substack{j=1 \\ i \neq j}}^{s} J_{jiL} - J_{iG1} + J_{iG2}, \tag{23.1}$$

where J_{jiL} is the rate of mass exchange between j-th and i-th components in the liquid phase; J_{iG1} and J_{iG2} are the rates of evaporation and condensation during phase transitions at the interface; $\mathbf{w}_{iL}$ is the relative velocity of i-th component in the solution.

Assume that there is no mass exchange between components in the solution and no condensation. Then $J_{jiL} = 0$ and $J_{G2} = 0$ and Eq. (23.1) simplifies to

$$\frac{\partial \rho_{iL}}{\partial t} + \nabla \cdot (\rho_{iL} \mathbf{w}_{iL}) = -J_{iG1}. \tag{23.2}$$

Separation of Multiphase, Multicomponent Systems. E. G. Sinaiski and E. J. Lapiga

ISBN: 978-3-527-40612-8

The gaseous phase consists of small-sized spherical bubbles with low volume concentration, so possible interaction between them can be neglected due to their compactness. Bubble radii can range from the minimum radius R_n of a germ to the maximum radius R_m that can be attained by a bubble that was born at the bottom of the liquid and has risen to the surface.

The state of the gaseous phase at each moment of time t at any point P of the volume can be characterized by a continuous distribution of bubbles over the mass of i-th component: $n_i(m_i, t, P)$. This distribution must obey the kinetic equation that looks like

$$\frac{\partial n_i}{\partial t} + \nabla \cdot (n_i \mathbf{u}_G) + \frac{\partial}{\partial m_i}\left(n_i \frac{dm_i}{dt}\right) = I_n, \tag{23.3}$$

if we neglect breakup and coagulation of bubbles. Here I_n is the rate of bubble formation due to nucleation (see Section 14.2), and u_G is the bubble's velocity.

Multiply both parts of Eq. (23.3) by m_i and integrate over m_i from 0 to ∞. The outcome is

$$\frac{\partial M_{iG}}{\partial t} + \nabla \cdot \mathbf{Q}_i = \int_0^\infty n_i \left(\frac{dm_i}{dt}\right) dm_i + \int_0^\infty I_n m_i \, dm_i, \tag{23.4}$$

where M_{iG} is the mass of i-th component in a unit volume of mixture; $\mathbf{Q}_i = \int_0^\infty m_i n_i u_G \, dm_i$ is the mass flux of i-th component through the surface enclosing the unit volume. Since the mass of i-th component changes due to its phase transition from the liquid to the gaseous phase,

$$J_{iG1} = \frac{\partial M_{iG}}{\partial t} + \nabla \cdot \mathbf{Q}_i. \tag{23.5}$$

The relative velocity $\mathbf{w}_{iL}$ of i-th component in the solution is equal to

$$\mathbf{w}_{iL} = -\frac{\rho_L}{\rho_{iL}} D_{iL} \nabla\left(\frac{\rho_{iL}}{\rho_L}\right), \quad \rho_L = \rho_s + \sum_i \rho_{iL}, \tag{23.6}$$

where D_{iL} is the coefficient of binary diffusion in the solution.

We shall restrict our analysis to an extremely diluted solution. Then $\sum_i \rho_{iL} \ll \rho_s$, $\rho \approx \rho_s$ and Eq. (23.6) may be rewritten in the form

$$\mathbf{w}_{iL} = -\frac{D_{iL}}{\rho_{iL}} \nabla \rho_{iL}. \tag{23.7}$$

Thus, Eq. (23.2) becomes

$$\frac{\partial \rho_{iL}}{\partial t} = D_{iL} \Delta(\rho_{iL}) - \int_0^\infty n_i \left(\frac{dm_i}{dt}\right) dm_i - \int_0^\infty I_n m_i \, dm_i. \tag{23.8}$$

23.1
Differential Separation of a Binary Mixture

Let the solution consist of two components, one of transits into the gaseous phase when the pressure above the surface of the mixture diminishes. The pressure is kept at constant value. Consider a thin layer of mixture with volume V and height H. Assume that all parameters depend on time and on a single coordinate x whose direction coincides with the direction of gravity. The value $x = 0$ corresponds to the upper surface of the layer. The assumptions just made allow us to write Eq. (23.8) as

$$\frac{\partial \rho_{iL}}{\partial t} = D_{iL}\frac{\partial^2 \rho_{iL}}{\partial x^2} - \int_0^\infty n_1\left(\frac{dm_1}{dt}\right) dm_1 - \int_0^\infty I_n m_1\, dm_1. \tag{23.9}$$

By implication, dm_1/dt is the rate of mass change for component 1 in the bubble, therefore

$$\frac{dm_1}{dt} = u\frac{dm_1}{\partial x} = -I_D, \tag{23.10}$$

where $u = \alpha V^{2/3}$ is the lift velocity of the bubble; I_D is the diffusion flux of component 1 through the bubble surface; V is the bubble volume.

Expressions for the diffusion flux for various cases were presented in the previous chapter. To solve Eq. (23.9), we must know the mass distribution of bubbles. Let $N_n(x_0, t)$ be the number of bubbles (germs) per unit volume per unit time formed at the depth x_0 at the moment t. It is obvious that the number of germs decreases in time and with the reduction of depth x_0 as the mixture is being depleted. The flux of bubbles through a fixed unit cross section is conserved, therefore

$$un_1\, dm_1 = N_n(x_0, t)\, dx_0. \tag{23.11}$$

Appearing in the left-hand side is the flux through the cross section of a vertical column at the depth x at the moment t, and in the right-hand side – the same quantity at the depth $x_0 > x$. Suppose the bubble growth formula $m_1 = m_1(x, x_0, m_n)$, where m_n is the mass of a germ, is known to us. This formula tells us the mass that the bubble nucleated at the depth $x_0 > x$ acquires when it rises to the depth x. For a fixed x, we can write $dm_1 = (\partial m_1/\partial x_0)\, dx_0$, in which case the expression (23.11) gives us

$$un = \frac{N_n(x_0, t)}{\partial m_1/\partial x_0}. \tag{23.12}$$

Let the sizes and the masses of incipient bubbles be identical, constant in time and over the height of the layer. Then

$$I_n = N_n\delta(m_1 - m_n). \tag{23.13}$$

and we get after substituting expressions (23.11) and (23.13) into (23.13),

$$\frac{\partial \rho_{1L}}{\partial t} = D_{1L}\frac{\partial^2 \rho_{1L}}{\partial x^2} - N_n m_n - \int_x^H \frac{I_D N_n}{u}\, dx_0. \tag{23.14}$$

Take for I_D the expression (22.7), which is valid for bubbles whose radius exceeds the critical radius R_{cr}. At a sufficiently high supersaturation of mixture, the bubble rapidly grows in size, and its radius quickly reaches the critical value, justifying our assumption. Suppose that the change of concentration of the dissolved component is predominantly caused by its ablation by bubbles. Then we can neglect the first term in the right-hand side of Eq. (23.14). Besides, if the layer is thin enough or if the bubble is growing quickly, we can assume the density of gas in the bubble to be constant. The assumptions just made allow us to rewrite Eq. (23.14) in the form

$$\frac{\partial \rho_{1L}}{\partial t} = -\rho_G V_n N_n - \frac{8D_{1L}}{\alpha^{2/3}}(\rho_{1L} - \rho_{1L0})\int_x^H N_n\, dx_0,$$
$$\rho_{1L}(0, x) = \rho_{1L0}. \tag{23.15}$$

The number N_n of nucleated bubbles per unit time per unit volume could be estimated from the well-known expression [6] for a pure liquid,

$$N_n = Z_1 \exp\left(-\frac{l}{kT}\right)\sqrt{6\Sigma(3-b)\pi m}\, \exp\left(-\frac{16\pi\Sigma^2}{3kT(\Delta p)^2}\right). \tag{23.16}$$

Here Z_1 is the number of liquid molecules in 1 cm^3; l is the molecular heat of evaporation at temperature T; k is Boltzmann's constant; Σ is the coefficient of surface tension of the liquid; $b = \Delta p/p_0$, $\Delta p = p_0 - p_\infty$; m is the mass of a molecule.

From the form of the dependence of N_n on Δp, it follows that on a small interval of Δp, N_n increases abruptly from 0 to $N_{nm} \sim Z_1\sqrt{18\Sigma m_n e^{-l/kT}}$. The dependence (23.16) can be approximated on this interval by the expression $N_n = C(\Delta p)^\Lambda$ ($\Lambda \geq 1$). The value of the exponent Λ depends on the physical constants entering Eq. (23.16) that describe the properties of the liquid. Since $\Delta p \sim \rho_{1L} - \rho_{1p}$, where ρ_{1p} is the equilibrium (at pressure p_∞) value of the mass concentration of the dissolved component, it is possible to represent N_n as

$$N_n = N_0\left(\frac{\rho_{1L} - \rho_{1p}}{\rho_{1L0} - \rho_{1p}}\right)^\Lambda, \tag{23.17}$$

where N_0 is the value of N_n at the initial moment of time.

Introducing dimensionless variables

$$\xi = \frac{x}{H}, \quad \tau = \frac{8N_0 D_{1L}^{2/3} Ht}{\alpha^{2/3}}, \quad \bar{\rho}_{1L} = \rho_{1L}\frac{M_s B_1}{M_1 \rho_s p_\infty},$$

$$\bar{\rho}_{1g} = \frac{1}{1 - p_\infty / B_1}, \quad \Delta\bar{\rho}_1 = \frac{\rho_{1L} - \rho_{1p}}{\rho_{1L0} - \rho_{1p}}, \quad y_0 = \frac{6D_{1L}^{2/3} ATH}{\pi R_0^3 \alpha^{2/3} B_1 M_s}.$$

we can bring Eq. (23.15) to the form

$$\frac{\partial \bar{\rho}_1}{\partial \tau} = -\frac{(\Delta\bar{\rho}_1)^\Lambda}{y_0} - (\bar{\rho}_{1L} - \bar{\rho}_{1p}) \int_\xi^1 (\Delta\bar{\rho}_1)^\Lambda \, d\xi, \quad \bar{\rho}_{1L}(\xi, 0) = \bar{\rho}_{1L0}. \tag{23.18}$$

Since we are only considering an infinitely diluted solution, $\rho_L = \rho_s$, and the density may be taken as constant.

The integral equation (23.18) is solved by the method of successive approximations, taking $\bar{\rho}_{1L0}$ as zeroth approximation, while the other approximations are found from the formula

$$\frac{\partial \bar{\rho}_1}{\partial \tau} = -\frac{(\Delta\bar{\rho}_1)^\Lambda}{y_0} - (\bar{\rho}_{1L}^{(j-1)} - \bar{\rho}_{1g}) \int_\xi^1 (\Delta\bar{\rho}^{(j-1)})^\Lambda \, d\xi, \quad (j \geq 1). \tag{23.19}$$

The parameter $y_0 \gg 1$, therefore the first term in the right-hand side is small. So, we eventually get the formula

$$\Delta\bar{\rho}_1 = \exp\{\Lambda^{-1}(Ei(-z) - \ln(z) - C)\}, \quad z = \Lambda(1 - \xi)\tau. \tag{23.20}$$

which is accurate to the third approximation.

Numerical solution of Eq. (23.19) has shown a good agreement with the approximate solution (23.20).

The dependence of relative supersaturation $\Delta\bar{\rho}$ of the solution on time at various layer depths x/H is shown in Fig. 23.1 for $\Lambda = 3$. An increase of Λ corresponds to a faster decrease of the number of incipient bubbles with the reduction of $\Delta\bar{\rho}$ and results in a noticeable slowing of the degassing process. Concentration of the dissolved gas is distributed over the depth of the mixture non-uniformly and decreases with the increase of depth. The reason for this is that the number of bubbles in a unit volume is larger at smaller depths, because newly-formed bubbles are joined by the bubbles lifting from the underlying layers. This results in a more intensive depletion of the upper layers of the mixture. In the region adjoining to the bottom, nucleating bubbles account for most of the depletion. Therefore, for this region, it is possible to leave only the first term in the right-hand side of Eq. (23.18):

$$\frac{\partial \bar{\rho}_1}{\partial \tau} = \frac{(\Delta\bar{\rho})^\Lambda}{y_0}. \tag{23.21}$$

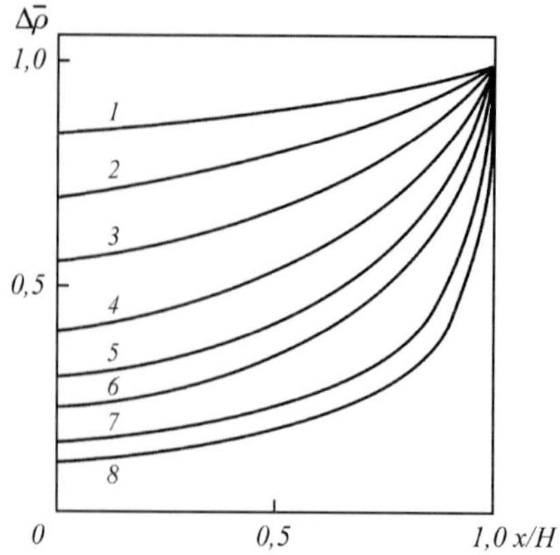

Fig. 23.1 Relative supersaturation $\Delta\bar{\rho}$ of the solution vs. time τ at various depths x/H of the layer: *1* – 0.2; *2* – 0.5; *3* – 1; *4* – 2; *5* – 3.5; *6* – 5; *7* – 10; *8* – 15.

The solution of this equation is

$$\Delta\bar{\rho}_1 = \begin{cases} (1 + (\Lambda - 1)\tau/\gamma_0(\bar{\rho}_{1L0} - \bar{\rho}_{1p}))^{1-\Lambda} & \text{at } \Lambda \neq 1, \\ \exp(-\tau/\gamma_0) & \text{at } \Lambda \neq 1. \end{cases} \tag{23.22}$$

It is worth to note that the characteristic time of change of concentration in the vicinity of the bottom is γ_0 times greater than in the near-surface layer. It means that equilibrium is established much faster at the surface of the mixture than at the bottom.

Denote by J_G the flux of gas from a unit surface area of the mixture. Then the balance-of-mass condition for a column of mixture of a unit cross section gives us:

$$J_0 = -0.5H\frac{d\bar{\rho}_{1L}}{dt}. \tag{23.23}$$

The efficiency of separation of a given volume of mixture can be characterized by the parameter η, equal to the ratio of the amount of gas released in time t to the total amount of gas that can be released by the time the equilibrium is established. One can easily derive:

$$\eta = 1 - \int_0^1 \Delta\bar{\rho}_1\, d\xi. \tag{23.24}$$

The dependence $\eta(\tau)$ is shown in Fig. 23.2. The efficiency of separation grows quickly at first, and then, in the course of mixture depletion, η slowly tends to 1.

The obtained dependence makes it possible to estimate the characteristic relaxation time of the mixture. One can see that at $\tau \geq 10$, the value of η practically

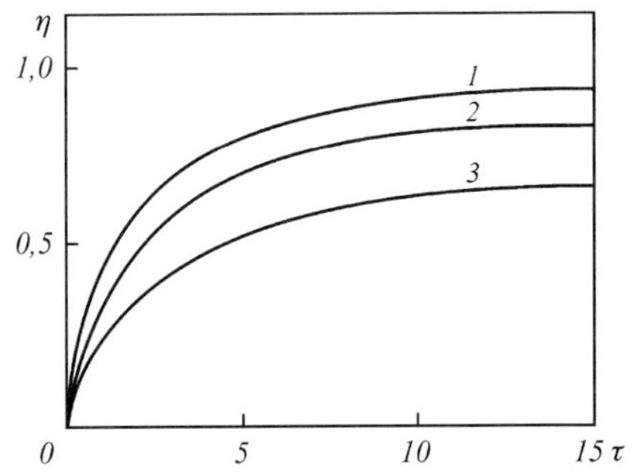

Fig. 23.2 η vs. τ for different values of Λ: 1 – 1; 2 – 2; 3 – 3.

does not change. This means that the characteristic relaxation time is estimated by $\tau_{eq} \sim 10$.

Finally, let us discuss how the layer depth H influences the efficiency of separation. Since $t \sim H^{-1}$, the characteristic relaxation time is inversely proportional to the layer depth. It means that for the same volume of the mixture, equilibrium will establish faster if we spread the mixture in a thicker layer (τ_{eq} will be smaller for the layer of thickness H_1 than for the layer of thickness H_2, where $H_1 > H_2$). But it is necessary to remember that we cannot set the layer depth to be arbitrarily large, as this can violate the assumption that the time it takes for a bubble to reach the surface is shorter than the characteristic relaxation time.

23.2 Contact Separation of a Binary Mixture

Consider now the case when the emanating gas is accumulated above the surface of the mixture in a volume of height h. Retain all assumptions made in the previous paragraph and in addition assume that as the bubbles rise up, the pressure above the mixture changes insignificantly. Now we must add to Eq. (23.15), describing the change of mass concentration of gas dissolved in the liquid, another equation describing the change of pressure above the surface of the mixture. Denote as $J_G(t)$ the gas flux through a unit surface of mixture at a moment t, and as $p_{\infty 0}$ – the initial pressure above the surface. Then the pressure above the surface at any moment $t > 0$ (under the condition that gas is ideal) is equal to

$$p_\infty(t) = p_\infty^{(0)} + \frac{AT}{M_1 h}\int_0^t J_G(t)\,dt. \tag{23.25}$$

At $t \to \infty$ the pressure approaches its equilibrium value:

$$p_\infty^{eq}(t) = p_\infty^{(0)} + \frac{AT}{M_1 h}\int_0^\infty J_G(t)\,dt. \tag{23.26}$$

For the intensity of bubble formation we will use, as before, the expression (23.17).

Now, introduce the following dimensionless variables:

$$\xi = \frac{x}{H}, \quad \tau = \frac{8N_0 D_{1L}^{2/3} Ht}{\alpha^{2/3}}, \quad \bar{\rho}_{1L} = \rho_{1L}\frac{M_s}{M_1\rho_s}, \quad \bar{p}_\infty = \frac{p_\infty}{p_\infty^{(0)}}, \quad \bar{\rho}_{1p} = \frac{\bar{p}_\infty}{1-\bar{p}_\infty},$$

$$\Delta\bar{\rho}_1 = \frac{\rho_{1L}-\rho_{1p}}{\rho_{1L0}-\rho_{1p}}, \quad \gamma_0 = \frac{6D_{1L}^{2/3} ATH}{\pi R_0^3 \alpha^{2/3} B_1 M_s}, \quad K_1 = \frac{B_1}{p_{\infty 0}^{(0)}}, \quad G = \frac{ATH\rho_s}{2p_\infty^{(0)} M_s h}.$$

Then the system of equations describing the process of contact degassing becomes:

$$\frac{\partial(\Delta\bar{\rho}_1)}{\partial\tau} = -\frac{K_1\bar{p}_\infty(\Delta\bar{\rho}_1)^\Lambda}{\gamma_0} - \Delta\bar{\rho}_1\int_0^t (\Delta\bar{\rho}_1)^\Lambda\, d\xi - \frac{K_1-1}{\bar{\rho}_{10}(K_1-1)}\frac{d}{dt}\left(\frac{\bar{p}_\infty}{K_1-\bar{p}_\infty}\right),$$

$$\bar{p}_\infty = 1 + G\int_0^t \Delta\bar{\rho}\, d\xi, \tag{23.27}$$

$$\Delta\bar{\rho}_1(\xi, 0) = 1, \quad \bar{p}_\infty(0) = 1.$$

Shown in Figs. 23.3 and 23.4 are some results of numerical solution of Eqs. (23.27): change in time of separation efficiency for various values of parameter G describing the ratio of liquid-gas-cap volumes, and dependence of equilibrium pressure in the gas cap on parameters K_1 and G. Separation efficiency of contact degassing is lower than that of differential degassing. The heavier the dissolved component (i.e. the smaller the parameter K_1), the greater is this difference.

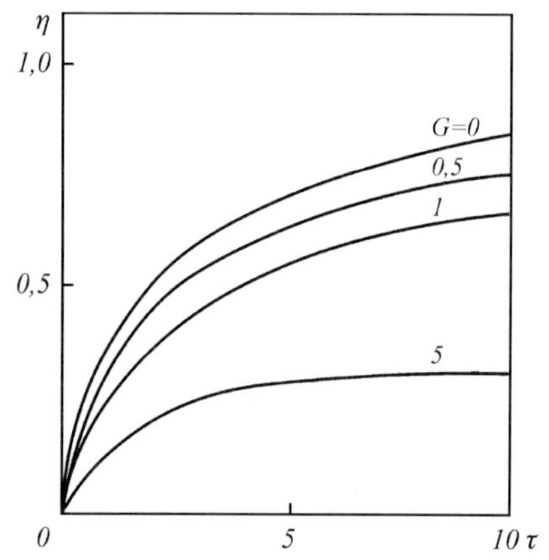

Fig. 23.3 Separation efficiency η vs. τ for various values of G ($K_1 = 10$; $\Lambda = 1$).

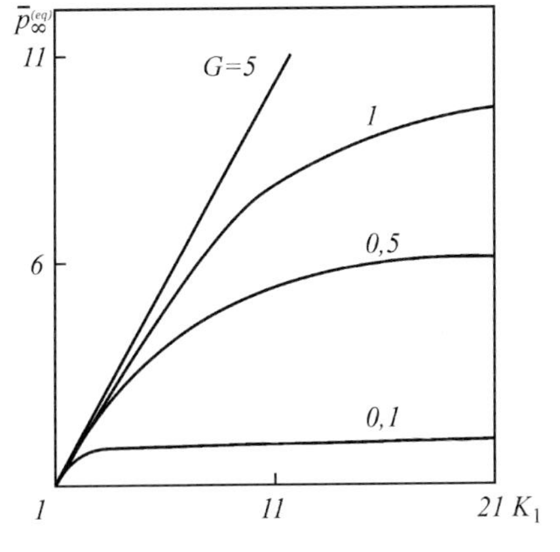

Fig. 23.4 Equilibrium pressure above the mixture surface vs. parameter K_1 for various values of G.

23.3 Differential Separation of Multi-Componenet Mixtures

Consider the process of differential degassing of a quiescent volume of a multi-component liquid-gas mixture, in which the disperse phase consists of spherical bubbles of equal size. Assume that the volume concentration of bubbles is small and their mutual interaction can be neglected. The problem reduces to the analysis of mass exchange for a unit bubble of initial radius R_0 placed in a multi-component solution. Assume that the process is isothermal; the pressure above the mixture surface is maintained at a constant value.

The distribution of component concentrations in the solution is described by the diffusion equation:

$$\frac{\partial \rho_{iL}}{\partial t} + u\frac{\partial \rho_{iL}}{\partial r} = D_{iL}\frac{1}{r^2}\frac{\partial}{\partial r}r^2\frac{\partial \rho_{iL}}{\partial r}, \quad (i = 1, 2, \ldots, n), \tag{23.28}$$

where u is the velocity of the bubble's boundary, equal to

$$u = \dot{R}\left(\frac{R}{r}\right)^2, \tag{23.29}$$

r is the distance from the bubble center, R is the bubble radius.

The condition of dynamic balance of the bubble surface with due consideration of inertial, viscous, and capillary forces is given by the equation

$$R\ddot{R} + \frac{3}{2}\dot{R}^2 = \frac{1}{\rho_L}\left(p_G - \frac{2\Sigma}{R} - \frac{4\mu_L}{R}\dot{R} - p_\infty\right), \tag{23.30}$$

where ρ_L is the density of the liquid; Σ is the coefficient of surface tension; μ_L is the liquid viscosity; p_G is the pressure inside the bubble; p_∞ is the pressure far away from the bubble.

Assuming that all parameters of the gas in the bubble are distributed uniformly but depend on time, we have the following equations:

- balance-of-mass of i-th component in the bubble:

$$\frac{dm_{iG}}{dt} = J_{Di} = 4\pi R^2 \left(\frac{\partial \rho_{iL}}{\partial r}\right)_{r=R}, \qquad m_{iG} = \frac{4}{3} 4\pi R^3 \rho_{iG}; \tag{23.31}$$

- gas state equation, under the assumption that the gas is ideal:

$$p_G = \sum_i p_{iG} = \sum_i \frac{AT_{iG}}{M_i}, \tag{23.32}$$

where A is the gas constant, T – the absolute temperature, M – the molecular mass of i-th component;

- the equation expressing Henry's law, under the assumption that the solution is extremely diluted:

$$p_{iG} = B_i \frac{\rho_{iw}/M_i}{\rho_s/M_s + \sum_j \rho_{jw}/M_j}, \tag{23.33}$$

where B_i is Henry's constant, ρ_{iw} is the mass concentration of i-th component in the solution at the bubble surface, ρ_s is the mass concentration of the solvent, and M_s is the molecular mass of the solvent.

To these equations, it is necessary to add the following initial and boundary conditions:

$$\begin{aligned}
&\rho_{iL}(r,0) = \rho_{iL0}, \quad \rho_{iL}(R,t) = \rho_{iwL}, \quad \rho_{iL}(\infty,t) = \rho_{i\infty L}, \\
&R(0) = R_0, \quad \dot{R}(0) = 0, \quad \rho_{iG}(0) = \rho_{iG0}.
\end{aligned} \tag{23.34}$$

If component concentrations in the bulk of the solution are constant, then the number of equations (23.28)–(23.34) would suffice to solve the formulated problem. In fact, $\rho_{i\infty L}$ changes with time. To derive the appropriate equation describing the change of ρ_{iL} with time, it is necessary to consider the mass balance of a unit volume of liquid-gas mixture under the condition that the volume concentration of bubbles is small ($W \ll 1$). At the moment t, the masses of i-th component in the liquid and gaseous phases, respectively, are equal to

$$m_{iL} = (1 - W)\rho_{i\infty L}, \quad m_{iG} = W\rho_{iG},$$

and the volume concentration of bubbles is

$$W = \frac{4}{3}\pi R^3 N = W_0 \left(\frac{R}{R_0}\right)^3, \tag{23.35}$$

where N is number of bubbles per unit volume of the mixture.

The mass of i-th component Δm_i that transfers from the liquid phase into the gas is equal to

$$\Delta m_i = m_{i0}\left(1 - \frac{(1-W)\rho_{i\infty L}}{(1-W_0)\rho_{iL0}}\right).$$

Now the balance-of-mass condition for i-th component in the solution can be written as

$$\frac{d\rho_{i\infty L}}{dt} = -NJ_D = -\frac{3W_0 D_{iL}}{R_0}\left(\frac{R}{R_0}\right)^3 \left(\frac{\partial \rho_{iL}}{\partial r}\right)_{r=R}, \tag{23.36}$$

$$\rho_{i\infty L}(0) = \rho_{iL0}.$$

Notice that the *constraint*, that is, variation of $\rho_{i\infty L}$ with volume content of bubbles has not been taken into account in a correct way, because the boundary condition (23.34) for ρ_{iL} far away from the drop should be formulated not at $r \to \infty$, but at $r = R_z$, where R_z is the radius of a cell, equal to the average distance between bubbles. Therefore the formulated problem should be treated merely as the first approximation to the goal of making a proper account of the constraint.

The condition that the solution is extremely diluted allows us to take $\rho_L = \rho_s$ and $M_L = M_s$.

Introduce the following dimensionless variables:

$$\bar{\rho}_i = \frac{\rho_{iL} - \rho_{iLw}}{\rho_{iL0} - \rho_{iLw}}, \quad \bar{R} = \frac{R}{R_0}, \quad \tau = \frac{tD_{1L}}{R_0^2}, \quad x = \frac{r-R}{R_0},$$

$$y = 1 - \frac{S}{S+x}, \quad \bar{\rho}_{iw} = \frac{\rho_{iwL}}{\rho_{iL0}}, \quad \bar{\rho}_{i\infty} = \frac{\rho_{i\infty L}}{\rho_{iL0}}, \quad d_i = \frac{D_{iL}}{D_{1L}},$$

$$q = \frac{2\Sigma}{R_0 p_\infty}, \quad \alpha_i = \frac{B_i M_s \rho_{iL0}}{M_i \rho_s p_\infty}, \quad \beta = \frac{\rho_s D_{1L}^2}{p_\infty R_0^2}, \quad \gamma = \frac{4\mu_L D_{1L}}{p_\infty R_0^2}, \quad \varepsilon_i = \frac{3AT\rho_L}{B_i M_L}.$$

Introduction of the dimensionless coordinate y instead of r results in that the external area $R \le r < \infty$ transforms into a strip $0 \le at < 1$. Such a transformation is convenient for the numerical solution procedure.

In new variables, Eqs. (23.28)–(23.34) and (23.36) become:

$$\frac{\partial \bar{\rho}_i}{\partial \tau} + \left\{\frac{(1-y)^2}{S}\frac{d\bar{R}}{d\tau} + \left(\frac{\bar{R}(1+y)}{Sy + \bar{R}(1+y)}\right)^2 \frac{(1-y)^2}{S}\frac{d\bar{R}}{d\tau} - \frac{2d_i(1-y)^2}{S[\bar{R}(1-y)+Sy]}\right\}\frac{\partial \bar{\rho}_i}{\partial y}$$

$$= \frac{d_i(1-y)^2}{S^2}\frac{\partial}{\partial y}\left((1-y)^2 \frac{\partial \bar{\rho}_i}{\partial y}\right) + \left(\frac{\bar{\rho}_{iL} - 1}{1 - \bar{\rho}_{iw}}\right)\frac{d\bar{\rho}_{iw}}{d\tau}, \tag{23.37}$$

$$\beta\left\{\frac{d^2\bar{R}}{d\tau^2}\bar{R}+\frac{3}{2}\left(\frac{d\bar{R}}{d\tau}\right)^2\right\}=\sum_i \alpha_i p_{iw}-1-\frac{q}{\bar{R}}+\frac{\gamma}{\bar{R}}\frac{d\bar{R}}{d\tau},$$

$$\frac{d(\bar{R}^3\bar{p}_{iw})}{d\tau}=\frac{d_i\varepsilon_i\bar{R}^2}{S}\left(\frac{d\bar{\rho}_i}{dy}\right)_{y=0}(1-\bar{\rho}_{iw}),$$

$$\frac{d\bar{\rho}_{i\infty L}}{d\tau}=-\frac{3W_0 d_i}{S}(1-\bar{\rho}_{iw})\bar{R}^2\left(\frac{\partial\bar{\rho}_i}{\partial y}\right)_{y=0}.$$

Eqs. (23.37) should be solved together with the following conditions:

$$\bar{\rho}_i(y,0)=1,\quad \bar{R}(0)=1,\quad \dot{\bar{R}}=0,\quad \bar{\rho}_{i\infty}(0)=1,$$
$$\bar{\rho}_i(0,\tau)=0,\quad \bar{\rho}_i(1,\tau)=\bar{\rho}_{i\infty}. \tag{23.38}$$

Solving them, we find the distributions $\bar{\rho}_i(y,\tau)$; $\bar{\rho}_{iw}(\tau)$; $\bar{R}(\tau)$, and $\bar{\rho}_{i\infty}(\tau)$, whereupon we can determine the dimensionless mass of i-th component that transfers from the liquid into the gaseous phase during the time interval τ:

$$\Delta\bar{m}_i=\frac{\Delta m_i}{m_{i0}}=1-\frac{\bar{\rho}_{i\infty}(1-W_0\bar{R}^3)}{1-W_0}.$$

A transition to molar fractions can be accomplished by using the formulas

$$x_i=\frac{\rho_{iL}M_L}{\rho_L M_i},\qquad y_i=\frac{\rho_{iG}/M_i}{\sum_j \rho_{jG}/M_j}.$$

Sometimes the initial composition of the solution is given by weight percentages Π_i. Then molar fractions of components in the bulk of the solution are expressed in terms of weight percentages as follows:

$$x_{i\infty}=\frac{\Pi_i M_L}{M_i\left(100-\sum_j \Pi_j\right)}.$$

When doing numerical integration of Eqs. (23.37), the main problem is to assign to the bubble an initial composition y_{i0}, because this composition is not known beforehand. Calculations show, however, that for any choice of initial composition, the final composition will be established very quickly and remains unchanged afterwards.

Consider an example of numerical solution of the problem of a five-component mixture at the following values of parameters: $d_i=1$; $q=0.1$; $\gamma=4\cdot 10^{-8}$; $\beta=10^{-10}$; $\alpha_1=0.187$; $\alpha_2=0.33$; $\alpha_3=0.649$; $\alpha_4=0.168$; $\alpha_5=0.364$; $\varepsilon_1=2.54$; $\varepsilon_2=25.4$; $\varepsilon_3=31.7$; $\varepsilon_4=42.3$; $\varepsilon_5=63.5$. Two variants of bubble initial composition are presented in Table 23.1.

The following notations are used in the table: y_{i0} – initial molar fractions of components in the bubble; $y_i^{(eq)}$ – the corresponding equilibrium values; Π_{i0} – the initial weight percentages of components in the solution.

Table 23.1

Component		C_1	C_2	C_3	i-C_4	n-C_4
I	y_{i0}	0.15	0.25	0.4	0.1	0.1
II	y_{i0}	0.025	0.15	0.3	0.225	0.3
	$y_i^{(eq)}$	0.05	0.186	0.385	0.115	0.264
	Π_{i0} (%)	0.01	0.33	1.19	0.54	2.43

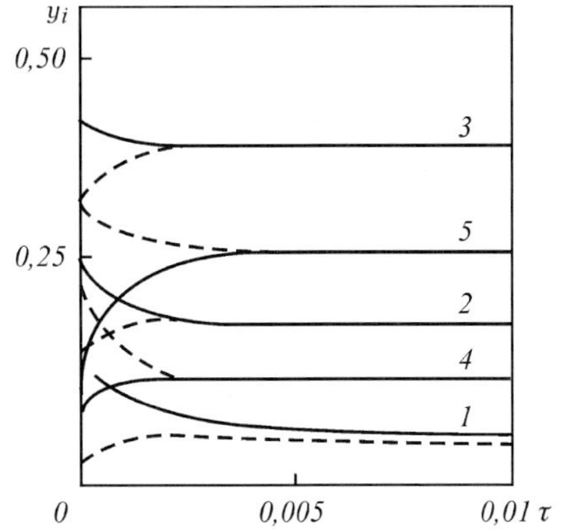

Fig. 23.5 Molar fractions of components in the bubble vs. time: 1–5 – component indices.

Molar composition of the bubble is established very quickly (see Fig. 23.5). The continuous line shows dependences $y_i(\tau)$ for the first variant, and dotted line – for the second one.

In the gaseous phase, equilibrium is established much faster than in the liquid phase. As an example, we consider three variants of initial solution composition, presented in Table 23.2.

Fig. 23.6 shows the bubble radius as a function of time. Switching back to dimensional time, we get $t = 0.1\ \tau$ for the characteristic values $D_1 \sim 10^{-9}$ m^2/s and $R_0 \sim 10^{-5}$ m $= 10$ µm.

Table 23.2

Component		C_1	C_2	C_3	i-C_4	n-C_5
I	Π_i	0.01	0.33	1.19	0.54	2.43
II	Π_i	0.015	0.35	1.25	0.6	2.5
III	Π_i	0.01	0.03	1.1	0.5	2.25

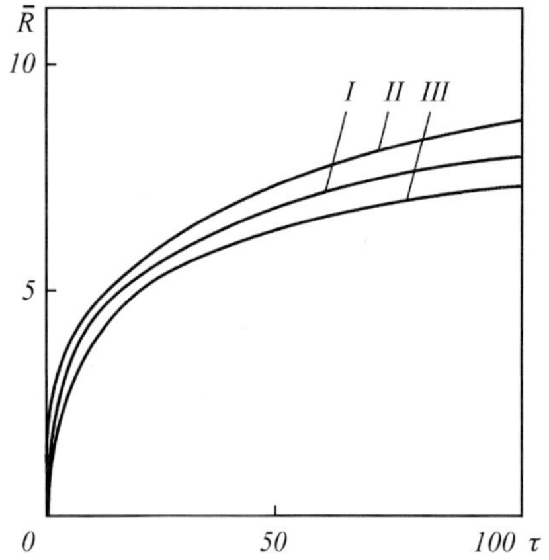

Fig. 23.6 Bubble radius vs. time: *I–III* – variants of initial composition of the liquid (see Tab. 23.2).

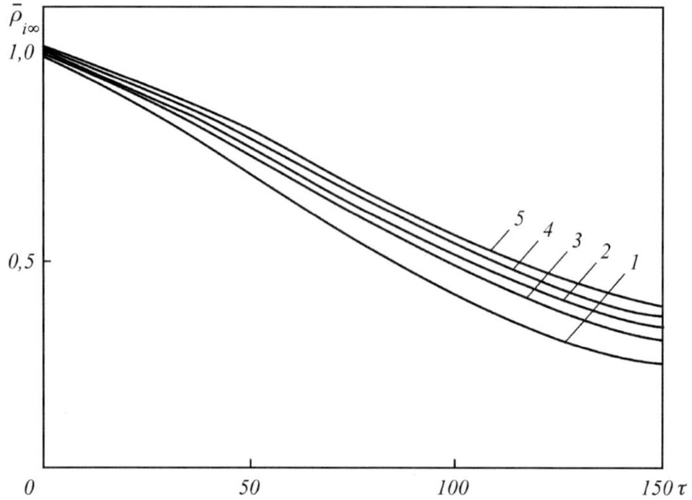

Fig. 23.7 Mass concentrations of liquid phase components vs. dimensionless time: *1–5* – component indices.

Therefore the characteristic time of bubble growth is estimated as $t \sim 10$ s. Shown in Fig. 23.7 is the change in time of mass concentrations of components in the liquid phase, divided by the initial values given in the first row (variant I) of Tab. 23.2. The initial volume concentration of the gaseous phase was $W_0 = 3 \cdot 10^{-4}$. Volume concentration of bubbles grows with time (Fig. 23.8). It increases to $W \sim 0.1$ in $\tau \sim 200$ ($t \sim 20$ s).

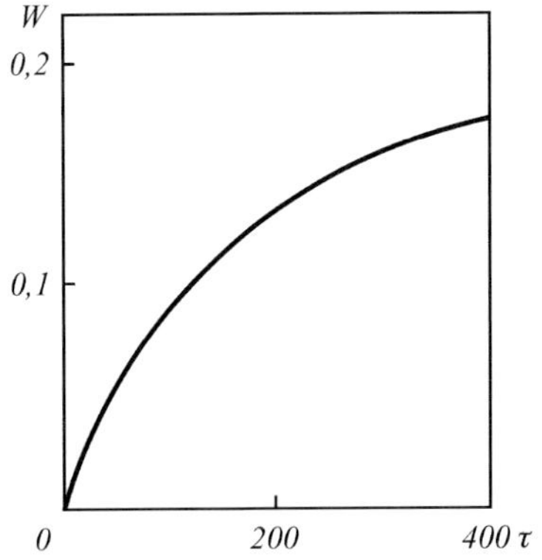

Fig. 23.8 Volume concentration of the gaseous phase vs. time.

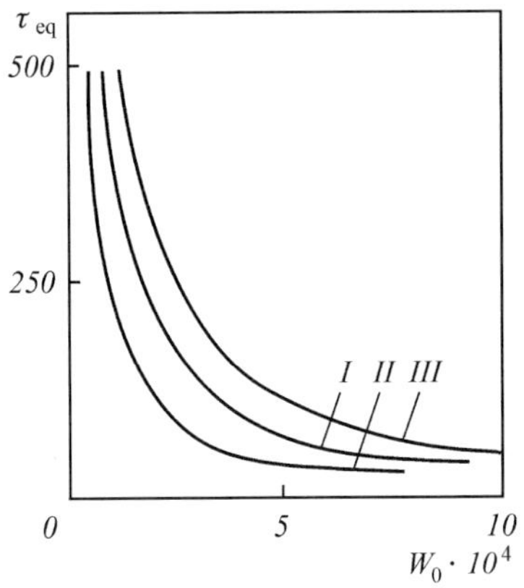

Fig. 23.9 Characteristic relaxation time τ_{eq} vs. initial volume concentration of the gaseous phase: *I–III* are variants of initial composition of the liquid phase.

Of greatest interest is the time τ_{eq} it takes for the system to come to equilibrium. Fig. 23.9 shows the dependence of τ_{eq} on the initial volume concentration of bubbles W_0. Thus, we have $t \sim 60$ s for $W_0 \sim 10^{-4}$, and $\tau_{eq} \sim 5$–10 s for $W_0 \sim 5 \cdot 10^{-4}$.

One should be aware that the results above should only be considered as estimations, because for real hydrocarbon mixtures, especially at high pressures, one has to use the equation of state for a real mixture. Besides, the extremely diluted solution approximation may not correspond to the real picture. The method of solution can be easily generalized for the case of a real liquid-gas mixture.

23.4
Separation of a Moving Layer

The problem of separation of a moving layer of liquid-gas mixture is of interest in research of separation efficiency of oil separators. The gaseous phase consists of bubbles brought by the flow from the supply pipeline, and of bubbles nucleated in the separator as a result of sharp pressure decrease. The majority of oil separators are equipped with a special device for preliminary removal of gas prior to separation, therefore only a small part of gas discharged in the supply pipeline enters the separator. Based on this assumption, we shall assume that the volume of the disperse phase in the mixture that moves inside the separator is small, so the interaction of bubbles can be neglected in the first approximation. Interaction between bubbles will be accounted for later on. If the mixture is sufficiently depleted of dissolved components, which can be the case, for example, at the last stage in the scheme of differential separation, then the bubbles present in the volume are predominantly of the first type. For a mixture enriched with dissolved components, for example, at the first stage of separation, bubbles of both types are present in the volume. In view of this fact, two cases are possible. In the first case, bubbles brought by the flow do not grow, but simply rise up to the surface and quit the volume. In contrast, in the second case, there are bubbles of two sorts in the mixture, which are lifting and simultaneously increasing in size as the dissolved components diffuse into them. Consider the two cases in succession.

Consider a layer of mixture of length L and height H, moving on an inclined plane (Fig. 23.10). The radius distribution of bubbles $n_0(R)$ at the entrance is given. Denote by $n_1(R)$ the radius distribution at the exit. To estimate the efficiency of separation of bubbles from the liquid, introduce the transfer function

$$\Phi(R) = \frac{n_1(R)}{n_0(R)}. \tag{23.39}$$

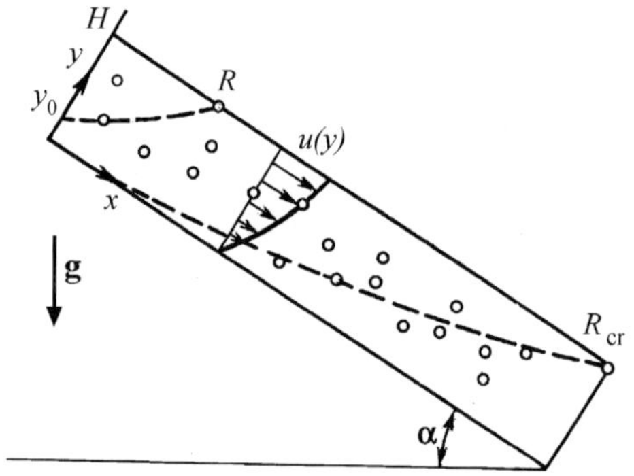

Fig. 23.10 Scheme of mixture flow on an inclined plane.

If the transfer function is known, then for any given distribution at the entrance, it is possible to find the distribution at the exit.

Suppose the mixture layer is sufficiently thin so that $H/L \ll 1$, with both H and L remaining constant. Then the profile of longitudinal velocity u can be determined from the equation of motion of a thin film:

$$\mu_L \frac{d^2u}{dy^2} = -\rho_L g \sin\alpha. \tag{23.40}$$

Integration of this equation, with the imposed conditions $u = 0$ at $y = 0$, $du/dy = 0$ at $y = H$ and the condition of mass flow rate Q conservation, gives

$$u = \frac{H^2\rho_L g \sin\alpha}{\mu_L}\left(-0.5\frac{y^2}{H^2} + \frac{y}{H}\right), \quad H = \left(\frac{3\mu_L Q}{\rho_L g \sin\alpha}\right)^{1/2}. \tag{23.41}$$

The velocity, averaged over the height of the layer, is equal to

$$\bar{u} = \frac{H^2\rho_L g \sin\alpha}{3\mu_L}. \tag{23.42}$$

Trajectories of non-inertial motion of a bubble of radius R in the layer, provided that it moves as a solid particle with Stokes velocity, are described by the equations

$$\frac{dx}{dt} = u - \frac{2R^2\rho_L g \sin\alpha}{9\mu_L}, \quad \frac{dy}{dt} = \frac{2R^2\rho_L g \cos\alpha}{9\mu_L}, \tag{23.43}$$

$$x(0) = 0, \quad y(0) = y_0,$$

which, in dimensionless variables

$$X = \frac{x}{H}, \quad Y = \frac{y}{H}, \quad \bar{R} = \frac{R}{R_0}, \quad v = \frac{\mu_L u}{H^2\rho_L g}, \quad \beta^2 = \frac{9H^2}{2R_0^2}, \quad X_L = \frac{L}{H},$$

have the solution

$$R^2 X \cos\alpha = \beta^2 \sin\alpha\left(0.5(Y^2 - Y_0^2) - \frac{Y^3 - Y_0^3}{3}\right) + R^2(Y - Y_0)\sin\alpha. \tag{23.44}$$

The distance from the entrance, at which the bubble will reach the surface, can be obtained from the solution (23.44) at $Y = 1$:

$$X_{cr} = \frac{\beta^2 \tan\alpha}{R}\left(0.5(1 - Y_0^2) - \frac{1 - Y_0^3}{3}\right) + (1 - Y_0)\tan\alpha. \tag{23.45}$$

Now, it is easy to find the critical radius of the bubble, i.e. the radius at which a bubble that was initially at the entrance at the point $(0, 0)$ comes up to the surface at the point (L, H):

$$\bar{R}_{cr} = \left(\frac{\beta^2 \tan \alpha}{6(X_L - \tan \alpha)}\right)^{1/2}. \tag{23.46}$$

Since $X_L \gg 1$ and $\alpha \sim 6$–10°, this gives us:

$$\bar{R}_{cr} \approx \left(\frac{\pi \beta^2 \alpha}{1080 X_L}\right)^{1/2}. \tag{23.47}$$

If at the entrance all bubbles have different sizes, then all bubbles with $R > R_{cr}$ will manage to rise to the surface and escape from the volume during their residence time in the layer, while a part of bubbles with $R < R_{cr}$ will be carried away by the flow. If the distribution of bubbles at the entrance is uniform over the cross section, then the ratio between the number of bubbles of radius R carried away by the flow, and the number of bubbles of the same radius at the entrance, will be equal to the transfer function $\Phi(R)$, which, in its turn, is equal to Y_{0cr}, where Y_{0cr} is such value of Y_0 at the entrance that the bubble that started from the point $(0, Y_{0cr})$ will get to the surface at the point $(X_L, 1)$. The value of Y_{0cr} can be found from the cubic equation (23.44), into which we must plug $X = X_L$ and $Y_0 = Y_{0cr}$. Since the solution looks bulky, let us estimate it by replacing the parabolic velocity profile (23.41) with the uniform profile $u = \bar{u}$ (see 23.42).

Then

$$Y_{0cr} = 1 - \frac{3R^2 X_L}{\beta^2 \tan \alpha} = 1 - \frac{R^2}{R_{cr}^2}, \quad R_{cr} = \left(\frac{9\mu_L Q}{2\rho_L g L}\right)^{1/2}. \tag{23.48}$$

Comparing the expressions (23.47) and (23.48) for the critical radius, we see that replacement of the parabolic velocity profile with a uniform one increases R_{cr} by the factor of $\sqrt{2}$.

The transfer function, by virtue of (23.48), is equal to

$$\Phi(R) = \begin{cases} 1 - (R/R_{cr})^2 & \text{at } R < R_{cr}, \\ 0 & \text{at } R \geq R_{cr}. \end{cases} \tag{23.49}$$

It is now possible to determine the mass of the gas carried away by the flow. The masses of the gas contained in the bubbles at the entrance and at the exit are, respectively,

$$M_0 = \rho_G \int_0^\infty V n_0(V)\, dV, \quad M_1 = \rho_G \int_0^\infty \Phi(R) V n_0(V)\, dV.$$

The separation efficiency is

$$\eta = 1 - \frac{M_1}{M_0} = 1 - \frac{\displaystyle\int_0^\infty \Phi(R) V n_0(V)\, dV}{\displaystyle\int_0^\infty V n_0(V)\, dV}$$

$$= 1 - \frac{\displaystyle\int_0^{V_{cr}} (1 - R^2/R_{cr}^2) V n_0(V)\, dV}{\displaystyle\int_0^\infty V n_0(V)\, dV}, \tag{23.50}$$

where $V_{cr} = 4\pi R_{cr}^3/3$.

As an example, suppose a liquid-phase mixture approaches the entrance, whose volume distribution of bubbles is given by a gamma distribution,

$$n_0(V) = \frac{N_0^2(k+1)}{W_0 k!} \left(\frac{V}{V_0}\right)^k \exp\left(-\frac{V}{V_0}\right), \tag{23.51}$$

where N_0 and W_0 are the number and volume concentrations of bubbles; k and V_0 are parameters of the distribution, related to the average volume V_{av} and variance σ^2 by the formulas

$$V_{av} = V_0(k+1), \quad \sigma^2 = V_{av}^2(k+1)^{-1}.$$

The distribution (23.51) for various values of k is shown in Fig. 23.11.

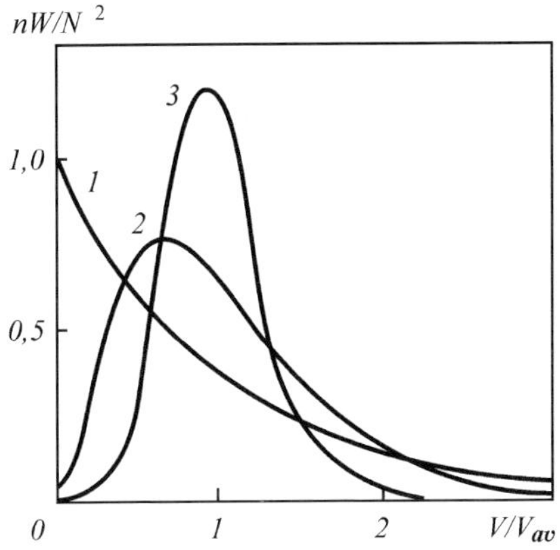

Fig. 23.11 Gamma distribution of bubbles over volumes for different values of the distribution parameter k: 1 – 0; 2 – 2; 3 – 15.

A substitution of (23.51) into the expression (23.50) gives

$$\eta = 1 - \frac{1}{(k+1)!}\left(\gamma(k+2,z) + \frac{1}{z^{2/3}(k+1)!}\gamma(k+8/3,z)\right), \tag{23.52}$$

where $z = V_{cr}(k+1)/V_{av}$ and $\gamma(x, y)$ is the reduced gamma function.

The expression (23.52) gives the dependence of the efficiency coefficient of separation on distribution parameters k and σ^2 at the entrance, and also on the geometrical and hydrodynamic parameters of the flow appearing in z. Note that the parameter z depends on the residence time t of the mixture on the inclined plane:

$$z = (k+1)\left(\frac{t}{t_{av}}\right)^{3/2}, \quad t_{av} = \frac{Lu_{av}}{\bar{u}H}, \quad u_{av} = \frac{2\Delta\rho g R_{av}^2}{9\mu_L}.$$

The dependence of η on dimensionless time $\tau = t/t_{av}$ is shown in Fig. 23.12. Note that k $= \infty$ corresponds to a monodisperse distribution of bubbles over volumes.

Consider now the second case, when a supersatutared mixture flows on an inclined plane. The difference from the case considered above is that bubbles can grow in size according to the expression (22.14) for a binary mixture:

$$\bar{R} = \frac{R}{R_0} = \left(1 + (\bar{\rho}_{1L} - \bar{\rho}_{1wL})(Y - Y_0)\gamma_H\right)^{1/3}, \tag{23.53}$$

where $\bar{\rho}_{1L}$, $\bar{\rho}_{1wL}$ are dimensionless parameters introduced in Section 23.3, $\gamma_H = 6D_{1L}^{2/3}\rho_L ATH/\pi R_0^2\alpha^{2/3}B_1M_L$ and $\alpha = 2g\Delta\rho/9\mu_L$. There is also the possibility of nucleation of bubbles, which occurs with intensity N_n.

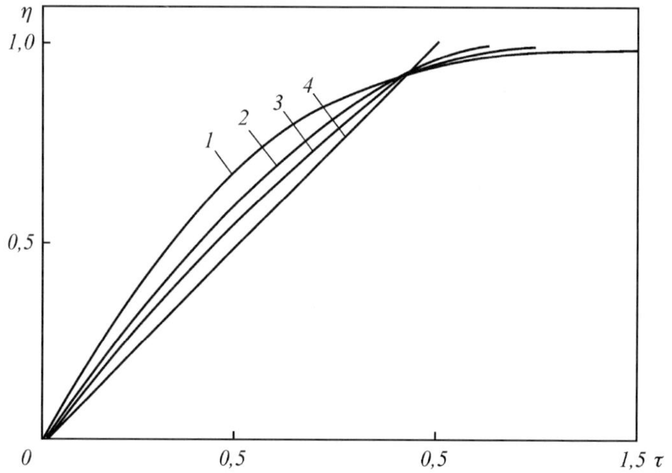

Fig. 23.12 Coefficient of efficiency of separation η vs. time τ for different values of k: 1 – 0; 2 – 2; 3 – 15; 4 – ∞.

Begin with the case when the number of incipient bubbles is small and their contribution to the separation process can be neglected. Substituting the expression (23.53) into (23.41), we obtain:

$$X = \frac{3\beta^2 \bar{u}}{\gamma_H(\rho_{1L} - \rho_{1wL})}\left(\left(1 + (\bar{\rho}_{1L} - \bar{\rho}_{1wL})(Y - Y_0)\gamma_H\right)^{1/3} - 1\right). \tag{23.54}$$

Taking $Y = 1$, $X = X_L = L/H$ and $Y_0 = 0$, we can determine from (23.54) the critical radius $R_{0\text{cr}}$, and then the transfer function:

$$\Phi(R_0) = \begin{cases} Y_{0cr} & \text{at } R_0 < R_{cr}, \\ 0 & \text{at } R_0 \geq R_{cr}, \end{cases} \tag{23.55}$$

where

$$Y_{0cr} = 1 - \frac{1}{F}\left(\frac{R_0}{R_{0cr}}\right)^3 \left\{ \frac{R_0}{R_{0cr}} \left(\left(1 + F\left(\frac{R_0}{R_{0cr}}\right)^3 \right)^{1/3} - 1 \right) - 1 \right\},$$

$$F = \frac{6(\rho_{1L} - \rho_{1wL}) D_{1L}^{2/3} \rho_L ATH}{\pi\alpha^{2/3} B_1 M_L R_{0\alpha v}^3}.$$

Here R_{0av} is the average radius of bubbles at the entrance and R_{0cr} is the critical radius, which can be found from the equation

$$S\frac{R_{0\alpha v}}{R_{0cr}} = \left(1 + F\left(\frac{R_0}{R_{0cr}}\right)^3\right)^{1/3} - 1, \quad S = \frac{2FR_{0\alpha v}^2 X_L}{27\bar{u}H^2}.$$

Suppose that together with growth of bubbles brought in from the outside, there occurs nucleation of the new bubbles of radius R_n, at the rate N_n. It is easy to see that a bubble nucleated at the depth Y_0, will, at the end of the layer, have the radius $R_1 = R_n(1 + X_L F/\beta^2 tg\alpha)$. In order for this to happen, the bubble must travel the distance $X_1 = (R_1/R_n - 1)\beta^2 tg\alpha/F$. Thus, in order for the bubble of initial radius R_n to reach the surface of the layer, it should be nucleated at a distance no less than X_1 from the end of the layer. Out of all the bubbles, only those will reach the exit, for whom the inequality $Y_0 < Y_{0cr}$ is satisfied.

As an example, consider the lifting of bubbles whose distribution at the entrance is given by a gamma-distribution of the kind (23.51).

In Fig. 23.13, distributions of bubbles at the exit for $k = 2$ are shown. The following parameter values were used in the calculation: $\bar{\mathbf{u}} = 0.1$ m/s; $H = 0.01$ m; $D_{iL} = 10^{-9}$ m^2/s; $T_n = 10^9$ m^{-3}s^{-1}, and $R_{0av} = 10^{-5}$ m. The length of the layer was taken as a variable. One can see that significant lengths (~25–50 m) are necessary for efficient separation on the inclined layer. In practice, such lengths are achieved by providing the separator with a system of shelves connected in parallel or in series. In such a separator, in order to improve separation efficiency, either

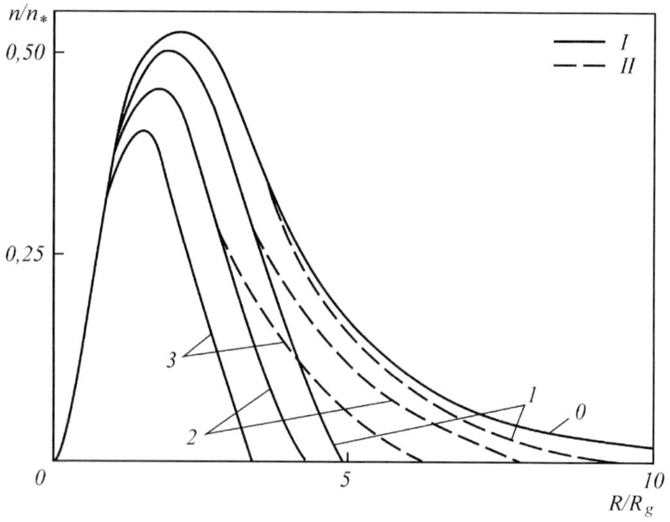

Fig. 23.13 The radius distribution of bubbles as the mixture leaves the layer for different lengths of the layer L, m: *0* – initial distribution; *1* – 10; *2* – 50; *3* – 100; *I*, *II* – with and without due consideration for diffusion growth of bubbles.

the flow rate of the mixture should be distributed uniformly over the shelves, or, if the flow rate is fixed, the shelf's length should be increased. Fig. 23.14 shows the dependence of separation efficiency on the shelf's length for the mixture moving on an inclined shelf.

Thus, separation of free gas in a thin layer moving on inclined shelves is efficient enough, while to achieve efficient separation of dissolved gas, it is necessary to increase the layer's thickness.

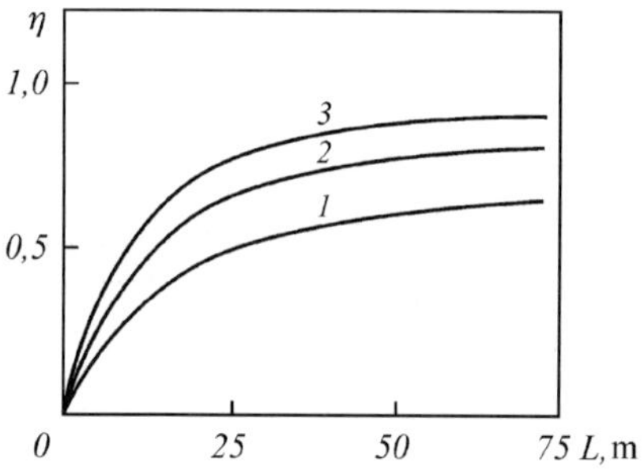

Fig. 23.14 Efficiency vs. length of the inclined shelf for separation of gas dissolved in a liquid ($H = 0.1$ m; $\bar{u} = 0.1$ m/s), for different values of Λ: *1* – 4; *2* – 1; *3* – 0.

24
Separation with Due Regard of Hinderness of Floating Bubbles

Consider the process of liquid-gas mixture separation in a gravitational horizontal separator. The mixture enters the separator from the supply pipeline equipped with a device for preliminary separation of free gas. A study of dispersivity of liquid-gas flow at the entrance is presented in [7]. As is shown in this work, the preliminary separation of large bubbles results in that the bubbles entering the separator have diameters ranging within a narrow interval from D_1 to D_2, with $D_2 \approx 3D_1$, and with the variance $\sigma^2 \sim 0.003$. The distribution of bubbles over diameters D at the entrance can be approximated in the interval $(D_1; D_2)$ by a uniform (in the simplest case) distribution,

$$n_0(D) = \frac{N_0}{(D_2 - D_1)} \tag{24.1}$$

or the normal one,

$$n_0(D) = \frac{N_0}{\sqrt{2\pi}\sigma} \exp\left(-\frac{(D - D_{av})^2}{2\sigma^2}\right), \tag{24.2}$$

where $N_0 = W_0/V_{av}$ is the number concentration of bubbles at the entrance; D_{av} is the mean diameter of bubbles at the entrance; D_1 and D_2 are the smallest and the largest diameter of bubbles; W_0 is the volume concentration of bubbles at the entrance.

Despite the preliminary separation of gas in the supply pipeline, the liquid-gas mixture enters the separator with sufficiently high concentration of the gaseous phase ($W_0 \sim 10^{-2}$–10^{-1}). High values of W_0 result in the necessity to take into account the constrained (hindered) character of bubble floating. As was experimentally shown in [7], the velocity of bubble's ascent, with the constraint taken into account, is well approximated by the expression

$$U = U_{st} f(W), \quad f(W) = (1 - W)^{\beta}, \tag{24.3}$$

where the exponent index β depends on the dimensionless parameter $K = \Sigma^{3/2}/g^{1/2}\nu_L^2\rho_L^{3/2}$ as follows:

$$\beta = 4.9 + 3.4 \cdot 10^{-5} K. \tag{24.4}$$

Separation of Multiphase, Multicomponent Systems. E. G. Sinaiski and E. J. Lapiga

ISBN: 978-3-527-40612-8

For the characteristic parameter values $\Sigma = 10^{-2}$ N/m; $g = 10$ m/s^2; $\nu_L = 10^{-5}$ m/s^2; $\rho_L = 10^3$ kg/m^3, we have $K \sim 10^2$. However, decrease of oil viscosity to $\nu_L = 10^{-6}$ m^2/s, and increase of surface tension up to $\Sigma \sim 5 \cdot 10^{-2}$ N/m results in the value $K \sim 10^5$. Hence, for highly viscous oil, $\beta \approx 5$, and for low-viscous oil, the parameter K influences the bubble's lift velocity.

Gravitational separators can operate either in the regime of periodic pump-out or in the flow regime. In the first case the separator is filled with liquid-gas mixture, the mixture undergoes sedimentation for some time, until most of the gas bubbles float to the surface, and then the separator is emptied. In this case one is interested in the optimal sedimentation time, which depends on the required degree of separation. In the second case the liquid-gas mixture flows continuously through the separator. The primary goal is to determine the residence time of mixture in the device that is necessary for the required degree of separation. Consider these two regimes in more detail.

24.1
Separation in the Periodic Pump-out Regine

The process of separation in the periodic pump-out regime can be modeled by the following problem. At the initial moment of time, there is a volume of height H, filled by a liquid-gas mixture with a given distribution $n_0(D)$ of bubbles over diameters, which is supposed to be uniform over the height of the layer. Then the distribution of bubbles at $t > 0$ depends on z, t and D, where z is the vertical coordinate directed against gravity, and is described by the kinetic equation, which in the absence of coagulation and diffusion growth of bubbles, takes the form

$$\frac{\partial n}{\partial t} + U_s \frac{\partial}{\partial z}(f(W)n) = 0. \tag{24.5}$$

Let U_s be the Stokes velocity of free motion of a bubble with completely "frozen" surface, which under the obvious condition $\rho_G \ll \rho_L$ equals

$$U_s = \frac{1}{18}\frac{gD^2}{\nu_L}. \tag{24.6}$$

To these equations, one must add the expression (24.3) for $f(W)$ and the relation for the volume concentration of bubbles:

$$W = \int_{D_1}^{D_2} \frac{\pi}{6} D^3 n(z,t,D)\, dD. \tag{24.7}$$

The initial conditions are:

$$n(z,0,D) = n_0(D), \quad W(z,0) = W_0. \tag{24.8}$$

As a boundary condition, we should take the condition of absence of bubbles at the bottom wall of the separator:

$$n(0, t, D) = 0. \tag{24.9}$$

For simplicity, take the uniform distribution (24.1) for $n_0(D)$. Then,

$$W_0 = \frac{\pi N_0}{24(D_2 - D_1)}(D_2^4 - D_1^4). \tag{24.10}$$

The solution of Eq. (24.5) can be obtained by method of characteristics. Introduce the dimensionless variable $\xi = z/U_s t$ and seek the solution of Eqs. (24.5)–(24.10) in the form $n = \eta(D, \xi)$ and $W = \zeta(\xi)$. Then η and ζ should satisfy the following equations:

$$-\xi\eta' + (\mathrm{f}\eta)' = 0, \quad \mathrm{f}(\zeta) = (1 - \zeta)^{\beta}. \tag{24.11}$$

$$\zeta = \int_{D_1}^{D_2} \frac{\pi}{6} D^3 \eta(\xi, D)\, dD, \tag{24.12}$$

$$\eta(0, D) = 0, \quad \eta(\infty, D) = n_0. \tag{24.13}$$

Since the diameters of bubbles range from D_1 to D_2, at $t > 0$ there are three regions in the considered volume (Fig. 24.1), divided by flat surfaces of contact discontinuities (these surfaces are already familiar to us from the similar problem of suspension sedimentation (Section 8.5)). Consider a plane (z, t) (Fig. 24.2) and straight lines

$$\frac{z}{U_s t} = \xi_1 = 1, \quad \frac{z}{U_s t} = \xi_2 = \frac{D_2^2}{D_1^2} f(W_0) \tag{24.14}$$

in this plane.

The region $\xi < \xi_1$ corresponds to a single-phase area I, which is free from bubbles. In it, $n = 0$ and $W = 0$. The region $\xi_2 < \xi < \infty$ (area III adjoining the top interface) is the zone of constant gas concentration. Here we have $\eta = \eta_0$. In the third intermediate region II for any value ξ_* satisfying the inequality $\xi_1 \le \xi_* \le \xi_2$, the following conditions are satisfied:

$$\xi^* = \xi(\xi^*) = \int_{D_1}^{D_*} \frac{\pi}{6} D^3 n_0\, dD = \frac{\pi n_0}{24}(D_*^4 - D_1^4), \quad \xi^* = \frac{D_*^2}{D_1^2} \mathrm{f}(\xi^*).$$

Eliminating D_*/D_1 from these equations, one obtains the dependence of volume concentration of bubbles $W = \zeta$ on z and t:

$$f(W) = \frac{\pi N_0 D_1^4}{24}\left(\frac{\xi^2}{f(W)} - 1\right). \tag{24.15}$$

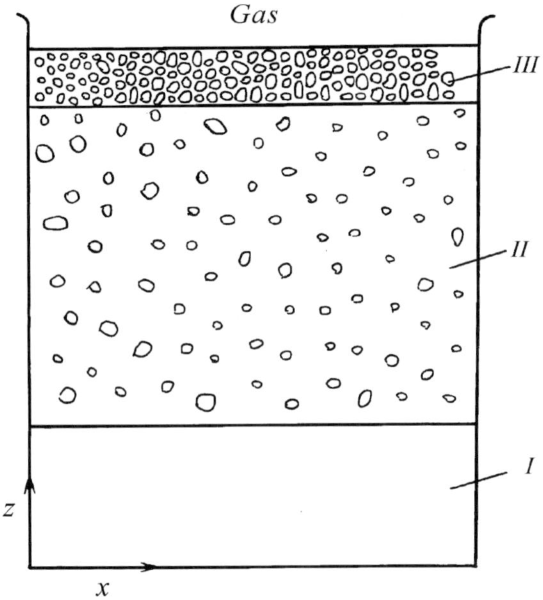

Fig. 24.1 Three regions arising during separation of liquid-gas mixture: I – pure liquid; II – intermediate region; III – zone of constant gas concentration.

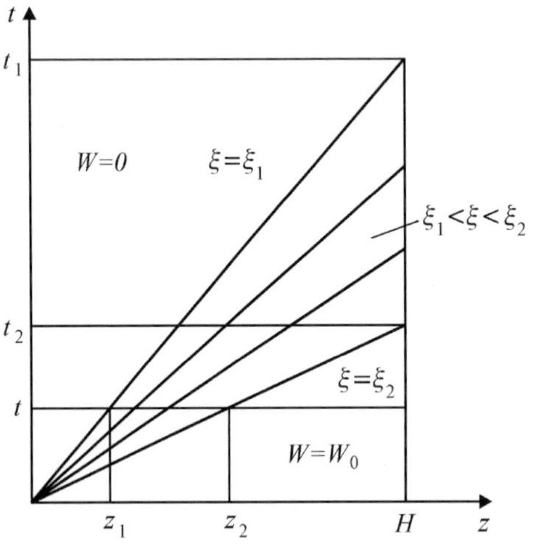

Fig. 24.2 Characteristics of Eq. (24.5): $z_1 = gD_1^2t/18\nu_L$; $z_2 = gD_2^2tf(W_0)/18g\nu_L$.

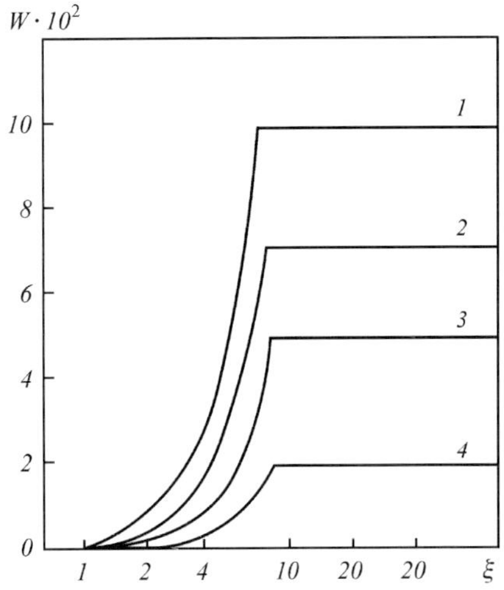

Fig. 24.3 Dependence of the gas content W on ξ for different values of the initial gas content W_0: $1 - 10^{-3}$; $2 - 7 \cdot 10^{-2}$; $3 - 5 \cdot 10^{-2}$; $4 - 2 \cdot 10^{-2}$.

Similarly, eliminating N_0 from (24.15) with the use of formula (24.10), one obtains:

$$z = \frac{gD_1^2}{18\nu_L}\left(\frac{W}{W_0}\left(\frac{D_2^4}{D_1^4} - 1\right) + 1\right)^{1/2} tf(W). \tag{24.16}$$

Depicted in Fig. 24.3 is the dependence of W on ξ for various values of the initial volume concentration W_0.

It is now possible to determine the fraction of the gaseous phase remaining in the volume at a moment of time $0 < t < t_2$:

$$Q_1 = \frac{1}{H}\left\{\int_{gD_1^2t/18\nu_L}^{(gD_2^2t/18\nu_L)f(W_0)} W\,dz + \int_{(gD_2^2t)/(18\nu_L)f(W_0)}^{H} W_0\,dz\right\}, \tag{24.17}$$

at a moment of time $t_2 \le t \le t_1$:

$$Q_1 = \frac{1}{H}\int_{gD_1^2t/18\nu_L}^{H} W\,dz, \tag{24.18}$$

and, finally, at $t > t_1$:

$$Q_1 = 0. \tag{24.19}$$

The time $t_1 = 18\nu_L H/gD_1^2$ can be very long. Therefore in practice some ablation of the gas together with the liquid is allowed. Usually the residual share of gas (residual gas content) is given. This enables us to find from the adduced expressions for Q_1 the necessary time to get given gas content.

24.2
Separation in the Flow Regime

Consider the emersion of bubbles in a liquid moving with a constant longitudinal velocity U in a flat channel of height H. The equation for the diameter distribution of bubbles $n(x, z, D)$, where x is the longitudinal coordinate, is

$$U\frac{\partial n}{\partial x} + U_s\frac{\partial}{\partial z}(f(W)n) = 0. \tag{24.20}$$

To this equation it is necessary to add the equations

$$U_s = \frac{1}{18}\frac{gD^2}{\nu_L}, \tag{24.21}$$

$$W = \int_{D_1}^{D_2} \frac{\pi}{6} D^3 n(x, z, D)\, dD. \tag{24.22}$$

The boundary conditions are

$$n(0, z, D) = n_0(D), \quad n(x, 0, D) = 0. \tag{24.23}$$

Let us consider only the simplest case when there is a uniform distribution of bubbles at the entrance.

The solution of the problem can be obtained in the same manner as in Section 24. Divide the area $0 \le x \le L$, $0 \le y \le H$ into three regions (Fig. 24.4):

the undisturbed region I:

$$z/x > \eta_1 = gD_2^2 f(W_0)/18\nu_L U, \quad n = n_0,\ W = W_0;$$

the region III, completely clear of gas bubbles:

$$z/x < \eta_1 = gD_1^2/18\nu_L U, \quad n = 0;\ W = 0.$$

the intermediate region II:

$$gD_1^2/18\nu_L U \le z/x \le gD_2^2 f(W_0)/18\nu_L U.$$

In this region the solution is similar to (24.16) and has the form

$$z = \frac{gD_1^2}{18\nu_L U}\left\{\frac{W}{W_0}\left(\frac{D_2^4}{D_1^4} - 1\right) + 1\right\}^{1/2} xf(W). \tag{24.24}$$

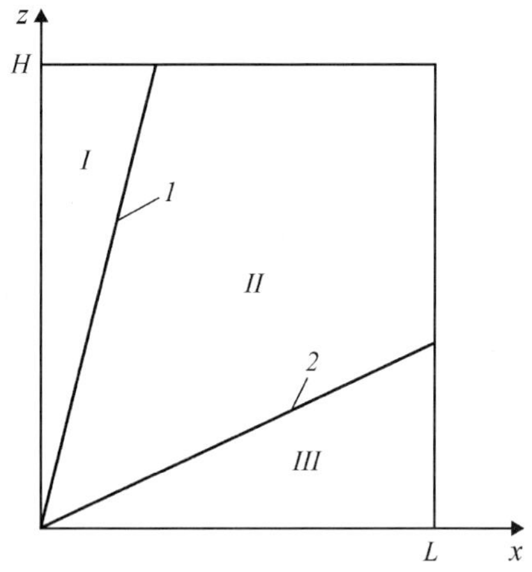

Fig. 24.4 Characteristics of Eq. (24.20): $1 - \eta = \eta_2 = gD_2^2 f(W_0)/18U\nu_L$; $2 - \eta = \eta_1 = gD_1^2 t/18U\nu_L$; $I - W = W_0(\eta > \eta_2)$; $III - W = 0(\eta < \eta_1)$; $II - W = W(\eta = z/x)(\eta_1 < \eta < \eta_2)$.

Eq. (24.24) allows us to find the minimum length of the separator, which at a given height and flow rate provides the required residual gas content of the mixture at the exit.

Using the expressions (24.17) to (24.19), in which t should be replaced by x/U, it is possible to determine the residual gas content at the exit. The volume flow rate of gas Q_G carried away by the mixture from the separator is

$$Q_G = U \int_{LU_s/U}^{H} W(L, z)\, dz. \tag{24.25}$$

The corresponding value at the entrance to the separator is equal to

$$Q_{G0} = UHW_0. \tag{24.26}$$

The efficiency of separation is determined by the fraction of the gas separated from the mixture in the separator

$$\eta = 1 - \frac{G_G}{Q_{G0}}. \tag{24.27}$$

Thus, having set the value of η, it is possible to find from (24.27) and (24.25) the geometrical sizes of the separator (L or H).

25
Coagulation of Bubbles in a Viscous Liquid

Hitherto, our study of bubble-growth dynamics has hinged on the assumption that the bubbles do not interact with one another. Such assumption is valid at small volume concentrations of bubbles ($W \ll 1$), since the average distance between them in this case is $L_{av} \sim R_{av}/\sqrt[3]{W} \gg R_{av}$. For example, for $W \sim 10^{-3}$, the average distance between bubbles is estimated as $L_{av} \sim 10R_{av}$, while an increase of W up to the value $\sim 10^{-1}$ results in the reduction of L_{av} down to $3R_{av}$. If besides the bubbles increase in sizes via diffusion growth, then the average distance between the bubbles becomes even smaller with time. Hence, during the growth of gas bubbles – either under the mixture supersaturation conditions, or with a drop of pressure, as in the motion of liquid-gas mixture along vertical pipes (wells), their volume concentration can increase considerably, which results in the necessity to take into account bubble interaction, in particular, their coagulation.

The dispersiveness of a liquid-gas mixture can be characterized by a continuous volume distribution of bubbles $n(V, t, P)$ at the moment of time t at the point P of the considered volume. The distribution n satisfies the kinetic equation, which, with due regard of the diffusion growth and coagulation, and neglecting the breakup of bubbles, takes the form:

$$\frac{\partial n}{\partial t} + \nabla \cdot (\mathbf{u}n) + \frac{\partial}{\partial V}\left(n\left(\frac{dV}{dt}\right)_D\right) = I_k, \tag{25.1}$$

$$I_k = \frac{1}{2}\int_0^V K(\omega, V-\omega)n(V-\omega, t, P)n(\omega, t, P)\,d\omega$$

$$- n(V, t, P)\int_0^\infty K(V, \omega)n(\omega, t, P)\,d\omega.$$

Eq. (25.1) represents the balance for the number of bubbles of volume V in space (V, t, P). The term with $(dV/dt)_D$ here represents the rate of change of bubble volume due to the diffusion growth or pressure drop in the system as the bubbles rise in a mixture column. Figuring in the right part is the collision term characterizing the change of the distribution due to bubble coagulation with regard

Separation of Multiphase, Multicomponent Systems. E. G. Sinaiski and E. J. Lapiga

ISBN: 978-3-527-40612-8

only to pair interactions. The interaction mechanism between the bubbles is determined by the coagulation constant $K(\omega, V)$, which is equal to the bubble collision frequency evaluated for the volumes ω and V and the unit concentration. The basic property of the coagulation constant is its symmetry with respect to volumes $K(\omega, V) = K(V, \omega)$.

The second term in the left-hand side is the convective transfer of bubbles, which is equal to the velocity of a bubble of volume V. We assume that bubbles are small enough, so that $u = \alpha V^{2/3}$, and $\alpha = 2\Delta\rho g/3\mu_L$ for bubbles with completely "frozen" surface, whereas for bubbles with free surface we have $\alpha = g\Delta\rho/3\mu_L$. It should be noted that the constrained character of bubble motion is not taken into account, otherwise α will depend on the volume concentration of bubbles W. One of such dependences has been suggested in [8] by introducing the correction factor $\lambda(W) = (1 - W)^n$ into α.

If the expressions $(dV/dt)_D$ and $K(\omega, V)$ are known, then, having solved Eq. (25.1), one can find the number and volume concentrations of bubbles, N and W, and also their average volume $V_{av} = W/N$. The expressions for $(dV/dt)_D$ were obtained in Section 22. In the present chapter, the expressions for $K(\omega, V)$ and some simple solutions of the kinetic equation will be obtained.

25.1 Coagulation of Bubbles in a Laminar Flow

In a laminar flow, the coagulation constant K is proportional to the capture cross section of bubbles of volume ω by a bubble of volume V, while in a turbulent flow, it is proportional to the flux of bubbles of volume ω towards the test bubble of volume V.

Consider the motion of bubbles in a laminar flow. Their interactions are caused, on the one hand, by the difference of their velocities in the liquid due to their various sizes, and on the other hand, by the forces of molecular interaction. Due to the difference in sizes, they begin to approach each other at rather large (in comparison with radiuses of bubbles) distances. At small distances, there emerge forces of resistance, which impede their mutual approach. At the same distances, Van der Waals attractive forces come into play, stimulating efficient capture of bubbles. Note that in the case of bubbles with completely frozen surface approaching one another, the force of hydrodynamic resistance at small gaps δ between their surfaces is singular $F_h \sim \delta^{-1}$, and therefore their collision would be impossible without Van der Waals forces. For the approaching bubbles with free surface, $F_h \sim \delta^{-1/2}$. In contrast to the first case, this singularity is integrable; therefore bubbles with free surface can collide even with no account taken of the molecular attraction force. It should be noted when the gaps are small, the bubble surfaces in the contact region can be appreciably flattened. However, if the sizes of bubbles are small enough, then their approach at such distances is controlled by the force of molecular attraction, and the deformation of the surface will not noticeably influence their approach velocity.

Let us restrict ourselves to the case of fast coagulation, assuming that each collision of bubbles results in their coalescence. The study of mutual approach of bubbles in a laminar flow is based on the analysis of trajectories of their relative motion. The equations of non-inertial motion of a bubble of radius a relative to a bubble of radius b in the quasi-stationary approximation are:

$$\boldsymbol{F}_h + \boldsymbol{F}_m = 0, \quad \frac{d\boldsymbol{r}_a}{dt} = \boldsymbol{u}_a, \tag{25.2}$$

where $\boldsymbol{F}_h$ is the hydrodynamic force acting on the bubble; $\boldsymbol{F}_m$ is the force of molecular interaction; $\boldsymbol{r}_a$ is the radius-vector of the center of the bubble of radius a in the coordinate system attached to the center of the second bubble of radius b; $\boldsymbol{u}_a$ is the velocity of the bubble of radius a relative to the bubble of radius b.

Consider the interaction of bubbles with free surface. Since the velocity of a bubble has two components directed along the line of centers and perpendicular to it, then the hydrodynamic force $\boldsymbol{F}_h$ has also two components. At relatively large distances between bubbles, the force $\boldsymbol{F}_h$ is not much different from Stokes force – $4\pi\mu_L a\boldsymbol{U}$, where $\boldsymbol{U}$ is the velocity of the flow far from the drops. At small distances, the second component still does not significantly deviate from its Stokesean counterpart, while the component directed along the line of centers, differs essentially from a Stokesean component. Therefore $\boldsymbol{F}_h$ could be presented as

$$\boldsymbol{F}_h = \boldsymbol{F}_{he} + \boldsymbol{F}_{hs}, \quad \boldsymbol{F}_{hs} = -4\pi\mu_L a(\boldsymbol{u}_a - \boldsymbol{U}). \tag{25.3}$$

Here $\boldsymbol{F}_{hs}$ is the Stokesean force; $\boldsymbol{F}_{he}$ is the force of the central approach of bubbles, which differs from the Stokesean force and takes into account the influence of the bubble of radius b.

All the forces and, consequently, the trajectories of bubble motion lie in the plane that passes through the line of centers and is parallel to the flow velocity at the infinity.

Having set the initial position of a bubble with respect to the other one, it is possible to determine the family of trajectories, part of which terminate at the surface of the bubble of radius b, while the others pass it by. Call the trajectory dividing these two families the limiting trajectory. On a large distance from the bubble of radius b, they form a cylindrical surface of circular cross section. Call the area of this section the cross section of capture of bubbles of radius a by the bubble of radius $b > a$.

Assume that a small bubble of radius a does not deform the velocity field of the liquid flowing around the large bubble. Then

$$U_r = U_\infty\left(\frac{b}{r} - 1\right)\cos\theta, \quad U_\theta = U_\infty\left(1 - \frac{b}{2r}\right)\sin\theta. \tag{25.4}$$

For a simple estimation of force F_{he} at small gaps δ between surfaces of approaching bubbles we use the corresponding expression for a bubble near a flat free surface [9]:

$$F_{he} = 1.2\pi\mu_L a u_{an}\left(\frac{a}{\delta}\right)^{1/2}, \tag{25.5}$$

where u_{an} is the component of bubble velocity normal to the free surface.

In case of mutual approach of two bubbles of different radii a and b, the formula (25.5) should be corrected:

$$F_{he} = 1.2\pi\mu_L a u_{an}\left(\frac{a}{\delta}\right)^{1/2}\left(\frac{b}{a+b}\right)^{1/2}. \tag{25.6}$$

For narrow clearances between the bubbles, the force of molecular attraction can be presented as

$$F_m = \Gamma\frac{ab}{\delta^2(a+b)}, \tag{25.7}$$

where Γ is Hamaker's constant having the order of $\Gamma \sim 10^{-19}$.

Now, in view of expressions for the acting forces, the equations of motion for a bubble of radius a can be written as

$$\begin{aligned} &4\pi\mu_L a\left(1 + 0.3\delta^{-1/2}\left(\frac{ab^3}{(a+b)^3}\right)^{1/2}\right)\frac{dr}{dt} \\ &\quad = 4\pi\mu_L a U_\infty\left(\frac{b}{r} - 1\right)\cos\theta - \frac{\Gamma ab}{\delta^2(a+b)}, \\ &4\pi\mu_L a r\frac{d\theta}{dt} = 4\pi\mu_L a U_\infty\left(1 - \frac{b}{2r}\right)\sin\theta. \end{aligned} \tag{25.8}$$

Let us introduce the following dimensionless variables:

$$R = \frac{r}{a}, \quad \tau = \frac{U_\infty t}{b}, \quad k = \frac{a}{b}, \quad S_A = \frac{\Gamma}{4\pi\mu_L U_\infty b^2}.$$

Then Eqs. (25.8) are transformed to

$$\begin{aligned} &1 + 0.3\left(\frac{k}{(k+1)^3(R-k-1)}\right)^{1/2}\frac{dR}{d\tau} \\ &\quad = \left(\frac{1}{R} - 1\right)\cos\theta - \frac{S_A}{(k+1)(R-k-1)^2}, \\ &R\frac{d\theta}{d\tau} = \left(1 - \frac{1}{2R}\right)\sin\theta, \\ &R = R_0, \quad \theta = \theta_0 \quad \text{at } \tau = 0. \end{aligned} \tag{25.9}$$

If the viscous resistance and molecular interactions are absent, then the trajectories coincide with streamlines

$$(R(R-1))^{1/2}\cos\theta = C, \tag{25.10}$$

where C is the constant of integration.

In the considered case, the critical trajectory leads to contact of bubbles at $\theta = \pi/2$ and $R = 1+k$. This condition enables us to find $C = \sqrt{k(k+1)}$. It is now possible to find the dimensionless radius of capture cross section

$$l_1 = \lim_{R\to\infty} R\sin\theta = \sqrt{k(k+1)}. \tag{25.11}$$

Without regard for Van der Waals forces ($S_A = 0$), but taking into account viscous resistance, the solution can be found in quadratures:

$$\sqrt{R(R+1)}\cos\theta = C\exp\left[-\frac{0.3}{(k+1)^2}\left(\sqrt{k}\arctan\sqrt{\frac{R-k-1}{k+1}} + \sqrt{k+1}\arctan\sqrt{\frac{R-k-1}{k+1}}\right)\right]. \tag{25.12}$$

Using (25.12), as in the previous case, it is easy to find the dimensionless radius of the capture cross section:

$$l_2 = \sqrt{k(k+1)}\exp\left[-\frac{0.15\pi}{(k+1)^2}(\sqrt{k}+\sqrt{k+1})\right]. \tag{25.13}$$

It follows from expressions (25.11) and (25.12) that the proper account of viscous resistance reduces the capture cross section. In the range $0 < k < 0.6$ the ratio of capture cross sections lies in the interval $0.59 < l_2/l_1 < 0.67$. Thus, if we take into account viscous forces, but not Van der Waals forces, the capture cross section radius will be reduced almost by the factor of 1.6, and consequently, the frequency of coagulation – by the factor of 2.6.

In the general case, where we account for both viscous and molecular forces, Eq. (25.9) should be integrated numerically. Typical trajectories are shown in Fig. 25.1. Fig. 25.2 shows the dependence of the dimensionless capture cross section radius on the ratio k of bubble radii for various values of parameter of molecular interaction S_A. The dependencies of l_1 and l_2 on k are also given in the same figure for comparison.

In the case of gravitational emersion of bubbles in a quiescent liquid, $U_\infty = g\Delta p/3\mu_L$ and $S_A = \Gamma/4\pi\mu_L b^2 U_\infty = 3\Gamma/4\pi\rho_L g b^4$, where $\Delta p = \rho_L - \rho_G \approx \rho_L$. For example, for the values $\rho_L = 10^3$ kg/m^3, $g = 10$ m/s^2, and $\Gamma = 10^{-19}$ J, we have $S_A = 10^{-23}/4b^4$. In particular, for bubbles of radius $b = 10^{-4}$ m, $S_A = 10^{-7}$.

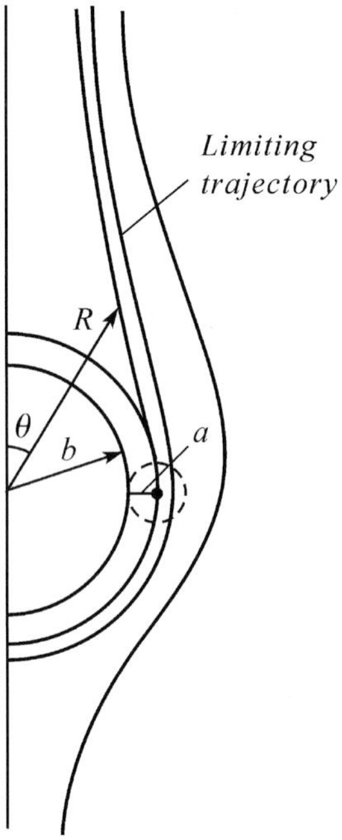

Fig. 25.1 Trajectories of motion of a small bubble relative to a large one ($k = 0.2$; $S_A = 10^{-2}$).

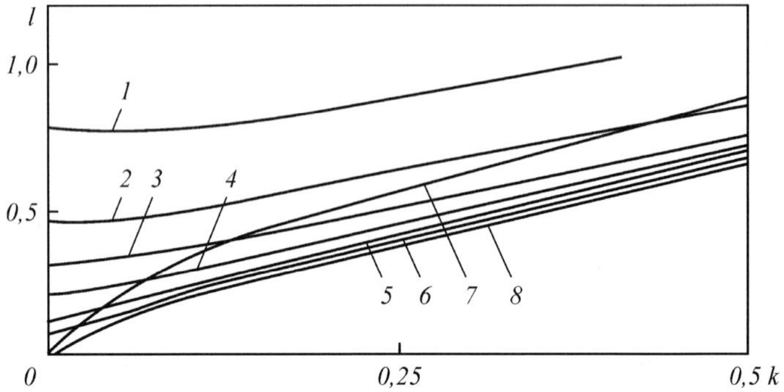

Fig. 25.2 Capture section radius l vs. k for different values of S_A: 1 – 10^{-2}; 2 – 10^{-3}; 3 – 10^{-4}; 4 – 10^{-5}; 5 – 10^{-6}; 6 – 10^{-7}; 7 – l_1.

The coagulation kernel is equal to

$$K = \pi b^2 l^2 |u_b - u_a| = \pi \rho_L g b^4 |1 - k^2| l^2 / 3\mu_L. \tag{25.14}$$

It is now possible to determine K for various models of the collision.

1. Viscous and molecular forces are neglected. In this case, from (25.11) and (25.14), there follows:

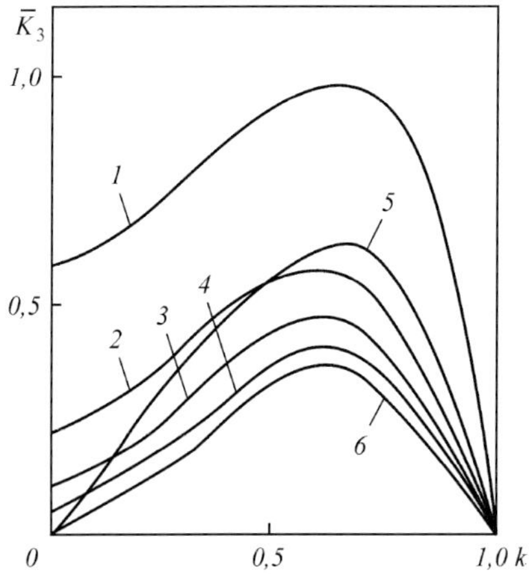

Fig. 25.3 Dependence of the coagulation kernel $\overline{K}_3$ on k for different values of S_A: 1 – 10^{-2}; 2 – 10^{-3}; 3 – 10^{-4}; 4 – 10^{-5}; 5 – $\overline{K}_1$; 6 – $\overline{K}_2$.

$$\overline{K}_1 = \frac{3\mu_L K}{\pi \rho_L b^4 g} = |1 - k^2|(1 + k)k. \tag{25.15}$$

The quantity $\overline{K}_1$ reaches its maximum value $\overline{K}_{1m} = 0.62$ at $k = 0.65$ (Fig. 25.3). It means that the probability of collision is greatest for the bubbles with the ratio of radii $a/b = 0.65$.

2. The viscous force is accounted for, but the van der Waals force is neglected. In this case, (25.13) and (25.14) give us

$$\overline{K}_2 = |1 - k^2|(1 + k)k \exp\left[-\frac{0.15\pi}{(k+1)^2}(\sqrt{k} + \sqrt{k+1})\right]. \tag{25.16}$$

The maximum value $\overline{K}_2 = 0.33$ is achieved at $k = 0.68$ (Fig. 25.3). In other words, consideration of the viscous force gives us the maximum probability of collision that is almost two times smaller than in the case when the viscous force is neglected, with the maximum value being achieved practically at the same value of k.

3. Both viscous and molecular forces are present. In this case

$$\overline{K}_3 = |1 - k^2| l_3^2. \tag{25.17}$$

The dependence of $\overline{K}_3$ on k and S_A is shown in Fig. 25.3. The comparison with cases 1 and 2 shows that accounting for the molecular force is essential at $S_A > 10^{-4}$. At smaller values of S_A, it is possible to use the expression (25.16).

25.2
Coagulation of Bubbles in a Turbulent Flow

For the turbulent flow regime, the flux of bubbles of volume ω toward the test bubble of volume V can be considered as a diffusion flux with the effective coefficient of diffusion D_T. Consider a bubble of volume V, placed in a turbulent flow of liquid containing bubbles of volume ω with the number concentration n. Assuming that the process is stationary and spherically symmetric, we have the following equation describing the distribution $n(r)$:

$$\frac{1}{r^2}\frac{d}{dr}r^2\left(D_T\frac{dn}{dr}-\frac{F_m n}{H(r)}\right)=0. \tag{25.18}$$

Here F_m is the force of molecular interaction between bubbles; $H(r)$ is the coefficient of hydrodynamic resistance in the expression

$$\boldsymbol{F}_h=-H(r)\boldsymbol{u}. \tag{25.19}$$

In a turbulent flow (see Section 11.3),

$$D_T=\left(\frac{H_0}{H}\right)^2. \tag{25.20}$$

where $D_0(r)$ is the coefficient of turbulent diffusion for non-hindered motion of bubbles ($H=H_0$) and $H_0=4\pi\mu_L a$. The expression for resistance H is:

$$H=4\pi\mu_L a\left(1+0.3\delta^{-1/2}\left(\frac{ab^3}{(a+b)^3}\right)^{1/2}\right). \tag{25.21}$$

In a turbulent flow, turbulent pulsations with different amplitudes are superimposed on the averaged motion. These pulsations are characterized by both the velocity u_λ and the distance (known as the scale of pulsations), at which the velocity of pulsations undergoes an appreciable change. Each of them can be associated with Reynolds number $\mathrm{Re}_\lambda=u_\lambda\lambda/\nu_L$. For large pulsations, $\lambda\sim L$, where L is the characteristic linear scale of the region where viscous forces do not have a noticeable influence, while for small pulsations, they can dominate. The scale λ_0 for which $\mathrm{Re}_{\lambda_0}=1$, is called the internal scale of turbulence. Viscous forces are important for pulsations with $\lambda<\lambda_0$. Under the action of pulsations, bubbles can move chaotically. In [10], it is shown that the form of D_0 depends on the type of pulsations that cause bubbles to perform random motion:

$$D_0=\begin{cases}\nu_L\lambda^2/\lambda_0^2\sim(\varepsilon/\nu_L)^{1/2}\lambda^2 & \text{at } \lambda\le\lambda_0,\\ \varepsilon\lambda^{4/3} & \text{at } \lambda>\lambda_0.\end{cases} \tag{25.22}$$

Let us find out which pulsations can force two bubbles with radii a and b to approach each other. For exactness, suppose that $a < b$. Pulsations with $\lambda \gg b$ will simply transfer both bubbles, keeping them away from each other, while pulsations with $\lambda \leq b$ will move the two bubbles closer together. In the limiting case $a \ll b$, the mutual approach of bubbles occurs under the influence of pulsations of the scale $\lambda \sim r - b - a$, and in the other case, $a \sim b$ – under the influence of the scale $\lambda \sim r + a$. Combining the two cases, we obtain

$$D_0 \sim \frac{v_L}{\lambda_0^2}\left(r - b + \frac{ab(a+b)}{a^2 + b^2 - ab}\right)^2. \tag{25.23}$$

Now, using (25.20), (25.21) and (25.23), we find:

$$D_T \sim \frac{v_L}{\lambda_0^2}\left(\frac{r - b + ab(a+b)/(a^2+b^2-ab)}{1 + 0.3(ab^3)^{1/2}/(r-a-b)^{1/2}/(a+b)^{3/2}}\right)^2. \tag{25.24}$$

The collision frequency of a bubble of radius b with bubbles of radius a is equal to the diffusion flux:

$$J_T = 4\pi b^2\left(D_T\frac{dn}{dr}\right)_{r=a+b}. \tag{25.25}$$

Eq. (25.18) will be integrated with the conditions

$$n = 0 \quad \text{at } r = a + b, \qquad n \to n_0 \quad \text{at } r \to \infty.$$

Note that, as opposed to the coagulation of bubbles in a laminar flow, coagulation in a turbulent flow is impossible in the absence of molecular forces, since $D_T \sim \delta$ at $\delta = r - a - b \to 0$. Therefore the integral term entering the solution of the diffusion equation has a nonintegrable singularity that can be eliminated only by the introduction of a molecular attraction force growing as δ^{-2} at $\delta \to 0$.

The solution of Eq. (25.18) gives the following expression for the diffusion flux:

$$J_T = 4\pi n_0\left\{\int_{a+b}^{\infty}\exp\left(-\int_r^{\infty}\frac{F_m(\xi)}{H(\xi)D_T(\xi)}d\xi\right)\frac{dr}{r^2 D_T(r)}\right\}^{-1}. \tag{25.26}$$

Switching to dimensionless variables,

$$x = \frac{r-a-b}{b}, \quad k = \frac{a}{b}, \quad \bar{D}_T = \frac{\lambda_0^2 D_T}{v_L b^2}, \quad S_m = \frac{\lambda_0^2 \Gamma}{4\pi\rho_L b^2 v_L^2},$$

we get:

$$\bar{D}_T = \left\{\frac{x + k + k(k+1)/(k^2-k+1)}{1 + 0.3k^{1/2}/x^{1/2}(1+k)^{3/2}}\right\}^2$$

and

$$J_T = \frac{4\pi n_0 v_L b^3}{\lambda_0^2}\left\{\{\int_0^\infty \exp\left(-\frac{S_m}{(k+1)}\int_x^\infty\left(1+\frac{0.3\sqrt{k}}{\sqrt{\xi}(1+k)^{3/2}}\right)\right.\right.$$
$$\left.\times\,\xi^{-2}\left(\xi+k+\frac{k(k+1)}{k^2-k+1}\right)^{-2}d\xi\right)\left(\left(1+\frac{0.3\sqrt{k}}{\sqrt{x}(1+k)^{3/2}}\right)\right.$$
$$\left.\left.\times\,(x+1+k)^{-1}\left(x+\frac{k(k^2+2)}{k^2-k+1}\right)^{-1}\right)^2 dx\right\}^{-1}. \qquad (25.27)$$

Fig. 25.4 shows the dependence of $\bar{J}_T = J_T\lambda_0^2/4\pi n_0 v_L b^3$ on the ratio k of bubble radii for different values of the parameter of molecular interaction S_m. There are two noteworthy peculiarities of bubble coagulation in turbulent flow: the flux has a weak dependence on S_m, and reaches the maximum value at $k = 1$. This means that the collision probability is highest for the bubbles of identical sizes.

Now, compare the collision frequencies of bubbles in laminar and turbulent flows. For J_T, we can use the expression (25.27), and for J_l,

$$J_l = \frac{\pi g b^4 n_0 \bar{K}}{3 v_L},$$

where $\bar{K}$ is determined by the expression (25.17).

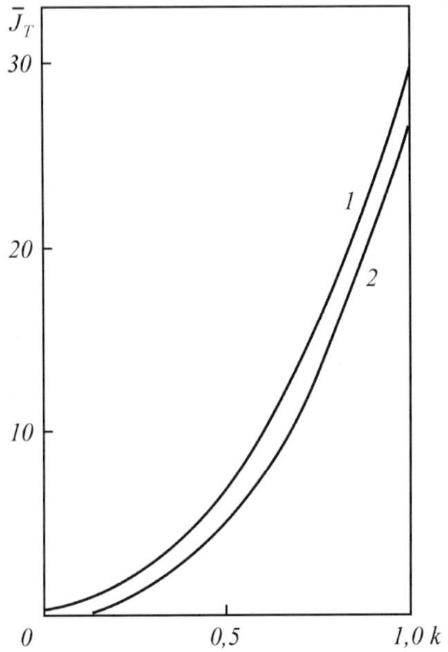

Fig. 25.4 Flux $\bar{J}_T$ vs. k for different values of S_m: 1 – 10^{-2}; 2 – 10^{-7}.

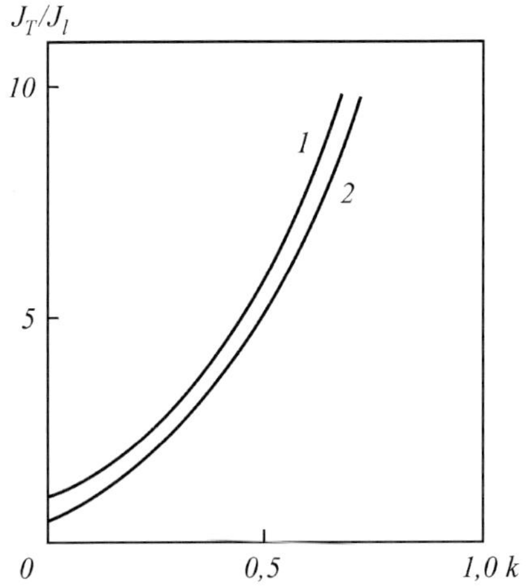

Fig. 25.5 Ratio J_T/J_l vs. k for different values of S_A ($S_m = 10S_A$): $1 - 10^{-7}$; $2 - 10^{-5}$.

For $\lambda_0 = 10^{-3}$ m, $b = 10^{-4}$ m, and $\nu_L = 10^{-5}$ m^2/s, we have $S_m = 10S_A$. The dependence of J_T/J_l on k for given values of parameters is presented in Fig. 25.5. One can see that for bubbles whose sizes differ by a lot ($k \ll 1$), $J_T \approx J_l$, while for bubbles whose sizes are about equal ($k \sim 1$), $J_T \gg J_l$.

25.3 Kinetics of Bubble Coagulation

The expressions for frequencies of bubble collision in laminar and turbulent flow which derived in the previous paragraphs make it possible to find the kernels of coagulation $K(\omega, V)$ and then proceed to solve the kinetic equation (25.1). Because the solution, generally speaking, presents significant mathematical difficulties, we shall only consider some simple special cases.

Let the process of coagulation progress in such a manner that the volume distribution of bubbles depends only on time t. In addition, we assume that the integration of bubbles occurs only due to their coagulation. Then the kinetic equation will take the form:

$$\frac{dn}{dt} = I_k, \quad n(V, 0) = n_0(V), \tag{25.28}$$

$$I_k = \frac{1}{2}\int_0^V K(\omega, V-\omega)n(V-\omega, t)n(\omega, t)\,d\omega - n(V, t)\int_0^\infty K(\omega, V)n(\omega, t)\,d\omega.$$

We will solve this equation by the method of moments, using the parametrical representation and making a further assumption that the distribution initially looks like a gamma distribution and later on, still remains in this class of distributions. The method of moments was thoroughly analyzed in Section 11.1. Limiting ourselves to the first two moments m_0 and m_l, we obtain the following equations for these moments:

$$\frac{dm_0}{dt} + m_0 \frac{V_0'}{V_0} = -\frac{1}{2}\int_0^\infty \int_0^\infty K(\omega, V)n(V,t)n(\omega,t)\, dV\, d\omega,$$
$$\frac{dm_1}{dt} + 2m_1 \frac{V_0'}{V_0} = 0, \quad V_0 = \frac{V_{av}}{k+1}. \tag{25.29}$$

Here V_0 and k are parameters of the gamma distribution; V_{av} is the average volume of bubbles.

At the initial moment,

$$m_0(0) = \frac{N_0(k+1)}{V_{av0}}, \quad m_1(0) = \frac{W_0(k+1)^2}{V_{av0}^2},$$

where N_0 and W_0 are the initial values for the number and volume concentrations of bubbles; V_{av0} is the initial average volume of bubbles.

From the second equation (25.29), it follows that the volume concentration of bubbles W remains constant, and from the first equation it follows that

$$\frac{dn}{dt} = -\frac{1}{2}\frac{(k+1)^2}{[\Gamma(k+1)]^2}\frac{N^4(t)}{W_0^2} V_0^2 \varphi(V_{av}), \tag{25.30}$$
$$\varphi(V_{av}) = -\frac{1}{2}\int_0^\infty \int_0^\infty K(V_0 x, V_0 y) x^k y^k e^{-x} e^{-y}(V,t)\, dx\, dy.$$

Consider now some models of bubble interaction.

1. The model assumes the interaction of bubbles with no forces of viscous resistance or molecular attraction, in other words, gravitational separation of bubbles is assumed.

The expression for the coagulation kernel is determined by the formula (25.25), which, when written in terms of volumes instead of radii, will become

$$K(\omega, V) = \left(\frac{3}{4\pi}\right)^{2/3} \frac{\pi \rho_L g}{3\mu_L} V^{2/3} |V^{2/3} - \omega^{2/3}| \frac{\omega^{1/3}}{V^{1/3}} \left(1 + \frac{\omega^{1/3}}{V^{1/3}}\right). \tag{25.31}$$

Substituting (25.31) into (25.30), we obtain:

$$\frac{dN}{dt} = -\left(\frac{3}{4\pi}\right)^{2/3} \frac{\pi \rho_L g}{3\mu_L} W_0^{4/3} \lambda_k N^{2/3}, \quad N(0) = N_0, \tag{25.32}$$

where

$$\lambda_k = \frac{3}{4}\frac{\Gamma(2k+10/3)}{[\Gamma(k+1)]^2(k+1)^{4/3}2^{2k+10/3}}\left\{\frac{1}{k+4/3}\Phi\left(1;2k+\frac{10}{3};k+\frac{7}{3};\frac{1}{2}\right)\right.$$
$$+\frac{1}{k+5/3}\Phi\left(1;2k+\frac{10}{3};k+\frac{8}{3};\frac{1}{2}\right)-\frac{1}{k+2}\Phi\left(1;2k+\frac{10}{3};k+\frac{9}{3};\frac{1}{2}\right)$$
$$\left.-\frac{1}{k+7/3}\Phi\left(1;2k+\frac{10}{3};k+\frac{9}{3};\frac{1}{3}\right)\right\},$$

$\Gamma(x)$ is the full gamma function; $\Phi(\alpha;\beta;\gamma;z)$ is the hypergeometric function. The λ_k values are: $\lambda_0 = 0.4639$; $\lambda_1 = 0.515$; $\lambda_2 = 0.5648$; $\lambda_3 = 0.6095$; $\lambda_4 = 0.6893$.

The solution of Eq. (25.32) is

$$N(\tau) = N_0\left(1-\frac{\lambda_k W_0 \tau}{3}\right)^3, \quad \tau = \left(\frac{3}{4\pi}\right)^{2/3}\frac{\rho_L g V_{av0}^{1/3} t}{3\mu_L}. \tag{25.33}$$

The number of bubbles decreases with time. The average volume and variance of the distribution grow with time according to the law

$$V_{av} = V_{av0}\left(1-\frac{\lambda_k W_0 \tau}{3}\right)^{-3}, \quad \sigma^2 = \sigma_0^2\left(1-\frac{\lambda_k W_0 \tau}{3}\right)^{-6}. \tag{25.34}$$

The characteristic time of coagulation is estimated as

$$t_k \sim 3\left(\frac{4\pi}{3}\right)^{2/3}\frac{\nu_L}{g V_{av0}^{1/3}\lambda_k W_0}. \tag{25.35}$$

Now it is possible to estimate the time it takes for the average volume of a bubble to increase by a given number of times. As an example, consider the initial distribution of bubbles at $k = 2$. Then the time during which the bubble volume will double is equal to $\tau \sim 7/4\lambda_2 W_0 = 3.13/W_0$, or, in the dimensional form, $t \sim 10\nu_L/W_0 g R_{av0}$. For the values $\nu_L = 5\cdot 10^{-6}$ m^2/s, $W_0 = 10^{-2}$, and $R_{av0} = 10^{-5}$ m, we have $t \sim 50$ s.

2. This model assumes interaction with the viscous resistance force present, but with no Van der Waals force.

In this case, the expression for the coagulation kernel is determined by the formula (25.16), which differs from (25.15) by the factor

$$g(k) = \exp\left\{-\frac{0.15\pi}{(k+1)^2}(\sqrt{k}+\sqrt{k+1})\right\}.$$

In the interval $0 \le k \le 1$, this function has a minimum at $k = 1/15$ and reaches its maximum value at $k = 0$, and the difference between the maximum and the minimum value is insignificant. Therefore, we can make an estimate by taking $g(k) \approx 0.33$. Repeating the calculations made above, we obtain an equation of the kind (25.32), in which, instead of λ_k, it is necessary to take $\bar{\lambda}_k = 0.33\lambda_k$. Because the coagulation time (25.35) contains λ_k in the denominator, it will increase approximately by a factor of three as compared to the previous case.

References

1 Gupallo Yu.P., Poljanin A.D., Rjasanzev Ju.S. Mass exchange of reacting particles with the flow. Moscow: Nauka, 1985. (in Russian).
2 Rulev N.N. Hydrodynamics of a lifting bubble // Colloid J. 1980. V.42. N 2. P. 252–263. (in Russian).
3 Plesset M.S., Zwick S.A. A nonsteady Heat Diffusion Problem with spherical Symmetry // J. Appl. Phys. 1952. V.23. N 1. P. 95–99.
4 Fridrichsberg D.A. A handbook on colloid chemistry. Leningrad: Chemistry, 1974. (in Russian).
5 Goncharov V.K., Klementeva N.Yu. A study of the influence of surfactant films on the dissolution of a bubble moving in sea-water // Proceedings Rus. Acad. Sci. Physics of atmosphere and ocean. 1995. V.31. N 5. P. 705–712. (in Russian).
6 Volmer M. Kinetik der phasenbilgung. Dresden und Leipzig: T. Steinkopff, 1939.
7 Shlikova M.P. Experimental analysis of hydrodynamic processes accompanying separation of gas-liquid mixtures. Ph.D. thesis. Moscow: Moscow Institute of Oil and Gas, 1986. (in Russian).
8 Wallis G.B. One-dimensional two-phase flow. New York McGraw-Hill, 1969.
9 Waholder E., Weihs D. Slow motion of fluid sphere in the vicinity of another sphere or a plane boundary // Chem. Eng. Sci. 1972. V.27. N 10. P. 1817–1827.
10 Levich V.G. Physico-chemical hydrodynamics. Englwood Cliffs, N.J. Prentice-Hall, 1962.

Author Index

Separation of Multiphase, Multicomponent Systems. E. G. Sinaiski and E. J. Lapiga

ISBN: 978-3-527-40612-8

Subject Index

Separation of Multiphase, Multicomponent Systems. E. G. Sinaiski and E. J. Lapiga

ISBN: 978-3-527-40612-8

Related Titles

Schmelzer, J. W. P.

Nucleation Theory and Applications

2005
ISBN 978-3-527-40469-8

Senatskommission zur Beurteilung von Stoffen in der Landwirtschaft (ed.)

Transient Phenomena in Multiphase and Multicomponent Systems

Research Report

2000
ISBN 978-3-527-27149-8